Trends in Mathematics is a series devoted to the publication of volumes arising from conferences and lecture series focusing on a particular topic from any area of mathematics. Its aim is to make current developments available to the community as rapidly as possible without compromise to quality and to archive these for reference.

Proposals for volumes can be sent to the Mathematics Editor at either

Birkhäuser Verlag
P.O. Box 133
CH-4010 Basel
Switzerland

or

Birkhäuser Boston Inc.
675 Massachusetts Avenue
Cambridge, MA 02139
USA

Material submitted for publication must be screened and prepared as follows:

All contributions should undergo a reviewing process similar to that carried out by journals and be checked for correct use of language which, as a rule, is English. Articles without proofs, or which do not contain any significantly new results, should be rejected. High quality survey papers, however, are welcome.

We expect the organizers to deliver manuscripts in a form that is essentially ready for direct reproduction. Any version of $\TeX$ is acceptable, but the entire collection of files must be in one particular dialect of $\TeX$ and unified according to simple instructions available from Birkhäuser.

Furthermore, in order to guarantee the timely appearance of the proceedings it is essential that the final version of the entire material be submitted no later than one year after the conference. The total number of pages should not exceed 350. The first-mentioned author of each article will receive 25 free offprints. To the participants of the congress the book will be offered at a special rate.

Mathematics and Computer Science II

Algorithms, Trees, Combinatorics and Probabilities

Brigitte Chauvin
Philippe Flajolet
Danièle Gardy
Abdelkader Mokkadem
Editors

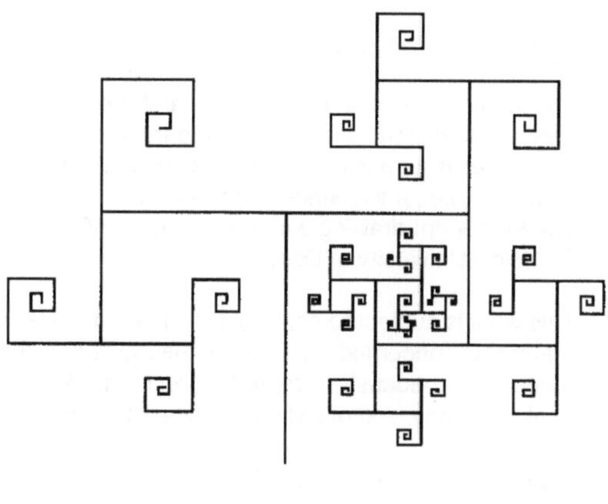

Springer Basel AG

Editors' addresses:

Brigitte Chauvin
Université de Versailles-St. Quentin
Département de Mathématiques
Bâtiment Fermat
45 avenue des Etats-Unis
78035 Versailles Cedex
France
e-mail: chauvin@math.uvsq.fr

Danièle Gardy
Université de Versailles-St-Quentin
PRISM
Bâtiment Descartes
45 avenue des Etats-Unis
78035 Versailles Cedex
France
e-mail: gardy@prism.uvsq.fr

Philippe Flajolet
INRIA Rocquencourt
78153 Le Chesnay
France
e-mail: Philippe.Flajolet@inria.fr

Abdelkader Mokkadem
Université de Versailles-St-Quentin
Département de Mathématiques
Bâtiment Fermat
45 avenue des Etats-Unis
78035 Versailles Cedex
France
e-mail: mokkadem@math.uvsq.fr

2000 Mathematical Subject Classification 68M20, 68P30, 68Q25, 68Rxx, 68W20, 90B15

A CIP catalogue record for this book is available from the Library of Congress, Washington D.C., USA

Deutsche Bibliothek - Cataloging-in-Publication Data

Mathematics and computer science II: algorithms, trees, combinatorics and probabilities.
Brigitte Chauvin ... ed.. - Basel ; Boston ; Berlin : Birkhäuser, 2002
(Trends in mathematics)
ISBN 978-3-0348-9475-3 ISBN 978-3-0348-8211-8 (eBook)
DOI 10.1007/978-3-0348-8211-8

ISBN 978-3-0348-9475-3

The logo on the cover is a binary search tree in which the directions of child nodes alternate between horizontal and vertical, and the edge lengths decrease as 1 over the square root of 2. The tree is a Weyl tree, which means that it is a binary search tree constructed from a Weyl sequence, i.e., a sequence (na) mod 1, $n = 1,2,...$, where a is an irrational real number.
The PostScript drawing was generated by Michel Dekking and Peter van der Wal from the Technical University of Delft.

Originally published by Birkhäuser Verlag, Basel - Boston - Berlin in 2002
Softcover reprint of the hardcover 1st edition 2002

ISBN 978-3-0348-9475-3

9 8 7 6 5 4 3 2 1

www.birkhauser-science.com

Contents

Foreword

These are the Proceedings of the International Colloquium of Mathematics and Computer Science held at the University of Versailles-St-Quentin, September 18-20, 2002. This colloquium is the second one in a now regularly established series following the first venue in September 2000 in Versailles. The present issue is centered around Combinatorics, Random Graphs and Networks, Algorithms Analysis and Trees, Branching Processes and Trees, Applied Random Combinatorics.

The contributions have been carefully reviewed for their scientific quality and originality by the Scientific Committee chaired by P. Flajolet and composed by P. Chassaing, B. Chauvin , M. Drmota, J. Fill, P. Flajolet, A. Frieze, D. Gardy, S. Janson, C. Krattenthaler, G. Louchard, A. Mokkadem, R. Pemantle, P. Robert, J. Spencer , B. Ycart. We do thank them for their impressive work.

We also thank the invited speakers: D. Aldous, L. Devroye, S. Janson, M. Krivelevich, B. Pittel, H. Prodinger, M. Steele, the authors of submitted papers and the participants for their contribution to the success of the conference.

A. Baffert and C. Ducoin deserve special thanks for their kind and efficient contribution to the material preparation of the colloquium.

Finally, we express our acknowledgements to the laboratory of Mathematics (LAMA), the laboratory of Computer Science (PRISM), the University of Versailles-St-Quentin, the Centre National de la Recherche Scientifique (CNRS) and the Institut National de Recherche en Informatique et Automatique (INRIA) for providing generous financial and material support.

The Organizing Committee
B. Chauvin
S. Corteel
D. Gardy
J.F. Marckert
A. Mokkadem
B. Ycart

Preface

These colloquium proceedings address problems at the interface between mathematics and computer science, with special emphasis on discrete probabilistic models and their relation to algorithms. Combinatorial and probabilistic properties of random graphs and networks, random trees and branching processes, as well as random walks are central. Applications are to be found in analysis of algorithms and data structures, the major application field, but also in statistical theory, information theory, and mathematical logic. This colloquium is the second one in a now regularly established series, following the first venue in September 2000 in Versailles. The book features a collection of original refereed contributions supplemented by survey articles written by the invited speakers, L. Devroye, S. Janson, M. Krivelevich, B. Pittel, H. Prodinger, and M.Steele. As the field is burgeoning with applications at the frontier of several scientific disciplines, authors have been asked to provide a perspective on the various subfields involved.

Combinatorics. The starting point of many studies of random discrete models is combinatorics, which often provides us with exact representations in terms of counting generating functions. Arquès and Micheli develop the combinatorial enumerative theory of maps with special attention to colouring problems. Banderier presents a synthetic theory of walks with returns over the half integer line, which is motivated by fast random generation of combinatorial structures. Bousquet-Mélou shows us an explicitly solvable model of walks in the quarter-plane that originates with basic queueing theory questions and gives rise to elegant combinatorial developments. Brlek, Duchi, Pergola and Pinzani enrich the theory of "ECO-systems" now recognized to provide a unifying framework for many problems of combinatorial random generation. Krikun and Malyshev finely characterize the boundary of a random triangulation of the disk by means of combinatorial-analytic methods. Labelle, Lamathe, and Leroux successfully apply the theory of species to tree-like arrangements of cells and derive an original combination of explicit and asymptotic counting results.

Random Graphs and Networks. Following Erdős and Rényi's pioneering work around 1960, random graph models have been the subject of intense study for four decades. Baert, Ravelomanana, and Thimonier base a novel analysis of triangle free graphs on breadth-first search and its associated stochastic properties in the line of Spencer's approach. Chassaing and Schaeffer solve a long-standing open questions: What is the diameter of a random map? Their result is achieved by an examplary combination of bijective and probabilistic methods. Coppersmith, Gamarnik, and Sviridenko characterize the diameter of a random graph with long range interactions—such problems are of interest for percolation models but also in relation to the geometry of the web. Devroye, McDiarmid, and Reed analyse the emergence of giant components in two graph models that are similarly motivated by our desire to understand the "graph of the web". Krivelevich's invited lecture surveys random graph colouring: the problem is *NP*-complete in the worst-case, but the perspective changes dramatically when one switches from the pessimistic worst-case scenario to the more realistic average-case analysis. Le Bars demonstrates the usefulness of probabilistic inequalities in the analysis of some threshold phenomena of logic. Palaysi explores combinatorial and algorithmic aspects of wavelength assignment in certain graphs representing interconnection networks. Last but not least, Steele's invited lecture reviews and revisits the celebrated problem of minimal spanning trees in graphs with random edge weights; his text gives for the first time surprisingly explicit formulæ out of which quantitative estimates can be derived.

Analysis of Algorithms and Trees. Trees are perhaps the most important structure of computer science. In particular, they appear as data structures in an amazing variety of domains, like textual data processing, data compression, fast retrieval of information, symbolic computation, and so on. Bourdon and Vallée exhibit versatile criteria informing us on conditions under which a complex pattern is or isn't likely to occur; their analysis is based on an original interplay of combinatorial and dynamical systems methods. The invited lecture by Devroye and Neininger develops an original analysis of a new structure, the suffix search tree that is a hybrid of two of the most important data structures, the suffix trie and the binary search tree. In particular, their study contributes significantly to our understanding of basic data structures when these are subjected to correlated data. Gittenberger proposes an approach via generating functions to the analysis of strata of nodes in random trees, which has applications to breadth-first search traversal. Hwang and Steyaert offer a definitive analytic treatment of the heap structure whose importance devolves from its widespread use in priority queue management and near-optimal sorting. Jacquet and Szpankowski are able to characterize the redundancy of Markov sources of order r by an ingenious combination of combinatorial and analytic methods. Quickfind, which is one of the most spectacular algorithms known for basic order statistics, is thoroughly analysed by Martínez, Panario, and Viola: their contribution even results in an eminently practical discussion of cut-off points for optimal performance. Nguyen-The's paper show that the study of random combinatorial trees is intimately related to the performance of basic algorithms of symbolic manipulation, in particular the formal simplification of expressions. The invited paper of Prodinger offers a unified analytic perspective on digit statistics in a wide variety of number representation systems. Beyond its pure number-theoretic aspects, this study finds numerous applications in parsing and compiling (register allocation), sorting networks, the design of adder circuits, mergesort, interpolation search, and even branching fractals.

Branching Processes and Trees. Branching processes constitute the probabilistic counterpart of the combinatorial theory of trees. Fayolle and Krikun provide ergodicity conditions for a model of random trees that is akin to binary search trees, but where evolution involves random insertions and deletions. The invited lecture of Janson tackles problems motivated by exhaustive search, along the lines of research by Ruskey and Knuth (see the forthcoming volume, *Combinatorial Algorithms*). In particular, Janson proves the existence of a limit distribution for the number of ideals in random trees and does so by a clever adaptation of contraction methods, using the size-biased branching process. Menshikov and Petritis report on recent results concerning random walks in a random environment on trees and their relationship to multiplicative chaos. Pittel's invited lecture revisits the loop erased random walk: this is a biased model of self-avoidance that is currently witnessing a number of spectacular developments. This article derives very precise estimates of limiting distributions that are also of interest for generating random spanning trees of the complete graph. Rösler, Topchii, and Vatutin finely characterize convergence rates of weighted branching processes and detect there the occurrence of stable laws. Vatutin and Dyakonova develop informative limit theorems for a critical branching process in a random environment.

Applied random combinatorics. Random combinatorics interacts with many other areas of science. Huillet and Porzo re-examine a version of the parking problem that is known to have numerous applications in computer science (hashing algorithms, resource allocation), combinatorial optimization, as well as statistical mechanics and adsorption models. Mossel and O'Donnell examine the sensitivity to input noise of Boolean functions; their results have implications in learning theory, complexity theory, neural networks, and even (the authors argue) the American election system. Teytaud develops a set of new results in learning theory basing himself on ergodicity properties. Trouvé and Yu establish upper bounds on the number of questions a user asks in the case of hierarchically structured databases. Weiermann characterizes 0-1 laws in ordinal theory via analytic combinatorics; his paper also serves as a valuable introduction to random combinatorics in its relation to finite model theory and logic. The last contribution to this book, by Zhang and Golin, develops explicit formulæ for the number of spanning trees in structured graphs having a highly regular shape.

Altogether papers assembled in this volume offer snapshots of current research. At the same time, they illustrate the numerous ramifications of the theory of random discrete structures throughout mathematics and computer science. Many of them, in particular invited lectures, include carefully crafted surveys of their field. We thus hope that the book may serve both as a reference text and as a smooth introduction to many fascinating aspects of this melting pot of continuous and discrete mathematics.

Enjoy!

Brigitte Chauvin,
Philippe Flajolet,
Danièle Gardy,
A. Mokkadem

PART I

Combinatorics

Trends in Mathematics, © 2002 Birkhäuser Verlag Basel/Switzerland

n-Colored Maps and Multilabel n-Colored Trees

Didier Arquès, Anne Micheli

ABSTRACT: *New topological operations are introduced in order to recover in another way the generalized Dyck equation for the generating function of n-colored maps presented in a former paper, by decomposing maps topologically and bijectively. Applying repeatedly the operations which allowed to reveal the generalized Dyck equation to the successive transformed maps, a one-to-one correspondence is obtained between n-colored maps on any surface and n-colored trees whose vertices can be labelled with several labels. This bijection provides us with a coding of these maps.*

1 Introduction

The enumerative study of maps starts in 1962 with W.T. Tutte [15, 16], who enumerates the number of rooted planar maps with n edges. Maps can also be described as combinatorials objects [12]. In 1975, R. Cori [7] studies planar maps in this perspective and extends these results with A. Machi [8] to orientable maps. In particular, R. Cori and B. Vauquelin determine a bijection between planar maps and well labelled trees [9], which leads to a code of these maps. These results were extended to maps of genus g and well labelled g-trees [13] and a code for maps of genus g by words product of a shuffle of Dyck words with constraints and of a sequence of integers was then obtained. Many studies can be found on maps of a strictly positive genus, orientable or not, as for example [1, 6, 3].
The study of rooted maps independently of their genus begins with T.R.S. Walsh and A. Lehman [17]. They give a recursive relation on the number of rooted maps with respect to the number of edges, which does not lead to an explicit enumeration formula of these maps. In 1990, D.M. Jackson and T.I. Visentin [11] use an algebraic approach and obtain a closed formula for the generating functions of orientable rooted maps with respcet to the number of edges and vertices.
More recently, D. Arquès and J.F. Béraud [2] determine a functional equation satisfied by the generating functions of rooted maps with respect to the number of edges and vertices and express the solution in a continued fraction form. This continued fraction reveals an interesting bijection, since it also enumerates connected fixed-point free involutions [14].

Topological operations applied to a map such as the removal or the addition of an edge, the fusion of two vertices, modify sometimes the genus of the map. These operations can not therefore be carried out in a systematic way when one works with fixed genus. However, these elementary operations make it possible to find new functional equations on maps studied independently of genus and to establish bijections between families of maps.

In Section 2, we recall general definitions on maps and n-colored maps. New topological operations are introduced in Section 3, in order to establish in Section 4, a bijection between n-colored maps of arbitrary genus, and n-colored maps of arbitrary genus with a root bridge, in which a subset of their vertices has

been selected. This bijection provides us with a new proof of the generalized Dyck equation on orientable n-colored rooted maps obtained formerly [4, 5]. This equation was obtained by an analytic resolution of a differential equation satisfied by the generating function of n-colored rooted maps and led to a solution in a multi-continued fraction form. We here present a new proof of this equation, without any computation over the generating function, by transcription of the presented bijection. P. Flajolet [10] moreover showed that many continued fractions having integer coefficients can be explained in a purely combinatorial way, and here is an example of his assertion on a multi-continued fraction with integer coefficients.
In Section 5, we give a bijection between orientable n-colored rooted maps and a family of n-colored trees whose vertices can be labelled by several labels according to certain rules, which is deduced from the one presented in Section 4 by successive applications of this bijection. The bijection enables us to determine a new language coding n-colored maps.

2 Definitions

Let us recall some definitions used afterwards (for further details, see for example [7, 8]).

A topological *map* C in an orientable surface Σ of $\mathbb{R}^3$ is a partition of Σ in three finite sets of cells: the set of *vertices* of C, which is a finite set of dots; the set of *edges* of C, which is a finite set of open Jordan arcs, pairwise disjoint, whose extremities are vertices; the set of *faces* of C. Each face is simply connected and its border is the union of vertices and edges.

The *genus* of the map C is the genus of Σ. A cell is *incident* to another cell if one is contained in the boundary of the other. A *bridge* is an edge incident on both sides to the same face. We call *half-edge* an oriented edge of the map.

Let B be the set of half-edges of the map. With each half-edge, one can associate its *initial vertex*, its *final vertex* and its underlying edge. α (resp. σ) is the permutation in B associating to each half-edge b its opposite half-edge (resp. the first half-edge met when turning round the initial vertex of b in the positive way of the surface). The cycles of α (resp. σ) represent the edges (resp. the vertices) of the map. The cycles of $\bar{\sigma} = \sigma \circ \alpha$ are the oriented borders of the faces of the map. (B, σ, α) is the combinatorial definition of the topological orientable map associated C.

A map $C = (B, \sigma, \alpha)$ is *rooted* if a half-edge $\tilde{b}$ is distinguished. The half-edge $\tilde{b}$ is called the *root half-edge* of C, and its initial vertex is the *root vertex*. C is then defined as the triplet $(\sigma, \alpha, \tilde{b})$. Face $\bar{\sigma}^*(\tilde{b})$ is called the *exterior face* of C. By convention, the one vertex map (one vertex, no edge) is said to be rooted.

Two orientable maps of the same genus are isomorphic if there is a homeomorphism of the surfaces, preserving its orientation, mapping vertices, edges and faces of one map onto vertices, edges and faces respectively of the other map. An isomorphism class of orientable rooted maps will simply be called a rooted map.

Definition 2.1. **n-colored map** An orientable rooted *n-colored map* ($n > 1$) is a rooted map, where a maximum of n colors are used to color the vertices and such that each edge is incident to two vertices of different colors.

The property "n-colored" is compatible with the equivalence relation whose classes are the rooted maps.

Let $\mathcal{M}$ be the set of orientable rooted maps, $\mathcal{M}_{n,i}$ the set of orientable rooted n-colored maps, whose root vertex is of color i, $\mathcal{I}_{n,i}$ the subset of $\mathcal{M}_{n,i}$ of maps with a root bridge, and for any map $I \in \mathcal{I}_{n,i}$, $Right(I)$ (resp. $Left(I)$) the maximal submap of I incident to the root vertex (resp. the final vertex of $\tilde{b}$) such that the root half-edge $\tilde{b}$ (resp. $\alpha(\tilde{b})$) of I does not belong to $Right(I)$ (resp. $Left(I)$) (see Figure 4.1). Let $\{p_i\}$ be the one vertex map of $\mathcal{M}_{n,i}$.

3 Preliminaries

In Section 3.1, we describe two algorithms of half-edges and vertices numbering of a map. Numbering induces an order relation on half-edges and vertices that allow us to define in Section 3.2, new topological operations on maps. These operations will be useful to prove Theorem 4.1. These two operations are reciprocal and they are interesting since the derivation operation allows to gather in one vertex a subset of vertices of a same color of a map, and the integration operation allows to get back this subset of vertices.

3.1 Order relations in a rooted map

Order relations on half-edges and vertices of a map are introduced in this Section. We show a map traversal algorithm. Half-edges are numbered beginning with the root half-edge and in the order of their appearance in the oriented circuit given by the algorithm (see map C in Figure 3.1). Half-edges are then naturally ordered by their number.

The root half-edge $\tilde{b}$ gets number 0, then the other half-edges of its face, $\bar{\sigma}^*(\tilde{b})$, are numbered. Afterwards while there still are numberless half-edges:

- Among numbered half-edges, the smallest half-edge b is chosen with a numberless opposite half-edge.
- Along the face $\bar{\sigma}^*(\alpha(b))$, beginning with $\alpha(b)$, half-edges are numbered.

Definition 3.1. Order relation on vertices. Let C be a rooted map and s_1, s_2 two vertices of C. The vertex s_1 is smaller than s_2 if the smallest half-edge of s_1 is smaller than the smallest half-edge of s_2.

Vertices are numbered by this order relation. Number 1 is affected to the root vertex and other vertices are numbered in an ascending order (see numbers in bold on map C of Figure 3.1).

A map is *ordered* when its half-edges and vertices are numbered by the algorithms given above.

Definition 3.2. Path and subpath of a map. The *path* of an ordered map C corresponds to the ascending ordered sequence of the half-edges of C, starting from its root half-edge. A *subpath* of C is defined as an increasing subsequence of ordered and successive half-edges of C.

Property 3.3. On the smallest half-edges of a face and of a vertex of an ordered map. *The smallest half-edge b_s of a vertex s different from the root vertex, of an ordered map $C = (\sigma, \alpha, \tilde{b})$, is not the smallest half-edge of its face*

$\bar{\sigma}^*(b_s)$. *The smallest half-edge b_f of a face f different from the exterior face, of an ordered map C, is not the smallest half-edge of its initial vertex.*

Proof: If b_s belongs to the exterior face of C, as s is different from the root vertex, we have $\tilde{b} < b_s$ and b_s cannot be the smallest half-edge of its face.
If b_s does not belong to the exterior face of C, half-edges of face $\bar{\sigma}^*(\alpha(b_s))$ have been numbered before b_s (see the algorithm above). Thus, $\alpha(b_s)$ is smaller than b_s. Then $\bar{\sigma}(\alpha(b_s)) = \sigma(b_s)$, which belongs to vertex s, is smaller than b_s $\diamond$

3.2 Topological and bijective operations on maps

In 3.2.1 we define the *derivation* operation that gather a subset of vertices of a map and the root vertex of a second map, in one vertex. These vertices can be recovered by applying the inverse operation, called *integration* operation and defined in 3.2.2, which uses the order properties on a map to get back all the gathered vertices. These operations are the main tools used in the proof of Theorem 4.1.

Let us denote by $\mathcal{M}_2$ the subset of maps of $\mathcal{M}$ which have at least two distinct vertices.

3.2.1 Derivation of maps

In this section we define a derived map of a pair of maps (C, R) of $\mathcal{M}_2 \times \mathcal{M}$ with respect to certain vertices of C. To derive a pair of maps with respect to vertices $s_1, \ldots, s_m$ of C means to collect these vertices in one vertex while respecting an order and afterwards to glue this vertex to the root vertex of R, as described in definition below.

Figure 3.1. *Derived map with respect to vertices* **3** *and* **5** *of a pair of maps.*

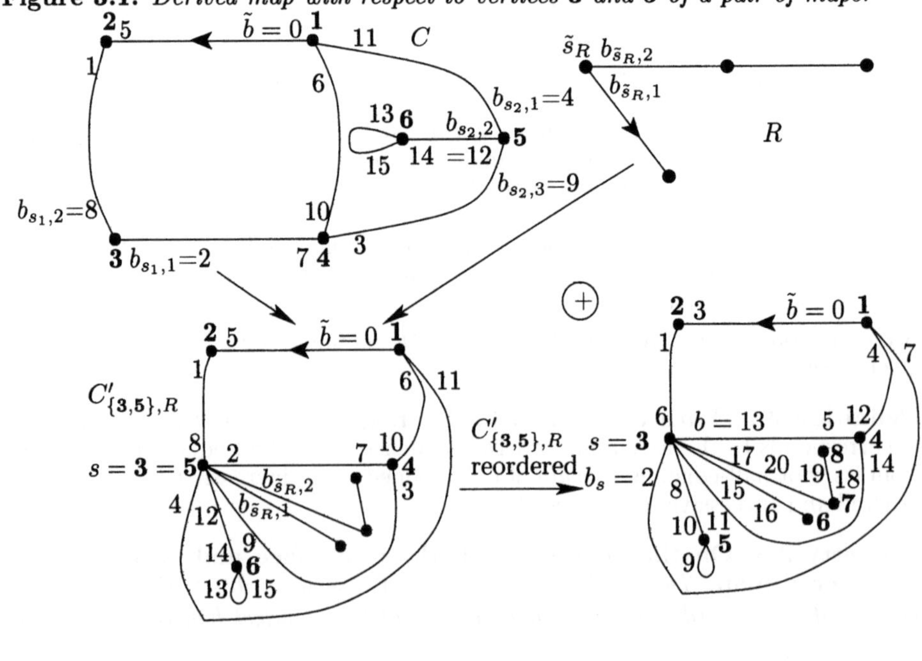

Definition 3.4. Derived map. Let $C = (\sigma, \alpha, \tilde{b})$ be a map of $\mathcal{M}_2$, with root vertex $\tilde{s}_C$ and $R = (\sigma_R, \alpha)$ be a map of $\mathcal{M}$, with root vertex $\tilde{s}_R$ and if $R \neq \{p\}$, let $(b_{\tilde{s}_R,1}, b_{\tilde{s}_R,2}, \ldots, b_{\tilde{s}_R,l_{\tilde{s}_R}})$ be the half-edges of $\tilde{s}_R$ and $b_{\tilde{s}_R,1}$ be the root half-edge of R. Let $\mathscr{S} = \{s_1, \ldots, s_m\}$ be a set of m distinct vertices of C such that $\tilde{s}_C < s_1 < s_2 < \ldots < s_m$. Forall i in $[1,m]$, let $(b_{s_i,1}, \ldots, b_{s_i,l_{s_i}}) = \sigma^*(b_{s_i,1})$, be the half-edges of initial vertex s_i, in which $b_{s_i,1}$ is the smallest half-edge of s_i.

The *derived map* $C'_{\mathscr{S},R} = (\sigma', \alpha, \tilde{b})$ of $(C, \mathscr{S}, R)$ is then the map obtained from C and R after the gathering in a unique vertex s, of the vertices of $\mathscr{S} \bigcup \{\tilde{s}_R\}$ in the following way (see Figure 3.1):

$$s = (\underbrace{b_{s_1,1}, \ldots, b_{s_1,l_{s_1}}}_{s_1}, \underbrace{b_{s_2,1}, \ldots, b_{s_2,l_{s_2}}}_{s_2}, \ldots, \underbrace{b_{s_m,1}, \ldots, b_{s_m,l_{s_m}}}_{s_m}, \underbrace{b_{\tilde{s}_R,1}, \ldots, b_{\tilde{s}_R,l_{\tilde{s}_R}}}_{\tilde{s}_R}) =$$

$\sigma'^*(b_{s_1,1})$. In terms of permutation, it means: $\sigma' = \tau_{1R}\tau_{1m}\ldots\tau_{12}\sigma = \gamma\sigma$ with $\tau_{1i} = (b_{s_1,1}b_{s_i,1})$, $\tau_{1R} = (b_{s_1,1}b_{\tilde{s}_R,1})$ and $\gamma = (b_{s_1,1}\ldots b_{s_m,1}b_{\tilde{s}_R,1})$.

Property 3.5. Orders of $C'_{\mathscr{S},R}$

1. *In the ordered map $C'_{\mathscr{S},R}$, if $R \neq \{p\}$, $b_{\tilde{s}_R,1}$ is the smallest half-edge among the half-edges of R (see Figure 3.1 in which $b_{\tilde{s}_R,1} = 15$ and $b_{\tilde{s}_R,l_{\tilde{s}_R}} = b_{\tilde{s}_R,2} =$ 17).*

2. *The subpaths from $\tilde{b}$ to $\alpha(b_{s_1,l_{s_1}})$ are identical in C and $C'_{\mathscr{S},R}$ (see Figure 3.1, $\alpha(b_{s_1,l_{s_1}}) = 1$ in C and in $C'_{\{3,5\},R}$).*

Proof :

1. By construction, R is recovered if in $C'_{\mathscr{S},R}$, the subset of half-edges belonging also to R, i.e. $\{b_{\tilde{s}_R,1}, \ldots, b_{\tilde{s}_R,l_{\tilde{s}_R}}\}$, is unglued from vertex s. Thus in the traversal of $C'_{\mathscr{S},R}$, starting from its root half-edge, $\tilde{b}$, to reach any half-edge of R, one has to pass through s. It implies that there exists i, $1 \leq i \leq l_{\tilde{s}_R}$ such that $b_{\tilde{s}_R,i}$ is the smallest half-edge of the half-edges of R in $C'_{\mathscr{S},R}$. If $l_{\tilde{s}_R} > 1$, let us prove that $b_{\tilde{s}_R,1}$ is the smallest half-edge of the half-edges of R in $C'_{\mathscr{S},R}$.
$b_{\tilde{s}_R,i}$ cannot be the smallest half-edge of its face, $\overline{\sigma'}^*(b_{\tilde{s}_R,i})$, otherwise $\alpha(b_{\tilde{s}_R,i})$, which belongs to R and which has been previously numbered to the face $\overline{\sigma'}^*(b_{\tilde{s}_R,i})$, is smaller than $b_{\tilde{s}_R,i}$.
If $i > 1$, $b_{\tilde{s}_R,i} = \sigma'(b_{\tilde{s}_R,i-1}) = \overline{\sigma'}(\alpha(b_{\tilde{s}_R,i-1}))$, so that $\alpha(b_{\tilde{s}_R,i-1})$, which belongs to R, is smaller than $b_{\tilde{s}_R,i}$ (as $b_{\tilde{s}_R,i}$ is not the smallest half-edge of its face), which contradicts definition of $b_{\tilde{s}_R,i}$. Thus $i = 1$.

2. In C, $s_1 < s_2 < \ldots < s_m$ implies that $b_{s_1,1} < b_{s_2,1} < \ldots < b_{s_m,1}$.
Furthermore forall i in $[1,m]$, $\bar{\sigma}(\alpha(b_{s_i,l_{s_i}})) = b_{s_i,1}$ and $b_{s_i,1}$ is not the smallest half-edge of its face (see Property 3.3), so that $\alpha(b_{s_i,l_{s_i}})$ precedes $b_{s_i,1}$ in the ordered map C.
One then has in C, $\tilde{b} < \alpha(b_{s_1,l_{s_1}}) < b_{s_1,1} < \alpha(b_{s_2,l_{s_2}}) < b_{s_2,1} < \ldots < \alpha(b_{s_m,l_{s_m}}) < b_{s_m,1}$.
Thus in C, the subpath from $\tilde{b}$ to $\alpha(b_{s_1,l_{s_1}})$ does not go through any half-edge $\alpha(b_{s_i,l_{s_i}})$.
If one proves that in $C'_{\mathscr{S},R}$, the subpath from $\tilde{b}$ to $\alpha(b_{s_1,l_{s_1}})$ does not go

through $\alpha(b_{\tilde{s}_R,l_{\tilde{s}_R}})$, then one will conclude from what precedes that in $C'_{\mathscr{S},R}$, the subpath from $\tilde{b}$ to $\alpha(b_{s_1,l_{s_1}})$ does not go through any of the half-edge $\alpha(b_{s_i,l_{s_i}})$. It means that the subpath from $\tilde{b}$ to $\alpha(b_{s_1,l_{s_1}})$ in $C'_{\mathscr{S},R}$ is unchanged as $\overline{\sigma'}(a) = \begin{cases} b_{s_{i+1},1} & \text{if } a = \alpha(b_{s_i,l_{s_i}}) \ \forall 1 \le i < m \\ b_{\tilde{s}_R,1} & \text{if } a = \alpha(b_{s_m,l_{s_m}}) \\ b_{s_1,1} & \text{if } a = \alpha(b_{\tilde{s}_R,l_{\tilde{s}_R}}) \\ \bar{\sigma}(a) & \text{if } a \in C, a \neq b_{s_i,l_{s_i}} \ \forall 1 \le i \le m \\ \overline{\sigma_R}(a) & \text{if } a \in R, a \neq b_{\tilde{s}_R,l_{\tilde{s}_R}} \end{cases}$.

Let us then prove that the subpath of $C'_{\mathscr{S},R}$ from $\tilde{b}$ to $\alpha(b_{s_1,l_{s_1}})$ does not go through the half-edge $\alpha(b_{\tilde{s}_R,l_{\tilde{s}_R}})$.
Since $\overline{\sigma'}(\alpha(b_{s_m,l_{s_m}})) = b_{\tilde{s}_R,1}$ and $b_{\tilde{s}_R,1}$ is not the smallest half-edge of its face (see item 1 of this proof), $\alpha(b_{s_m,l_{s_m}})$ precedes $b_{\tilde{s}_R,1}$ in the path of $C'_{\mathscr{S},R}$.
Furthermore, from Property 3.5.1, $b_{\tilde{s}_R,1} < \alpha(b_{\tilde{s}_R,l_{\tilde{s}_R}})$ as $\alpha(b_{\tilde{s}_R,l_{\tilde{s}_R}}) \in R$.
Thus in $C'_{\mathscr{S},R}$, $\alpha(b_{s_1,l_{s_1}}) < \alpha(b_{\tilde{s}_R,l_{\tilde{s}_R}})$ ◇

The following technical lemma gives us the way to recover vertices $s_1, \ldots, s_m, \tilde{s}_R$, which compose vertex s, as will be shown in Lemma 3.10.
Notations of Definition 3.4 are used here.

Lemma 3.6. *In $C'_{\mathscr{S},R}$, $\sigma'(b_{s_1,l_{s_1}}) = \begin{cases} b_{s_2,1} & \textit{if } m > 1 \\ b_{\tilde{s}_R,1} & \textit{if } R \neq \{p\} \textit{ and } m = 1 \\ b_{s_1,1} & \textit{if } R = \{p\} \textit{ and } m = 1 \end{cases}$ is the smallest half-edge among half-edges of vertex s.*

Proof :

1. If $R = \{p\}$ and $m = 1$ then $C = C'_{\mathscr{S},R}$, $s = s_1$ and thus, $\sigma'(b_{s_1,l_{s_1}}) = b_{s_1,1}$ is the smallest half-edge among the half-edges of s.

2. Let us assume that $R \neq \{p\}$ or $m \neq 1$. Let $\hat{b}$ be the smallest half-edge of face $\overline{\sigma}^*(b_{s_1,1})$ in C.

   (a) In C, $b_{s_1,1}$ is the smallest half-edge of vertex s_1. From Property 3.3, as $s_1 \neq \tilde{s}_C$, $b_{s_1,1}$ is not the smallest half-edge of its face. Thus, there exists $j > 0$ such that $\bar{\sigma}^j(\hat{b}) = b_{s_1,1}$.

   (b) Let us prove at last Lemma 3.6, that is: $\sigma'(b_{s_1,l_{s_1}})$ is the smallest half-edge of s in $C'_{\mathscr{S},R}$ (see Figure 3.1, $\sigma'(b_{s_1,l_{s_1}}) = b_s$ in $C'_{\{3,5\},R}$).
   From Property 3.5.2, one knows that the subpath from $\tilde{b}$ to $\alpha(b_{s_1,l_{s_1}})$ in $C'_{\mathscr{S},R}$ is identical to the one in C. Thus $\alpha(b_{s_1,l_{s_1}}) = \bar{\sigma}^{j-1}(\hat{b}) = \overline{\sigma'}^{j-1}(\hat{b})$.
   Furthermore, in C, the subpath from $\tilde{b}$ to $\alpha(b_{s_1,l_{s_1}})$ does not go through s as $b_{s_1,1}$ is the smallest half-edge of the half-edges of s in C and $\alpha(b_{s_1,l_{s_1}})$ is smaller than $b_{s_1,1}$ in C (see the proof of Property 3.5.2). It is the same in $C'_{\mathscr{S},R}$.
   Thus $\sigma'(b_{s_1,l_{s_1}}) = \overline{\sigma'}(\alpha(b_{s_1,l_{s_1}}))$ is the smallest half-edge of s in $C'_{\mathscr{S},R}$ ◇

$b_{s_1,1}$ is the smallest half-edge of $\mathscr{S}$ in C. Its predecessor in the path of C, is the half-edge $\alpha(b_{s_1,l_{s_1}})$ as $b_{s_1,1}$ is not the smallest half-edge of its face (see Property

3.3). In map $C'_{\mathcal{S},R}$, built from C and R by gluing together vertices of C and the root vertex of R in one vertex s, the successor of $\alpha(b_{s_1,l_{s_1}})$ becomes $b_{s_2,1}$, which then is the smallest half-edge of s in $C'_{\mathcal{S},R}$ reordered. If $b_{s_1,1}$ has been marked, one gets back thus vertex s_1 which is detached from s, then recursively vertices $s_2, \ldots, s_m$. Thus the pair of initial maps can be recovered from its derived map. A formal definition of this inverse operation, which will be called integration, is given in the next Section.

3.2.2 Integration of a map

A topological operation of opening of a vertex into two vertices is introduced in order to define the integration of a map, which consists in the splitting of a vertex into several vertices. It will then be seen that to recover a pair of maps (C, R) and the subset of vertices of C if its derived map is known, one has to integrate this last map.

Definition 3.7. Topological operation of opening of a map with respect to a half-edge. Let $C = (\sigma, \alpha, \tilde{b})$ be a map and b a half-edge of C. Let b_s be the smallest half-edge of a vertex $s = \sigma^*(b)$. The *opening* of C with respect to b consists in the splitting of the vertex s into two vertices s_1 and s_2 in the following way:

$$s = (b, \ldots, \sigma^{-1}(b_s), b_s, \ldots, \sigma^{-1}(b)) \rightarrow s_1 = (b, \ldots, \sigma^{-1}(b_s))$$

and $s_2 = (b_s, \ldots, \sigma^{-1}(b))$.
It means that the following permutation $\widehat{\sigma_b}$ is applied to the half-edges of C: $\widehat{\sigma_b} = \tau\sigma$ with $\tau = (bb_s)$.
The result of the opening of C with respect to b is a map or a pair of maps:

(i) If $b_s \neq b$ and if the group generated by $(\widehat{\sigma_b}, \alpha, \tilde{b})$ acts transitively on the set of half-edges of C (i.e. $(\widehat{\sigma_b}, \alpha, \tilde{b})$ generates a map and not two disconnected maps), then a new map $\widehat{C_b} = (\widehat{\sigma_b}, \alpha, \tilde{b})$ is defined.

(ii) Otherwise a pair of maps $(\widehat{C_b}, D)$, $\widehat{C_b} = (\widehat{\sigma_b}, \alpha, \tilde{b})$, $D = (\widehat{\sigma_b}, \alpha, b_s)$, is obtained, D being the map $\{p\}$ if $b_s = b$.

Remark 3.8. If $s \neq \tilde{s}$, $\widehat{C_b} \in \mathcal{M}_2$.

Figure 3.2. *Integration of map $C'_{\{3,5\},R}$ reordered of Figure 3.1 with respect to the half-edge $b = 13$: a pair of maps $(\widehat{C_b}, D)$ of $\mathcal{M}_2 \times \mathcal{M}$ is obtained.*

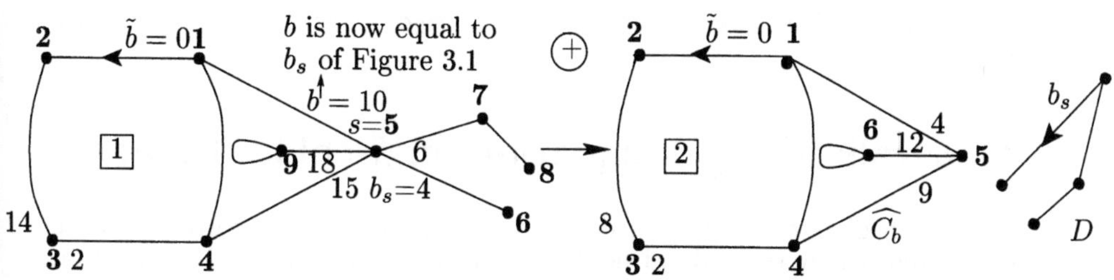

Definition 3.9. Integration of a map. Let $C = (\sigma, \alpha, \tilde{b})$ be a map of $\mathcal{M}_2$, of root vertex $\tilde{s}$. Let $s \neq \tilde{s}$ be a vertex of C and $b \in s$. Let $\mathscr{S} = \emptyset$.

It will be said that a map C is *integrated* with respect to a half-edge b, when the operation of the opening of C is recursively applied until case (ii) of Definition 3.7 is reached, that is:

- Let b_s be the smallest half-edge of $\sigma^*(b)$. C is opened with respect to b (see Definition 3.7).
- If this operation gives a map $\widehat{C_b}$ (see Figure 3.2, drawing $\boxed{1}$), the vertex obtained after the opening, incident to b (the other obtained vertex is incident to b_s), is added to $\mathscr{S}$ and the opening operation starts again with $C \leftarrow \widehat{C_b}$ and $b \leftarrow b_s$.
- Otherwise, a pair of maps of $\mathcal{M}_2 \times \mathcal{M}$, $(\widehat{C_b}, D)$ is obtained (see Figure 3.2, drawing $\boxed{2}$), and also a set of vertices of $\widehat{C_b}$, $\mathscr{S}$ with the added vertex of $\widehat{C_b}$ which was split from the root vertex of D (vertex of $\widehat{C_b}$ to which b belongs).

Lemma 3.10. *Let $C'_{\mathscr{S},R}$, be the derived map of a pair of maps (C, R) of $\mathcal{M}_2 \times \mathcal{M}$ with respect to a set of vertices $\mathscr{S}$ of C. Let us denote by b ($= b_{s_1,1}$ of Definition 3.4) the smallest half-edge of $\mathscr{S}$ in C. Integration of $C'_{\mathscr{S},R}$ with respect to b gives $(C, \mathscr{S}, R)$.*

Proof : With notations of Definitions 3.4 and 3.9, the map $C'_{\mathscr{S},R} = (\sigma', \alpha, \tilde{b})$ is integrated with respect to the half-edge $b_{s_1,1}$: $b = b_{s_1,1}$ and $b_s = b_{s_2,1}$ (from Lemma 3.6). The opening operation of vertex s unglues vertex s_1 from s, and gives the map $\widehat{(C'_{\mathscr{S},R})}_b = (\widehat{\sigma'}_b, \alpha, \tilde{b})$:

$$s = (\underset{\substack{\uparrow \\ b}}{b_{s_1,1}}, \ldots, b_{s_1,l_{s_1}}, \underset{\substack{\uparrow \\ b_s}}{b_{s_2,1}}, \ldots, b_{s_2,l_{s_2}}, \ldots, b_{s_m,1}, \ldots, b_{s_m,l_{s_m}}, b_{\tilde{s}_R,1}, \ldots, b_{\tilde{s}_R,l_{\tilde{s}_R}})$$

Two vertices are obtained, a vertex $s_1 = (b_{s_1,1}, \ldots, b_{s_1,l_{s_1}})$ and a vertex $s = (b_{s_2,1}, \ldots, b_{s_2,l_{s_2}}, \ldots, b_{s_m,1}, \ldots, b_{s_m,l_{s_m}}, b_{\tilde{s}_R,1}, \ldots, b_{\tilde{s}_R,l_{\tilde{s}_R}})$. One has: $\widehat{\sigma'}_b = \tau_{12}\sigma'$. Thus, $(\sigma_2 = \widehat{\sigma'}_b, \alpha, \tilde{b}) = C'_{\{s_2,\ldots,s_m\},R}$ and $\mathscr{S} = \{s_1\}$. One successively obtains maps $C'_{\{s_i,\ldots,s_m\},R} = (\sigma_i = \tau_{i-1i}\sigma_{i-1}, \alpha, \tilde{b})$ for $\tau_{i-1i} = (b_{s_{i-1},1} b_{s_i,1})$, and $\mathscr{S} = \{s_1, \ldots, s_{i-1}\}$, with $3 \leq i \leq m$. Applying for the last time to $C'_{\{s_m\},R}$ the topological operation of opening of $s = (b_{s_m,1}, \ldots, b_{s_m,l_{s_m}}, b_{\tilde{s}_R,1}, \ldots, b_{\tilde{s}_R,l_{\tilde{s}_R}})$, two disconnected maps, $C = (\sigma, \alpha, \tilde{b})$ and $R = (\sigma, \alpha, b_{\tilde{s}_R,1})$, are recovered and also $\mathscr{S} = \{s_1, \ldots, s_m\}$. One has: $\sigma = \tau_{Rm}\tau_{mm-1}\ldots\tau_{12}\sigma' = \delta\sigma'$ with $\delta = \gamma^{-1}$ (see Definition 3.4) ◇

4 Generalized Dyck equation on n-colored maps

The well-known Dyck equation on trees, is based on a one-to-one correspondence between rooted planar trees $\mathcal{A}$, without the one vertex tree, and $\mathcal{A}^2$. In Section

4.1, an equation generalizing the Dyck equation to n-colored rooted maps of arbitrary genus, is given. This equation is equivalent to an equation on sets which is determined. A proof of the equation on sets is given in 4.2. Topological operations introduced in Section 3.2 will be used for this proof.

4.1 Generalized Dyck equations

The equation on sets is given as a bijection between the set of n-colored rooted maps of arbitrary genus, $\mathcal{M}_{n,i}$, and the set of pairs of maps of $\bigcup_{j=1,j\neq i}^{n} \mathcal{M}_{n,j}\times\mathcal{M}_{n,i}$, where in one of these maps a subset (possibly empty) of its vertices of color i is selected. Equation (1) is then a translation with generating functions of this bijection.

For any map M of $\mathcal{M}_{n,i}$, let us denote by $\mathcal{V}_{i,M}$ the set of vertices of color i of M and $\mathcal{P}(\mathcal{V}_{i,M})$ the set of all subsets of $\mathcal{V}_{i,M}$.

Theorem 4.1. $\mathcal{M}_{n,i} \leftrightarrow \{p_i\} \bigcup_{j=1,j\neq i}^{n} \left[\bigcup_{M\in\mathcal{M}_{n,j}} M \times \mathcal{P}(\mathcal{V}_{i,M}) \right] \times \mathcal{M}_{n,i}$

The proof of this theorem is given in Section 4.2.

Let $\mathcal{I}_n$ be the set $\{1,\ldots,n\}$. Let $M_{n,i}$ (resp. M_i), $i \in \mathcal{I}_n$, be the generating function of maps of $\mathcal{M}_{n,i}$, enumerated by vertices (resp. vertices of color $j \in \mathcal{I}_n$) and half-edges whose initial vertex is of color $j \in \mathcal{I}_n$. Let c_i, $i \in \mathcal{I}_n$, be the variable whose exponent represents the number of half-edges with initial vertex of color i. Let y be the variable whose exponent represents the number of vertices of the map. Henceforth we will write $M_{n,i}$ for $M_{n,i}(y, c_1, \ldots, c_n)$ and $M_i(u)$ for $M_i(u; c_1, \ldots, c_n)$ with $u = (u_j)_{1\le j\le n}$.

Corollary 4.2. Generalized Dyck equation:

$$M_{n,i} = y + c_i M_{n,i} \sum_{j=1,j\neq i}^{n} c_j M_j(v) \quad \textit{with } v = (v_j)_{1\le j\le n} = (y + \delta_{ij})_{1\le j\le n}. \qquad (1)$$

4.2 Proof of Theorem 4.1

A bijection between maps of $\mathcal{M}_{n,i}$, different from the one vertex map and $\mathcal{P}_{n,i} \times \mathcal{M}_{n,i}$, with $\mathcal{P}_{n,i} = \bigcup_{j=1,j\neq i}^{n} \left[\bigcup_{M\in\mathcal{M}_{n,j}} M \times \mathcal{P}(\mathcal{V}_{i,M}) \right]$, is described, which means between maps of $\mathcal{M}_{n,i}$ and maps of $\mathcal{I}_{n,i}$ in which for each map I of $\mathcal{I}_{n,i}$, a set $\mathscr{S}_i$ of vertices of color i of $Left(I)$, has been selected. As a matter of fact $\mathcal{I}_{n,i}$ is in one-to-one correspondence with $\bigcup_{j=1,j\neq i}^{n} \mathcal{M}_{n,j} \times \mathcal{M}_{n,i}$, as to each map I of $\mathcal{I}_{n,i}$, a pair of maps of $\mathcal{M}_{n,j} \times \mathcal{M}_{n,i}$, $(Left(I), Right(I))$, can be associated, and $j \neq i$ as a half-edge of a n-colored map is incident to two vertices of distinct colors. Furthermore the set of pairs $(Left(I), \mathscr{S}_i)$ is the set $\mathcal{P}_{n,i}$.

Lemma 4.3. Bijection of theorem 4.1. *There is a one-to-one correspondence between $\mathcal{M}_{n,i}$ and the set of pairs $(I, \mathscr{S}_i)$, in which I is a map of $\mathcal{I}_{n,i}$ and $\mathscr{S}_i$ a set of vertices of color i of $Left(I)$, possibly empty.*

Proof : Integration of a map with respect to a half-edge b of initial vertex of color i allows to recover a pair of maps as well as a set of vertices of color i of one of the obtained maps. Thus when a derived map I' is obtained, to have the possibility of going back, one has to memorize the half-edge b. To do this, if the root vertex of I' is only incident to the root half-edge and is of color i, then it is sufficient to glue the root half-edge just before b in order to obtain a map M of $\mathcal{M}_{n,i}$.
Starting with a map I of $\mathcal{I}_{n,i}$ (see Figure 4.1), in which a set $\mathscr{S}_i$ of vertices of color i of $Left(I)$ has been selected, to obtain a map $M = (\sigma_M, \alpha, \tilde{b})$ of $\mathcal{M}_{n,i}$, one has to (see Figure 4.2):

- apply the derivation operation to $(I_L, \mathscr{S}_i, Right(I)) = I'$, with I_L, the map I without $Right(I)$, with the same root half-edge than I,

- if $\mathscr{S}_i \neq \emptyset$, to glue the root vertex of I' in the following way $\sigma_M(\tilde{b}) = b_{s_1,1}$, with $b_{s_1,1}$ is the smallest half-edge of $\mathscr{S}_i$ in I ◇

Figure 4.1. *Map of $\mathcal{I}_{n,i}$*

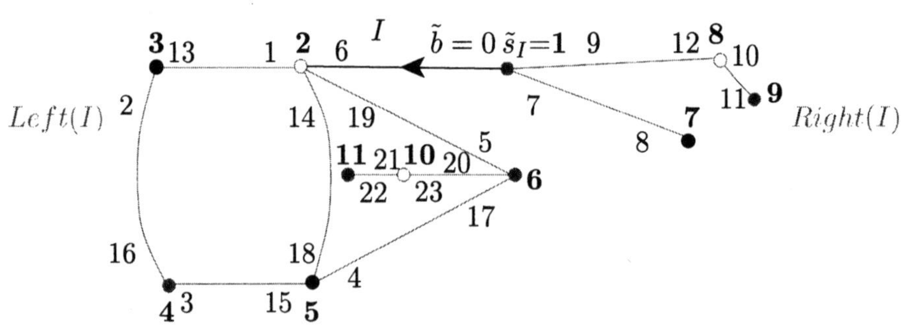

Figure 4.2. *A map $I' = (I_L)'_{\{\mathbf{4,6}\},Right(I)}$ and a map M of $\mathcal{M}_{n,i}$ are obtained (I' and M have not been reordered)*

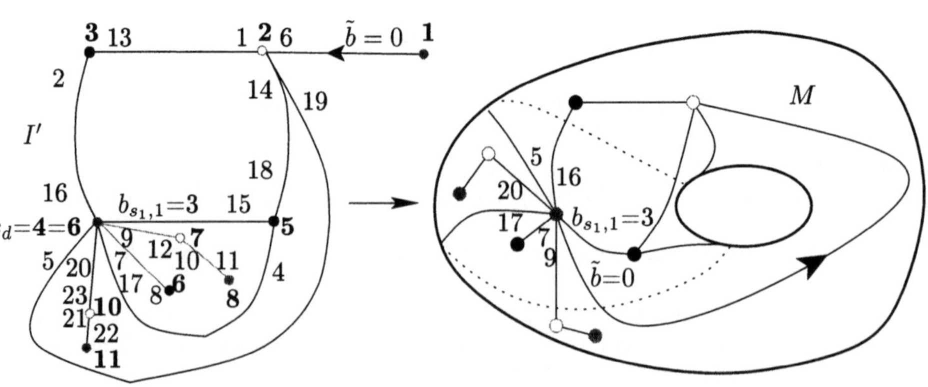

5 Bijection between maps of arbitrary genus and multilabel n-colored trees

The operation that allowed to prove Theorem 4.1 transforms a map of $\mathcal{M}_{n,i}$ into a map with a root bridge in which a subset of its vertices of color i has been selected. If this operation is iterated on the successive submaps incident to the two vertices incident to the bridge, and if the subset of vertices associated with each map is labelled (one distinct label for each subset), the initial map is transformed into a n-colored tree whose vertices can be labelled with several labels, following repartition rules. One then obtains what we will call a multilabel n-colored tree.
In Section 5.1, we give the definition of a multilabel n-colored tree and in Section 5.2, we give the Theorem 5.3 explaining the one-to-one correspondence between maps of $\mathcal{M}_{n,i}$ and multilabel n-colored trees. This bijection leads to a coding of maps by words of a language, as shown in Section 5.3.

5.1 Multilabel n-colored trees

We give definitions of a multilabel n-colored tree. Order relations given in Section 3.1 are applied to multilabel n-colored trees. An order on half-edges and vertices is thus established in a classical in-depth descent of the tree. Let us notice that the smallest half-edge of a vertex is also its left son in the tree structure, since a tree has only one face.

Figure 5.1. *Illustration of Definition 5.1*

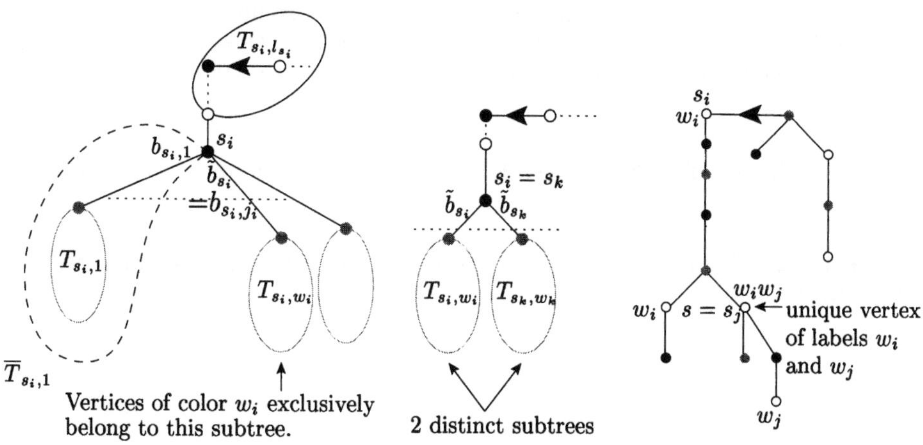

Illustration of the rule 2a Illustration of the rule 2b Illustration of the rule 3

Definition 5.1. Multilabel n-colored tree. Let $T = (\sigma, \alpha, \tilde{b})$ be a n-colored rooted tree. Let $\mathcal{W} = \{w_1, \ldots, w_p\}$ be a set of p distinct labels, eventually empty ($p \geq 0$). Each vertex of T can have 0 to p labels in $\mathcal{W}$. Forall i in $[1,p]$, let us denote by s_i, the smallest vertex of T of label w_i.

T is a *multilabel n-colored tree* if T complies with the following rules (see Figure 5.1):

1. each label of W is assigned to at least two distinct vertices of T and the vertices of a same label must be of a same color;

2. let $(b_{s_i,1}, \ldots, b_{s_i,l_{s_i}}) = \sigma^*(b_{s_i,1})$ be the half-edges of initial vertex s_i, where $b_{s_i,1}$ is the smallest half-edge of $\sigma^*(b_{s_i,1})$, i.e. the left son of s_i. $b_{s_i,j}, 1 \leq j < l_{s_i}$ are the half-edges, sons of s_i, and $b_{s_i,l_{s_i}}$ is the half-edge which goes up towards the father of s_i. Let $T_{s_i,j}$, be the subtree of T incident to the final vertex of $b_{s_i,j}$, rooted in $\bar{\sigma}(b_{s_i,j})$ and $\overline{T}_{s_i,j}$, the tree composed of $T_{s_i,j}$ and of the half-edge $b_{s_i,j}$ which is its root half-edge. Then:

   (a) there is a single j_i such that in T, w_i is assigned to s_i and exclusively vertices of T_{s_i,j_i}. Let us denote this subtree by $T_{s_i,w_i} = T_{s_i,j_i}$, its root half-edge by $\tilde{b}_{s_i} = b_{s_i,j_i}$ and $\overline{T}_{s_i,w_i} = \overline{T}_{s_i,j_i}$;

   (b) forall k in $[1,p]$, $k \neq i$, if $s_i = s_k$ then $\overline{T}_{s_i,w_i} \cap \overline{T}_{s_k,w_k} = \emptyset$.

3. For all distinct labels w_i and w_j, if there is a vertex s of labels w_i and w_j where s_i is smaller than s_j, then $s = s_j$ and s is the only vertex of label w_j which is also of label w_i.

We will be say that two multilabel n-colored trees are isomorphous if one can be obtained from the other by a permutation on its labels. A class of isomorphism of multilabel n-colored trees will simply be called multilabel n-colored tree.

Let $\mathcal{T}_{n,i}$ be the set of multilabel n-colored trees, with a root vertex of color i.

Remark 5.2. If T is a multilabel n-colored tree with m vertices and p distinct labels, then $p < m$.

5.2 Bijection between $\mathcal{M}_{n,i}$ and $\mathcal{T}_{n,i}$

Theorem 5.3. *The set of rooted n-colored maps with root vertex of color i and p edges is in bijection with the family of multilabel n-colored trees with root vertex of color i and p edges.*

Figure 5.2. *The multilabel n-colored tree associated with the map M of Figure 4.2*

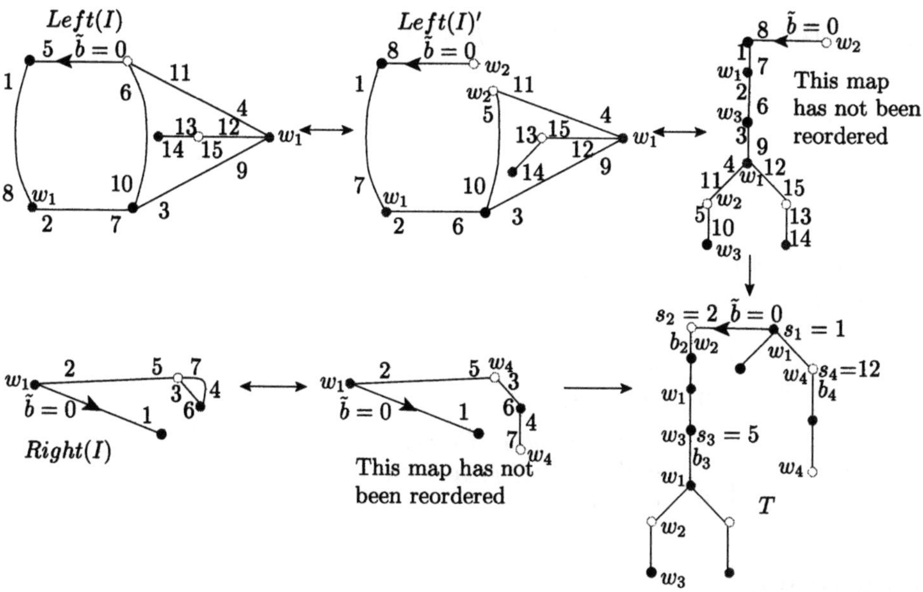

5.3 Application: a language coding maps of arbitrary genus

In this Section, we present a language coding rooted n-colored maps. The equation defining this language is a generalization of the well-known equation on Dyck words. In fact this language codes multilabel n-colored trees and thus by bijection rooted n-colored maps.

In order to clarify the significance of each letter of the alphabet of the language that we present, we need to give a definition.

Definition 5.4. Twin labels. Two labels w and w' of a tree of $\mathcal{T}_{n,i}$ are *twin* if there is a vertex of T labelled by these two labels or if there is a subsequence of labels of T, $w_1 = w, w_2, \ldots, w_n = w'$ such that forall j in $[1, n[$, w_j and w_{j+1} label the same vertex. One then defines classes of equivalence of labels, where two labels are in the same class if they are twin.

Let us denote by e (resp. $\bar{e}$) the variable coding a half-edge, whose opposite half-edge is not coded (resp. is coded), v_i the variable coding a vertex of color i in case of maps, and in case of multilabel n-colored trees a vertex of color i without any label or the smallest vertex of color i of a same or twin label. Let y_j, $j \geq 1$, be the variable coding a vertex of label w_j (with $w_j \neq w_k$ if $j \neq k$) of a multilabel n-colored tree. In a rooted n-colored map, y_j, codes the half-edges belonging to a subset of the set of half-edges of initial vertex s_j, for a given vertex s_j of arity strictly superior to 1 (s_j can be equal to s_k if $j \neq k$).
Let us denote by $\vec{v} = (v_j)_{1 \leq j \leq n}$ and $\vec{v}_{q,p} = (v_j + y_p \delta_{jq})_{1 \leq j \leq n}$.

Theorem 5.5. *The set of rooted n-colored maps with root vertex of color i is coded by the language $L_{\infty,i} = \lim\limits_{p \to \infty} L_{p,i}$, where $L_{p,i}$ represents the language coding maps* *of $\mathcal{M}_{n,i}$ with at most p edges and is defined in the following way:*

$$\begin{aligned} L_{p,i}(\vec{v}, y_1, \ldots, y_p, e, \bar{e}) &= v_i + e \sum_{j=1, j\neq i}^{n} L_{p-1,j}(\vec{v}_{i,p}, y_1, \ldots, y_{p-1}, e, \bar{e}) \qquad (2) \\ & \qquad \bar{e}\ L_{p-1,i}(\vec{v}, y_1, \ldots, y_{p-1}, e, \bar{e})\ (1 - \epsilon_p + y_p \epsilon_p)\delta_{e,p} \\ L_0(\vec{v}, e, \bar{e}) &= v_i \qquad (3) \end{aligned}$$

where for every word m_1 of $L_{p-1,j}(\vec{v}_{i,p}, y_1, \ldots, y_{p-1}, e, \bar{e})$ and m_2 of $L_{p-1,i}(\vec{v}, y_1, \ldots, y_{n-1}, e, \bar{e})$:

$$\epsilon_p = \begin{cases} 1 & \text{if } y_p \in m_1 \\ 0 & \text{otherwise} \end{cases}$$

$$\delta_{e,p} = \begin{cases} 1 & \text{if (the number of occurrences of } e \text{ in } em_1\bar{e}m_2) \leq p \\ & \text{and } \nexists 1 \leq k \leq p / y_k \in m_1 \text{ and } y_k \in m_2 \\ 0 & \text{otherwise} \end{cases}$$

References

[1] D. Arquès, Relations fonctionnelles et dénombrement des cartes pointées sur le tore, J. Combin. Theory Ser. B 43 (1987) 253-274.

[2] Arquès, D. and Béraud, J.F., "Rooted maps on orientable surfaces, Riccati's equation and continued fractions." *Discrete Math.* **215**(2000), 1–12.

[3] D. Arquès and A. Giorgetti, "Enumération des cartes pointées de genre quelcquonque en fonction des nombres de sommets et de faces", J. Combin. Theory Ser. B 77 (1999) 1-24.

[4] Arquès, D. and Micheli, A., "Enumeration of multi-colored rooted maps", *Actes du Lacim 2000* **27**(2000), 31-42.

[5] Arquès, D. and Micheli, A., "Enumeration of multi-colored rooted maps", *Discrete Math.*, to be published.

[6] E.A. Bender and E.R. Canfield, "The number of rooted maps on an orientable surface", J. Combin. Theory Ser. B, 53 (1991), 293-299.

[7] Cori, R., "Un code pour les graphes planaires et ses applications", *Astérisque* **27**(1975).

[8] Cori, R. and A. Machi, "Maps, hypermaps and their automorphisms: a survey I, II, III", Expo. Math. **10** (1992), 403-427, 429-447, 449-467.

[9] R. Cori et B. Vauquelin, "Planar maps are well labelled trees", Canad. J. Math. **33 (5)** (1981), 1023-1042.

[10] P. Flajolet, "Combinatorial aspects of continued fractions", Discrete Math. **32** (1980) 125-161.

[11] D.M. Jackson et T.I. Visentin, "A character theoretic approach to embeddings of rooted maps in an orientable surface of given genus", Trans. Amer. Math. Soc. **322** (1990), 343-363.

[12] A. Jacques, "Sur le genre d'une paire de substitutions", C. R. Acad. Sci. Paris **267** (1968), 625-627.

[13] M. Marcus et B. Vauquelin, "Un codage des cartes de genre quelconque", Actes de FPSAC'95, Marne-La-Vallée (1995), 399-415.

[14] P. Ossona de Mendez et P. Rosenstiehl, "Connected permutations and hypermaps", les cahiers du CAMS 183, available at http://www.ehess.fr/centres/cams/papers.

[15] W.T. Tutte, "A census of planar maps", Canad. J. Math. 15 (1963) 249-271.

[16] W.T. Tutte, "On the enumeration of planar maps", Bull. Amer. Math. Soc. 74 (1968) 64-74.

[17] T.R.S. Walsh et A.B. Lehman, "Counting rooted maps by genus. I", J. Combin. Theory Ser. B **13** (1972), 192-218.

Didier Arquès

Université de Marne-La-Vallée
5 Boulevard Descartes, Champs sur Marne
77454 Marne la Vallée Cedex 2
arques@univ-mlv.fr

Anne Micheli

Université de Marne-La-Vallée
5 Boulevard Descartes, Champs sur Marne
77454 Marne la Vallée Cedex 2
micheli@univ-mlv.fr

Trends in Mathematics, © 2002 Birkhäuser Verlag Basel/Switzerland

Limit Laws for Basic Parameters of Lattice Paths with Unbounded Jumps

Cyril Banderier

ABSTRACT: *This paper establishes the asymptotics of a class of random walks on* $\mathbb{N}$ *with regular but unbounded jumps and studies several basic parameters (returns to zero for meanders, bridges, excursions, final altitude for meanders). All these results are generic (obtained by the kernel method for the combinatorial part and by singularity analysis for the asymptotic part).*

This paper completes the article [3] which was only dealing with the combinatorics (enumeration and bijections) of walks with unbounded jumps (the so-called "factorial walks"), which play an important rôle for uniform random generation of some combinatorial objects. We fully parallelize the analytical approach from [4] which was dealing with walks with bounded jumps only.

1 Introduction

Our main motivation for analyzing a class of walks with unbounded jumps comes from the fact that several classes of combinatorial objects can be generated via the so-called "generating trees". Enumerating these trees (and predicting the number of nodes at a given depth) allows uniform random generation. The concept of generating trees has been used from various points of view and has been introduced in the literature by Chung, Graham, Hoggatt and Kleiman [11] to examine the reduced Baxter permutations. This technique has been successively applied to other classes of permutations and the main references on the subject are due to West [14, 25, 26], then followed by the Florentine school [6, 7, 16, 19, 20, 22, 23] and other authors [3, 12, 18]. A generating tree is a rooted labeled tree (labels are integers) with the property that if v_1 and v_2 are any two nodes with the same label then, for each label ℓ, v_1 and v_2 have exactly the same number of children with label ℓ. To specify a generating tree it therefore suffices to specify: 1) the label of the root; 2) a set of rules explaining how to derive from the label of a parent the labels of all of its children. Points 1) and 2) define what we call a *rewriting rule*. Any random walk in the generating tree can also be seen as a lattice path (random walk on the integers, with an infinite number of possible jumps). The regularity of the rewriting rules determines the "solvability" (combinatorially speaking) of the corresponding random walk process.

Few years ago, Pinzani and al. [6] exhibited several cases of factorial-like rewriting rules for which the generating functions were algebraic. This was calling for a general solution of the factorial-like rewriting rules case. This problem was solved in [3], by establishing a link between the generating trees and families of lattice paths with unbounded jumps (with respect to a given rewriting rule, the number of nodes with label k at depth n in the tree is the number of walks of length n ending at altitude k); then, the corresponding generating functions for walks are always algebraic and are made explicit via the kernel method (we give more details in Section 3). The asymptotic properties of such walks were remaining open.

The article [4] and an important part of the PhD thesis [2] are dedicated to the analysis of several parameters of discrete random walks on $\mathbb{Z}$ or $\mathbb{N}$ with *bounded* jumps. For this case (but not for the case of unbounded jumps), a context-free grammar approach is also possible (as the jumps are bounded and thus can be encoded by a finite alphabet). However this language theory approach (which was previously the main one considered in combinatorics) reveals almost nothing about the shape of the generating function and is even less talkative about the asymptotics. An orthogonal approach (the kernel method) has the merit of giving a direct access to the generating functions and their asymptotics.

A natural question is: can the same approach be the winning one for the study of walks with *unbounded* jumps? We show here that the answer is clearly: yes, for a quite general family of walks! What follows is a slightly modified copy/paste of [4] which gives however some new original results for generating functions and asymptotics of walks with unbounded jumps.

2 Lattice paths and generating functions

This section presents the varieties of lattice paths to be studied as well as their companion generating functions (in the same terms as in [4]).

Definition 2.1. *Fix a set of vectors of* $\mathbb{Z}\times\mathbb{Z}$, $\mathcal{S}=\{(x_1,y_1),\ldots\}$. *(*$\mathcal{S}$ *can be finite or not). A* lattice path *or* walk *relative to* $\mathcal{S}$ *is a sequence* $v=(v_1,\ldots,v_n)$ *such that each* v_j *is in* $\mathcal{S}$. *The geometric realization of a lattice path* $v=(v_1,\ldots,v_n)$ *is the sequence of points* $(P_0,P_1,\ldots,P_n)$ *such that* $P_0=(0,0)$ *and* $\overrightarrow{P_{j-1}P_j}=v_j$. *The quantity* n *is referred to as the* size *of the path.*

In what follows, we focus our attention to a class of infinite sets $\mathcal{S}$ and we shall identify a lattice path with the polygonal line admitting $P_0,\ldots,P_n$ as vertices. The elements of $\mathcal{S}$ are called *steps* or *jumps*, and we also refer to the vectors $\overrightarrow{P_{j-1}P_j}=v_j$ as the steps of a particular path.

Various constraints will be imposed on paths. In particular we restrict attention throughout this paper to *directed paths* defined by the fact that if (i,j) lies in $\mathcal{S}$, then necessarily one should have $i>0$. In other words, a step always entails progress along the horizontal axis and the geometric realization of the path naturally lives in the half plane $\mathbb{N}\times\mathbb{Z}$. (This constraint implies that the paths studied can be treated essentially as 1-dimensional objects.) The following conditionings are to be considered (Figure 1).

Definition 2.2. *A* bridge *is a path whose end-point* P_n *lies on the* x*-axis. A* meander *is a path that lies in the quarter plane* $\mathbb{N}\times\mathbb{N}$. *An* excursion *is a path that is at the same time a meander and a bridge; it thus connects the origin to a point lying on the* x*-axis and involves no point with negative* y*–coordinate.*

A family of paths is said to be factorial *if each allowed step in* $\mathcal{S}$ *(Definition 2.1) is of the form* $(1,-y)$ *for any* $y\geq 1$ *or of the form* $(1,j)$ *with* $j\in\mathcal{J}$ *a given finite subset of* $\mathbb{Z}$. *We thus simply note* $\mathcal{S}=\{\mathbb{Z}_{<0},\mathcal{J}\}$.

In the factorial case the size of a path coincides with its span along the horizontal direction, that is, its *length.* The terminology of bridges, meanders, and excursions is chosen to be consistent with the standard one adopted in Brownian motion

walks	ending anywhere	ending in 0
unconstrained (on $\mathbb{Z}$)	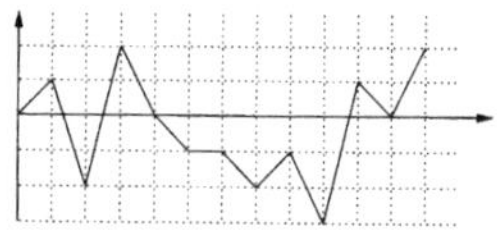 walk ($\mathcal{W}$) $W(z,u) \equiv \sum_{k\in\mathbb{Z}} W_k(z)\, u^k$ $= \frac{1}{1 - z\,Q(u)}$ $W_n = +\infty$	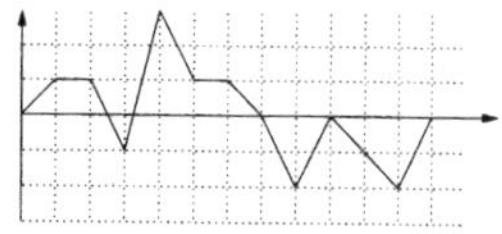 bridge ($\mathcal{B}$) $B(z) \equiv W_0(z) = z \sum_{i=0}^{b} \frac{u_i'(z)}{u_i(z)}$ $B_n \sim \beta_0 \frac{Q(\tau)^n}{\sqrt{2\pi n}}$
constrained (on $\mathbb{N}$)	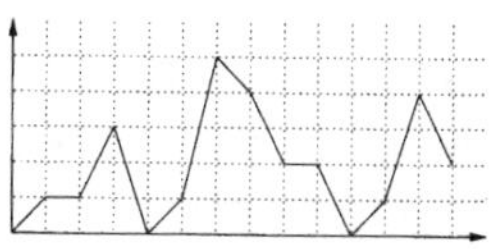 meander ($\mathcal{M}$) $M(z) \equiv \sum_{k\geq 0} F_k(z)$ $= -\frac{1}{z} \prod_{i=0}^{b} (1 - u_i(z))$ $M_n \sim \mu_0 \frac{Q(\tau)^n}{2\sqrt{\pi n^3}}$	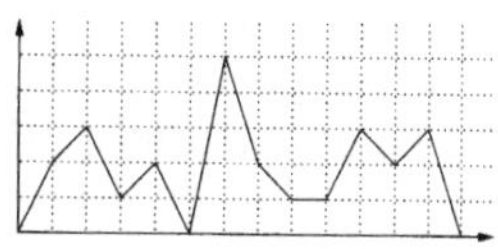 excursion ($\mathcal{E}$) $E(z) \equiv F_0(z) = \frac{(-1)^{b+1}}{z p_{-b}} \prod_{i=0}^{b} u_i(z)$ $E_n \sim \epsilon_0 \frac{Q(\tau)^n}{2\sqrt{\pi n^3}}$

Figure 1: The four types of paths with unbounded jumps: walks, bridges, meanders, and excursions. We give the corresponding generating functions and the asymptotics of their coefficients. (*N.B.*: there is an infinite number of unconstrained walks as jumps are unbounded.)

theory; see, e.g., [24]. A factorial walk is simply a walk for which there is, at each step, not only a finite amount of "bounded" jumps below or above the actual position but also the possibility to go anywhere below the actual position.

The main objective of this paper is to enumerate exactly as well as asymptotically paths, bridges, and meanders, this with special attention to factorial families. Once the set of steps is fixed, we let $\mathcal{W}$ and $\mathcal{B}$ denote the set of paths and bridges respectively ($\mathcal{W}$ being reminiscent of "walk"); we denote by $\mathcal{M}$ and $\mathcal{E}$ the set of meanders and excursions.

Given a class $\mathcal{C}$ of paths, we let $\mathcal{C}_n$ denote the subclass of paths that have size n, and, whenever appropriate, $\mathcal{C}_{n,k} \subset \mathcal{C}_n$ those that have final vertical abscissa (also known as "final altitude") equal to k. With the convention of using standard fonts to denote cardinalities of the corresponding sets (themselves in calligraphic style), $C_n = \text{card}(\mathcal{C}_n)$ and $C_{n,k} = \text{card}(\mathcal{C}_{n,k})$, the corresponding (ordinary) *generating functions* are then

$$C(z) := \sum_{n\in\mathbb{N}} C_n z^n, \qquad C(z,u) = \sum_{k\in\mathbb{Z}} C_k(z)\, u^k = \sum_{n\in\mathbb{N},k\in\mathbb{Z}} C_{n,k}\, u^k z^n = \sum_{n\in\mathbb{N}} c_n(u)\, z^n.$$

This paper is entirely devoted to characterizing these generating functions: they are either rational functions (W) or algebraic functions (B, M, E)[1]. As we shall see, a strong algebraic decomposition prevails which, as opposed to other approaches, renders the calculation of the generating functions effective. Even more importantly, the decomposability of generating functions makes it possible to extract their singular structure, and in turn solve the corresponding asymptotic enumeration problems in a wholly satisfactory fashion.

Weighted paths. For several applications, it is useful to associate *weights* to single steps. In this case, the set of steps $\mathcal{S}$ is coupled with a system of weights $\Pi = \{w_i\}_{i\in\mathbb{Z}}$, with $w_i > 0$ the weight associated to $(1,i) \in \mathcal{S}$; the weight of a path is then defined as the *product* of the weights of its individual steps. Then the quantity C_n, still referred to as *number of paths* (of size n), represents the sum of the weights of all paths of size n. Such weighted paths cover several situations of interest: (i) combinatorial paths in the standard sense above when each $w_i = 1$; (ii) paths with coloured steps, e.g., $w_i = 2$ means that the corresponding step $(1,i)$ has two possible coloured incarnations (say blue and red); (iii) $\sum w_i = 1$ corresponds to a probabilistic model of paths where, at each stage, step $(1,i)$ is chosen with probability w_i.

3 Functional equation and the kernel method

In this section, we characterize the generating functions of the four types of directed paths (unconstrained, bridges, meanders, and excursions). It will be seen that a specific algebraic curve, the "characteristic curve" plays a central rôle.

Definition 3.1. *Let $\mathcal{S} = \{\mathbb{Z}_{<0}, \mathcal{J}\}$ be a factorial set of jumps, with $\Pi = \{w_i\}_{i\in\mathbb{Z}}$ the corresponding system of weights ($w_i \equiv 1$ in the unweighted case). The* characteristic series *of $\mathcal{S}$ is defined as the Laurent series*[2]

$$Q(u) := \sum_{i\in\mathbb{Z}_{<0}\cup\mathcal{J}} w_i\, u^i\,.$$

Let $b = -\min \mathcal{J} \cup \{0\}$ and $a = \max \mathcal{J}$ be the two extreme vertical amplitudes of any jump of $\mathcal{J}$, and assume throughout $a > 0, b \geq 0$. We restrict now attention to the unweighted case (but with possibly coloured jumps in $\mathcal{J}$, see the paragraph "weighted paths" in Section 2). The characteristic series can be then rewritten as

$$Q(u) = \sum_{i=-\infty}^{-1} u^i + P(u) \equiv P(u) - \frac{1}{1-u}, \quad \textit{where } P(u) := \sum_{j=-b}^{a} p_j u^j \quad (p_j \in \mathbb{N}). \quad (1)$$

[1]The attentive reader should have understood that this does not stand for the acronym of a well-known Belgian theorem (Brownian Motion Everywhere)!

[2]By Laurent series, we mean objects like $\sum_{k=m}^{+\infty} g_k u^k$ ($m \in \mathbb{Z}$) *or* $\sum_{k=-\infty}^{m} g_k u^k$. The reader can check that our generating functions are holomorphic/meromorphic functions; they can be expanded at 0 or at infinity, and so they can be seen as belonging either to $\mathbb{C}[[\frac{1}{u}]][u]$ or $\mathbb{C}[\frac{1}{u}][[u]]$.

So p_j can be seen as the multiplicity of the jump $(1, j)$. The kernel *is defined by*

$$K(z,u) := (1-u)u^b - z(u^b(1-u)P(u) - u^b). \tag{2}$$

The characteristic curve *of the lattice paths determined by $\mathcal{S}$ is the plane algebraic curve defined by the* kernel equation

$$1 - z\,Q(u) = 0, \qquad \textit{or equivalently} \quad K(z,u) = 0. \tag{3}$$

As we shall see the characteristic equation plays a central rôle, the second form being the entire version (that is, a form without negative powers).

Proposition 3.2. *The kernel equation* (3) *admits $a+b+1$ roots in u : $b+1$ roots $u_0(z), \ldots, u_b(z)$ finite for $z \sim 0$ and a large roots $v_1(z), \ldots, v_a(z)$ infinite for $z \sim 0$.*

Proof : This polynomial has degree $a+b+1$ in u, and hence, admits $a+b+1$ solutions, which are algebraic functions of z. The classical theory of algebraic functions and the Newton polygon construction enable us to expand the solutions near any point as Puiseux series (that is, series involving fractional exponents; see [13]). The $a+b+1$ solutions, expanded around 0, can be classified as follows:

- the "unit" branch, denoted by u_0, is a power series in z with constant term 1;
- b "small" branches, denoted by $u_1, \ldots, u_b$, are power series in $z^{1/b}$ whose first nonzero term is $\zeta z^{1/b}$, with $\zeta^b + 1 = 0$;
- a "large" branches, denoted by $v_1, \ldots, v_a$, are Laurent series in $z^{1/a}$ whose first nonzero term is $\zeta z^{-1/a}$, with $\zeta^a + 1 = 0$.

In particular, all the roots are distinct. □

Formulae (4) and (5) in the following theorem were first derived in [3]:

Theorem 3.3 (Excursions and meanders). *The generating function $F(z,u)$ for factorial walks starting from 0 is algebraic; it is given by* (8)*, where $u_0, \ldots, u_b$ (resp. $v_1, \ldots, v_a$) are the finite (resp. infinite) solutions at $z = 0$ of the equation $K(z,u) = 0$ and the kernel K is defined by* (2)*. In particular, the generating function for all walks, irrespective of their endpoint, is*

$$M(z) = F(z,1) = -\frac{1}{z}\prod_{i=0}^{b}(1-u_i), \tag{4}$$

and the generating function for excursions, *i.e., walks ending at 0, is, for $b < 0$:*

$$E(z) = F(z,0) = \frac{(-1)^{b+1}}{z p_{-b}}\prod_{i=0}^{b} u_i\,. \tag{5}$$

(For $b = 0$, the relation becomes $F(z,0) = \frac{u_0}{1+(1-p_0)z}$.)

More generally, the generating function for meanders ending at altitude k is

$$F_k(z) = \frac{1}{z p_a}\sum_{i=1}^{a}\frac{v_i^{-k-1}}{\prod_{j\neq i} v_j - v_i}\,. \tag{6}$$

Proof: The allowed jumps imply that from position k (encoded by u^k), one can go to the position encoded[3] by $u^0 + u^1 + \ldots + u^{k-1} + \{u^{\geq 0}\}P(u)u^k = \frac{u^k - 1}{u-1} + \{u^{\geq 0}\}P(u)u^k$, as this is a linear mapping, this leads to the recurrence on the $f_n(u)$'s (the polynomials encoding the possible walk positions at time n):

$$f_{n+1}(u) = \frac{f_n(u) - f_n(1)}{u-1} + \{u^{\geq 0}\}P(u)f_n(u)$$

and equivalently to the following equality

$$\begin{aligned} F(z,u) &= \sum_{n\geq 0} f_n(u)z^n \\ &= 1 + z\left(\frac{F(z,1) - F(z,u)}{1-u} + P(u)F(z,u) - \{u^{<0}\}[P(u)F(z,u)]\right). \end{aligned}$$

Thus, $F(z,u)$ satisfies the following functional equation:

$$F(z,u)\left(1 + \frac{z}{1-u} - zP(u)\right) = 1 + \frac{zF(z,1)}{1-u} - z\sum_{k=0}^{b-1} r_k(u)F_k(z)\,, \tag{7}$$

where $r_k(u)$ is a Laurent polynomials whose exponents belong to $[k-b, -1]$:

$$r_k(u) := \{u^{<0}\}\left(P(u)u^k\right) \equiv \sum_{j=-b}^{-k-1} p_j u^{j+k}.$$

Now comes the second ingredient of the proof, the so called "kernel method". The right-hand side of (7), once multiplied by $u^b(1-u)$, is

$$R(z,u) = u^b(1-u)\left(1 + \frac{z}{1-u}F(z,1) - z\sum_{k=0}^{b-1} r_k(u)F_k(z)\right).$$

By construction, it is a ***polynomial*** in u of degree $b+1$ and leading coefficient -1. Hence, it admits $b+1$ roots, which depend on z. Replacing u by the series $u_0, u_1, \ldots, u_b$ in Eq. (7) shows that these series are exactly the $b+1$ roots of R, so that

$$R(z,u) = -\prod_{i=0}^{b}(u - u_i).$$

Let $p_a := [u^a]P(u)$ be the multiplicity of the largest forward jump.

Then the coefficient of u^{a+b+1} in $K(z,u)$ is $p_a z$, and we can write

$$K(z,u) = p_a z \prod_{i=0}^{b}(u - u_i)\prod_{i=1}^{a}(u - v_i).$$

[3]We make use of the conventional notations for coefficients of entire and Laurent series: $[z^n]\sum_n f_n z^n := f_n$ and $\{u^{\geq 0}\}g(u)$ is the sum of the monomials of $g(u)$ with a nonnegative exponent.

Finally, as $K(z,u)F(z,u) = R(z,u)$, we obtain

$$F(z,u) = \frac{-\prod_{i=0}^{b}(u-u_i)}{u^b(1-u)+zu^b-zu^b(1-u)P(u)} = -\frac{1}{p_a z\prod_{i=1}^{a}(u-v_i)}. \tag{8}$$

Setting $u = 1$ and $u = 0$ gives formulae (4) and (5) and a partial fraction decomposition of the rightmost part of (8) gives (6). □

The "kernel method" has been part of the folklore of combinatorialists for some time and is related to the what is known as "the quadratic method" in enumeration of planar maps [10]. Earlier references (see [17] Ex. 2.2.1.11 for Dyck paths, [21, Sec. 15.4] for a pebbling game) were dealing with the case of a single unknown in the right part of (7). The kernel method in its more general version was developed by Banderier, Bousquet-Mélou, Flajolet, Petkovšek [1, 2, 3, 4, 9]. A somewhat similar idea (involving a reduction to a Riemann–Hilbert problem) was used in [15] for a queuing theory application.

Theorem 3.4 (Bridges). *The bivariate generating function of paths (with z marking size and u marking final altitude) relative to a simple set of steps $\mathcal{S}$ with characteristic series $Q(u)$ is a rational function. It is given by*

$$W(z,u) = \frac{1}{1-zQ(u)}. \tag{9}$$

The generating function of bridges is an algebraic function given by

$$B(z) = z\sum_{j=0}^{b}\frac{u_j'(z)}{u_j(z)} = z\frac{d}{dz}\log\left(u_0(z)\cdots u_b(z)\right), \tag{10}$$

where the expressions involve all the small branches $u_0,\ldots,u_b$ of the characteristic curve (3). *Generally, the generating function W_k of paths terminating at altitude k is, for $-\infty < k < b$,*

$$W_k(z) = z\sum_{j=0}^{b}\frac{u_j'(z)}{u_j(z)^{k+1}} = -\frac{z}{k}\frac{d}{dz}\left(\sum_{j=0}^{b}u_j(z)^{-k}\right), \tag{11}$$

and for $-a < k < +\infty$,

$$W_k(z) = -z\sum_{j=1}^{a}\frac{v_j'(z)}{v_j(z)^{k+1}} = \frac{z}{k}\frac{d}{dz}\left(\sum_{j=1}^{a}v_j(z)^{-k}\right), \tag{12}$$

where $v_1,\ldots,v_a$ are the large branches.

(For W_0, the second form in (11) and (12) is to be taken in the limit sense $k \to 0$.)

Proof: The proof of an identity similar to (10) for walks with bounded jumps is given in [4] and holds verbatim for walks with unbounded jumps: Consider a bridge and let m (with $m \le 0$) be the minimal altitude of any vertex. Any nonempty bridge β decomposes uniquely into a walk φ_1 of size ≥ 1 from 0 to m that only reaches level m at its right end, followed by an excursion ε, followed by a path φ_2 of size ≥ 0 from m to 0 that only touches level m at its beginning. By rearrangement, one can write $\beta = \varepsilon \cdot (\varphi_2|\varphi_1)$, where the gluing of $\varphi_2\varphi_1$ is an arch (that is, an excursion which reaches 0 only at its beginning and its end) and the bar keeps track of where the splitting should occur. This links bridges and excursions:

$$\overbrace{B(z)-1}^{\text{bridges}} = \overbrace{E(z)}^{\text{excursions}} \cdot \overbrace{\left(z\frac{d}{dz}A(z)\right)}^{\text{split arches}}, \tag{13}$$

as $E(z) = 1/(1-A(z))$ ($A(z)$ stands for the generating function of arches), this is equivalent to

$$B(z) - 1 = E(z) \cdot z\frac{d}{dz}\left(1 - \frac{1}{E(z)}\right) = z\frac{E'(z)}{E(z)},$$

using Formula (5) for $E(z)$ gives the identity (10).

This reinforces the discussion of [4] about ubiquitous Spitzer, Andersen-like relations and here also, this gives the possibility of analysing the number of times a bridge attains its minimum or maximum value by adapting the decomposition (13).

Set $w_n(u) = [z^n]W(z,u)$, the Laurent series that describes the possible altitudes and the number of ways to reach them in n steps. We have $w_0(u) = 1$, $w_1(u) = Q(u)$, and $w_{n+1}(u) = Q(u)w_n(u)$, so that $w_n(u) = Q(u)^n$ for all n. The determination of $W(z,u)$ in (9) follows from

$$\sum_{n\geq 0} Q(u)^n z^n = \frac{1}{1 - zQ(u)}.$$

Observe that the resulting series is entire in z but of the Laurent type in u (it involves arbitrary negative powers of u).

For positive $Q(u)$, the radius of convergence of $W(z,u)$ viewed as a function of z is exactly $1/Q(u)$. Also, by the link between $E(z)$ and $B(z)$ (see above), the radius of convergence of $B(z)$ as a function of z is $\rho = 1/Q(\tau)$, the radius of convergence of $E(z)$ ($\tau > 1$, as it is proven in the next section). Consider now $|z| < r$, where $r := \frac{\rho}{2}$ and then follow the scheme of the proof from [4]. □

4 Asymptotics

Lemma 4.1. *Let $Q(u) = P(u) - 1/(1-u)$ be the rational series associated to the jumps a factorial walk. Then, there exists a unique number τ, called the* structural constant, *such that $Q'(\tau) = 0$, $\tau > 1$. The* structural radius *is by definition the quantity*

$$\rho := \frac{1}{Q(\tau)}.$$

The following domination amongst the roots holds

$$|u_i(z)| < u_0(z) \leq v_1(z) < |v_j(z)| \quad \forall |z| \leq \rho \text{ for } i = 1, \ldots, b \text{ and } j = 2, \ldots, a. \tag{14}$$

Proof: Differentiating twice Q as given in (1), we see that $Q''(x) > 0$ for all $x > 1$. Thus, the real function $x \mapsto Q(x)$ is strictly convex on $[1, +\infty]$. Since it satisfies $Q(1^+) = Q(+\infty) = +\infty$, it must have a unique positive minimum attained at some τ, and $Q'(\tau) = 0$.

As Q is aperiodic, a strong version of the triangular inequality gives

$$Q(v_1) = \frac{1}{z} = |Q(v_i)| < Q(|v_i|)$$

since Q is strictly increasing on the interval $[1, +\infty]$ and since $|v_i| > \tau > 1$ belongs to this interval for $z \in [0, \rho]$, one has the three last inequalities of (14); a duality argument gives the first inequality of (14). □

As one of the referee pointed out, the structural constant τ is such that the jumps with law $\frac{w_j \tau^j}{Q(\tau)}$ are centered. Similarly, the factoriality assumption results in steps which can be seen as a mixture of a geometric probability law and a finitely supported one.

Theorem 4.2. *The asymptotics for the number of bridges, meanders, excursions is given by*

$$B_n \sim \beta_0 \frac{Q(\tau)^n}{\sqrt{2\pi n}} (1 + \frac{\beta_1}{n} + \frac{\beta_2}{n^2} + \ldots), \quad \beta_0 = \frac{1}{\tau}\sqrt{\frac{Q(\tau)}{Q''(\tau)}},$$

$$M_n \sim \mu_0 \frac{Q(\tau)^n}{2\sqrt{\pi n^3}} (1 + \frac{\mu_1}{n} + \frac{\mu_2}{n^2} + \ldots), \quad \mu_0 = \widetilde{U}(\rho)\sqrt{\frac{2Q^3(\tau)}{Q''(\tau)}},$$

$$E_n \sim \epsilon_0 \frac{Q(\tau)^n}{2\sqrt{\pi n^3}} (1 + \frac{\epsilon_1}{n} + \frac{\epsilon_2}{n^2} + \ldots), \quad \epsilon_0 = U(\rho)(-1)^b\sqrt{\frac{2Q^3(\tau)}{Q''(\tau)}},$$

where $U(\rho) = u_1(\rho) \ldots u_b(\rho)$ *and* $\widetilde{U}(\rho) = (1 - u_1(\rho)) \ldots (1 - u_b(\rho))$.

Proof: Here again, the approach used in [4] is the winning one. A saddle point method gives

$$\begin{aligned} B_n &= \frac{1}{2i\pi} \int_{|u|=\tau} Q(u)^n \frac{du}{u} \\ &\sim \frac{1}{2i\pi} \int_{\tau e^{-i\epsilon}}^{\tau e^{+i\epsilon}} \exp\left(n \left(\log Q(\tau) + \frac{1}{2}\frac{Q''(\tau)}{Q(\tau)}(u-\tau)^2 + O((u-\tau)^3)\right)\right) \frac{du}{u} \\ &\sim \frac{Q(\tau)^n}{2\pi\tau} \int_{-\infty}^{+\infty} e^{-nht^2/2}\, dt = \frac{Q(\tau)^n}{\tau\sqrt{2\pi n h}}, \qquad h = \frac{Q''(\tau)}{Q(\tau)}. \end{aligned}$$

The approximation is valid as $Q(\tau)$ dominates on the circle of integration (this can be seen by the Laurent series expression of $Q(u)$).

Contrary to what is observed for the bounded jumps case, it may happen that the small roots cross for $|z| < \rho$ (but their product remains analytic). We follow the scheme of proof from [4] which uses the link between $B(z)$ and $E(z)$. One has, by local inversion of the kernel equation,

$$u_0(z) = \tau - \sqrt{2\frac{Q(\tau)}{Q''(\tau)}}\sqrt{1 - z/\rho} + \cdots \qquad (z \to \rho^-). \tag{15}$$

Then the only possible behaviour compatible with the above asymptotics for B_n is that $U(z) := u_1(z)\dots u_b(z)$ is analytical for $|z| < \rho$; the same hold for $\widetilde{U}(z) := (1-u_1(z))\dots(1-u_b(z))$.
Singularity analysis on the following expressions then gives the asymptotic expansions from the theorem

$$E_n \sim [z^n]U(\rho)\frac{(-1)^{b+1}}{z}\sqrt{2\frac{Q(\tau)}{Q''(\tau)}}\sqrt{1-z/\rho},$$

$$M_n \sim [z^n]\widetilde{U}(\rho)\frac{-1}{z}\sqrt{2\frac{Q(\tau)}{Q''(\tau)}}\sqrt{1-z/\rho}.$$

□

5 Returns to zero

Theorem 5.1 (Excursions). *The number of returns to zero of an excursion with unbounded jumps is asymptotically the sum of two independent geometric laws. The average is* $2E(\rho)-1+O\left(\frac{1}{n}\right)$ *returns to zero, with a variance* $2E(\rho)(E(\rho)-1)+O\left(\frac{1}{n}\right)$.

Proof: An excursion is a sequence of arches, so $E(z)=\frac{1}{1-A(z)}$ and $A(z)=1-\frac{1}{E(z)}$ for $E(z)$ and $A(z)$ generating functions of excursions and arches respectively. We note $F(z,u,t)$ the generating functions with respect to their length, final altitude, number of returns to zero. Thus, one has

$$F(z,0,t)=\sum_{n,j} f_{nj}(0)t^j z^n = \frac{1}{1-tA} = \frac{1}{1-t\left(1-\frac{1}{E}\right)},$$

where $f_{nj}(0)$ stands for the number of excursions of length n with j returns to 0. Then, all the moments can be made explicit as the m-th derivatives in t of $F(z,0,t)$ are computable ($\partial_t^m F(z,0,t) = m!\frac{(1-E^{-1})^m}{(1-t(1-E^{-1}))^{m+1}}$) and simplify when $t=1:\ \partial_t^m F(z,0,1) = m!E(z)(E(z)-1)^m$.
Thus, the average number of returns to zero is

$$\mu_n = \frac{[z^n]\partial_t F(z,0,1)}{[z^n]F(z,0,1)} = \frac{[z^n]E(z)^2}{[z^n]E(z)} - 1 = 2e_0 - 1 + O\left(\frac{1}{n}\right)$$

as $E(z) = e_0 - e_1\sqrt{\rho - z} + \cdots$ and the variance is given by

$$\sigma_n^2 = \frac{f_n''(1)}{f_n(1)} + \mu_n - \mu_n^2 = \frac{[z^n]2E(z) - 4E(z)^2 + 2E(z)^3}{[z^n]E(z)} + \mu_n - \mu_n^2$$

$$= 6e_0^2 - 8e_0 + 2 + \mu_n - \mu_n^2 = 2e_0(e_0-1) + O\left(\frac{1}{n}\right).$$

The number of excursions of length n with j returns to zero is given by

$$f_{nj}(0) = [z^n]\left(1 - \frac{1}{E(z)}\right)^j = [z^n](1-\frac{1}{e_0})^j - \frac{e_1 j(1-e_0^{-1})^{j-1}\sqrt{r-z}}{e_0^2} + O(r-z).$$

Consequently, the probability to get asymptotically j returns to zero is $\pi_j = f_{nj}(0)/f_n \to \frac{j(1-e_0^{-1})^{j-1}}{e_0^2}$ for $n \to +\infty$, and $\sum_{j\geq 0}\pi_j = 1$ for any e_0. The probability generating function is $x\,(\frac{1}{e_0}\frac{1}{1-x(1-e_0^{-1})})^2$ and one has so a discrete limit law which is asymptotically the sum of two independent geometric laws of parameters $1-1/e_0$. □

Perhaps it can seem strange than a walk with a infinite negative drift has such a small average number of returns to zero[4], the explanation of this "paradox" is that most of the walks have much more returns, but their probabilities are very low, decreasing exponentially (so, like for Zeno's paradox, the sum is finite).

Theorem 5.2 (Meanders). *The average number of returns to zero of a meander with unbounded jumps follows a discrete limit law of a geometrical type.*

Proof: Equation (8) gives $F(z,u)$, the bivariate generating function for meanders (length, final altitude). Taking into account the number of returns to zero (via another variable t) leads to

$$F(z,u,t) = \sum_{n,j\geq 0} f_{nj}(u)t^j z^n = \frac{1}{1-t(1-1/E(z))}\frac{F(z,u)}{E(z)}.$$

This reflects the fact that a meander is a sequence of arches, followed by a prefix (i.e. a left part) of an arch, so $M(z) = \frac{1}{1-A(z)}M^+(z)$ and that a prefix of arch (note $M^+(z,u)$ their generating function) times an excursion gives a meander, so $M^+(z,u) = F(z,u)/E(z)$. The number $f_{nj}(1)$ of meanders of length n with j returns to zero is then given by

$$f_{nj}(1) = [z^n](1-1/E(z))^j\frac{M(z)}{E(z)}.$$

Notice that

$$\frac{(1-1/E(z))^j}{E(z)} \sim \frac{(1-\frac{1}{e_0})^j}{e_0} + \left(-\frac{j}{e_0} + (1-\frac{1}{e_0})\right)\left(1-\frac{1}{e_0}\right)^{j-1}\frac{e_1}{e_0^2}\sqrt{\rho - z} + \ldots$$

Multiplying by the behaviour of $M(z) = F(z,1) = m_0 + m_1\sqrt{\rho - z}$ around $z = \rho$ gives

$$f_{nj}(1) \sim \left(m_1\frac{(1-\frac{1}{e_0})^j}{e_0} + m_0\left(-\frac{j}{e_0} + (1-\frac{1}{e_0})\right)\left(1-\frac{1}{e_0}\right)^{j-1}\frac{e_1}{e_0^2}\right)[z^n]\sqrt{\rho - z}.$$

[4]One referee pointed out that a similar result was known in a special case of bridge, cf. Proposition 2.2 page 101 of [8].

So $f_{nj}(1)/f_n(1) \to (\frac{1}{e_0}+\frac{m_0}{m_1}\frac{e_1}{e_0^2})(1-\frac{1}{e_0})^j - \frac{m_0 e_1 j}{m_1 e_0^3}\left(1-\frac{1}{e_0}\right)^{j-1}$ for $n \to +\infty$. Asymptotics of moments is also easily computable from

$$\partial_t F(z,u,1) = F(z,u)(E(z)-1) \text{ and } \partial_t^2 F(z,u,1) = 2(E(z)-1)^2 F(z,u).$$

Average and variance are $O(1)$. □

Theorem 5.3 (Bridges). *The number of returns to zero of a bridge with unbounded jumps is asymptotically the sum of two independent geometric laws. The average is* $2B(\rho) - 1 + O\left(\frac{1}{n}\right)$ *returns to zero, with a variance* $2B(\rho)(B(\rho)-1) + O\left(\frac{1}{n}\right)$.

Proof : We can play the same game as above:

$$W_k(z,t) = \frac{1}{1-t\left(1-\frac{1}{B(z)}\right)}\frac{W_k(z)}{B(z)},$$

The number of walks w_{nj} of length n ending at altitude k with j returns to zero is then given by

$$w_{nj}(1) = [z^n](1-1/B(z))^j \frac{W_k(z)}{B(z)}.$$

□

6 Final altitude of a meander.

The *final altitude* of a path is the abscissa of its end point. The random variable associated to finite altitude when taken over the set of all meanders of length n is denoted by X_n, and it satisfies

$$\Pr(X_n = k) = \frac{[z^n u^k]F(z,u)}{[z^n]F(z,1)}.$$

We state:

Theorem 6.1 (Meanders). *The final altitude of a random meander of size n admits a discrete limit distribution characterized in terms of the large branches:*

$$\lim_{n\to\infty} \Pr(X_n = k) = [u^k]\,\varpi(u), \quad \textit{where} \quad \varpi(u) = \frac{(1-\tau)^2}{(u-\tau)^2}\prod_{\ell\geq 2}\frac{1-v_\ell(\rho)}{u-v_\ell(\rho)}.$$

The limiting distribution admits an explicit form

$$[u^k]\,\varpi(u) = \tau^{-k}(c_0 + c_1 k) + \sum_{\ell\geq 2} c_\ell v_\ell(\rho)^{-k},$$

for a set of constants c_j that can be made explicit by a partial fraction expansion of $\varpi(u)$.

Proof: Similarly to [4], one directly shows that the probability generating function of X_n at u converges pointwise to a limit that precisely equals $\varpi(u)$, the convergence holding for $u \in (0,1)$. By the fundamental continuity theorem for probability generating functions, this entails convergence in law of the corresponding discrete distributions.

We now fix a value of u taken arbitrarily in $(0,1)$ and treated as a parameter. The probability generating function of X_n is

$$\frac{[z^n]F(z,u)}{[z^n]F(z,1)},$$

where $F(z,u)$ is given by Theorem 3.3. We know from the proof of Theorem 4.2 that $\tau = v_1(\rho)$ satisfies $\tau > 1$ while the radius of convergence of $F(z,1)$ coincides with the structural radius ρ. Then, the quantity

$$\widetilde{V}(z,u) = \prod_{\ell \geq 2}^{a} \frac{1}{u - v_\ell(z)}$$

is analytic in the closed disk $|z| \leq \rho$: being a symmetric function of the nonprincipal large branches, it has no algebraic singularity there; given the already known domination relations between the large branches (Lemma 4.1), the denominators cannot vanish.

It then suffices to analyse the factor containing the principal large branch v_1. This factor has a branch point at ρ, where

$$\frac{1}{u - v_1(z)} \sim \frac{1}{u - \tau} + \frac{1}{(u-\tau)^2}\sqrt{2\frac{Q(\tau)}{Q''(\tau)}}\sqrt{1 - z/\rho},$$

as follows directly from (15) and the fact that v_1 is conjugate to u_0 at $z = \rho$. Singularity analysis then gives instantly the fact that, for some nonzero constant C,

$$[z^n]F(z,u) \sim C\rho^{-n}n^{-3/2}\Omega(u), \quad \text{where} \quad \Omega(u) = \frac{1}{(u-\tau)^2}\widetilde{V}(\rho,u),$$

and the result follows after normalization by $[z^n]F(z,1)$.

□

7 Variations...

All the above theorems hold with a slightly more general model of walks, for which all the backward unbounded jumps are coloured (say, there is m colors). The only modification is that the roots are then the roots of the kernel $K(z,u) = (1-u)u^b - z(u^b(1-u)P(u) - mu^b)$. The analysis for the F_k's and W_k's is more delicate as it involves a better "individual" knowledge of the small and large roots.

Some more general models of walks were considered in [5], there is still some algebraic generating functions but their asymptotic properties remain to be established, this seems quite difficult as there is no clear simple closed form formula (in terms of the roots of the kernel) in the general case.

Acknowledgements. This work was partially supported by the Future and Emerging Technologies programme of the EU under contract number IST-1999-14186 (ALCOM-FT), by the INRIA postdoctoral programme and by the Max-Planck Institut.

References

[1] Cyril Banderier. Combinatoire analytique : application aux marches aléatoires. *Mémoire de DEA, Université Paris VII*, 1998.

[2] Cyril Banderier. *Combinatoire analytique des chemins et des cartes.* PhD thesis, Université de Paris 6, 2001.

[3] Cyril Banderier, Mireille Bousquet-Mélou, Alain Denise, Philippe Flajolet, Danièle Gardy, and Dominique Gouyou-Beauchamps. Generating functions for generating trees. *Discrete Mathematics*, 246(1-3):29–55, 2002.

[4] Cyril Banderier and Philippe Flajolet. Basic analytic combinatorics of directed lattice paths. *Theoretical Computer Science*, 281(1-2):37–80, 2002.

[5] Cyril Banderier and Donatella Merlini. Lattice paths with an infinite set of jumps. *Proceedings of Formal power series and algebraic combinatorics (Melbourne, 2002)*, 2001.

[6] Elena Barcucci, Alberto Del Lungo, Elisa Pergola, and Renzo Pinzani. A methodology for plane tree enumeration. *Discrete Mathematics*, 180(1-3):45–64, 1998.

[7] Elena Barcucci, Alberto Del Lungo, Elisa Pergola, and Renzo Pinzani. Permutations avoiding an increasing number of length-increasing forbidden subsequences. *Discrete Mathematics & Theoretical Computer Science. DMTCS. An Electronic Journal*, 4(1):31–44 (electronic), 2000.

[8] Philippe Bougerol and Thierry Jeulin. Brownian bridge on hyperbolic spaces and on homogeneous trees. *Probability Theory and Related Fields*, 115(1):95–120, 1999.

[9] Mireille Bousquet-Mélou and Marko Petkovšek. Linear recurrences with constant coefficients: the multivariate case. *Discrete Mathematics*, 225(1-3):51–75, 2000.

[10] W. G. Brown and W. T. Tutte. On the enumeration of rooted non-separable planar maps. *Canad. J. Math.*, 16:572–577, 1964.

[11] F. R. K. Chung, R. L. Graham, V. E. Hoggatt, and M. Kleiman. The number of Baxter permutations. *Journal of Combinatorial Theory, Series A*, 24:382–394, 1978.

[12] Sylvie Corteel. Séries génératrices exponentielles pour les eco-systèmes signés. In *FPSAC'00*. Springer, June 2000.

[13] J. Dieudonné. *Infinitesimal calculus.* Hermann, Paris, 1971. Appendix 3.

[14] S. Dulucq, S. Gire, and J. West. Permutations with forbidden subsequences and nonseparable planar maps. *Discrete Mathematics*, 153(1-3):85–103, 1996. Proceedings of the 5th Conference on Formal Power Series and Algebraic Combinatorics (Florence, 1993).

[15] Guy Fayolle and Roudolf Iasnogorodski. Two coupled processors: the reduction to a Riemann-Hilbert problem. *Zeitschrift für Wahrscheinlichkeitstheorie und Verwandte Gebiete*, 47(3):325–351, 1979.

[16] O. Guibert and E. Pergola. Enumeration of vexillary involutions which are equal to their mirror/complement. *Discrete Mathematics*, 224(1-3):281–287, 2000.

[17] Donald Ervin Knuth. *The Art of Computer Programming*, volume 1. Addison-Wesley, third edition, 1997.

[18] Darla Kremer. Permutations with forbidden subsequences and a generalized Schröder number. *Discrete Mathematics*, 218(1-3):121–130, 2000.

[19] D. Merlini and M. C. Verri. Generating trees and proper Riordan Arrays. *Discrete Mathematics*, 218:167–183, 2000.

[20] Donatella Merlini, Renzo Sprugnoli, and Maria Cecilia Verri. An algebra for generating trees. In *Mathematics and computer science (Versailles, 2000)*, pages 127–139, Basel, 2000. Birkhäuser.

[21] A. M. Odlyzko. *Asymptotic enumeration methods*, volume II. Elsevier, 1995. In Handbook of Combinatorics, (R. Graham, M. Grötschel, and L. Lovász, eds.).

[22] Elisa Pergola, Renzo Pinzani, and Simone Rinaldi. Towards an algebra of succession rules. In *Mathematics and computer science (Versailles, 2000)*, pages 141–152, Basel, 2000. Birkhäuser.

[23] Elisa Pergola and Robert A. Sulanke. Schröder triangles, paths, and parallelogram polyominoes. *Journal of Integer Sequences*, 1:Article 98.1.7, 1998.

[24] Jim Pitman. Brownian motion, bridge, excursion, and meander characterized by sampling at independent uniform times. *Electronic Journal of Probability*, 4:no. 11, 33 pp. (electronic), 1999.

[25] J. West. Generating trees and the Catalan and Schröder numbers. *Discrete Mathematics*, 146:247–262, 1995.

[26] J. West. Generating trees and forbidden subsequences. *Discrete Mathematics*, 157:363–374, 1996.

Cyril Banderier

INRIA-Rocquencourt
F-78153 Le Chesnay (France)
cyril.banderier@inria.fr

Counting Walks in the Quarter Plane

Mireille Bousquet-Mélou

ABSTRACT: *We study planar walks that start from a given point (i_0, j_0), take their steps in a finite set $\mathfrak{S}$, and are confined in the first quadrant $x \geq 0, y \geq 0$. Their enumeration can be attacked in a systematic way: the generating function $Q(x, y; t)$ that counts them by their length (variable t) and the coordinates of their endpoint (variables x, y) satisfies a linear functional equation encoding the step-by-step description of walks. For instance, for the square lattice walks starting from the origin, this equation reads*

$$(xy - t(x + y + x^2y + xy^2))\, Q(x, y; t) = xy - xtQ(x, 0; t) - ytQ(0, y; t).$$

The central question addressed in this paper is the nature *of the series $Q(x, y; t)$. When is it algebraic? When is it D-finite (or holonomic)? Can these properties be derived from the functional equation itself?*

Our first result is a new proof of an old theorem due to Kreweras, according to which one of these walk models has, for mysterious reasons, an algebraic generating function. Then, we provide a new proof of a holonomy criterion recently proved by M. Petkovšek and the author. In both cases, we work directly from the functional equation.

1 Walks in the quarter plane

The enumeration of lattice walks is one of the most venerable topics in enumerative combinatorics, which has numerous applications in probability [16, 30, 39]. These walks take their steps in a finite subset $\mathfrak{S}$ of $\mathbb{Z}^d$, and might be constrained in various ways. One can only cite a small percentage of the relevant litterature, which dates back at least to the next-to-last century [1, 20, 27, 33, 34]. Many recent publications show that the topic is still active [4, 6, 12, 22, 24, 35, 36].

After the solution of many explicit problems, certain patterns have emerged, and a more recent trend consists in developing methods that are valid for generic sets of steps. A special attention is being paid to the *nature* of the generating function of the walks under consideration. For instance, the generating function for unconstrained walks on the line $\mathbb{Z}$ is rational, while the generating function for walks constrained to stay in the half-line $\mathbb{N}$ is always algebraic [3]. This result has often been described in terms of *partially directed* 2-dimensional walks confined in a quadrant (or *generalized Dyck walks* [14, 21, 28, 29]), but is, essentially, of a 1-dimensional nature.

Similar questions can be addressed for *real* 2-dimensional walks. Again, the generating function for unconstrained walks starting from a given point is clearly rational. Moreover, the argument used for 1-dimensional walks confined in $\mathbb{N}$ can be recycled to prove that the generating function for the walks that stay in the half-plane $x \geq 0$ is always algebraic. What about doubly-restricted walks, that is, walks that are confined in the quadrant $x \geq 0, y \geq 0$?

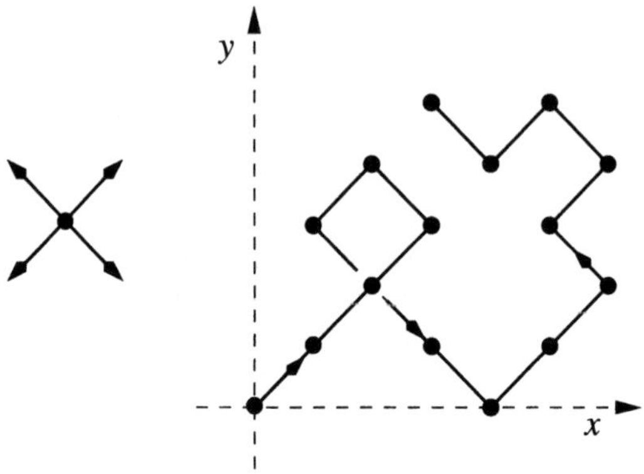

Figure 1: A walk on the diagonal square lattice confined in the first quadrant.

A rapid inspection of the most standard cases suggests that these walks might have always a D-finite generating function[1]. The simplest example is probably that of the diagonal square lattice, where the steps are North-East, South-East, North-West and South-West (Figure 1): by projecting the walks on the x- and y-axes, we obtain two decoupled prefixes of Dyck paths, so that the length generating function for walks that start from the origin and stay in the first quadrant is

$$\sum_{n\geq 0} \binom{n}{\lfloor n/2 \rfloor}^2 t^n,$$

a D-finite series. For the ordinary square lattice (with North, East, South and West steps), the generating function is

$$\sum_{m,n\geq 0} \binom{m+n}{m} \binom{m}{\lfloor m/2 \rfloor} \binom{n}{\lfloor n/2 \rfloor} t^{m+n} = \sum_{n\geq 0} \binom{n}{\lfloor n/2 \rfloor} \binom{n+1}{\lceil n/2 \rceil} t^n,$$

another D-finite series. The first expression comes from the fact that these walks are shuffles of two prefixes of Dyck walks, and the Chu-Vandermonde identity transforms it into the second simpler expression [25].

In both cases, the number of n-step walks grows asymptotically like $4^n/n$, which prevents the generating function from being algebraic (see [17] for the possible asymptotic behaviours of coefficients of algebraic series).

The two above results can be refined by taking into account the coordinates of the endpoint: if $a_{i,j}(n)$ denotes the number of n-step walks of length n ending at (i, j), then we have, for the diagonal square lattice:

$$\sum_{i,j,n\geq 0} a_{i,j}(n) x^i y^j t^n = \sum_{i,j,n\geq 0} \frac{(i+1)(j+1)}{(n+1)^2} \binom{n+1}{\frac{n-i}{2}} \binom{n+1}{\frac{n-j}{2}} x^i y^j t^n,$$

[1]A series $F(t)$ is D-finite (or *holonomic*) if it satisfies a linear differential equation with polynomial coefficients in t. Any algebraic series is D-finite.

where the binomial coefficient $\binom{n}{(n-i)/2}$ is zero unless $0 \le i \le n$ and $i \equiv n \bmod 2$. Similarly, for the ordinary square lattice,

$$\sum_{i,j,n\ge 0} a_{i,j}(n)x^i y^j t^n = \sum_{i,j,n\ge 0} \frac{(i+1)(j+1)}{(n+1)(n+2)} \binom{n+2}{\frac{n+i-j+2}{2}} \binom{n+2}{\frac{n-i-j}{2}} x^i y^j t^n. \quad (1)$$

These two series can be seen to be D-finite in their three variables.

This holonomy, however, is not the rule: as proved in [11], walks that start from $(1,1)$, take their steps in $\mathfrak{S} = \{(2,-1),(-1,2)\}$ and always stay in the first quadrant have a non-D-finite length generating function. The same holds for the subclass of walks ending on the x-axis. These walks are sometimes called ***knight's walks***.

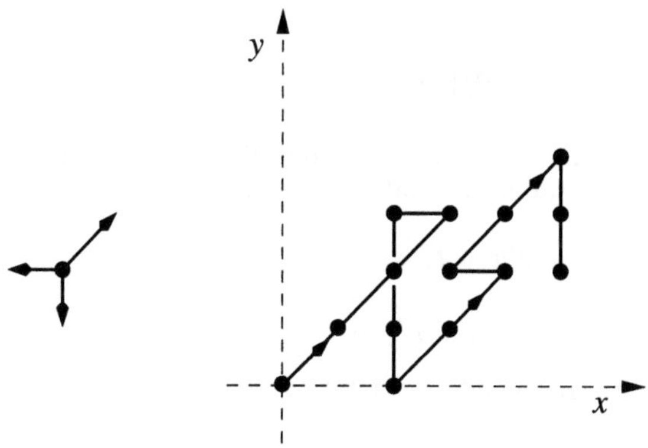

Figure 2: Kreweras' walks in a quadrant.

At the other end of the hierarchy, another walk model displays a mysteriously simple algebraic generating function: when the starting point is $(0,0)$, and the allowed steps South, West and North-East (Figure 2), the number of walks of length $3n+2i$ ending at the point $(i,0)$ is

$$\frac{4^n(2i+1)}{(n+i+1)(2n+2i+1)} \binom{2i}{i} \binom{3n+2i}{n}. \quad (2)$$

This result was first proved by Kreweras in 1965 [27, Chap. 3], and then rederived by Niederhausen [35] and Gessel [20]. It is, however, not well-understood, for two reasons:

– no direct proof of (2) is known, even when $i=0$. The number of walks ending at the origin is closely related to the number of non-separable planar maps, to the number of *cubic* non-separable maps [37, 38, 41, 42], and to the number of two-stack sortable permutations [5, 43, 44]. All available proofs of (2) are rather long and complicated. Moreover, in all of them, the result is *checked* rather than *derived*.

– most importantly, the three-variate generating function for these walks can be shown to be algebraic [20], but none of the proofs explain combinatorially this algebraicity.

All problems of walks confined in a quadrant can be attacked by writing a functional equation for their three-variate generating function, and it is this uniform approach that we discuss here. This functional equation simply encodes the step-by-step construction of the walks. For instance, for square lattice walks, we can write

$$Q(x,y;t) := \sum_{i,j,n\geq 0} a_{i,j}(n)x^i y^j t^n$$

$$= 1+t(x+y)Q(x,y;t)+t\,\frac{Q(x,y;t)-Q(0,y;t)}{x}+t\,\frac{Q(x,y;t)-Q(x,0;t)}{y},$$

that is,

$$\left(xy - t(x+y+x^2y+xy^2)\right) Q(x,y;t) = xy - xtQ(x,0;t) - ytQ(0,y;t), \quad (3)$$

and the solution of this equation, given by (1), is D-finite (but transcendental). Similarly, for the diagonal square lattice, we have

$$\left(xy - t(1+x^2)(1+y^2)\right) Q(x,y;t) =$$

$$xy - t(1+x^2)Q(x,0;t) - t(1+y^2)Q(0,y;t) + tQ(0,0;t),$$

with again a D-finite transcendental solution, while for Kreweras' algebraic model, we obtain

$$\left(xy - t(x+y+x^2y^2)\right) Q(x,y;t) = xy - xtQ(x,0;t) - ytQ(0,y;t). \quad (4)$$

Finally, the equation that governs the non-holonomic model of [11] is

$$\left(xy - t(x^3+y^3)\right) Q(x,y;t) = x^2y^2 - tx^3Q(x,0;t) - ty^3Q(0,y;t).$$

The general theme of this paper is the following: the above equations completely solve, in some sense, the problem of enumerating the walks. But they are not the kind of solution one likes, especially if the numbers are simple, or if the generating function is actually algebraic! How can one derive these simple solutions from the functional equations? And what is the essential difference between, say, Eqs. (3) and (4), that makes one series transcendental, and the other algebraic?

We shall answer some of these questions. Our main result is a new proof of (2), which we believe to be simpler than the three previous ones. It has, at least, one nice feature: we *derive the algebraicity from the equation* without having to guess the formula first. Then, we give a new proof of a (refinement of) a holonomy criterion that was proved combinatorially in [11]: if the set of steps $\mathfrak{S}$ is symmetric with respect to the y-axis and satisfies a *small horizontal variations* condition, then the generating function for walks with steps in $\mathfrak{S}$, starting from any given point (i_0, j_0), is D-finite. This result covers the two above D-finite transcendental cases, but not Kreweras' model... We finally survey some perspectives of this work.

Let us conclude this section with a few more formal definitions on walks and power series.

Let $\mathfrak{S}$ be a finite subset of $\mathbb{Z}^2$. A walk with steps in $\mathfrak{S}$ is a finite sequence $w = (w_0, w_1, \ldots, w_n)$ of vertices of $\mathbb{Z}^2$ such that $w_i - w_{i-1} \in \mathfrak{S}$ for $1 \leq i \leq n$. The number of steps, n, is the *length* of w. The starting point of w is w_0, and its endpoint is w_n. The *complete generating function* for a set $\mathfrak{A}$ of walks starting from a given point $w_0 = (i_0, j_0)$ is the series

$$A(x, y; t) = \sum_{n\geq 0} t^n \sum_{i,j\in\mathbb{Z}} a_{i,j}(n) x^i y^j,$$

where $a_{i,j}(n)$ is the number of walks of $\mathfrak{A}$ that have length n and end at (i, j). This series is a formal power series in t whose coefficients are polynomials in $x, y, 1/x, 1/y$. We shall often denote $\bar{x} = 1/x$ and $\bar{y} = 1/y$.

Given a ring $\mathbb{L}$ and k variables $x_1, \ldots, x_k$, we denote by $\mathbb{L}[x_1, \ldots, x_k]$ the ring of polynomials in $x_1, \ldots, x_k$ with coefficients in $\mathbb{L}$, and by $\mathbb{L}[[x_1, \ldots, x_k]]$ the ring of formal power series in $x_1, \ldots, x_k$ with coefficients in $\mathbb{L}$. If $\mathbb{L}$ is a field, we denote by $\mathbb{L}(x_1, \ldots, x_k)$ the field of rational functions in $x_1, \ldots, x_k$ with coefficients in $\mathbb{L}$.

Assume $\mathbb{L}$ is a field. A series F in $\mathbb{L}[[x_1, \ldots, x_k]]$ is *rational* if there exist polynomials P and Q in $\mathbb{L}[x_1, \ldots, x_k]$, with $Q \neq 0$, such that $QF = P$. It is *algebraic* (over the field $\mathbb{L}(x_1, \ldots, x_k)$) if there exists a non-trivial polynomial P with coefficients in $\mathbb{L}$ such that $P(F, x_1, \ldots, x_k) = 0$. The sum and product of algebraic series is algebraic.

The series F is *D-finite* (or *holonomic*) if the partial derivatives of F span a finite dimensional vector space over the field $\mathbb{L}(x_1, \ldots, x_k)$ (this vector space is a subspace of the fraction field of $\mathbb{L}[[x_1, \ldots, x_k]]$); see [40] for the one-variable case, and [31, 32] otherwise. In other words, for $1 \leq i \leq k$, the series F satisfies a non-trivial partial differential equation of the form

$$\sum_{\ell=0}^{d_i} P_{\ell,i} \frac{\partial^\ell F}{\partial x_i^\ell} = 0,$$

where $P_{\ell,i}$ is a polynomial in the x_j. Any algebraic series is holonomic. The sum and product of two holonomic series are still holonomic. The specializations of a holonomic series (obtained by giving values from $\mathbb{L}$ to some of the variables) are holonomic, if well-defined. Moreover, if F is an *algebraic* series and $G(t)$ is a holonomic series of one variable, then the substitution $G(F)$ (if well-defined) is holonomic [32, Prop. 2.3].

2 A new proof of Kreweras' result

Consider walks that start from $(0, 0)$, are made of South, West and North-East steps, and always stay in the first quadrant (Figure 2). Let $a_{i,j}(n)$ be the number of n-step walks of this type ending at (i, j). We denote by $Q(x, y; t)$ the complete generating function of these walks:

$$Q(x, y; t) := \sum_{i,j,n\geq 0} a_{i,j}(n) x^i y^j t^n.$$

Constructing the walks step by step yields the following equation:

$$\left(xy - t(x + y + x^2y^2)\right) Q(x, y; t) = xy - xtQ(x, 0; t) - ytQ(0, y; t). \tag{5}$$

We shall often denote, for short, $Q(x, y; t)$ by $Q(x, y)$. Let us also denote the series $xtQ(x, 0; t)$ by $R(x; t)$ or even $R(x)$. Using the symmetry of the problem in x and y, the above equation becomes:

$$\left(xy - t(x + y + x^2y^2)\right) Q(x, y) = xy - R(x) - R(y). \tag{6}$$

This equation is equivalent to a recurrence relation defining the numbers $a_{i,j}(n)$ by induction on n. Hence, it defines completely the series $Q(x, y; t)$. Still, the characterization of this series we have in mind is of a different nature:

Theorem 2.1. *Let $X \equiv X(t)$ be the power series in t defined by*

$$X = t(2 + X^3).$$

Then the generating function for Kreweras' walks ending on the x-axis is

$$Q(x, 0; t) = \frac{1}{tx}\left(\frac{1}{2t} - \frac{1}{x} - \left(\frac{1}{X} - \frac{1}{x}\right)\sqrt{1 - xX^2}\right).$$

Consequently, the length generating function for walks ending at $(i, 0)$ is

$$[x^i]Q(x, 0; t) = \frac{X^{2i+1}}{2.4^i\, t}\left(C_i - \frac{C_{i+1}X^3}{4}\right),$$

where $C_i = \binom{2i}{i}/(i+1)$ is the i-th Catalan number. The Lagrange inversion formula gives the number of such walks of length $3n + 2i$ as

$$a_{i,0}(3n + 2i) = \frac{4^n(2i + 1)}{(n + i + 1)(2n + 2i + 1)}\binom{2i}{i}\binom{3n + 2i}{n}.$$

The aim of this section is to derive Theorem 2.1 from the functional equation (5).

Note. Kreweras also gave a closed form expression for the number of walks containing exactly p West steps, q South steps, and r North-East steps, that is, for walks of length $m = p + q + r$ ending at $(i, j) = (r - p, r - q)$:

$$a_{r-p,r-q}(p+q+r) = \binom{p + q + r}{p, q, r}\left(1 - \frac{p + q}{r + 1}\right)$$
$$+ \sum_{h=1}^{p}\sum_{k=1}^{q} \frac{(-1)^{h+k}}{(h + k)(h + k - 1)}\binom{h + k}{h}\binom{2h + 2k - 2}{2h - 1}\binom{p + q + r}{p - h, q - k, r + h + k}.$$

The functional equation (5), combined with the expression of $Q(x, 0)$ given in Theorem 2.1, gives an alternative expression for this number, still in the form of a double sum:

$$a_{r-p,r-q}(p+q+r) = \binom{p + q + r}{p, q, r} - \sum_{i\ge 0}\sum_{n\ge 0} \frac{4^n(2i + 1)}{(n + i + 1)(2n + 2i + 1)}\binom{2i}{i}\binom{3n + 2i}{n}$$

$$\times \left(\binom{p+q+r-3n-2i-1}{p-n, q-n-i-1, r-n-i} + \binom{p+q+r-3n-2i-1}{p-n-i-1, q-n, r-n-i} \right).$$

This expression has a straightforward combinatorial explanation (all walks, except those that cross the x- or y-axis). But none of these formulas specialize to the above simple expression of $a_{i,0}(m)$ when $q = r$...

2.1 The obstinate kernel method

The kernel method is basically the only tool we have to attack Equation (6). This method had been around since, at least, the 70's, and is currently the subject of a certain revival (see the references in [2, 3, 10]). It consists in coupling the variables x and y so as to cancel the ***kernel*** $K(x, y) = xy - t(x + y + x^2y^2)$. This should give the "missing" information about the series $R(x)$.

As a polynomial in y, this kernel has two roots

$$\begin{aligned} Y_0(x) &= \frac{1 - t\bar{x} - \sqrt{(1 - t\bar{x})^2 - 4t^2x}}{2tx} = t + \bar{x}t^2 + O(t^3), \\ Y_1(x) &= \frac{1 - t\bar{x} + \sqrt{(1 - t\bar{x})^2 - 4t^2x}}{2tx} = \frac{\bar{x}}{t} - \bar{x}^2 - t - \bar{x}t^2 + O(t^3). \end{aligned}$$

The elementary symmetric functions of the Y_i are

$$Y_0 + Y_1 = \frac{\bar{x}}{t} - \bar{x}^2 \quad \text{and} \quad Y_0Y_1 = \bar{x}. \tag{7}$$

The fact that they are polynomials in $\bar{x} = 1/x$ will play a very important role below.

Only the first root can be substituted for y in (6) (the term $Q(x, Y_1; t)$ is not a well-defined power series in t). We thus obtain a functional equation for $R(x)$:

$$R(x) + R(Y_0) = xY_0. \tag{8}$$

It can be shown that this equation – once restated in terms of $Q(x, 0)$ – defines uniquely $Q(x, 0; t)$ as a formal power series in t with polynomial coefficients in x. Equation (8) is the standard result of the kernel method.

Still, we want to apply here the ***obstinate*** kernel method. That is, we shall not content ourselves with Eq. (8), but we shall go on producing pairs (X, Y) that cancel the kernel and use the information they provide on the series $R(x)$. This obstination was inspired by the book [15] by Fayolle, Iasnogorodski and Malyshev, and more precisely by Section 2.4 of this book, where one possible way to obtain such pairs is described (even though the analytic context is different). We give here an alternative construction that actually provides the same pairs.

Let $(X, Y) \neq (0, 0)$ be a pair of Laurent series in t with coefficients in a field $\mathbb{K}$ such that $K(X, Y) = 0$. We define $\Phi(X, Y) = (X', Y)$, where $X' = (XY)^{-1}$ is ***the other solution*** of $K(x, Y) = 0$, seen as a polynomial in x. Similarly, we define $\Psi(X, Y) = (X, Y')$, where $Y' = (XY)^{-1}$ is the other solution of $K(X, y) = 0$. Note that Φ and Ψ are involutions. Let us examine their action on the pair (x, Y_0). We obtain the diagram of Figure 3.

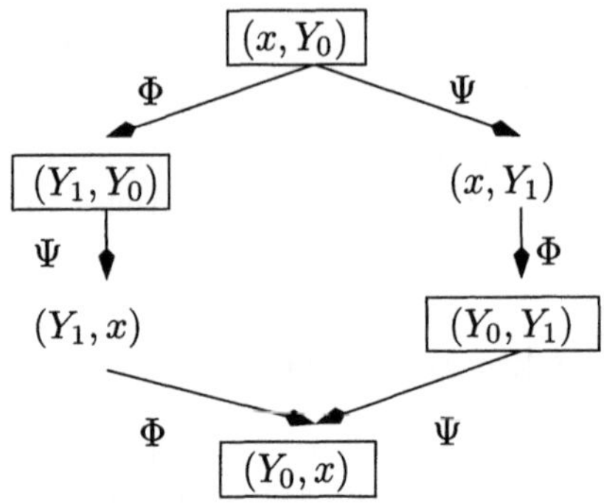

Figure 3: The orbit of (x, Y_0) under the action of Φ and Ψ.

All these pairs of power series cancel the kernel, and we have framed the ones that can be legally substituted[2] in the main functional equation (6). We thus obtain *two* equations for the unknown series $R(x)$:

$$R(x) + R(Y_0) = xY_0, \tag{9}$$

$$R(Y_0) + R(Y_1) = Y_0Y_1 = \bar{x}. \tag{10}$$

Remark. Let p, q, r be three nonnegative numbers such that $p+q+r=1$. Take $x = (pr)^{1/3}q^{-2/3}$, $y = (qr)^{1/3}p^{-2/3}$, and $t = (pqr)^{1/3}$. Then $K(x, y; t) = 0$, so that $R(x) + R(y) = xy$. This equation can be given a probabilistic interpretation by considering random walks that make a North-East step with (small) probability r and a South (resp. West) step with probability p (resp. q). This probabilistic argument, and the equation it implies, is the starting point of Gessel's solution of Kreweras problem [20, Eq. (21)].

2.2 Symmetric functions of Y_0 and Y_1

After the kernel method, the next tool of our approach is the extraction of the positive part of power series. More precisely, let $S(x; t)$ be a power series in t whose coefficients are Laurent polynomials in x:

$$S(x; t) = \sum_{n \geq 0} t^n \sum_{i \in \mathbb{Z}} s_i(n) x^i,$$

where for each $n \geq 0$, only finitely many coefficients $s_i(n)$ are non-zero. We define the positive part of this series by

$$S^+(x; t) := \sum_{n \geq 0} t^n \sum_{i \in \mathbb{N}} s_i(n) x^i.$$

This is where the values of the symmetric functions of Y_0 and Y_1 become crucial: the fact that they only involve negative powers of x (see (7)) will simplify the extraction of the positive part of certain equations.

[2]The fact that the series $Q(Y_0, Y_1; t)$ and $Q(0, Y_1; t)$ *are* well-defined is not immediate, and depends strongly on the three steps taken by the walks.

Lemma 2.2. *Let $F(u, v; t)$ be a power series in t with coefficients in $\mathbb{C}[u, v]$, such that $F(u, v; t) = F(v, u; t)$. Then the series $F(Y_0, Y_1; t)$, if well-defined, is a power series in t with polynomial coefficients in $\bar{x}$. Moreover, the constant term of this series, taken with respect to $\bar{x}$, is $F(0, 0; t)$.*

Proof. All symmetric polynomials of u and v are polynomials in $u + v$ and uv with complex coefficients. ■

We now want to form a symmetric function of Y_0 and Y_1, starting from the equations (9–10). The first one reads

$$R(Y_0) - xY_0 = -R(x).$$

By combining both equations, we then obtain the companion expression:

$$R(Y_1) - xY_1 = R(x) + 2\bar{x} - 1/t.$$

Taking the product[3] of these two equations gives

$$(R(Y_0) - xY_0)(R(Y_1) - xY_1) = -R(x)(R(x) + 2\bar{x} - 1/t).$$

The extraction of the positive part of this identity is made possible by Lemma 2.2. Given that $R(x; t) = xtQ(x, 0; t)$, one obtains:

$$x = -t^2x^2Q(x, 0)^2 + (x - 2t)Q(x, 0) + 2tQ(0, 0),$$

that is,

$$t^2x^2Q(x, 0)^2 + (2t - x)Q(x, 0) - 2tQ(0, 0) + x = 0. \tag{11}$$

2.3 The quadratic method

Equation (11) – which begs for a combinatorial explanation – is typical of the equations obtained when enumerating planar maps, and the rest of the proof will be routine to all maps lovers. This equation can be solved using the so-called *quadratic method*, which was first invented by Brown [13]. The formulation we use here is different both from Brown's original presentation and from the one in Goulden and Jackson's book [23]. This new formulation is convenient for generalizing the method to equations of higher degree with more unknowns [7].

Equation (11) can be written as

$$P(Q(x), Q(0), t, x) = 0, \tag{12}$$

where $P(u, v, t, x) = t^2x^2u^2 + (2t - x)u - 2tv + x$, and $Q(x, 0)$ has been abbreviated in $Q(x)$. Differentiating this equation with respect to x, we find

$$\frac{\partial P}{\partial u}(Q(x), Q(0), t, x)\frac{\partial Q}{\partial x}(x) + \frac{\partial P}{\partial x}(Q(x), Q(0), t, x) = 0.$$

[3] An alternative derivation of Kreweras' result, obtained by considering the divided difference $(R(Y_0) - xY_0 - R(Y_1) + xY_1)/(Y_0 - Y_1)$, will be discussed on the complete version of this paper.

Hence, if there exists a power series in t, denoted $X(t) \equiv X$, such that

$$\frac{\partial P}{\partial u}(Q(X), Q(0), t, X) = 0, \tag{13}$$

then one also has

$$\frac{\partial P}{\partial x}(Q(X), Q(0), t, X) = 0, \tag{14}$$

and we thus obtain a system of three polynomial equations, namely Eq. (12) written for $x = X$, Eqs. (13) and (14), that relate the three unknown series $Q(X)$, $Q(0)$ and X. This puts us in a good position to write an algebraic equation defining $Q(0) = Q(0, 0; t)$.

Let us now work out the details of this program: Eq. (13) reads

$$X = 2t^2X^2Q(X) + 2t,$$

and since the right-hand side is a multiple of t, it should be clear that this equation defines a unique power series $X(t)$. The system of three equations reads

$$\begin{cases} t^2X^2Q(X)^2 + (2t - X)Q(X) - 2tQ(0) + X = 0, \\ 2t^2X^2Q(X) + 2t - X = 0, \\ 2t^2XQ(X)^2 - Q(X) + 1 = 0. \end{cases}$$

Eliminating $Q(X)$ between the last two equations yields $X = t(2 + X^3)$, so that the series X is the parameter introduced in Theorem 2.1. Going on with the elimination, we finally obtain

$$Q(0, 0; t) = \frac{X}{2t}\left(1 - \frac{X^3}{4}\right),$$

and the expression of $Q(x, 0; t)$ follows from (11).

■

3 A holonomy criterion

Using functional equations, we can recover, and actually refine, a holonomy criterion that was recently proved combinatorially [11]. Let $\mathfrak{S}$ be a finite subset of $\mathbb{Z}^2$. We say that $\mathfrak{S}$ is symmetric with respect to the y-axis if

$$(i, j) \in \mathfrak{S} \Rightarrow (-i, j) \in \mathfrak{S}.$$

We say that $\mathfrak{S}$ has small horizontal variations if

$$(i, j) \in \mathfrak{S} \Rightarrow |i| \leq 1.$$

The usual square lattice steps satisfy these two conditions. So do the steps of the diagonal square lattice (Figure 1).

Theorem 3.1. *Let $\mathfrak{S}$ be a finite subset of $\mathbb{Z}^2$ that is symmetric with respect to the y-axis and has small horizontal variations. Let $(i_0, j_0) \in \mathbb{N}^2$. Then the complete generating function $Q(x, y; t)$ for walks that start from (i_0, j_0), take their steps in $\mathfrak{S}$ and stay in the first quadrant is D-finite.*

A combinatorial argument proving the holonomy of $Q(1, 1; t)$ is presented in [11].

3.1 Example

Before we embark on the proof of this theorem, let us see the principle of the proof at work on a simple example: square lattice walks confined in a quadrant. The functional equation satisfied by their complete generating function is

$$\left(xy - t(x + y + x^2y + xy^2)\right) Q(x,y) =$$
$$xy - xtQ(x,0) - ytQ(0,y) = xy - R(x) - R(y), \qquad (15)$$

where, as in Kreweras' example, we denote by $R(x)$ the series $txQ(x,0)$. The kernel $K(x,y) = xy - t(x + y + x^2y + xy^2)$, considered as a polynomial in y, has two roots:

$$Y_0(x) = \frac{1 - t(x+\bar{x}) - \sqrt{(1-t(x+\bar{x}))^2 - 4t^2}}{2t} = t + (x+\bar{x})t^2 + O(t^3),$$

$$Y_1(x) = \frac{1 - t(x+\bar{x}) + \sqrt{(1-t(x+\bar{x}))^2 - 4t^2}}{2t} = \frac{1}{t} - x - \bar{x} - t - (x+\bar{x})t^2 + O(t^3).$$

The elementary symmetric functions of the Y_i are

$$Y_0 + Y_1 = \frac{1}{t} - x - \bar{x} \quad \text{and} \quad Y_0Y_1 = 1.$$

Observe that they are no longer polynomials in $\bar{x} = 1/x$.
If, as above, we apply to the pair (x, Y_0) the transformations Φ and Ψ, we obtain a very simple diagram:

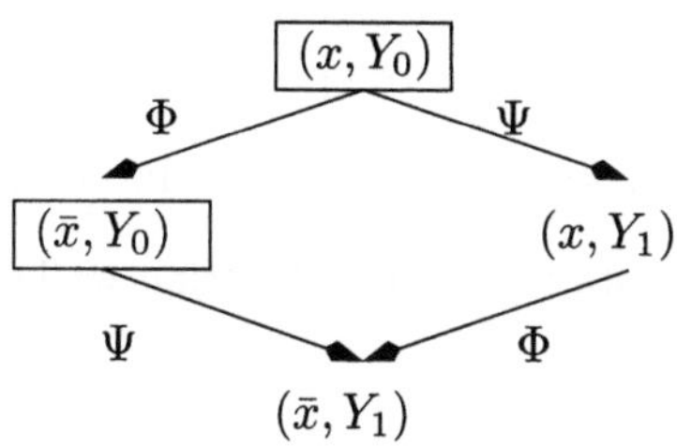

From the two pairs that can be substituted for (x,y) in Equation (15), we derive the following system:

$$\begin{aligned} R(x) + R(Y_0) &= xY_0, \\ R(\bar{x}) + R(Y_0) &= \bar{x}Y_0. \end{aligned}$$

From here, the method has to diverge from what we did in Kreweras' case. Eliminating $R(Y_0)$ between the two equations gives

$$R(x) - R(\bar{x}) = (x - \bar{x})Y_0. \qquad (16)$$

Since $R(0) = 0$, extracting the positive part of this identity gives $R(x)$ as *the positive part of an algebraic series*. It is known that the positive part of a D-finite series is always D-finite [31]. In particular, the series $R(x)$ is D-finite. The same holds for $Q(x,0)$, and, by (15), for $Q(x,y)$.

This argument is enough for proving the holonomy of the series, but, given the simplicity of this model, we can proceed with explicit calculations. Given the polynomial equation defining Y_0,

$$Y_0 = t(1 + \bar{x}Y_0 + xY_0 + Y_0^2) = t(1 + \bar{x}Y_0)(1 + xY_0),$$

the Lagrange inversion formula yields the following expression for Y_0:

$$Y_0 = \sum_{m\geq 0}\sum_{i\in\mathbb{Z}} \frac{x^i\, t^{2m+|i|+1}}{2m+|i|+1}\binom{2m+|i|+1}{m+|i|}\binom{2m+|i|+1}{m}.$$

Since $R(0) = 0$, extracting the positive part in the identity (16) now gives, after some reductions,

$$R(x) = txQ(x,0) = \sum_{m\geq 0}\sum_{i\geq 0} \frac{x^{i+1}t^{2m+i+1}(i+1)}{(2m+i+1)(2m+i+2)}\binom{2m+i+2}{m+i+1}\binom{2m+i+2}{m}.$$

This naturally fits with the general expression (1).

3.2 Proof of Theorem 3.1.

We define two Laurent polynomials in y by

$$P_0(y) := \sum_{(0,j)\in\mathfrak{S}} y^j \quad \text{and} \quad P_1(y) := \sum_{(1,j)\in\mathfrak{S}} y^j.$$

Let $-p$ be the largest down move; more precisely,

$$p = \max(0, \{-j : (i,j) \in \mathfrak{S} \text{ for some } i\}).$$

The functional equation obtained by constructing walks step-by-step reads:

$$K(x,y)Q(x,y) =$$

$$x^{1+i_0}y^{p+j_0} - ty^pP_1(y)Q(0,y) - t\sum_{(i,-j)\in\mathfrak{S}}\sum_{m=0}^{j-1}\left(Q_m(x) - \delta_{i,1}Q_m(0)\right)x^{1-i}y^{p+m-j} \tag{17}$$

where

$$K(x,y) = xy^p\left(1 - tP_0(y) - t(x+\bar{x})P_1(y)\right)$$

is the kernel of the equation, and $Q_m(x)$ stand for the coefficient of y^m in $Q(x,y)$. All the series involved in this equation also depend on the variable t, but it is omitted for the sake of briefness. For instance, $K(x,y)$ stands for $K(x,y;t)$.

As above, we shall use the kernel method – plus another argument – to solve the above functional equation. The polynomial $K(x,y)$, seen as a polynomial in y, admits a number of roots, which are Puiseux series in t with coefficients in an algebraic closure of $\mathbb{Q}(x)$. Moreover, all these roots are distinct. As $K(x,y;0) =$

xy^p, exactly p of these roots, say $Y_1, \ldots, Y_p$, vanish at $t = 0$. This property guarantees that these p series can be substituted for y in (17), which yields

$$x^{1+i_0} Y^{p+j_0} = tY^p P_1(Y) Q(0,Y) + t \sum_{(i,-j)\in\mathfrak{S}} \sum_{m=0}^{j-1} \left(Q_m(x) - \delta_{i,1} Q_m(0)\right) x^{1-i}\, Y^{p+m-j} \tag{18}$$

for any $Y = Y_1, \ldots, Y_p$.

Given the symmetry of K in x and $\bar{x}$, each of the Y_i is invariant by the transformation $x \to 1/x$. Replacing x by $\bar{x}$ in the above equation gives, for any $Y = Y_1, \ldots, Y_p$,

$$\bar{x}^{1+i_0} Y^{p+j_0} = tY^p P_1(Y) Q(0,Y) + t \sum_{(i,-j)\in\mathfrak{S}} \sum_{m=0}^{j-1} \left(Q_m(\bar{x}) - \delta_{i,1} Q_m(0)\right) x^{i-1} Y^{p+m-j}. \tag{19}$$

We now combine (18) and (19) to eliminate $Q(0,Y)$:

$$(x^{1+i_0} - \bar{x}^{1+i_0}) Y^{p+j_0} = t \sum_{(i,-j)\in\mathfrak{S}} \sum_{m=0}^{j-1} \left(x^{1-i} Q_m(x) - x^{i-1} Q_m(\bar{x})\right) Y^{p+m-j}$$

for any $Y = Y_1, \ldots, Y_p$. This is the generalization of Eq. (16). The right-hand side of the above equation is a polynomial P in Y, of degree at most $p-1$. We know its value at p points, namely $Y_1, \ldots, Y_p$. The Lagrange interpolation formula implies that these p values completely determine the polynomial. As the left-hand side of the equation is algebraic, then each of the coefficients of P is also algebraic. That is,

$$t \sum_{(i,-j)\in\mathfrak{S}} \sum_{m=0}^{j-1} \left(x^{1-i} Q_m(x) - x^{i-1} Q_m(\bar{x})\right) y^{p+m-j} = \sum_{m=0}^{p-1} A_m(x) y^m,$$

where each of the A_m is an algebraic series. Let us extract the positive part of this identity. Given that i can only be 0, 1 or -1, we obtain

$$t \sum_{(i,-j)\in\mathfrak{S}} \sum_{m=0}^{j-1} \left(x^{1-i} Q_m(x) - \delta_{i,1} Q_m(0)\right) y^{p+m-j} = \sum_{m=0}^{p-1} H_m(x) y^m$$

where $H_m(x) := A_m^+(x)$ is the positive part of $A_m(x)$. Again, this series can be shown to be D-finite. Going back to the original functional equation (17), this gives

$$K(x,y) Q(x,y) = x^{1+i_0} y^{p+j_0} - ty^p P_1(y) Q(0,y) - \sum_{m=0}^{p-1} H_m(x) y^m.$$

Let us finally[4] consider the kernel $K(x,y)$ as a polynomial in x. One of its roots, denoted below X, is a formal power series in t that vanishes at $t = 0$. Replacing

[4] In the square lattice case, the symmetry of the model in x and y makes this step unnecessary: once the holonomy of $Q(x,0)$ is proved, the holonomy of $Q(x,y)$ follows.

x by this root allows us to express $Q(0,y)$ as a D-finite series:

$$ty^p P_1(y)Q(0,y) = X^{1+i_0}y^{p+j_0} - \sum_{m=0}^{p-1} H_m(X)y^m.$$

The functional equation finally reads

$$K(x,y)Q(x,y) = (x^{1+i_0} - X^{1+i_0})y^{p+j_0} - \sum_{m=0}^{p-1}(H_m(x) - H_m(X))y^m.$$

Since the substitution of an algebraic series into a D-finite one gives another D-finite series, this equation shows that $Q(x,y)$ is D-finite.

■

4 Further comments, and perspectives

We first give some asymptotic estimates for the number of n-step walks in the quadrant, for various sets of steps. Then, a number of research directions, which I have started to explore, or would like to explore in the coming months, are presented. All of them are motivated by the new proof of Kreweras' formula given in Section 2.

4.1 Asymptotic estimates

Following a suggestion of one of the referees, the table below summarizes the asymptotic behaviour on the number of n-step walks in the first quadrant, for the four models mentioned in the introduction, with three different conditions: the endpoint is fixed, the endpoint lies on the x-axis, the endpoint is free. This is all the more relevant that the argument proving the transcendence of the square lattice case is based on asymptotic estimates. The results for the two versions of the square lattice can be obtained directly from the formulas given in the introduction. The results for Kreweras' walks are derived from Theorem 2.1 and the functional equation (5), by analysing the singularities of the series [18]. The last series of results is derived from [11] using, again, an analysis of the singularities of the generating functions.

Model	Specific endpoint	Endpoint on the x-axis	Free endpoint	Nature of the series
Ordinary or diagonal sq. lattice	$\frac{4^n}{n^3}$	$\frac{4^n}{n^2}$	$\frac{4^n}{n}$	D-finite transcendental
Kreweras' walks	$\frac{3^n}{n^{5/2}}$	$\frac{3^n}{n^{7/4}}$	$\frac{3^n}{n^{3/4}}$	algebraic
Knight walks	0	$\frac{1}{n^{3/2}}\left(\frac{3}{4^{1/3}}\right)^n$	2^n	not D-finite

4.2 Other starting points

It was observed by Gessel in [20] that the method he used to prove Kreweras' result was hard to implement for a starting point different from the origin. The reason of this difficulty is that, unlike the method presented here, Gessel's approach *checks* the known expression of the generating function, but does not *construct* it. I am confident that the new approach of Section 2 can be used to solve such questions. If the starting point does not lie on the main diagonal, the $x - y$ symmetry is lost; the diagram of Figure 3 now gives *four* different equations between the *two* unknown functions $Q(x, 0)$ and $Q(0, y)$.

4.3 Other algebraic walk models

A close examination of the ingredients that make the proof of Section 2 work might help to construct other walk models which, for non-obvious reasons, would have an algebraic generating function. Note that for some degenerate sets of steps, like those of Figure 4, the quadrant condition is equivalent to a half-plane condition and thus yields an algebraic series.

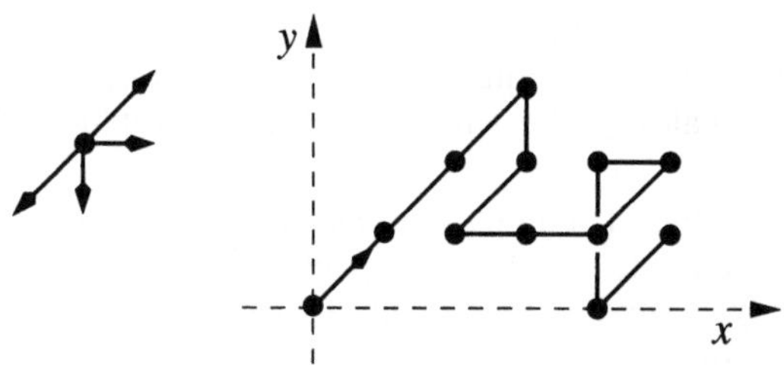

Figure 4: A degenerate set of steps.

I have started a systematic exploration of walks with few steps and only one up step: the non-trivial algebraic cases do not seem to be legion! However, I met in this exploration one model that seems to yield nice numbers (with a D-finite generating function) and for which the method of Section 2 "almost" works. I then realized that the same problem had been communicated to me, under a slightly different form, by Ira Gessel, a few months ago. I plan to explore this model further.

4.4 Other equations

Any combinatorial problem that seems to have an algebraic generating function and for which a linear functional equation with two "catalytic" variables (in the terminology of Zeilberger [45]) is available is now likely to be attacked via the method of Section 2. These conditions might seem very restrictive, but there is at least one such problem! The *vexillary* involutions, which were conjectured in 1995 to be counted by Motzkin numbers, satisfy the following equation:

$$\left(1+\frac{t^2x^2y}{1-x}+\frac{t^2y}{1-y}\right)F(x,y;t)=$$

$$\frac{t^2x^2y^2}{(1-ty)(1-txy)}+t\left(1+\frac{ty}{1-y}\right)F(xy,1;t)+\frac{t^2x^2y}{1-x}F(1,y;t).$$

The conjecture was recently proved via a difficult combinatorial construction [26]. I have been able to apply successfully the method of Section 2 to this equation [9].

4.5 Random walks in the quarter plane

Random walks in the quarter plane are naturally studied in probability. Given a Markov chain on the first quadrant, a central question is the determination of an/the invariant measure $(p_{i,j})_{i,j\geq 0}$. The invariance is equivalent to a linear equation satisfied by the series $P(x,y)=\sum p_{i,j}x^iy^j$, in which the variables x and y are "catalytic". A whole recent book is devoted to the solution of this equation in the case where the walk has small horizontal and vertical variations [15]. This book contains *one* example for which the series $P(x,y)$ is algebraic: no surprise, the steps of the corresponding walk are exactly Kreweras' steps... This result is actually due to Flatto and Hahn [19]. The equation satisfied by the series $P(x,y)$ does not work exactly like the equations for complete generating functions like $Q(x,y;t)$: roughly speaking, the third variable t is replaced by the additional constraint $P(x,y)=1$.

Very recently, I have found a new, simpler proof of Flatto and Hahn's result (at least, in the symmetric case). The principle is the same as in Section 2. One can either study a version of the enumeration problem in which each walk is weighted by its probability (so that the invariant distribution is obtained as a limit distribution), or directly adapt the method to the context of the series $P(x,y)$. With both approaches, one remains, from the beginning to the end of the proof, in the field of algebraic series [8]. This offer a significant shortcut to Flatto and Hahn's proof, which is based on non-trivial complex analysis, and uses a parametrisation of the roots of the kernel by elliptic functions, which are *not* algebraic.

Acknowledgements. To my shame, I must recall that, in the lecture that I gave at FPSAC'01 in Phoenix, I mentioned (part of) Kreweras' result as a conjecture. I am very grateful to Ira Gessel who enlightened my ignorance by giving me the right references. I also thank the anonymous referees for their very valuable comments.

References

[1] D. André, Solution directe du problème résolu par M. Bertrand, *C. R. Acad. Sci. Paris* **105** (1887) 436–437.

[2] C. Banderier, M. Bousquet-Mélou, A. Denise, D. Gardy, D. Gouyou-Beauchamps and P. Flajolet, Generating functions for generating trees, *Discrete Math.* **246** No. 1-3 (2002) 29–55.

[3] C. Banderier and P. Flajolet, Basic analytic combinatorics of directed lattice paths, to appear in *Theoret. Comput. Sci.*

[4] E. Barcucci, E. Pergola, R. Pinzani and S. Rinaldi, A bijection for some paths on the slit plane, *Adv. in Appl. Math.* **26** No. 2 (2001) 89–96.

[5] M. Bousquet-Mélou, Multi-statistic enumeration of two-stack sortable permutations, *Electron. J. Combin.* **5** (1998) R21.

[6] M. Bousquet-Mélou, Walks on the slit plane: other approaches, *Adv. in Appl. Math.* **27** No. 2-3 (2001) 243–288.

[7] M. Bousquet-Mélou, On (some) functional equations arising in enumerative combinatorics, in *Proceedings of FPSAC'01*, Arizona State University, pp. 83–89.

[8] M. Bousquet-Mélou, Walks in the quarter plane: an algebraic example, in preparation.

[9] M. Bousquet-Mélou, The number of vexillary involutions, in preparation.

[10] M. Bousquet-Mélou and M. Petkovšek, Linear recurrences with constant coefficients: the multivariate case, *Discrete Math.* **225** (2000) 51–75.

[11] M. Bousquet-Mélou and M. Petkovšek, Walks confined in a quadrant are not always D-finite, submitted.

[12] M. Bousquet-Mélou and G. Schaeffer, Walks on the slit plane, preprint 2000, `arXiv:math.CO/0012230`. To appear in *Proba. Theory Related Fields.*

[13] W. G. Brown, On the existence of square roots in certain rings of power series, *Math. Annalen* **158** (1965) 82-89.

[14] P. Duchon, On the enumeration and generation of generalized Dyck words, *Discrete Math.* **225** (2000) 121–135.

[15] G. Fayolle, R. Iasnogorodski and V. Malyshev, *Random walks in the quarter plane*, Vol. 40 of *Applications of Mathematics: Stochastic Modelling and Applied Probability*, Springer-Verlag, Berlin Heidelberg, 1999.

[16] W. Feller, *An introduction to probability theory and its applications*, Vol.1, John Wiley and Sons, New York, 1950.

[17] P. Flajolet, Analytic models and ambiguity of context-free languages, *Theoret. Comput. Sci.* **49** (1987) 283–309.

[18] P. Flajolet and A. Odlyzko, Singularity analysis of generating functions, *SIAM J. Discrete Math.* **3** No. 2 (1990) 216–240.

[19] L. Flatto and S. Hahn, Two parallel queues created by arrivals with two demands I. *SIAM J. Appl. Math.* **44** (1984) 1041–1053.

[20] I. M. Gessel, A probabilistic method for lattice path enumeration, *J. Stat. Planning Infer.* **14** (1986) 49–58.

[21] I. M. Gessel, A factorization for formal Laurent series and lattice path enumeration, *J. Combin. Theory. Ser. A* **28** (1980) 321–337.

[22] I. M. Gessel and D. Zeilberger, Random walk in a Weyl chamber, *Proc. Amer. Math. Soc.* **115** (1992) 27-31.

[23] I. P. Goulden and D. M. Jackson, *Combinatorial enumeration,* John Wiley and Sons, 1983.

[24] D. J. Grabiner, Random walk in an alcove of an affine Weyl group, and non-colliding random walks on an interval, in *Proceedings of FPSAC 2001*, Tempe, Arizona, pp. 217–225.

[25] R. Guy, B. E. Sagan and C. Krattenthaler, Lattice paths, reflections, and dimension-changing bijections, *Ars Combin.* **34** (1992) 3–15.

[26] O. Guibert, E. Pergola and R. Pinzani, Vexillary involutions are enumerated by Motzkin numbers, *Ann. Comb.* **5** (2001) 1–22.

[27] G. Kreweras, Sur une classe de problèmes liés au treillis des partitions d'entiers, *Cahiers du B.U.R.O.* **6** (1965) 5–105.

[28] J. Labelle, Langages de Dyck généralisés, *Ann. Sci. Math. Québec* **17** (1993) 53–64.

[29] J. Labelle and Y.-N. Yeh, Generalized Dyck paths, *Discrete Math.* **82** (1990) 1–6.

[30] G. F. Lawler, *Intersections of random walks*, Probabilities and its applications, Birkhäuser Boston, 1991.

[31] L. Lipshitz, The diagonal of a D-finite power series is D-finite, *J. Algebra* **113** (1988) 373–378.

[32] L. Lipshitz, D-finite power series, *J. Algebra* **122** (1989) 353–373.

[33] S. G. Mohanty, *Lattice path counting and applications*, Academic Press, 1979.

[34] T. V. Narayana, A partial order and its applications to probability theory, *Sankhya* **21** (1959) 91–98.

[35] H. Niederhausen, The ballot problem with three candidates, *European J. Combin.* **4** (1983) 161-167.

[36] H. Niederhausen, Lattice paths between diagonal boundaries, *Electron. J. Combin.* **5** (1998) R30.

[37] D. Poulalhon and G. Schaeffer, A bijection for loopless triangulations of a polygon with interior points, FPSAC 2002.

[38] G. Schaeffer, Random sampling of large planar maps and convex polyhedra, in *Proceedings of the 31th annual ACM Symposium on the Theory of Computing* (STOC'99), Atlanta, ACM press, 1999.

[39] F. Spitzer, *Principles of random walk*, The University Series in Higher Mathematics, Van Nostrand Company, Princeton, 1964.

[40] R. P. Stanley, Differentiably finite power series, *European J. Combin.* **1** (1980) 175–188.

[41] W. T. Tutte, A census of planar triangulations, *Canad. J. Math.* **14** (1962) 21–38.

[42] W. T. Tutte, A census of planar maps, *Canad. J. Math.* **15** (1963) 249–271.

[43] J. West, Sorting twice through a stack, *Theoret. Comput. Sci.* **117** (1993) 303–313.

[44] D. Zeilberger, A proof of Julian West's conjecture that the number of two-stack-sortable permutations of length n is $2(3n)!/((n+1)!(2n+1)!)$, *Discrete Math.* **102** (1992) 85–93.

[45] D. Zeilberger, The umbral transfer-matrix method: I. Foundations, *J. Combin. Theory Ser. A* **91** (2000) 451–463.

Mireille Bousquet-Mélou

CNRS, LaBRI, Université Bordeaux 1
351 cours de la Libération
33405 Talence Cedex, France
bousquet@labri.fr

Trends in Mathematics, © 2002 Birkhäuser Verlag Basel/Switzerland

Bijective Construction of Equivalent Eco-systems

Srečko Brlek, Enrica Duchi, Elisa Pergola, Renzo Pinzani

ABSTRACT: *First, we explicit an infinite family of equivalent succession rules parametrized by a positive integer α, for which two specializations lead to the equivalence of rules defining the Catalan and Schröder numbers. Then, from an ECO-system for Dyck paths, we easily derive an ECO-system for complete binary trees, by using a widely known bijection between these objects. We also give a similar construction in the less easy case of Schröder paths and Schröder trees which generalizes the previous one.*

1 Introduction

The concept of succession rule was introduced in [4] by Chung et al. in the study of Baxter permutations. Later West [12], Gire [6] and Guibert [7] used succession rules for the enumeration of permutations with forbidden sequences. More recently, this concept was deepened by Barcucci et al. [2] as a fundamental tool for ECO method, which is a method for constructing and enumerating combinatorial objects. In particular, let $\mathcal{O}$ be a class of combinatorial objects and p a parameter on $\mathcal{O}$ such that $\mathcal{O}_n = \{O \in \mathcal{O} \mid p(O) = n\}$ is finite. ECO method provides, by means of an operator ϑ, a construction for the class $\mathcal{O}$ with respect to the parameter p. If ϑ is an operator on $\mathcal{O}$ satisfying the following conditions:

$$\begin{array}{rl} \text{(i)} & \textit{for each } O' \in \mathcal{O}_{n+1}, \textit{ there exists } O \in \mathcal{O}_n \textit{ such that } O' \in \vartheta(O), \\ \text{(ii)} & \textit{for each } O, O' \in \mathcal{O}_n \textit{ such that } O \neq O', \textit{ then } \vartheta(O) \cap \vartheta(O') = \emptyset, \end{array} \qquad \text{(E)}$$

then the family of sets $\mathcal{F}_{n+1} = \{\vartheta(O) : O \in \mathcal{O}_n\}$ is a partition of $\mathcal{O}_{n+1}$.
Note that many different operators may exist on a class $\mathcal{O}$. Consequently, when an operator ϑ is fixed on $\mathcal{O}$, we will denote it by $\vartheta_{\mathcal{O}}$, and the ECO-pair by $(\mathcal{O}, \vartheta_{\mathcal{O}})$. The subscript will be omitted when no confusion arises. The conditions (E) above state that the construction of each object $O' \in \mathcal{O}_{n+1}$ is obtained from one and only one object $O \in \mathcal{O}_n$. This construction can be described by a *generating tree* [2, 4], a rooted tree whose vertices are objects of $\mathcal{O}$. The objects having the same value of p lie at the same level, and the sons of an object are the objects produced from it by using ϑ. A generating tree can be sometimes described by means of a succession rule of the form:

$$\Lambda = \begin{cases} (a) \\ (k) \rightsquigarrow (e_1)(e_2)\dots(e_k), \end{cases} \qquad (1)$$

where $a, k, e_i \in \mathbb{N}$, meaning that the root object has a sons, and the k objects $O'_1, \dots, O'_k$, produced by an object O are such that $|\vartheta(O'_i)| = e_i$, $1 \leq i \leq k$. A succession rule Ω of type (1) defines a sequence $\{f_n\}_n$ of positive integers, where f_n is the number of nodes at level n of the generating tree of Ω.

[1]The first author was supported by NSERC (Canada), and GNCS - Istituto Nazionale di Alta Matematica (Italia)

Given an ECO-system $(\mathcal{O}, p, \vartheta_{\mathcal{O}}, \Lambda)$ and a bijection $\Phi : \mathcal{O} \longrightarrow \mathcal{V}$ between two classes of combinatorial objects, it is always possible to map formally $\vartheta_{\mathcal{O}}$ on the class $\mathcal{V}$ along the bijection Φ. Indeed, let $O \in \mathcal{O}$ then we define the ECO-system $(\mathcal{V}, p', \vartheta_{\mathcal{V}}, \Lambda)$ by

$$\Phi(O') \in \vartheta_{\mathcal{V}}(\Phi(O)) \iff O' \in \vartheta_{\mathcal{O}}(O).$$

This means that the generating tree $T_{\mathcal{O}}$ for the class $\mathcal{O}$ is mapped on the generating tree $T_{\mathcal{V}}$, where each node of $T_{\mathcal{V}}$ contains the image of the corresponding node of $T_{\mathcal{O}}$. It is clear that in this case the same succession rule is obtained, but the problem of describing the operator $\vartheta_{\mathcal{V}}$, independently from Φ, remains and is not easy in general. In Section 4, we describe explicitly this construction on two examples. Firstly, we carry out the description in the easy case of the bijection between Dyck paths and complete binary trees, and, secondly, in the less easy case of the bijection between Schröder paths and Schröder trees.

Two succession rules Λ and Λ' are *equivalent* (written $\Lambda \sim \Lambda'$) if they define the same number sequence [9]. The problem of determining classes of equivalent succession rules, is still open. In section 5, by using both a combinatorial and a generating function approach, we prove that $\Omega_\alpha \sim \Omega'_\alpha$, where Ω_α and Ω'_α are defined as follows. Let $\alpha \in \mathbb{N}^+$,

$$\Omega_\alpha = \begin{cases} (\alpha) \\ (\alpha) \rightsquigarrow (\alpha+1)^\alpha \\ (k) \rightsquigarrow (\alpha+1)(\alpha+2)\dots(k-1)(k)(k+1)^\alpha, \end{cases}$$

and

$$\Omega'_\alpha = \begin{cases} (\alpha) \\ (\alpha) \rightsquigarrow (\alpha)^{\alpha-1}(2\alpha) \\ (2k\alpha) \rightsquigarrow (\alpha)^{k\alpha}(2\alpha)^{\alpha-1}(4\alpha)^\alpha(6\alpha)^\alpha\dots(2(k-1)\alpha)^\alpha(2k\alpha)^\alpha(2(k+1)\alpha). \end{cases}$$

These succession rules are related to the well known classical rules for Catalan and Schröder numbers.

2 Some classical combinatorial structures

In the plane $\mathbb{Z} \times \mathbb{Z}$, we consider lattice paths using three step types: *rise steps* $(1,1)$, *fall steps* $(1,-1)$ and *k-length horizontal steps* $(k,0)$ (briefly, k-horizontal steps).

Definition 2.1. A *generalized Motzkin path* is a sequence of rise, fall and k-horizontal steps, running from $(0,0)$ to $(n,0)$, and remaining weakly above the x-axis.

These paths have been extensively studied, an account of which can be found in [11] for instance. They include many classical lattice paths, and, among others, Dyck, Motzkin and Schröder paths correspond respectively to the cases $k = 0$, $k = 1$ and $k = 2$. A path remaining strictly above the x-axis except for $(0,0)$ and $(n,0)$ is called *elevated.* A *coloured generalized Motzkin path* is a generalized Motzkin path for which the horizontal steps can have more than one colour. We give now the classical ECO construction for Dyck and Schröder paths.

Let $\mathcal{D}$ be the class of Dyck paths, and let $D \in \mathcal{D}$. Then, $\vartheta(D)$ is the set of Dyck paths obtained by adding a peak on each point of the last sequence of D's fall steps. The rule associated to this construction is the classical rule for Catalan numbers:

$$\Omega_1 = \begin{cases} (1) \\ (1) \rightsquigarrow (2) \\ (k) \rightsquigarrow (2)\dots(k)(k+1) \end{cases}$$

Let $\mathcal{S}$ be the class of Schröder paths and let $S \in \mathcal{S}$. The set $\vartheta(S)$ contains the Schröder paths obtained from S by inserting a horizontal step at the end of S, or by inserting both a rise step in each point of the last sequence of fall and horizontal steps, and a fall one at the end of S. The rule

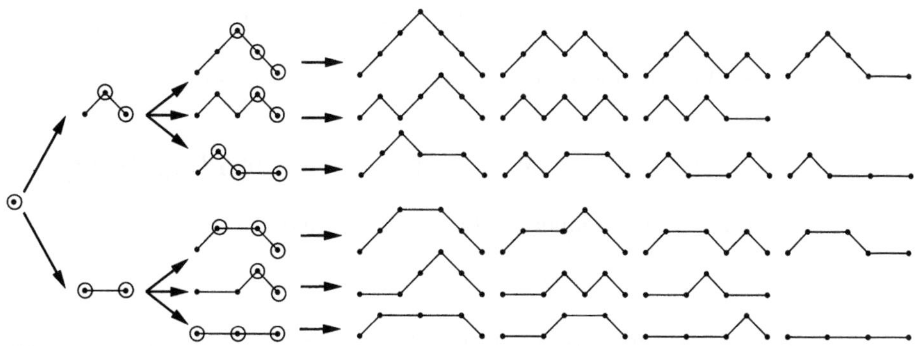

Figure 1: Classical ECO construction for Schröder paths.

$$\Omega_2 = \begin{cases} (2) \\ (2) \rightsquigarrow (3)(3) \\ (k) \rightsquigarrow (3)\dots(k)(k+1)^2 \end{cases}$$

associated to this construction is the classical rule for Schröder numbers (Fig. 1).

Definition 2.2. A *Schröder tree* is either a leaf or a list $(r, A_1, \dots, A_m)$, where $m \geq 2$, and such that each A_i is a Schröder tree.

The class $\mathcal{T}$ of Schröder trees contains planar trees whose internal nodes have degree at least two, and are enumerated by little Schröder numbers (i.e. the half of Schröder numbers) according to the number of their leaves [8]. As a particular case, the class $\mathcal{B}$ of *complete binary trees*, i. e. binary trees whose nodes have degree 0 or 2, is a subclass of $\mathcal{T}$.

3 A construction for Dyck and Schröder paths

The specialisations $\alpha = 1$ and $\alpha = 2$ of Ω'_α yield two new succession rules defining, respectively, Catalan and Schröder numbers,

$$\Omega'_1 = \begin{cases} (1) \\ (1) \rightsquigarrow (2) \\ (2k) \rightsquigarrow (1)^k(4)(6)\dots(2(k-1))(2k)(2(k+1)) \end{cases}$$

$$\Omega'_2 = \begin{cases} (2) \\ (2) \rightsquigarrow (2)(4) \\ (4k) \rightsquigarrow (2)^{2k}(4)(8)^2(12)^2\dots(4(k-1))^2(4k)^2(4(k+1)), \end{cases}$$

for which we are able to describe the corresponding constructions.

3.1 A construction for Dyck paths corresponding to Ω'_1

Each Dyck path D factors uniquely in blocks of elevated Dyck paths,

$$D = D_1D_2\dots D_k,$$

and, D is said of *even type* (respectively *odd type*) if $k = 2j$ for some j (resp. $k = 2j+1$). The last sequence of fall steps, or *last descent*, of D is denoted $\ell_d(D)$ and satisfies

$$\ell_d(D) = \ell_d(D_k).$$

Let $\mathbf{P}(D)$ be the set of points of $\ell_d(D)$, excepting the point at level 0. The set of Dyck paths having length $2n$ is denoted by $\mathcal{D}_n$, and the operator

$$\vartheta_\mathcal{D} : \mathcal{D}_n \longrightarrow 2^{\mathcal{D}_{n+1}}$$

is defined as follows:

D1. If D is of even type, then $\vartheta_\mathcal{D}(D)$ contains a single Dyck path, obtained by glueing a peak of height 1 at the end of D (see Fig. 2(D1)).

D2. If D is of odd type, then $\vartheta_\mathcal{D}(D)$ is the set of Dyck paths obtained from D by performing on each $A \in \mathbf{P}(D)$ the following actions:

(a) insert a peak;

(b) let A' be the leftmost point such that $A'A$ is a Dyck path; remove the subpath $A'A$ from D, elevate it by 1, and glue it at the end of D (see Fig. 2(D2)).

This construction yields the succession rule Ω'_1.

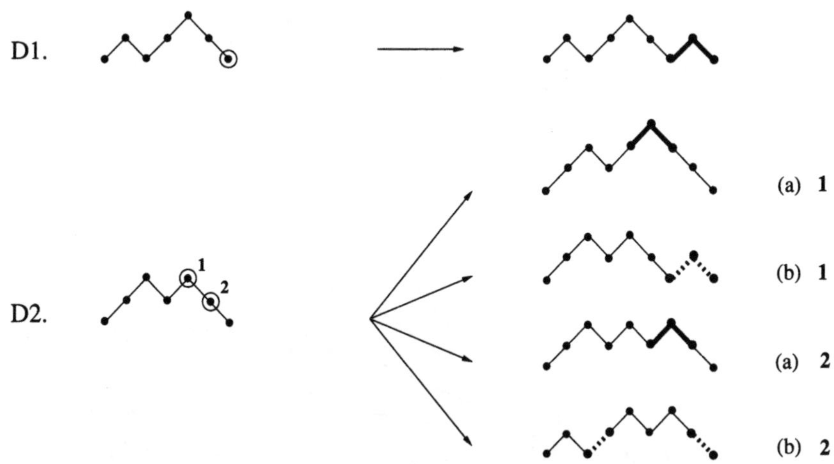

Figure 2: The construction for Dyck paths according to the rule Ω_1'.

3.2 A construction for Schröder paths corresponding to Ω_2'

We give now a similar construction for Schröder paths. Each Schröder path S factors uniquely,

$$S = S_1 S_2 \dots S_k,$$

where $S_i, 1 \leq i \leq k$, is either elevated or a horizontal step on the x-axis. The path S is said of *even type* (respectively *odd type*) if the number of elevated factors following the rightmost horizontal step is even (resp. odd). The *last descent* $\ell_d(S)$ of S is the last run of fall steps, and $\mathbf{P}(S)$ is the set of its points, excepted the last point on the x-axis.

The set of Schröder paths having length $2n$ is denoted $\mathcal{S}_n$, and the operator

$$\vartheta_S : \mathcal{S}_n \longrightarrow 2^{\mathcal{S}_{n+1}}$$

is defined by the following rules:

S1. If S is of even type, then $\vartheta_S(S)$ contains two Schröder paths, obtained respectively by glueing at the end of S, either a peak of height 1, resulting in an odd type path, or a horizontal step, resulting in an even type path (Fig. 3(S1)).

S2. If S is of odd type, then $\vartheta_S(S)$ is the set obtained by performing the following actions on every point $A \in \mathbf{P}(S)$ (Fig. 3(S2)):

(a) insert a peak of height 1 or a horizontal step;

(b) let A' be the leftmost point such that $A'A$ is a Schröder path. Then cut $A'A$, elevate it by 1, and glue it at the end of S;

(c) let A'' be the first left point such that $A''A$ is a Schröder path; if $A''A$ is not empty, then replace it by a horizontal step and glue $A''A$ at the end of S; if $A''A$ is empty then glue a horizontal step at the end of S. In this way we obtain an even type path.

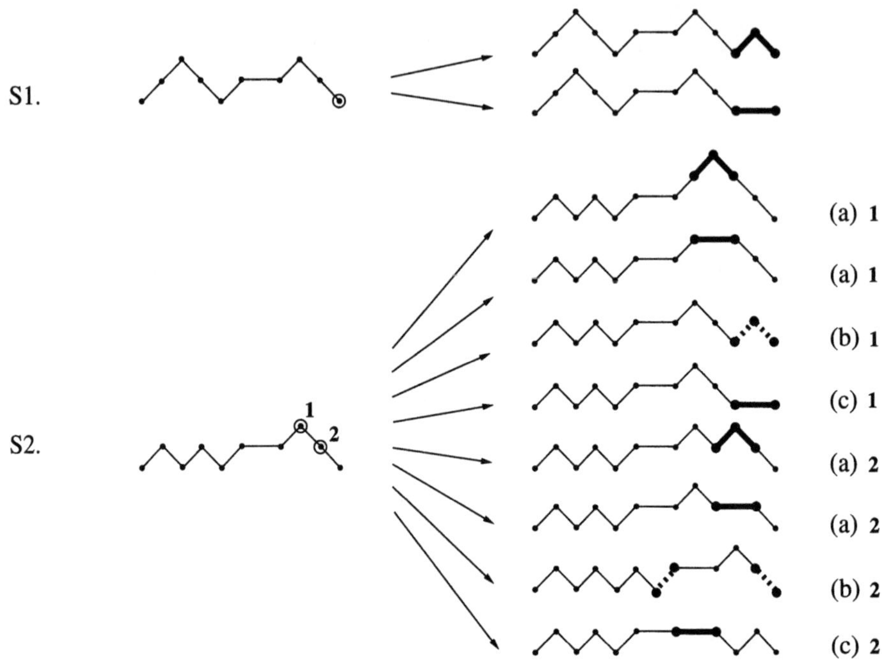

Figure 3: The construction for Schröder paths corresponding to the rule Ω_2'.

The previous construction for Schröder paths, can be easily extended to Schröder α-coloured paths by using α-coloured horizontal steps. It leads to the succession rule $\Omega_{\alpha+1}'$, with $\alpha \geq 2$. For instance, when horizontal steps of two colours are used, we obtain Schröder bi-coloured paths associated to the succession rule Ω_3'. Moreover, if we use α–coloured horizontal steps in the classical ECO construction for Schröder paths we obtain α–coloured Schröder paths to which the rule $\Omega_{\alpha+1}$, $\alpha \geq 2$, is associated. So we have proved the equivalence between Ω_α and Ω_α' in a combinatorial way.

4 A new construction for the classes $\mathcal{B}$ and $\mathcal{T}$

In this section we show how to transport an operator ϑ along a bijection, and we provide a description that is independent from the bijection in two classes of trees.

The nodes of a planar tree T can be totally ordered by means of the prefix traversal, and indexed increasingly by the integers, so that, given two nodes x_i and x_j,

$$x_i < x_j \iff i < j.$$

Accordingly, the maximum of two nodes is defined by

$$\max(x_i, x_j) = x_j \iff i < j.$$

Also, the total order allows to define notions like *first, last, successor, predecessor, etc.*, consequently, for every node p of T, we denote by (see Fig. 5 and 7):

- $\ell_i(T), \ell_l(T), \ell_s(T)$ the last, respectively, internal node, leaf, internal sibling;
- $f(p)$ the set of leaves following p;
- father(p) the father of p;
- succ(p) the successor of p;

A common abuse of notation identifies a tree with the name of its root, and, consequently subtrees as nodes. The total order extends to the the class $\mathcal{F}$ of *forests*, whose objects are lists of trees, in the obvious way, making all the above definitions relevant for forests as well.

For convenience we denote the tree consisting of a single point by "$\bullet$", and define the "tree" and "raise" constructors

$$\text{tree}, \text{raise} : \mathcal{F} \longrightarrow \mathcal{T}$$

respectively, by

$$\text{tree}(T_1, T_2, \ldots, T_k) = (\bullet, T_1, T_2, \ldots, T_k),$$

and (see Fig. 4),

$$\text{raise}(T_1, T_2, \ldots, T_k) = \text{tree}(T_1, T_2, \ldots, T_k, \bullet).$$

A useful operation on trees is the substitution. Given two trees $T_1, T_2 \in \mathcal{T}$, the

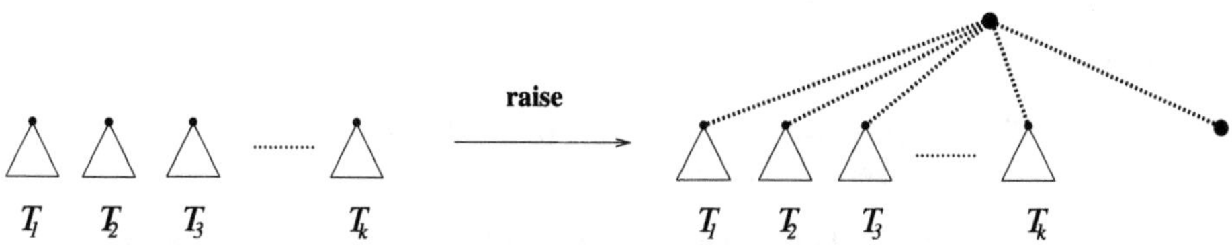

Figure 4: The raise constructor.

substituting of T_2 by T_1 ($T_2 \leftarrow T_1$) is denoted

$$\text{subs}(T_1, T_2).$$

Moreover, we say that T is of *even type* (resp. *odd type*) if the length of its rightmost branch is even (resp. odd).

>From here on, we consider this total order on two subclasses of planar trees, namely, the class $\mathcal{B}$ of complete binary trees and the class $\mathcal{T}$ of Schröder trees. The parameter p considered on these two classes of combinatorial objects is the number of leaves.

There is a well-known bijection between Dyck paths and complete binary trees,

$$\Psi : \mathcal{D} \longrightarrow \mathcal{B}$$

(for instance, see [10] and Fig. 5). For $D \in \mathcal{D}$ and $B = \Psi(D)$, define

$$\mathbf{P}(B) = f(\ell_i(B)) \setminus \{\ell_l(B)\},$$

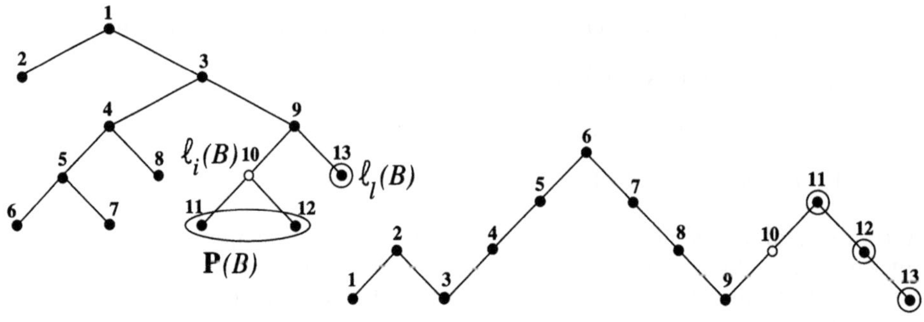

Figure 5: A complete binary tree B in $\mathcal{B}_7$, and the corresponding Dyck path.

and observe that the number of elevated Dyck paths in D corresponds to the length of the right branch of B. Moreover, we have the underlying set bijection on nodes

$$\begin{aligned} f(\ell_i(B)) &= \Psi(\ell_d(D)); \\ \mathbf{P}(B) &= \Psi(\mathbf{P}(D)). \end{aligned}$$

These observations lead to an almost direct translation of the operator $\vartheta_{\mathcal{D}}$. Indeed, let $\mathcal{B}_n$ be the set of binary trees having n leaves, and let $B \in \mathcal{B}_n$, then the operator

$$\vartheta_{\mathcal{B}} : \mathcal{B}_n \longrightarrow 2^{\mathcal{B}_{n+1}}$$

is defined as follows (see Fig. 6):

B1. if B is of even type then add two sons to $\ell_l(B)$, i.e. $\vartheta_{\mathcal{B}}(B) = \text{subs}(\text{raise}(\bullet), \ell_l(B))$;

B2. if B is of odd type then $\vartheta_{\mathcal{B}}(B)$ is the set of complete binary trees obtained by performing on each leaf $A \in \mathbf{P}(B)$ the following actions:

(a) $\text{subs}(\text{raise}(\bullet), A)$;

(b) let A' be the largest complete binary subtree of B such that $A = \ell_l(A')$; then, do $\text{subs}(\text{raise}(A'), \ell_l(B))$ and $\text{subs}(\bullet, A')$.

Clearly, $\vartheta_{\mathcal{D}}$ and $\vartheta_{\mathcal{B}}$ share the same succession rule Ω_1'.

4.1 A construction for Schröder trees

Let $\mathcal{S}'$ be the class of Schröder paths, without horizontal steps at level 0, and let $\vartheta_{\mathcal{S}'}$ be the restriction of $\vartheta_{\mathcal{S}}$ to $\mathcal{S}'$. That is

$$\vartheta_{\mathcal{S}'}(\mathcal{S}_n') = \vartheta_{\mathcal{S}}(\mathcal{S}_n) \cap \mathcal{S}_{n+1}' \,, \; \forall n \geq 1.$$

As for Dyck paths, we show how to transport the operator $\vartheta_{\mathcal{S}'}$ along the bijection [8](see Fig.7)

$$\Psi' : \mathcal{S}' \longrightarrow \mathcal{T}.$$

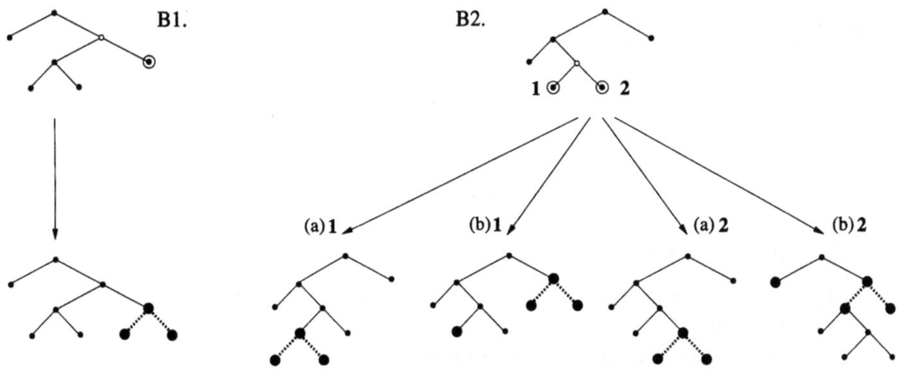

Figure 6: The construction for complete binary trees.

This bijection provides a simple interpretation of the required parameters. Indeed, a rise (resp. fall) step of S corresponds to a leftmost (resp. rightmost) sibling of T, and the horizontal steps of S correspond to the internal siblings of T, that is, those siblings strictly between the leftmost one and the rightmost one. The last run of fall steps $\ell_d(S)$ corresponds to, either the leaves following the last internal node $\ell_i(T)$, or, the last internal sibling $\ell_s(T)$ and its successors, whichever occurs the last. Therefore, define

$$z = \max(\text{succ}(\ell_i(T)), \ell_s(T)),$$

($z = 14$ in Fig. 7), and set

$$\mathbf{P}(T) = \Psi'(\mathbf{P}(S)) = \{z\} \cup f(z) \setminus \{\ell_l(T)\}.$$

Observe that this generalizes the corresponding definition in the class $\mathcal{B}$.

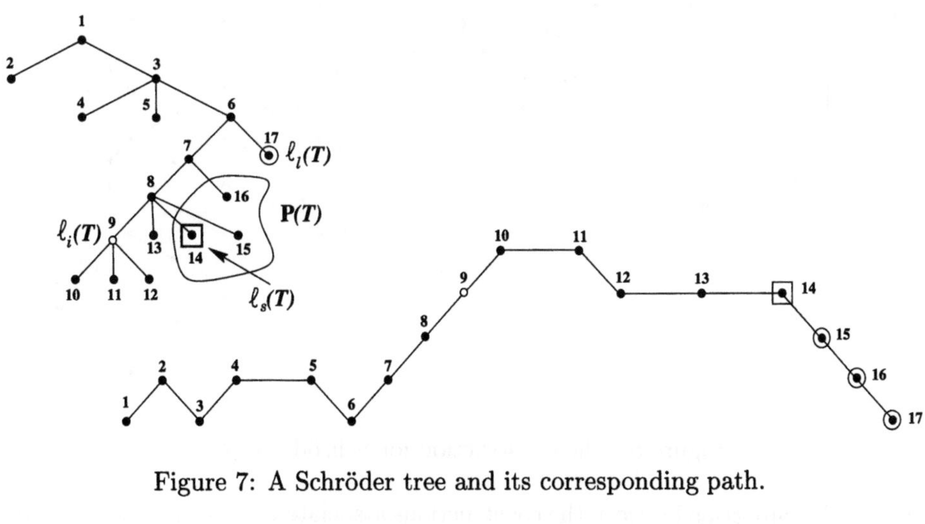

Figure 7: A Schröder tree and its corresponding path.

Let $\mathcal{T}_n$ be the set of Schröder trees having n-leaves. The operator

$$\vartheta_{\mathcal{T}} : \mathcal{T}_n \longrightarrow 2^{\mathcal{T}_{n+1}}$$

is defined as follows (see Fig. 8):

ST1. If T is of even type, then $\vartheta_{\mathcal{T}}(T) = \text{subs}(\text{raise}(\bullet), \ell_l(T))$ (see Fig. 8(ST1)).

ST2. If T is of odd type, then $\vartheta_{\mathcal{T}}(T)$ is obtained by performing on each point $A \in \mathbf{P}(T)$ the following actions (see Fig. 8(ST2)):

(a) $\text{subs}(\text{raise}(\bullet), A)$, or add a left brother to $succ(A)$;

(b) let A' be the largest Schröder subforest of T, such that $A = \ell_l(A')$; then, do
$\text{subs}(\text{raise}(A'), \ell_l(T))$ and $\text{subs}(\bullet, A')$;

(c) if $A \neq z$, let A'' be the tree having father(A) for root; then, do $\text{subs}(A'', \ell_l(T))$, $\text{subs}(\bullet, A'')$, and add a right brother to A'' .

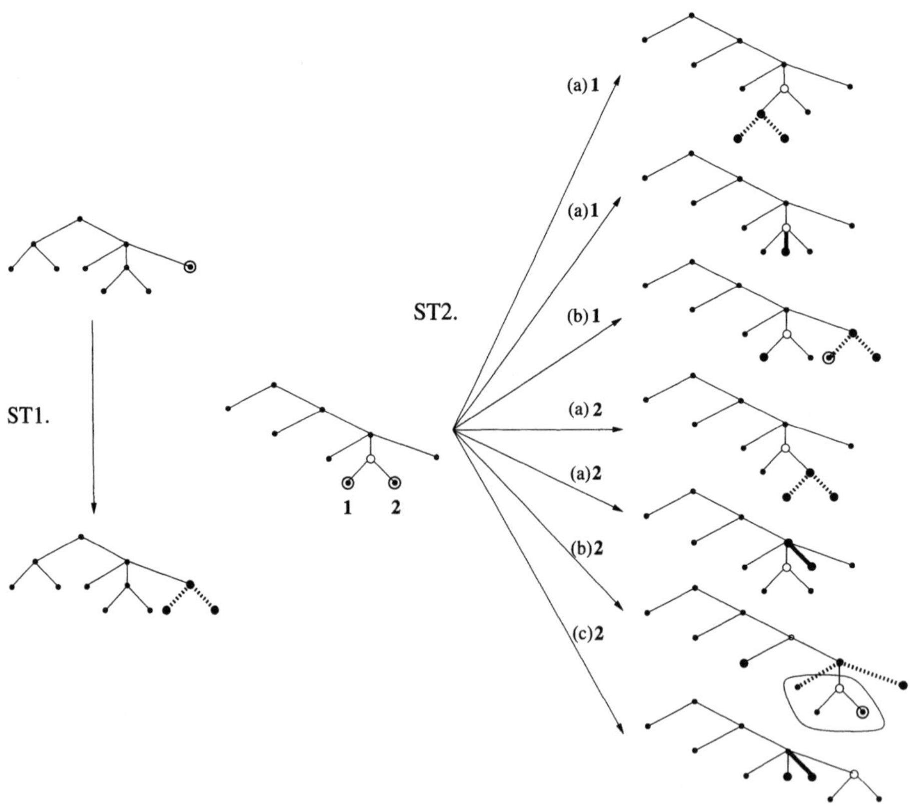

Figure 8: The construction for Schröder trees.

A careful comparison between the constructions associated to the operators $\vartheta_{\mathcal{T}}$ and $\vartheta_{\mathcal{S}}$ shows some differences. Indeed, since we are concerned with the restriction $\vartheta_{\mathcal{S}'}$, it was necessary to avoid the cases that generate a Schröder path with a horizontal step at level 0. This occurs precisely when the node z is treated.

5 Equivalence between two succession rules

We show now that the two succession rules Ω_α and Ω'_α defined in Section 1 have the same generating function. The computation is based on the *kernel method*, which was succesfully used for similar computations in [1, 3].

The bivariate generating function $F(x,y)$ counts the structures which satisfy Ω_α according to their size and the value of the associated label. Obviously, we suppose the size of the structure represented by the root of the generating tree being equal to 0. Therefore, we have:

$$F(x,y)\left(1+\frac{xy}{1-y}-xy\alpha\right)=y^\alpha+\frac{xy^{\alpha+1}}{1-y}F(x,1).$$

If

$$1+\frac{xy}{1-y}-xy\alpha=0, \tag{2}$$

then

$$y^\alpha+\frac{xy^{\alpha+1}}{1-y}F(x,1)=0.$$

The solution of the equation (2) is:

$$y_0(x)=\frac{x(\alpha-1)+1-\sqrt{(x(1-\alpha)-1)^2-4x\alpha}}{2x\alpha}$$

so, the generating function for Ω_α is:

$$\begin{aligned}F(x,1) &= \frac{y_0(x)-1}{xy_0(x)}\\ &= \frac{x(1-\alpha)+1-\sqrt{(x(1-\alpha)-1)^2-4x\alpha}}{2x}.\end{aligned}$$

In an analogous way we determine the generating function $G(x,y)$ arising from Ω'_α. After some computations we get:

$$G(x,y)=B_1(x,y)+B_2(x,y), \tag{3}$$

where

$$B_1(x,y)=y^\alpha+x(\alpha-1)y^\alpha B_1(x,1)+xy^\alpha\alpha\frac{\partial}{\partial y}B_2(x,y)\Big|_{y=1}$$

and

$$B_2(x,y)=xyB_1(x,1)+xy(\alpha-1)B_2(x,1)+\frac{xy^2\alpha}{1-y}B_2(x,1)-\frac{xy\alpha}{1-y}B_2(x,y)+xyB_2(x,y),$$

which simplify into:

$$B_1(x,1)=\frac{1+B_2(x,1)}{1-x\alpha+2x}$$

and

$$B_2(x,1) = \frac{y_0(x) - 1}{\alpha - x\alpha^2 + 3x\alpha - 2x - xy_0(x)\alpha + 2xy_0(x)}$$

where

$$y_0(x) = \frac{1 - x(\alpha - 1) - \sqrt{(x(\alpha - 1) - 1)^2 - 4x}}{2x}.$$

Substituting these values in (3), we have $F(x,1) = G(x,1)$, that is Ω_α and Ω'_α are equivalent.

6 Concluding remarks

The constructions we provided in this paper are natural because, in a sense, they commute. Indeed let $\pi_{\mathcal{D}}$ and $\pi_{\mathcal{B}}$ be the projections

$$\pi_{\mathcal{D}} : \mathcal{S} \longrightarrow \mathcal{D}; \quad \text{and} \quad \pi_{\mathcal{B}} : \mathcal{T} \longrightarrow \mathcal{B};$$

which erase, respectively, the horizontal steps and the internal siblings. The following diagram

$$\begin{array}{ccc} \mathcal{S}' & \xrightarrow{\Psi'} & \mathcal{T} \\ \pi_{\mathcal{D}} \downarrow & & \downarrow \pi_{\mathcal{B}} \\ \mathcal{D} & \xrightarrow{\Phi} & \mathcal{B} \end{array}$$

commutes, and the ECO-operators also commute. We believe that the problem of characterizing the natural bijections between objects (allowing the translation of ECO-operators) is a problem that is worth investigating.

References

[1] C. Banderier, M. Bousquet-Mélou, A. Denise, P. Flajolet, D. Gardy, and D. Gouyou-Beauchamps, (2002) *Generating functions for generating trees*, Discrete Mathematics, 246 (1-3) 29–55.

[2] E. Barcucci, A. Del Lungo, E. Pergola, R. Pinzani, (1999) *ECO: a methodology for the Enumeration of Combinatorial Objects*, Journal of Difference Equations and Applications, Vol.5, 435–490.

[3] M. Bousquet-Mélou, M. Petkovšek, (2000) *Linear Recurrences with Constant Coefficients: the Multivariate Case*, Discrete Mathematics 225 (1-3)51–75.

[4] F. R. K. Chung, R. L. Graham, V. E. Hoggatt, M. Kleimann, (1978) *The number of Baxter permutations*, Journal of Combinatorial Theory Ser. A, 24, 382–394.

[5] S. Corteel, (2000) Séries génératrices exponentielles pour les eco-systèmes signés, in D. Krob, A.A. Mikhalev, A.V. Mikhalev (Eds.), *Proceedings of the 12-th International Conference on Formal Power Series and Algebraic Combinatorics, Moscow, Russia*, Springer, 655–666.

[6] S. Gire, (1993) *Arbres, permutations à motifs exclus et cartes planaires: quelques problèmes algorithmiques et combinatoires*, Thèse de l'université de Bordeaux I.

[7] O. Guibert, (1995) *Combinatoire des permutations à motifs exclus en liaison avec mots, cartes planaires et tableaux de Young*, Thèse de l'université de Bordeaux I.

[8] J.G. Penaud, E. Pergola, R. Pinzani, O. Roques, (2001) *Chemins de Schröder et hiérarchies aléatoires*, Theoretical Computer Science, 255, 345–361.

[9] E. Pergola, R. Pinzani, S. Rinaldi, (2000) *A set of well-defined operations on succession rules*, Proceedings of MCS, Versailles,**1**, 141–152.

[10] R. P. Stanley, (1986) *Enumerative Combinatorics, Vol. I*, Wadworth & Brooks/Cole, Monterey, Cal.

[11] R. A. Sulanke, (2000) *Moments of Generalized Motzkin Paths*, J. of Integer Sequences, Vol. 3, article 00.1.1.

[12] J. West, (1995) *Generating trees and the Catalan and Schröder numbers*, Discrete Mathematics, 146, 247–262.

Srečko Brlek

LaCIM, Université du Québec à Montréal
CP 8888, Succ. Centre-Ville, Montréal (QC)
Canada H3C 3P8
brlek@lacim.uqam.ca

Enrica Duchi, Elisa Pergola, Renzo Pinzani

Dipartimento di Sistemi e Informatica
Via Lombroso 6/17, 50134 Firenze, Italy
{duchi, elisa, pinzani}@dsi.unifi.it

Trends in Mathematics, © 2002 Birkhäuser Verlag Basel/Switzerland

Random Boundary of a Planar Map

Maxim Krikun, Vadim Malyshev

ABSTRACT: *We consider the probability distribution P_N on the class of near-triangulations T of the disk with N triangles, where each T is assumed to have the weight y^m, $m = m_N = m_N(T)$ is the number of boundary edges of T. We find the limiting distribution of the random variable $m_N(T)$ as $N \to \infty$: in the critical point $y = y_{cr} = 6^{-\frac{1}{2}}$ the random variables $N^{-\frac{1}{2}} m_N$ converge to a non-gaussian distribution, for $y > y_{cr}$ for some constant c the random variables $N^{-\frac{1}{2}}(m_N - cN)$ converge to a gaussian distribution.*

1 Introduction

Enumeration of maps is an important part of the art of combinatorics. It started in sixties with the papers by W. Tutte. He invented powerful "deleting a rooted edge" and analytic "quadratic" methods, that have been exploited and developped in hundreds of subsequent papers, until nowadays. Unfortunately since then, no essentially new analytic methods for enumeration of maps appeared in combinatorics itself. This lack of essentially new ideas was compensated by two breakthroughs in other fields of mathematics and physics, where maps played an important role. One breakthrough occured in theoretical physics in eighties. Maps provided a discrete approximation to the string theory and two-dimensional quantum gravity. To deal with maps new powerful matrix methods were invented. Second one was initiated by A. Grothendieck in his program devoted to algebraic geometry and Galois theory. Some connections between these two breakthroughs were understood in nineties as having essential physical interpretation. We do not give references here, see a detailed introduction and references in [5]. For several reasons enumerative combinatorics of maps has been developping all this period in a stand alone way.

We study here some probabilistic problems for maps. Enumeration of maps deals in fact with the uniform distribution on some finite class $\mathcal{A}$ of maps. If this class has $|\mathcal{A}|$ elements then the probability of each map T is $P(T) = |\mathcal{A}|^{-1}$. In physics one is interested in the probability when maps $T \in \mathcal{A}$ have non-negative weights $w(T)$, the weights have a special Gibbs form, derived from physics. We use one below. Then the probabilities are $P(T) = Z^{-1}w(T)$, where $Z = \sum_{T\in\mathcal{A}} w(T)$ is called a partition function. We hope that rigorous probability approach can establish interconnections between differents applications of maps clearer.

As a particular case of probability for maps, we consider classes $\mathcal{T}_0(N,m)$ of rooted maps of a disk, called rooted near-triangulations in [2], with N triangles and m edges on the boundary. Enumeration problem for the number $C_0(N,m) = |\mathcal{T}_0(N,m)|$ was completely solved by Tutte [1], see also [2]. We remind that this class of maps is defined by the following restrictions: the boundary of each cell consists exactly of three edges, moreover the map is assumed to be nonseparable, thus multiple edges are allowed but no loops.

In this paper we consider the probability distribution P_N on a class $\mathcal{T}_0(N) = \cup_{m=2}^{\infty}\mathcal{T}_0(N,m)$ of maps with fixed N but variable boundary length, given by the formula

$$P_N(T) = Z_N^{-1} y^{m(T)}.$$

Here y is a positive parameter, that corresponds to $y = e^{-\mu/2}$ according to [4], and $m(T) = m_N(T)$ is the number of the boundary edges of the triangulation T. We will be interested with asymptotic propeties of the random variable $m_N = m_N(T)$. Its distribution is given by

$$P_N(m_N = m) = Z_N^{-1} y^m C_0(N, m), m \geq 2$$

where we use the normalization factor (canonical partition function)

$$Z_N(y) = \sum_{T:F(T)=N} \exp(-\frac{\mu}{2} m(T)) = \sum_{m=2}^{\infty} y^m C_0(N, m)$$

Note that N and m are always of one parity, because $m + 3N$ equals twice the number of edges, consequently $P_N(m_N = m) = 0$ if $N + m$ is odd.

In [4] relations with quantum gravity are explained, and several equivalent definitions of the distribution P_N are given, showing its naturalness, also in [4] the phase transition phenomena for m_N is described.

Here we essentially strengthen the results of section 4.2 of [4] and get explicit expressions for the limiting distributions for all three phases. Moreover, complex analytic methods we use here are quite different from [4], where the explicit combinatorial formula for $C_0(N, m)$ by Tutte was used. The method used here seems to be more adequate also in more general situations.

In the subcritical region a finite limit of m_N exists. In the critical point and the supercritical region by choosing an appropriate scaling we get a limiting distribution, which is non-gaussian or gaussian correspondingly. This is summarized in the following three theorems.

Here and further the critical parameter value is $y_{cr} \equiv \frac{1}{\sqrt{6}}$.

Theorem 1.1 (subcritical). *If $y < y_{cr}$ then for any $z, |z| < 1$, the generating function of $(m_N - 2)$,*

$$f_N(z) = \sum_{m=2}^{\infty} (m-2) P_N(m_N = m) z^{m-2},$$

for even N tends as $N \to \infty$ to

$$f_{even}(z) = \frac{(1 - \sqrt{6}yz)^{-3/2} + (1 + \sqrt{6}yz)^{-3/2}}{(1 - \sqrt{6}y)^{-3/2} + (1 + \sqrt{6}y)^{-3/2}},$$

and for odd N to

$$f_{odd}(z) = \frac{(1 - \sqrt{6}yz)^{-3/2} - (1 + \sqrt{6}yz)^{-3/2}}{(1 - \sqrt{6}y)^{-3/2} - (1 + \sqrt{6}y)^{-3/2}}.$$

Theorem 1.2 (critical). *If $y = y_{cr}$ then $\xi_N = \frac{m_N}{\sqrt{N}}$ tends in probability to the random variable ξ with the density*

$$p_\xi(x) = \frac{2}{3^{\frac{3}{2}}} \sqrt{x} e^{-\frac{x^2}{2}}, \quad x \geq 0.$$

Theorem 1.3 (supercritical). *If $y > y_{cr}$ then*

$$Em_N = c_1 N(1 + O(\frac{1}{N})), \qquad \frac{m_N - Em_N}{\sqrt{N}} \xrightarrow{Pr} \mathcal{N}(0, \sigma^2),$$

where

$$c_1 = \frac{24y^3 + 8y - (12y^2+1)\sqrt{4y^2+2}}{\sqrt{4y^2+2}(1 + 4y^2 - 2y\sqrt{4y^2+2})},$$

$$\sigma^2 = 4y\frac{32y^4 + 16y^2 + 1 - (16y^3 + 4y)\sqrt{4y^2+2}}{(2y^2+1)\sqrt{4y^2+2}(1 + 4y^2 - 2y\sqrt{4y^2+2})^2}.$$

2 The generating function

It is known [1, 2] that the generating function

$$U_0(x,y) = \sum_{N=0}^{\infty}\sum_{m=2}^{\infty} C_0(N,m)x^N y^{m-2} \tag{1}$$

is analytic in $(0,0)$ and satisfies the following equation (in a neighborhood of $(0,0)$)

$$U_0(x,y) = xy^{-1}(U_0(x,y) - U_0(x,0)) + xyU_0^2(x,y) + 1, \tag{2}$$

which also can be rewritten as

$$(2xy^2U_0(x,y) + x - y)^2 = (x-y)^2 - 4xy^3 + 4x^2y^2S(x), \tag{3}$$

where $S(x) = U_0(x,0)$. We will need some analytic techniques which slightly differs from the original method by Tutte.

Denote by $D(x,y)$ the righthand side of (3) and consider the analytic set $\mathcal{D} = \{(x,y) : D(x,y) = 0\}$ in a small neighbourhood of $(0,0)$. This set is not empty as it contains the point $(0,0)$, and it defines the branch of the function $y = y(x)$ such that $y(x) = x + O(x^2)$ in a neighbourhood of $x = 0$, we denote it further mostly by $h(x)$. In particular, it will be shown that $h(x)$ and $S(x)$ are algebraic functions. Because $D(x,y)$ is a square of an analytic function, we have two equations valid at the points of $\mathcal{D}$

$$D(x,y) = 0, \qquad \frac{\partial D(x,y)}{\partial y} = 0$$

or

$$4x^2y^2S(x) + (x-y)^2 - 4xy^3 = 0, \tag{4}$$

$$8x^2yS(x) - 2(x-y) - 12xy^2 = 0.$$

One can exclude the function $S(x)$ by multiplying the second equation (4) on $\frac{y}{2}$ and subtracting it from the first equation, then

$$y = x + 2y^3 \tag{5}$$

or

$$y = \frac{x}{1-2y^2}. \tag{6}$$

We have exposed the quadratic method belonging to Tutte. Now we have to get more information about analytic properties of the solution.

By the theorem on implicit functions equation (6) gives the unique function $h(x) = y(x)$, analytic for small x with $h(0) = 0$. It is evident from (6) that the convergence radius of $h(x)$ is finite, and its series have nonnegative coefficients. Moreover, $y(x)$ is an algebraic function satisfying the equation $y^3+py+q = 0$ with $p = -\frac{1}{2}, q = \frac{x}{2}$. The polynomial $f(y) = y^3+py+q$ has multiple roots only when $f = f'_y = 0$, which gives $x_\pm = \pm\sqrt{\frac{2}{27}}$. These roots are double roots because $f''_y \neq 0$ at these points. From $f'_y = 3y^2 - \frac{1}{2} = 0$ and $f = 0$ it follows that $y(x_\pm) = \pm\frac{1}{\sqrt{6}}$. From (6) it also follows that $x(-y) = -x(y)$ and thus $y(x)$ is odd. It follows that $y(x)$ has both $x_\pm = \pm\sqrt{\frac{2}{27}}$ as its singular points.

From (4) we know $S(x)$ explicitely. The unique branch $y(x) = h(x)$, defined by equation (6), is related to the unique branch of $S(x)$ by the equation

$$S(x) = \frac{1-3h^2(x)}{(1-2h^2(x))^2} = x^{-2}h^2(1-3h^2) \tag{7}$$

that is obtained by substituting $x = h - 2h^3$ to the first equation (4).

We know that $S(x)$ has positive coefficients, that is why $x_+ = \sqrt{\frac{2}{27}}$ should be among its first singularities. Then $x_- = -\sqrt{\frac{2}{27}}$ should also be a singularity of both $h(x)$ and $S(x)$. We proved also that the generating functions are algebraic. The principal part of the singularity at the root x_+ is $h(x) = A(x-x_+)^{d+\frac{1}{2}}$ for some integer d (as the sigularity is algebraic and the root is a double root). As $y_+ = h(x_+)$ is finite then $d \geq 0$. At the same time $h'(x) = \frac{1}{1-6h^2(x)}$ that is ∞ for $x = x_+$. It follows that $d = 0$. For $S(x)$ we have the same type of singularity $A(x-x_+)^{d+\frac{1}{2}}$ but here $d = 1$ as $S(x_+)$ and $S'(x_+)$ are finite but $S''(x_+)$ is infinite. As $y = h(x)$ is a double root of the main equation, we have by substituting (7) to (3)

$$\begin{aligned} D(x,y) &= 4y^2h^2(1-3^2h^2) + (h(1-2h^2)-y)^2 - 4y^3h^2(1-2h^2) \\ &= (y-h)^2\left(\frac{x^2}{h^2} - 4xy\right) \end{aligned} \tag{8}$$

Remember that $D(x,y) = (2xy^2U_0(x,y) + x - y)^2$, so

$$U_0(x,y) = \frac{-(x-y) + (h-y)\sqrt{d(x,y)}}{2xy^2}, \qquad d(x,y) = \frac{x^2}{h^2} - 4xy. \tag{9}$$

In the last equality we have chosen the sign appropriately, that is the sign + should be chosen so that for $x = y > 0$ the value $U_0(x,y)$ were positive.

Singularities of $U_0(x,y)$ Let us prove that for any fixed $y \in (0, y_{cr})$ the minimal singularities of $U_0(x,y)$ (as a function of x) coincide with the minimal singularities of $h(x)$ that is with $x_\pm = \pm\sqrt{\frac{2}{27}}$. Consider the right hand side of (9). All singularities of $U_0(x,y)$ that do depend on y are described by the equation $d(x,y) = 0$, which is equivalent to $\frac{4h^2(x)}{x} = y^{-1}$. The series of the function $\frac{4h^2(x)}{x}$ has all coefficients nonnegative, that's why for $|y| < y_{cr}$

$$\max_{|x|\le x_+} \left|\frac{4h^2(x)}{x}\right| = \sqrt{6} = y_{cr}^{-1} < |y^{-1}|.$$

Thus for $y < y_{cr}$ the minimal singularities are at $x_\pm$. Moreover, the equation

$$\frac{x^2}{h^2} = 4xy \tag{10}$$

becomes, as $x = h - 2h^3$,

$$\frac{h-2h^3}{h^2} - 4y = 0.$$

Its solutions are

$$h_{1,2} = -y \pm \frac{1}{2}\sqrt{4y^2+2}, \qquad x_{1,2} = 2y + 8y^3 \mp 4y^2\sqrt{4y^2+2}.$$

In particular this means that for every real y the solution of (10) is real too. As we are interested only in $y > 0$, a minimal singularity is unique and is given by choosing minus in the latter equation,

$$x_1(y) = 2y + 8y^3 - 4y^2\sqrt{4y^2+2}. \tag{11}$$

For each $y \ge \frac{1}{\sqrt{6}}$ this gives $x_1(y) \le x_{cr} = \sqrt{\frac{2}{27}}$, equalities are achieved simultaneously. This can be easy checked by plotting a graph of $(h-2h^3)/h^2$ and using the fact that the function $h(x)$ is strictly increasing, we omit this construction.

3 Subcritical region

The canonical partition function is the coefficient in the expansion

$$U_0(x,y) = \sum_{N=0}^{\infty} Z_N(y)x^N.$$

$U_0(x,y)$ is algebraic, and we will prove that for any fixed $y, 0 < y < y_{cr}$, in the vicinity of $x_\pm$

$$U_0(x,y) = f_{\pm,0}(x,y) + f_{\pm,1}(x,y)(1-\frac{x}{x_\pm})^{\frac{3}{2}}$$

where for fixed y the functions $f_{\pm,0}, f_{\pm,1}$ are analytic near $x_\pm$ correspondingly, the values of $f_{\pm,1}$ at $x_\pm$ are nonzero, namely

$$f_{+,1}(x_+,y) = \frac{6^{\frac{3}{4}}3}{(1-\sqrt{6}y)^{3/2}}, \quad f_{-,1}(x_-,y) = \frac{6^{\frac{3}{4}}3}{(1+\sqrt{6}y)^{3/2}}.$$

Expand $h(x)$ near $x_+ = \sqrt{\frac{2}{27}}$ in $t = x_+ - x$

$$h(x) = \frac{1}{\sqrt{6}} - \frac{1}{\sqrt[4]{6}}t^{1/2} - \frac{1}{6}t - \frac{5\sqrt[4]{6}}{72}t^{3/2} + O(t^2) \quad . \tag{12}$$

Substitute (12) together with $x = x_+ - t$ to the expression (9) for $U_0(x,y)$ and expand in $t^{1/2}$

$$U_0(x_+ - t, y) = a_+(y) + b_+(y)t + \frac{6^{3/4}3}{(1-\sqrt{6}y)^{3/2}}t^{3/2} + O(t^2).$$

Similary we find

$$U_0(x_- + t, y) = a_-(y) + b_-(y)t + \frac{6^{3/4}3}{(1+\sqrt{6}y)^{3/2}}t^{3/2} + O(t^2),$$

Then as $N \to \infty$

$$Z_N(y) \sim 6^{3/4}3\Big(\frac{1}{(1-\sqrt{6}y)^{3/2}} + \frac{(-1)^N}{(1+\sqrt{6}y)^{3/2}}\Big)[x^N](x_+ - x)^{\frac{3}{2}} \tag{13}$$

This is known under different names (for example, as Darboux theorem in [3]). However, it can be proved elementarily, using the following expansion for $a = \frac{3}{2}$

$$t^a = (x_0 - x)^a = \sum_{N=0}^{\infty} \frac{\Gamma(N-a)}{N!\Gamma(-a)} x_0^{a-N} x^N \tag{14}$$

where $[x^N]F(x)$ stands for the N-th coefficient in the $F(x)$ power series. Secondly, subtracting this main term and proving that the rest is asymptotically negligible. In fact, (13) should be read as two separate equations,

$$Z_N(y) \sim 6^{3/4}3\Big(\frac{1}{(1-\sqrt{6}y)^{3/2}} \pm \frac{1}{(1+\sqrt{6}y)^{3/2}}\Big)[x^N](x_+ - x)^{\frac{3}{2}},$$

with a plus sign standing for even values of N and a minus sign for odd.

Finally for given y the generating function for $m_N - 2$ is obtained from the partition function $Z_N(y)$ by normalization, that is

$$f_N(z) = \sum_{m=2}^{\infty} P\{m_N = m\} z^{m-2} = \frac{Z_N(yz)}{Z_N(y)},$$

and after taking limits in N (by even an odd values separately) we come to the assertion of Theorem 1.1.

4 Critical point

In a critical point the expectation of m_N has no finite limit. To describe the limiting distribution we shall calculate the asymptotics (as $N \to \infty$) of the factorial

moments of m_N and find the appropriate scaling. That is we have to study the singularities of all the partial derivatives $\frac{\partial^n}{\partial y^n}U_0(x,y)$ at $y = y_{cr}$, as we have done in the previous section for $U_0(x,y)$ only.
From the previous analysis we know that for $y = y_{cr}$ the singularity defined by $d(x,y) = 0$ is among the minimal ones. According to (11) is equal to $\sqrt{\frac{2}{27}}$ and coinsides to x_+ singularity of $h(x)$, so there are two minimal singularities at points x_+ and x_-.

Lemma 4.1. *Put $t = x - x_0$. Then there exist functions $\varphi_{n,i}(t) = \varphi_{n,i}(t,y), i = 0,1,2,$ analytic in the vicinity of $t = 0$ such that*

$$U_0^{(n)}(x,y_{cr}) = \varphi_{n,0}(t) + \varphi_{n,1}(t)\,t^{3/4-n/2} + \varphi_{n,2}(t)\,t^{5/4-n/2}, \quad \varphi_{n,1}(0) \neq 0$$

Proof. Instead of calculating the y-derivatives of $U_0(x,y)$ we calculate them for $2xy^2U_0(x,y)$, which is much simpler, but keeps all information on $C_0(N,m)$. We have

$$xy^2U_0(x,y) = y - x + (h-y)\sqrt{4x}\Big(\frac{x}{4h^2} - y\Big)^{1/2}, \quad x \geq 0,$$

$$xy^2U_0(x,y) = y - x + (h-y)\sqrt{-4x}\Big(-\frac{x}{4h^2} + y\Big)^{1/2}, \quad x \leq 0.$$

To get the derivatives put $y = y_{cr} + u$ and consider the formal series in u:

$$\begin{aligned} 2xy^2U_0(x,y)\Big|_{y=y_{cr}+u} &= (y_{cr} - x) + u + \Big((h - y_{cr}) - u\Big) \\ &\times\sqrt{4x}\sum_{n=0}^{\infty}\frac{\Gamma(n-\frac{1}{2})}{n!\Gamma(-\frac{1}{2})}\Big(\frac{x}{4h^2} - y_{cr}\Big)^{1/2-n}u^n, \quad x \geq 0, \end{aligned}$$

$$\begin{aligned} 2xy^2U_0(x,y)\Big|_{y=y_{cr}+u} &= (y_{cr} - x) + u + \Big((h - y_{cr}) - u\Big) \\ &\times\sqrt{-4x}\sum_{n=0}^{\infty}\frac{\Gamma(n-\frac{1}{2})}{n!\Gamma(-\frac{1}{2})}\Big(-\frac{x}{4h^2} + y_{cr}\Big)^{1/2-n}(-u)^n, \quad x \leq 0. \end{aligned}$$

For $n > 1$, $x \geq 0$ the n-the coefficient (we denote it $[u^n]$) is equal to

$$\begin{aligned} [u^n]\Big(2xy^2U_0(x,y)\Big|_{y=y_0+u}\Big) &= \sqrt{4x}\frac{\Gamma(n-\frac{3}{2})}{(n-1)!\Gamma(-\frac{1}{2})}\Big(\frac{x}{4h^2} - y_{cr}\Big)^{3/2-n} \\ &\quad -(h-y_{cr})\sqrt{4x}\frac{\Gamma(n-\frac{1}{2})}{n!\Gamma(-\frac{1}{2})}\Big(\frac{x}{4h^2} - y_{cr}\Big)^{1/2-n} \\ &= \sqrt{4x}\frac{\Gamma(n-\frac{3}{2})}{(n-1)!\Gamma(-\frac{1}{2})}\Big(\frac{x}{4h^2} - y_{cr}\Big)^{3/2-n} \\ &\quad \times\Big(1 - (h-y_{cr})\frac{n-\frac{3}{2}}{n}\Big(\frac{x}{4h^2} - y_{cr}\Big)^{-1}\Big), \end{aligned} \tag{15}$$

and similarly for $n > 1$, $x \le 0$

$$\begin{aligned}
[u^n]\Big(2xy^2U_0(x,y)\Big|_{y=y_0+u}\Big) &= \sqrt{-4x}\frac{\Gamma(n-\frac{3}{2})}{(n-1)!\Gamma(-\frac{1}{2})}\Big(-\frac{x}{4h^2}+y_{cr}\Big)^{3/2-n}(-1)^{n-1} \\
&\quad -(h-y_{cr})\sqrt{4x}\frac{\Gamma(n-\frac{1}{2})}{n!\Gamma(-\frac{1}{2})}\Big(-\frac{x}{4h^2}+y_{cr}\Big)^{1/2-n}(-1)^n \\
&= \sqrt{4x}\frac{\Gamma(n-\frac{3}{2})}{(n-1)!\Gamma(-\frac{1}{2})}\Big(-\frac{x}{4h^2}+y_{cr}\Big)^{3/2-n}(-1)^{n-1} \\
&\quad \times\Big(1+(h-y_{cr})\frac{n-\frac{3}{2}}{n}\Big(-\frac{x}{4h^2}+y_{cr}\Big)^{-1}\Big). \qquad (16)
\end{aligned}$$

Next we need the following auxiliary expansions

$$(h-y_{cr})\Big(\frac{x}{4h^2}-y_{cr}\Big)^{-1}\Big|_{x=x_+-t} = -\frac{1}{2}+\frac{3}{8}6^{1/4}t^{1/2}+O(t),$$

$$\Big(\frac{x}{4h^2}-y_{cr}\Big)\Big|_{x=x_+-t} = \frac{1}{3}6^{3/4}t^{1/2}+O(t),$$

$$(h-y_{cr})\Big(-\frac{x}{4h^2}+y_{cr}\Big)^{-1}\Big|_{x=x_-+t} = -1+\frac{3}{2}6^{1/4}t^{1/2}+O(t),$$

$$\Big(-\frac{x}{4h^2}+y_{cr}\Big)\Big|_{x=x_-+t} = \frac{1}{3}\sqrt{6}+\frac{1}{3}6^{3/4}t^{1/2}+O(t).$$

(note that the second one has no constant term). Using these expansions we obtain from (15) and (16) the behaviour of the $U_0(x,y)$ derivatives near $x_\pm$, namely

$$\frac{\partial^n}{\partial y^n}U_0(x,y)\Big|_{x=x_+-t} = \text{const } t^{3/4-n/2}(1+O(t^{1/2})),$$

$$\frac{\partial^n}{\partial y^n}U_0(x,y)\Big|_{x=x_-+t} = \text{const } + O(t^{1/2}).$$

Lemma is proved.

The factorial moments of m_N are

$$M_1(N) \sim 2^{-2}3^2\frac{\Gamma(-\frac{3}{4})}{\Gamma(-\frac{1}{4})}N^{\frac{1}{2}}, \qquad M_2(N) \sim 2^{-4}3^4\frac{-\Gamma(-\frac{3}{4})}{\Gamma(\frac{1}{4})}N,$$

$$M_n(N) = \frac{[x^N]U_n}{[x^N]U_0} \sim 2^{-2n}3^{n+1}(2n-1)(2n-5)!!\frac{-\Gamma(-\frac{3}{4})}{\Gamma(\frac{n}{2}-\frac{3}{4})}N^{n/2},$$

Consequently the moments of a random variable $\xi = \lim_{N\to\infty} m_N/\sqrt{N}$ are

$$E\xi = 3(3/4)\frac{\Gamma(\frac{3}{4})}{\Gamma(-\frac{1}{4})}, \qquad E\xi^2 = 3(3/4)^2\frac{-\Gamma(-\frac{3}{4})}{\Gamma(\frac{1}{4})} = \frac{9}{4},$$

$$E\xi^n = 2^{-2n}3^{n+1}(2n-1)(2n-5)!!\frac{-\Gamma(-\frac{3}{4})}{\Gamma(\frac{n}{2}-\frac{3}{4})} = \frac{\Gamma(\frac{n}{2}+\frac{3}{4})3^n}{\Gamma(\frac{3}{4})}.$$

The moment generating function for ξ^2 is uniquely defined by this sequence (by classical uniqueness criteria, see sections VII.3 and VIII.6(b) of [6]), as they grow slower than $C^n n!$ for some C) and is equal to

$$\varphi_{\xi^2}(s) = \sum_{n=0}^{\infty} E\xi^{2n}\frac{(-s)^n}{n!} = (1+9s)^{-3/4}$$

Using the Laplace transform we get the density of ξ^2

$$p_{\xi^2}(t) = 3^{-\frac{3}{2}}\frac{1}{\Gamma(\frac{3}{4})}e^{-\frac{t}{9}}t^{-1/4}$$

5 Supercritical region

We shall prove that $Em_N \sim cN$ and all the semiinvariants (coefficients in the Taylor expansion of the logarithm of the generating function) of m_N are of order N. Then it follows that the semiinvariants of order greater than two of a scaled random variable $(m_N - Em_N/\sqrt{N}$ tend to zero as $N \to \infty$, which means the imiting distribution is uniquely defined by its moments (see above), and moreover it is gaussian (as the log of its generating function is a quadratic polynomial).

The semiinvariants of m_N are given by the formula

$$s_k(N) = (\frac{\partial}{\partial\lambda})^k \ln\varphi_N(\lambda)|_{\lambda=0}, \quad k \geq 1,$$

where

$$\varphi_N(t) = Ee^{\lambda m_N} = \frac{[x^N]U_0(x, ye^\lambda)}{[x^N]U_0(x,y)}$$

ined thing is the characteristic function of m_N.

We saw that for fixed $y > y_{cr}$ the minimal singularity of $U_0(x,y)$ (as the function of x) is unique and is given by (11). The expansion of $U_0(x,y)$ (as the function of x) at the singular point $x_{cr}(y)$ is

$$U_0(x,y) = a(y) + b(y)(x_{cr}(y)-x)^{1/2} + O(|x_{cr}(y)-x|)$$

for some constants $a(y), b(y)$. Then

$$[x^N]U_0(x,y) \sim b(y)[x^N](x - x_{cr}(y))^{1/2} = b(y)\frac{\Gamma(N-\frac{1}{2})}{N!\Gamma(-\frac{1}{2})}x_{cr}(y)^{\frac{1}{2}-N},$$

$$\ln\varphi_N(t) = \ln[x^N]U_0(x,ye^\lambda) - \ln[x^N]U_0(x,y) \sim N\Big(-\ln x_{cr}(ye^\lambda) + \ln x_{cr}(y)\Big).$$

It follows that all semiinvariants of m_N are $O(N)$.

6 Some remarks

Equivalent presentations of the model The factor $y^m = \exp(-\frac{\mu}{2}m)$ is quite natural: it is derived from the Hilbert-Einstein action in two-dimensional pure quantum gravity, see introductory exposition in [5]. The case $y = 1$ that could be natural for combinatorics seems to have no special interest for physics, where the critical point is of most interest. We could assign weights to maps as $\exp(-\mu L(T)$, where $L(T)$ is the number of all edges of the map T. This would give the same probability distribution because of the formula $|L(T)| = \frac{3N}{2} + \frac{m(T)}{2}$.

Second kind phase transition The free energy for this model is defined as

$$F(\mu) = \lim_{N\to\infty} \frac{1}{N} \log Z_{0,N}, Z_{0,N} = \sum_T \exp\{-\mu L(T)\}$$

The next theorem gives an explicit formula for the free energy, it corrects a calculational mistake in the corresponding result in [4]. It shows also that the phase transtion is a second order phase transition, as in the critical point the free energy is differentiable but not twice differentiable.

Theorem 6.1. *The free energy is equal to* $-\frac{3}{2}\mu + \ln\left(\sqrt{\frac{27}{2}}\right)$ *if* $y \le y_{cr}$ *and is equal to* $-\frac{3}{2}\mu + \ln x_{cr}(y)$ *if* $y > y_{cr}$.

Proof. It easily follows from the proofs in the preceding sections. We have

$$Z_{0,N} = \sum_T \exp\{-\mu L(T)\} = \sum_T \exp\{-\frac{\mu}{2}(3N+m)\} = \exp\{-\frac{3}{2}\mu N\}[x^N]U_0(x, e^{-\mu/2}),$$

$$\frac{1}{N} \log Z_{0,N} = -\frac{3}{2}\mu + \frac{1}{N} \log\left([x^N]U_0(1, e^{-\mu/2})\right)$$

Put $y = e^{-\mu/2}$. Following section 3, as $y < y_{cr}$:

$$[x^N]U_0(1, e^{-\mu/2}) = f(y)[x^N](x_0 - x)^{3/2},$$

$$\frac{1}{N} \log Z_{0,N} \to -\frac{3}{2}\mu + \ln x_0 = -\frac{3}{2}\mu + \ln\left(\sqrt{\frac{27}{2}}\right).$$

When $y = y_{cr}$:

$$[x^N]U_0(1, e^{-\mu/2}) = f(y)[x^N](x_0 - x)^{3/4}$$

$$\frac{1}{N} \log Z_{0,N} \to -\frac{3}{2}\mu + \ln\left(\sqrt{\frac{27}{2}}\right).$$

Following section 5, as $y > y_{cr}$:

$$[x^N]U_0(1, e^{-\mu/2}) = b(y)[x^N](x_{cr}(y) - x)^{1/2},$$

$x_{cr}(y) = 2y + 8y^3 - 4y^2\sqrt{4y^2 + 2}$ being defined as in (11) we get

$$\frac{1}{N} \log Z_{0,N} \to -\frac{3}{2}\mu + \ln x_{cr}(y)$$

Further problems The similar problem for two holes in the sphere could be the next solvable problem, that is consider a ring (or cylinder) with two boundaries of lengths m_1, m_2. Joint distribution of these two random variables is to be found. Not that if for one boundary there is the combinatorial formula for $C_0(N, m)$

$$C_0(N,m) = \frac{2^{j+2}(2m+3j-1)!(2m-3)!}{(j+1)!(2m+2j)!((m-2)!)^2}$$

by Tutte (used in [4]). Nothing similar is known for the number $C_0(N, m_1, m_2)$ of rooted near triangulations of a ring with N triangles and the lengths m_1, m_2 of the boundaries, where only analytic methods can be of use.

References

[1] W. Tutte, (1962) *A Census of Planar Triangulations.* Canad. J. of Math., 14, 21-38.

[2] I. Goulden, D. Jackson, (1983) *Combinatorial Enumeration.* John Wiley.

[3] P. Henrici, (1977) *Applied and computational complex analysis*, v.2. John Wiley.

[4] V. Malyshev, (2001) *Gibbs and Quantum discrete spaces.* Russian Math. Reviews., No. 5, 117-172.

[5] V. Malyshev, 2001 *Combinatorics and Probability of Maps.* In "Proceedings of the NATO Advanced Study Institute on Asymptotic Combinatorics in Mathematical Physics". Kluwer (to appear).

[6] W. Feller, *An introduction to probability theory and its applications*, Vol. II. John Wiley, 2nd edition.

Maxim Krikun

Moscow State University
Vorobievy Gory
Moscow, Russia
krikun@lbss.math.msu.su

Vadim Malyshev

I.N.R.I.A.
B.P. 105, Rocquencourt
78153, Le Chesnay Cedex
Vadim.Malyshev@inria.fr

Trends in Mathematics, © 2002 Birkhäuser Verlag Basel/Switzerland

Enumération des 2-arbres k-gonaux

Gilbert Labelle, Cédric Lamathe, Pierre Leroux

RÉSUMÉ : *Dans ce travail*[1], *nous généralisons les 2-arbres en remplaçant les triangles par des quadrilatères, des pentagones ou des polygones à k côtés (k-gones), où $k \geq 3$ est fixe. Cette généralisation, aux 2-arbres k-gonaux, est naturelle et est étroitement liée dans le cas planaire aux arbres cellulaires. Notre objectif est le dénombrement, étiqueté et non étiqueté, des 2-arbres k-gonaux selon le nombre n de k-gones. Nous donnons des formules explicites dans le cas étiqueté, et, dans le cas non étiqueté, des formules de récurrence et des formules asymptotiques.*

ABSTRACT: *In this paper*[1], *we generalize 2-trees by replacing triangles by quadrilaterals, pentagons or k-sided polygons (k-gons), where $k \geq 3$ is given. This generalization, to k-gonal 2-trees, is natural and is closely related, in the planar case, to some specializations of the cell-growth problem. Our goal is the enumeration, labelled and unlabelled, of k-gonal 2-trees according to the number n of k-gons. We give explicit formulas in the labelled case, and, in the unlabelled case, recursive and asymptotic formulas.*

1 Introduction

L'espèce des arbres bidimensionnels, ou 2-arbres, a été bien étudiée dans la littérature. Voir par exemple [4] et [2, 3]. Essentiellement, un 2-arbre est un graphe simple connexe constitué de triangles qui sont liés entre eux par les arêtes de manière arborescente, c'est-à-dire sans former de cycles (de triangles). Dans [5], Harary et al. ont énuméré une variante des arbres cellulaires (relié au "cell-growth problem"), à savoir des 2-arbres k-gonaux plans et planaires[2], dans lesquels les triangles ont été remplacés par des quadrilatères, des pentagones ou des polygones à k côtés (k-gones), où $k \geq 3$ est fixe. De tels 2-arbres, bâtis sur des k-gones, sont appelés 2-arbres k-gonaux. Cette généralisation apparaît naturellement et le but de ce travail est l'énumération des 2-arbres k-gonaux libres, c'est-à-dire vus comme graphes simples, sans question de planarité. La figure 1 a) propose un exemple de 2-arbres k-gonal, dans le cas où $k = 4$.

Nous disons qu'un 2-arbre k-gonal est *orienté* si ses arêtes sont orientées de façon telle que chaque k-gone forme un cycle orienté, voir la figure 1 b). Notons par $\mathcal{a}$ et par $\mathcal{a}_o$ les espèces des 2-arbres k-gonaux et des 2-arbres k-gonaux orientés respectivement. Pour ces deux espèces, nous utilisons les symboles $-$, $\diamond$ et $\underline{\diamond}$ en exposant pour indiquer que les structures ont été pointées en une arête, en un polygone, et en un polygone muni d'une arête distinguée, respectivement.

Notre objectif est le dénombrement, étiqueté et non étiqueté, des 2-arbres k-gonaux selon le nombre n de k-gones. Nous donnons des formules explicites dans le cas étiqueté, et dans le cas non étiqueté, des formules de récurrence et des formules asymptotiques. Pour cela, nous adaptons l'approche de Fowler et al. dans [2, 3]

---

[1] Avec l'appui du FCAR (Québec) et du CRSNG (Canada)

[2] Au sens où toutes les faces, à part la face externe, sont des k-gones

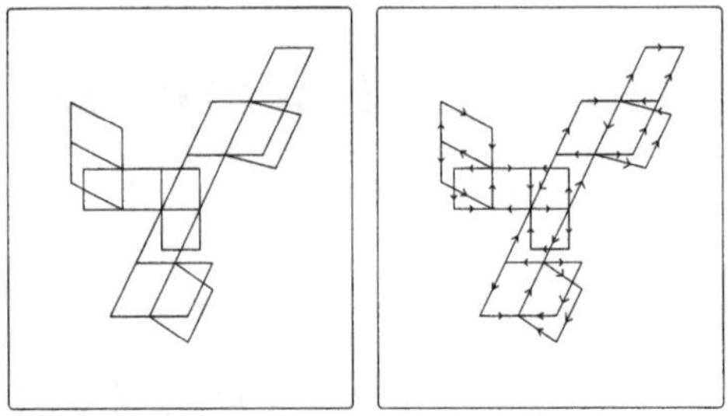
a) b)

Figure 1: Un 2-arbre 4-gonal non orienté et orienté

qui correspond au cas $k = 3$. En particulier, les 2-arbres sont étiquetés aux k-gones. La principale difficulté à cette extension vient, comme on le verra, du cas où k est pair.

Les deux premières étapes sont assez directes. Il s'agit d'étendre le théorème de dissymétrie au cas k-gonal et de caractériser l'espèce $B = \mathfrak{a}^{\to}$ des 2-arbres k-gonaux munis d'une arête distinguée et orientée, à l'aide d'une équation fonctionnelle de type lagrangien. Le premier résultat est une extension immédiate du cas $k = 3$ et la démonstration est omise.

Théorème 1.1. THÉORÈME DE DISSYMÉTRIE. *Les espèces $\mathfrak{a}$ et $\mathfrak{a}_o$ des 2-arbres k-gonaux orientés et non orientés respectivement satisfont les isomorphismes d'espèces suivants :*

$$\mathfrak{a}_o^{-} + \mathfrak{a}_o^{\diamond} = \mathfrak{a}_o + \mathfrak{a}_o^{\underline{\diamond}}, \tag{1}$$

$$\mathfrak{a}^{-} + \mathfrak{a}^{\diamond} = \mathfrak{a} + \mathfrak{a}^{\underline{\diamond}}. \tag{2}$$

Dans la prochaine section, nous caractérisons l'espèce $B = \mathfrak{a}^{\to}$ et nous en donnons ses propriétés. Par la suite, nous exprimons les diverses espèces pointées qui apparaissent dans le théorème de dissymétrie en fonction de l'espèce B et nous en déduisons les résultats énumératifs désirés pour les espèces $\mathfrak{a}_o$ et $\mathfrak{a}$. Le cas orienté, plus simple, est traité d'abord, dans la section 3. Le cas non orienté, suit, dans la section 4, en distinguant les deux cas de parité de k, pour le dénombrement non étiqueté. Enfin, les résultats asymptotiques sont présentés dans la section 5.

2 L'espèce $B = \mathfrak{a}^{\to}$

L'espèce $B = \mathfrak{a}^{\to}$ joue un rôle fondamental dans l'étude des 2-arbres k-gonaux.

Théorème 2.1. *L'espèce $B = \mathfrak{a}^{\to}$ des 2-arbres k-gonaux pointés en une arête orientée satisfait l'équation (isomorphisme) fonctionnelle suivante :*

$$B = E(XB^{k-1}), \tag{3}$$

où E représente l'espèce des ensembles.

Preuve. On décompose une $\mathfrak{a}^{\to}$-structure en un ensemble de *pages*, c'est-à-dire en sous-graphes maximaux qui partagent un seul k-gone avec l'arête distinguée.

Pour chaque page, l'orientation de l'arête pointée permet alors de définir un ordre et une orientation sur les $k-1$ arêtes restantes du polygone possédant cette arête, selon la figure 2 a) pour le cas impair, et b) pour le cas pair. Ces arêtes étant orientées, on peut alors y accrocher des B-structures. On en déduit alors l'équation (3). ∎

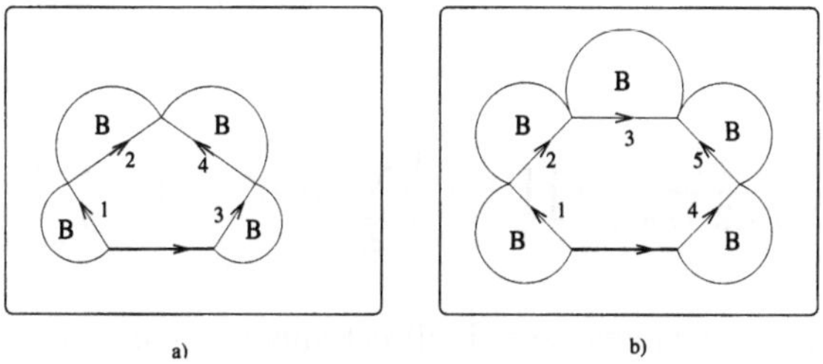

Figure 2: Une page orientée a) $k = 5$ b) $k = 6$

On peut relier simplement l'espèce $B = \mathfrak{a}^{\rightarrow}$ à celle des arborescences (arbres enracinés), A, caractérisée par l'équation fonctionnelle $A = XE(A)$, où X est ici l'espèce des sommets. En effet de (3), on déduit successivement

$$(k-1)XB^{k-1} = (k-1)XE((k-1)XB^{k-1}), \tag{4}$$

sachant que $E^m(X) = E(mX)$, et, par unicité,

$$(k-1)XB^{k-1} = A((k-1)X). \tag{5}$$

Finalement, on obtient l'expression suivante pour l'espèce B en fonction de l'espèce des arborescences :

Proposition 2.2. *L'espèce $B = \mathfrak{a}^{\rightarrow}$ des 2-arbres k-gonaux pointés en une arête orientée vérifie*

$$B = \sqrt[k-1]{\frac{A((k-1)X)}{(k-1)X}}. \tag{6}$$

Proposition 2.3. *Les nombres $a_n^{\rightarrow}$, $a_{n_1,n_2,\ldots}^{\rightarrow}$, et $b_n = \widetilde{a}_n^{\rightarrow}$ de 2-arbres k-gonaux pointés en une arête orientée et ayant n k-gones, respectivement étiquetés, laissés fixes par une permutation de $\mathbb{S}_n$ de type cyclique $1^{n_1}2^{n_2}\ldots$, et non étiquetés, satisfont les relations suivantes :*

$$a_n^{\rightarrow} = ((k-1)n+1)^{n-1} = m^{n-1}, \tag{7}$$

où $m = (k-1)n+1$ est le nombre d'arêtes,

$$a_{n_1,n_2,\ldots}^{\rightarrow} = \prod_{i=1}^{\infty}(1+(k-1)\sum_{d|i} dn_d)^{n_i-1}(1+(k-1)\sum_{\substack{d|i\\ d<i}} dn_d). \tag{8}$$

et

$$b_n = \frac{1}{n}\sum_{1\leq j\leq n}\sum_{\alpha}(|\alpha|+1)b_{\alpha_1}b_{\alpha_2}\ldots b_{\alpha_{k-1}}b_{n-j}, \qquad b_0 = 1, \tag{9}$$

la deuxième somme étant prise sur les $(k-1)$-uplets d'entiers $\alpha = (\alpha_1, \alpha_2, \ldots, \alpha_{k-1})$ tels que $|\alpha|+1$ divise l'entier j, où $|\alpha| = \alpha_1 + \alpha_2 + \cdots + \alpha_{k-1}$.

Preuve. Les formules (7) et (8) s'obtiennent en spécialisant avec $\mu = (k-1)^{-1}$ les formules suivantes, données par Fowler et al. dans [2, 3],

$$\left(\frac{A(x)}{x}\right)^{\mu} = \sum_{n\geq 0} \mu(\mu+n)^{n-1}\frac{x^n}{n!}, \tag{10}$$

$$Z_{\left(\frac{A(X/\mu)}{X/\mu}\right)^{\mu}} =$$

$$\sum_{n_1,n_2,\ldots} \frac{x_1^{n_1}x_2^{n_2}\ldots}{1^{n_1}n_1!2^{n_2}n_2!\ldots}\prod_{i=1}^{\infty}(1+\frac{1}{\mu}\sum_{d|i} dn_d)^{n_i-1}(1+\frac{1}{\mu}\sum_{d|i,d<i} dn_d). \tag{11}$$

La formule (7) peut également se voir directement par une adaptation de la bijection de Prüfer. Pour obtenir la récurrence (9), il suffit de prendre la dérivée logarithmique de l'équation

$$\widetilde{B}(x) = \exp\left(\sum_{i\geq 1}\frac{x^i\widetilde{B}^{k-1}(x^i)}{i}\right), \tag{12}$$

où $\widetilde{B}(x) = \sum_{n\geq 0} b_n x^n$, qui découle de la relation (3). ∎

La suite des nombres $\{b_n\}$, pour $k = 2,3,4,5$, est répertoriée dans l'encyclopédie des suites d'entiers [11] et l'équation (3), dans l'encyclopédie des structures combinatoires [6]. Le comportement asymptotique des nombres b_n est analysé, notamment en fonction de k, dans la section 5.

3 Cas orienté

Commençons par déterminer les espèces pointées qui apparaissent dans le théorème de dissymétrie. Ces relations sont assez immédiates et la démonstration est laissée au lecteur.

Proposition 3.1. *Les espèces a_o^{-}, $a_o^{\diamond}$, et $a_o^{\underline{\diamond}}$ sont caractérisées par les isomorphismes suivants*

$$a_o^{-} = B, \quad a_o^{\diamond} = XC_k(B), \quad a_o^{\underline{\diamond}} = XB^k, \tag{13}$$

où $B = a^{\rightarrow}$ et C_k représente l'espèce des cycles (orientés) de longueur k.

Le théorème de dissymétrie permet d'exprimer la série génératrice ordinaire $\widetilde{a}_o(x)$ des 2-arbres k-gonaux orientés non étiquetés, en termes des espèces pointées,

$$\widetilde{a}_o(x) = \widetilde{a}_o^{\rightarrow}(x) + \widetilde{a}_o^{\diamond}(x) - \widetilde{a}_o^{\underline{\diamond}}(x), \tag{14}$$

et par la proposition 3.1, nous pouvons alors exprimer $\widetilde{a}_o(x)$ en fonction de $\widetilde{B}(x) = \widetilde{a}^{\rightarrow}(x)$.

Proposition 3.2. *La série génératrice ordinaire $\widetilde{a}_o(x)$ de l'espèce des 2-arbres k-gonaux orientés non étiquetés est donnée par l'expression*

$$\widetilde{a}_o(x) = \widetilde{B}(x) + \frac{x}{k} \sum_{\substack{d|k \\ d>1}} \phi(d) \widetilde{B}^{\frac{k}{d}}(x^d) - \frac{k-1}{k} x \widetilde{B}^k(x). \tag{15}$$

Corollaire 3.3. *Les nombres $a_{o,n}$ et $\widetilde{a}_{o,n}$ de 2-arbres k-gonaux orientés étiquetés et non étiquetés, sur n k-gones sont donnés par*

$$a_{o,n} = ((k-1)n+1)^{n-2} = m^{n-2}, \quad n \geq 2, \tag{16}$$

$$\widetilde{a}_{o,n} = b_n - \frac{k-1}{k} b_{n-1}^{(k)} + \frac{1}{k} \sum_{\substack{d|k \\ d>1}} \phi(d) b_{\frac{n-1}{d}}^{(\frac{k}{d})}, \tag{17}$$

où $b_i^{(j)} = \sum_{i_1+\cdots+i_j=i} b_{i_1} b_{i_2} \ldots b_{i_j}$, représente le coefficient de x^i dans la série $\widetilde{B}^j(x)$, avec $b_r^{(j)} = 0$ si r est non entier ou négatif.

Preuve. Pour le cas étiqueté, il suffit de remarquer que $a_n^{\rightarrow} = m a_{o,n}$. Dans le cas non étiqueté, l'équation (17) s'obtient directement de (15). ∎

4 Cas non orienté

Dans le cas non orienté, le nombre a_n de 2-arbres k-gonaux étiquetés sur n polygones satisfait $2a_n = a_{o,n}+1$, puisque le seul 2-arbre k-gonal orienté étiqueté laissé fixe par changement d'orientation pour un nombre de polygones donné, est celui dont les polygones partagent tous une arête commune. On obtient

Proposition 4.1. *Le nombre a_n de 2-arbres k-gonaux étiquetés sur n polygones est donné par*

$$a_n = \frac{1}{2}\left(m^{n-2}+1\right), \qquad n \geq 2, \tag{18}$$

où $m = (k-1)n+1$.

Pour le dénombrement non étiqueté des 2-arbres k-gonaux (non orientés), nous allons considérer certaines espèces quotients de la forme $F/\mathbb{Z}_2$, où F est une espèce de structures "orientées" et $\mathbb{Z}_2 = \{1, \tau\}$, est un groupe dont l'action de τ sur les F-structures est de renverser l'orientation. Une structure d'une telle espèce quotient consiste alors en une orbite $\{s, \tau \cdot s\}$ de F-structures selon l'action de $\mathbb{Z}_2$.

Par exemple, les diverses espèces pointées de 2-arbres k-gonaux , a^-, $a^\diamond$ et $a^{\underline{\diamond}}$, s'expriment comme espèces quotients des espèces de 2-arbres k-gonaux orientés correspondantes :

$$a^- = \frac{a^{\rightarrow}}{\mathbb{Z}_2}, \quad a^\diamond = \frac{a_o^\diamond}{\mathbb{Z}_2} = \frac{XC_k(B)}{\mathbb{Z}_2}, \quad a^{\underline{\diamond}} = \frac{a_o^{\underline{\diamond}}}{\mathbb{Z}_2} = \frac{XB^k}{\mathbb{Z}_2}. \tag{19}$$

Pour le dénombrement non étiqueté de telles espèces quotients, on utilise la formule suivante qui est évidente :

$$(F/\mathbb{Z}_2)^{\sim}(x) = \frac{1}{2}(\widetilde{F}(x) + \widetilde{F}_{\tau}(x)), \tag{20}$$

où $\widetilde{F}_{\tau}(x) = \sum_{n\geq 0} |\mathrm{Fix}_{\widetilde{F}_n}(\tau)| x^n$ est la série génératrice des F-structures non étiquetées laissées fixes par l'action de τ, c'est-à-dire par changement d'orientation. Toutefois, le calcul de ces séries $\widetilde{F}_{\tau}(x)$ est assez complexe et il est avantageux de différencier en deux cas selon la parité de k.

4.1 Cas k impair

On peut remarquer, en observant les figures 2 a) et b), que dans tout k-gone contenant l'arête pointée (mais non orientée), d'une $\mathfrak{a}^{-}$-structure, il est possible d'orienter les $k-1$ autres arêtes, dans la direction s'éloignant de l'arête pointée comme dans la figure 2 a), lorsque k est impair, mais qu'il restera une arête ambiguë si k est pair. Ce phénomène permet d'introduire des espèces squelettes, lorsque k est impair, en analogie avec l'approche de Fowler et al. [2, 3] où $k = 3$. Ce sont les espèces à deux sortes $Q(X,Y)$, $S(X,Y)$ et $U(X,Y)$, où X représente la sorte des k-gones et Y celle des arêtes orientées, définies par les figures 3 a), b) et c), où $k = 5$. En analogie avec le cas $k = 3$, on a les propositions suivantes.

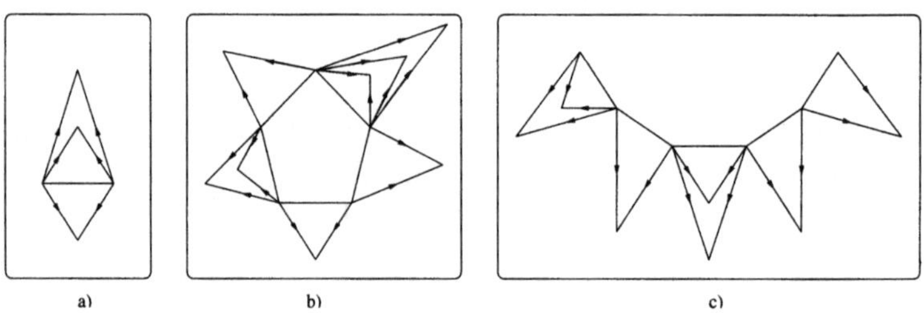

Figure 3: Espèces squelettes a) $Q(X,Y)$, b) $S(X,Y)$ et c) $U(X,Y)$

Proposition 4.2. *Les espèces squelettes Q, S et U admettent des expressions en termes d'espèces quotients :*

$$Q(X,Y) = E(XY^2)/\mathbb{Z}_2, \quad S(X,Y) = C_k(E(XY^2))/\mathbb{Z}_2, \quad U(X,Y) = (E(XY^2))^k/\mathbb{Z}_2. \tag{21}$$

Proposition 4.3. *Lorsque k est impair, $k \geq 3$, on a les expressions suivantes pour les espèces pointées de 2-arbres k-gonaux , où $B = \mathfrak{a}^{\rightarrow}$:*

$$\mathfrak{a}^{-} = Q(X, B^{\frac{k-1}{2}}), \quad \mathfrak{a}^{\diamond} = S(X, B^{\frac{k-1}{2}}), \quad \mathfrak{a}^{\underline{\diamond}} = U(X, B^{\frac{k-1}{2}}). \tag{22}$$

Dans le but d'obtenir des formules d'énumération, il faut préalablement calculer les séries indicatrices de cycles des espèces Q, S et U.

Proposition 4.4. *Les séries indicatrices de cycles des espèces $Q(X,Y)$, $S(X,Y)$ et $U(X,Y)$ sont données par la formule*

$$Z_Q = \frac{1}{2}\Big(Z_{E(XY^2)} + q\Big), \tag{23}$$

$$Z_S = \frac{1}{2}\Big(Z_{C_k(E(XY^2))} + q\cdot(p_2\circ Z_{E(XY^2)})^{\frac{k-1}{2}}\Big), \tag{24}$$

$$Z_U = \frac{1}{2}\Big(Z_{(E(XY^2))^k} + q\cdot(p_2\circ Z_{E(XY^2)})^{\frac{k-1}{2}}\Big), \tag{25}$$

où $q = h\circ(x_1y_2 + p_2\circ(x_1\frac{y_1^2-y_2}{2}))$, p_2 représente la fonction somme de puissances de degré deux, h la fonction symétrique homogène et $\circ$, la composition pléthystique.

Preuve. La formule (23) et la méthode utilisée se trouvent dans [2, 3]. Il s'agit de dénombrer les $F(X,Y)$-structures colorées non étiquetées laissées fixes par τ. Dans le cas de S, on doit laisser fixe une $C_k(E(XY^2))$-structure colorée. Pour cela le cycle de base de longueur k doit posséder au moins un axe de symétrie passant par le milieu d'un des côtés. On peut voir que lorsqu'une telle structure possède plusieurs axes de symétrie, le choix d'un axe est arbitraire. De part et d'autre de l'axe de symétrie, chaque $E(XY^2)$-structure colorée doit avoir son image miroir; ce qui contribue pour un terme de $(p_2\circ Z_{E(XY^2)})^{\frac{k-1}{2}}$. Ensuite, la structure attachée à l'arête distinguée doit être globalement laissée fixe, ce qui donne le facteur q. Le raisonnement est très similaire pour l'espèce U. ∎

Combinant le théorème de dissymétrie, les équations (23), (24), (25) et les lois de substitution de la théorie des espèces, on obtient les séries génératrices des types de l'espèce des 2-arbres k-gonaux .

Proposition 4.5. *Soit $k \geq 3$ impair. La série génératrice ordinaire $\widetilde{a}(x)$ des 2-arbres k-gonaux non étiquetés est donnée par*

$$\widetilde{a}(x) = \frac{1}{2}\Big(\widetilde{a}_o(x) + \exp\big(\sum_{i\geq 1}\frac{1}{2i}(2x^i\widetilde{B}^{\frac{k-1}{2}}(x^{2i}) + x^{2i}\widetilde{B}^{k-1}(x^{2i}) - x^{2i}\widetilde{B}^{\frac{(k-1)}{2}}(x^{4i}))\Big). \tag{26}$$

Corollaire 4.6. *Pour $k \geq 3$ impair, le nombre $\widetilde{a}_n$ de 2-arbres k-gonaux non étiquetés sur n k-gones satisfait la récurrence suivante*

$$\widetilde{a}_n = \frac{1}{2n}\sum_{j=1}^{n}\Big(\sum_{l|j} l\omega_l\Big)\Big(\widetilde{a}_{n-j} - \frac{1}{2}\widetilde{a}_{o,n-j}\Big) + \frac{1}{2}\widetilde{a}_{o,n}, \quad \widetilde{a}_k[0] = 1, \tag{27}$$

où, pour tout $n \geq 1$,

$$\omega_n = 2b_{\frac{n-1}{2}}^{(\frac{k-1}{2})} + b_{\frac{n-2}{2}}^{(k-1)} + b_{\frac{n-2}{4}}^{(\frac{k-1}{2})}, \tag{28}$$

et $b_i^{(j)}$ est défini au corollaire 3.3.

4.2 Cas k pair

Le cas où k est pair est plus délicat. Dans le but d'exprimer les séries génératrices ordinaires des types des trois espèces $\mathscr{a}^{-}$, $\mathscr{a}^{\diamond}$ et $\mathscr{a}^{\hat{\diamond}}$, nous appliquons la formule (20) aux formules (19). Pour l'espèce $\mathscr{a}^{-}$, on a

$$\widetilde{\mathscr{a}}^{-}(x) = \frac{1}{2}(\widetilde{\mathscr{a}}^{\rightarrow}(x) + \widetilde{\mathscr{a}}_{\tau}^{\rightarrow}(x)), \tag{29}$$

où $\widetilde{\mathscr{a}}_{\tau}^{\rightarrow}(x) = \sum_{n\geq 0} |\mathrm{Fix}_{\mathscr{a}_n^{\rightarrow}}(\tau)| x^n$ est la série génératrice des 2-arbres k-gonaux pointés en une arête orientée, non étiquetés, laissés fixes par changement d'orientation. Il faut donc calculer $\widetilde{\mathscr{a}}_{\tau}^{\rightarrow}(x)$. Pour cela, introduisons quelques espèces auxiliaires. La première, notée $\mathscr{a}_{\mathrm{TS}}$, est l'espèce des 2-arbres k-gonaux pointés en une arête orientée et dont toutes les pages attachées autour de cette arête sont verticalement symétriques, sans symétries croisées (voir plus loin); on dira *totalement symétriques*. On peut caractériser cette espèce par l'équation fonctionnelle

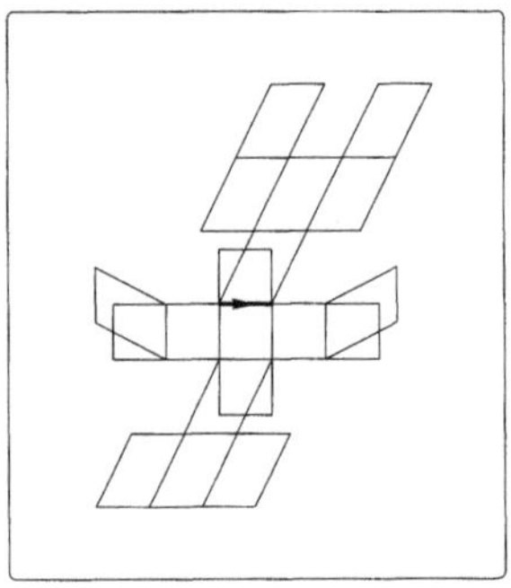

Figure 4: Une structure de l'espèce $\mathscr{a}_{\mathrm{TS}}$

suivante (voir figure 4),

$$\mathscr{a}_{\mathrm{TS}} = E(X \cdot X^2_{=} < B^{\frac{k-2}{2}} > \cdot \mathscr{a}_{\mathrm{TS}}) = E(P_{\mathrm{TS}}), \tag{30}$$

où $X^2_{=} < F >$ représente l'espèce des couples de F-structures isomorphes et P_{TS} est l'espèce des ***pages totalement symétriques.*** Cette équation se traduit au niveau des séries génératrices des types par

$$\widetilde{\mathscr{a}}_{\mathrm{TS}}(x) = \exp\left(\sum_{i\geq 1} \frac{1}{i} x^i \widetilde{B}^{\frac{k-2}{2}}(x^{2i}) \widetilde{\mathscr{a}}_{\mathrm{TS}}(x^i)\right). \tag{31}$$

Proposition 4.7. *Les nombres $\beta_n = |\widetilde{\mathscr{a}}_{\mathrm{TS}}[n]|$, de $\mathscr{a}_{\mathrm{TS}}$-structures non étiquetées sur n polygones satisfont la récurrence*

$$\beta_n = \frac{1}{n}\sum_{i=1}^{n}\Big(\sum_{d|i} d\omega_d\Big)\beta_{n-i}, \ n\geq 1 \quad \beta_0 = 1, \tag{32}$$

où

$$\omega_n = \sum_{\substack{i+j=n-1 \\ i \ pair}} b_{\frac{i}{2}}^{(\frac{k-2}{2})} \beta_j.$$

Preuve. Il suffit de prendre la dérivée logarithmique de l'expression (31). ∎

Passons maintenant à l'introduction des deux espèces P_{CR} et P_{M}, des ***paires de pages croisées*** et des ***pages mixtes***. Une paire de pages *croisées* est, par définition, une paire de pages orientées (des $\mathfrak{a}^{\to}$-structures comportant une seule page) de la forme $\{s, \tau\cdot s\}$ avec s et $\tau\cdot s$ non isomorphes. La figure 5 a) montre une structure de cette espèce. Une page *mixte* est une page symétrique possédant une (ou plusieurs) symétrie de type croisée. Une telle structure est dessinée en figure 5 b). On peut alors exprimer ces deux espèces l'une en fonction de l'autre, comme suit

$$P_{\mathrm{CR}} = \Phi_2 < XB^{k-1} - (P_{\mathrm{TS}} + P_{\mathrm{M}}) >, \tag{33}$$
$$P_{\mathrm{M}} = X \cdot X^2_{=} < B^{\frac{k-2}{2}} > \cdot \mathfrak{a}_{\mathrm{TS}} \cdot E_+(P_{\mathrm{CR}} + P_{\mathrm{M}}), \tag{34}$$

où $\Phi_2 < F >$ représente l'espèce des paires de F-structures de la forme $\{s, \tau \cdot s\}$ et E_+ est l'espèce des ensembles non vides. Passant aux séries génératrices des types, il vient

$$\widetilde{P}_{\mathrm{CR}}(x) = \frac{1}{2}(x^2\widetilde{B}^{k-1}(x^2) - \widetilde{P}_{\mathrm{TS}}(x^2) - \widetilde{P}_{\mathrm{M}}(x^2)), \tag{35}$$
$$\widetilde{P}_{\mathrm{M}}(x) = x\widetilde{B}^{\frac{k-2}{2}}(x^2)\widetilde{\mathfrak{a}}_{\mathrm{TS}}(x)\left(\exp\left(\sum_{i\geq 1}\frac{1}{i}(\widetilde{P}_{\mathrm{CR}}(x^i) + \widetilde{P}_{\mathrm{M}}(x^i))\right) - 1\right). \tag{36}$$

Après manipulations et la prise de la dérivée logarithmique de (36), on obtient les nombres $\widetilde{P}_{\mathrm{CR},n}$ et $\widetilde{P}_{\mathrm{M},n}$ de pages croisées et mixtes respectivement sur n polygones

$$\widetilde{P}_{\mathrm{CR},n} = b^{(k-1)}_{\frac{n-2}{2}} - \widetilde{P}_{\mathrm{TS},\frac{n}{2}} - \widetilde{P}_{\mathrm{M},\frac{n}{2}}, \tag{37}$$
$$\widetilde{P}_{\mathrm{M},n} = \sum_{i=1}^{n}\left(\sum_{d|i}\varepsilon_d\right)c_{n-i} + f_n, \tag{38}$$

où

$$\varepsilon_n = \frac{k-2}{2}b^{(k-1)}_{n-1} + \widetilde{P}_{\mathrm{TS},n} + \widetilde{P}_{\mathrm{CR},n} + \widetilde{P}_{\mathrm{M},n}, \tag{39}$$
$$c_n = \widetilde{P}_{\mathrm{M},n} + \sum_{i+j=n-1} b^{(\frac{k-2}{2})}_{\frac{i}{2}}\widetilde{a}_{\mathrm{TS},j}, \tag{40}$$
$$f_n = \sum_{i+j=n-1} b^{(\frac{k-2}{2})}_{\frac{i}{2}}\widetilde{a}_{\mathrm{TS},j} + 2\sum_{i+j+l=n-2} b^{(\frac{k-4}{2})}_{\frac{i}{2}} j b_{\frac{j}{2}}\widetilde{a}_{\mathrm{TS},l} + \sum_{i+j=n-1} j b^{(\frac{k-2}{2})}_{\frac{i}{2}}\widetilde{a}_{\mathrm{TS},j}. \tag{41}$$

Notons par $\widetilde{\mathfrak{a}}_{\mathrm{S}}(x)$ la série génératrice des $\mathfrak{a}^{\to}$-structures non étiquetées symétriques. On a alors (voir figure 6)

$$\widetilde{\mathfrak{a}}_{\mathrm{S}}(x) = E(P_{\mathrm{TS}} + P_{\mathrm{CR}} + P_{\mathrm{M}})^{\sim}(x), \tag{42}$$
$$= \exp\left(\sum_{i>1}\frac{1}{i}(\widetilde{P}_{\mathrm{TS}}(x^i) + \widetilde{P}_{\mathrm{CR}}(x^i) + \widetilde{P}_{\mathrm{M}}(x^i))\right). \tag{43}$$

On en déduit alors une récurrence pour le nombre $\alpha_n = \widetilde{a}_{\mathrm{S},n}$ de 2-arbres k-gonaux pointés en une arête laissés fixes par changement d'orientation.

$$\alpha_n = \frac{1}{n}\sum_{i=1}^{n}\Big(\sum_{d|i} d\omega_d\Big)\alpha_{n-i}, \quad \alpha_0 = 1, \tag{44}$$

où

$$\omega_k = \widetilde{P}_{\mathrm{TS},k} + \widetilde{P}_{\mathrm{CR},k} + \widetilde{P}_{\mathrm{M},k}.$$

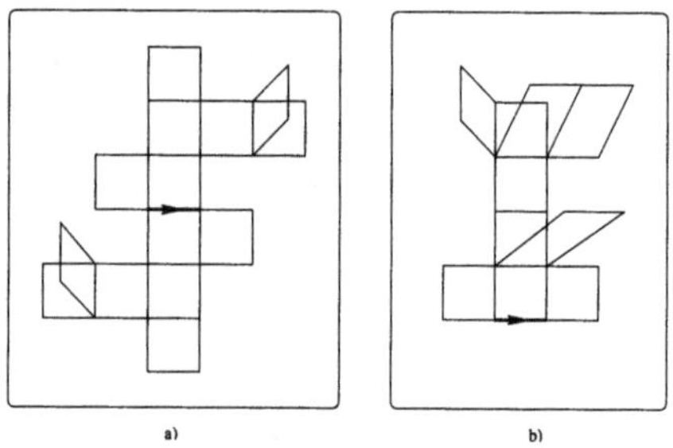

Figure 5: Une paire de pages croisées et une page mixte

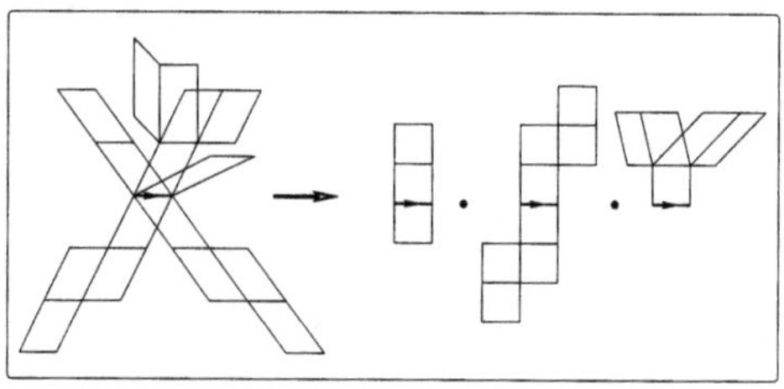

Figure 6: Décomposition d'une $a^{\rightarrow}$-structure fixée sous τ

Proposition 4.8. *Si k est un entier pair, $k \geq 4$, alors le nombre de 2-arbres k-gonaux pointés en une arête (non orientée) sur n k-gones est donné par*

$$\widetilde{a}^{-}_{n} = \frac{1}{2}(b_n + \alpha_n). \tag{45}$$

Passons maintenant à l'espèce $a^{\overset{\diamond}{-}}$ des 2-arbres k-gonaux pointés en un k-gone possédant une arête distinguée. On trouve

$$\widetilde{a}^{\overset{\diamond}{-}}(x) = \frac{1}{2}\Big(\widetilde{a}_o^{\overset{\diamond}{-}}(x) + \widetilde{a}_{o,\tau}^{\overset{\diamond}{-}}(x)\Big), \quad \text{où} \quad \widetilde{a}_{o,\tau}^{\overset{\diamond}{-}}(x) = x\widetilde{a}_{\mathrm{S}}^{2}(x)\widetilde{B}^{\frac{k-2}{2}}(x^2), \tag{46}$$

puisque une $a_o^{\overset{\diamond}{-}}$-structure non étiquetée τ-symétrique possède un axe de symétrie qui est, en fait, la médiatrice de l'arête distinguée dans le polygone pointé, et, qui est donc aussi naturellement la médiatrice de l'arête opposée à celle pointée.

Les structures attachées à ces deux arêtes sont donc symétriques, d'où le terme $(\widetilde{\mathfrak{a}}_{\mathrm{S}}(x))^2$; ensuite, de part et d'autre de l'axe, les B-structures que l'on y attache doivent s'échanger par paire, soit une contribution d'un facteur $\widetilde{B}(x^2)$ pour chacune des $\frac{k-2}{2}$ paires. On en déduit alors une expression du nombre de $\mathfrak{a}^{\underline{\diamond}}$-structures non étiquetées $\widetilde{\mathfrak{a}}_n^{\underline{\diamond}}$,

$$\widetilde{\mathfrak{a}}_n^{\underline{\diamond}} = \frac{1}{2}\left(\widetilde{\mathfrak{a}}_{o,n}^{\underline{\diamond}} + \sum_{i+j=n-1} \alpha_i^{(2)} \cdot b_{\frac{j}{2}}^{(\frac{k-2}{2})}\right), \tag{47}$$

où $\alpha_i^{(2)} = [x^i]\widetilde{\mathfrak{a}}_S^2(x)$.

Procédons de façon similaire pour l'espèce $\mathfrak{a}^{\diamond}$, des 2-arbres k-gonaux pointés en un polygone. Une nouvelle fois, nous utilisons la relation (20), qui donne

$$\widetilde{\mathfrak{a}}^{\diamond}(x) = \frac{1}{2}\left(\widetilde{\mathfrak{a}}_o^{\diamond}(x) + \widetilde{\mathfrak{a}}_{o,\tau}^{\diamond}(x)\right). \tag{48}$$

Remarquons d'abord que pour qu'une $\mathfrak{a}_o^{\diamond}$-structure soit laissée fixe par changement d'orientation, elle doit comporter au moins un axe de symétrie, qui peut être de deux types :

1. un axe passant par le milieu de deux arêtes opposées, ou
2. un axe passant par deux sommets opposés,

du polygone pointé. Le dénombrement se fait en orientant d'abord l'axe de symétrie. On trouve

$$\widetilde{\mathfrak{a}}_{o,\tau}^{\diamond}(x) = \frac{x}{2}\widetilde{\mathfrak{a}}_{\mathrm{S}}^2(x)\widetilde{B}^{\frac{k-2}{2}}(x^2) + \frac{x}{2}\widetilde{B}^{\frac{k}{2}}(x^2), \tag{49}$$

où le premier terme correspond à une symétrie de type 1, et le deuxième, de type 2. Les structures qui possèdent les deux symétries sont précisément celles qui sont comptées une demi fois dans chacun des deux termes. Le théorème de dissymétrie donne donc, pour $k \geq 4$ pair,

$$\begin{aligned}\widetilde{\mathfrak{a}}(x) &= \frac{1}{2}\widetilde{\mathfrak{a}}_o(x) + \frac{1}{2}\widetilde{\mathfrak{a}}_{\mathrm{S}}(x) + \frac{1}{2}\widetilde{\mathfrak{a}}_{o,\tau}^{\diamond}(x) - \frac{1}{2}\widetilde{\mathfrak{a}}_{o,\tau}^{\underline{\diamond}}(x),\\ &= \frac{1}{2}\widetilde{\mathfrak{a}}_o(x) + \frac{1}{2}\widetilde{\mathfrak{a}}_{\mathrm{S}}(x) + \frac{x}{4}(\widetilde{B}^{\frac{k}{2}}(x^2) - \widetilde{\mathfrak{a}}_{\mathrm{S}}^2(x)\widetilde{B}^{\frac{k-2}{2}}(x^2)),\end{aligned} \tag{50}$$

où $\widetilde{\mathfrak{a}}_o(x)$ est donné par (15) et $\widetilde{\mathfrak{a}}_{\mathrm{S}}(x)$ par (43).

Théorème 4.9. *Si $k \geq 4$ est pair, le nombre de 2-arbres k-gonaux non étiquetés sur n k-gones est donné par*

$$\widetilde{\mathfrak{a}}_n = \frac{1}{2}\widetilde{\mathfrak{a}}_{o,n} + \frac{1}{2}\alpha_n + \frac{1}{4}b_{\frac{n-1}{2}}^{(\frac{k}{2})} - \frac{1}{4}\sum_{i+j=n-1} \alpha_i^{(2)} \cdot b_{\frac{j}{2}}^{(\frac{k-2}{2})}, \tag{51}$$

avec

$$b_l^{(m)} = [x^l]\widetilde{B}^m(x), \quad \alpha_i^{(2)} = [x^i]\widetilde{\mathfrak{a}}_{\mathrm{S}}^2(x).$$

5 Dénombrement asymptotique

Grâce au théorème de dissymétrie et aux diverses équations combinatoires qui lui sont associées, le dénombrement asymptotique des 2-arbres k-gonaux (étiquetés ou non) dépend essentiellement de celui des B-structures où B est l'espèce auxiliaire caractérisée par l'équation combinatoire (3). Dans le cas étiqueté, la situation est triviale puisque l'on dispose des formules closes simples (7), (16) et (18). Dans le cas non étiqueté, la situation est vraiment plus délicate puisque la série $\widetilde{B}(x)$ est caractérisée par l'équation fonctionnelle complexe (12).

Voici quelques notations préliminaires à l'énoncé du résultat principal de la présente section. Si $\lambda = (\lambda_1 \geq \lambda_2 \geq \ldots \geq \lambda_\nu)$ est un partage d'un entier n en ν parts, on écrit $\lambda \vdash n$, $n = |\lambda|$, $\nu = l(\lambda)$, $m_i(\lambda) = |\{j : \lambda_j = i\}|$ = nombre de parts de taille i dans λ. De plus, on pose

$$\sigma_i(\lambda) = \sum_{d|i} dm_d(\lambda), \quad \sigma_i^*(\lambda) = \sum_{d|i, d<i} dm_d(\lambda) \tag{52}$$

$$\widehat{\lambda} = 1 + |\lambda| + l(\lambda), \quad \widehat{z}(\lambda) = 2^{m_1(\lambda)} m_1(\lambda)! 3^{m_2(\lambda)} m_2(\lambda)! \ldots . \tag{53}$$

On a le résultat suivant.

Proposition 5.1. *Posons $p = k - 1$ et $\widetilde{B}(x) = \sum b_n(p)x^n$. Alors*

i) $b_n(p)$ est un polynôme en p de degré $n-1$, $n \geq 1$,

ii) il existe des constantes α_p et β_p telles que

$$b_n(p) \sim \alpha_p \beta_p^n n^{-\frac{3}{2}}, \qquad \text{pour } n \to \infty. \tag{54}$$

De plus, $\alpha_p = \alpha(\xi_p) = \dfrac{1}{\sqrt{2\pi}} \dfrac{1}{(p\xi_p)^{\frac{1}{p}} p} \left(1 + \dfrac{p\xi_p \omega'(\xi_p)}{\omega(\xi_p)}\right)^{\frac{1}{2}}$ et $\beta_p = \dfrac{1}{\xi_p}$, où ξ_p est la plus petite racine de l'équation

$$\xi = \frac{1}{ep} \omega^{-p}(\xi), \tag{55}$$

où $\omega(x)$ est la série (absolument convergente au voisinage de ξ_p) donnée par (58). On a le développement convergent

$$\xi_p = \sum_{n=1}^{\infty} \frac{c_n}{p^n}, \tag{56}$$

où les coefficients c_n sont des constantes, indépendantes de p, données explicitement par

$$c_n = \sum_{\lambda \vdash n} \frac{e^{-\widehat{\lambda}}}{\widehat{\lambda} \widehat{z}(\lambda)} \prod_{i \geq 1} (\sigma_i(\lambda) - \widehat{\lambda})^{m_i(\lambda)-1} (\sigma_i^*(\lambda) - \widehat{\lambda}), \tag{57}$$

lorsque λ parcourt l'ensemble des partages de n.

Preuve. La partie i) de l'énoncé découle immédiatement de la formule explicite (8). Pour la partie ii) qui affirme l'existence des constantes α_n et β_n, on s'inspire de l'approche de Fowler et al. pour les 2-arbres ($k = 3$) en utilisant le théorème classique de Bender. Posons, pour simplifier $b(x) = \widetilde{B}(x)$. Alors, grâce à (12), $y = b(x)$ satisfait la relation

$$y = e^{xy^p}\omega(x), \quad \text{où} \quad \omega(x) = e^{\frac{1}{2}x^2 b^p(x^2) + \frac{1}{3}x^3 b^p(x^3) + \cdots} \tag{58}$$

Par le théorème de Bender, appliqué à la fonction $f(x,y) = y - e^{xy^p}\omega(x)$, on doit chercher un couple (ξ_p, τ_p) solution du système

$$f(x,y) = 0 \qquad \text{et} \qquad f_y(x,y) = 0. \tag{59}$$

Ceci équivaut à dire que ξ_p est solution de (55) et que $p\xi_p\tau_p^p = 1$. Les formules explicites (56) et (57) s'obtiennent en appliquant préalablement l'inversion de Lagrange à l'équation $\xi = zR(\xi)$ où $z = \frac{1}{ep}$ et $R(t) = \omega^{-p}(t)$, pour obtenir

$$\xi_p = \xi = \sum_{n\geq 0} \frac{a_n}{n!}\left(\frac{1}{ep}\right)^n, \quad \frac{a_n}{n!} = \frac{1}{n}[t^{n-1}]\omega^{-np}(t). \tag{60}$$

Ensuite, pour évaluer explicitement $\omega^{-np}(x)$, on utilise la version de Labelle [7] de la formule d'inversion de Good pour les séries indicatrices en tenant compte de (6) et en remarquant que

$$\omega^{-np}(x) = e^{-n(\frac{x_2}{2} + \frac{x_3}{3} + \cdots)} \circ Z_A(x_1, x_2, \ldots)|_{x_i := px^i}, \tag{61}$$

où $A = XE(A)$ est l'espèce des arborescences. ∎

Dans le cas orienté non pointé, une méthode similaire basée sur l'équation (15), mène à

$$\widetilde{a}_{o,n} \sim \overline{\alpha}_p \beta_p^n n^{-\frac{5}{2}}, \quad \text{où} \quad \overline{\alpha}_p = 2\pi p (p\xi_p)^{\frac{2}{p}} \alpha_p^3. \tag{62}$$

Enfin, une analyse fine de la formule (51) montre que

$$\widetilde{a}_n \sim \frac{1}{2}\widetilde{a}_{o,n}. \tag{63}$$

La table 1 donne, à 20 décimales, les constantes ξ_p, α_p et $\beta_p = \frac{1}{\xi_p}$ pour $p = 1, \ldots, 5$.

p	ξ_p	α_p	β_p
1	0.3383218568 9920769520	1.3003121246 8216843599	2.955765285651994974715
2	0.177099522303285617693	0.349261381742311443973	5.646542616232949712893
3	0.119674100436145452060	0.191997258649948899321	8.356026879295995368276
4	0.090334539604383047938	0.131073637348549764379	11.069962877759326312419
5	0.072539192528125499910	0.099178841365021748147	13.785651110084685198930

Table 1 : Valeurs numériques de ξ_p, α_p et β_p, $p = 1, \ldots, 5$.

Voici les premières valeurs des constantes universelles c_n apparaissant dans (56), pour $n = 1, \ldots 5$.

$$c_1 = \frac{1}{e} = 0.36787944117144232160, \tag{64}$$

$$c_2 = -\frac{1}{2}\frac{1}{e^3} = -0.024893534183931971 49, \tag{65}$$

$$c_3 = \frac{1}{8}\frac{1}{e^5} - \frac{1}{3}\frac{1}{e^4} = -0.0052629695880257 1004, \tag{66}$$

$$c_4 = -\frac{1}{48}\frac{1}{e^7} + \frac{1}{e^6} - \frac{1}{4}\frac{1}{e^5} = 0.00077526788594593923, \tag{67}$$

$$c_5 = \frac{1}{384}\frac{1}{e^9} - \frac{4}{3}\frac{1}{e^8} + \frac{49}{72}\frac{1}{e^7} - \frac{1}{5}\frac{1}{e^6} = 0.00032212622183609932. \tag{68}$$

Remarque 5.1. *Les calculs de cette section sont également valables pour le cas où $k = 2$ et $p = 1$, correspondant aux arborescences ordinaires (de Cayley) définies par l'équation $A = XE(A)$. Dans ce cas, la constante de croissance $\beta = \beta_1$, dans* (54), *est connue sous le nom de constante d'Otter (voir* [10]*). Il est intéressant de noter que cette constante prend la forme explicite $\beta = \frac{1}{\xi_1}$, avec*

$$\xi_1 = \sum_{n \geq 1} c_n. \tag{69}$$

Il est à noter que lorsque $k = 3$, nous retrouvons les résultats asymptotiques obtenus par Fowler et al. dans [2, 3].

References

[1] F. Bergeron, G. Labelle, and P. Leroux, *Combinatorial Species and tree-like structures*, Encyclopedia of Mathematics and it's Applications, vol. 67, Cambridge University Press, (1998).

[2] T. Fowler, I. Gessel, G. Labelle, P. Leroux, *Specifying 2-trees*, Proceedings FPSAC'01, Moscou, 26-30 juin 2000, 202–213.

[3] T. Fowler, I. Gessel, G. Labelle, P. Leroux, *The Specification of 2-trees*, Advances in Applied Mathematics, 28, 145–168, (2002).

[4] F. Harary and E. Palmer, *Graphical Enumeration*, Academic Press, New York, (1973).

[5] F. Harary, E. Palmer and R. Read, *On the cell-growth problem for arbitrary polygons*, Discrete Mathematics, 11, 371–389, (1975).

[6] INRIA, *Encyclopedia of combinatorial structures.* http://algo.inria.fr/encyclopedia/index.html.

[7] G. Labelle, *Some new computational methods in the theory of species*, Combinatoire énumérative, Proceedings, Montréal, Québec, Lectures Notes in Mathematics, vol. 1234, Springer-Verlag, New-York/Berlin, 160–176, (1985).

[8] G. Labelle, C. Lamathe and P. Leroux, *Développement moléculaire de l'espèce des 2-arbres planaires*, Proceedings GASCom01, 41–46, (2001).

[9] G. Labelle, C. Lamathe and P. Leroux, *A classification of plane and planar 2-trees*, preprint CO/0202052, submitted.

[10] R. Otter, *The number of trees*, Annals of Mathematics, 49, 583–599, (1948).

[11] N. J. A. Sloane and S. Plouffe, *The Encyclopedia of Integer Sequences*, Academic Press, San Diego, (1995).

Gilbert Labelle, Cédric Lamathe, Pierre Leroux

LaCIM
Université du Québec à Montréal
Case Postale 8888, succursale centre-ville
H3C 3P8 Montréal
{gilbert, lamathe, leroux}@math.uqam.ca

[illegible]

PART II

Random Graphs and Networks

Trends in Mathematics, © 2002 Birkhäuser Verlag Basel/Switzerland

Breadth First Search, Triangle-Free Graphs and Brownian Motion

Anne-Elisabeth Baert, Vlady Ravelomanana, Loÿs Thimonier

ABSTRACT: *One major problem in the enumeration of random graphs concerns triangle-free graphs. In this paper, we study Breadth First Search processes and the associated queues to compute, in terms of Wright's constants, the number of triangle-free graphs . Next, we prove that this number is equivalent to the number of connected labelled graphs by using arguments of the Brownian excursion type.*

1 Introduction

In computer science, trees are combinatorial structures evolving with time, and both the algorithm to traverse them (here *Breadth First Search*) and the underlying data structure (namely a queue; more generally for graph algorithms a priority queue) imply a wealth of interesting properties and other algorithms, as remarked in Sedgewick's book [15]. *Depth First Search* and *Breadth First Search* processes have been studied in a lot of published works, one can cite for example Aldous [1, 2, 3], Chassaing *et al.* [6, 7], Drmota and Gittenberger [8], Marckert and Mokkadem [12] ....

We consider here simple labelled connected graphs, i.e., graphs with labelled vertices, undirected edges and without self-loops or multiple edges. Throughout this paper, a $(n, n+k)$ graph is one having n vertices and $n+k$ edges; k is then called its *excess*. Denote by $c(n, n+k)$ the number of connected graphs on n vertices and $n+k$ edges, usually expressed with Wright's constants. Wright and Bender *et al.* [18, 19, 4] gave exact and asymptotic formulae for the numbers $c(n, n+k)$ by means of enumerative and analytic approaches. As an alternative method, Spencer [16] surprisingly used Breadth First Search and developped formulae for $c(n, n+k)$ in terms of appropriate expectations.

Harary and Palmer [9] pinpointed that one of the major problem in graphical enumeration concerns the triangle-free graphs. Since Breadth First Search is convenient to examine the connections between vertices at the same level in a tree, we exploit this fact to enumerate connected triangle-free graphs.

In this paper, we recall shortly the Breadth First Search method to traverse rooted labelled trees and graphs, then we examine Breadth Fisrt Search Trees to precise some enumeration in the associated queue. We use Breadth First Search processes to show that the number of triangle-free graphs with excess k can be expressed in terms of Wright's constants. Then, using Breadth First Walk and Brownian excursion, we show that the number of triangle-free graphs with n vertices and $n+k$ edges, is equivalent to the number of connected labelled $(n, n+k)$-graphs, as $n \to +\infty$.

2 The Breadth First Search method

2.1 Breadth First Search on random trees

Let T be a tree with $\{1, \cdots, n\}$ as vertex set, rooted on the first vertice 1. The well-known Cayley's formula [5] gives n^{n-1} for the number of such trees. All adjacency lists being in numerical order, the tree can be obtained through Breadth First Search (BFS).

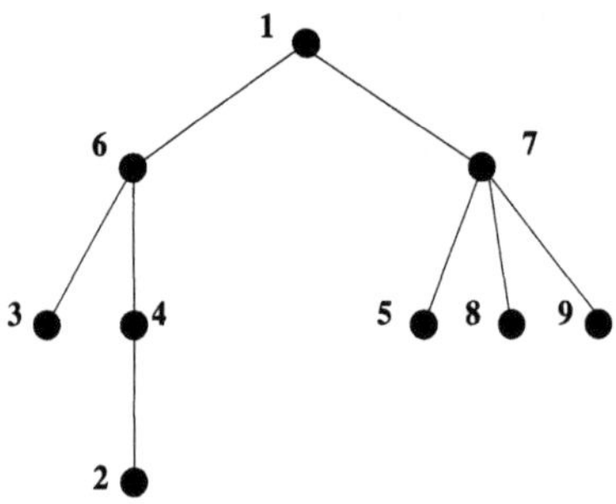

Figure 1: Tree with 9 vertices

Steps	**Edges**	Queue
1	**1**	6, 7
2	**6**	7, 3, 4
3	**7**	3, 4, 5, 8, 9
4	**3**	4, 5, 8, 9
5	**4**	5, 8, 9, 2
6	**5**	8, 9, 2
7	**8**	9, 2
8	**9**	2
9	**2**	

Figure 2: Associated queue

Recall this method: a queue Q is initialized with 1 (the root of the labelled tree), and the BFS ends when the queue is empty. At each step $i\,(i \geq 1)$ a vertex x is taken at the head of the queue and removed from Q; all new adjacent vertices to x are added to Q. Such vertices are said to be *found* by x. Let $q_i^{(n)}$ be the size of Q after the ith vertex (to remove) is processed and let x_i be the number of vertices found by this i^{th} vertex. We have

$$q_0^{(n)} = 1 \quad \text{and for} \quad 1 \leq i \leq n\,,\; q_i^{(n)} = q_{i-1}^{(n)} + x_i - 1\,. \tag{1}$$

Remark that -1 corresponds to removing the ith vertex, $q_n^{(n)} = 0$ (empty queue), and the total number of vertices is equal to:

$$\sum_{i=1}^{n} x_i = n - 1\,. \tag{2}$$

For random simulation of such trees of a given type $(x_1, \cdots, x_n)$, we consider independent random variables $X_1, \cdots, X_n$ and the random variables $Q_i^{(n)}$ defined recursively by: $Q_0^{(n)} = 1$, $Q_i^{(n)} = Q_{i-1}^{(n)} + X_i - 1$. We have $Q_n^{(n)} = 0$ if and only if $X_1 + \cdots + X_n = n - 1$. The constraints

$$Q_n^{(n)} = 0 \quad \text{and} \quad Q_i^{(n)} \geq 0 \quad \text{for} \quad i < n\,, \tag{3}$$

are necessary and sufficient for a tree to be connected or for the queue to become empty only after step n.

2.2 Breadth First Search on random graphs

Let G^+ be a connected graph on vertex set $\{1, \cdots, n\}$. Once traversed by BFS, one obtains a tree T called a *Breadth First Search Tree* (BFST).W want to find the number of graphs that split BFST T, for a given type T. BFSTs have nice properties: for instance, the edges of G^+ are of three types: some edges are in T, some connect two vertices at the same level of T, and the remaining ones connect two vertices on two adjacents levels. It is not possible for an edge to skip a level. Finally, the maximum number $M^{(n)}$ of edges that can be added in a given tree T in order to build a graph with the same BFST, satisfies:

$$M^{(n)} = \sum_{t=1}^{n-1} \left(Q_t^{(n)} - 1\right). \tag{4}$$

Proposition 2.1. *Let $M^{(n)}$ be the number of edges that can be added in a given tree T in order to build a graph with the same BFST:*

$$M^{(n)} = \binom{n}{2} - (n-1) - \sum_{t=1}^{n} (t-1) X_t . \tag{5}$$

Proof: This follows directly from the recursive definition of $Q_t^{(n)}$. Indeed, one has $Q_t^{(n)} = \sum_{j=1}^{t} X_j - (t-1)$, and thus:

$$\sum_{t=1}^{n-1} Q_t^{(n)} = \frac{(n-1)(n+2)}{2} - \sum_{t=1}^{n} t X_t .$$

□

$N = \binom{n}{2}$ is the total number of edges in a complete graph with n vertices and $(n-1)$ is the number of vertices in a tree. $(t-1)x_t$ is the number of edges we can add between the vertices found by the tth vertice processed, i.e., x_t and the vertices processed before t, i.e., the first one to the $(t-1)^{th}$.

2.3 Results on the queue

Let $\left(G_l^{(n)}\right)_{l\geq 0}$ be the number of vertices at distance l from the root. $\left(G_l^{(n)}\right)_{l\geq 0}$ is called the *profile* of the tree and satisifies:

$$\begin{aligned} G_0^{(n)} &= 1, \quad G_1^{(n)} = q_1^{(n)} \quad \text{and} \\ G_l^{(n)} &= q^{(n)}_{\sum_{i=0}^{l-1} G_i^{(n)}} . \end{aligned} \tag{6}$$

At time $t = \sum_{i=0}^{l-1} G_i^{(n)}$ the last vertex at distance $l-1$ from the root is processed and all the vertices at distance l are in the queue. Therefore, $M^{(n)}$ can be classified according to the three different types of edges.

Proposition 2.2. *Let $M^{(n)}$ be the number of edges that can be added in a given tree T in order to build a graph with the same BFST. The total number of edges in $M^{(n)}$ at level l is:*

$$\binom{G_l^{(n)}}{2}.$$

In particular, at level l the number of edges between two vertices with the same "father" is :

$$\sum_{t=\sum_{i=0}^{l-2} G_i^{(n)}+1}^{\sum_{i=0}^{l-1} G_i^{(n)}} \binom{x_t}{2}. \tag{7}$$

Between level l and level $l+1$, the number of edges is:

$$\sum_{t=1}^{G_l^{(n)}-1} \left(G_l^{(n)} - t\right) . x_{\left(\sum_{i=0}^{l-1} G_i^{(n)}+t\right)}. \tag{8}$$

Proof : By induction on l, and using the fact that between

$$t = \sum_{i=0}^{l-2} G_i^{(n)} + 1 \text{ and } t = \sum_{i=0}^{l-1} G_i^{(n)}$$

all the vertices at level $l-1$ are processed.

□

3 Enumeration of triangle-free graphs

3.1 Triangle-free graphs with excess at most 2

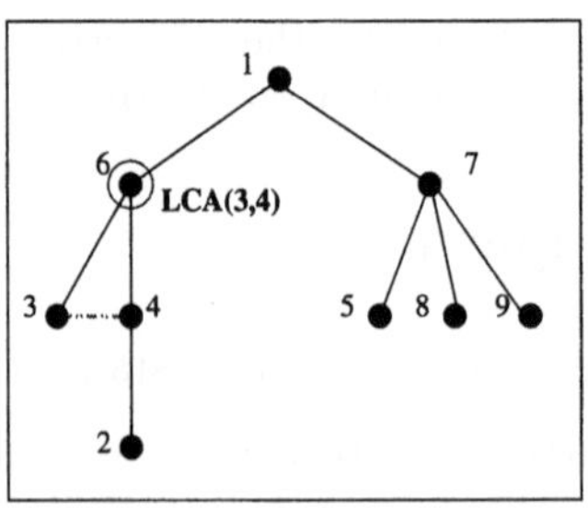

Figure 3: Triangle graph with excess 0

Let T^* be a connected triangle-free graph with excess 0 on vertex set $\{1, \cdots, n\}$ traverse by BFS method. Any two vertices in a tree have one or more common ancestors. The *Lowest Common Ancestor* of a set of vertices is the one that is closest to the vertices. We denote by $LCA(t_i, t_i')$ the Lowest Common Ancestors of

t_i and t'_i. When an edge is added at random to T, the only way to have a triangle is to connect two vertices at the same level and for which their Lowest Common Ancestor is at distance 1.

Similarly to $M^{(n)}$ in (4), let $R^{(n)}$ be the number of edges which can be added to T in order to obtain a triangle. The number of sons of the tth vertex processed is x_t, hence $R^{(n)}$ is given by:

$$R^{(n)} = \sum_{t=1}^{n-1} \binom{X_t}{2}. \tag{9}$$

Thus from a tree T, $M^{(n)} - R^{(n)}$ triangle-free graphs with excess 0 can be constructed. $M^{(n)}$ and $R^{(n)}$ are random variables under the conditioned distribution of the X_t and we have the following

Theorem 3.1. *The number $f^*(n,n)$ of triangle-free graphs with excess 0 satisfies*

$$\frac{f^*(n,n)}{c(n,n-1)} = E\big(M^{(n)} - R^{(n)}\big). \tag{10}$$

We denote by (t_i, t'_i) the ith edge added between t_i and t'_i. Recall that the considered graphs have undirected edges and labelled vertices; we can suppose without loss of generality that t_i is processed before t'_i.

Definition 3.2. *Let $d(t_i, t'_i)$ be the distance between t_i and t'_i, and $(t_i \wedge t'_i)$ the first vertex processed. Let*

$$\delta_i = \begin{cases} 0 & \text{if } d(LCA(t_i,t'_i),t_i) = 1 \\ 1 & \text{otherwise} \end{cases}$$

$$\Delta_{i,j} = \begin{cases} 1 & \text{if } (t_i, t'_i, t_j, t'_j \text{ are on two adjacent levels) and } (t_i = t'_j \text{ or } t_j = t'_i) \\ 0 & \text{otherwise}. \end{cases}$$

Let $\phi(t_0) = 0$ and for $1 \le i \le k$, $\phi(t_i) = x_{t_i} - \delta_i$.

$\Delta_{i,j} = 1$ if there is a *path* between three vertices in two adjacent levels.

Theorem 3.3. *Let $f^*(n, n+k)$ be the number of triangle-free graphs with excess k at most 2, and let $V^{(n)} = M^{(n)} - R^{(n)}$ and $\Gamma^{(u)}_{i,j} = \sum_{i=0}^{u}(\phi(t_i) + \sum_{j=1,j>i}^{u} \Delta_{i,j})$.*

$$\frac{f^*(n,n+k)}{c(n,n-1)} = \frac{1}{(k+1)!} E\left[\prod_{u=0}^{k}\left(V^{(n)} - u - \Gamma^{(u)}_{i,j}\right)\right]. \tag{11}$$

Proof : To obtain a triangle-free graph with excess 0, the first edge (t_1, t'_1) must be chosen amongst the $M^{(n)} - R^{(n)}$ possible edges. Then, a second edge (t_2, t'_2), is added in order to obtain triangle-free graphs with excess 1. As described in figure 4, two situations appear depending on the choice of the first edge. The case where the first edge added connects two vertices and one of them is at distance one of their Lowest Common Ancestors, and the case where the distance of one of the two vertices is at least two. For the first case, since x_{t_1} is the number of vertices

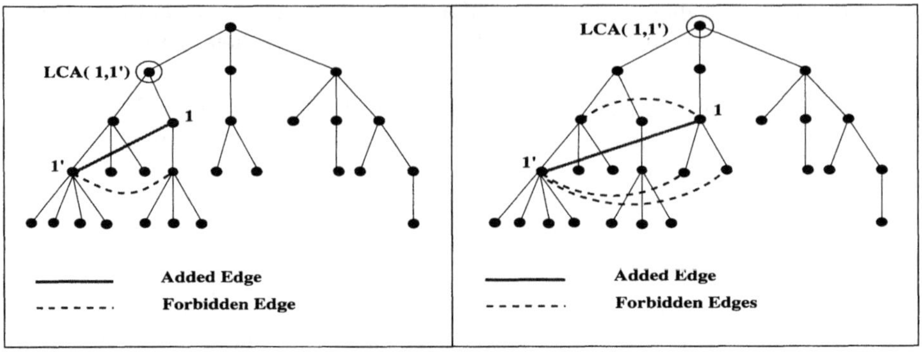

Figure 4: Two possible cases to add the second edge

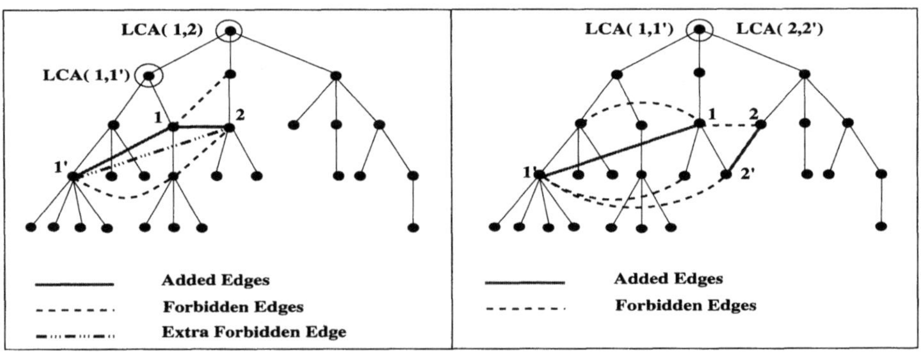

Figure 5: Two cases to add the third edge

found by t_1, there is $(M^{(n)} - R^{(n)} - 1 - x_{t_1})$ possible choices for the second edge whereas in the second case, this number becomes $(M^{(n)} - R^{(n)} - 1 - x_{t_1} - 1)$.

At least, we have to add the last edge, (t_3, t_3') in order to obtain a Triangle- Free graph with excess 2. When this edge is added, (t_1, t_1') and (t_2, t_2') can have a common vertex. Suppose that $(t_1 = t_2')$ hence $\Delta_{1,2} = 1$, if (t_1', t_2) is added to the graph t_1, t_2, t_2' is a triangle (as seen in figure 5): this sort of edge must be avoided. If the second added edge has no common vertex with the first one, we can proceed for the third edge as for the second one.

□

3.2 Triangle-free graphs with excess k ≥ 3

Theorem 3.4. *Let $f^*(n, n+k)$ be the number of triangle-free graphs with excess k for $k \geq 3$ fixed, and let $V^{(n)} = M^{(n)} - R^{(n)}$ and $\Gamma_{i,j}^{(u)} = \sum_{i=0}^{u}(\phi(t_i) + \sum_{j=0, j>i}^{u} \Delta_{i,j})$.*

$$\frac{1}{(k+1)!}E\left[\prod_{u=0}^{k}\left(V^{(n)} - u - \Gamma_{i,j}^{(u)}\right)\right] \leq \frac{f^*(n,n+k)}{c(n,n-1)} \leq E\left[\binom{V^{(n)}}{k+1}\right] \quad (12)$$

Proof : A connected triangle-free graph with excess k is obtained by adding $k+1$ edges to the tree T. If these edges do not share some common vertices, the triangle-free graph has the same evolution as previously described.

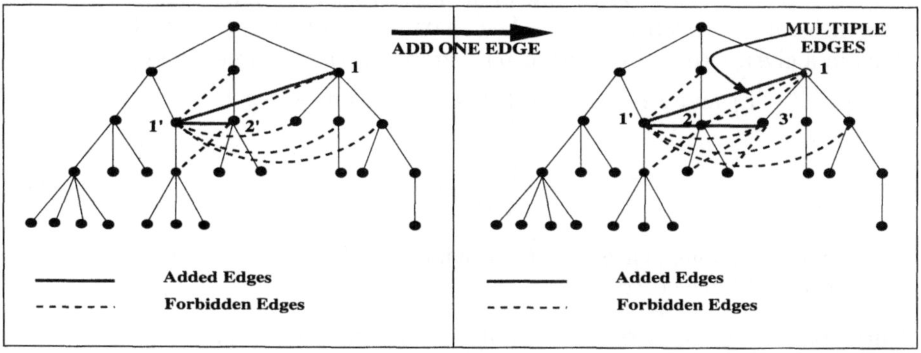

Figure 6: Example of path of length four

However, as depicted in figure 6, four edges can form a path and some edges can be counted twice or more. Consequently, putting all these repetitions in

$$\sum_{i=0}^{u}\left(\phi(t_i) + \sum_{\substack{j=0 \\ j>i}}^{u} \Delta_{i,j}\right)$$

leads to the lower bound given by (12).
Obviously, $\binom{M^{(n)}-R^{(n)}}{k+1}$ is a natural upper bound.

□

4 Brownian motion and k excess graphs

4.1 Breadth First Walk and Brownian excursion

We follow here the way suggested by Spencer [16]. We consider Poisson independent random variables, $X_1, \ldots, X_n$, with mean 1, the sequence $Q_i^{(n)}$ is conditioned to satisfy (3). $Q_1^{(n)}, \cdots, Q_n^{(n)}$ can be considered as a *random walk* whose step size has distribution $Y_i = Q_i^{(n)} - Q_{i-1}^{(n)} = X_i - 1$. X_i are Poisson with mean 1; Y_i are Poisson 1 minus 1. It is well-known that:

$$\left(\frac{Q_{\lfloor nt \rfloor}^{(n)}}{\sqrt{n}}\right)_{0 \le t \le 1} \xrightarrow{weakly} (e(t))_{0 \le t \le 1}, \tag{13}$$

where e is a normalized *Brownian excursion.* Let

$$L = \int_0^1 e(s) ds$$

which may be interpreted as the mean distance from the origin in this conditioned 1-dimensional Brownian motion. The maximum number of edges that can be added in a given tree T (in order to buid graphs with the same BFST), is $M^{(n)} = \sum_{t=1}^{n} \left(Q_t^{(n)} - 1\right)$ and we obtain in distribution, as $n \to +\infty$:

$$\frac{M^{(n)}}{n^{3/2}} \to L\,. \tag{14}$$

In [7], similar arguments are used to provide tight bounds for the moments of the width of rooted labelled trees.

Proposition 4.1. *In distribution as $n \to +\infty$, and for any fixed $k \ge 0$:*

$$E\left[\left(\frac{M^{(n)}}{n^{\frac{3}{2}}}\right)^{k+1}\right] \sim E\left[L^{k+1}\right]. \tag{15}$$

Proof: Following the works of Chassaing and Marckert [7], we use connections between Breadth First Search random walks and *empirical processes.* Let $(U_i)_{1 \le i \le n}$ be a sequence of n i.i.d random variables uniformly distributed on $[0, 1]$. Denote by $F_n(t)$ the *empirical distribution function* of $(U_i)_{1 \le i \le n}$ defined for $t \in [0, 1]$ by

$$F_n(t) = \frac{\text{card}\{i \in \{1, \ldots, n\}, U_i \le t\}}{n}.$$

The empirical distribution function converges towards the distribution function, $F(t) = t$, of the uniform law. The speed of convergence is revealed by the *empirical process*

$$\alpha_n(t) = \sqrt{n}\,(F_n(t) - F(t))\,,$$

and $(\alpha_n(t)_{0 \le t \le 1})$ converges in distribution to the *Brownian bridge* $(b(t))_{0 \le t \le 1}$. The theorem of "strong convergence" of Komlos, Major and Tusnady [10] shows the existence on the same probability space as $(U_i)_{i \ge 1}$ of a sequence of *Brownian bridges* $(b_n)_{n \ge 1}$ that approximate closely the sequence $(\alpha_n)_{n \ge 1}$, i.e., if a, M, μ are positive absolute constants, for all n and x:

$$\Pr\left(\sup_{0 \le t \le 1} |\sqrt{n}\,(\alpha_n(t) - b_n(t))| \ge A \log n + x\right) \le M e^{-\mu x}\,. \tag{16}$$

The relation

$$\alpha_n\left(\frac{T(n)}{n+1}\right) = \min_{0\le j\le n} \alpha_n\left(\frac{j}{n+1}\right),$$

defines a unique number $T(n)$ between 0 and n and $T(n)$ is uniformly distributed on $\{0,1,\cdots,n\}$. Using the relations established between parking functions and empirical processes in [7], we denote by

$$w_j^{(n)} = \frac{n-j}{n+1} + \sqrt{n}\left[\alpha_n\left\{\frac{j+1+T(n)}{n+1}\right\} - \alpha_n\left(\frac{T(n)}{n+1}\right)\right] \quad \text{for } 0\le j\le n. \tag{17}$$

We can write [7, Prop. 4.6 p.13]:

$$\left(Q_j^{(n)}\right)_{0\le j\le n} \overset{law}{=} \left(w_j^{(n)}\right)_{0\le j\le n}. \tag{18}$$

We can establish the following relation using (17)

$$\begin{aligned}\left(\sum_{j=1}^{n} w_j\right)^{k+1} &= \left(\sum_{j=1}^{n}\frac{n-j}{n+1} + \sqrt{n}\left[\alpha_n\left(\left\{\frac{j+1+T(n)}{n+1}\right\}\right) - \alpha_n\left(\frac{T(n)}{n+1}\right)\right]\right)^{k+1}\\ &= \left(\sum_{j=1}^{n}\frac{n-j}{n+1} + \sqrt{n}\left[b_n\left(\frac{T(n)}{n+1}\right) - \alpha_n\left(\frac{T(n)}{n+1}\right)\right]\right.\\ &\quad + \sqrt{n}\left[\alpha_n\left(\left\{\frac{j+1+T(n)}{n+1}\right\}\right) - b_n\left(\left\{\frac{j+1+T(n)}{n+1}\right\}\right)\right]\\ &\quad \left.+ \sqrt{n}\left[b_n\left(\left\{\frac{j+1+T(n)}{n+1}\right\}\right) - b_n\left(\frac{T(n)}{n+1}\right)\right]\right)^{k+1}.\end{aligned} \tag{19}$$

We recall

Theorem 4.2 (Vervaat, 1979). *Let $b=(b(t))_{0\le t\le 1}$ be a brownian bridge, and let T be the almost surely unique point such that $b(T)=\min_{0\le t\le 1} b(t)$. Then T is uniform and $e=(e(t))_{0\le t\le 1}$, defined by $e(t)=b(\{T+t\})-b(T)$ is a normalized Brownian excursion independent of T.*

According to Skorohod representation theorem with $(b_n)_n \ge 1$ comes a sequence of Brownian excursions $(e_n)_{n\ge 1}$. One can deduce from (16),(19) and (13), (18) the proposition.

□

4.2 Triangle-free graphs and Brownian motion

We can now express the number of triangle free graphs in terms of a certain restricted Brownian motion:

Theorem 4.3. *For any fixed $k\ge 0$, and when n tends to $+\infty$:*

$$\frac{f^*(n,n+k)}{c(n,n-1)} \sim n^{\frac{3(k+1)}{2}} E\left[\frac{L^{k+1}}{(k+1)!}\right]. \tag{20}$$

To prove this theorem we need the following

Proposition 4.4. *For any fixed $k \geq 0$, we have:*

$$E\left[\left(\frac{M^{(n)} - R^{(n)}}{n^{\frac{3}{2}}}\right)^{k+1}\right] \sim E\left[\left(\frac{M^{(n)}}{n^{\frac{3}{2}}}\right)^{k+1}\right]. \tag{21}$$

Proof : Marckert and Mokkadem in [12] show how to reduce the study on discrete excursions to the same study on (non-conditioned) random walks; if there is a moderate deviations principles for a functional of a centered random walk, there exists an upper bound for the analogous principle on the associated excursion. The number $R^{(n)}$ of edges which can be added to T in order to obtain a triangle is given by $R^{(n)} = \sum_{t=1}^{n-1} \frac{X_t(X_t-1)}{2}$; the triangular inequality once applied to $M^{(n)} - R^{(n)}$ leads to:

$$\left| \left\| \frac{M^{(n)} - R^{(n)}}{n^{3/2}} \right\|_{k+1} - \left\| \frac{M^{(n)}}{n^{3/2}} \right\|_{k+1} \right| \leq \left\| \frac{R^{(n)}}{n^{3/2}} \right\|_{k+1}, \tag{22}$$

Thanks to "the conditioning argument" [12, 3.2 pp.8]: $E\left(\sum_{t=1}^{n-1} \frac{X_t(X_t-1)}{2}\right) = \frac{n-1}{2}$, and there exits $\alpha \geq 0$ such that $E\left(\exp \alpha \left(\sum_{t=1}^{n-1} \frac{X_t(X_t-1)}{2}\right)\right) < +\infty$. According to Petrov's theorem [13], for $0 < \gamma < 1$, there exists $c > 0$ such that

$$\Pr\left(\left|R^{(n)} - nc\right| \geq n^{\frac{1}{2}+\gamma}\right) \leq \exp\left(-n^{\gamma}\right). \tag{23}$$

Consequently, one can decompose the right term of (22) in two parts:

$$\left[E\left(\left(\frac{R^{(n)}}{n^{3/2}}\right)^{k+1} \mathbf{1}_{|R^{(n)}-nc| \leq n^{1/2+\gamma}} + \left(\frac{R^{(n)}}{n^{3/2}}\right)^{k+1} \mathbf{1}_{|R^{(n)}-nc| \geq n^{1/2+\gamma}}\right)\right]^{\frac{1}{k+1}}.$$

Now, each term of the previous sum can be bounded, according to (23):

$$\left(\frac{R^{(n)}}{n^{3/2}}\right)^{k+1} \mathbf{1}_{|R^{(n)}-nc| \leq n^{1/2+\gamma}} \leq \left(\frac{nc + n^{1/2+\gamma}}{n^{3/2}}\right)^{k+1},$$

and using the fact that $R^{(n)} \leq n^3$:

$$E\left[\left(\frac{R^{(n)}}{n^{3/2}}\right)^{k+1} \mathbf{1}_{|R^{(n)}-nc| \geq n^{1/2+\gamma}}\right] \leq \exp(-n^{\gamma}) \left(\frac{n^3}{n^{3/2}}\right)^{k+1}.$$

Thus, as $n \to +\infty$,

$$\left\| \frac{R^{(n)}}{n^{3/2}} \right\|_{k+1} \to 0,$$

and so,

$$\left\| \frac{M^{(n)} - R^{(n)}}{n^{3/2}} \right\|_{k+1} \sim \left\| \frac{M^{(n)}}{n^{3/2}} \right\|_{k+1}.$$

□

Proof of Theorem 4.3: For $k \geq 3$, the inequality (12)of the theorem 3.4 gives two bounds for $\frac{f^*(n,n+k)}{c(n,n-1)}$. The lower one, is asymptotically equivalent to

$$E\left[\frac{\left(M^{(n)} - R^{(n)}\right)^{k+1}}{(k+1)!}\right].$$

The upper one is the same than the lower one, as for any fixed k and when n tends to $+\infty$

$$E\left[\binom{M^{(n)} - R^{(n)}}{k+1}\right] \sim E\left[\frac{(M^{(n)} - R^{(n)})^{k+1}}{(k+1)!}\right].$$

Obviously, one can deduce from Proposition 4.4 that for any fixed $k \geq 0$ and when $n \to +\infty$:

$$\frac{f^*(n,n+k)}{c(n,n-1)} \sim \frac{1}{(k+1)!} E\left[\left(M^{(n)}\right)^{k+1}\right],$$

and from Proposition 4.1 comes the end of the proof.

□

Theorem 4.5. *The number of triangle-free graphs with excess k is asymptotically equivalent to the number of graphs with excess k.*

Proof: Theorem 3.2 in [16] allows us to write when $n \to +\infty$:

$$\frac{c(n,n+k)}{c(n,n-1)} \sim \frac{n^{\frac{3(k+1)}{2}}}{(k+1)!} E\left[L^{k+1}\right]$$

which combined with the previous theorem leads to the theorem.

□

5 Conclusion

In this paper, we addressed the problem of counting constrained graphs. We showed that the Breadth First Search processes and Brownian excursion perfom well with connected triangle-free $(n, n+k)$ graphs and permit us to estimate their numbers whenever n tends to ∞ but k is fixed. Although, the work presented here suggests that the analysis can be significantly more complex (especially if $k \equiv k(n)$), however the methods still have potential and connections between dynamic algorithms and their underlying structures can be deepened.

References

[1] D. Aldous, (1991) *The continuum random tree. II: An overview.*, In M.T. Barlow and N.H. Bingham, editors, Stochastic Analysis, pages 23–70. Cambridge University Press, 23-70.

[2] D. Aldous,(1993) *The continuum random tree. III*, Ann. of Probab., 21, 248-289.

[3] D. Aldous, (1998) *Brownian excursion conditioned on its local time*, Elect. Comm. in Probab., 3, 79-90.

[4] E. A. Bender, E. R. Canfield and B. D. McKay, (1990) *The Asymptotic Number of labelled Connected Graphs with a Given Number of Vertices and Edges*, Random Structures & Algorithms 1, 127-169.

[5] A. Cayley, (1889) *A theorem on trees*, Q. J. Pure Appl. Math., 23, 376-378.

[6] P. Chassaing, J.F. Marckert, M. Yor, (2000) *The height and width of simple trees*, Mathematics and Computer Science: Algorithms, Trees, Combinatorics and Probabilities, Ed. D. Gardy, A. Mokkadem. Birkhauser, 17-30.

[7] P. Chassaing, J.F. Marckert, (2001), *Parking functions, empirical processe, and the width of rooted labelled trees*, Elect. Journ. Comb, 8.

[8] M. Drmota, B. Gittenberger,(1997) *On the profile of random tree*, Random Structures & Algorithms 10, No. 4, 421-451.

[9] F. Harary and E. Palmer, (1973) *Graphical Enumeration*, Academic Press.

[10] J. Komloś, P. Major and G. Tusnády, (1976) *An approximation of partial sums of independent RV's and the sample DF.II*, Z. Wahrscheinlichkeitstheorie verw. Gebiete, 34, 33-58.

[11] G. Louchard,(1984) *The Brownian excurion: a numerical analysis*, Comput. Math. Appl. 10, 413-417.

[12] J.F. Marckert, A. Mokkadem, (2001) *The depth first processes of Galton-Watson trees converge to the same Brownian excursion*, Preprint UVSQ 56-01.

[13] V. V. Petrov, (1975) *Sums of independent random variables*, Springer Berlin.

[14] R. Sedgewick et P. Flajolet, (1996) *An Introduction to the Analysis of Algorithms*, Addison-Wesley.

[15] R. Sedgewick, (1988) *Algorithms*, Addison Wesley.

[16] J. Spencer, (1997) *Enumerating graphs and Brownian Motion*, Commun. Pure Appl. Math. 60, No. 3, 291-294.

[17] L. Takacs, (1993) *Limit distributions for queues and random rooted trees*, J. Appl. Math. Stoch. Ana. 6, No.3, 189-216.

[18] E. Maitland Wright, (1977) *The number of connected sparsely edged graphs*, Journal of Graph Theory 1, 317-330.

[19] E. Maitland Wright, (1980) *The number of connected sparsely edged graphs III, Asymptotic results* , Journal of Graph Theory 4, 393-407.

Anne-Elisabeth Baert
LaRIA, Université de Picardie Jules-Verne,
5, rue du Moulin Neuf,
80000 Amiens, France,
baert@laria.u-picardie.fr

Vlady Ravelomanana
LIPN, Université de Paris-Nord,
99, av. J.B. Clement,
93430 Villetaneuse, France,
vlad@lipn.univ-paris13.fr

Loÿs Thimonier
LaRIA, Université de Picardie Jules-Verne,
5, rue du Moulin Neuf,
80000 Amiens, France,
thimon@laria.u-picardie.fr

Trends in Mathematics, © 2002 Birkhäuser Verlag Basel/Switzerland

Random Planar Lattices and Integrated SuperBrownian Excursion

Philippe Chassaing, Gilles Schaeffer

ABSTRACT: *In this extended abstract, a surprising connection is described between a specific brand of random lattices, namely planar quadrangulations, and Aldous' Integrated SuperBrownian Excursion (ISE). As a consequence, the radius* r_n *of a random quadrangulation with* n *faces is shown to converge, up to scaling, to the width* $r = R - L$ *of the support of the one-dimensional ISE, or precisely:*

$$n^{-1/4} r_n \xrightarrow{\text{law}} (8/9)^{1/4}\, r.$$

More generally the distribution of distances to a random vertex in a random quadrangulation is described in its scaled limit by the random measure ISE shifted to set the minimum of its support in zero.

1 Introduction

From a distant perspective, this article uncovers a surprising, and hopefully deep, relation between two famous models: *random planar maps*, as studied in combinatorics and quantum physics, and *Brownian snakes*, as studied in probability theory and statistical physics. More precisely, our results connect some distance-related functionals of *random quadrangulations* with functionals of Aldous' *Integrated SuperBrownian Excursion* (ISE) in dimension one.

In this extended abstract, most proofs are omitted. A detailed manuscript is available from the authors or as `arXiv:math.CO/0205226`.

Quadrangulations On the one hand, quadrangulations are finite plane graphs with four-regular faces (see Figure 1 and Section 2 for precise definitions). Random quadrangulations, like random triangulations, random polyhedra, or the ϕ^4-models of physics, are instances of a general family of random lattices that has received considerable attention both in combinatorics (under the name *random planar maps*, following Tutte's terminology [27]) and in physics (under the name *Euclidean two-dimensional discretised quantum geometry*, or simply *dynamical triangulations* or *fluid lattices* [2, 7, 13]).

Many probabilistic properties of random planar maps have been studied, that are *local properties* like vertex or face degrees [12], or $0-1$ laws for properties expressible in first order logic [6]. Other well documented families of properties are related to connectedness and constant size separators [5], also known as branchings into baby universes [16]. In this article we consider another fundamental aspect of the geometry of random maps, namely *global properties of distances*. The *profile* $(H_k^n)_{k\geq 0}$ and *radius* r_n of a random quadrangulation with n faces are defined in analogy with the classical profile and height of trees: H_k^n is the number of vertices at distance k from a basepoint, while r_n is the maximal distance reached. The

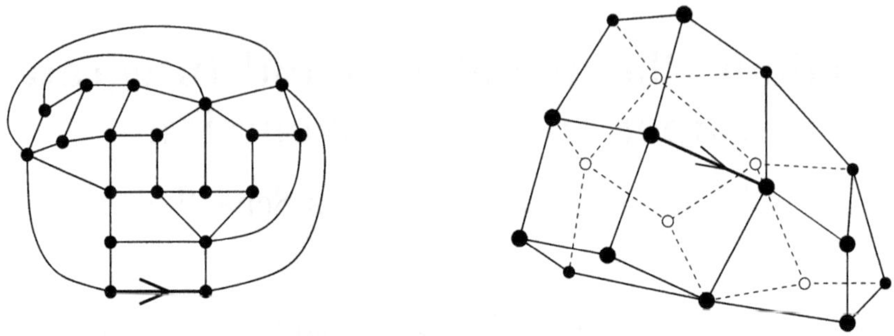

Figure 1: Random quadrangulations, in planar or spherical representation.

profile was studied (with triangulations instead of quadrangulations) by physicists Watabiki, Ambjørn et *al.* [3, 28] who gave a consistency argument proving that the only possible scaling for the profile is $k \sim n^{1/4}$, a property which reads in their terminology *the internal Hausdorff dimension is 4*. Independently the conjecture that $\mathbb{E}(r_n) \sim cn^{1/4}$ was proposed by Schaeffer [24].

Integrated SuperBrownian Excursion On the other hand, ISE was introduced by Aldous as a model of random distributions of masses [1]. He considers random embedded discrete trees as obtained by the following two steps: first an abstract tree t, say a Cayley tree with n nodes, is taken from the uniform distribution and each edge of t is given length 1; then t is embedded in the regular lattice on $\mathbb{Z}^d$, with the root at the origin, and edges of the tree randomly mapped on edges of the lattice. Assigning masses to leaves of the tree t yield a random distribution of mass on $\mathbb{Z}^d$. Upon scaling the lattice to $n^{-1/4}\mathbb{Z}^d$, these random distributions of mass admit, for n going to infinity, a continuum limit $\mathcal{J}$ which is a random probability measure on $\mathbb{R}^d$ called ISE.

Derbez and Slade proved that ISE describes in dimension larger than eight the continuum limit of a model of lattice trees [11], while Hara and Slade obtained the same limit for the incipient infinite cluster in percolation in dimension larger than six [14]. As opposed to these works, we shall consider ISE in dimension one and our embedded discrete trees should be thought of as folded on a line. The support of ISE is then a random interval (L, R) of $\mathbb{R}$ that contains the origin.

From quadrangulations to ISE The purpose of this paper is to draw a relation between, on the one hand, random quadrangulations, and, on the other hand, Aldous' ISE: upon proper scaling, the profile of a random quadrangulations is described in the limit by ISE translated to have support $(0, R-L)$. This relation implies in particular that the radius r_n of random quadrangulations, again upon scaling, weakly converges to the width of the support of ISE in one dimension, that is the continuous random variable $r = R-L$. We shall indeed prove (Corollary 5.4) that

$$n^{-1/4} r_n \xrightarrow{\text{law}} (8/9)^{1/4}\, r,$$

as well as the convergence of moments. While this proves the conjecture $\mathbb{E}(r_n) \sim cn^{1/4}$, the value of the constant c remains unknown because, as mentioned by

Aldous [1], little is known on R or $R - L$.

The path from quadrangulations to ISE consists of three main steps, the first two of combinatorial nature and the last with a more probabilistic flavor. Our first step, Theorem 3.1, revisits a correspondence of Cori and Vauquelin [9] between planar maps and some *well labelled trees*, that can be viewed as plane trees embedded in the positive half-line. Thanks to an alternative construction [24, Ch. 7], we show that under this correspondence the profile can be mapped to the mass distribution on the half-line. In particular, the radius r_n of a random quadrangulation is equal in law to the maximal label μ_n of a random well labelled tree.

Safe for the positivity condition, well labelled trees would be constructed exactly according to Aldous' prescription for embedded discrete trees. Well labelled trees are thus to Aldous' embedded trees what the Brownian excursion is to the Brownian bridge, and we seek an analogue of Vervaat's relation. At the discrete level a classical elegant explanation of such relations is based on Dvoretsky and Motzkin's cyclic shifts and cycle lemma. Our second combinatorial step, Theorem 4.4, consists in the adaptation of these ideas to embedded trees. More precisely, via the *conjugation of tree principle* of [24, Chap. 2], we bound the discrepancy between the mass distribution of our conditioned trees on the positive half-line and a translated mass distribution of freely embedded trees. In particular we construct a coupling between well labelled trees and freely embedded trees such that the largest label μ_n, and thus the radius r_n, is coupled to the width of the support (L_n, R_n) of random freely embedded trees:

$$|r_n - (R_n - L_n)| \leq 3.$$

Since our freely embedded trees are constructed according to Aldous' prescription, one could expect to be able to conclude directly. However two obstacles still need to be bypassed at this point.

Contour walks and Brownian snakes. The first obstacle is that the construction of ISE as a continuum limit of mass distributions supported by embedded discrete trees was only outlined in Aldous' original paper. The original mathematical definition is by embedding a continuum random tree (CRT), which amounts to exchanging the embedding and the continuum limit. But Borgs et *al.* proved that indeed ISE is the limit of mass distributions supported by embedded Cayley trees [8] and their proof could certainly be adapted to other simple classes of trees and in particular to our embedded plane trees.

The second, more important, obstacle is that weak convergence of probability measures is not adequate to our purpose, since we are interested in particular in convergence of the width of the support, *which is not a continuous functional on the space of measures.* In order to circumvent this difficulty, we turn to the description of ISE in terms of superprocesses: ISE can be constructed from the Brownian snake with lifetime e, the standard Brownian excursion [1, 18].

From the discrete point of view, we consider the encoding of an embedded plane tree by a pair of contour walks (x_k, y_k), that encode respectively the height of the node visited at time k and its position on the line. Our last result, Theorem 5.2, is the weak convergence, upon proper scaling, of this pair of walks to the Brownian snake with lifetime e:

$$\left(e^{(n)}(s), \hat{W}^{(n)}(s)\right) \xrightarrow{\text{law}} \left(e(s), \hat{W}_s\right).$$

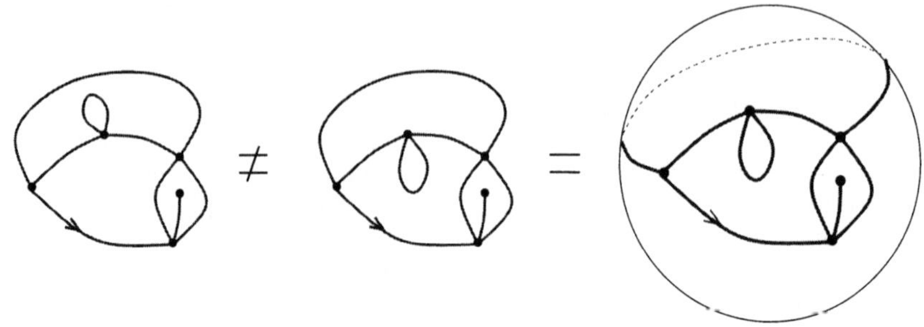

Figure 2: Two distinct planar maps, and a spherical representation of the second.

As $R = \sup_s \hat{W}_s$ and $L = \inf_s \hat{W}_s$ this convergence, together with some deviation bounds obtained in the proof allows us to conclude on the radius. (A similar weak convergence was independently proved by Marckert and Mokkadem [21] but without the deviation bounds we need here.)

More generally the joint convergence of the minimum and the mass distribution of discrete embedded trees implies that, upon scaling, the label distribution of well labelled trees converges to ISE translated to have the minimum of its support at the origin. The same then holds for the profile of random quadrangulations.

Dynamical triangulations and a Continuum Random Map Although we concentrate in this article on the radius and profile of random quadrangulations, our derivation suggests a much tighter link between random quadrangulations and ISE. We conjecture that a Continuum Random Map (CRM) can be built from ISE that would describe the continuum limit of scaled random quadrangulations, in a similar way as the CRT describes the continuum limit of scaled random discrete trees. From the point of view of physics, the resulting CRM would describe in the limit the geometry of scaled dynamical triangulations as studied in discretised two-dimensional Euclidean pure quantum geometries [2, 7, 13]. We plan to discuss this connection further in future work.

2 The combinatorial models of random lattice

2.1 Planar maps and quadrangulations

A *planar map* is a proper embedding (without edge crossings) of a connected graph in the plane. Loops and multiple edges are *a priori* allowed. A planar map is *rooted* if there is a *root*, *i.e.* a distinguished edge on the border of the infinite face, which is oriented counterclockwise. The origin of the root is called the *root vertex*. Two rooted planar maps are considered identical if there exists an homeomorphism *of the plane* that sends one map onto the other (roots included).

The difference between planar graphs and planar maps is that the cyclic order of edges around vertices matters in maps, as illustrated by Figure 2. Observe that planar maps can be equivalently defined on the sphere. In particular Euler's characteristic formula applies and provides a relation between the numbers n of

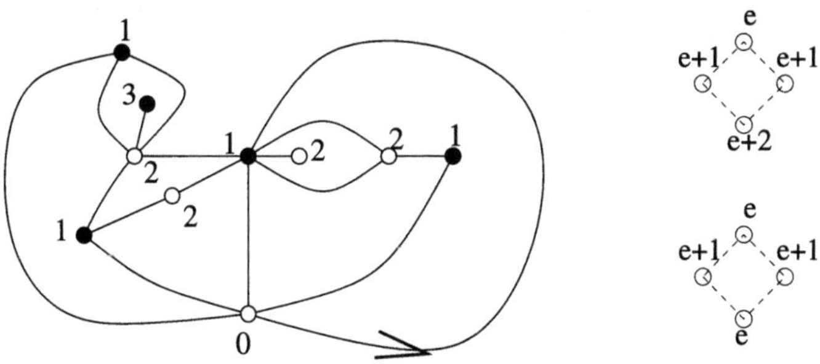

Figure 3: Labelling by distance from the root vertex and the two possible configurations of labels (top: a simple face; bottom: a confluent face).

edges, f of faces and v of vertices of any planar map: $f + v = n + 2$.

The *degree* of a face or of a vertex of a map is its number of incidence of edges. A planar map is a *quadrangulation* if all faces have degree four. All (planar) quadrangulations are *bipartite*: their vertices can be colored in black or white so that the root is white and any edge joins two vertices with different colors. In particular a quadrangulation contains no loop but may contain multiple edges. See Figures 1 and 3 for examples of quadrangulations.

Let $\mathcal{Q}_n$ denote the set of rooted quadrangulations with n faces. A quadrangulation with n faces has $2n$ edges (because of the degree constraint) and $n+2$ vertices (applying Euler's formula). The number of rooted quadrangulations with n faces was obtained by W.T. Tutte [27]:

$$|\mathcal{Q}_n| \;=\; \frac{2}{n+2}\,\frac{3^n}{n+1}\binom{2n}{n}. \tag{1}$$

Various alternative proofs of this result have been obtained (see *e.g.* [7, 9, 4, 24]). Our treatment will indirectly provide another proof, related to [9, 24].

2.2 Random planar lattices

Let L_n be a random variable with uniform distribution on $\mathcal{Q}_n$. Formally, L_n is the $\mathcal{Q}_n$-valued random variable such that for all $Q \in \mathcal{Q}_n$

$$\Pr(L_n = Q) \;=\; \frac{1}{|\mathcal{Q}_n|} \;=\; \frac{1}{\frac{2}{n+2}\frac{3^n}{n+1}\binom{2n}{n}}.$$

The random variable L_n is our *random planar lattice*. To explain this terminology, taken from physics, observe that locally the usual planar square lattice is a planar map whose faces and vertices all have degree four. Our random planar lattice corresponds to a relaxation of the constraint on vertices.

Classical variants of this definition are obtained by replacing quadrangulations with n faces by triangulations with $2n$ triangles, or by (vertex-)4-regular maps with n vertices, or by all planar maps with n edges, *etc.* All these random planar lattices have been considered both in combinatorics (see [5] and references therein) and in mathematical physics (see [2] and references therein; in the physics literature, definitions are usually phrased using "symmetry weights" instead of rooted objects, but this is strictly equivalent to the combinatorial definition). Although details of local topology vary between families, most probabilistic properties are believed to be "universal", that is qualitatively analogue for all "reasonable" families. Observe also that random maps in classical families have exponentially small probability to be symmetric, so that all results hold as well as in the model of uniform unrooted maps [23].

In this article we focus on quadrangulations because of their combinatorial relation, detailed in Section 3, to well labelled trees.

2.3 The profile of a map

The distance $d(x, y)$ between two vertices x and y of a map is the minimal number of edges on a path from x to y (in other terms all edges have abstract length 1).

The *profile* of a rooted map M is the sequence $(H_k)_{k\geq 1}$, where $H_k \equiv H_k^{[M]}$ is the number of vertices at distance k of the root vertex v_0. We shall also consider the cumulated profile $\widehat{H}_k^{[M]} = \sum_{\ell=1}^{k} H_\ell^{[M]}$. By construction the support of the profile of a rooted map is an interval *i.e.* $\{k \mid H_k > 0\} = [1, r]$ where r is the *radius* of the map (sometimes also called *eccentricity*). The radius r is closely related to the *diameter*, that is the largest distance between two vertices of a map: in particular $r \leq d \leq 2r$. The quadrangulation of Figure 3 has radius 3.

The *profile of the random planar lattice* L_n is the random variable $(H_k^{(n)})_{k\geq 1}$ that is defined by taking the profile $(H_k^{[L_n]})_{k\geq 1}$ of an instance of L_n, while $(\widehat{H}_k^{(n)})_{k\geq 1}$ denotes the *cumulated* profile of L_n. Similarly the radius of a random planar lattice is a positive integer valued random variable r_n.

3 Encoding the profile with well labelled trees

3.1 Well labelled trees and the encoding result

A *plane tree* is a rooted planar map without cycle (and thus with only one face). Equivalently plane trees can be recursively defined as follows:

- the smallest tree is made of a single vertex,
- any other tree is a non-empty sequence of subtrees attached to a root.

In other term, each vertex has a possibly empty sequence of sons, and each vertex but the root has a father. The number of plane trees with n edges is the well known Catalan number $C(2n) = \frac{1}{n+1}\binom{2n}{n}$.

A plane tree is *well labelled* if all its vertices have positive integral labels, the labels of two adjacent vertices differ at most by one, and the label of the root vertex is one. Let $\mathcal{W}_n$ denote the set of well labelled trees with n edges.

The *label distribution* of a well labelled tree T is the sequence $(\lambda_k)_{k\geq 1} \equiv (\lambda_k^{[T]})_{k\geq 1}$ where $\lambda_k^{[T]}$ is the number of vertices with label k in the tree T. The cumulated

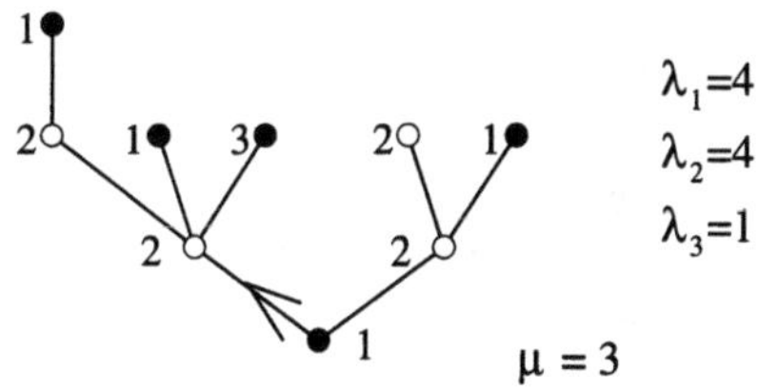

Figure 4: A well labelled tree with its label distribution.

label distribution is defined by $\widehat{\lambda}_k^{[T]} = \sum_{\ell=1}^{k} \lambda_\ell^{[T]}$. By construction the support of the label distribution is an interval: there exists an integer μ such that $\{k \mid \lambda_k > 0\} = [1, \mu]$. This integer μ is the maximal label of the tree. These definitions are illustrated by Figure 4.

The following theorem will serve us to reduce the study of the profile of quadrangulations to the study of the label distribution of well labelled trees.

Theorem 3.1 (Schaeffer [24]). *There exists a bijection $\mathcal{T}$ between rooted quadrangulations with n faces and well labelled trees with n edges, such that the profile $(H_k^{[Q]})_{k\geq 1}$ of a quadrangulation Q is mapped onto the label distribution $(\lambda_k^{[T]})_{k\geq 1}$ of the tree $T = \mathcal{T}(Q)$.*

Theorem 3.1 and Tutte's formula (1) imply that the number of well labelled trees with n edges equals

$$|\mathcal{W}_n| = \frac{2}{n+2}\frac{3^n}{n+1}\binom{2n}{n}. \qquad (2)$$

This result was proved already by Cori and Vauquelin [9], who introduced well labelled trees to give an encoding of all planar maps with n edges. Because of a classical bijection between the latter maps and quadrangulations with n faces, their result is equivalent to the first part of Theorem 3.1. Their bijection has been extended to bipartite maps by Arquès [4] and to higher genus maps by Marcus and Vauquelin [20]. All these constructions were recursive and based on encodings of maps with permutations (also known as rotation systems).

However, our interest in well labelled trees lies in the relation between the profile and the label distribution, which does not appear in Cori and Vauquelin's bijection. The bijection we use here is much simpler and immediately leads to the second part of Theorem 3.1. This approach was extended to non separable maps by Jacquard [15] and to higher genus by Marcus and Schaeffer [19].

We postpone to Section 4 the discussion of the interesting form of Formula (2) and its relation to Catalan's numbers. Instead the rest of this part is concerned with the proof of Theorem 3.1, which goes in three steps. First some properties of distances in quadrangulations are indicated (Section 3.2). This allows in a second step to define the encoding, as a mapping $\mathcal{T}$ from quadrangulations to well labelled trees (Section 3.3). The proof that this encoding is correct is omitted here.

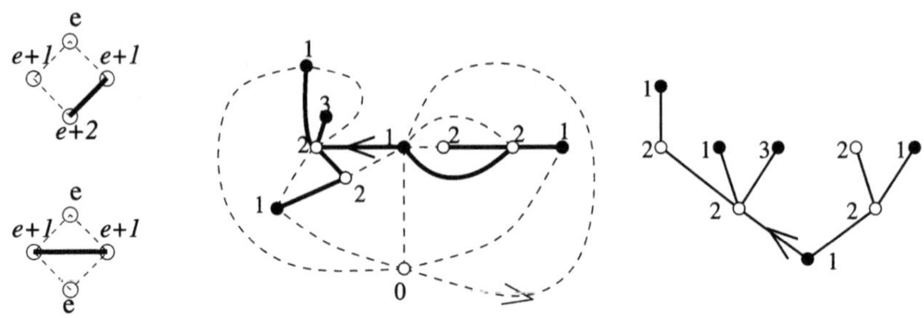

Figure 5: The rules of selection of edges and an example.

3.2 Properties of distances in a quadrangulation

Let Q be a rooted quadrangulation and denote v_0 its root vertex. The labelling ϕ of the map Q is defined by $\phi(x) = d(x, v_0)$ for each vertex x, where $d(x, y)$ denote the distance in Q (cf. Figure 3). Observe that in the number of label k in the labelling of the map Q is precisely the number of vertices at distance k of v_0, that is $H_k^{[Q]} = |\{x \mid \phi(x) = k\}|$. This labelling satisfies the following immediate properties:

Proposition 3.2. *If x and y are joined by an edge, $|\phi(x) - \phi(y)| = 1$. Indeed the quadrangulation being bipartite, a vertex x is white if and only if $\phi(x)$ is even, black if and only if $\phi(x)$ is odd.*

Proposition 3.3. *Around a face, four vertices appear: a black x_1, a white y_1, a black x_2 and a white y_2. These vertices satisfy at least one of the two equalities $\phi(x_1) = \phi(x_2)$ or $\phi(y_1) = \phi(y_2)$ (cf. Figure 3).*

A face will be said *simple* when only one equality is satisfied and *confluent* otherwise (see Figure 3). It should be noted that one may have $x_1 = x_2$ or $y_1 = y_2$.

3.3 Construction of the encoding $\mathcal{T}$

Let Q be a rooted quadrangulation with its distance labelling. The map Q' is obtained by dividing all confluent faces Q into two triangular faces by an edge joining the two vertices with maximal label. Let us now define a subset $\mathcal{T}(Q)$ of edges of Q' by two selection rules:

- In each confluent face of Q, the edge that was added to form Q' is selected.
- For each simple face f of Q, an edge e is selected: let v be the vertex with maximal label in f, then e is the edge leaving v with f on its left.

These two selection rules are illustrated by Figure 5. The first selected edge around the endpoint of the root of Q is taken to be the root of $\mathcal{T}(Q)$.

Proposition 3.4. *The mapping $\mathcal{T}$ sends a quadrangulation Q with n faces on a well labelled tree with n edges.*

More precisely, $\mathcal{T}$ is the bijection of Theorem 3.1.

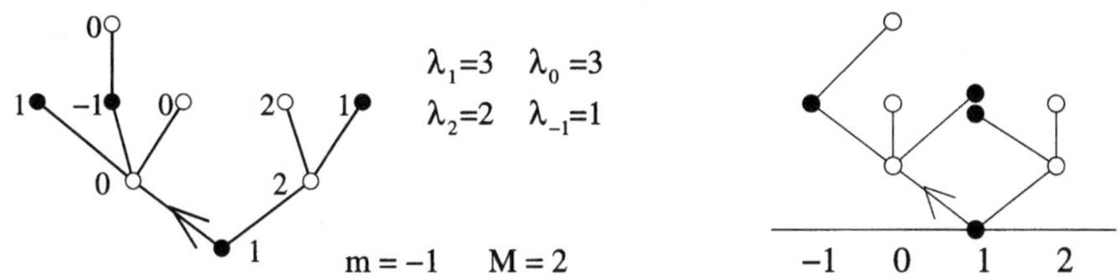

Figure 6: An unconstrained well labelled tree with its label distribution and a representation of the embedding on the line (the plane order structure of the tree is lost in the latter representation).

4 Well labelled and embedded trees

4.1 Unconstrained well labelled trees as embedded trees

Formula (2) for the number of well labelled trees with n edges,

$$|\mathcal{W}_n| = \frac{2}{n+2}\frac{3^n}{n+1}\binom{2n}{n} = \frac{2}{n+2}\cdot 3^n \cdot C(2n),$$

is remarkably simple and yet not immediately clear from definition. Indeed, even though $C(2n)$ is known to be the number of plane trees, the positivity of labels makes it difficult to count labellings that make a plane tree well labelled.

It is thus natural to work first without this positivity condition: define a plane tree to be an *unconstrained well labelled tree* if its vertices have integral labels, the labels of two adjacent vertices differ at most by one, and the label of the root vertex is one. Let $\mathcal{U}_n$ denote the set of unconstrained well labelled trees with n edges.

The labelling of a labelled tree can be recovered uniquely from the label of its root and the variations of labels along all edges. We shall denote $\kappa(\epsilon) \in \{-1,0,1\}$ the variation of labels along the edge ϵ when it is traversed away from the root. Since there is no positivity condition on the labels of unconstrained well labelled trees, all $\kappa(\epsilon)$ can be set independently and the number of labellings of a plane tree that yield an unconstrained well labelled tree is just 3^n. That is,

$$|\mathcal{U}_n| = \frac{3^n}{n+1}\binom{2n}{n} = 3^n \cdot C(2n).$$

The definition of label distribution extends to unconstrained well labelled trees. For $U \in \mathcal{U}_n$ let $(\lambda_k)_{m<k<M} \equiv (\lambda_k^{[U]})_{k\in\mathbb{Z}}$ be the number of vertices with label k in the tree U. The label distribution of U is supported by an interval $[m, M]$ with $m \leq 1 \leq M$. The cumulated label distribution is defined with respect to the minimum label m by $\widehat{\lambda}_k^{[U]} = \sum_{\ell=1}^{k} \lambda_{m+\ell-1}^{[U]}$. These definitions are illustrated by Figure 6.

Observe moreover that similar unconstrained labellings have been considered by D. Aldous [1] with the following interpretation (we restrict to our special one-dimensional case). The tree is folded on the lattice $\mathbb{Z}$ with the root set at position

1 and each edge mapped on an elementary vector (here +1, 0, or −1). The label of a node then describes its position on the line and, upon counting the number of nodes at position j, a mass distribution is obtained. More precisely, with our notations, Aldous' discrete mass distribution associated to a tree $U \in \mathcal{U}_n$ is just the empirical measure of labels

$$\mathcal{J}^{[U]} = \frac{1}{n} \sum_{k \in \mathbb{Z}} \lambda_k^{[U]} \delta_k,$$

where δ_k denote the dirac mass at k.

In view of this interpretation and for concision's sake, let us rename *unconstrained well labelled trees* and call them instead *embedded trees.*

4.2 Random trees and random quadrangulations

Let W_n and U_n be random variables with uniform distribution on $\mathcal{W}_n$ and $\mathcal{U}_n$. More precisely,

$$\Pr(W_n = W) = \frac{1}{\frac{2}{n+2}\frac{3^n}{n+1}\binom{2n}{n}}, \quad \text{and} \quad \Pr(U_n = U) = \frac{1}{\frac{3^n}{n+1}\binom{2n}{n}},$$

for all $W \in \mathcal{W}_n$ and $U \in \mathcal{U}_n$.

The label distribution of the corresponding random trees are two random variables that we shall denote $(\lambda_k^{(n)})_{k\geq 1} \equiv (\lambda_k^{[W_n]})_{k\geq 1}$ for random well labelled trees, and $(\Lambda_k^{(n)})_{k\in\mathbb{Z}} \equiv (\lambda_k^{[U_n]})_{k\in\mathbb{Z}}$ for random embedded trees. For random well labelled trees we also use the notation μ_n for the maximal label, and for random embedded trees the notations m_n and M_n for the minimal and maximal label respectively. Finally cumulated profiles $\widehat{\lambda}_k^{(n)} = \sum_{\ell=1}^k \lambda_\ell^{[W_n]}$ and $\widehat{\Lambda}_k^{(n)} = \sum_{\ell=1}^k \lambda_{m_n+\ell-1}^{[U_n]}$ are defined accordingly (the minimum m_n in $\widehat{\Lambda}_k^{(n)}$ is understood for the same realisation U_n).

At this point we are given three random variables: random quadrangulations L_n, random well labelled trees W_n and random embedded trees U_n. On the one hand, according to Theorem 3.1, random quadrangulations "are" random well labelled trees, as illustrated by the next corollary.

Corollary 4.1. *The label distribution of random well labelled trees has the same distribution as the profile of quadrangulations:*

$$(\lambda_k^{(n)})_{k\geq 1} \overset{law}{=} (H_k^{(n)})_{k\geq 1}.$$

In particular $r_n = \mu_n$.

On the other hand, random embedded trees seem to be a simple variant of well labelled trees that has the great advantage to be defined in accordance with Aldous' prescription for discrete embedded trees. This leads us to study more precisely the relation between W_n and U_n. By definition, $\mathcal{W}_n \subset \mathcal{U}_n$, and according to Tutte's formula (2),

$$|\mathcal{W}_n| = \frac{2}{n+2} \cdot |\mathcal{U}_n|. \tag{3}$$

For combinatorists, this relation could be reminiscent of the relation between the number of Dyck walks and the number of bilatere Dyck walks (see [26, Ch. 5]).
Equivalently, from a more probabilistic point of view, the relation reads

$$\Pr(U_n \in \mathcal{W}_n) = \frac{2}{n+2},$$

and random well labelled trees are random embedded trees conditioned to positivity. This is exactly similar to Kemperman's formula for the probability that a simple symmetric walk on $\mathbb{Z}$ starting from $k > 0$ and ending at 0 after n steps remains positive until the last step (see [22]).

4.3 How to lift the positivity condition for labelled trees

In view of Relation (3), it is tempting to look for a cyclic shift argument in the spirit of the classical combinatorial argument for Łukasiewicz words. This idea to consider cyclic shifts originates in Dvoretsky and Motzkin's work and was used by Raney to prove Lagrange inversion formula and by Takács to prove and extend Kemperman's formula for random walks (see [26, Ch. 5] and [22] for historical references). From the probabilistic point of view this approach should be compared to Vervaat's relation between the Brownian excursion and the Brownian bridge and their local times relatively to the minimum.
It is indeed possible to adapt the cyclic shift idea to draw a relation between well labelled and embedded trees. More precisely, in the complete article we prove the following theorem.

Theorem 4.2. *There exists a partition of $\mathcal{U}_n = \bigcup_{C\in\mathcal{C}_n} C$ into disjoint* conjugacy classes *each of size at most $n+2$ and such that in each class $C \in \mathcal{C}_n$*

- *well labelled trees are fairly represented:*

$$2 \cdot |C| = (n+2) \cdot |C \cap \mathcal{W}_n|,$$

- *and for any $W \in \mathcal{W}_n \cap C$, $U \in C$ and $k \geq 1$,*

$$\widehat{\Lambda}_{k-2}(U) \leq \widehat{\lambda}_k(W) \leq \widehat{\Lambda}_{k+2}(U).$$

Corollary 4.3 (Cori-Vauquelin, 1981). *The number of well labelled trees with n edges, (which is also the number of quadrangulations with n faces), is*

$$|\mathcal{W}_n| = \frac{2}{n+2} \cdot |\mathcal{U}_n| = \frac{2}{n+2} \cdot \frac{3^n}{n+1}\binom{2n}{n}.$$

The proof of Theorem 4.2 relies on an encoding of plane trees in terms of another family of trees, called *blossom trees*, and on the *conjugation of trees* principle which is an analogue of the cycle lemma for blossom trees. This principle was introduced in [24] in order to give a direct combinatorial proof of Corollary 4.3 based on the cycle lemma. However that proof did not rely on well labelled trees and does not provide the link to the profile.
Theorem 4.2 admits the following probabilistic restatement.

Theorem 4.4. *There is a coupling* (W_n, U_n) *(*i.e. *a distribution on* $\mathcal{W}_n \times \mathcal{U}_n$ *such that the marginals are* W_n *and* U_n *as previously defined) such that the induced joint distribution* $(\lambda^{(n)}, \Lambda^{(n)})$ *satisfies for all* k

$$\widehat{\Lambda}^{(n)}_{k-2} \le \widehat{\lambda}^{(n)}_{k} \le \widehat{\Lambda}^{(n)}_{k+2},$$

and in particular

$$|\mu_n - (M_n - m_n)| \le 3.$$

Proof. [Proof of Theorem 4.4] The distribution on $\mathcal{W}_n \times \mathcal{U}_n$ is immediately obtained from the partition $\mathcal{U}_n = \bigcup_{C \in \mathcal{C}_n} C$ as follows: for any (W, U) in $\mathcal{W}_n \times \mathcal{U}_n$, let

$$\Pr((W_n, U_n) = (W, U)) = \begin{cases} \frac{1}{2|\mathcal{U}_n|} & \text{if } U, W \text{ are both in } C \text{ with } |C \cap \mathcal{W}_n| = 2, \\ \frac{1}{|\mathcal{U}_n|} & \text{if } U, W \text{ are both in } C \text{ with } |C \cap \mathcal{W}_n| = 1, \\ 0 & \text{if } U \in C_1 \text{ and } W \in C_2 \text{ with } C_1 \ne C_2. \end{cases}$$

In view of the first part of Theorem 4.2, the marginals are uniformly distributed. The second part of Theorem 4.2 gives the two inequalities. □

5 Quadrangulations, Brownian snake and ISE

5.1 Encoding embedded trees by pairs of contour walks

Let $\bar{\mathcal{U}}_n$ be the set of embedded trees with root label zero instead of one. These trees, that are simply obtained from trees of $\mathcal{U}_n$ by shifting all labels down by one, will be more convenient for our purpose.

Let U be an embedded tree of $\bar{\mathcal{U}}_n$ and consider the following traversal of U, where traversing an edge takes unit time:

- At time $t = 0$, the traversal arrives at the root.
- If the traversal reaches at time t a vertex v_t having k sons for the ℓth time with $\ell \le k$, its next step is toward the ℓth son of v_t.
- If the traversal reaches at time t a vertex v_t having k sons for the $(k+1)$th time, its next step is back toward the father of v_t.

This traversal is called the *contour traversal* because, as exemplified by Figure 7, it turns around the tree. In particular every edge is traversed twice (first away from and then toward the root) and the complete traversal takes $2n$ steps. The *contour pair* of U is then defined by the height (*i.e.* distance to the root in the abstract tree), $E^{[U]}(t)$ and label $V^{[U]}(t)$ of vertex v_t traversed at time $t = 0, \dots, 2n$. (The path E is often called the *Dyck path* associated to the tree U [26, Ch. 5], or the *contour process* in [18, Ch. I.3].)

The following proposition is immediate from the definition of contour pairs.

Proposition 5.1. *The contour pair construction is a one-to-one correspondence between* $\bar{\mathcal{U}}_n$ *(or* $\mathcal{U}_n$*) and the set* $\mathcal{EV}_{2n}$ *of pairs of walks of length* $2n$ *such that:*

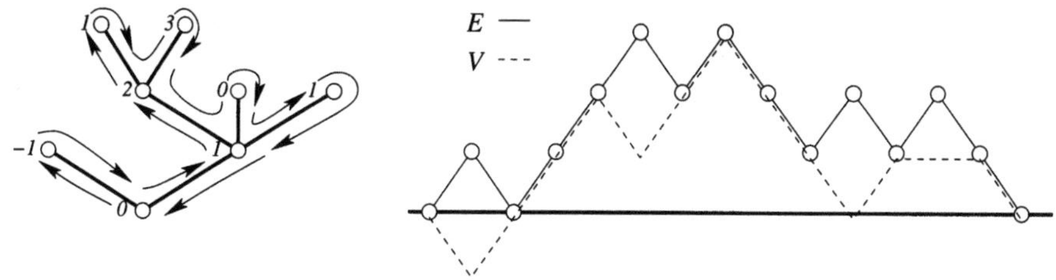

Figure 7: Contour traversal and contour pair (E, V) of a tree.

- *the walk E is an excursion with increment ± 1 or Dyck path, that is $E(0) = E(2n) = 0$, $|E(t) - E(t+1)| = 1$ and $E(t) \geq 0$ for all $t = 0, \ldots, 2n-1$;*
- *the walk V is a bridge with increment $\{-1, 0, 1\}$ or bilatere Motzkin path, that is $V(0) = V(2n) = 0$ and $(V(t) - V(t+1)) \in \{-1, 0, 1\}$ for all t;*
- *and the consistency condition hold:*
  $(E(t) = E(t')$ and $E(s) \geq E(t)$ for all $t < s < t') \Rightarrow V(t) = V(t')$.

The excursion E alone determines a unique unlabelled rooted plane tree, while the walk V describes one of the 3^n labelling of the tree encoded by E. Recall that for an embedded tree U, $\kappa(\epsilon) \in \{-1, 0, 1\}$ denotes the variation along edge ϵ when traversed away from the root. In particular if ϵ is traversed for the first time between time t and $t+1$ and for again between t' and $t'+1$, then

$$\kappa(\epsilon) = V(t+1) - V(t) = V(t') - V(t'+1).$$

This local condition is equivalent to the consistency condition of Proposition 5.1.

5.2 Random trees as random contour pairs

Endow now $\bar{\mathcal{U}}_n$ with the uniform distribution and let $(E^{(n)}, V^{(n)}) \equiv (E^{[U_n]}, V^{[U_n]})$ denote the contour pair of the random tree U_n. According to Proposition 5.1, the random contour pair $(E^{(n)}, V^{(n)})$ is uniformly distributed on $\mathcal{EV}_{2n}$ and E_n is uniformly distributed on $\mathcal{E}_{2n}$, the set of Dyck walks of length $2n$. More precisely, for all $(E, V) \in \mathcal{EV}_{2n}$,

$$\Pr((E^{(n)}, V^{(n)}) = (E, V)) = \frac{1}{\frac{3^n}{n+1}\binom{2n}{n}}, \qquad \Pr(E^{(n)} = E) = \frac{1}{\frac{1}{n+1}\binom{2n}{n}}.$$

In order to state convergence results, let us now defined scaled version of these random walks: given a random tree U_n and its contour pair $(E^{(n)}, V^{(n)})$, let

$$e^{(n)} = \left(\frac{E^{(n)}(\lfloor 2ns \rfloor)}{\sqrt{2n}}\right)_{0 \leq s \leq 1} \quad \text{and} \quad \hat{W}^{(n)} = \left(\frac{V^{(n)}(\lfloor 2ns \rfloor)}{(8n/9)^{1/4}}\right)_{0 \leq s \leq 1}.$$

The random variables $e^{(n)}$ and $\hat{W}^{(n)}$ take their values in the Skorohod space $D([0,1], \mathbb{R})$ of càdlàg real functions (right continuous with left limits).

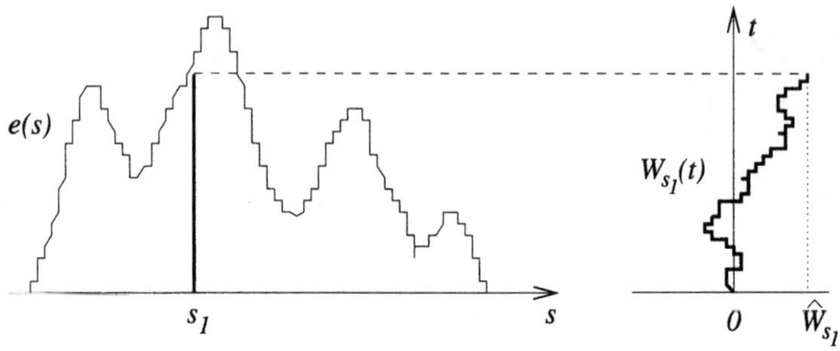

Figure 8: Spacial extension of the snake at time s_1.

As was proved by Kaigh [17], the scaled version $e^{(n)}$ of the contour process converges weakly to the normalised Brownian excursion e. Our aim is to state an analogous result for the random variable

$$X^{(n)} \equiv \left(e^{(n)}, \hat{W}^{(n)}\right),$$

that takes its value in the Skorohod space $D([0,1], \mathbb{R}^2)$.

5.3 A Brownian snake

Let e be the normalised Brownian excursion and

$$W = (W_s(t))_{0 \leq s \leq 1,\ 0 \leq t \leq e(s)}$$

be the Brownian snake with lifetime e, as studied previously in [1, 8, 10, 11, 18, 25]. More precisely, the process W can be defined as follows:

- for all $0 \leq s \leq 1$, $t \to W_s(t)$ is a standard Brownian motion defined for $0 \leq t \leq e(s)$ (see Figure 8);
- the application $s \to W_s(.)$ is a path-valued Markov process with transition function satisfying: for $s_1 < s_2$, and for $m = \inf_{s_1 \leq u \leq s_2} e(u)$, conditionally given $W_{s_1}(.)$ (see Figure 9),
  - on the one hand we have that

$$(W_{s_1}(t))_{0 \leq t \leq m} = (W_{s_2}(t))_{0 \leq t \leq m},$$

  - and on the other hand $(W_{s_2}(m+t))_{0 \leq t \leq e(s_2)-m}$ is a standard Brownian motion starting from $W_{s_2}(m)$, independent of $W_{s_1}(.)$.

The Brownian snake can be viewed as a branching Brownian motion, or as an embedded continuum random tree (see [1]). More precisely the excursion e can be thought of as the contour walk obtained by contour traversal of a continuum random tree, while the snake $W_s(\cdot)$ at times s describes the embedding of the branch to the root at time s.

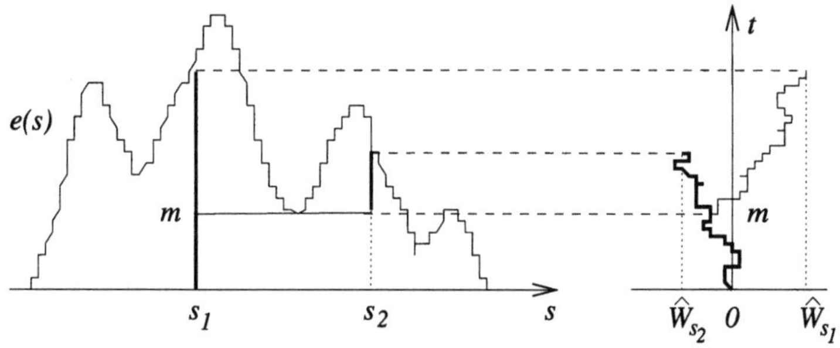

Figure 9: Consistency of the snake between times s_1 and s_2.

Instead of considering the full Brownian snake $W_s(t)$ we shall concentrate, as we did in the discrete case, on its description by a contour pair (or "head of the snake" description) $X = (X_s)_{0\leq s\leq 1}$, defined by (see also Figure 10)

$$\hat{W}_s = W_s(e(s)), \quad X_s = \left(e(s), \hat{W}_s\right), \qquad \text{for } 0 \leq s \leq 1.$$

In complete analogy with the discrete case, the full Brownian snake can be reconstructed from its contour pair description since $W_s(t) = \hat{W}_{\sigma(s,t)}$ where $\sigma(s,t) = \sup\{s' \leq s \mid e(s') = t\}$. However we need only and shall content with results in terms of X (see [21] for a complete discussion of the relation between the full snake and its contour description).

5.4 Integrated SuperBrownian Excursion

Let $\mathcal{J}_n$ denote the empirical measure of labels of a random embedded tree:

$$\mathcal{J}_n = \frac{1}{n}\sum_k \Lambda_k^{(n)}\ \delta_k.$$

Following Aldous [1], for any simple family of trees like our embedded trees, $\mathcal{J}_n$ is expected to converge upon scaling to a random mass distribution $\mathcal{J}$ supported by a random interval $0 \in [L,R] \subset \mathbb{R}$. This random measure $\mathcal{J}$ is called Integrated SuperBrownian Excursion (ISE) by Aldous, in view of its relation to W through

$$\int g\, d\mathcal{J} = \int_0^1 g\left(\hat{W}_s\right)\, ds, \tag{4}$$

for any measurable test function g, see [18, Ch. IV.6]. In [8] the convergence of $\mathcal{J}_n$ to $\mathcal{J}$ is proved for random embedded Cayley trees. Although these trees are not exactly our random embedded *plane* trees, the proof could easily be adapted. According to Corollary 4.1 and Theorem 4.4, the radius r_n is given by the width of the support of $\mathcal{J}_n$. However the weak convergence of $\mathcal{J}_n$ to $\mathcal{J}$, as obtained in [8] is not sufficient for our purpose since $r = R - L$, the width of the support of $\mathcal{J}$, is not a continuous functional of the measure $\mathcal{J}$.

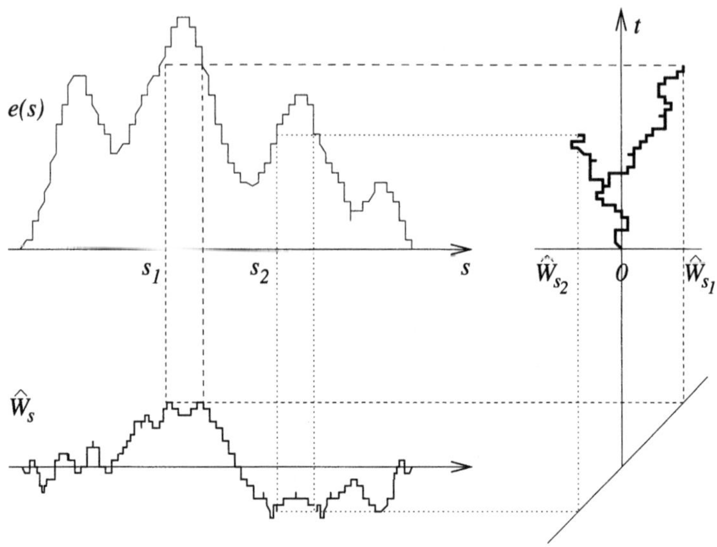

Figure 10: The contour description $(e, \hat{W}_s)$: the excursion e encodes the extension of the snake, the second walk describes the horizontal position of its head.

5.5 Convergence of snakes

Instead of weak convergence of $\mathcal{J}_n$ to $\mathcal{J}$, we shall prove in the extended version the following stronger result.

Theorem 5.2. *The scaled contour pair $X^{(n)}$ converges weakly to X in $D([0,1], \mathbb{R}^2)$.*

This theorem establishes weak convergence of the scaled contour (or head of the snake) description of embedded trees to the head of the snake description of the Brownian snake with lifetime e. We moreover obtain a deviation bound for the maximal extension of the snake $\hat{W}_s^{(n)}$.

Proposition 5.3. *There exists $y_0 > 0$ such that for all $y > y_0$ and n,*

$$\mathbb{P}\left(\sup_{0\leq s\leq 1} \hat{W}_s^{(n)} > (8/9)^{1/4} y\right) \leq e^{-y}.$$

Theorem 5.2 was independently obtained by Marckert and Mokkadem [21]. They extend the convergence result to the explicit full description $(W_s(t))_{s,t}$ but their alternative proof does not provide the exponential bound of Proposition 5.3.

5.6 The radius of quadrangulations and the width of ISE

According to Corollary 4.1 and to Theorem 4.4, the radius r_n of the quadrangulation corresponding to U_n satisfies

$$\left|(8/9)^{1/4}\left(\sup_{0\leq s\leq 1} \hat{W}_s^{(n)} - \inf_{0\leq s\leq 1} \hat{W}_s^{(n)}\right) - n^{-1/4} r_n\right| \leq 3n^{-1/4}.$$

Theorem 5.2 and Proposition 5.3 thus prove the conjecture $\mathbb{E}(r_n) = \Theta(n^{1/4})$ and lead to a much more precise characterization:

Corollary 5.4. *The random variable $n^{-1/4}\, r_n$ converges weakly to $(8/9)^{1/4}\, r$, in which*

$$r = \sup_{0\le s\le 1} \hat{W}_s - \inf_{0\le s\le 1} \hat{W}_s.$$

Furthermore, convergence of all moments holds true.

In view of Relation (4) the random variable r is also the width of ISE process $\mathcal{J}$.

5.7 The profile and a CRM

Actually, Theorems 3.1 and 4.4 suggest that not only the scaled radius but the full scaled profile converges (at least in distribution) to the ISE mass distribution. More precisely, define the distribution function $F(x)$ of the translated ISE by

$$W_{\min} = \inf_{0\le s\le 1} \hat{W}_s, \qquad F(x) = \mathcal{J}\left((-\infty, W_{\min} + x]\right) = \mathcal{J}\left([W_{\min}, W_{\min} + x)]\right),$$

and the scaled distribution function of the profile of random quadrangulations by

$$F_n(x) = \frac{1}{n+1}\, \widehat{\lambda}^{(n)}_{\lfloor (8n/9)^{1/4} x\rfloor} = \frac{1}{n+1}\, \widehat{H}^{(n)}_{\lfloor (8n/9)^{1/4} x\rfloor}.$$

where $\widehat{\lambda}^{(n)}_k$ is the cumulated distribution of labels of a random well labelled tree (as defined in Section 3) and $\widehat{H}^{(n)}_k$ is the cumulated profile of a random quadrangulation (as defined in Section 2).

Then the following is a corollary of Theorems 3.1, 4.4, 5.2 and Corollary 5.4.

Corollary 5.5. *The scaled profile F_n converges weakly to F in $D([0, +\infty), \mathbb{R})$.*

A natural conjecture is that there is a continuum analogue to Theorem 3.1 that allows to define from ISE a Continuum Random Map (CRM), such that the properties of scaled distances in random quadrangulations (distances between arbitrary pairs of points, not only with respect to a basepoint) would be described by the properties of distance in the CRM. In view of the interpretation of random quadrangulations as 2d Euclidean pure quantum geometries, this CRM might be considered as a natural candidate model of continuum 2d pure quantum geometry. We plan to discuss this connection further in a subsequent paper.

References

[1] D. J. Aldous. Tree-based models for random distribution of mass. *J. Statist. Phys.*, 73(3-4):625–641, 1993.

[2] J. Ambjørn, B. Durhuus and T. Jónsson. *Quantum gravity, a statistical field theory approach.* Cambridge Monographs on Mathematical Physics, 1997.

[3] J. Ambjørn and Y. Watabiki, Scaling in Quantum Gravity. *Nucl. Phys. B*, 445:129–144, 1995.

[4] D. Arquès. Les hypercartes planaires sont des arbres très bien étiquetés. *Discrete Math.*, 58:11–24, 1986.

[5] C. Banderier, P. Flajolet, G. Schaeffer and M. Soria. Random Maps, Coalescing Saddles, Singularity Analysis, and Airy Phenomena. *Random Struct. & Algorithms,* 19:194-246, 2001.

[6] E. A. Bender, K. J. Compton and L. B. Richmond. 0–1 laws for maps. *Random Struct. & Algorithms*, 14:215–237, 1999.

[7] E. Brezin, C. Itzykson, G. Parisi and J.-B. Zuber. Planar Diagrams. *Comm. Math. Phys.*, 59:35–47, 1978.

[8] C. Borgs, J. Chayes, R. van der Hofstad and G. Slade. Mean-field lattice trees. On combinatorics and statistical mechanics. *Ann. Comb.*, 3(2-4):205–221, 1999.

[9] R. Cori and B. Vauquelin. Planar maps are well labeled trees. *Canad. J. Math.*, 33(5):1023–1042, 1981.

[10] A. Dembo and O. Zeitouni. Large deviations for random distribution of mass. *Random discrete structures (Minneapolis, MN, 1993)*, 45–53, IMA Vol. Math. Appl., 76, Springer, New York, 1996.

[11] E. Derbez and G. Slade. The scaling limit of lattice trees in high dimensions. *Comm. Math. Phys.*, 193(1):69–104, 1998.

[12] Z. Gao and N. C. Wormald. The Distribution of the Maximum Vertex Degree in Random Planar Maps. *J. Combinat. Theory, Ser. A*, 89:201–230, 2000.

[13] D. Gross, T. Piran and S. Weinberg. *Two dimensional quantum gravity and random surfaces.* World Scientific, 1992.

[14] T. Hara and G. Slade. The incipient infinite cluster in high-dimensional percolations. *Electron. Res. Announc. Amer. Math. Soc.*, 4:48-55, 1998.

[15] B. Jacquard. *Cartes et Arbres : énumération, génération et dessins.* PhD thesis, École Polytechnique, 1997, Palaiseau.

[16] S. Jain and S. D. Mathur. World sheet geometry and baby universes in 2-d quantum gravity. *Phys. Lett. B*, 305:208–213, 1993.

[17] W. D. Kaigh. An invariance principle for random walk conditioned by a late return to zero. *Ann. Prob.*, 4(1):115–121, 1976.

[18] J.-F. Le Gall, *Spatial branching processes, random snakes and partial differential equations.* Lectures in Mathematics ETH Zürich, Birkhäuser Verlag, Basel, 1999.

[19] M. Marcus and G. Schaeffer. Une bijection simple pour les cartes orientables. Manuscript, 10pp, july 2001.

[20] M. Marcus and B. Vauquelin. Un codage des cartes de genre quelconque. In B. Leclerc and J.Y. Thibon, ed., *Séries formelles et combinatoire algébrique, 7ème colloque*, 399–416. Université de Marne-la-Vallée, 1995.

[21] J.-F. Marckert and A. Mokkadem. States spaces of the snake and of its tour – Convergence of the discrete snake. Manuscript, 2002.

[22] J. Pitman. Enumerations of Trees and Forests related to Branching Processes and Random Walks. Tech. Report 482, Dept. Statistics, University of California at Berkeley.

[23] L. B. Richmond and N. C. Wormald. Almost all maps are asymmetric. *J. of Combinat. Theory, Ser. B*, 63(1):1–7, 1995.

[24] G. Schaeffer. *Conjugaison d'arbres et cartes combinatoires aléatoires.* PhD. thesis, Université Bordeaux I, 1998, Bordeaux.

[25] L. Serlet. A large deviation principle for the Brownian snake, *Stochastic Process. Appl.*, 67(1):101–115, 1997.

[26] R. Stanley. *Enumerative Combinatorics,* volume II, Cambridge Series in Advanced Mathematics, 1999.

[27] W.T. Tutte. A census of planar maps. *Canad. J. Math.*, 15:249–271, 1963.

[28] Y. Watabiki. Construction of noncritical string field theory by transfert matrix formalism in dynamical triangulations. *Nucl. Phys. B*, 441:119–166, 1995.

Philippe Chassaing
Université Henri Poincaré
B.P. 239
54506 Vandoeuvre
chassain@iecn.u-nancy.fr
Gilles Schaeffer
CNRS – LORIA
B.P. 239
54506 Vandoeuvre
Gilles.Schaeffer@loria.fr

Trends in Mathematics, © 2002 Birkhäuser Verlag Basel/Switzerland

The Diameter of a Long-Range Percolation Graph

Don Coppersmith, David Gamarnik, Maxim Sviridenko

ABSTRACT: *e consider the following long-range percolation model: an undirected graph with the node set $\{0, 1, \ldots, N\}^d$, has edges $(\mathbf{x}, \mathbf{y})$ selected with probability $\approx \beta/||\mathbf{x} - \mathbf{y}||^s$ if $||\mathbf{x} - \mathbf{y}|| > 1$, and with probability 1 if $||\mathbf{x} - \mathbf{y}|| = 1$, for some parameters $\beta, s > 0$. This model was introduced by Benjamini and Berger [2], who obtained bounds on the diameter of this graph for the one-dimensional case $d = 1$ and for various values of s, but left cases $s = 1, 2$ open. We show that, with high probability, the diameter of this graph is $\Theta(\log N/\log\log N)$ when $s = d$, and, for some constants $0 < \eta_1 < \eta_2 < 1$, it is at most N^{η_2} when $s = 2d$, and is at least N^{η_1} when $d = 1, s = 2, \beta < 1$ or when $s > 2d$. We also provide a simple proof that the diameter is at most $\log^{O(1)} N$ with high probability, when $d < s < 2d$, established previously in [2].*

1 Introduction

Long-range percolation is a model in which any two elements x, y of some (finite or countable) metric space are connected by edges with some probability, inverse proportional to the distance between the points. The motivation for studying this model is dual. First, it naturally extends a classical percolation models on a lattice, by adding edges between non-adjacent nodes with some positive probability. The questions of existence of infinite components were considered specifically by Schulman [8], Aizenman and Newman [1] and Newman and Schulman [7], where the metric space is $\mathcal{Z}$ and edges $(i, j) \in \mathcal{Z}^2$ are selected with probability $\beta/|i - j|^s$ for some parameters β, s. Existence of such an infinite component with positive probability usually implies its existence with probability one, by appealing to Kolmogorov's $0 - 1$ law. It was shown in [7] and in [1] respectively, that percolation occurs if $s = 2, \beta > 1$ and (suitably defined) short range probability is high enough, and does not occur if $s = 2, \beta \leq 1$, for any value of the short range probability.

The second motivation for studying long-range percolation is modelling social networks, initiated by Watts and Strogatz [9]. They considered a random graph model on integer points of a circle, in which neighboring nodes are always connected by an edge, and, in addition, each node is connected to a constant number of other nodes uniformly chosen from a circle. Their motivation was a famous experiment conducted by Milgram [6], which essentially studied the diameter of the "social acquaintances" network and introduced the notion of "six degrees of separation". Watts and Strogatz argued that their graph provides a good model for different types of networks, not only social networks (world wide web, power grids), and showed that the diameter of their random graph is much smaller than the size of the graph. This model was elaborated later by Kleinberg [5], who considered a model similar to a long-range percolation model on a two-dimensional grid, although the work was concerned mostly with algorithmic questions of constructing simple decentralized algorithms for finding short paths between the nodes.

The present paper is motivated by a recent work by Benjamini and Berger [2]. They consider a one-dimensional long-range percolation model in which the nodes are elements of a finite circle $\{0, 1, \ldots, N\}$. An edge (i, j) exists with probability one if $|i - j| = 1$, and with probability $1 - \exp(-\beta/|i - j|^s)$ otherwise, for some parameters β, s, here the distance $|\cdot|$ is taken with respect to a circle. Since for large $|i - j|$, $1 - \exp(-\beta/|i - j|^s) \approx \beta/|i - j|^s$, this model is closely related to the infinite percolation model on $\mathcal{Z}$, with an important distinction, however. The graph is finite and, since neighboring nodes are connected with probability one, the graph is connected. Thus, the percolation question is irrelevant as such; rather, as in models of "social networks", the diameter of the graph is of interest. It is shown in [2] that the diameter of the circle graph above is, with high probability, a constant, when $s < 1$; is $O(\log^\delta N)$, for some $\delta > 1$, when $1 < s < 2$; and is linear $\Theta(N)$, when $s > 2$. These results apply immediately to a graph on an interval $\{0, 1, \ldots, N\}$. A multidimensional version of this problem with a graph on a node set $\{0, 1, \ldots, N\}^d$ was also considered by Benjamini, et al in [3], who showed that the diameter is $\lceil d/(d - s) \rceil$ when $s < d$. The critical cases $s = 1, 2$ were left open in [2] and the authors conjectured that the diameter is $\Theta(\log N)$ when $s = 1$, and $\Theta(N^\eta)$ for some $0 < \eta < 1$, when $s = 2$. In addition, the authors conjectured that, for the case $1 < s < 2$, $\Theta(\log^\delta N)$ is also a lower bound for some $\delta > 1$. In other words, the system experiences a phase transition at $s = 1$ and $s = 2$. Recently Biskup [4] proved that for the case $1 < s < 2$ the diameter is indeed $\Theta(\log^\delta N)$ for some constant δ which Biskup computes explicitly.

In this work we consider a multidimensional version of the problem. Our graph has a node set $\{0, 1, \ldots, N\}^d$ and edges are selected randomly using a long-range percolation $\beta/||\mathbf{x} - \mathbf{y}||^s$ law. We obtain upper and lower bounds on the diameter for the regimes $s = d, d < s < 2d, s = 2d$ and $s > 2d$. This corresponds to regimes $s = 1, 1 < s < 2, s = 2, s > 2$ for the one-dimensional case. We show that, with high probability, for $s = d$, the diameter of this graph is $\Theta(\log N / \log\log N)$; for $d < s < 2d$ the diameter is at most $\log^\delta N$ for some constant $\delta > 1$; and for $s = 2d$, the diameter is at most N^{η_2}, for some constant $0 < \eta_2 < 1$. We also prove a lower bound $N^{\eta_1}, \eta_1 < 1$ on the diameter, which holds with high probability but only when $d \geq 1, s > 2d$ or $d = 1, s = 2, \beta < 1$. We do not have lower bounds for other cases. Note that our lower bound for $s > 2d$ is weaker than known linear lower bound when $d = 1$. We conjecture that the linear lower bound holds for general dimensions. Our results, when applied to the one-dimensional case, support bounds conjectured in [2] for the case $s = 2$ and disprove it for the case $s = 1$. As we mentioned above, the upper bound $\log^\delta N$ for the case $d < s < 2d$ was proven in [2] for the one-dimensional case. It was pointed to the authors that the proof extends to a multidimensional case as well. We provide here an alternative proof which seems simpler. Summarizing the results of present paper and of [2], the diameter of the long-range percolation graph in one-dimensional case experiences a phase transition at $s = 1, 2$ and has a qualitatively different values for $s < 1$; $s = 1$; $1 < s < 2$; $s = 2$ and $\beta < 1$; $s > 2$. Whether the same holds true for general dimensions (whether $s = d, s = 2d$ are the only critical values) remains to be seen. Our results only partially support this conjecture.

2 Model and the main result

Our model is a random graph $G = G(N)$ on a node set $[N]_d \equiv \{0, 1, \ldots, N\}^d$ - integral points of the d-dimensional cube with side length N. Let $||\mathbf{x}||$ denote an $\mathbf{L_1}$ norm in the space $\mathcal{Z}^d$. That is $||\mathbf{x}|| = \sum_{i=1}^d |x_i|$. Nodes $\mathbf{x}, \mathbf{y} \in [N]_d$ are connected with probability 1 if $||\mathbf{x} - \mathbf{y}|| = 1$, and, otherwise, with probability $1 - \exp(-\frac{\beta}{||\mathbf{x}-\mathbf{y}||^s})$, where $\beta > 0, s > 0$ are some fixed parameters. Let $D(N)$ denote the (random) diameter of the graph $G(N)$, and let $P(N)$ denote the (random) length of a shortest path between nodes $\mathbf{0} \equiv (0, \ldots, 0)$ and $\mathbf{N} = (N, \ldots, N)$. For any $\mathbf{x}, \mathbf{y} \in [N]_d$ let also $P(\mathbf{x}, \mathbf{y})$ denote the length of a shortest path between nodes $\mathbf{x}, \mathbf{y}$ in the graph $G(N)$. Our main result is as follows.

Theorem 2.1. *There exist constants $C_1, C_2, C_s > 0, \delta > 1$, $0 < \eta_1 < \eta_2 < 1$, which in general depend on s, β and on dimension d such that*

1. $\lim_{N\to\infty} \text{Prob}\{D(N) \geq N^{\psi}\} = 1$, *for any $s > 2d, \psi < \frac{s-2d}{s-d-1}$.*

2. $\lim_{N\to\infty} \text{Prob}\{D(N) \leq N^{\eta_2}\} = 1$, *for $s = 2d$ and* $\lim_{N\to\infty} \text{Prob}\{D(N) \geq N^{\eta_1}\} = 1$, *for $d = 1, s = 2, \beta < 1$.*

3. $\lim_{N\to\infty} \text{Prob}\{C_s \log N \leq D(N) \leq \log^{\delta} N\} = 1$, *for $d < s < 2d$.*

4. $\lim_{N\to\infty} \text{Prob}\{\frac{C_1 \log N}{\log \log N} \leq D(N) \leq \frac{C_2 \log N}{\log \log N}\} = 1$, *for $s = d$.*

As we mentioned above, it was shown in [3] that the diameter is, with high probability, $\lceil d/(d-s) \rceil$, when $s < d$. Also part 3 of the theorem above was proven by Benjamini and Berger in [2] for the one-dimensional case. They also pointed out to the authors that their proof holds for a multidimensional case as well. We provide here a simpler proof. Throughout the paper we use standard notations $f = O(g), f = \Omega(g), f = \Theta(g), f = o(g)$, which mean respectively that for two functions $f(N), g(N)$, $f(N) \leq C_1 g(N), f(N) \geq C_2 g(N), C_3 g(N) \leq f(N) \leq C_4 g(N), f(N)/g(N) \to 0$, for some constants $C_i, i = 1, 2, 3, 4$ which in general depend on β, s, but do not depend on N. Also, throughout the paper $[n]_d$ denotes an integral cube $\{0, 1, \ldots, n\}^d$ for any nonnegative integer n. The logarithmic function is always assumed to be with the base e.

3 Case $s > 2d$. Lower bound

In this section we show that,with high probability, the diameter of the graph $G(N)$ is at least essentially $N^{\frac{s-2d}{s-d-1}}$. As we noted, for the one dimensional case $d = 1$ this is weaker than the existing linear lower bound $\Omega(N)$ ([2]).

Proof of Theorem 2.1, Part 1: We fix a constant $\psi < \frac{s-2d}{s-d-1}$. For any $k > N^{1-\psi}$ let $L(k)$ be the total number of edges between pairs of points at distance exactly k. We will now show that if $\psi < (s-2d)/(s-d-1)$ then $\sum_{k > N^{1-\psi}} kL(k) \leq dN/2$, with high probability. Since $||\mathbf{N}|| = dN$, then this would imply that, with high probability, any path between $\mathbf{0}$ and $\mathbf{N}$ would contain at least $dN/(2N^{1-\psi}) = (dN^{\psi})/2$ edges and the proof would be completed. For a fixed pair of nodes $\mathbf{x}, \mathbf{y}$ at a distance k, the probability that the edge between them exist is $1-\exp(-\beta/k^s) \le \beta/k^s$, where we use $\exp(-\beta x) \ge 1-\beta x$ for all $0 \le x \le 1$. For a fixed node $\mathbf{x}$ there are $\Theta(k^{d-1})$ nodes $\mathbf{y}$ which are at distance k from $\mathbf{x}$; also there are N^d choices for the node $\mathbf{x}$. Combining $E[L(k)] = O(N^d k^{d-1}(\beta/k^s))$. Then

$$\sum_{k>N^{1-\psi}} kE[L(k)] = O(\beta N^d \sum_{k>N^{1-\psi}} k^{d-s}) = O(N^d N^{(1-\psi)(d-s+1)}).$$

For the given choice of ψ, we have $d+(1-\psi)(d-s+1) < 1$ and the value above is $o(N)$. Using Markov's inequality, we obtain

$$\mathrm{Prob}\{\sum_{k>N^{1-\psi}} kL(k) > N/2\} \le \frac{o(N)}{(N/2)} = o(1).$$

□

4 Case $s = 2d$.

4.1 Upper bound

In this subsection we prove that when $s = 2d$, there exists a constant $0 < \eta < 1$, which depends on β and d, such that with high probability $D(N) \le N^\eta$. To this end we first establish an upper bound on $\max_{\mathbf{x},\mathbf{y}\in[N]_d} E[P(\mathbf{x},\mathbf{y})]$ and then use this bound to obtain a polynomially small bound on $\mathrm{Prob}\{D(N) > N^\eta\}$ for some constant $\eta < 1$.

Proof of Theorem 2.1, Part 2:
We first assume that N is a power of $3 : N = 3^m$, for some integer $m > 0$, and then consider the general case. For any fixed integer n let

$$R(n) = \max_{\mathbf{x},\mathbf{y}\in[n]_d} E[P(\mathbf{x},\mathbf{y})].$$

That is, $R(n)$ is the maximum over expected lengths of shortest paths between all the pairs of points in the cube $[n]_d$. We obtain an upper bound on $R(N)$ by relating it to $R(N/3)$. Divide the cube $[N]_d$ into 3^d subcubes of the type $I_{i_1\ldots i_d} \equiv \prod_{j=1}^d [i_j \frac{N}{3}, (i_j+1)\frac{N}{3}], 0 \le i_j \le 2$. Each cube has a side length $N/3$ (which is integer since N is a power of three). We say that two such cubes are neighboring if they have at least a common node. For example $[0, N/3]^d$ and $[N/3, 2N/3]^d$ are neighboring through a corner node $(N/3, \ldots, N/3)$. We now fix a pair of points $\mathbf{x}, \mathbf{y} \in [N]_d$ and estimate $P(\mathbf{x},\mathbf{y})$ by considering two cases.

1. $\mathbf{x}, \mathbf{y}$ belong to the same subcube $I = I_{i_1\ldots i_d}$. The length of a shortest path between these two points using edges of $[N]_d$ is not bigger than the length of the shortest path between same points but using only edges of the subcube I. Therefore $E[P(\mathbf{x},\mathbf{y})] \le R(N/3)$.

2. $\mathbf{x}, \mathbf{y}$ belong to different subcubes I, I'. Let $\mathcal{E} = \mathcal{E}(I, I')$ be the event "there exists at least one edge between some nodes $\mathbf{v} \in I, \mathbf{v}' \in I'$". The probability

that $\mathcal{E}$ occurs is at least $1-\exp(-\beta(\frac{N}{3}+1)^{2d}/(dN)^{2d})$ since there are $(\frac{N}{3}+1)^d$ nodes in each cube, and the largest possible distance between them is dN. In particular, $\text{Prob}\{\mathcal{E}\}$ is not smaller than a certain constant $\delta > 0$, independent of N. We now estimate $E[P(\mathbf{x},\mathbf{y})]$ conditioned on $\mathcal{E}$ and $\bar{\mathcal{E}}$. Given that $\mathcal{E}$ occurs, select an edge $(\mathbf{v},\mathbf{v}')$ between the cubes I, I'. Then

$$E[P(\mathbf{x},\mathbf{y})|\mathcal{E}] \le E[P(\mathbf{x},\mathbf{v})|\mathcal{E}] + E[P(\mathbf{v}',\mathbf{y})|\mathcal{E}] + 1$$

Note, however, that edges within each cube I, I' are selected independently from edges between cubes and specifically are independent from the event $\mathcal{E}$. Therefore, since $\mathbf{x},\mathbf{v}$ belong to the same cube, $E[P(\mathbf{x},\mathbf{v})|\mathcal{E}] \le R(N/3)$. Similarly, $E[P(\mathbf{v}',\mathbf{y})|\mathcal{E}] \le R(N/3)$. We conclude $E[P(\mathbf{x},\mathbf{y})|\mathcal{E}] \le 2R(N/3)+1$. Now, suppose $\mathcal{E}$ does not occur. Select a cube I'' which is a neighboring cube for cubes I, I' (it is easy to see that such a cube exists). Specifically, let $\mathbf{z}(\mathbf{z}')$ be the nodes shared by cubes I and I'' (I' and I''). Then arguing as above $E[P(\mathbf{x},\mathbf{y})|\bar{\mathcal{E}}] \le E[P(\mathbf{x},\mathbf{z})|\bar{\mathcal{E}}] + E[P(\mathbf{z},\mathbf{z}')|\bar{\mathcal{E}}] + E[P(\mathbf{z}',\mathbf{y})|\bar{\mathcal{E}}] \le 3R(N/3)$. Combining, we obtain

$$\begin{aligned} E[P(\mathbf{x},\mathbf{y})] &\le (2R(N/3)+1)\text{Prob}\{\mathcal{E}\} + 3R(N/3)(1-\text{Prob}\{\mathcal{E}\}) = \\ &(3-\text{Prob}\{\mathcal{E}\})R(N/3) + \text{Prob}\{\mathcal{E}\} \le (3-\delta)R(N/3)+1. \end{aligned}$$

We conclude, $R(N) = \max_{\mathbf{x},\mathbf{y}\in[N]_d} E[P(\mathbf{x},\mathbf{y})] \le (3-\delta)R(N/3)+1$. Applying this bound $m-1 = \log N/\log 3 - 1$ times, we obtain

$$R(N) \le (3-\delta)^{m-1}R(3) + \sum_{i=0}^{m-2}(3-\delta)^i = O((3-\delta)^m) = O(N^{\frac{\log(3-\delta)}{\log 3}}),$$

Note, $\alpha \equiv \log(3-\delta)/\log 3 < 1$. We obtain $R(N) = O(N^\alpha)$ for some $\alpha < 1$.

In order to generalize the bound for all N, it is tempting to argue that $R(N) \le R(3^m)$ as long as $N \le 3^m$. This would require proving a seemingly obvious statement that $R(n)$ is a non-decreasing function of n. While this is most likely correct, proving it does not seem to be trivial. Instead, we proceed as follows. Let m be such that $3^m \le N < 3^{m+1}$. We cover the cube $[N]_d$ with 3^d cubes $I_i, i = 1,\ldots,3^d$ with side length 3^m, with a possible overlapping. Specifically, $I_i \subset [N]_d$ and $\cup_i I_i = [N]_d$. Let $\mathbf{x},\mathbf{y} \in [N]_d$ be arbitrary. Find cubes $I_{i_1}, I_{i_2}, I_{i_3}$ such that $\mathbf{x} \in I_{i_1}, \mathbf{y} \in I_{i_3}$ and $I_{i_1} \cap I_{i_2} \ne \emptyset, I_{i_2} \cap I_{i_3} \ne \emptyset$. Let $\mathbf{z}_1, \mathbf{z}_2$ be some nodes lying in these intersections. Then $E[P(\mathbf{x},\mathbf{y})] \le E[P(\mathbf{x},\mathbf{z}_1)] + E[P(\mathbf{z}_1,\mathbf{z}_2)] + E[P(\mathbf{z}_2,\mathbf{y})] = O((3^m)^\alpha)$, where the last equality follows since pairs $(\mathbf{x},\mathbf{z}_1), (\mathbf{z}_1,\mathbf{z}_2), (\mathbf{z}_2,\mathbf{y})$ lie within cubes $I_{i_1}, I_{i_2}, I_{i_3}$ respectively and each of them has a side length 3^m. But $3^m \le N$. We conclude $E[P(\mathbf{x},\mathbf{y})] = O(N^\alpha)$ and $R(N) = \max_{\mathbf{x},\mathbf{y}} E[P(\mathbf{x},\mathbf{y})] = O(N^\alpha)$.

We now finish the proof of part 2, upper bound, by obtaining a similar bound on the diameter $D(N)$. Fix an arbitrary $0 < \epsilon, \gamma < 1$ such that $\alpha+\epsilon < 1$ and $\epsilon - d(1-\gamma) > 0$. Divide the cube $[N]_d$ into equal subcubes $I_{i_1\ldots i_d} = \prod_{j=1}^d [i_j N^\gamma, (i_j+1)N^\gamma]$, $0 \le i_j \le N^{1-\gamma}$, each with side length N^γ. The total number of subcubes is $N^{d(1-\gamma)}$. Fix any such cube I and let $\mathbf{x}(I)$ be its lower corner (the node with smallest possible coordinates). We showed above $E[P(\mathbf{0},\mathbf{x}(I))] \le O(N^\alpha)$, from which, using Markov inequality,

$$\text{Prob}\{P(\mathbf{0},\mathbf{x}(I)) \ge N^{\alpha+\epsilon}\} = O(\frac{N^\alpha}{N^{\alpha+\epsilon}}) = O(\frac{1}{N^\epsilon}).$$

Then

$$\text{Prob}\{\max_I P(\mathbf{0}, \mathbf{x}(I)) \geq N^{\alpha+\epsilon}\} = O(\frac{N^{d(1-\gamma)}}{N^\epsilon}) = O(\frac{1}{N^{\epsilon-d(1-\gamma)}}).$$

On the other hand for every cube I and every $\mathbf{x} \in I$ we have trivially, $P(\mathbf{x}, \mathbf{x}(I)) \leq dN^\gamma$. Since $D(N) \leq 2\sup_{\mathbf{x}\in[N]_d} P(0, \mathbf{x})$, then

$$\text{Prob}\{D(N) \geq 2(dN^\gamma + N^{\alpha+\epsilon})\} = O(\frac{1}{N^{\epsilon-d(1-\gamma)}}) = o(1).$$

We take $\eta = \max\{\gamma, \alpha+\epsilon\} < 1$ and obtain $\text{Prob}\{D(N) \geq 4dN^\eta\} = o(1)$. This completes the proof of the upper bound. □

4.2 Lower bound

The proof of the lower bound for the one-dimensional case $d = 1, s = 2, \beta < 1$ is similar to the proof for the case $s > 2$, from [2] and uses the notion of a cut point. We first show that $E[D(N)] \geq N^\eta$ for a certain constant $0 < \eta < 1$, for large N. Then we show that this bound holds with high probability. Given a node $1 \leq i \leq N-1$, we call it a cut node if there are no edges which go across i. Namely, i is a cut point if edges (j,k) do not exist for all $j < i < k$. The probability that i is a cut node is $\exp(-\beta\sum_{j<i<k}\frac{1}{|j-k|^2}) \geq \exp(-\beta\sum_{1\leq n\leq N}\frac{n-1}{n^2}) = \Theta(\frac{1}{N^\beta})$. Then the expected number of cuts is $\Omega(N^{1-\beta})$ (which will be helpful to us only if $\beta < 1$). But the shortest path $P(N)$ and as a result the diameter $D(N)$ are not smaller than the number of cuts. Taking $\eta < 1-\beta$, we obtain the bound $E[D(N)] \geq N^\eta$ for large N.

We now complete the proof, by showing that the lower bound holds with high probability. Divide the interval $[N]$ into $N^{\frac{2}{3}}$ intervals $I_1, I_2, \ldots, I_{N^{\frac{2}{3}}}$ each of length $N^{\frac{1}{3}}$. For each interval I_i and each $x \in I_i$, we say that x is a local cut point if it is a cut point with respect to just the graph induced by vertices from I_i. We showed above that the expected number of local cut points in the interval I_i is at least $|I_i|^\eta = N^{\frac{\eta}{3}}$, for any $\eta < 1-\beta$ and for all i. Let $C(I_i)$ be the number of local cut points in the interval I_i. We now show that, with high probability, at least one of the intervals has at least $(1/2)N^{\frac{\eta}{3}}$ local cut points. Note $\{C(I_i)\}_{1\leq i\leq N^{\frac{2}{3}}}$ are independent from each other. We have $E[C(I_i)] \geq N^{\frac{\eta}{3}}$. Also $\text{Var}(C(I_i)) \leq |I_i|^2 = N^{\frac{2}{3}}$. Applying Chebyshev's inequality, we have

$$\text{Prob}\{\sum_i C(I_i)/N^{\frac{2}{3}} < \frac{1}{2}N^{\frac{\eta}{3}}\} \leq \frac{\text{Var}(\text{I}_\text{i})}{\frac{1}{2}N^{\frac{\eta}{3}}N^{\frac{2}{3}}} = O(\frac{1}{N^{\frac{\eta}{3}}}),$$

Therefore, with high probability, at least one of the intervals contains at least $(1/2)N^{\frac{\eta}{3}}$ local cut points. We denote this interval by I_{i^*}. Let us estimate the number of edges between I_{i^*} and $[N] \setminus I_{i^*}$. Note that in defining interval I_{i^*} with many local cut points, we only considered edges within intervals I_i. Note also, that for each $k \geq 1$ there are at most $2k$ edges of length k between I_{i^*} and its complement. Then, the expected number of edges between I_{i^*} and $[N] \setminus I_{i^*}$ is at most

$$\sum_{k=1}^N 2k(1-\exp(-\frac{\beta}{k^2})) + O(1) = O(\sum_{k=1}^N \frac{\beta}{k}) = O(\log N),$$

where we use $\exp(-\beta x) \geq 1 - \beta x$ for all $x \in [0,1]$. Using Markov's inequality, the probability that the number of edges between I_{i^*} and its complement is bigger than $\log^2 N$ is at most $O(1/\log N)$. We conclude that with high probability there are at most $\log^2 N$ edges between I_{i^*} and its complement. Since the number of local cuts in I_{i^*} is $\Omega(N^{\frac{\eta}{3}})$ then there are two local cuts i_1, i_2, such that the interval $[i_1, i_2]$ contains at least $\Omega(N^{\frac{\eta}{3}}/\log^2 N) = \Omega(N^{\frac{\eta}{4}})$ local cuts and no outside edges are connected to nodes in interval $[i_1, i_2]$. Let the number of local cuts in $[i_1, i_2]$ be L. We take the $(1/3)L$-th and the $(2/3)L$-th local cut in this interval. By construction, the shortest path between these local cuts is at least $(1/3)L = \Omega(N^{\frac{\eta}{4}})$. We conclude, $D(N) = \Omega(N^{\frac{\eta}{4}})$, with high probability. □

5 Case $d < s < 2d$.

The lower bound $D(N) \geq C_s \log N$ was proven to hold with high probability in [2] for the case $d = 1$, using branching theory and the fact that for each node, the expected number of its neighbors is a constant. The proof extends easily to all dimensions d. We now focus on an upper bound. Our proof is similar to the one in [2] and is based on renormalization technique, although our analysis is simpler.

Proof of Theorem 2.1, Part 3: We have $d < s < 2d$. Let us fix $\alpha < 1$ such that $2d\alpha > s$. Split the cube $[N]_d$ into equal subcubes $I_{i_1 \dots i_d} \equiv \prod_{j=1}^d [i_j \lceil N^\alpha \rceil, (i_j + 1)\lceil N^\alpha \rceil - 1]$ with side length $\lceil N^\alpha \rceil$. If $N/\lceil N^\alpha \rceil$ is not an integer then we make the cubes containing nodes $(\dots, N, \dots)$ overlap partially with some other cubes. In the following we drop the rounding $\lceil \cdot \rceil$ for simplicity, the argument still holds. Consider the following event $\mathcal{E}_1$: "there exist two cubes I, I' such that no edge exists between points $\mathbf{x} \in I$ and $\mathbf{y} \in I'$". Each resulting cube $I = I_{i_1 \dots i_d}$ we split further into subcubes with side length N^{α^2}. We consider the event $\mathcal{E}_2$: "there exist a cube I with side length N^α and its two subcubes I_1, I_2 with side length N^{α^2}, such that no edge exists between points in I_1 and I_2". We continue this process m times, obtaining in the end cubes with side length N^{α^m}. Assume that none of the events $\mathcal{E}_1, \mathcal{E}_2, \dots, \mathcal{E}_m$ occurs. We claim that then the diameter of our original graph is at most $2^{m+1} N^{\alpha^m}$. In fact, since event $\mathcal{E}_1$ does not occur any two points $\mathbf{x}, \mathbf{y} \in [N]_d$ are connected by a path with length at most $2\bar{D}(N^\alpha) + 1$, where $\bar{D}(N^\alpha)$ is the (random) largest diameter of the cubes $I_{i_1 \dots i_d}$ with side length N^α. Similarly, since event $\mathcal{E}_2$ does not occur, $\bar{D}(N^\alpha) \leq 2\bar{D}(N^{\alpha^2}) + 1$, where $\bar{D}(N^{\alpha^2})$ is the largest diameter of the subcubes with side length N^{α^2}, obtained in second stage. In the end we obtain that the diameter of our graph satisfies $D(N) \leq 2^m \bar{D}(N^{\alpha^m}) + 2^m \leq 2^{m+1} d N^{\alpha^m}$, since trivially, $D(N^{\alpha^m}) \leq d N^{\alpha^m}$. We now show that for a certain value of m, which depends on N, this upper bound on the diameter $D(N)$ is at most $\log^\delta N$ for some constant $\delta > 1$ and simultaneously, the probability $\text{Prob}\{\wedge_{r=1}^m \bar{\mathcal{E}}_r\} \to 1$, as $N \to \infty$. For a given cube with side length $N^{\alpha^{r-1}}$ and its two given subcubes with side length N^{α^r}, the probability that no edges exist between these two subcubes is at most $\exp(-\beta N^{2d\alpha^r}/(dN)^{s\alpha^{r-1}}) = \exp(-\Theta(N^{\alpha^{r-1}(2d\alpha - s)}))$, since there are $N^{2d\alpha^r}$ pairs of points considered and the largest distance among any two of them is $dN^{\alpha^{r-1}}$. Since there are at most N^{2d} pairs of such subcubes, then the probability of the event $\mathcal{E}_r$ is bounded above by

$N^{2d}\exp(-\Theta(N^{\alpha^{r-1}(2d\alpha-s)}))$. We conclude

$$\mathrm{Prob}\{\vee_{r=1}^{m}\mathcal{E}_r\} \le \sum_{r=1}^{m} N^{2d}e^{-\Theta(N^{\alpha^{r-1}(2d\alpha-s)})} \le mN^{2d}e^{-\Theta(N^{\alpha^{m}(2d\alpha-s)})}.$$

Let us fix a large constant C and take

$$m = \frac{\log\log N - \log\log\log N + \log(2d\alpha - s) - \log C}{\log\frac{1}{\alpha}} = O(\log\log N).$$

A straightforward computation shows that for this value of m,

$$\mathrm{Prob}\{\vee_{r=1}^{m}\bar{\mathcal{E}}_r\} = O(e^{-\Theta(\log^{C} N)}). \tag{1}$$

On the other hand, we showed above that, conditioned on event $\wedge_r\bar{\mathcal{E}}_r$, we have $D(N) = O(2^m dN^{\alpha^m})$. For our choice of m a simple calculation shows that $\alpha^m \log N = O(\log\log N)$ or $N^{\alpha^m} = \log^{O(1)} N$. Also, since $m = O(\log\log N)$, then $2^m = O(\log^{O(1)} N)$. This completes the proof. □

In the course of the proof we established the following bound which follows immediately from (1).

Corollary 5.1. *For any constant C, there exists a constant $\delta > 1$ such that*

$$\mathrm{Prob}\{D(N) > \log^{\delta} N\} \le O(e^{-\Theta(\log^{C} N)}).$$

6 Case $s = d$.

Proof of Theorem 2.1, Part 4: We first prove a lower bound. We show that $D(N) \ge (d-\epsilon)\log N/\log\log N$ with high probability, for any constant $0 < \epsilon < 1$. Observe, that, for any $1 < k \le N$ and for each node $\mathbf{x} \in [N]_d$, there are $\Theta(k^{d-1})$ nodes at distance k from $\mathbf{x}$. Each such node is connected to $\mathbf{x}$ with probability $1 - \exp(-\beta/k^d) \le \beta/k^d$. (We used $\exp(-\beta x) \ge 1 - \beta x$ for all $x \in [0,1]$). Then the expected number of nodes connected to $\mathbf{x}$ by an edge is at most $O(1) + O(\sum_{1\le k\le dN}(k^{d-1}/k^d)) = O(\log N)$. Then, the total expected number of nodes which are reachable from $\mathbf{x}$ by paths with length $\le m$ is at most $c^m \log^m N$, for some constant c. We denote the number of such nodes $B(m)$. Using Markov's inequality

$$\mathrm{Prob}\{B(m) \ge N^d\} \le \frac{E[B(m)]}{N^d} \le \frac{c^m \log^m N}{N^d} \to 0$$

if $m = (d-\epsilon)\log N/\log\log N$. Therefore, with probability tending to one, the diameter $D(N)$ is $\Omega(\log N/\log\log N)$.

We now focus on a more difficult part – the upper bound. The proof is fairly technical, but is based on a simple observation which we present now. We have already noted that any fixed node $\mathbf{z}$, in particular, node $\mathbf{N} = (N, N, \ldots, N)$, has in expectation $\Theta(\log N)$ neighbors. We will show later in the formal proof that this actually holds with high probability. Consider a subcube $I = [0, N/\log^c N]^d$ for a certain constant c. Let $\mathbf{y}$ be a neighbor of $\mathbf{x}$. The probability that $\mathbf{y}$ has no neighbors in I is at most $\exp(-\beta N^d/(d^d N^d \log^{cd} N))$, since the largest possible distance is dN and the number of nodes in I is $N^d/\log^{cd} N$. Then probability that none of the $\Theta(\log N)$ neighbors of $\mathbf{N}$ is connected to some node of I by a path of length $\leq$ two is at most $\exp(-\beta N^d \log N/(d^d N^d \log^{cd} N)) = \exp(-\Theta(\log^{1-cd} N))$. If $c < 1/d$ then this quantity converges to zero. Therefore, with high probability $\mathbf{N}$ is connected to some node $\mathbf{X}_1 \in I$ by a path of length 2. Applying this argument for $\mathbf{X}_1$ we find a node $\mathbf{X}_2$ which is connected to $\mathbf{X}_1$ by a path of length two and such that all the coordinates of $\mathbf{X}_2$ are at most $N/\log^{2c} N$. Continuing m times we will obtain that $\mathbf{N}$ is connected by a path of length $O(m)$ to some node $\mathbf{X}_m$ with all the coordinates $\leq N/\log^{cm} N$. Taking $m = O(\log N/\log\log N)$ we will obtain that, with high probability, $\mathbf{N}$ is connected to $\mathbf{0}$ by a path of length $\leq O(m)$. We now formalize this intuitive argument.

We fix an arbitrary node $\mathbf{z}_0 \in [N]_d$. Consider all the paths $(\mathbf{x}, \mathbf{y}, \mathbf{z}_0)$ with length two, which end in node $\mathbf{z}_0$. That is edges $(\mathbf{x}, \mathbf{y}), (\mathbf{y}, \mathbf{z}_0)$ exist. Let $\mathbf{X}_1 = \text{argmin}||\mathbf{x}||$, where the minimum is taken over all such paths. In other words, $\mathbf{X}_1$ is the smallest, in norm, node connected to z_0 via a path of length at most 2. Note, $\mathbf{X}_1$ is random and $||\mathbf{X}_1|| \leq ||\mathbf{z}_0||$, as $\mathbf{z}_0$ is connected to itself by a path of length two. Similarly, let $\mathbf{X}_2 < \mathbf{X}_1$ be the smallest, in norm, node, connected to $\mathbf{X}_1$ via a path of length 2. We continue this procedure for m (to be defined later) steps and obtain a (random) node $\mathbf{X}_m$.

Lemma 6.1.
For any constantly large integer c, if $m = (2d+2)\cdot 2^{c+1}\log N/\log\log N$, then the bound $||\mathbf{X}_m|| \leq \exp((\log N)^{d/2^c})$ holds with probability at least $1 - 1/N^{2d}$.

Before we prove the lemma, let us show how it is used to prove the result. We invoke part 3 of Theorem 2.1, which we proved in the previous section. Choose a constant integer c such that $2^c/d \geq 2\delta$, where $\delta > 1$ is a constant from part 3 of Theorem 2.1. Applying part 3 of Theorem 2.1, the diameter of the cube $[\exp((\log N)^{d/2^c})]_d$ is at most $((\log N)^{d/2^c})^\delta \leq \log^{\frac{1}{2}} N = o(\log N/\log\log N)$ with high probability. In particular $\sup_{\mathbf{x}:||\mathbf{x}||\leq \exp((\log N)^{d/2^c})} P(0,\mathbf{x}) = o(\log N/\log\log N)$ with high probability. By the conclusion of the lemma, with probability at least $1 - O(1/N^{2d})$, each fixed node $z_0 \in [N]_d$ is connected to some node $\mathbf{X}_m$ with $||\mathbf{X}_m|| \leq \exp((\log N)^{d/2^c})$ by a path of length $m = O(\log N/\log\log N)$. Then, with probability at least $1 - O(1/N^d)$, all the nodes $z_0 \in [N]_d$ are connected to some corresponding nodes $\mathbf{X}_m \in [\exp((\log N)^{d/2^c})]_d$ by a path of length $O(\log N/\log\log N)$. Combining, we obtain that $\sup_{\mathbf{z}_0 \in [N]_d} P(0, \mathbf{z}_0) = O(\log N/\log\log N)$ with probability at least $1 - o(1)$. But $D(N) \leq 2\sup_{\mathbf{z}_0\in[N]_d} P(0,\mathbf{z}_0)$. □

Proof of Lemma 6.1: We fix a node $\mathbf{x}$ with $||\mathbf{x}|| \leq ||\mathbf{z}_0||$, fix $1 \leq r \leq m$ and consider $\mathbf{X}_r$ conditioned on event $\mathbf{X}_{r-1} = \mathbf{x}$ (assume $\mathbf{X}_0 = \mathbf{z}_0$). Our goal for the remaining part is the following

Lemma 6.2. *If $||\mathbf{x}|| > \exp((\log N)^{\frac{d}{2^c}})$, then*

$$E\Big[||\mathbf{X}_r||\Big|\mathbf{X}_{r-1} = \mathbf{x}\Big] \leq O\Big(\frac{||\mathbf{x}||}{(\log N)^{1/2^{c+1}}}\Big). \tag{2}$$

In other words, at each step $r = 1, 2, \ldots, m$, the expected value of $||\mathbf{X}_r||$ decreases by a factor of $O(\frac{1}{(\log N)^{1/2^{c+1}}})$, provided that $||\mathbf{X}_{r-1}||$ is still bigger than

$\exp((\log N)^{\frac{d}{2^c}})$.

Proof. Let $B(\mathbf{x})$ be the total number of nodes which are connected to $\mathbf{X}_{r-1} = \mathbf{x}$ and which have a norm smaller than $||\mathbf{x}||$. Note, that for each such node $\mathbf{y}$, $||\mathbf{y} - \mathbf{x}|| \leq ||\mathbf{y}|| + ||\mathbf{x}|| < 2||\mathbf{x}||$. We first show that with probability at least $1 - O(\frac{1}{(\log N)^{d/2^c}})$, the equality $B(\mathbf{x}) = \Omega(\log ||\mathbf{x}||)$ holds. For any fixed $k \leq ||\mathbf{x}||$ there are $\Theta(k^{d-1})$ nodes $\mathbf{y}$ which for which $||\mathbf{y} - \mathbf{x}|| = k$ and $||\mathbf{y}|| < ||\mathbf{x}||$. Each such node is connected by an edge to $\mathbf{x}$ with probability $1 - \exp(-\beta/k^d)$. Then

$$E[B(\mathbf{x})] = \sum_{0 \leq k \leq ||\mathbf{x}||} (1 - \exp(-\frac{\beta\Theta(k^{d-1})}{k^d})) = \Theta(\log ||\mathbf{x}||).$$

Let $c_1 < c_2$ be constants, such that $c_1 \log ||\mathbf{x}|| \leq E[B(\mathbf{x})] \leq c_2 \log ||\mathbf{x}||$. We now estimate the second moment

$$E[B^2(\mathbf{x})] = E[B(\mathbf{x})] +$$

$$\sum_{\mathbf{y}_1 \neq \mathbf{y}_2, ||\mathbf{y}_1||, ||\mathbf{y}_2|| < ||\mathbf{x}||} (1 - \exp(-\frac{\beta}{||\mathbf{y}_1 - \mathbf{x}||}))(1 - \exp(-\frac{\beta}{||\mathbf{y}_2 - \mathbf{x}||})) \leq$$

$$E[B(\mathbf{x})] + \sum_{||\mathbf{y}_1||, ||\mathbf{y}_2|| < ||\mathbf{x}||} (1 - \exp(-\frac{\beta}{||\mathbf{y}_1 - \mathbf{x}||}))(1 - \exp(-\frac{\beta}{||\mathbf{y}_2 - \mathbf{x}||})) =$$

$$E[B(\mathbf{x})] + (E[B(\mathbf{x})])^2.$$

It follows, $\mathrm{Var}(B(\mathbf{x})) \leq E[B(\mathbf{x})]$. Using Chebyshev's inequality,

$$\mathrm{Prob}\{B(\mathbf{x}) \leq (1/2)c_1 \log ||\mathbf{x}||\} \leq \mathrm{Prob}\{|B(\mathbf{x}) - E[B(\mathbf{x})]| \geq (1/2)c_1 \log ||\mathbf{x}||\} \leq$$

$$\frac{\mathrm{Var}(B(\mathbf{x}))}{(1/4)c_1^2 \log^2 ||\mathbf{x}||} \leq \frac{c_2 \log ||\mathbf{x}||}{(1/4)c_1^2 \log^2 ||\mathbf{x}||} = O(\frac{1}{\log ||\mathbf{x}||}) \leq O(\frac{1}{(\log N)^{d/2^c}}), \quad (3)$$

where the last inequality follows from the assumption $||\mathbf{x}|| > \exp((\log N)^{d/2^c})$ of the lemma. Let

$$V(\mathbf{x}) = \{\mathbf{z} : ||\mathbf{z}|| \leq \frac{||\mathbf{x}||}{(\log N)^{1/2^{c+1}}}\}.$$

In particular, $|V(\mathbf{x})| = \Theta(||\mathbf{x}||^d/(\log N)^{d/2^{c+1}})$. Suppose $\mathbf{y}, ||\mathbf{y}|| < ||\mathbf{x}||$ is any node connected by an edge to $\mathbf{x}$ (if any exist). Note that the distance between $\mathbf{y}$ and any node in $V(\mathbf{x})$ is smaller than $3||\mathbf{x}||$. Then, the probability that $\mathbf{y}$ has no nodes in $V(\mathbf{x})$ connected to it by an edge is at most

$$\exp(-\frac{\beta\Theta(||\mathbf{x}||^d)}{(\log N)^{d/2^{c+1}}||\mathbf{x}||^d}) = \exp(-\frac{\Theta(1)}{(\log N)^{d/2^{c+1}}}).$$

By (3), with probability at least $1 - O(\frac{d}{(\log N)^{1/2^c}})$, $\mathbf{x}$ has $\Omega(\log ||\mathbf{x}||)$ nodes $\mathbf{y}, ||\mathbf{y}|| < ||\mathbf{x}||$ connected to it. Conditioned on this event, the probability that no node in $V(\mathbf{x})$ is connected to $\mathbf{x}$ by a path of length two is at most $\exp(-\frac{\Omega(\log ||\mathbf{x}||)}{(\log N)^{d/2^{c+1}}})$. By assumption, $||\mathbf{x}|| > \exp((\log N)^{\frac{d}{2^c}})$ or $\log ||\mathbf{x}|| > (\log N)^{\frac{d}{2^c}}$, using which,

$\exp(-\frac{\Omega(\log ||\mathbf{x}||)}{(\log N)^{d/2^{c+1}}}) \leq \exp(-\Omega((\log N)^{d/2^{c+1}}))$. It follows, that the probability that no node in $V(\mathbf{x})$ is connected to $\mathbf{x}$ by a path of length two, is at most

$$O(\frac{1}{(\log N)^{d/2^c}}) + \exp(-\Omega((\log N)^{d/2^{c+1}})) = O(\frac{1}{(\log N)^{d/2^c}}).$$

Summarizing, conditioned on $\mathbf{X}_{r-1} = \mathbf{x}$, the bound $||\mathbf{X}_r|| \leq \frac{||\mathbf{x}||}{(\log N)^{1/2^{c+1}}}$ holds with probability at least $1 - O(\frac{d}{(\log N)^{1/2^c}})$. On the other hand, with probability one $||\mathbf{X}_r|| \leq ||\mathbf{X}_{r-1}||$. We conclude

$$E\Big[||\mathbf{X}_r||\Big|\mathbf{X}_{r-1} = \mathbf{x}\Big] \leq \frac{||\mathbf{x}||}{(\log N)^{1/2^{c+1}}} + O(\frac{||\mathbf{x}||}{(\log N)^{d/2^c}}) = O(\frac{||\mathbf{x}||}{(\log N)^{1/2^{c+1}}}).$$

This completes the proof of Lemma 6.2. □

We now complete the proof of Lemma 6.1. Note, that for any $2 \leq r \leq m$, $E[\mathbf{X}_r|\mathbf{X}_{r-1}, \mathbf{X}_{r-2}, \ldots, \mathbf{X}_1] = E[\mathbf{X}_r|\mathbf{X}_{r-1}]$. We denote $\exp((\log N)^{d/2^c})$ by $\alpha(N)$. We have,

$$\mathrm{Prob}\{||\mathbf{X}_m|| > \alpha(N)\} =$$

$$\sum_{\alpha(N)<||\mathbf{x}_m||\leq||\mathbf{x}_{m-1}||<||\mathbf{z}_0||} \mathrm{Prob}\{\mathbf{X}_m = \mathbf{x}_m|\mathbf{X}_{m-1} = \mathbf{x}_{m-1}\}\mathrm{Prob}\{\mathbf{X}_{m-1} = \mathbf{x}_{m-1}\} \leq$$

$$\sum_{\alpha(N)<||\mathbf{x}_m||\leq||\mathbf{x}_{m-1}||<||\mathbf{z}_0||} ||\mathbf{x}_m||\mathrm{Prob}\{\mathbf{X}_m = \mathbf{x}_m|\mathbf{X}_{m-1} = \mathbf{x}_{m-1}\}\mathrm{Prob}\{\mathbf{X}_{m-1} = \mathbf{x}_{m-1}\} \leq$$

$$\sum_{\alpha(N)<||\mathbf{x}_{m-1}||<||\mathbf{z}_0||} E\Big[||\mathbf{X}_m||\Big|\mathbf{X}_{m-1} = \mathbf{x}_{m-1}\Big]\mathrm{Prob}\{\mathbf{X}_{m-1} = \mathbf{x}_{m-1}\}.$$

But, using bound (2) of Lemma 6.2, we have

$$E\Big[||\mathbf{X}_m||\Big|\mathbf{X}_{m-1} = \mathbf{x}_{m-1}] \leq O(||\mathbf{x}_{m-1}||/(\log N)^{1/2^{c+1}}),$$

as long as $||\mathbf{x}_{m-1}|| > \alpha(N)$. We obtain

$$\mathrm{Prob}\{||\mathbf{X}_m|| > \alpha(N)\}$$

$$\leq O(\frac{1}{(\log N)^{1/2^{c+1}}}) \sum_{\alpha(N)<||\mathbf{x}_{m-1}||<||\mathbf{z}_0||} ||\mathbf{x}_{m-1}||\mathrm{Prob}\{\mathbf{X}_{m-1} = \mathbf{x}_{m-1}\}$$

$$= O(\frac{1}{(\log N)^{1/2^{c+1}}}) \sum_{\alpha(N)<||\mathbf{x}_{m-1}||\leq||\mathbf{x}_{m-2}||<||\mathbf{z}_0||}$$

$$||\mathbf{x}_{m-1}||\mathrm{Prob}\{\mathbf{X}_{m-1} = \mathbf{x}_{m-1}|\mathbf{X}_{m-2} = \mathbf{x}_{m-2}\}\mathrm{Prob}\{\mathbf{X}_{m-2} = \mathbf{x}_{m-2}\}$$

$$\leq O(\frac{1}{(\log N)^{1/2^{c+1}}}) \sum_{\alpha(N)<||\mathbf{x}_{m-2}||<||\mathbf{z}_0||} E\Big[||\mathbf{X}_{m-1}||\Big|\mathbf{X}_{m-2}=\mathbf{x}_{m-2}\Big]\mathrm{Prob}\{\mathbf{X}_{m-2}=\mathbf{x}_{m-2}\}$$

$$\leq \Big(O(\frac{1}{(\log N)^{1/2^{c+1}}})\Big)^2 \sum_{\alpha(N)<||\mathbf{x}_{m-2}||<||\mathbf{z}_0||} ||\mathbf{x}_{m-2}||\mathrm{Prob}\{X_{m-2} = \mathbf{x}_{m-2}\},$$

where in the last inequality we used bound (2) of Lemma 6.2 again. Continuing this conditioning argument $m-1$ times, we obtain that for some constant C

$$\text{Prob}\{||\mathbf{X}_m|| > \alpha(N)\} \le \frac{C^{m-1}}{(\log N)^{\frac{m-1}{2^c+1}}}||\mathbf{z}_0|| \le \frac{(\log N)^{\frac{1}{2^c+1}} C^m}{(\log N)^{\frac{m}{2^c+1}}} dN.$$

But, by assumption of the lemma, $m = (2d+2)\cdot 2^{c+1}\log N/\log\log N$, from which $(\log N)^{\frac{1}{2^c+1}} C^m = o(N)$ and $\text{Prob}\{||\mathbf{X}_m|| > \alpha(N)\} \le 1/N^{2d}$ for large N. □

7 Concluding remarks and open questions

We considered a long-range percolation model on an graph with a node set $\{0,1,\dots,N\}^d$. Answering some open questions raised by Benjamini and Berger in [2], we showed that if two nodes at a distance r are connected by an edge with probability $\approx \beta/r^s$, then, with high probability, the diameter of this graph is $\Theta(\frac{\log N}{\log\log N})$ when $s=d$, and is at most N^{η} for some value $\eta<1$, when $s=2d$. We also proved a lower bound $N^{\eta'}, \eta'<1$ on the diameter for the cases $d=1, s=2, \beta<1$ and $s>2d, d\ge 1$. Note that for the case $d=1, s>2$ our bound is weaker than known linear lower bound $\Omega(N)$ established in [2]. We conjecture that this linear lower bound holds for all dimensions d as long as $s>2d$. Other unanswered regimes are lower bounds for $s=2d$ and $d=1, s=2, \beta>1$. It would also be interesting to compute the limits $\frac{D(N)}{(\log N/\log\log N)} \to C$ and $\log D(N)/\log N \to \eta$ or even show that these limits actually exist when $s=d, 2d$ respectively.

Acknowledgments. We wish to thank I. Benjamini and N. Berger for clarifying their work and identifying several errors in an earlier version of this paper.

References

[1] M. Aizenman and C. M. Newman. Discontinuity of the percolation density in one dimensional $1/|x-y|^2$ percolation models. *Commun. Math. Phys.*, 107:611–647, 1986.

[2] I. Benjamini and N. Berger. The diameter of long-range percolation clusters on finite cycles. *Random Structures and Algorithms*, 19(2):102–111, 2001.

[3] I. Benjamini, H. Kesten, Y. Peres, and O. Schramm. The geometry of the uniform spanning forest: transitions in dimensions 4, 8, 12, ... *In preparation*, 2000.

[4] M. Biskup. *Work in Preparation*, 2001.

[5] J. Kleinberg. The small-world phenomenon: An algorithmic perspective. Cornell Computer Science Technical Report 99-1776, 1999.

[6] S. Milgram. The small world problem. *Psychology Today*, 1(61), 1967.

[7] C. M. Newman and L. S. Schulman. One dimensional $1/|j-i|^s$ percolation models: the existence of a transition for $s\le 2$. *Commun. Math. Phys.*, 180:483–504, 1986.

[8] L. S. Schulman. Long-range percolation in one dimension. *J.Phys. A*, 16(17):L639–L641, 1983.

[9] D. Watts and S. Strogatz. Collective dynamics of small-world networks. *Nature*, 393(440), 1998.

Don Coppersmith

IBM T.J. Watson Research Center
PO Box 218, Yorktown Heights, NY 10598, USA
dcopper@us.ibm.com

David Gamarnik

IBM T.J. Watson Research Center
PO Box 218, Yorktown Heights, NY 10598, USA
gamarnik@watson.ibm.com

Maxim Sviridenko

IBM T.J. Watson Research Center
PO Box 218, Yorktown Heights, NY 10598, USA
sviri@us.ibm.com

Trends in Mathematics, © 2002 Birkhäuser Verlag Basel/Switzerland

Giant Components for Two Expanding Graph Processes

Luc Devroye, Colin McDiarmid, Bruce Reed

ABSTRACT: *We discuss the emergence of giant components in two random graph models (one directed, one undirected). Our study of these models was motivated by an interest in finding a random model of the Internet.*

1 Introduction

The hyperlinks between the pages of the internet yield a directed graph whose vertices are the web pages and whose arcs correspond to the hyperlinks themselves.

This directed graph and the undirected graph underlying it have been intensely studied (see Adamic and Huberman 1999, Broder et al, 2000, Kleinberg et al. 1999) as an understanding of its structure could be useful in designing searching engines or identifying communities on the web. Researchers are also attempting to build random models of the web (see Barabási Albert and Jeong 1999, Cooper and Frieze 2001, Kumar et al. 1999).

As pointed out in (Kumar et al. 1999), standard random graph models do not accurately represent the web for two reasons. The first is that the web has more vertices of high degree than an average graph. The second is that the web expands as pages get added over time, and a page is more likely to link to those which were present when it was added.

Indeed, this expansion is to some extent responsible for the existence of high degree vertices, as old pages tend to have high degree. However, the function a page serves is also important in determining its degree. For example, the home page for Google has very high degree.

Researchers (see Aiello Chung and Lu 2000, Strogatz and Watts 1999) have applied the techniques of (Molloy and Reed 1995), to study the connectivity properties of graphs whose degree sequence is similar to that of the undirected graph underlying the web. However, less attention has been devoted to developing models which reflect the time dependency inherent in the internet graph. In this paper we study the threshold for the existence of a giant component in two expanding graph processes.

Although the analysis of our processes was motivated by attempts to model the internet, we present the results for their intrinsic interest. Indeed other time-dependent random processes will obviously provide better models of the web graph, yielding e.g. a degree sequence like that of the web graph (see Aiello Chung and Lu 2002, Barabási Albert and Jeong 1999).

2 The Models

We are interested in the following random process UGROW with parameter a constant p, with $0 < p \leq 1$, for constructing an undirected graph.

0. Initialize with the single vertex 1.

1. For $i = 2, ..., n$ add vertex i and with probability p add an edge between i and a vertex chosen uniformly at random from $1, ..., i-1$.

We are interested in the following random process DGROW for constructing a directed graph. Again the parameters $0 \leq p_{down}, p_{extra} \leq 1$ are constants, and all choices are independent.

0. Initialize with the single vertex 1.

1. For $i = 2, ..., n$ add vertex i and with probability p_{down} add an arc from i to a vertex chosen uniformly at random from $1, ..., i-1$.

2. For each ordered pair (i, j) of vertices, add an arc from i to j with probability $\frac{p_{extra}}{n-1}$.

3 The Results

Obviously, if $p = 1$ in UGROW then the algorithm produces a spanning tree of G. We prove:

Theorem 3.1. *Let M_n be the maximum order (number of nodes) of a component of the n-node graph constructed by UGROW. Then the expected value of M_n satisfies $\mathbf{E}(M_n) = \Theta(n^p)$; and for any $\epsilon > 0$ there are positive constants c_1 and c_2 such that*

$$\mathbf{P}(c_1 n^p \leq M_n \leq c_2 n^p) > 1 - \varepsilon$$

for all n.

In the directed case, we are interested in whether or not there is a 'giant' strong component, that is one with $\Omega(n)$ vertices.

Theorem 3.2. *If $p_{down} + p_{extra} \leq 1$ then the digraph constructed by DGROW almost surely has no giant strong component.*

Theorem 3.3. *If $p_{down} + p_{extra} > 1$ then the digraph constructed by DGROW almost surely has a giant strong component.*

Theorem 3.1 is a consequence of much finer results on the output of UGROW. We discuss these results in the next section and then turn to the directed case. We close the paper with some concluding remarks.

4 Analyzing UGROW

We orient each edge of the random graph we obtain to point to its endpoint of smaller index. The directed graph obtained is a random forest, consisting of a number of trees which is distributed like $1 + B(n-1, 1-p)$, where $B(n,p)$ denotes a binomial random variable with parameters n and p. Let N_i denote the order (number of vertices) of the subtree in the forest rooted at node i, and let $M_n = \max(N_1, \ldots, N_n)$ be the maximal order of a subtree. We will show the following.

Lemma 4.1. *For fixed k,*

$$\frac{N_k}{n^p} \to \mathcal{Z}(k,p)$$

in distribution, and $\mathcal{Z}(k,p)$ is a random variable with ℓ-th moment

$$\frac{\Gamma(\ell+1)\Gamma(k)}{\Gamma(k+p\ell)} .$$

Note that for $p = 1$, these are the moments of the beta $(1, k-1)$ distribution when $k > 1$. For $k = 1$, $\mathcal{Z}(k,1) \equiv 1$.

Lemma 4.2. *For all $\ell \geq 0$, and all $1 \leq k \leq n$,*

$$\mathbf{E}\{N_k(N_k+1)\cdots(N_k+\ell)\} \leq (\ell+1)! \left(\frac{n}{k}\right)^{p(\ell+1)} e^{p(\ell+1)/k} .$$

Lemma 4.3. *For $t > 0$,*

$$\mathbf{P}\left\{\frac{M_n}{n^p} \geq t\right\} \leq \frac{\Gamma(2+2/p)e^2\pi^2/6}{t^{2/p}} .$$

Note that Lemma 4.3 may be generalized to bounds of the form $C(a,p)/t^a$ for any $a > 0$ and some constants $C(a,p) > 0$. The order n^p for M_n is actually achieved in all cases in a probabilistic sense: for all $t > 0$, we have,

$$\mathbf{P}\left\{\frac{M_n}{n^p} \leq t\right\} \leq \mathbf{P}\left\{\frac{N_1}{n^p} \leq t\right\} = \mathbf{P}\{\mathcal{Z}(k,p) \leq t\} + o(1)$$

But $\mathbf{P}\{\mathcal{Z}(k,p) \leq t\}$ tends to zero as $t \downarrow 0$:

Lemma 4.4. *For all $p \in (0,1)$, and all $k \geq 1$, $\mathcal{Z}(k,p)$ is a continuous random variable. In particular,*

$$\lim_{t \downarrow 0} \mathbf{P}\{\mathcal{Z}(k,p) \leq t\} = 0 .$$

The forest we are studying is somewhat related to uniform random recursive trees. A uniform random recursive tree (or URRT) on n nodes is a tree recursively constructed by letting the i-th node pick its parent uniformly and at random from among the first $i-1$ nodes. This corresponds to $p = 1$ in our model. A uniform random recursive dag (or URRD) on n nodes starts this process only at node $m+1$, so that the first m nodes are roots. Furthermore, the i-th node picks r nodes uniformly from among the first $i-1$ nodes to be its "parents", thus creating a directed acyclic graph. Na and Rapoport (1970), Moon (1974), Gastwirth (1977), Meir and Moon (1978), Najock and Heyde (1982), Dondajewski and Szymański (1982), Gastwirth and Bhattacharya (1984), Devroye (1987, 1988), Szymański (1987, 1990), Mahmoud (1992), Mahmoud and Smythe (1991), Pittel (1994), and Devroye and Lu (1995) have studied the URRT in some detail. A URRT of course is just a URRD with $m = 1$. Dags model expression trees in which the symbols are the roots and the mathematical operators correspond to internal nodes. They also model PERT networks, and represent partial orders in general.

There is also a Pólya urn model view for our process. In Pólya urns (Pólya, 1931), one starts with a fixed finite number of urns, each having a given number of balls. An urn is picked with probability proportional to the size of the urn, and a ball is added to that urn. An urn in our setting is of course a tree in the forest. It was shown by Pólya and others (Defays, 1974, Athreya, 1969; for a survey, see Johnson and Kotz, 1977) that the proportions of the balls in the urns tends almost surely to a Dirichlet random vector. The urn occupancies are thus not concentrated in the sense that the proportion of balls in the first urn does not tend in probability to a constant. This lack of concentration is also apparent from the results below. In fact, the moment method proof of Lemma 4.4 is mimicked after the standard proof of the beta limit law for the proportion of balls in the first urn in Pólya's urn model. However, our limit law for each subtree size is not beta! In fact, the subtrees have sizes that are roughly $(n/k)^p$. Theorem 3.1 shows that the maximal tree size is $O(n^p)$ in probability.

PROOF OF LEMMA 4.1.

Consider the following process started at node k. Let $X_k = 1$, and for $j > k$, let X_j denote the size of the subtree rooted at k when j nodes have been processed. When the j-th node is processed, note that that subtree grows by one with probability $pX_{j-1}/(j-1)$. Clearly, $X_n = N_k$. For fixed $\ell \geq 0$, it takes a moment to verify the following relationship for the $(\ell+1)$-st increasing factorial moment:

$$\begin{aligned}
&\mathbf{E}\{X_{j+1}(X_{j+1}+1)\cdots(X_{j+1}+\ell)\} \\
&\quad = \mathbf{E}\{X_j(X_j+1)\cdots(X_j+\ell)\} \\
&\quad\quad +(\ell+1)\mathbf{E}\left\{(X_j+1)\cdots(X_j+\ell)\times\frac{pX_j}{j}\right\} \\
&\quad = \mathbf{E}\{X_j(X_j+1)\cdots(X_j+\ell)\}\times\left(1+\frac{p(\ell+1)}{j}\right).
\end{aligned}$$

>From this, we have without further work,

$$\begin{aligned}
\mathbf{E}\{X_n(X_n+1)\cdots(X_n+\ell)\} &= (\ell+1)!\prod_{j=k+1}^{n}\left(1+\frac{p(\ell+1)}{j-1}\right) \\
&= (\ell+1)!\frac{\Gamma(n+p(\ell+1))\Gamma(k)}{\Gamma(k+p(\ell+1))\Gamma(n)}.
\end{aligned}$$

For fixed k and l, we note that the right-hand side is asymptotic to

$$n^{p(\ell+1)}\frac{\Gamma(\ell+2)\Gamma(k)}{\Gamma(k+p(\ell+1))}.$$

Thus,

$$\lim_{n\to\infty} \frac{\mathbf{E}\{X_n(X_n+1)\cdots(X_n+\ell)\}}{n^{p(\ell+1)}} = \frac{\Gamma(\ell+2)\Gamma(k)}{\Gamma(k+p(\ell+1))} .$$

The limit of $\mathbf{E}\left\{(X_n/n^p)^{(\ell+1)}\right\}$ is identical. Carleman's condition applied to the limiting moments shows that these are the moments of a distribution that is uniquely determined by its moments. We call the limiting distribution $\mathcal{Z}(k,p)$. This proves Lemma 4.1.

PROOF OF LEMMA 4.2.

From the proof of Lemma 4.1, we recall

$$\begin{aligned}
&\mathbf{E}\{X_n(X_n+1)\cdots(X_n+\ell)\} \\
&\quad = (\ell+1)! \prod_{j=k+1}^{n} \left(1+\frac{p(\ell+1)}{j-1}\right) \\
&\quad \le (\ell+1)! \exp\left(\sum_{j=k}^{n-1} \frac{p(\ell+1)}{j}\right) \\
&\quad \le (\ell+1)! \exp\left(p(\ell+1)(\log(n/k)+1/k)\right) \\
&\quad \le (\ell+1)! \left(\frac{n}{k}\right)^{p(\ell+1)} e^{p(\ell+1)/k} .
\end{aligned}$$

PROOF OF LEMMA 4.3.

For $t<1$, there is nothing to prove, so assume $t \ge 1$. Let $X_n = N_k$. By Markov's inequality,

$$\begin{aligned}
\mathbf{P}\{X_n \ge tn^p\} &\le \mathbf{P}\left\{X_n(X_n+1)\cdots(X_n+\ell) \ge t^{\ell+1}n^{p(\ell+1)}\right\} \\
&\le \frac{\mathbf{E}\{X_n(X_n+1)\cdots(X_n+\ell)\}}{t^{\ell+1}n^{p(\ell+1)}} \\
&\le (\ell+1)! \left(\frac{e^{1/k}}{t^{1/p}k}\right)^{p(\ell+1)}
\end{aligned}$$

uniformly over all n.

In particular, if we set $\ell = \lceil 2/p \rceil - 1$, then $p(\ell+1) \ge 2$. Thus,

$$\mathbf{P}\{N_k \ge tn^p\} \le \Gamma(2+2/p) \left(\frac{e}{t^{1/p}k}\right)^2 .$$

From this, we deduce by Boole's inequality,

$$\begin{aligned}
\mathbf{P}\{\max(N_1,\ldots,N_n) \ge tn^p\} &\le \sum_{k=1}^{n} \Gamma(2+2/p) \left(\frac{e}{t^{1/p}k}\right)^2 \\
&\le \frac{\Gamma(2+2/p)e^2\pi^2/6}{t^{2/p}} .
\end{aligned}$$

PROOF OF LEMMA 4.4.
The random variable $\mathcal{Z}(k,p)$ has characteristic function given by

$$\begin{aligned}\varphi(t) &= \mathbf{E}\left\{e^{it\mathcal{Z}(k,p)}\right\} \\ &= \sum_{r=0}^{\infty} \frac{(it)^r}{r!} \mathbf{E}\left\{(\mathcal{Z}(k,p))^r\right\} \\ &= \Gamma(k) \sum_{r=0}^{\infty} \frac{(it)^r}{\Gamma(k+pr)} \\ &= \Gamma(k) \mathcal{M}_{p,k}(it) ,\end{aligned}$$

where $\mathcal{M}_{p,k}(z) = \sum_{r=0}^{\infty} z^r / \Gamma(k+pr)$ is the Mittag-Leffler function with parameters p and k. This function is of semiexponential type, analytic on the positive complex halfspace (Henrici, 1986, p. 333), and thus, $\mathcal{M}_{p,k}(z) \to 0$ if $z \to \infty$ along the imaginary axis. Thus, $|\varphi(t)| \to 0$ as $|t| \to \infty$, and thus, $\mathcal{Z}(k,p)$ is a continuous random variable. As $\mathcal{Z}(k,p)$ has no atoms, it has no atom at zero, and thus, $\mathbf{P}\{\mathcal{Z}(k,p) \leq t\} = o(1)$ as $t \downarrow 0$.

5 Analyzing DGROW

We let $D = D(n, p_{down}, p_{extra})$ be the random digraph constructed by DGROW. For each vertex v, we let $F(v)$ be the set of vertices which can be reached by a directed path From v in D. We let $T(v)$ be the set of vertices from which there is a directed path To v in D. We note that the strong component containing v is exactly $T(v) \cap F(v)$. Our approach is to model the construction of $F(v)$ for each vertex using a branching process.

The expected number of arcs out of vertex i in D is essentially $p_{down} + p_{extra}$. (More precisely, for $i > 1$ this expected value is $p_{down} + p_{extra} - \frac{p_{down} p_{extra}}{n-1}$ whilst for $i = 1$ it is p_{extra}). If this value is at most 1 then it is not hard to show that almost surely the maximum size of a strong component is $o(n)$, as we now see.

Proof of Theorem 3.2 The outdegree of a node is stochastically at most the sum of independent random variables $B(n-1, \frac{p_{extra}}{n-1})$ and $B(1, p_{down})$. Let X_n, $X_n^{(1)}, X_n^{(2)}, \ldots$ be independent random variables with this distribution. Note that X_n is a sum of n Poisson trials, and $\mathbf{E}(X_n) = p_{down} + p_{extra}$. We consider a Galton-Watson branching process in which the family sizes are distributed like X_n. Let $R = R_n$ be the random tree constructed by this process. Clearly, for any node v, $\mathbf{P}(|F(v)| \geq k) \leq \mathbf{P}(|R| \geq k)$.

Consider first the case when $p_{down} + p_{extra} = 1 - \epsilon$ for some $\epsilon > 0$, so that the expected number of offspring in our branching process is $1-\epsilon$. Now

$$\mathbf{P}(|R|>k) = \mathbf{P}(\sum_{i=1}^{j}(X_n^{(i)}-1)\geq 0 \quad \forall j=1,\ldots,k)$$
$$\leq \mathbf{P}(\sum_{i=1}^{k} X_n^{(i)} \geq k).$$

But $\sum_{i=1}^{k} X_n^{(i)}$ is a sum of nk Poisson trials with (total) mean $\mu = (p_{down} + p_{extra})k = (1-\epsilon)k$. Hence

$$\mathbf{P}(|R|>k) \leq \mathbf{P}(\sum_{i=1}^{k} X_n^{(i)} \geq (1+\frac{\epsilon}{1-\epsilon})\mu) \leq e^{-\frac{\epsilon^2}{2}k}$$

by standard bounds. But this last term is $o(1/n)$ for $k \geq (3/\epsilon^2)\log n$, and so in this case each component of D almost surely has $O(\log n)$ nodes.

Now consider the case $p_{down} + p_{extra} = 1$, when the expected number of offspring in our process equals 1. We need to be a little more careful. Note first that, if v is in a strong component of D with at least k nodes then $|F(v)| \geq k$. Thus,

$$\begin{aligned}
&\mathbf{P}(\textit{some strong component has } \geq k \textit{ nodes})\\
&\leq \mathbf{E}(\# \textit{ of strong components with } \geq k \textit{ nodes})\\
&\leq \frac{1}{k}\,\mathbf{E}(\# \textit{ of nodes in strong components with } \geq k \textit{ nodes})\\
&\leq \frac{n}{k}\,\mathbf{P}(|R|\geq k).
\end{aligned}$$

We may assume that $p_{down} < 1$, since otherwise $p_{extra} = 0$ and D has only trivial strong components. Note that

$$var(X_n - 1) = p_{extra}(1-\frac{p_{extra}}{n-1}) + p_{down}(1-p_{down}) \rightarrow 1 - p_{down}^2 > 0$$

as $n \rightarrow \infty$, and $\mathbf{E}(|X_n - 1|^3) = O(1)$. Hence by the Berry-Esseen theorem, there is a constant c such that for all n and k we have

$$\mathbf{P}(X_n^{(1)} + \cdots + X_n^{(k)} = k-1) \leq ck^{-\frac{1}{2}}.$$

It follows (see Dwass (1969)) that, for the tree R corresponding to the X_n distribution, we have

$$\mathbf{P}(|R| = k) = \mathbf{P}(X_n^{(1)} + \cdots + X_n^{(k)} = k-1)/k \leq ck^{-\frac{3}{2}},$$

and so $\mathbf{P}(|R| \geq k) = O(k^{-\frac{1}{2}})$. Thus

$$\mathbf{P}(\textit{some strong component has } \geq k \textit{ nodes}) \leq \frac{n}{k}\,\mathbf{P}(|R|\geq k) = O(nk^{-\frac{3}{2}}),$$

and this last term is $o(1)$ if $k = \omega(n)n^{\frac{2}{3}}$. Thus each component of D almost surely has $O(\omega(n)n^{\frac{2}{3}})$ vertices. This completes the proof of Theorem 3.2.

Conversely, if $p_{down} + p_{extra} > 1$ then we almost surely have a giant strong component, as we now show.

Proof of Theorem 3.3

For our branching process analysis to work, we need a final 'post-processing' stage. We will reserve a constant proportion of the extra arcs to be added in this stage. That is, for some constant $p_{final} > 0$, we add an arc from i to j with probability $\frac{(p_{extra}-p_{final})(1-\frac{p_{final}}{n-1})^{-1}}{n-1}$ in step 2 (which completes the first stage), and then with probability $\frac{p_{final}}{n-1}$ in the new final stage. We use our branching process analysis to show that before this final stage we have:

Proposition 5.1. *For some $\epsilon = \epsilon(p_{down}, p_{extra}) > 0$ there are almost surely at least ϵn vertices in the set $A_\epsilon = \{v : |F(v)| \geq \epsilon n\}$.*

Proposition 5.2. *For some $\delta = \delta(p_{down}, p_{extra}) > 0$ there are almost surely at least δn vertices in the set $B_\delta = \{v : |T(v)| \geq \delta n\}$.*

It is an easy matter to show that

Proposition 5.3. *For any $\delta, \epsilon > 0$, almost surely for every $u \in A_\epsilon$ and $v \in B_\delta$ there are at least $\frac{\delta\epsilon p_{final}^2 n}{2}$ vertices w for which we add both an arc from $F(u)$ to w and an arc from w to $T(v)$ in the final stage.*

Proposition 5.4. *If $|A_\epsilon||B_\delta| > n\log n$ holds for some $\delta, \epsilon > 0$, then almost surely there is an arc xy with $x \in B_\delta$ and $y \in A_\epsilon$.*

Combining these last two results we see that if $|A_\epsilon||B_\delta| > n\log n$ holds for some $\delta, \epsilon > 0$, then almost surely there is an $x \in B_\delta$ such that the strong component containing x has at least $\frac{\delta\epsilon p_{final}^2 n}{2}$ vertices. So to prove the theorem we need only prove Propositions 5.1 and 5.2.

Now, since the sum of the sizes of the $F(v)$ equals the sum of the sizes of the $T(v)$, if Proposition 5.1 holds for some $\epsilon > 0$ then an easy averaging argument shows that Proposition 5.2 holds for $\delta = \frac{\epsilon^2}{2}$. So, in fact we need only prove Proposition 5.1.

Before doing so, we specify our choice of p_{final}. We recall that in step 2, instead of adding an arc from i to j with probability $\frac{p_{extra}}{n-1}$, we add the arc with probability $\frac{p'_{extra}}{n-1}$ for $p'_{extra} = (p_{extra} - p_{final})(1 - \frac{p_{final}}{n-1})^{-1}$. Now, no matter how small we make p_{final}, Propositions 5.3 and 5.4 will still hold so by decreasing p_{final} we can make p'_{extra} arbitrarily close to p_{extra}. In particular, we want to ensure that $p_{down} + p'_{extra} > 1$. It turns out that choosing $p_{final} = \frac{p_{down}+p_{extra}-1}{2}$ ensures this is true.

Thus, the expected number of arcs out of a vertex in step 2 exceeds 1. From now on then, we may ignore p_{final} and the final stage, and just assume that $p_{down}+p_{extra} > 1$. It remains to prove (the cleaned-up version of) Proposition 5.1. However, if we try to analyze growing $F(v)$ using a simple branching process we soon run into difficulties because the 'down' arcs make it highly likely we pick vertices with low indices and so the expected outdegree of a low index vertex outside of the already picked vertices rapidly becomes less than 1.

Instead, we will think of a step in the branching process as consisting of starting with a vertex i , exposing all the 'extra' arcs out of i and then exposing the set of vertices reachable from these vertices by 'down' arcs. Now clearly, the expected number of vertices reachable from j by down arcs is essentially $1+p_{down}+p_{down}^2+p_{down}^3\cdots = \frac{1}{1-p_{down}}$ (this isn't quite true if j is small e.g. if $j=1$ this value will be 0 but if e.g. $j>\sqrt{n}$ then this value will be $\frac{1}{1-p_{down}} - o(1)$). So, the expected number of vertices reachable from i in a step is essentially $\frac{p_{extra}}{1-p_{down}}$, which exceeds one.

In order to avoid the complications due to low index vertices, we actually only consider arcs which go to vertices of reasonably high index. Furthermore, we only consider arcs from i added in Step 1 which go to vertices whose index is reasonably high in terms of i. Forthwith the details.

For a given (p_{extra}, p_{down}), we choose $\epsilon_1, \epsilon_2 > 0$ and $C \geq 1$ so that setting $p^* = \sum_{i=0}^{C}((1-\epsilon_2)p_{down})^i$ we have:

$$(1-\epsilon_1)p_{extra}\, p^* > 1.$$

This is possible since the inequality holds if $\epsilon_1 = \epsilon_2 = 0$ and $C = \infty$ (in which case we have $p^* = (1-p_{down})^{-1}$), and we are free to choose the ϵ's as small as we like and C as large as we like.

We set $\epsilon_3 = \epsilon_1 - \frac{\epsilon_2}{2}$ and $\epsilon_4 = \epsilon_3(\frac{\epsilon_2}{2})^{C+2}$. We will restrict our attention to the subgraph D' of D consisting of those arcs (i,j) with $j > \epsilon_4 n$.

To begin, we obtain for each node v, a lower bound on the size of the random set $F'(v)$, consisting of those vertices which can be reached from v along a path P which satisfies:

(a) for any 'extra' arc (i,j) of P added in Step 2 we have $j > \frac{\epsilon_3 n}{2}$,

(b) for any 'down' arc (i,j) of P added in Step 1 we have $j > \frac{\epsilon_2 i}{2}$, and

(c) any set of $C+1$ consecutive arcs of P contains at least one which was added in Step 2.

We will grow $F'(v)$ iteratively. In each iteration we will explore from some vertex i in $F'(v)$ by exposing all the extra arcs out of i which satisfy (a) and go to new vertices, and then exposing the set of new vertices reachable from these vertices by paths of up to C down arcs which satisfy (b). We begin with $F'(v) = \{v\}$, and continue until either there are no unexplored vertices of $F'(v)$ or $|F'(v)| \geq \epsilon_4 n$. Thus throughout the process, there are at most $(\epsilon_3 + \epsilon_4)n \leq \epsilon_1 n$ vertices which are either already known to be in $F'(v)$ or which have indices less than ϵ_3n. In the same vein, from any vertex i, there are at most $\frac{\epsilon_2 i}{2} + \epsilon_4 n$ vertices which have indices less than $\frac{\epsilon_2 i}{2}$ or are already in $F'(v)$. If $i \geq (\frac{\epsilon_2}{2})^C \epsilon_3 n$ then this is less than $\epsilon_2 i$.

Consider the corresponding search tree, while it contains less than $\epsilon_4 n$ nodes. The distribution of the number of (new) children of a node v is stochastically at least the distribution D_n defined as follows. Take a sum of $B((1-\epsilon_1)n, \frac{p_{extra}}{n-1})$ independent random variables Y, where each of these random variables Y takes values in $\{0,1,\ldots,C\}$ and satisfies:

$$\text{for } 0 \leq i \leq C, \;\; \mathbf{P}(Y \geq i) \geq ((1-\epsilon_2)p_{down})^i.$$

Further, our choices of ϵ_1, ϵ_2 and C ensure that (for n sufficiently large) this distribution is stochastically at least a fixed distribution D^* taking a bounded set of values $\{0, 1, \ldots, b\}$ and having mean > 1, where we take a sum of a truncated Poisson number of independent random variables like Y above.

Thus the probability that $|F'(v)| \geq \epsilon_4 n$ is at least the probability that the Galton Watson branching process with family size distribution D^* constructs a tree with at least $\epsilon_4 n$ nodes.

Consider such a Galton Watson branching process. Let its generation sizes be $Z_0 = 1, Z_2, \ldots$ and let $|R|$ be the total number of descendants. We need two facts:

$$\mathbf{P}(|R| = \infty) = \epsilon_5 > 0,$$

and

$$\mathbf{P}(|R| = \infty \| R| \geq \omega \log n) = 1 - o(\frac{1}{n}).$$

Let S be the set of nodes v such that $|F'(v)| \geq \epsilon_4 n$, and let S' be the set of nodes v such that $|F'(v)| \geq \omega(n) \log n$. Then $S \subseteq A_{\epsilon_4}$, and from the above we have $S' \subseteq S$ a.s., and $\mathbf{E}(|S'|) \geq \epsilon_5 n$.

To complete the proof we show that $|S'|$ is concentrated around its expected value, using the second moment method. Having exposed the set W of the first up to $\omega(n) \log n$ vertices of $F'(u)$ we explore, it is quite likely that for some other vertex v, when we expose the first up to $\omega(n) \log n$ vertices of $F'(v)$ we will not touch W. Thus, $\mathbf{P}(v \in S' | u \in S') = \mathbf{P}(v \in S')(1 + o(1))$, which is enough to apply the second moment method. Thus $|S'| \geq \frac{1}{2}\epsilon_5 n$ a.s., and hence $|A_\epsilon| \geq \epsilon n$ a.s., where $\epsilon = \min(\epsilon_4, \frac{1}{2}\epsilon_5)$. Thus the proof is complete.

6 Concluding Remarks

We could attempt to compute the probability that D has no giant component when $p_{down} + p_{extra} > 1$: we believe it is exponentially small in n and our technique may perhaps be pushed to yield this.

We could also imagine a random process where each vertex throws up to k edges back according to probabilities $p_1, \ldots, p_k$.

Finally, in Kim et al. 2002, results are given for the undirected model in which at iteration i, we add one of the $\binom{i}{2}$ possible edges with endpoints in $\{1, .., i\}$ with probability p. The authors show that the threshold for having a giant component is $p = \frac{1}{8}$ and determine bounds on the size of the largest component in the subcritical case.

References

Adamic L. and Huberman B., "Growth Dynamics of the World Wide Web," *Nature*, vol. 401, p. 131, 1999.

Aiello W. Chung F. and Lu L., "Random Evolution in Massive Graphs," *to appear in the Handbook on Massive Data Sets*, 2002.

Aiello W. Chung F. and Lu L., "A random graph model for massive graphs," *Proceedings of the 32nd Annual ACM Symposium on Theory of Computing*, p. 171-180, 2000.

K. B. Athreya, "On a characteristic property of Pólya's urn," *Stud. Sci. Math. Hung.*, vol. 4, pp. 31–35, 1969.

Barabási A. Albert R. and Jeong H., "Scale-free characteristics of random networks: the topology of the world wide web," *Physics A.*, vol. 272, pp. 173-187, 1999.

Broder A. Kumar R. Maghoul F. Raghavan P. Rajagopalan S. Stata R. Tompkins A. and Wiener J. "Graph Structure in the Web," *Computer Networks*, vol. 33, pp. 309-321, 2000.

Cooper C. and Frieze A., "A General Model for Web Graphs," *Proceedings of ESA 2001*, 2001.

D. Defays, "Etude du comportement asymptotique de schémas d'urnes," *Bull. Soc. Roy. Sci. Liège*, vol. 43, pp. 26–34, 1974.

L. Devroye, "Branching processes in the analysis of the heights of trees," *Acta Informatica*, vol. 24, pp. 277–298, 1987.

L. Devroye, "Applications of the theory of records in the study of random trees," *Acta Informatica*, vol. 26, pp. 123–130, 1988.

L. Devroye and J. Lu, "The strong convergence of maximal degrees in uniform random recursive trees and dags," *Random Structures and Algorithms*, vol. 6, pp. 1–14, 1995.

M. Dondajewski and J. Szymański, "On the distribution of vertex-degrees in a strata of a random recursive tree," *Bulletin de l'Académie Polonaise des Sciences, Série des Sciences Mathématiques*, vol. 30, pp. 205–209, 1982.

M. Dwass, "The total progeny in a branching process," *Journal of Applied Probability*, vol. 6, pp. 682–686, 1969.

J. L. Gastwirth, "A probability model of a pyramid scheme," *The American Statistician*, vol. 31, pp. 79–82, 1977.

J. L. Gastwirth and P. K. Bhattacharya, "Two probability models of pyramid or chain letter schemes demonstrating that their promotional claims are unreliable," *Operations Research*, vol. 32, pp. 527–536, 1984.

P. Henrici, *Applied and Computational Complex Analysis Volume 2*, vol. 1, John Wiley, New York, 1986.

N. L. Johnson and S. Kotz, *Urn Models and Their Application*, John Wiley, New York, N.Y., 1977.

Kim J. Kohayakawa Y. McDiarmid C. Reed B. Spencer J. Vu V., "An expanding graph process," *in preparation* , 2002.

Kleinberg J. Kumar S. Raghavan P. Rajagopalan S. and Tomkins A., "The web as a graph: Measurements, models and methods," *Proceedings of the International Conference on Combinatorics and Computing*, 1999.

Kumar S. Raghavan P. Rajagopalan S. and Tomkins A., "Extracting Large-scale Knowledge Bases from the Web," *Proceedings of the 25th VLDB conference*, 1999.

H. M. Mahmoud and R. T. Smythe, "On the distribution of leaves in rooted subtrees of recursive trees," *Annals of Applied Probability*, vol. 1, pp. 406–418, 1991.

H. M. Mahmoud, *Evolution of Random Search Trees*, John Wiley, New York, 1992.

A. Meir and J. W. Moon, "Path edge-covering constants for certain families of trees," *Utilitas Mathematica*, vol. 14, pp. 313–333, 1978.

Molloy M. and Reed B., "A critical point for random graphs with a given degree sequence," *Random Structures and Algorithms* vol. 6, pp. 161-179, 1995.

J. W. Moon, "On the maximum degree in a random tree," *Michigan Mathematical Journal*, vol. 15, pp. 429–432, 1968.

J. W. Moon, "The distance between nodes in recursive trees," *London Mathematical Society Lecture Notes*, vol. 13, pp. 125–132, Cambridge University Press, London, 1974.

H. S. Na and A. Rapoport, "Distribution of nodes of a tree by degree," *Mathematical Biosciences*, vol. 6, pp. 313–329, 1970.

D. Najock and C. C. Heyde, "On the number of terminal vertices in certain random trees with an application to stemma construction in philology," *Journal of Applied Probability*, vol. 19, pp. 675–680, 1982.

B. Pittel, "Note on the heights of random recursive trees and random m-ary search tres," *Random Structures and Algorithms*, vol. 5, pp. 337–347, 1994.

G. Pólya, "Sur quelques points de la théorie de probabilité," *Ann. Inst. Henri Poincaré*, vol. 1, pp. 117–161, 1931.

Strogatz S. and Watts D., "Collective Behaviour of Small World Networks," *Nature*, vol. 393, 1998.

J. Szymański, "On a nonuniform random recursive tree," *Annals of Discrete Mathematics*, vol. 33, pp. 297–306, 1987.

J. Szymański, "On the maximum degree and the height of a random recursive tree," in: *Random Graphs 87*, (edited by M. Karoński, J. Jaworski and A. Ruciński), pp. 313–324, John Wiley, Chichester, 1990.

Luc Devroye

McGill University
Montreal, Canada
luc@cs.mcgill.ca

Colin McDiarmid

Oxford University
Oxford, UK
cmcd@stats.ox.ac.uk

Bruce Reed

McGill University
Montreal, Canada
breed@cs.mcgill.ca

Trends in Mathematics, © 2002 Birkhäuser Verlag Basel/Switzerland

Coloring Random Graphs – an Algorithmic Perspective

Michael Krivelevich

1 Introduction

Algorithmic Graph Coloring and Random Graphs have long become one of the most prominent branches of Combinatorics and Combinatorial Optimization. It is thus very natural to expect that their mixture will produce quite many very attractive, diverse and challenging problems. And indeed, the last thirty or so years witnessed rapid growth of the field of Algorithmic Random Graph Coloring, with many researchers working in this area and bringing there their experience from different directions of Combinatorics, Probability and Computer Science. One of the most distinctive features of this field is indeed the diversity of tools and approaches used to tackle its central problems.

This survey is not intended to be a very detailed, monograph-like coverage of Algorithmic Random Graph Coloring. Instead, our aim is to acquaint the reader with several of the main problems in the field and to show several of the approaches that proved most fruitful in attacking those problems. We do not and we simply cannot provide all details of the proofs, referring the (hopefully) enthusiastic reader to the papers where those proofs are presented in full, or to his/her previous experience in Random Graphs, which should be sufficient to recover many sketched arguments. But of course, the best way to become fluent in this field is to learn from the best, most influential papers, and above all, to engage in independent research, which will undoubtedly bring new and exciting results.

2 Graph coloring is hard

Graph coloring ([21]) has long been one of the central notions in Graph Theory and Combinatorial Optimization. Great many diverse problems can be formulated in terms of finding a coloring of a given graph in a small number of colors or calculating, exactly or approximately, the chromatic number of the graph. Unfortunately, it turns out that these computational problems are very hard. Karp proved already in 1972 [25] that is it NP-complete to decide, for any fixed $k \geq 3$, whether a given graph G is k-colorable. Recent results show that one should not even hope to obtain an efficient algorithm which approximates the chromatic number within a non-trivial approximation ratio. Specifically, Feige and Kilian proved [12] that, unless $coRP = NP$, there is no approximation algorithm for the chromatic number whose approximation ratio over graphs on n vertices is less than $n^{1-\epsilon}$, for any fixed $\epsilon > 0$. Coloring 3-colorable graphs in four colors is NP-complete as well ([28], [18]). Altogether, the situation does not appear particularly encouraging, from both theoretical and practical points of view.

3 The chromatic number of random graphs

The picture changes dramatically when one switches from the somewhat pessimistic worst case scenario to possibly more applicable in practice average case analysis. Already the term "average case analysis" suggests that there should be some underlying probability distribution, whose ground set is composed of graphs, and whose purpose is to help to measure the typical or average performance of various coloring algorithms. Usually this underlying probability space is composed of graphs with the same number n of vertices.

It appears that the most natural and interesting definition of the probability measure on graphs on n vertices is the so called *binomial random graph* $G(n,p)$. This is the probability space whose elements are all labeled graphs $G = (V,E)$ with vertex set $V = \{1,\ldots,n\}$, where each pair of vertices $(i,j) : 1 \leq i < j \leq n$ is chosen to be an edge of G independently and with probability p, in general p may be a function of the number of vertices n: $p = p(n)$. Thus $G(n,p)$ can be viewed as a product probability space, formed by $\binom{n}{2}$ i.i.d. Bernoulli random variables with parameter p. The probability of an individual graph G on n vertices in $G(n,p)$ is easily seen to be $Pr[G] = p^{|E(G)|}(1-p)^{\binom{n}{2}-|E(G)|}$. The special case $p = 0.5$ occupies a very prominent position in the study of random graphs, as for this case the probabilities of every pair (i,j) to be an edge or be a non-edge are equal, resulting in the uniform distribution on the set on all labeled graphs on n vertices: $Pr[G] = 2^{-\binom{n}{2}}$. Therefore studying asymptotic properties of the random graph $G(n,0.5)$ is in a sense equivalent to counting graphs on n vertices with specified properties.

Usually asymptotic properties of random graphs $G(n,p)$ are of interest. For this reason we will assume that the number of vertices n tends to infinity. Also, for a graph property A (where "graph property" means just a family of graphs closed under isomorphism), we say that A holds *almost surely*, or a.s. for brevity, in $G(n,p)$, if the probability that a random graph G, drawn according to the distribution $G(n,p)$, possesses A tends to 1 as n tends to infinity. With some abuse of notation we will use $G(n,p)$ both for the probability distribution on graphs on n vertices and for a graph G drawn from this distribution.

The theory of random graphs is one of the most rapidly developing areas of Combinatorics, with already thousands papers devoted to the subject. We certainly do not intend to cover it here, instead referring the reader to recent monographs [20] and [9]. We will however represent the state of the art of one aspect of random graphs, relevant to the subject of this survey – the asymptotic behavior of the chromatic number of random graphs.

To begin with, consider the most important case $p = 0.5$. For an integer k, let

$$f(k) = \binom{n}{k}\left(\frac{1}{2}\right)^{\binom{k}{2}} . \tag{1}$$

Obviously, $f(k)$ is just the expectation of the number of independent sets of size k in $G(n,0.5)$. When $f(k) = o(1)$, this expectation tends to zero as n grows, and applying Markov's inequality we get immediately that a.s. $G(n,0.5)$ does not contain an independent set of size k. We thus set

$$k_0 = \max\{k : f(k) \geq 1\} . \tag{2}$$

Applying standard asymptotic estimates on the binomial coefficients, one can easily solve asymptotically the above equation for k_0, getting $k_0 = (1-o(1))2\log_2 n$. Hence a.s. $\alpha(G(n,0.5)) \leq (1-o(1))2\log_2 n$. Providing a matching lower bound on the independence number of $G(n,0.5)$ requires more effort, but this can be done as follows [41]. Let X_k be a random variable, counting the number of independent sets of size k in $G(n,0.5)$. Clearly, $E[X_k] = f(k)$. If k is chosen around k_0, using direct calculations one can show that $VAR[X_k] = o(E[X_k]^2)$. Thus Chebyshev's inequality applies, and we get that X_k is concentrated around its mean. In particular, when $f(k) \to \infty$, we derive that a.s. $X_k \geq 1$, which means exactly that $\alpha(G) \geq k$. To satisfy the former condition it is enough to choose $k = k_0$ or $k = k_0 + 1$. Altogether we get that a.s. $\alpha(G(n,0.5)) = (1-o(1))2\log_2 n$.

As for every graph G, $\chi(G) \geq |V(G)|/\alpha(G)$, the above asymptotic (upper) bound on $\alpha(G(n,0.5))$ supplies an easy asymptotic bound for the chromatic number – a.s. $\chi(G(n,0.5)) \geq (1+o(1))n/(2\log_2 n)$. Providing a matching upper bound for the chromatic number was one of the major open questions in the theory of random graphs for about quarter of a century until Béla Bollobás [8] discovered a very inspiring proof of the following theorem.

Theorem 3.1. *Almost surely in the probability space* $G(n,0.5)$, $\chi(G) \leq (1+o(1))\frac{n}{2\log_2 n}$.

Proof. Set $m = n/\log^2 n$ and define

$$k_1 = \max\left\{k : \binom{m}{k}\left(\frac{1}{2}\right)^{\binom{k}{2}} \geq n^3\right\}.$$

The parameter k_1 is chosen so as to guarantee that the expected number of independent subsets of size k_1 in a *fixed* subset $V_0 \subseteq V$ of m vertices is at least n^3. An asymptotic computation very similar to the one mentioned above for k_0 shows that still $k_1 = (1-o(1))2\log_2 n$.

The theorem will easily follows from the following lemma.

Lemma 3.2. *Almost surely in* $G(n,0.5)$, *every subset* V_0 *of* m *vertices contains an independent set of size* k_1.

We will return to the proof of the lemma shortly, but let us see first how it implies Theorem 3.1. Assume that a graph G on n vertices satisfies the conclusion of the lemma. We will prove then that $\chi(G) < n/k_1 + m = (1+o(1))n/(2\log_2 n)$. To show this, we will act in a rather typical for existential coloring arguments way, coloring the graph G by excavation. As long as G contains at least m uncolored vertices, there exists an independent set I of size k_1 in G, all of whose vertices are still uncolored. We then color I by a fresh color and discard all of its vertices from the graph. Clearly this procedure is repeated at most n/k_1 times. Once less than m vertices are left uncolored, we can color each one of them in a new and separate color, resulting in less than m additional colors. The total number of colors used by the above argument is less than $n/k_1 + m$, as promised.

Of course, the above derivation was pretty easy, so the crux of the proof of Theorem 3.1 lies in proving Lemma 3.2. In order to prove Lemma 3.2, fix a subset $V_0 \subset V$ of cardinality $|V_0| = m$. The subgraph of $G(n,0.5)$, induced by V_0, behaves like a random graph $G(m, 0.5)$. Let Y be the number of independent sets of size k_1 inside V_0. Then

$$E[Y] = \binom{m}{k}\left(\frac{1}{2}\right)^{\binom{k}{2}} \geq n^3 ,$$

due to the definition of k_1. To prove the lemma it is enough to prove that

$$Pr[Y = 0] \ll \frac{1}{\binom{n}{m}} , \tag{3}$$

as then by the simple Union bound almost surely every subset of size m in $G(n, 0.5)$ contains an independent set of the required size.

Notice that (3) is a typical large deviation statement – one needs to show that the probability that a random variable (Y) deviates from its expectation ($E[Y] \geq n^3$) by a large quantity (n^3) is exponentially small. However, this task, rather standard and accessible by now, was very challenging fifteen years ago! The main contribution of Bollobás was first to switch from Y to another random variable Z, easier to tackle and such that the positivity of Z implies the positivity of Y, and then to provide an exponential bound for $Pr[Z = 0]$, using martingales – a very novel by then tool for combinatorialists. Currently, there are at least three alternative proofs of Lemma 3.2, based on three different large deviation techniques – the one of Bollobás through martingales, a proof through the so called generalized Janson Inequality, and a proof using the Talagrand concentration of measure inequality. Since all three of them require some technical calculations, we prefer not to present any of them here, instead suggesting the reader to consult [5], where all three tools are discussed in great details.

Bollobas' argument works also for smaller values of $p(n)$ down to $p(n) = n^{-a}$ for a small positive constant a. Later, Łuczak [37] was able to establish the asymptotic value of the chromatic number of $G(n, p)$ for all values of $p(n)$ down to $p(n) \geq C/n$ for a large enough constant $C > 0$:

Theorem 3.3. *There exists C_0 such that for every $p = p(n)$ satisfying $C_0/n \leq p \leq \log^{-7} n$ a.s. in $G(n, p)$*

$$\frac{np}{2\log(np) - 2\log\log(np) + 1} \leq \chi(G) \leq \frac{np}{2\log(np) - 40\log\log(np)} .$$

Łuczak's argument is quite challenging technically and relies heavily on the so called *expose-and-merge approach* invented by Matula [42]. We will not discuss it here. For future reference we summarize the above discussion by noting that the chromatic number of $G(n, p)$ is almost surely $(1+o(1))n\log_2(1(1-p))/(2\log_2 n)$ for a constant edge probability p, and $(1 + o(1))np/(2\ln(np))$ for $C/n \leq p(n) \leq o(1))$, where the $o(1)$ term tends to 0 as np tends to infinity.

The above described results of Bollobás and Łuczak have settled the most important problem in random graph coloring – the asymptotic value of the chromatic number of a random graph. Still many quite significant and attractive problems in this area remain unsolved, for example, the concentration of the chromatic number of random graphs ([45], [38], [2]), list coloring ([3], [29], [33]), thresholds for non-k-colorability for a fixed value of $k \geq 3$ (see a recent survey of Molloy [43]), to mention just a few. And of course, there are many algorithmic problems related to random graph coloring, some of them to be addressed later in this survey.

4 The greedy algorithm for coloring random graphs

The *greedy algorithm*, sometimes also called the first fit algorithm for reasons to become evident immediately, is probably the simplest imaginable algorithm for graph coloring. The greedy algorithm proceeds as follows: given a graph $G = (V, E)$ on n vertices, one first fixes some order $(v_1, \ldots, v_n)$ of the vertices of G, and then scans the vertices of G according to the chosen order, each time coloring a current vertex v_i in the first available color, not used by any already colored neighbor $v_j, j < i$, of v_i. Of course, the resulting number of colors may depend not only on the graph G itself, but also on the chosen order of its vertices. A distinctive feature of the greedy algorithm is that it is essentially an *online* algorithm as the color of a vertex is determined by the edges from the vertex to already seen vertices, and once the color is chosen it will remain unchanged (see [27] for a survey on online graph coloring). This fact makes the analysis of the performance of the greedy algorithm on random graphs $G(n,p)$ quite accessible, as we can generate the random graph as the algorithm proceeds, using the so called *vertex exposure* mechanism – once the algorithm reaches vertex i, a p-Bernoulli coin is flipped for each pair $(j, i), 1 \leq j < i$, to decide whether this pair is an edge of $G(n,p)$, and then a color of i is chosen based on the results of coin flips and a coloring of vertices $1, \ldots, i-1$.

The greedy algorithm turns out to be quite successful for most graphs, using about twice as many colors as the chromatic number of a graph – a remarkable achievement taking into account its simplicity and also the hardness results for graph coloring mentioned above!

Theorem 4.1. *[16] Almost all graphs on n vertices are colored by the greedy algorithm in at most $n/(\log_2 - 3\log_2\log_2 n)$ colors.*

Proof. Let $k = \lfloor \frac{n}{\log_2 n - 3\log_2\log_2 n} \rfloor$. We assume that the greedy algorithm colors the vertices of G according to their natural order $1, \ldots, n$. Denote by $\chi_g(G)$ the number of colors used by the greedy algorithm to color G. In the probability space $G(n, 0.5)$ define A_i to be the event "Vertex i is the first to get color $k+1$". Then clearly the event "$\chi_g(G) > k$" is the union of the events $A_i, 1 \leq i \leq n$, which are pairwise disjoint, and thus:

$$Pr[\chi_g(G) > k] = Pr[\bigcup_{i=1}^{n} A_i] = \sum_{i=1}^{n} Pr[A_i]\,,$$

and the theorem will follow if we will prove $Pr[A_i] = o(1/n)$ for all i.

Consider vertex i of G. The probability $Pr[A_i]$ obviously depends only on the coloring of preceding vertices $1, \ldots, i-1$. Moreover, we can assume that exactly k colors have been used by the algorithm to color those vertices, for otherwise $Pr[A_i] = 0$. So we fix a k-coloring $(C_1, \ldots, C_k)$ of $\{1, \ldots, i-1\}$ and estimate the conditional probability $Pr[A_i|(C_1, \ldots, C_k)]$. In order to force vertex i to be colored in color $k+1$ at least one edge should connect i with each of the color classes $C_1, \ldots, C_k$. We therefore get:

$$Pr[A_i|(C_1,\ldots,C_k)] = \prod_{j=1}^{k}\left(1-\left(\frac{1}{2}\right)^{|C_j|}\right) \le \left(1-\left(\frac{1}{2}\right)^{\sum_{j=1}^{k}|C_j|/k}\right)^k$$
$$< \left(1-\left(\frac{1}{2}\right)^{\frac{n}{k}}\right)^k < e^{-\left(\frac{1}{2}\right)^{\frac{n}{k}}k}$$
$$\le e^{-\left(\frac{1}{2}\right)^{\log_2 n - 3\log_2\log_2 n}k} = e^{-\frac{k\log_2^3 n}{n}} = e^{-(1+o(1))\log_2^2 n} = o(1/n) ,$$

where the first inequality above follows from the convexity of $\log(1-(1/2)^x)$ for $x > 0$.

Grimmett and McDiarmid also showed in [16] that the upper bound on the greedy algorithm from Theorem 4.1 is asymptotically tight – almost every graph in $G(n, 0.5)$ will be colored by at least $(1+o(1))n/\log_2 n$ colors by the greedy algorithm. Moreover, almost surely *all* color classes produced by the greedy algorithm have size at most $(1+o(1))\log_2 n$. Another attractive feature of the greedy algorithm is its extreme robustness when applied to random graphs, as shown by the following result of McDiarmid [39]:

Theorem 4.2. *$Pr[\chi_g(G(n,0.5)) > (1+5\log_2\log_2 n/\log_2 n)n/\log_2 n] < 1/n^n$. Therefore, almost every graph on n vertices is such that no matter which order of the vertices is chosen, the greedy algorithm uses fewer than $(1+5\log_2\log_2 n/\log_2 n)n/\log_2 n$ colors.*

When the edge density becomes lower, the greedy algorithm becomes less competitive. Pittel and Weishaar show in [44] that when applied in the probability space $G(n, c/n)$, the greedy algorithm almost surely outputs a coloring with $(1+o(1))\log_2\log n$ colors, while the chromatic number of most of the graphs in this probability space is bounded from above by a constant $C = C(c)$ (see, e.g., Theorem 3.3). As explained in [44], it is quite easy to see why the number of colors used by the greedy algorithm in $G(n, c/n)$ is a.s. unbounded. To show this, define a sequence of trees T_k as follows: T_1 is a single vertex, and for $k \ge 2$ the tree T_k in obtained from T_{k-1} by joining each vertex of T_{k-1} with a new pendant vertex. A standard second moment argument shows that $G(n, c/n)$ contains a.s. $\Theta(n)$ connected components isomorphic to T_k, for each fixed k. Observe that if the vertices of T_k are colored from "outside in", k colors will be required. As the graph a.s. has so many copies of T_k, at least one of them will a.s. be ordered in this adversarial way. One can however reach essentially the same approximation ratio $2+o(1)$ like in the dense case by a simple modification of the greedy algorithm as suggested by Shamir and Upfal in [46]. The algorithm of Shamir and Upfal proceeds in two phases. The first phase is the usual greedy algorithm as described above. In the second phase, called the correction phase in [46], a subgraph of G spanned by the set V_0 of all vertices that received color higher than some predetermined quantity $K(n,p) = (1+o(1))np/\ln(np)$ is considered. This set is then shown to be a.s. colorable by a breadth-first search in a bounded number of colors.

Unfortunately, the greedy algorithm is not always as good as the typical behavior analysis suggests. It is quite easy to construct an example of a bipartite graph G on n vertices for which the greedy algorithm will use a linear in n number of colors for a certain ordering of the vertices of G. Choosing an ordering of vertices at random does not necessarily help much, as Kučera shows in [35] that for every positive $\epsilon > 0$ and all sufficiently large n there exists a graph G on n vertices with chromatic number at most n^ϵ, for which the greedy algorithm with a randomly chosen initial vertex ordering uses almost surely at least $(1-\epsilon)n/\log_2 n$ colors.

Let us return to the most basic case $p = 0.5$. We know already that almost surely in $G(n, 0.5)$, $\chi_g(G)/\chi(G) = 2 + o(1)$. Is there a better on average coloring algorithm than the greedy coloring? Specifically,

Research Problem 1. *Does there exist a polynomial time algorithm which colors most of the graphs on n vertices in* $(1-\epsilon)n/\log_2 n$ *colors, for some fixed* $\epsilon > 0$?

This is certainly one of the major problems in algorithmic random graph coloring. It is instructive to observe that any such algorithm would produce also an independent set of size $(1+\epsilon')\log_2 n$ (a largest independent set in such a coloring). Therefore the coloring problem is closely related to a quarter century old question of Karp, who asked [26] for a polynomial time algorithm for finding an independent set of size $(1+\epsilon)\log_2 n$ in almost all graphs on n vertices. As we mentioned already, the greedy algorithm almost surely does not provide such a large set. Jerrum proved in [22] that the Metropolis algorithm, which in this case is a random walk on independent sets of the graph biased towards larger independent sets, also requires almost surely a super-polynomial time to reach an independent set of size $(1+\epsilon)\log_2 n$. So apparently the problem of finding a large independent set in a typical graph is hard algorithmically, the fact which has been used even for cryptographic purposes [23]. An interesting fact is that it follows from the above mentioned expose-and-merge technique of Matula that an algorithm for finding a.s. an independent set of size $(1+\epsilon)\log_2 n$ in $G(n, 0.5)$ can be used as a subroutine in an algorithm for coloring a typical graph in $(1-\epsilon')n/\log_2 n$ colors.

A marginal improvement over the greedy algorithm (Theorem 4.1) has been achieved by Krivelevich and Sudakov [31], who provided a randomized polynomial time algorithm that colors almost every graph on n vertices is $n/(\log_2 n + c\sqrt{\log_2 n})$ colors for every positive constant c, thus outputting a coloring with color classes of average size $\log_2 n + c\sqrt{\log_2 n}$, compared to $\log_2 n - \Theta(\log\log n)$ of the greedy algorithm. Again, the critical task here is to find an independent set of size $\log_2 n + c\sqrt{\log n}$. This is achieved in [31] by running the greedy algorithm for finding an independent set I of size $|I| = \log_2 n - 2c\sqrt{\log n}$ in the first $n/2$ vertices of the graph. The set U of non-neighbors of I in the second half of the graph has then almost surely about $2^{2c\sqrt{\log_2 n}}$ vertices and contains inside an independent set I_1 of size $(1-o(1))2\log_2|U| = (1-o(1))4c\sqrt{\log_2 n}$, which can be found in polynomial time by checking exhaustively all $\binom{|U|}{4c\sqrt{\log_2 n}}$ subsets of U of the appropriate size. The union of I and I_1 forms a desired independent set.

There is (at least) one reason to believe that it would be hard to break the $\log_2 n + \Theta(\sqrt{\log n})$ barrier. Going back to the expectation (1) of the number of independent sets of size k, we can easily check that $f(k)$ is polynomially smaller than the total number of independent sets in $G(n, 0.5)$ only if $k \le \log_2 n + \Theta(\sqrt{\log n})$. This may indicate that finding independent sets of larger size may take superpolynomial time. Further discussion can be found in [31].

5 Approximating the chromatic number in expected polynomial time

As discussed above, the greedy algorithm is quite successful for most of graphs, providing a coloring that uses about twice as many colors as an optimal coloring. But there are some (very rare though) graphs for which the greedy coloring performs rather miserably. On the other hand, one cannot probably expect to design a coloring algorithm that beats significantly the trivial approximation ratio n for all graphs, due to the complexity results stated in Section 2. It is therefore desirable to provide a coloring algorithm with a guaranteed approximation ratio for *all* graphs on n vertices and with running time polynomial on *average* in n. We thus arrive naturally to the concept of algorithms with expected polynomial running time.

Given an algorithm A whose domain is the set of all graphs on n vertices, and a probability distribution $P[\cdot]$ on the same set, the *expected running time* of A is defined as $\sum_G Pr[G]R_A(G)$, where the sum runs over all graphs on n vertices, $Pr[G]$ is the probability of G according to the chosen probability measure, and $R_A(G)$ is the running time of A on G. Thus, while looking for an algorithm A whose expected running time is polynomial, we can allow A to spend a superpolynomial time on some graphs on n vertices, but it should be efficient on average.

Observe that if an algorithm A has expected polynomial running time with respect to the probability distribution $P(\cdot)$, then A is polynomial for almost all graphs according to P. Therefore, it is more difficult to develop algorithms with expected polynomial running time than algorithms that perform the same algorithmic task for almost all graphs.

Obviously the problem is quite sensitive to the choice of the underlying probability distribution. In this section we concentrate on the case where the distribution is chosen to be $G(n,p)$, the binomial random graph. There have been a few papers addressing different probability distributions, we will discuss some of them later.

Krivelevich and Vu prove in [32] the following result on the existence of an approximate coloring algorithm with expected polynomial running time:

Theorem 5.1. *For any constant $\epsilon > 0$ the following holds. If the edge probability $p(n)$ satisfies $n^{-1/2+\epsilon} \le p(n) \le 0.99$, then there exists a deterministic coloring algorithm, approximating the chromatic number $\chi(G)$ within a factor $O((np)^{1/2}/\log n)$ and having polynomial expected running time over $G(n,p)$.*

Thus in the most basic case $p = 0.5$ we get a coloring algorithm with approximation ratio $O(\sqrt{n}/\log n)$ – a considerable improvement over best known approximation algorithm for the worst case [19], whose approximation ratio is only $O(n/polylog(n))$. Note also that the approximation ratio decreases with the edge probability $p(n)$.

Before describing the basic idea of the algorithm of [32], we would like to say a few words about combinatorial ideas forming the core of its analysis. As is typically the case with developing algorithms whose expected running time is polynomial, we need to distinguish efficiently between "typical" graphs in the probability space $G(n,p)$, for which it is relatively easy to provide a good approximation algorithm, and "non-typical" ones, which are rare but may be hard for approximating a desired quantity. As these rare and possibly hard graphs have an exponentially small probability in $G(n,p)$, this gives us a possibility to spend an exponential time on each of them. This in turn enables to approximate the chromatic number within the desired factor even for these graphs.

A separation between typical and non-typical instances will be made based on the first eigenvalue of an auxiliary matrix, to be defined later. Then we will apply a large deviation result to show that this eigenvalue deviates from its expectation with exponentially small probability, and thus bad instances are indeed extremely rare. Thus our main tools will come from two seemingly unrelated fields - graph spectral techniques and large deviation inequalities.

We now describe a proof of a somewhat weaker version of Theorem 5.1 for the case $p = 0.5$. Recall that according to Theorem 4.2 the probability that the greedy algorithm fails to color a graph in $G(n, 0.5)$ in $1 + o(1))n/\log_2 n$ colors is less than n^{-n}. Therefore we can run the greedy algorithm, and then apply the exhaustive search for graphs colored by more than, say, $2n/\log_2 n$ colors.

In order to show that the greedy coloring is within $\tilde{O}(\sqrt{n})$ factor from an optimal coloring of a graph G, we need to bound from below the chromatic number of G. As $\chi(G) \geq n/\alpha(G)$, it is enough to certify that $\alpha(G) = \tilde{O}(\sqrt{n})$. We thus need an efficiently computable graph parameter that bounds from above the independence number of G. Given a graph $G = (V, E)$ with vertex set $V = \{1, \ldots, n\}$ define an n-by-n matrix $M = (m_{ij})$ as follows:

$$m_{ij} = \begin{cases} 1, & \text{if } i, j \text{ are non-adjacent in } G, \\ -1, & \text{otherwise ,} \end{cases} \tag{4}$$

Then M is a real symmetric matrix and has therefore n real eigenvalues $\lambda_1 \geq \lambda_2 \geq \ldots \geq \lambda_n$. The connection between $\alpha(G)$ and the first eigenvalue $\lambda_1(M(G))$ is given by the following lemma.

Lemma 5.2. *Let $M = M(G)$ be as defined in (4). Then $\lambda_1(M) \geq \alpha(G)$.*

Proof. Let $k = \alpha(G)$. Then M contains a k by k block of all 1's, indexed by the vertices of an independent set of size k. It follows from interlacing (see, e.g., Chapter 31 of [51]) that $\lambda_1(M) \geq \lambda_1(1_{k \times k}) = k$.

(In fact, $\lambda_1(M(G))$ is an upper bound not only for the independence number of G, but also for its Lovász theta-function [36].)

The spectrum of a real symmetric matrix can be efficiently calculated within any desired precision. So if we calculate $\lambda_1(M(G))$ and see that $\lambda_1(M(G)) = \tilde{O}(\sqrt{n})$, then we have a certificate of the desired lower bound for $\chi(G)$.

What is a typical value of $\lambda_1(M(G))$? Recall that G is distributed according to $G(n, 0.5)$, and therefore $M(G)$ is a random symmetric matrix, each of its entries above the main diagonal is independently 1 or -1 with probability 0.5. This enables us to apply known results on eigenvalues of random symmetric matrices. Füredi and Komlós proved in [14] that

$$E[\lambda_1(M)] = (1 + o(1))2\sqrt{n}\ .$$

This shows that typically $\lambda_1(M(G))$ is of order $\sqrt{n}$.

Now we need to estimate the probability that $\lambda_1(M(G))$ is significantly larger than its expectation. The desired estimate is provided by the following large deviation result from [32] (see also [4] for a more general result):

Lemma 5.3. *Let $M = (m_{ij})$ be an n-by-n random symmetric matrix with all entries bounded by 1 in their absolute values. Then for all $t > 0$,*

$$Pr[\lambda_1(M) > E[\lambda_1(M)] + t] \leq 2e^{-(1+o(1))t^2/32} .$$

This lemma is proven by applying the Talagrand concentration of measure inequality, see [32], [4] for details of the proof.

Plugging an estimate of Füredi and Komlös on $E[\lambda_1(M)]$ in the above lemma we conclude that $Pr[\lambda_1(M(G)) > 6\sqrt{n \log n}] < n^{-n}$.

Now we have at hand all necessary ingredients to formulate our coloring algorithm and analyze its performance.

Step 1. Run the greedy algorithm on G. Let C be the resulting coloring. If C uses more than $2n/\log_2 n$ colors, go to *Step 3*;

Step 2. Define $M = M(G)$ according to (4) and compute $\lambda_1(M)$. If $\lambda_1(M) \leq 6\sqrt{n \ln n}$, output C;

Step 3. Find an optimal coloring by the exhaustive search and output it.

Let us verify that the above algorithm approximates $\chi(G)$ within a factor of $O(\sqrt{n/\log n})$. If coloring C is output at Step 2, then $|C| \leq 2n/\log_2 n$ and $\chi(G) \geq n/\alpha(G) \geq n/\lambda_1(G) \geq O(\sqrt{n/\log n})$, implying $|C|/\chi(G) = O(\sqrt{n/\log n})$. Of course, if we ever get to Step 3 of the algorithm, an optimal coloring is output.

To estimate the expected running time, observe that Steps 1 and 2 take obviously a polynomial in n number of steps. The probability of getting to Step 3 is at most $O(n^{-n})$ as follows from the above discussion. As the complexity of Step 3 is $O(n^n poly(n))$ the desired expected running time estimate follows.

A more careful implementation of the same basic idea enables to shave off an extra logarithmic factor from the above described result. We refer the reader to [32] for details.

Research Problem 2. *Find a coloring algorithm with approximation ratio $o(\sqrt{n}/\log n)$ and polynomial expected running time over the probability space $G(n, 0.5)$.*

Research Problem 3. *Find coloring algorithms with good approximation ratios and polynomial expected running time in probability spaces $G(n, n^{-a})$ for $a > 0.5$.*

6 Deciding k-colorability in expected polynomial time

In this section we continue our coverage of coloring algorithms with expected polynomial running time in probability spaces $G(n,p)$. The subject here is algorithms for deciding k-colorability.

As stated in Section 2, deciding k-colorability in NP-complete for every fixed $k \geq 3$. Observe however that for absolute most of the graphs G on n vertices the answer to the question whether G is k-colorable is "No":

Proposition 6.1. *For every fixed positive integer k, the random graph $G(n, 0.5)$ is non-k-colorable with probability $1 - 2^{-\Theta(n^2)}$.*

Proof. If G has n vertices and is k-colorable then it contains an independent set of size at least n/k. The probability of the latter event in $G(n, 0.5)$ is at most:

$$\binom{n}{\frac{n}{k}} 2^{-\binom{\frac{n}{k}}{2}} < 2^n \cdot 2^{-\Theta(n^2)} = 2^{-\Theta(n^2)} .$$

The above proposition indicates that it should be probably easy to decide k-colorability quickly on average – the answer is "No" for vast majority of the graphs, and exceptional instances are very rare. And indeed, Wilf showed in [52] that the backtrack algorithm for deciding k-colorability in graphs on n vertices has a *constant* expected running time over $G(n, 0.5)$ as n approaches infinity. For instance, a backtrack search tree for 3-coloring a graph has an average of about 197 nodes only! The backtrack search tree of a graph G with vertex set $\{1, \ldots, n\}$ is the tree whose nodes are on levels $0, \ldots, n$, and in which there is a node on level i corresponding to every proper k-coloring of the subgraph of G induced by its first i vertices. A node v' at level i is connected by an edge to a node v'' at level $i+1$ if the colors of vertices $1, \ldots, i$ are the same at v' and v''. Level 0 contains a single root node, corresponding to the empty coloring.

In fact, it is quite easy to see why there exists an algorithm for deciding k-colorability in constant expected time. To show this, fix $t = C(k)n$ edge disjoint copies $K_1, \ldots, K_t$ of the complete graph K^{k+1} in the complete graph on n vertices, where $C(k)$ is a large enough constant (this is of course feasible, as in fact $\Theta(n^2)$ such copies can be found). The probability of each copy K_i to appear in the random graph $G(n, 0.5)$ is $2^{-\binom{k+1}{2}}$, which is a constant. The appearance of a copy of K^{k+1} is $G(n, 0.5)$ can serve as a certificate for non-k-colorability. Our algorithm scans chosen copies K_i looking for a clique on $k+1$ vertices. If such clique is found, the algorithm rejects the graph G. If no such copy is found, the algorithm decides k-colorability of G by performing the exhaustive search. The correctness of the above algorithm is immediate. To estimate the expected running time, observe the running time of the first phase is the truncated geometric distribution with parameter $p = 2^{-\binom{k+1}{2}} = \Theta(1)$ and has therefore a constant expectation. The probability of ever getting to the second phase can be made much smaller than k^n by choosing the constant $C(k)$ large enough, and hence the expected number of steps spent at the second phase is $o(1)$.

The problem becomes significantly harder as the edge probability $p(n)$ decreases. Bender and Wilf proved [6] that in this case the backtrack algorithm has expected running time $e^{\Theta(1/p)}$, i.e., becomes exponential in n. Also, one can easily show that for every fixed $k \geq 3$, if the edge probability p satisfies $p(n) = o(n^{-2/k})$, then a.s. every subgraph of $G(n, p)$ with a bounded number of vertices is k-colorable, and thus one cannot hope to find a certificate for non-k-colorability by performing local search only.

Here we present an algorithm from [30] for deciding k-colorability in expected polynomial time in $G(n, p)$ for every fixed $k \geq 3$, as long as $p(n) \geq C/n$, where $C = C(k) > 0$ is a sufficiently large constant. Our algorithm can be immediately extended for larger values of $p(n)$. Note that if C is sufficiently large, the random graph $G(n, p)$ is not k-colorable with probability $1 - e^{-\Theta(n)}$. Therefore the algorithm still rejects most of the graphs from $G(n, p)$. In order to be able to reject an input graph, the algorithm needs some graph parameter whose value can serve as a certificate for non-k-colorability. This parameter should be computable in polynomial time. The parameter we will use in our algorithm is the so called *vector chromatic number* of a graph [24]. Besides being computable in polynomial time, the vector chromatic number turns out to be extremely robust, and the probability that its value is small is exponentially small in n. This will enable us to invest exponential time in "exceptional" graphs, i.e. those with small vector chromatic number.

Let us now provide necessary background on the vector chromatic number. This concept has been introduced by Karger, Motwani and Sudan in [24]. Suppose we are given a graph $G = (V, E)$ with vertex set $V = \{1, \dots, n\}$. A *vector k-coloring* of G is an assignment of unit vectors $v_i \in R^n$ to the vertices of G so that for every edge $(i, j) \in E(G)$ the standard scalar product of the corresponding vectors v_i, v_j satisfies the inequality $(v_i, v_j) \leq -\frac{1}{k-1}$. The graph G is called *vector k-colorable* if such a vector k-coloring exists. Finally, the *vector chromatic number* of G, which we denote by $v\chi(G)$, is the minimal real $k \geq 1$ for which G is vector k-colorable.

Karger, Motwani and Sudan established the connection between the usual chromatic number of a graph, $\chi(G)$, and its vector chromatic number, $v\chi(G)$. Below we repeat some of their arguments and conclusions.

Lemma 6.2. *If $\chi(G) = k$, then G is vector k-colorable. Thus, $v\chi(G) \leq \chi(G)$.*

Proof. The statement will follow easily from the proposition below.

Proposition 6.3. *For every $k \leq n+1$, there exists a family $\{v_1, \dots, v_k\}$ of k unit vectors is R^n satisfying $(v_i, v_j) = -\frac{1}{k-1}$ for every $1 \leq i \neq j \leq k$.*

Proof. The existence of such a family can be proven by induction on n, as in [24]. Here we present an alternative proof.

Clearly it is enough to prove the proposition for the case $k = n + 1$ (if $k < n + 1$, find such a family in R^{k-1} and complete the found vectors by zeroes in the last $n-k+1$ coordinates to get the desired family). Define an n-by-n matrix $A = (a_{ij})$ by setting $a_{ii} = 1$ for $1 \leq i \leq n$, and $a_{ij} = -1/n$ for all $1 \leq i \neq j \leq n$. Then A is a symmetric positive definite matrix (the eigenvalues of A are $\lambda_1 = \ldots = \lambda_{n-1} = 1 + 1/n$, $\lambda_n = 1/n$). Therefore it follows from standard linear algebra results that there exists a family $\{v_1, \dots, v_n\}$ of n vectors in R^n so that $a_{ij} = (v_i, v_j)$ for all $1 \leq i, j \leq n$. In particular, $(v_i, v_i) = a_{ii} = 1$, so all members of this family are unit vectors. Also, $(v_i, v_j) = a_{ij} = -1/n$ for all $1 \leq i \neq j \leq n$. Set now $v_{n+1} = -(v_1 + \ldots + v_n)$. Then $(v_{n+1}, v_{n+1}) = (v_1 + \ldots + v_n, v_1 + \ldots + v_n) = n \cdot 1 + 2\binom{n}{2}(-1/n) = 1$, and v_{n+1} is a unit vector as well. Also, for all $1 \leq i \leq n$, $(v_i, v_{n+1}) = (v_i, -v_1 - \ldots - v_n) = -1 + (n-1)/n = -1/n$. Hence, $\{v_1, \dots, v_n, v_{n+1}\}$ forms the desired family.

Returning to the proof of the lemma, we argue as follows. Let $V = C_1 \cup \ldots \cup C_k$ be a k-coloring of G. Based on the above proposition, we can find a family $\{v_1, \dots, v_k\}$ of unit vectors in R^n so that $(v_i, v_j) = -1/(k-1)$ for all $1 \leq i \neq j \leq k$. Now, for each color class C_i, every vertex from C_i gets the vector v_i assigned to it. The obtained assignment is clearly a vector k-coloring of G.

The most important algorithmic feature of the vector chromatic number, noticed by Karger et al., is that it is polynomially computable. Formally, if a graph G on n vertices is vector k-colorable, then a vector $(k + \epsilon)$-coloring of the graph can be constructed in time polynomial in k, n and $\log \frac{1}{\epsilon}$. This is due to the fact that vector chromatic number can be represented as a solution of a Semidefinite Program and as such is polynomial time computable (see [17]).

Another concept, needed for the analysis of our algorithm and borrowed again from [24], is that of a semi-coloring. Given a graph $G = (V, E)$ on n vertices and an integer $1 \leq t \leq n$, a *semi-coloring* of G in t colors is a family $(C_1, \ldots, C_t)$, where each $C_i \subseteq V(G)$ is an independent set in G, the subsets C_i are pairwise disjoint, and $|\bigcup_{i=1}^t C_i| \geq \frac{n}{2}$.

Lemma 6.4. *[24] For any $k \geq 3$, there exist $c = c(k) > 0, n_0 = n_0(k) > 0$ so that the following holds. For any $n > n_0$ and for any graph G on n vertices and with $m > n$ edges, if $v\chi(G) \leq k$ then there exists a semi-coloring of G in t colors, where*

$$t \leq c\left(\frac{m}{n}\right)^{\frac{k-2}{k}} \ln^{1/2}\left(\frac{m}{n}\right) .$$

Thus the assumption that the vector chromatic number of G is small enables to claim the existence of many pairwise disjoint and large on average independent sets in G.

Now we formulate an algorithm for deciding k-colorability. As the reader will see immediately, the algorithm is extremely simple and in a sense just calculates the vector chromatic number of an input graph.

Input: An integer $k \geq 3$ and a graph $G = (V, E)$ on n vertices.

Step 1. Calculate the vector chromatic number $v\chi(G)$ of the input graph G;

Step 2. If $v\chi(G) > k$, output "G is not k-colorable";

Step 3. Otherwise, check exhaustively all k^n potential k-colorings of G. If a proper k-coloring of G is found, output "G is k-colorable", else output "G is not k-colorable".

The correctness of the algorithm is immediate from Lemma 6.2. Let us see what its expected running time is polynomial in $G(n, C/n)$ for $C = C(k)$ large enough. Steps 1 and 2 of the algorithm take polynomial time. Notice that we get to Step 3 only if $v\chi(G) \leq k$. At Step 3 we check exhaustively all k^n potential k-colorings of G, and checking each potential coloring costs us time polynomial in n. Therefore it takes at most $k^n poly(n)$ time to perform Step 3. Thus it is enough to prove the following lemma:

Lemma 6.5. *If $C = C(k) > 0$ is large enough, and G is distributed according to $G(n,p)$ with $p = C/n$, then*

$$Pr[v\chi(G) \leq k] \leq k^{-n} .$$

Proof. The proof is based on the following technical propositions about the probability space $G(n,p)$.

Proposition 6.6. *If $C > 0$ is large enough then*

$$Pr[|E(G)| \leq 2n^2p] \geq 1 - o(k^{-n}) .$$

Proposition 6.7. *For every fixed $c > 0$, $k \geq 3$, if $C > 0$ is large enough then the following is true in $G(n,p)$ with $p = C/n$. Let $t = c(2C)^{\frac{k-2}{k}} \ln^{1/2}(2C)$. Then*

$$Pr[G \text{ has a semi-coloring in } t \text{ colors}] = o(k^{-n}) \ .$$

Assuming the above two propositions hold, we prove now Lemma 6.5. By Proposition 6.6 we may assume that G has at most $2n^2p = 2Cn$ edges. If the vector chromatic number of such a graph is at most k, then by Proposition 6.4 G has a semi-coloring in $t = c(m/n)^{(k-2)/k)} \ln^{1/2}(m/n) \leq c(2C)^{(k-2)/k} \ln^{1/2}(2C)$ colors. However, by Proposition 6.7 this happens in $G(n,p)$ with probability $o(k^{-n})$.

Both Propositions 6.6 and 6.7 are proven by straightforward calculations, quite standard for the probability space $G(n,p)$. We omit the details here, referring the reader to [30].

Research Problem 4. *Find an algorithm for deciding k-colorability whose expected running time is polynomial over $G(n,p)$ for* any *value of the edge probability $p = p(n)$.*

Note that if $p(n) \leq c/n$ with $c = c(k) > 0$ sufficiently small, then $G(n,p)$ is k-colorable almost surely (see, e.g. [43]), and the algorithm should thus accept most of the input graphs. The problem becomes especially challenging when $p(n)$ is close to the threshold probability for non-k-colorability. This is due to the widespread belief that typical instances at the non-colorability threshold are computationally hard (see, e.g., [10] for a relevant discussion).

Research Problem 5. *Find an algorithm for deciding k-colorability in expected polynomial time in $G(n,p)$ when the parameter k is a growing function of n: $k = k(n)$.*

The apparent difficulty here lies in the fact that the vector chromatic number seems no longer be useful as Lemma 6.4 degenerates to a trivial statement already for $k \gg \log n$.

7 Coloring random k-colorable graphs

The somewhat contradictory title of this section should not confuse the reader – of course, a k-coloring of the input graph is not known to an algorithm, and the task it to recover it or to find some k-coloring.

We should first define models or probability spaces we will be working with. The first one, which we denote by $G(n,p,k)$ is formed as follows. The vertex set is a union of k disjoint subsets of $V_1, \ldots, V_k$ of size n each, and for every pair of vertices $u \in V_i, v \in V_j, i \neq j$, (u,v) is an edge of $G(n,p,k)$ independently and with probability $p = p(n)$. The second model $G_S(n,p,k)$, usually called the *semi-random model* is more complicated – first a random graph G is generated according to the distribution $G(n,p,k)$, and then the adversary can for every non-edge (u,v) of G, where u and v belong to different color classes, add this pair to the set of edges. Adding extra edges to the otherwise random graph $G(n,p,k)$ can spoil its random structure, making the task of recovering its k-coloring significantly more difficult. Of course, the resulting graphs in both models are guaranteed to be

k-colorable with proper coloring $(V_1, \ldots, V_k)$, but this k-coloring is not necessarily unique, and in fact the resulting graph can even have chromatic number smaller than k. It should be clear to the reader that those two models are not the most general ones, and quite a few other models of random k-colorable graphs exist (see, e.g., Section 1 of [49] for a detailed discussion). In this survey we will mostly restrict ourselves with the case of a constant $k \geq 3$, although some of the results we will describe work for growing $k = k(n)$ as well.

Just as in the previous sections, we will consider here two algorithmic tasks. The first task is to provide an algorithm that k-colors in polynomial time almost all graphs in a chosen probability space. The second, more challenging task is to give a k-coloring algorithm with expected polynomial running time.

Let us start with the random model $G(n, p, k)$ and the edge probability $p = 0.5$. This value of the edge probability is the most natural choice as most k-colorable graphs are easily seen to have a quadratic number of edges. Turner [50] proposed the following very simple algorithm for finding a.s. a k-coloring in this case. The algorithm of Turner starts with finding a clique $K = \{v_1, \ldots, v_k\}$ of size k in G and coloring its vertices arbitrary in k distinct colors. Then the algorithm repeatedly searches for an uncolored vertex v that has neighbors in exactly $k-1$ colors, and colors such a vertex in a unique available color. It is rather easy to see that for a constant k such a vertex can almost surely be found at each step. Turner proves in fact that the above algorithm works as long as the number of colors k satisfies $k \leq (1-\epsilon)\log_2 n$. The result of Turner has been strengthened by Dyer and Frieze [11], who proposed an algorithm for finding a k-coloring in $G(n, 0.5, k)$ in $O(n^2)$ expected time. As the number of edges in $G(n, 0.5, k)$ is almost surely quadratic in n, the algorithm of Dyer and Frieze colors this random graph in linear in the number of edges expected time.

An equally simple algorithm has been proposed by Kučera [34] for the case of $k \leq Cn/\log n$. (It could be helpful for the reader to note here that a formal statement of Kučera's result in his paper appears different; this is due to the fact that he considers the probability space of k-colorable graphs on n vertices, so if $k = \Theta(\sqrt{n/\log n})$ in his result, this approximately translates to $\Theta(n/\log n)$ colors in our setting). Observe that if vertices u, v belong to the same color class V_i of $G(n, p, k)$, then the number of their common neighbors is binomially distributed with parameters $(k-1)n, p^2)$, while if u and v come from distinct color classes $u \in V_i, v \in V_j, i \neq j$, the number of their common neighbors is again binomially distributed, but this time with parameters $(k-2), n, p^2)$. Therefore we expect a pair of vertices in the same color class to have more common neighbors than a pair from different color classes. Using standard bounds on the tails of the binomial distribution, one can easily show that for the case $p = 0.5$, $k \leq cn/\log n$, almost surely the number of common neighbors of any two vertices in the same color class is strictly larger than the number of common neighbors in different color classes. We can thus use the number of common neighbors to classify vertices into the same of different color classes. Technical details of the proof can be easily filled or alternatively found in [34].

Research Problem 6. *Find an algorithm that almost surely k-colors a random graph $G(n, 0.5, k)$ for $k \gg n$.*

However, as the edge probability $p = p(n)$ decreases, it becomes harder and harder to find a k-coloring in $G(n, p, k)$ even for fixed k. This should not be surprising – the more random edges we have, the more evident becomes the prefixed coloring scheme. Still, algorithms are known even for very sparse random graphs. The best achievement belongs to Alon and Kahale [1], who gave an algorithm for k-coloring $G(n,p,k)$ for $p \geq C/n$, where $C = c(k)$ is a large enough constant. Notice that if $p = C/n$, then the random graph has typically only a linear in n number of edges, and a linear number of vertices are isolated.

Let us describe briefly the main idea of the algorithm of [1] for the case of $k = 3$ colors. Denote $d = pn$, then d is the expected number of neighbors of every vertex $v \in V_i$ in every other color class. Let us assume for simplicity that every vertex v has indeed exactly d neighbors in every other color class in G. Consider the adjacency matrix $A = A(G)$ of G. Let $\lambda_1 \geq \ldots \geq \lambda_{3n-1} \geq \lambda_{3n}$ be the eigenvalues of A, and $e_1, \ldots, e_{3n-1}, e_{3n}$ be the corresponding orthonormal basis of eigenvectors. The largest eigenvalue of A is then $\lambda_1 = d$, and the spectrum of A is in the interval $[-d, d]$. Let F be the 2-dimensional subspace of all vectors $x = (x_v : v \in V)$ that are constant on every color class, and whose sum is zero: $\sum_{v \in V} x_v = 0$. A simple calculation shows that any non-zero vector from F is an eigenvector of A with eigenvalue $-d$. One can also show that almost surely the multiplicity of the eigenvalue $-d$ is two, and thus F is in fact the eigenspace of $-d$. Therefore any linear combination t of the vectors e_{3n-1} and e_{3n} (both can be efficiently computed) is constant on every color class. Now we find a non-zero linear combination t of e_{3n-1} and e_{3n}, whose median is zero, that is, the numbers of positive and negative components of t both do not exceed $3n/2$. (It is easy to see that such a combination always exists and can be found efficiently.) Normalizing such t to have it with l_2-norm $\sqrt{2n}$, we get a vector t' whose coordinates take values 0,1 or -1 depending on the color class. Defining now

$$\begin{aligned} V_1 &= \{v \in V : t'_v = 0\}; \\ V_2 &= \{v \in V : t'_v = 1\}; \\ V_3 &= \{v \in V : t'_v = -1\}, \end{aligned}$$

we get a proper coloring of G in three colors. The real algorithm of Alon and Kahale is of course much more complicated, it starts from defining an approximate coloring (V_1, V_2, V_3) according to the last two eigenvectors e_{3n-1}, e_{3n} as described above, and then refines it to get a proper coloring.

Very recently, McSherry [40] described a very general spectral algorithm, applicable to several partitioning problems in random graphs. This algorithm differs significantly from the one of Alon and Kahale, and when applied to the k-coloring problem works for the edge probability $p(n)$ down to $p(n) \geq c\log^3(n)/n$.

Research Problem 7. *Find an algorithm that k-colors almost every graph in $G(n, c/n, k)$ for a fixed $k \geq 3$, for all values of the constant $c > 0$.*

If the constant c in the above problem is small enough, the random graph $G(n, c/n, k)$ almost surely does not contain a subgraph with minimal degree k and hence can be easily colored by a greedy-type algorithm. The problem is most challenging for moderate values of c.

The expected time version of the problem has been considered by Subramanian in [48], who proposed a k-coloring algorithm with expected running time polynomial in n as long as $p(n) \geq n^{-a}$ and $a < 3/4$. It would be interesting to obtain coloring algorithms with polynomial expected time for smaller values of the edge probability $p(n)$.

Let us now switch to the semi-random model $G_S(n,p,k)$. This model has been considered by Blum and Spencer [7], who applied the so-called forced coloring approach. To explain this approach, assume that vertices u, v of G are fully connected to a common clique of size $k-1$. Then every proper k-coloring of G should put u, v in the same color class. Blum and Spencer implemented this approach as follows: (a) Given a graph $G = (V, E)$, define a new graph $G' = (V, F)$ where $(u, v) \in F$ if u and v are both adjacent to a common $(k-1)$-clique; (b) Find all connected components of G'. If G' contains exactly k connected components, then they coincide with color classes of the original graph G. Returning to the probability space $G_S(n,p,k)$, observe that adding edges to the random graph $G \sim G(n,p,k)$ can only add edges in the corresponding auxiliary graph G', and hence it is enough to show that already in $G(n,p,k)$ almost surely the graph G' has exactly k connected components. The expected number of copies of $K^{k+1} - e$ becomes linear in n for $p(n) = n^{\frac{2k}{(k-1)(k+2)}}$, and Blum and Spencer were able to show that increasing this probability a bit (say, by factor n^{ϵ} for any $\epsilon > 0$) suffices to get almost surely k connected components in G'. Later, Subramanian, Fürer and Veni Madhavan [49] extended the result of Blum and Spencer by giving a k-coloring algorithm for $G_S(n,p,k)$ for the same range of edge probabilities.

The best result for the semi-random coloring problem has been obtained by Feige and Kilian [13]:

Theorem 7.1. *For every constant k, there is a polynomial time algorithm that k-colors almost graphs in the probability space $G_S(n,p,k)$ for $p(n) \geq (1+\epsilon)k \ln n/n$.*

Feige and Kilian observed also that the above result is close to optimal, as given by the next theorem:

Theorem 7.2. *Let $\epsilon > 0$, $k \geq 3$ be constants and let $p(n) \leq (1-\epsilon)\ln n/n$. The unless $NP \subseteq BPP$, every random polynomial time algorithm will fail almost surely to k-color a semi-random graph from $G_S(n,k,p)$.*

We do not intend to cover a rather complicated algorithm of Feige and Kilian here. Very briefly, it starts by finding a large independent set in $G_S(n,p,k)$, using Semidefinite Programming in the spirit of Lemma 6.4. This independent set I is shown to belong almost entirely to one of the color classes V_i. Then I is purified to get rid of the vertices outside V_i, and then remaining vertices from V_i are found to recover one color class completely; the algorithm then proceeds to recovering the next color class and so on.

Finally we note that a random graph $G(n,p,k)$ can have in fact chromatic number less than k. Subramanian in [47] provides algorithms that color $G(n,p,k)$ and $G_S(n,p,k)$ in minimal possible number of colors in expected polynomial time, as long as $p(n) \geq n^{-\gamma(k)+\epsilon}$ for the random model and $p(n) \geq n^{-\alpha(k)+\epsilon}$ for the semi-random model, where $\gamma(k) = \frac{2k}{k^2-k+2}$ and $\alpha(k) = \frac{2k}{(k-1)(k+2)}$.

References

[1] N. Alon and N. Kahale, *A spectral technique for coloring random 3-colorable graphs*, Proc. of the 26th Annual ACM Symposium on Theory of Computing (STOC'94), 346–355.

[2] N. Alon and M. Krivelevich, *The concentration of the chromatic number of random graphs*, Combinatorica 17 (1997), 303–313.

[3] N. Alon, M. Krivelevich and B. Sudakov, *List coloring of random and pseudo-random graphs*, Combinatorica 19 (1999), 453–472.

[4] N. Alon, M. Krivelevich and V. H. Vu, *On the concentration of eigenvalues of random symmetric matrices*, Israel Journal of Mathematics, in press.

[5] N. Alon and J. Spencer, The probabilistic method, 2nd ed., Wiley, New York, 2000.

[6] E. Bender and H. Wilf, *A theoretical analysis of backtracking in the graph coloring problem*, J. Algorithms 6 (1985), 275–282.

[7] A. Blum and J. Spencer, *Coloring random and semirandom k-colorable graphs*, J. Algorithms 19 (1995), 204–234.

[8] B. Bollobás, *The chromatic number of random graphs*, Combinatorica 8 (1988), 49–55.

[9] B. Bollobás, Random graphs, 2nd ed., Cambridge Univ. Press, Cambridge, 2001.

[10] J. Culberson and I. Gent, *Well out of reach: why hard problems are hard*, Technical report APES-13-1999, APES Research Group, 1999. Available from: http://www.cs.strath.ac.uk/ apes/apesreports.html.

[11] M. Dyer and A. Frieze, *The solution of some random NP-hard problems in polynomial expected time*, J. Algorithms 10 (1989), 451–489.

[12] U. Feige and J. Kilian, *Zero knowledge and the chromatic number*, Proc. 11th IEEE Conf. Comput. Complexity, IEEE (1996), 278–287.

[13] U. Feige and J. Kilian, *Heuristics for semirandom graph problems*, J. Computer and System Sciences, to appear.

[14] Z. Füredi and J. Komlós, *The eigenvalues of random symmetric matrices*, Combinatorica 1 (1981), 233–241.

[15] M. Fürer, C. R. Subramanian and C. E. Veni Madhavan, *Coloring random graphs in polynomial expected time*, Algorithms and Comput. (Hong Kong 1993), Lecture Notes Comp. Sci. 762, Springer, Berlin, 1993, 31–37.

[16] G. Grimmett and C. McDiarmid, *On colouring random graphs*, Math. Proc. Cam. Phil. Soc. 77 (1975), 313–324.

[17] M. Grötschel, L. Lovász and A. Schrijver, Geometric algorithms and combinatorial optimization, Algorithms and Combinatorics 2, Springer Verlag, Berlin, 1993.

[18] V. Guruswami and S. Khanna, *On the hardness of 4-coloring a 3-colorable graph*, Proc. 15th Annual IEEE Conf. on Computational Complexity, IEEE Comput. Soc. Press, Los Alamitos, 2000, 188–197.

[19] M. M. Halldórsson, *A still better performance guarantee for approximate graph coloring*, Inform. Process. Letters 45 (1993), 19–23.

[20] S. Janson, T. Łuczak and A. Ruciński, Random graphs, Wiley, New York, 2000.

[21] T. Jensen and B. Toft, Graph coloring problems, Wiley, New York, 1995.

[22] M. Jerrum, *Large cliques elude the metropolis process*, Random Structures and Algorithms 3 (1992), 347–359.

[23] A. Juels and M. Peinado, Hiding cliques for cryptographic security, *Proc. of the Ninth Annual ACM-SIAM SODA*, ACM Press (1998), 678–684.

[24] D. Karger and R. Motwani and M. Sudan, *Approximate Graph coloring by semidefinite programming*, Journal of the ACM, 45 (1998), 246–265.

[25] R. Karp, Reducibility among combinatorial problems, in: *Complexity of computer computations* (E. Miller and J. W. Thatcher, eds.) Plenum Press, New York, 1972, 85–103.

[26] R. M. Karp, Probabilistic analysis of some combinatorial search problems, In: *Algorithms and Complexity: New Directions and Recent Results*, J. F. Traub, ed., Academic Press, New York, 1976, pp. 1–19.

[27] H. Kierstead, *Coloring graphs on-line*, Online algorithms (Schloss-Dagstuhl 1996), Lecture Notes in Computer Science 1442, Springer, Berlin, 1998, 281–305.

[28] S. Khanna, N. Linial and S. Safra, *On the hardness of approximating the chromatic number*, Combinatorica 20 (2000), 393–415.

[29] M. Krivelevich, *The choice number of dense random graphs*, Combinatorics, Probability and Computing 9 (2000), 19–26.

[30] M. Krivelevich, *Deciding k-colorability in expected polynomial time*, Information Processing Letters 81 (2002), 1–6.

[31] M. Krivelevich and B. Sudakov, *Coloring random graphs*, Inform. Process. Letters 67 (1998), 71–74.

[32] M. Krivelevich and V. H. Vu, *Approximating the independence number and the chromatic number in expected polynomial time*, Proc. 27th Int. Colloq. on Automata, Languages and Programming (ICALP'2000), Lecture Notes in Computer Science 1853, Springer, Berlin, 13–24.

[33] M. Krivelevich and V. H. Vu, *Choosability in random hypergraphs*, Journal of Combinatorial Theory Ser. B 83 (2001), 241–257.

[34] L. Kučera, *Graphs with small chromatic numbers are easy to color*, Inform. Process. Letters 30 (1989), 233–236.

[35] L. Kučera, *The greedy coloring is a bad probabilistic algorithm*, J. Algorithms 12 (1991), 674–684.

[36] L. Lovász, *On the Shannon capacity of a graph*, IEEE Trans. Inform. Theory 25 (1979), 1–7.

[37] T. Łuczak, *The chromatic number of random graphs*, Combinatorica 11 (1991), 45–54.

[38] T. Łuczak, *A note on the sharp concentration of the chromatic number of random graphs*, Combinatorica 11 (1991), 295–297.

[39] C. J. H. McDiarmid, *Colouring random graphs badly*, Graph Theory and Combinatorics (R. J. Wilson, Ed.), Pitman Research Notes in Mathematics, vol. 34, 76–86.

[40] F. McSherry, *Spectral partitioning of random graphs*, Proc. 42nd IEEE Symposium on Found. of Comp. Science (FOCS'01), IEEE Comp. Society, 529–537.

[41] D. W. Matula, *On the complete subgraph of a random graph*, Combinatory mathematics and its applications, Chapel Hill, North Carolina (1970), 356–369.

[42] D. Matula, *Expose-and-merge exploration and the chromatic number of a random graph*, Combinatorica 7 (1987), 275–284.

[43] M. Molloy, *Thresholds for colourability and satisfiability in random graphs and Boolean formulae*, in: Surveys in Combinatorics 2001, London Math. Soc. Lect. Note Ser. 288, Cambridge Univ. Press, Cambridge, 2001.

[44] B. Pittel and R. S. Weishaar, *On-line coloring of sparse random graphs and random trees*, J. Algorithms 23 (1997), 195–205.

[45] E. Shamir and J. Spencer, *Sharp concentration of the chromatic number of random graphs $G_{n,p}$*, Combinatorica 7 (1987), 124–129.

[46] E. Shamir and E. Upfal, *Sequential and distributed graph coloring algorithms with performance analysis in random graph spaces*, J. Algorithms 5 (1984), 488–501.

[47] C. R. Subramanian, *Minimum coloring k-colorable graphs in polynomial average time*, J. Algorithms 33 (1999), 112–123.

[48] C. R. Subramanian, *Algorithms for colouring random k-colourable graphs* Combinatorics, Probability and Computing 9 (2000), 45–77.

[49] C. R. Subramanian, M. Fürer and C. E. Veni Madhavan, *Algorithms for coloring semi-random graphs*, Random Struct. Algorithms 13 (1998), 125–158.

[50] J. S. Turner, *Almost all k-colorable graphs are easy to color*, J. Algorithms 9 (1988), 63–82.

[51] J. H. van Lint and R. M. Wilson, A course in combinatorics, Cambridge Univ. Press, Cambridge, 1992.

[52] H. Wilf, *Backtrack: a $O(1)$ expected time algorithm for the graph coloring problem*, Inform. Proc. Letters 18 (1984), 119–121.

Michael Krivelevich
Department of Mathematics
Faculty of Exact Sciences
Tel Aviv University
Tel Aviv, 69978
Israel.
krivelev@post.tau.ac.il

Trends in Mathematics, © 2002 Birkhäuser Verlag Basel/Switzerland

A Sharp Threshold for a Non-monotone Digraph Property

Jean-Marie Le Bars

ABSTRACT: *We define a non-monotone digraph property TOUR1, a variant of the digraph property KERNEL, which refines the notion of maximal tournament. First we prove that there is a constant $0 < \alpha < 1$ such that TOUR1 is asymptotically almost surely true in random digraphs with constant arc probability $p \leq \alpha$ and asymptotically almost surely false in random digraphs with constant arc probability $p > \alpha$. Then we concentrate our study on random digraphs with arc probability close to α and we obtain a sharp threshold.*

1 Introduction

The study of random digraphs began with the fundamental paper by Erdös and Rényi [6] where they studied the asymptotic probability of graph properties over random graphs with constant edge probability and gave applications of probabilistic method in combinatorics. Several books give a very complete presentation of this area, such that the books *Random graphs* of Bollobás [2], *The Probabilistic Method* of Alon and Spencer [1] and *Random graphs* of Janson, Luczak and Rucinski [9]. Studies may be conducted for digraphs. Let $\mathcal{H}(p(n))$ be the set of digraphs with n vertices for which any pair (a, b) of vertices gives rise to an arc with probability $p(n)$. Note that $p = \frac{1}{2}$ leads to the uniform distribution over digraphs. Let $\mathcal{P}$ be some digraph property. One denote by $\mu_n^p(\mathcal{P})$ the probability that a digraph H from $\mathcal{H}(p(n))$ satisfies $\mathcal{P}$. The limit of $\mu_n^p(\mathcal{P})$ – if it exists – is denoted by $\mu^p(\mathcal{P})$ and is called the asymptotic probability of $\mathcal{P}$ on $\mathcal{H}(p(n))$. We use the notations $\mu_n(\mathcal{P})$ and $\mu(\mathcal{P})$ for the uniform distribution. We will say that a digraph property $\mathcal{P}$ is *asymptotically almost surely true* when $\mu^p(\mathcal{P}) = 1$ and *asymptotically almost surely false* when $\mu^p(\mathcal{P}) = 0$. In many papers on random structures the phrase "almost surely" is used but we prefer the phrase "asymptotically almost surely" (abbreviated *a.a.s.*) to avoid confusions (this notation is choosen by Janson, Luczak and Rucinski in [9]).

The study of 0-1 laws give a well-known interaction between asymptotic probabilities and logic. Let $p = p(n)$ a fixed function. Consider the probability space $\mathcal{H}(p(n))$ and a class of properties $\mathcal{C}$. If each $\mathcal{P}$ in $\mathcal{C}$ is either *a.a.s* true ($\mu^p(\mathcal{P}) = 1$) or *a.a.s* false ($\mu^p(\mathcal{P}) = 0$), one has a 0-1 law for $\mathcal{C}$ in $\mathcal{H}(p(n))$. In many studies, $\mathcal{C}$ is the class of properties expressible in a logic.

In the studies of 0-1 laws, we fix the sequence $p = p(n)$ and we consider the asymptotic probability of a class of properties on random graphs with edge probability p. In most of the problems [2, 1, 9], we give a property $\mathcal{P}$ and we study the asymptotic behavior of the probability that a random graph has $\mathcal{P}$, where $p = p(n)$ varies. This is the case of the studies of threshold functions. The definitions of a threshold, a coarse threshold and a sharp threshold are usually made for monotones properties. Let $\mathcal{P}$ be a monotone property. Assume that $\mathcal{P}$ is an increasing property. A sequence $\hat{p} = \hat{p}(n)$ is called a threshold if for every sequence $p = p(n)$

$$\begin{aligned} \mu^p(\mathcal{P}) &= 0 \text{ if } p = o(\hat{p}), \\ &= 1 \text{ if } \hat{p} = o(p). \end{aligned}$$

Furthermore, a threshold $\hat{p}(n)$ is called a sharp threshold if

$$\begin{aligned} \mu^p(\mathcal{P}) &= 0 \text{ if } p \leq (1-\eta)\hat{p}(n), \\ &= 1 \text{ if } p \geq (1+\eta)\hat{p}(n) \end{aligned}$$

for every $\eta > 0$. For any $i \in [0,1]$, denote by $p_i = p_i(n)$ a sequence such that the $\mu^{p_i}(\mathcal{P}) = i$. Let $0 < \varepsilon < \frac{1}{2}$, we will denote by $\Delta(\varepsilon)$ the term $p_{1-\varepsilon}(n) - p_\varepsilon(n)$. It is already checked that the existence of a sharp threshold is equivalent to $\Delta(\varepsilon) = o(\hat{p}(n))$. Bollobàs and Thomason proved in the existence of threshold functions for all monotone set properties [3]. Friedgut and Kalai [7] gave a method to prove the existence of a sharp threshold for some monotones properties. For non-monotones properties one adopts a local version of the definition of a threshold. On the opposite of monotone properties, a non-monotone have not necessary exactly on threshold, it may have 0 or several thresholds. Besides, the thresholds $\hat{p} = \hat{p}(n)$ of monotones properties satisfies $\hat{p}(n) = o(1)$. Our purpose is to study thresholds $\hat{p}(n)$ of non-monotone properties which satisfy $\hat{p}(n) = \alpha + o(1)$, for a fixed $\alpha \in]0,1[$. The definition above is not interesting since there exists no sequence p such that $\hat{p} = o(p)$. We propose a more restricted definition of a threshold: let $\hat{p}(n) = \alpha + o(1)$ and $\mathcal{P}$ be a non-monotone property. $\hat{p}(n)$ will be called a threshold of $\mathcal{P}$ if it satisfies

$$\begin{aligned} \mu^p(\mathcal{P}) &= i \text{ if } p < \alpha + \varepsilon \\ &= j \text{ if } p > \alpha + \varepsilon, \end{aligned}$$

for every fixed $\varepsilon > 0$, where (i,j) is $(0,1)$ or $(1,0)$. Moreover, $\hat{p}(n)$ will be a sharp threshold if it satisfies $\Delta(\varepsilon) = o(1)$.

We introduce now some properties which are of the utmost interest from the point of view of asymptotic probabilities. First, we consider the digraph property KERNEL, studied in various areas. Given a directed graph $H = (V, A)$, a kernel U of H is a nonempty subset of V that satisfies the following

- for any $(a,b) \in U^2$, $(a,b) \notin A$ (U is a stable set).
- for any $a \notin U, \exists b \in U, (a,b) \in A$ (U is a dominating set).

H has the property KERNEL if it has at least one kernel. This property has been shown NP-complete by V. Chvàtal [4, 5]. Fernandez de la Vega [8] and Tomescu [14] obtained independently the result that the property KERNEL is *a.a.s* true over random digraphs from $\mathcal{H}(p)$, for every fixed $p \in]0,1[$. Before their result, there was no insight that the property was *a.a.s* true. This result comes from a frail balance between ***stability*** and ***dominance***, the two parts of the property KERNEL. Indeed the truth could have been different if the ratio of this two parts have been slightly different. Our idea was to change this ratio in order to break this balance. Kernel properties provide the best counterexamples of 0-1 laws in fragments of monadic existential second-order logic. In [11, 12] we consider such a variant –namely KERNEL$_2$– which has no asymptotic probability. KERNEL$_2$ is expressible in a small fragment of the logic above. It involves several types of arcs that change the ratio between stability and dominance. Recently we improve this result by proving that the 0-1 law fails for frame satisfiability of propositional modal logic, a very small fragment of the logic above [13]. It is established that the property KERNEL and its variants play a central role for the failure of 0-1 law for the fragments of the monadic second-order logic (see the survey of Kolaitis and Vardi [10]).

We define now a variant of the KERNEL. Given a directed graph $H = (V, A)$, a tournament1 U of H is a nonempty subset of V that satisfies the following

- for any $(a, b) \in U^2$, $(a, b) \in U$ and $(b, a) \notin U$ or $(a, b) \notin U$ and $(b, a) \in U$ (U is a tournament).
- for any $a \notin U, \exists b \in U, (a, b) \in A$ and $(b, a) \in A$ (U is neutralized).

H satisfies the property TOUR1 if it has at least one tournament1.

The main result of this paper is the following: there is a constant $0 < \alpha < 1$ such that TOUR1 has a sharp threshold. As we did to get counterexamples of 0-1 laws, we use the well-known first and second moment method to compute the asymptotic probabilities. We present before some results for KERNEL that we need to adapt for the property TOUR1.

2 The digraph property KERNEL

Given a directed graph $H = (V, A)$, recall that a kernel U of H is a nonempty subset of V that satisfies the following

- for any $(a, b) \in U^2$, $(a, b) \notin A$ (U is a stable set),
- for any $a \notin U, \exists b \in U, (a, b) \in A$ (U is a dominating set)

and H has the property KERNEL if it has at least one kernel.

2.1 Random variables

Let $\mathcal{P}$ be one of the properties stable set (stable), dominating set (dom) and kernel ($\mathcal{K}$) applied to subsets of V. Let $r \in \{1, \ldots, n\}$. We define (elementary) random variables $X^{\mathcal{P}}_{r,\mathcal{P}}$, for any subset U of V_n of cardinality r, $X^{\mathcal{P}}_{r,U} = 1$ if the property $\mathcal{P}$ holds for U and 0 otherwise. We denote by $X^{\mathcal{P}}_r$ the random variable equal to the number of subsets U of order r satisfying property $\mathcal{P}$. With the above definition, it is obvious that the random variables $X^{stable}_{r,U}$ and $X^{dom}_{r,U}$ are mutally independent, since they involve two disjoint sets of arcs. Furthermore, the properties in concern are monotone; each subset of a stable set is also a stable set, that is to say

$$\Pr\left(\cup_{r' \geq r} [X^{stable}_{r'} > 0]\right) = \Pr\left(X^{stable}_r > 0\right)$$

and each superset of a dominating set is also a dominating set, that is to say

$$\Pr\left(\cup_{r' \leq r} [X^{dom}_{r'} > 0]\right) = \Pr\left(X^{dom}_r > 0\right)$$

. For the computations, we use the following asymptotic notations: $f(n) = O(g(n))$ iff there are two positive constants c and N such that $|f(n)| \leq c|g(n)|$ for all $n \geq N$, $f(n) = o(g(n))$ iff $lim_{n \to +\infty} |\frac{f(n)}{g(n)}| = 0$ and $f(n) = \Theta(g(n))$ iff there are three positive constants c_1, c_2 and N such that $c_1|g(n)| \leq |f(n)| \leq c_2|g(n)|$, for all $n \geq N$.

2.2 First results

Tomescu and Fernandez de la Vega independently obtained very close results:

Theorem 2.1. *(Tomescu, [14])*
Let $H \in \mathcal{H}(n, \frac{1}{2})$ and $\beta = \log_2 n - \log_2 \log_2 n$. There exist two reals k_1 and k_2, where $k_1 \approx 1.43$ and $k_2 \approx 2.11$, such that for every $\varepsilon > 0$,

1. *there is a.a.s. a kernel of order r in H for every natural number r with $\beta - k_1 + \varepsilon < r < \beta + k_2 - \varepsilon$;*
2. *there is a.a.s. no kernel of order $< \beta - k_1 - \varepsilon$ or $> \beta + k_2 + \varepsilon$.*

Moreover, the number $K(H)$ of kernels of H satisfies

$$n^{0.913+o(1)} < K(H) < n^{1+o(1)} \text{ as } n \to +\infty.$$

Theorem 2.2. *(Fernandez de la Vega, [8])*
For any fixed p, $H \in \mathcal{H}(p(n))$ has a.a.s. a kernel of order $\lceil \beta \rceil$ for $\beta = \log_b n - \log_b \log_b n$, where $b = \frac{1}{1-p}$.

2.3 Tools: first and second moment methods

Let $r = r(n)$ be a sequence of positive integers. Let U be a subset of V_n, we denote by K applied to U the property that U is a kernel.

$$E\left[X_r^K\right] = \sum_{U \subset V_n, |U|=r} E\left[X_{r,U}^K\right] = \binom{n}{r} \Pr(U_r \text{ is a K}) = \binom{n}{r} p^{\frac{r(r-1)}{2}},$$

where $U_r \subset V_n$ denotes the set $\{0, ..., r-1\}$.
On one hand, if the sequence $r(n)$ satisfies $\lim_{n \to +\infty} E[X_r^K] = 0$, by using the first moment method $\Pr\left(X_r^K > 0\right) \leq E\left[X_r^K\right]$, we conclude $\lim_{n \to +\infty} \Pr\left(X_r^K > 0\right) = 0$. On the other hand, if the sequence $r(n)$ satisfies

$$\lim_{n \to +\infty} \frac{E\left[(X_r^K)^2\right] - E\left[X_r^K\right]^2}{E\left[X_r^K\right]^2} = 0,$$

by using the second moment method (Chebychev's inequality),

$$\Pr\left(X_r^K = 0\right) \leq \frac{E\left[(X_r^K)^2\right] - E\left[X_r^K\right]^2}{E\left[X_r^K\right]^2},$$

this implies

$$\lim_{n \to +\infty} \Pr\left(X_r^K > 0\right) = 1.$$

It is possible to extend the above theorems as follows

Theorem 2.3. *Let $p \in]0, 1[$, $H \in \mathcal{H}(p(n))$ and $\beta = \log_b n - \log_b \log_b n$. There exist two sequence of reals $min(n, p)$ and $Max(n, p)$ such that $min(n, p) < Max(n, p)$ and $min(n, p) = \beta - k_1$ and $Max(n, p) = \beta + k_2$, where $k_1 = k_1(p)$ and $k_2 = k_2(p)$ only depend on p so that the interval $I_\beta = [\beta - k_1(p), \beta + k_2(p)]$ satisfies:*

1. *there is a.a.s. a kernel of order r in H for every natural number r with $\beta - k_1 + \varepsilon < r < \beta + k_2 - \varepsilon$;*

2. *there is a.a.s. no kernel of order $< \beta - k_1 - \varepsilon$ or $> \beta + k_2 + \varepsilon$.*

2.4 Steps of the proof of Theorem 2.3

From now on, we will use the symbols n and p to denote respectively a natural number and a real $\in]0,1[$ and b still equals $\frac{1}{1-p}$. Then $\beta(n)$ denotes the real $\log_b n - \log_b \log_b n$.

Definition 2.4. *Let $r = r(n)$ be a sequence, we will denote by $c(n, r(n))$ the unique real c which satisfies*

$$n = e^c\, r(n)\, b^{r(n)}.$$

Definition 2.5. *Denote by f and g $\mathbb{R}$ into $\mathbb{R}$ the functions defined by $f(c) = c + 1 + \ln b$ and $g(c) = -e^c + c + 1 + \ln b$.*

Definition 2.6. *The definition of $min(n,p)$ and $Max(n,p)$ require a study of the function g. Since $g'(c) = 0$ iff $c = 0$ and g is strictly decreasing for $c > 0$ and strictly increasing for $c < 0$, the maximum of the function is $g(0) = \ln b$. Thus there are only two reals $0 < c_{Max} < c_{min}$, such that $g(c_{min}) = g(c_{Max}) = 0$. We denote by $Max(n,p)$ and $min(n,p)$ the sequences of reals defined by $n = e^{c_{min}}\, min\, b^{min}$ and $n = e^{c_{Max}}\, Max\, b^{Max}$.*

Observe that c_{min} and c_{Max} do not depend upon n, Claim 2.5 implies that the interval $[min, Max]$ contains a finite number of natural numbers. Let $m(n)$ defined by $m(n) = \lceil \beta(n) \rceil$ if $g(n, c(\lceil \beta(n) \rceil)) > g(n, c(\lfloor \beta(n) \rfloor))$ and $m(n) = \lfloor \beta(n) \rfloor$, otherwise. In other words, the maximum of the restriction of g on natural numbers is reached by $c(n, m(n))$.

Let $m'(n)$ denote the sequence $\lfloor \frac{3}{4}\, m(n) \rfloor$.

Definition 2.7. *An m-sequence r is a sequence such that $m'(n) \leq r(n) \leq 2\, m(n)$, for any $n \in \omega$.*

From now on, we denote by $r = r(n)$ a m-sequence and by $c(n, r(n))$ the real c which satisfies $n = e^c\, r(n)\, b^{r(n)}$.

The proof consists of several steps which requires the following lemmas:

Lemma 2.8. *There is a.a.s. no kernel of order less than or equal to $m'(n)$.*

Lemma 2.9. *Let $m' = m'(n)$ be the sequence $m'(n) = \lfloor \frac{3}{4}\, m(n) \rfloor$, for any $n \in \omega$. There is a.a.s. no kernels of order r, where $m'(n) < r < min(n,p)$.*

Lemma 2.10. *There is a.a.s. no kernels of order strictly greater than Max.*

Lemma 2.11. *Let r be a natural number in I_β. There is a.a.s. a kernel of order r.*

Let $\mathcal{P}^{stable}(r)$, $\mathcal{P}^{dom}(r,\nu)$ and $\mathcal{P}^{\mathcal{K}}(r))$ be the probabilities that a subset of V_n of order r will be respectively a stable set, a dominating set over a subset of V_n of order ν and a kernel. It is easily seen that we have $\mathcal{P}^{stable}(r) = b^{-r(r-1)}$ and $\mathcal{P}^{dom}(r,\nu) = (1-b^{-r})^{\nu}$. Since the events "$U_r$ is a stable set" and "U_r is a dominating set" are independent, we have $\mathcal{P}^{\mathcal{K}}(r) = b^{-r(r-1)}\,(1-b^{-r})^{n-r}$.
It follows that

$$E[X_r^{stable}] = \binom{n}{r}\, b^{-r(r-1)}$$

and

$$E[X_r^{dom}] = \binom{n}{r}\,(1-b^{-r})^{n-r}.$$

Hence

$$E[X_r^{\mathcal{K}}] = \binom{n}{r}\, b^{-r(r-1)}\,(1-b^{-r})^{n-r}.$$

Lemmas 2.8,2.9,2.10 require the following claims:

Claim 2.1.

$$\binom{n}{r} = \Theta\Big(\frac{n^r\, r^{-r}}{\sqrt{r}}\Big).$$

Claim 2.2. *The expectation of the number of stable sets of order r satisfies* $E\left[X_r^{stable}\right] = \Theta\big(\frac{e^{rf(c)}}{\sqrt{r}}\big)$.

Claim 2.3. *The probability that a subset of V of order r will be a dominating set satisfies* $\mathcal{P}^{dom}(r, n-r) = (1+o(1))\, e^{-r\, e^{c}}$.

Claim 2.4. *The expectation of the number of kernels of order r satisfies*

$$E\left[X_r^{\mathcal{K}}\right] = \Theta\Big(\frac{e^{r\, g(c)}}{\sqrt{r}}\Big).$$

Claim 2.5. *Let $r+1$ denote the sequence $r(n)+1$, then*

$$c(n, r(n)+1) = c(n, r(n)) - \ln b + o(1).$$

2.5 The use of the second moment method

We prove in this part Lemma 2.11. Assume that r is a sequence $r = m + k$, where k is a fixed integer k. We adapt with some modifications Fernandez de la Vega's proof which establishes this result for $r = \lceil \beta \rceil$.
By Claim 2.5,

$$-\frac{\ln b}{2} - k\, \ln b + o(1) \le c(n,m) \le \frac{\ln b}{2} - k \ln b + o(1) \qquad (1)$$

Let U be a subset of V_n and x be a vertex of V_n, U dominates x if there is a vertex y of U such that (x,y) is an arc of the digraph H. We turn to the evaluation of the variance of the random variable $X^{r,U}(H)$. Let U_1 and $U_2 \subset V_n = \{1,\ldots,n\}$ with $|U_1| = |U_2| = r$ and $U_1 \cap U_2 = i$ where $0 \le i \le r-1$. We shall require the probability, denoted by $R(i)$, that both U_1 and U_2 are kernels of H. Hence in particular $R(0) = \Big(\Pr(U_1 \textit{ is } \text{a kernel of } H)\Big)^2$. We have

$$R(i) = \Pr(U_1 \textit{ and } U_2 \textit{ stable}).\Pr(U_1 \textit{ and } U_2 \text{ dominating set } |U_1 \text{ and } U_2 \text{ stable}).$$

The first factor is equal to $q^{2r(r-1)-i(i-1)}$. The second factor is equal to the product $\mu^{n-2r+i}.\nu^{2r-2i}$ where μ denotes the probability that a vertex $x \in V_n \setminus (U_1 \cup U_2)$ is dominating by both U_1 and U_2 and ν denotes the probability that a vertex $y \in U_i \setminus U_j$, where $(i,j) = (1,2)$ or $(2,1)$, is dominated by U_j. Clearly

$$\nu = 1 - q^{r-i} \leq 1 - q^r \text{ and}$$

$$\begin{aligned}\mu &= \Pr(U_1 \cap U_2 \text{ dominates } x) \\ &\quad +(1 - \Pr(U_1 \cap U_2 \text{ dominates } x) \ \Pr(U_1 \setminus U_2 \text{ and } U_2 \setminus U_1 \text{ dominates } x) \\ &= 1 - q^i + q^i(1-q^{r-i})^2 = 1 - 2q^r + q^{2r-i} \\ &\leq (1-q^r)^2(1 + q^{2r-i}(1+3q^r)),\end{aligned}$$

for sufficiently large n, using the fact that $q^r \to 0$. Collecting the obtained upper bounds we get

$$R(i) \leq q^{2r(r-1)-i(i-1)}(1-q^r)^{2n-2r}(1+q^{2r-i}(1+3q^r))^{n-2r+i},$$

which gives

$$\frac{R(i)}{R(0)} \leq \frac{(1+q^{2r-i}(1+3q^r))^n}{q^{i(i-1)}}.$$

Let us write N for $X_r^{\mathcal{K}}$ where r is previously defined for a fixed k.

Lemma 2.12.

$$E\left[N(N-1)\right] = (1 + o(1))\ E\left[N\right]^2.$$

Proof. Recall that r is a sequence such that $r = m + k$, where k is a fixed integer k.

$$E\left[N(N-1)\right] = \binom{n}{r}\left[\binom{n-r}{r}R(0) + \sum_{i=1}^{r-1}\binom{r}{i}\binom{n-r}{r-i}R(i)\right].$$

$$\frac{E\left[N(N-1)\right]}{E\left[N\right]^2} \leq 1 + \sum_{i=1}^{r-1} h(i) \tag{2}$$

where

$$h(i) = \frac{\binom{r}{i}\binom{n-r}{r-i}R(i)}{\binom{n}{r}R(0)} \leq \frac{\binom{r}{i}\binom{n-r}{r-i}}{\binom{n}{r}}\frac{\Big(1+q^{2r-i}(1+3q^r)\Big)^n}{q^{i(i-1)}}. \tag{3}$$

We shall make use of the following bounds for the binomial coefficients:

$\binom{u}{v} \leq \left(\frac{u\,e}{v}\right)^v$ and $\binom{u}{v} \geq \delta v^{-\frac{1}{2}}\left(\frac{u\,e}{v}\right)^v$, for $v \leq u^{\frac{1}{2}}$, where δ is an absolute constant. Since

$$\binom{n}{r} = \frac{\binom{n}{r-i}\binom{n-r+i}{i}}{\binom{r}{i}}$$

we get

$$\frac{\binom{r}{i}\binom{n-r}{r-i}}{\binom{n}{r}} \leq \frac{\binom{r}{i}^2}{\binom{n-r+i}{i}} \leq \delta^{-1}\, i^{\frac{1}{2}} \left(\frac{r^2\, e}{i(n-r+i)}\right)^i \leq \delta'\, i^{\frac{1}{2}} \left(\frac{r^2\, e}{n\, i}\right)^i, \tag{4}$$

with $\delta' = 2\delta^{-1}$. Setting $j = r - i$, we get

$$\frac{\binom{r}{i}\binom{n-r}{r-i}}{\binom{n}{r}} = \frac{\binom{r}{j}\binom{n-r}{j}}{\binom{n}{r}} \leq \delta'\, r^{\frac{1}{2}} \left(\frac{r}{n\, e}\right)^r \left(\frac{r\, n\, e^2}{j^2}\right)^i, \tag{5}$$

Let ε denote an arbitrarily small positive constant.
We suppose firstly $i \leq i_0 = r - \lfloor (1+\varepsilon)\log_b\, \log_b n \rfloor$. Then, using $e^c = n\, r^{-1}\, q^r$, we get

$$q^i \geq (\log_b\, n)^{1+\varepsilon} \geq \frac{e^c r^{2+\varepsilon}(1-o(1))}{n}, \tag{6}$$

and $q^{2r-i} \leq \frac{e^c}{n\, r^{\varepsilon}}$. This implies $\left(1+q^{2r-i}(1+3q^r)\right)^n = 1+o(1)$. This gives, using (3) and (4) and by separating the cases $i \geq r - \frac{1+\varepsilon}{\ln q}\ln r + 1$ and $i < r - \frac{1+\varepsilon}{\ln q}\ln r + 1$,

$$h(i) \leq \delta' i^{\frac{1}{2}} \left(\frac{r^2 e}{n\, i\, q^{i-1}}\right)^i (1+o(1)) \leq \delta' \left(\frac{2e}{\sqrt{i}\, r^{\varepsilon}}\right)^i (1+o(1)),$$

which clearly implies

$$\sum_{i=1}^{i_0} h(i) \leq \frac{4\, e\, \delta'}{r^{\varepsilon}} = o(1). \tag{7}$$

It remains now to consider the case $i > i_0$, *i.e.* $1 \leq j \leq j_0 = \lfloor (1+\varepsilon)\log_b \log_b n \rfloor - 1$. Here we shall use (5) with $q^{i-1} \geq q^r\, e^c = rn^{-1}$ by applying (1), $q^{2r-i} = q^{r+j} = e^c\, rn^{-1}\, q^j$, obtaining

$$\begin{aligned} h(i) &\leq \delta'\, r^{\frac{1}{2}} \left(\frac{r}{n\, e}\right)^r \left(\frac{r\, n\, e^2}{j^2}\right)^j \left(\frac{n}{r}\right)^{r-j} e^{e^c\, r\, q^i(1+3q^r)} \\ &\leq \delta'\, r^{\frac{1}{2}} \left(\frac{r\, e}{j}\right)^{2j} e^{r(e^c\, q^j\, (1+3q^r)-1)} \end{aligned}$$

Fix $j' > 0$ the less natural number which satisfies $e^c\, q^j < q$, for every $j > j'$. Setting $i' = r - j'$. By computing the first bound of $h(i)$, we obtain

$$h(i) = o(1),$$

for every $i' \leq i \leq r-1$. Let $i < i'$, we have $h(i) \leq e^{r(q-1-\beta)}$ for sufficient large n, where β denotes an arbitrarily small positive constant. Combining with the previous result for $i' \leq i \leq r-1$, it implies

$$\sum_{i=i_0}^{r-1} h(i) = o(1). \tag{8}$$

Finally (2), (7) and (8) imply

$$E\left[N(N-1)\right] = (1+o(1))\, E\left[N\right]^2.$$

By using the second moment method (Chebychev's inequality), □

$$\Pr\left(N=0\right) \leq \frac{E\left[N^2\right] - E\left[N\right]^2}{E\left[N\right]^2} = (1+o(1))\frac{1}{E\left[N\right]}.$$

Since by definition of I_β we have $\lim_{n\to+\infty} E\left[N\right] = +\infty$, for any natural number $r \in I_\beta$, we conclude that there is *a.a.s.* a kernel of order r in H.

3 The property TOUR1

Given a directed graph $H = (V, A)$, recall that a tournament1 U of H is a nonempty subset of V that satisfies the following

- for any $(a,b) \in U^2$, $(a,b) \in U$ and $(b,a) \notin U$ or $(a,b) \notin U$ and $(b,a) \in U$ (U is a tournament).

- for any $a \notin U, \exists b \in U, (a,b) \in A$ and $(b,a) \in A$ (U is neutralized). H satisfies the property TOUR1 if it has at least one tournament1.

Let $\alpha = \frac{1}{3}\alpha' - \frac{2}{3}\frac{1}{\alpha'} - \frac{1}{3}$, where $\alpha' = (17+3\sqrt{33})^{\frac{1}{3}}$, $\alpha \approx 0.5437$. It is easy to show that α is the unique $p \in [0,1]$ solution of the equation $p^3 - p^2 - p + 1 = 0$.

Proposition 3.1. *The property TOUR1 has a threshold.*

Proof. It requires the following propositions,

Proposition 3.2. *TOUR1 is a.a.s. true in $\mathcal{H}(\alpha)$.*

Proposition 3.3. *Let a fixed $p \in]\alpha, 1[$. TOUR1 is a.a.s. true in $\mathcal{H}(p(n))$.*

Proposition 3.4. *Let a fixed $p \in]0, \alpha[$. TOUR1 is a.a.s. false in $\mathcal{H}(p(n))$.*

3.1 Study of the first and the second moment

Let $p = p(n)$ a sequence such that $0 < p < 1$. From now on, H will be denote a digraph in $\mathcal{H}(p(n))$. Let $P^T(r)$ the probability that a subset of V_n of cardinality r be a tournament. It is easily seen that

$$P^T(r) = q_T^{r(r-1)},$$

where $q_T = p(1-p)$.

Let U be a subset of V_n of cardinality r. We denote by $P^N(r)$ the probability that U is neutralized. It follows that

$$P^N(r) = (1 - q_N)^{n-r},$$

$q_N = 1 - p^2$.
Let U_1 and $U_2 \subset V_n = \{1, \dots, n\}$ with $|U_1| = |U_2| = r$ and $U_1 \cap U_2 = i$ where $0 \leq i \leq m-1$. Let $S(i)$ denote the probability that both U_1 and U_2 are tournament1's of $H \in \mathcal{H}(n,p)$. It follows $S(0) = \Big(\Pr(U_1 \; is \text{ a tournament of } H)\Big)^2$.
We have $S(i) = \Pr(U_1 \; and \; U_2 \; tournaments)$.
$\Pr(U_1 \; and \; U_2 \text{ neutralized } |U_1 \text{ and } U_2 \text{ tournaments})$.
The first factor is equal to $q_T^{2r(r-1)-i(i-1)}$. The second factor is equal to the product $\mu'^{n-2r+i}.\nu'^{r-i}$ where μ' denotes the probability that a vertex $x \in V_n \setminus (U_1 \cup U_2)$ neutralized both U_1 and U_2 and ν' denotes the probability that a vertex $y \in U_i \backslash U_j$, where $(i,j) = (1,2)$ or $(2,1)$, neutralizes U_j, *i.e.* there exists $b_j \in U_j$ such that (a_i, b_j) and (b_j, a_i) are arcs of H. Since we may have $a_1 = b_1$ and $a_2 = b_2$,

$$\nu' \leq (1 - q_N^{r-i})^2 \leq (1 - q_N^r)^2.$$

Besides,

$$\begin{aligned} \mu' &= \Pr(x \text{ neutralizes } U_1 \cap U_2) \\ &\quad +(1 - \Pr(x \text{ neutralizes } U_1 \cap U_2) \; \Pr(x \text{ neutralizes } U_1 \setminus U_2 \text{ and } U_2 \setminus U_1) \\ &= 1 - 2q_N^r + q_N^{2r-i} \end{aligned}$$

Hence

$$\begin{aligned} S(i) &\leq R(i)(1+\lambda)^{2m(m-1)-i(i-1)}, \\ S(0) &= R(0)(1+\lambda)^{2m(m-1)}, \\ \tfrac{S(i)}{S(0)} &\leq \tfrac{R(i)}{R(0)} \, (1+\lambda)^{-i(i-1)}. \end{aligned}$$

3.2 Proof of Proposition 3.2

We consider the particular case $p = \alpha$. It is easy to check that α is the unique value $p \in [0,1]$ such that $q_T = q_N$. The probability that a specific subset is a tournament in random digraphs with constant arc probability α or a kernel in random digraphs with constant arc probability $1 - \alpha^2$ are the same. Hence the expected number of tournament1's and kernels are the same. So, we get similar results from the first moment method. However, the correlations are slightly different. Let U_1 and $U_2 \subset V_n = \{1, \dots, n\}$ with $|U_1| = |U_2| = r$ and $U_1 \cap U_2 = i$ where $0 \leq i \leq m-1$. Let $S(i)$ (resp. $R(i)$) denote the probability that both U_1 and U_2 are tournament1s (resp. kernels) of $H \in \mathcal{H}(\alpha)$ (resp. $H' \in \mathcal{H}(1-\alpha^2)$). We keep the notations of Section 2.5. It follows

$$\frac{S(i)}{S(0)} \leq \frac{R(i)}{R(0)},$$

then the correlation is less than in H'. Hence, by using the second moment method, we deduce that H satisfies *a.a.s.* TOUR1. Furthermore the size of those tournaments are the same than the size of the kernels in H'.

3.3 Proof of Proposition 3.3

Recall that $m(n)$ denote $m(n) = \lceil\beta(n)\rceil$ if $g(n, c(\lceil\beta(n)\rceil)) > g(n, c(\lfloor\beta(n)\rfloor))$ and $m(n) = \lfloor\beta(n)\rfloor$, otherwise. In other words, the maximum of the restriction of g on natural numbers is reached by $c(n, m(n))$.

Fix $\alpha < p < 1$, it implies $q_T > q_N$. Let λ be the positive constant such that

$$q_T = (1+\lambda)\, q_N.$$

Let $X_r^{\mathcal{K},1-q_N}$ denote the random variable of the number of kernels in $H' = \mathcal{H}(1-q_N)$ and $X_r^{T_1}$ denote the random variable of the number of tournament1's of H.

Let $m = \log_{b_N} n \log_{b_N} \log_{b_N} n$. From now on, we will write N for $X_m^{\mathcal{K},1-q_N}$, M for $X_m^{T_1}$, q for q_N and b for b_N.

$$E[M] = E[N]\,(1+\lambda)^{m(m-1)},$$

Let U_1 and $U_2 \subset V_n = \{1, \dots, n\}$ with $|U_1| = |U_2| = r$ and $U_1 \cap U_2 = i$ where $0 \le i \le m-1$. Let $S(i)$ (resp. $R(i)$) denote the probability that both U_1 and U_2 are tournament1s (resp. kernels) of $H \in \mathcal{H}(p(n))$ (resp. $H' \in \mathcal{H}(1-q_N(n))$). Recall that $q_N = (1-p^2)$. It follows that

$$\frac{S(i)}{S(0)} = \frac{R(i)}{R(0)}\,(1+\lambda)^{m(m-1)} \le \frac{R(i)}{R(0)}.$$

Let $k(i)$ denote $\dfrac{\binom{m}{i}\binom{n-m}{m-i} S(i)}{\binom{n}{m} S(0)}$.

$$\frac{E[M(M-1)]}{E[M]^2} \le 1 + \sum_{i=1}^{m-1} k(i) \le 1 + \sum_{i=1}^{m-1} h(i) \le \frac{E[N(N-1)]}{E[N]^2}.$$

And the last term tends to 0 as $n \to +\infty$. Since $\lim_{n\to+\infty} E[M] = +\infty$, there is *a.a.s.* a tournament1 of order m. □

Proposition 3.4 $p \in]0, \alpha[$ implies $q_T < q$. Let r be a sequence such that $r(n) < n$ and λ the positive constant such that

$$q_T = (1-\lambda)\, q.$$

By Claim 2.2 we have

$$E[X_r^{stable}] = \Theta\left(\frac{e^{r f(c)}}{\sqrt{r}}\right)(1-\lambda)^{r(r-1)},$$

where $f(c) = c + 1 + \ln b$, $b = \frac{1}{q}$ and $n = e^c\, r\, b^r$. It follows that

$$E[M] = E[N]\,(1-\lambda)^{r(r-1)}.$$

Observe that in $\mathcal{H}(p)$ the expectation of TOUR1 is less than the expectation of KERNEL in $\mathcal{H}(1-q)$. We adapt the proof of Theorem 2.3 and it is sufficient to check that there is *a.a.s.* no tournament1 of order close to m.

4 Sharp threshold

Let $q_T = (1-\lambda)\, q$, where $\lambda = \frac{t}{\ln n}$ for an absolute constant t. It is easily seen that $(1-\lambda)^{m(m-1)} = \Theta\Big(e^{-\frac{m\,t}{\ln q}}\Big)$, hence

$$E\,[M] = \Theta\big(\frac{e^{m\,g(c)}}{\sqrt{m}}\; e^{-\frac{m\,t}{\ln q}}\big).$$

By definition of g, $0 < g(c) \leq \ln b$. Consequently, the sequence $\delta = \delta(n)$ defined by

$$g(c) = \delta \ln b$$

satisfies $-1 < \delta < 1$. Moreover, $e^{mg(c)} = b^{m\,\delta}$ and

$$-\ln b < c < \ln b.$$

Hence

$$E\,[M] = \Theta\Big(b^{m(\delta - \frac{t}{(\ln b)^2})}\; m^{-\frac{1}{2}}\Big). \tag{9}$$

Lemma 4.1. *One can divide the study into three cases:*

1. $t = (\ln b)^2 \delta$:
   $E\,[M] = \Theta\Big(m^{-\frac{1}{2}}\Big)$, *then* $\lim_{n\to+\infty} E\,[M] = 0$.

2. $t < (\ln b)^2 \delta$:
   $E\,[M] = \Theta\Big(b^{ml}\Big)$, *for some* $l > 0$, *then* $\lim_{n\to+\infty} E\,[M] = +\infty$.

3. $t > (\ln b)^2 \delta$:
   $E\,[M] = \Theta\Big(b^{-ml}\Big)$, *for some* $l > 0$, *then* $\lim_{n\to+\infty} E\,[M] = 0$.

As we did for KERNEL, we prove that the possible sizes of tournament1's are in a finite interval I which contains m. Furthermore by definition of g and m, the expectation of the number of tournaments1's of order $r \in I$ is less than the expectation of the number of tournaments1's of order m. Hence TOURS1 is *a.a.s.* false for $t \geq \delta$.

Lemma 4.2.

$$E\,[M(M-1)] = (1+o(1))\; E\,[M]^2.$$

Proof. The proof is similar as these of Lemma 2.11.

$$\frac{S(i)}{S(0)} \leq \frac{R(i)}{R(0)}\; (1-\lambda)^{-i(i-1)}.$$

Recall that $k(i) = \dfrac{\binom{m}{i}\binom{n-m}{m-i} S(i)}{\binom{n}{m}\; S(0)}$, hence $k(i) \leq h(i)\; (1-\lambda)^{-i(i-1)}$.

Let $i_0 = m - \lfloor (1+\varepsilon) \log_b \log_b n \rfloor$.
As we did for KERNEL, we separate the cases $i \le i_0$ and $i > i_0$.
$\underline{i \le i_0}$

$$k(i) \le \delta' \left(\frac{2e}{\sqrt{i}\, r^{\varepsilon}} (1-\lambda)^{-i+1} \right)^{i} (1+o(1)),$$

where δ' is an absolute constant.

$$(1-\lambda)^{-i+1} \le (1-\lambda)^{-m} = 0(1),$$

which implies $\sum_{i=1}^{i=i_0} k(i) \le \frac{4\, e\, \delta'}{r^{\varepsilon}} = o(1)$.
$\underline{i > i_0}$ $1 \le j \le j_0 = \lfloor (1+\varepsilon) \log_b \log_b n \rfloor - 1$. We adapt the result in 2.5. Since

$$(1-\lambda)^{-i+1} \le (1-\lambda)^{-m} < e^{-\frac{m\, t}{\ln q}},$$

$$k(i) \le \delta'\, m^{\frac{1}{2}} \left(\frac{m\, e}{j} \right)^{2j} e^{m(e^{c}\, q^{j}\, (1+3q^{m}) - 1 + \frac{t}{\ln q})}.$$

By computing the first bound of $h(i)$ for $i = r-1$, we obtain

$$k(r-1) = o(1).$$

Let $i < r-1$. By definition of t, we have $\frac{t}{\ln q} < -\ln q$. Then we show that

$$e^{c}\, q^{j} - 1 + \frac{t}{\ln q} < 0.$$

Hence

$$h(i) \le e^{m(q-1-\beta)}$$

for sufficient large n, where β denotes an arbitrarily small positive constant. Combining with the previous result for $i' \le i \le r-1$, it implies

$$\sum_{i=i_0}^{m-1} k(i) = o(1). \tag{10}$$

By (9) we deduce the lemma below: □

Lemma 4.3. *$E\,[M] = \theta(1)$ if and only if*

$$t = \delta + o(1).$$

Furthermore, if $E\,[M] = \theta(1)$, m is the unique possible size of a tournament1. It means that

$$\mu(TOUR1) = \Pr(M > 0),$$

provided the limit exists.
Let $0 \le p_i = p_i(n) \ge 1$ be a sequence such that $\mu^{p_i}(T) = i$. Fix $i \in [0,1[$, recall that $p_i = p_i(n)$ denote a sequence such that the $\mu^{p_i}(\mathcal{P}) = i$. We denote by t_i the corresponding value of t.

The first and the second moment method, imply that $E\ [M] = \theta(1)$, for every $i \in [0,1[$. By Lemma 4.3, $t_i = \delta + o(1)$. Besides, by definition of t_i, we have

$$p_i = \alpha - \Theta\Big(\frac{1}{\ln n}\Big), \tag{11}$$

Fix $\varepsilon > 0$. Recall that $\Delta(\varepsilon)$ denote the term $p_{1-\varepsilon}(n) - p_\varepsilon(n)$, (11) implies

$$\Delta(\varepsilon) = \Theta\Big(\frac{1}{\ln n}\Big),$$

which clearly prove the sharp thershold.

Acknowledgements

We thank Wenceslas Fernandez de la Vega for stimulating discussions, Brigitte Vallée for her drafting help and the anonymous referees for their constructive remarks.

References

[1] N. Alon and J. Spencer, The probabilistic method. *John Wiley ans Sons, New York*, 1992, 254 pp.

[2] B. Bollobas, Random Graphs, Second Edition, *Cambridge University Press*, 2001, 498 pp.

[3] B. Bollobás and A. Thomason, Threshold Functions. *Combinatorica*, 7, 1987

[4] V. Chvàtal. On the computational complexity of finding a kernel. Technical report, Report NO CRM-300, univ. de Montréal, 1973. Centre de Recherches Mathématiques.

[5] N. Creignou. *The class of problems that are linearly equivalent to satisfiability or a uniform method for proving NP-completeness.* Theoretical Computer Science 145:111-145, 1995.

[6] P. Erdös and A. Rényi. On the evolution of random graphs. *Pupl. Math. Inst. Hungar. Acad. Sci.*, 7:17–61, 1960.

[7] E. Friedgut and G. Kalai. Every monotone graph property has a sharp threshold. Proc. Amer. Math. Soc., 124:2993–3002, 1996

[8] W. Fernandez De La Vega. Kernel on random graphs. *Discrete Mathematics*, 82:213–217, 1990.

[9] S. Janson, T. Luczak and A. Rucinski, Random Graphs. *John Wiley ans Sons, New York*, 2000, 333 pp.

[10] Ph. G. Kolaitis and M.Y. Vardi. 0-1 laws for Fragments of Existential Second-Order Logic: A Survey. *Mathematical Foundations of Computer Science*, pages 84–98, 2000.

[11] J-M. Le Bars. Fragments of existential second-order logic without 0-1 laws. *In Proceedings of the 13th IEEE Symposium on Logic in Computer Science*, 1998.

[12] J-M. Le Bars. Counterexamples of the 0-1 law for fragments of existential second-order logic: an overview. *Bulletin of Symbolic Logic*, 9:67–82, 2000.

[13] J-M. Le Bars. The 0-1 law fails for frame satisfiability of propositional modal logic. *to appear at Symposium on Logic in Computer Science (LICS'2002), 2002* .

[14] I. Tomescu. Almost all digraphs have a kernel. *Discrete Mathematics*, 84:181-192, 1990.

Jean-Marie Le Bars

Université de Caen, Campus cote de nacre
14032 CAEN, France
lebars@info.unicaen.fr

Trends in Mathematics, © 2002 Birkhäuser Verlag Basel/Switzerland

Approximability of Paths Coloring Problem in Mesh and Torus Networks

Jérôme Palaysi

ABSTRACT: *In optical networks, the use of bandwidth can be optimized by a technique called "Wavelength Division Multiplexing" (WDM). In these networks, the data undergo some optical-electronic conversions which make them slow down. To solve this problem, the path was computed and set up before the data transmission: these networks are refered as all-optical networks. Signals can be transmitted through a same fiber link at the same time only if they have different wavelengths. We deal with particular networks families: meshes and toroidal meshes. Let a set of paths assigned to a set of connection requests. We try to find a feasible assignment of wavelengths (called "colors" in our model) to the paths. The goal is to minimize the number of wavelengths used.*

We show the existence of approximation algorithms for paths computed by a line-column routing, while the problem is shown to be no-APX when paths are computed by a free-routing, a shortest-path routing or a minimal load routing.

1 Introduction

In optical networks, links are optical fibers, each time a message reaches a router, it is converted from optical to electronic state and back again to optical state. These electronic switching are considered as bottlenecks for the network.

Contrary to optical networks which use expensive optoelectronic conversions, all-optical networks allocate to each communication request a physical path into the network, as for usual circuit switching; each router being set up, messages can stay in their optical state from start to end. The all-optical network commutation node that we study are Wavelength Routing Optical Cross-connect (WR-OXC) with Optical Add/Drop Multiplexer (OADM) [2] (cf. figure 1)

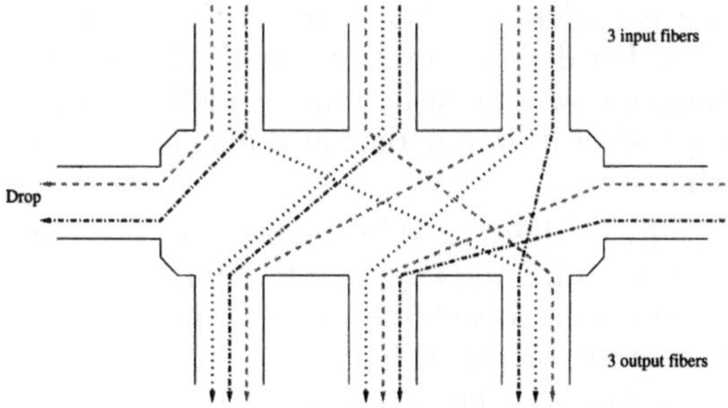

Figure 1: WR-OXC with OADM

Wavelength Division Multiplexing (WDM) is a well known technique [1] that proposes to share the huge bandwich of optical fiber by allocating one frequency to each communication. Several communications can simultaneously use the same fiber as long as their wavelength are different.

In this context, networks can be viewed as graphs, wether directed or not. Up to the concern of this paper, we can restrict ourselves to undirected graphs. Then the ALL-OPTICAL-ROUTING problem is defined as : given a graph and a family of requests (a request is a pair of nodes, some pairs may not be unique in the family), satisfy each request, that is, find a path in the graph linking its two nodes, and allocate one color to each path in such a way that no two paths using a common edge bear the same color.

1.1 Previous results

The last few years, several researchers have dealt with the ALL-OPTICAL-ROUTING problem. It is known to be polynomial in linear networks and trees with bounded degree, but NP-Hard in trees, rings or meshes [6]. This last result can be extended to other families of graphs such as tori and some graphs products with cycles [10]. However, approximation algorithms are known in trees and rings [6].

Some researchers dealt with particular requests collections giving, for instance, a polynomial algorithm for total exchange [2, 11] in the torus and an over one for the multicast [2] in any graph.

We deal with the problem of wavelengths assignment to a given paths collection, which we call the "PATH-COLORING" problem, also known as the "Fixed Path Coloring" [7]. This problem is obviously polynomial in linear networks and bounded degree tree since ALL-OPTICAL-ROUTING is polynomial in these graphs. As a direct consequence of a result of [6], it is NP-Hard but approximable in networks in the "shape" of tree. In the same way, it is NP-Hard and approximable in ring networks and a 2-approximation algorithm exists [12].

1.2 Meshes and tori

One of the mesh and toroidal mesh commun properties is that for a given number of nodes to be linked, the toroidal mesh and the mesh, which are naturally very high secure networks, will use at last only twice edges more than a non-secure tree.

In the all-optical networks case, torus and toroidal mesh have been largely studied [2, 4, 11] and are considered as real competitive solutions among current metropolitan topologies. For deflection routing methods [4], good results corroborate this idea. Furthermore, we can easily imagine deflection routing and routing by path and wavelength allocation together in the same network by dividing the fiber optical bandwith.

The toroidal mesh has already been used in the past, for example to make parallel computers (2D toroidal mesh for the Fujitsu AP 1000, 3D toroidal mesh for T3D and T3E of Cray), and the mesh as well (2D mesh for multiprocessor Intel Paragon and 3D mesh for Wavetracer computer Zephir).

We first define the **cartesian sum** [3] of two graphs[1] G and G' as the graph whose vertices are the ordered pairs (x, x') where x is a vertex of G and x' a vertex of G' and such that there is an edge from (x, x') to (y, y') if and only if $x = y$ and $\{x', y'\}$ is an edge of G', or $x' = y'$ and $\{x, y\}$ is an edge of G.

[1] Our graphs are always simple graphs with out loop

Definition 1.1. *For any integer interval $I = [a, b]$ and $J = [c, d]$, a*

1. **path (or linear network)** *P_I is the graph (I, E) where $E = \{\{i, i+1\} | a \leq i < b\}$;*

2. **mesh** *$M_{I \times J}$ is the cartesian sum of the two paths P_I and P_J;*

3. **ring** *R_I is the graph (I, E) where $E = \{\{i, i+1\} | a \leq i < b\} \cup \{a, b\}$;*

4. **torus (or toroidal mesh)** *$T_{I \times J}$ is the cartesian sum of the two rings R_I and R_J.*

As usual, we extend these definitions up to isomorphism. The reader is referred to [5] for more definitions on graphs. We take advantage of the representation of meshes (and tori) as grids (or extended grids) to use words such as "line" or "column" and expressions like "follow a line (or a column)" that have a simple and intuitive meaning on the figures.

1.3 Problems and contents of the paper

We deal mainly with the PATH-COLORING problem. In this problem, the paths were calculated and the algorithm must assign a color for each path. Paths being given, we often call "routing" the paths family himself. Sometimes, we will also use the term "routing" to designate the process that calculates a path for each connection request.

Definition 1.2 (PATH-COLORING). *is defined by:*

- **input** *: a graph G and a paths family R*

- **output** *: an assignment of color for any R path such that paths using the same edge are assigned different colors.*

- **goal** *: minimize the number of colors used, denoted $w(R)$*

This problem is equivalent to the vertices coloring problem of the conflicts graph:

Definition 1.3 (conflicts graph). *The conflicts graph associated to a paths family R is the undirected graph $G_R = (R, E)$ such that two paths of E are adjacent in G_R if and only if they share at least one edge.*

Definition 1.4 (load). *The load of an edge a, associated to a routing R, is the number of paths which contain this edge. We denote this by $\pi(a, R)$ or $\pi(a)$ if there is no ambiguity. The load of a routing R is the maximum load among all the edges of G and is referred to as $\pi(R)$.*

Fact : clearly, for any paths family R, $w(R) \geq \pi(R)$.

Here, problems are minimization problems for which every solution cost is positive. In these conditions:

Definition 1.5 (d-approximation, APX, No-APX). *An algorithm A is said to be a $d-$ approximation if A is polynomial and if for any instance x, we have $\frac{A(x)}{Opt(x)} \leq d$ where $A(x)$ is the cost obtained by the algorithm A on the instance x and $Opt(x)$ is the cost obtained by an optimal algorithm on the instance x. An APX problem is a problem for which a d-approximation exists for some real number d. Otherwise, this problem is said to be No-APX.*

For more details about the approximation theory, the reader is referred to [9].

In the next section, we show that the PATH-COLORING problem is NP-Hard and No-APX in the mesh and in the torus when the paths family can be any paths family (so called a free routing) and that it remains NP-Hard and No-APX when paths families are produced by a shortest paths routing or a routing that minimizes the load.

In section 3, we deal with line-column paths routing (also called line-column routing or simply lc-routing). A path is called a **line-column path** (or lc-path) when its edges belong to one line and one column of the mesh or the torus. We show approximation algorithms for coloring line-column paths routing in meshes or in tori.

2 No-approximable paths coloring problems

Theorem 2.1. *The paths coloring problem in a mesh or a torus is NP-Hard and NoAPX when the paths instance is a solution to*

1. *a free routing problem*
2. *a shortest paths routing problem*
3. *an optimal load routing problem*

Proof

1. This result comes from the graph vertices coloring problem complexity: NP-Hard and No-APX [9]. The reduction used polynomialy builds paths in a mesh (resp. torus) in order that the conflicts graph is isomorphic to the graph to be colored. This way, it becomes obvious that any paths coloring approximation algorithm in a mesh (resp. a torus) would also be a graph vertices coloring approximation algorithm in the general case.

   Figure 2 shows a graph, on the left hand side, and the paths built for the reduction, on the right hand side.

   Given a graph G with m edges and n vertices, we construct the mesh $M_{[0,n+m]\times[-n,n]}$. The vertices of G are treated in the order $x_0, x_1, x_2, ..., x_{n-1}$. We call $\delta_i = |\{x_j | \{x_i, x_j\} \in E \text{ and } j < i\}|$. The first vertex x_0 leads to a path beginning from the vertex $(0,0)$ to the vertex $(n+m, 0)$ through vertices $(j,0)$ with $1 \leq j < n+m$. This is the path P_0 on figure 2.

   Vertex x_i gives rise to a path beginning from vertex $(0,-i)$, going vertically to vertex $(\sum_{j=0}^{i-1} \delta_j + i, -i)$, next, horizontally to column i, crossing every path P_j for $0 \leq j < i$, dropping down to the next line along P_j whenever x_j

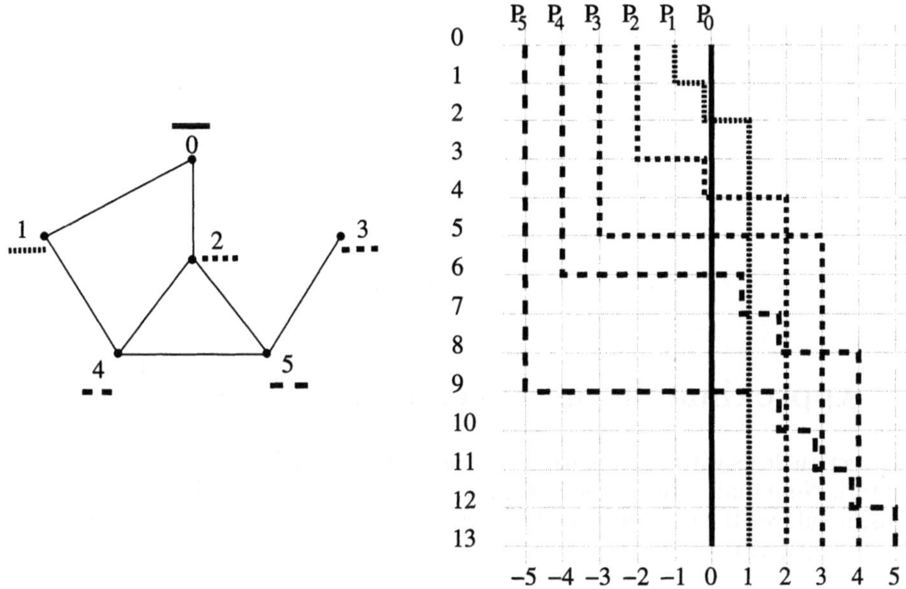

Figure 2: reduction from COLORING to PATH-COLORING in the torus and in the mesh

is adjacent to x_i. Once column i is reached, the path goes down to vertex $(n+m, i)$.

One can verify that the conflicts graph of this routing is isomorphic to the graph G. Furthermore, the size of the requested mesh (resp. torus) is polynomial.

2. For meshes, this result is a direct consequence of the previous reduction. For tori, we only have to perform this reduction on the torus $T_{[0,3(n+m)]\times[-n,3n]}$ (the torus is large enough so that the above construction yields paths which are actually shortest paths). The torus size stays obviously polynomial in the size of the graph to be colored.

3. **Fact** : for any family of paths, if these paths are shortest paths and if the load is the same on all the edges, then this routing is optimal for the load. Now, it is easy to give the scheme of proof for the third part of the theorem. One only has to extend the first reduction adding to the family path:

   - 2 paths of length 1 on each edge (of the mesh or torus) of load 0
   - 1 path of length 1 on each edge (of the mesh or torus) of load 1.

   The routing is an optimal routing for the load (the load is uniformally distributed in the mesh and the chosen paths are the shortest). If an algorithm finds a K-coloration for this routing then the graph G is obviously K-colorable. Besides, if an approximation paths coloring algorithm exists then an approximation graph vertices coloring algorithm exists: which is impossible. □

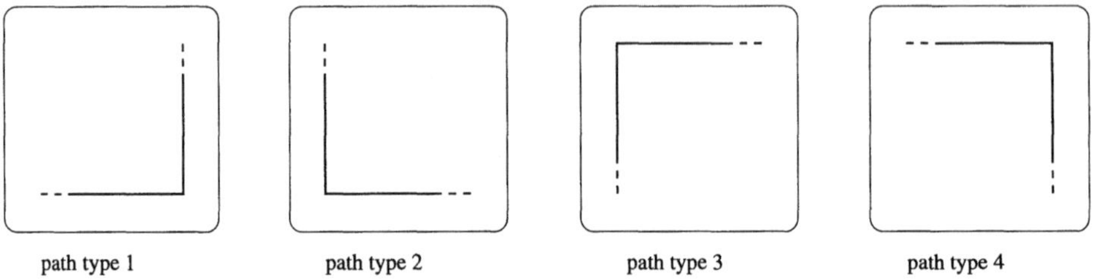

Figure 3: the different types of paths

3 Approximable paths coloring problems

This section deals with the lc-paths or lc-shortest-paths coloring problem in meshes and tori. Note that in the mesh, a lc-path is always a shortest path and that for the torus, it suffices to show that the lc-shortest-paths coloring problem is NP-Hard.

First let us recall :

Theorem 3.1. *[6] The paths coloring problem in the ring of size n, is NP-Hard, even if the length paths are at most $\lfloor \frac{n}{4} \rfloor$.*

Our first result stems from the above theorem.

Theorem 3.2. *The lc-paths coloring problem and the lc-shortest-paths coloring problem are NP-Hard in the mesh and in the torus.*

Proof : We first note that the paths coloring problem in a ring of size $4n$ with paths of length at most $4n$ is NP-Hard (hint : if the routers number on the ring is not a multiple of 4, one can add up to 3 nodes spaced out along the ring in such a way that every path length grows by at most one).

Taking advantage of a reduction used in [6], we map in a straightforward way the $4n$ routers of the ring to the $4n$ outside nodes of the mesh. Obviously, solving the path-coloring problem in the ring is polynomially equivalent to solving the lc-path-coloring problem on the mesh. A similar argument still holds for the torus, mapping the previous mesh onto a torus of size $(2n+1) \times (2n+1)$ to ensure that lc-paths are lc-shortest-paths indeed. □

Next results need preliminary definitions. In a mesh $M_{I \times J}$ or in a torus $T_{I \times J}$ of n lines and n columns ($I = [0, n-1]$ and $J = [0, n-1]$), we characterize a lc-path by a triplet (s, c, d) where s and d are the ends of the path and c the "corner" of the path[2]. We distinguish 4 types of paths (cf. figure 3). For instance, for a path of type 1, we go from s to c from left to right and from c to d from down to up.

In the torus, among paths of type 1 we define four sub-types of line-column paths (cf. figure 4):

- paths of type 1.1 are paths $((i,y),(i,j),(x,j))$ with $y \in [0, j[$ and $x \in [0, i[$
- paths of type 1.2 are paths $((i,y),(i,j),(x,j))$ with $y \in [0, j[$ and $x \in]i, n-1]$ (these paths cross the line 0)

[2] $c = (i,j)$,$s = (i,y)$,$d = (x,j)$, $y \neq j$ and $x \neq i$, where $i, j, x, y \in [0, n-1]$.

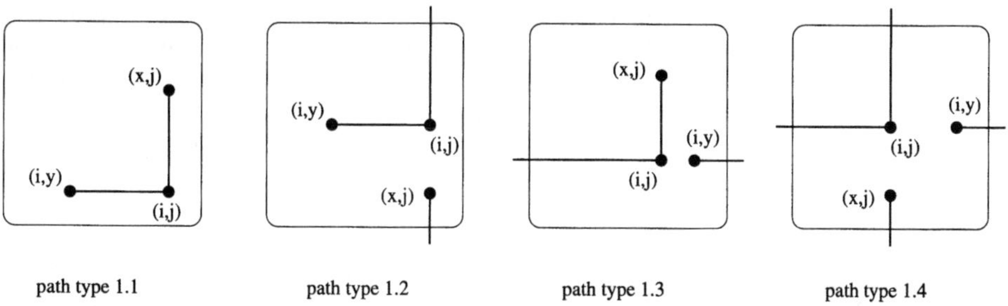

Figure 4: the different types of paths

- paths of type 1.3 are paths $((i,y),(i,j),(x,j))$ with $y \in]j,n-1]$ and $x \in [0,i[$ (these paths cross the column 0)
- paths of type 1.4 are paths $((i,y),(i,j),(x,j))$ with $y \in]j,n-1]$ and $x \in]i,n-1]$ (these paths cross the line 0 and the column 0)

Paths of type $k.1$, $k.2$, $k.3$ and $k.4$ are defined similary from paths of type k with $k \in \{2,3,4\}$.

Lemma 3.3. *Given any instance R of the lc-paths coloring problem and for any $k \in \{1,2,3,4\}$:*

1. *a paths family restricted to paths of type $k.1$, $k.2$ or $k.3$ is polynomialy colorable with 2π colors;*
2. *a paths family restricted to paths of type $k.4$ is polynomialy colorable with π colors.*

Proof :

We prove the lemma for $k = 1$ (other cases are similar). We partition the paths family of type 1 in the instance R into three classes :

- $V_{i,j}$ is the sub-family of paths (s,c,d) for which $c = (i,j)$
- $L_{i,j} = \cup_{k \in]j,n-1]} V_{i,k}$
- $C_{i,j} = \cup_{k \in]i,n-1]} V_{k,j}$

for all $i,j \in [0,n-1]$.

Our algorithm is a greedy algorithm. In all cases (1.1, 1.2, 1.3 and 1.4) the algorithm starts at coloring the paths in $V_{n-1,n-1}$. When the paths family $V_{i,j}$ is colored, it deals with the family $V_{i,j-1}$ or the family $V_{i-1,n-1}$ if $j-1<0$. It terminates when the last family: $V_{0,0}$ is colored.

1. Let us consider the sub-family 1.1. We prove by induction on the algorithm steps that 2π colors are enough. First, we note that 2π colors are enough for coloring $V_{n-1,n-1}$. Assume that no more than 2π colors have been used before coloring $V_{i,j}$. Let $r = |V_{i,j}|$. Let p denote the paths number of $L_{i,j}$

in conflict with the paths of $V_{i,j}$. These p paths and paths of $V_{i,j}$ share the edge $\{(i,j),(i,j-1)\}$. Obviously, $p+r \leq \pi$. Let q denote the paths number of $C_{i,j}$ in conflict with the paths of $V_{i,j}$. These q paths and paths of $V_{i,j}$ share the edge $\{(i,j),(i-1,j)\}$. Obviously, $q+r \leq \pi$. So, $p+q+2r \leq 2\pi$ and $r \leq 2r \leq 2\pi - p - q$, that is to say $|V_{i,j}|$ is less or equal to the number of free colors.

2. Proofs for sub-families 1.2 and 1.3 are similar.

3. Let us consider the sub-family 1.4. We build a bipartite[3] graph $B = (L, C, E)$ such that L is the lines set of the torus, C is the columns set of the torus and for any line-column path $p = (s, c, d)$ which $c = (i, j)$, one edge, denoted $e(p)$, exists in the bipartite graph between the line i and the column j. We can check that any coloring of the sub-family 1.4 induces an edge coloring of B (each edge $e(p)$ takes the color of p) such that two incident edges in B have two different colors. Inversely, an edges coloring of B such that two incident edges in B have two different colors induces a coloring of the sub-family 1.4. Indeed, two paths p_i and p_j are in conflict if and only if $e(p_i)$ and $e(p_j)$ are incident in B.

   We note that the edges of B are colorable with $\Delta(B)$ colors (($\Delta(B)$ is the max degree of B) using a polynomial algorithm [3]. At last, we can check that $\Delta(B)$ is equal to the 1.4 sub-family load. □

Now, let us define a l-or-c paths family as a family of paths whose nodes are on one single line (resp. column). Because interval graphs are known to be polynomialy $(\Delta - 1)$-colorable [8], no more than π colors are needed to color any l-or-c paths sub-family on the mesh. Because a 2-approximation exists for the paths coloring problem on a ring (that uses at most 2π colors [12]), the same result stands for any l-or-c paths sub-family coloring problem on the torus.

Theorem 3.4. *lc-paths coloring problem is 9-APX in meshes.*

Proof :

In a mesh, the paths are of type 1.1, 2.1, 3.1 or 4.1 or else of l-or-c type. In the worst case, for each type we use 2π colors expect for the last one which requires only π colors. If w^{algo} denotes the number of colors used by our algorithm:

$$\frac{w^{algo}}{w^*} \leq \frac{9\pi}{\pi} \leq 9$$

□

Theorem 3.5. *lc-paths coloring problem is 30-APX in tori.*

Proof :

In the torus, all types of line-column path exist plus the l-or-c type. Among which, $(4 \times 3) + 1 = 13$ of them need 2π colors at most and 4×1 of them need π colors at most. Therefore : we can color the paths with less than 30π colors:

$$\frac{w^{algo}}{w^*} \leq \frac{30\pi}{\pi} \leq 30$$

□

[3]that is not necessary a simple graph

Theorem 3.6. *shortest lc-path coloring problem is 18-APX in tori.*

Proof:
On one hand, l-or-c paths require 2π colors at most.
On the other hand, if the line-column paths are shortest paths in the torus, 4π colors are enough for each type of paths: 1,2,3 and 4. Indeed, let us see a torus $T_{[0,n-1]\times[0,n-1]}$ with equal and even number of lines and number of columns. We divide each family of type of path in four parts:

1. paths of type 1.a are paths $(s,(i,j),d)$ with $i \in [0, \frac{n}{2}-1]$ and $j \in [0, \frac{n}{2}-1]$
2. paths of type 1.b are paths $(s,(i,j),d)$ with $i \in [\frac{n}{2}, n-1]$ and $j \in [\frac{n}{2}, n-1]$
3. paths of type 1.c are paths $(s,(i,j),d)$ with $i \in [0, \frac{n}{2}-1]$ and $j \in [\frac{n}{2}, n-1]$
4. paths of type 1.d are paths $(s,(i,j),d)$ with $i \in [\frac{n}{2}, n-1]$ and $j \in [0, \frac{n}{2}-1]$

According to the lemma, the line-column paths of type 1.a are colorable polynomialy with 2π colors. Since the paths 1.a and 1.b are not in conflict, we can use the same 2π colors for the paths 1.b. In the same way, 2π colors are enough for coloring 1.c and 1.d. □

4 Conclusion

The paths coloring problem in a graph G is known to be NP-Hard and NoAPX in general.
When G is itself a path, the problem becomes polynomial. It remains NP-Hard when G is a tree or a ring but then becomes approximable.
The topologies of meshes and tori have been investigated in this paper. It turns out that, G being a mesh or a torus, the problem is NP-Hard and No-APX in general. Focussing on specific families of paths of interest, we proved APX or NoAPX results as summarized in the following table.

	mesh	**torus**
free routing, shortest paths routing or optimal load routing	NoAPX	NoAPX
lc-routing	APX (d=9)	APX (d=30 or d=18 if shortest paths)

Table 1: abstract table on the paths coloring problem in meshes and tori

Furthermore, as the load minimizing problem in a mesh and in a torus turns out to be APX when the routing is a lc-routing (work in preparation), this would yield an APX algorithm for the all-optical routing problem in meshes and tori when the routing is a lc-routing.

References

[1] Bruno Beauquier, Jean-Claude Bermond, Luisa Gargano, Pavol Hell, Stéphane Pérennes and Ugo Vaccaro. *Graph problems arising from wavelength-routing in all-optical networks*, research report, I.N.R.I.A., France, May 1997.

[2] Bruno Beauquier, *Communication dans les réseaux optiques par multiplexage en longueur d'onde*, PdD dissertation, University Nice-Sophia Antipolis, France, January 2000.

[3] Claude Berge. *Graphes et Hypergraphes*, theorem 12, page 238. Monographies Universitaires de Mathématiques. Dunod, Paris, France, 1970.

[4] Thierry Chich. *Optimisation du routage par déflexion dans les réseaux de télécommunications métropolitains.* PhD dissertation, École Normale Supérieure de Lyon, France, December 1997.

[5] Reinhard Diestel. *Graph Theory*, Graduate Texts in Mathematics, Springer, 1997.

[6] Thomas Erlebach and Kaus Jansen. *Call-scheduling in trees, rings and meshes.* In IEEE Computer Society Press, editor, Proceedings of the 30th Hawaii International Conference on System Sciences (HICSS-30), number 1, pages 221-222, 1997.

[7] Thomas Erlebach. *Scheduling Connections in Fast Networks.* PhD dissertation, Technical University of Munich, Germany, 1999.

[8] Fănică Gavril. *Algorithms for minimum coloring, maximum clique, minimum covering by cliques, and maximum independent set of a chordal graph.* SIAM Journal on Computing, 1:180-187, 1972.

[9] Dorit S. Hochbaum. *Approximation Algorithms for NP-Hard Problems.* PWS Publishing Company, 1997.

[10] Jérôme Palaysi, *Routage tout-optique et coloration de chaînes dans les réseaux en grilles et les réseaux toriques.* research report 00128, L.I.R.M.M., Montpellier, France, October 2000.

[11] Heiko Schröder, Ondrej Sýkora and Imrich Vrt'o. *Optical all-to-all communication for some product graphs.* In proceeding of the 24th Seminar on Current Trends in Theory and Practice of Informatics (SOF-SEM),1997.

[12] Alan Tucker, *Coloring a family of circular arcs.* SIAM Journal Appl. Math., 29(3):493-502, november 1975.

Jérôme Palaysi

Laboratoire Informatique Robotique Microélectronique Montpellier
161 rue Ada
34 392 Montpellier Cedex 5
France
palaysi@lirmm.fr

Trends in Mathematics, © 2002 Birkhäuser Verlag Basel/Switzerland

Minimal Spanning Trees for Graphs with Random Edge Lengths

J. Michael Steele

ABSTRACT: *The theory of the minimal spanning tree (MST) of a connected graph whose edges are assigned lengths according to independent identically distributed random variables is developed from two directions. First, it is shown how the Tutte polynomial for a connected graph can be used to provide an exact formula for the length of the minimal spanning tree under the model of uniformly distributed edge lengths. Second, it is shown how the theory of local weak convergence provides a systematic approach to the asymptotic theory of the length of the MST and related power sums. Consequences of these investigations include (1) the exact rational determination of the expected length of the MST for the complete graph K_n for $2 \le n \le 9$ and (2) refinements of the results of Penrose (1998) for the MST of the d-cube and results of Beveridge, Frieze, and McDiarmid (1998) and Frieze, Ruzinkó, and Thoma (2000) for graphs with modest expansion properties. In most cases, the results reviewed here have not reached their final form, and they should be viewed as part of work-in-progress.*

1 Introduction and Main Results

Consider a finite, connected, simple graph G with vertex set $v(G)$, and for each element of the edge set $e(G)$ let ξ_e denote a nonnegative random variable that one views as the length of the edge e. The random variables $\{\xi_e : e \in e(G)\}$ are assumed to be independent with a common distribution F, and the quantities that are of central concern here are the total length of the minimal spanning tree (MST) of G,

$$L_{\mathrm{MST}}(G) = \sum_{e \in G} \xi_e \mathbb{I}(\, e \in \mathrm{MST}(G)\,),$$

and the associated sums for power weighted edges

$$L^{\alpha}_{\mathrm{MST}}(G) = \sum_{e \in G} \xi_e^{\alpha} \mathbb{I}(\, e \in \mathrm{MST}(G)\,).$$

The first of these sums has been studied extensively since Frieze (1985) showed that for edge lengths with the uniform distribution on $[0,1]$ that one has

$$\mathbb{E}[L_{\mathrm{MST}}(K_n)] \to \zeta(3) = \sum_{k=1}^{\infty} \frac{1}{k^3} = 1.202\cdots \quad \text{as} \quad n \to \infty \tag{1}$$

where K_n is the complete graph on n vertices.

In particular, this result has now been refined or extended by numerous investigations. There are relaxations of the distributional assumption by Steele (1987), extensions to the bipartite MST expectations $\mathbb{E}[L_{\mathrm{MST}}(K_{n,n})]$ by Frieze and McDiarmid (1989), and even the development of a central limit theorem for $L_{\mathrm{MST}}(K_n)$

by Janson (1995). More recently, the basic limit (1) has been extended to larger classes of graphs, including an extension to the d-cube Q_d by Penrose (1998) and extensions to general classes of "modestly expansive" regular graphs by Beveridge, Frieze, and McDiarmid (1998) and Frieze, Ruzinkó, and Thoma (2000).

The path taken here diverges from this earlier work in several respects, but one key difference comes from the focus on exact calculations, rather than asymptotic relations. Specifically, we provide a formula for $\mathbb{E}[L_{\text{MST}}(G)]$ that permits one to determine the exact rational value of $\mathbb{E}[L_{\text{MST}}(G)]$ for many concrete choices of G.

We also pursue exact calculations for a certain infinite graph $\mathcal{T}$ that is in a sense the universal limit for any sequence of randomly rooted independently weighted finite graphs whose vertex degrees go to infinity. This calculation then permits us to provide a necessary and sufficient conditions for the determination of the asymptotic behavior of $\mathbb{E}[L^{\alpha}_{\text{MST}}(G_n)]$ for a large class of sequences of graphs.

After framing our main results more fully in the next few paragraphs, we turn to the proofs. In particular, Section 2 develops an exact formula for $\mathbb{E}[L_{\text{MST}}(G)]$ finite G, and then in Section 3 we calculate the expected length per vertex of a special subgraph of $\mathcal{T}$ that holds the key to many of the limit theorems for the MST. Section 4 then addresses some foundational results that connect calculations on $\mathcal{T}$ to calculations for sequences of finite randomly rooted graphs, and these results are subsequently applied to complete the proof of the basic limit theorem for $E[L^{\alpha}_{\text{MST}}(G_n)]$. Section 4 also examines a critical example that serves to illustrate the role of uniform integrability in the limit theory of the MST. The final section reviews some open problems and briefly speculates on the possibilities for further development.

A Formula for $\mathbb{E}[L_{\text{MST}}(G)]$

Theorem 1.1. *If G is a finite connected graph and the Tutte polynomial*[1] *of G is $T(G;x,y)$, then for independent edge lengths that are uniformly distributed on $[0,1]$, one has*

$$\mathbb{E}[L_{\text{MST}}(G)] = \int_0^1 \frac{(1-p)}{p} \frac{T_x\big(G;1/p,1/(1-p)\big)}{T\big(G;1/p,1/(1-p)\big)}\,dp, \tag{2}$$

where $T_x(x,y)$ denotes the partial derivative of $T(x,y)$ with respect to x.

We illustrate the efficacy of this formula by providing what we believe to be the first explicit computations for $\mathbb{E}[L_{\text{MST}}(K_n)]$ for finite values of n that go beyond the trivial $n=2$ and the easy $n=3$. Specifically, we use this formula to calculate $\mathbb{E}[L_{\text{MST}}(K_n)]$ for $2 \le n \le 9$, and these calculations lead to several compelling conjectures.

Asymptotic Consequences of an Exact Calculation

If d_n, $n=1,2,...$ is a sequence of integers such that $d_n \to \infty$ as $n \to \infty$, we say that the sequence of graphs G_n, $n=1,2,...$ is *nearly regular* provided that the maximum $\Delta(G_n)$ degree and the minimum degrees $\delta(G_n)$ satisfy the degree conditions

$$\Delta(G_n) \sim d_n \quad \text{and} \quad \delta(G_n) \sim d_n \quad \text{as } n \to \infty. \tag{3}$$

[1] Subsection 2.2 provides a brief but friendly development of the necessary background on the Tutte polynomial — beginning with its definition.

We also relax our assumption on $F(x) = P(\xi_e \le x)$, and instead of requiring the that the ξ_e be uniformly distributed on $[0,1]$ we only require

$$F(0) = 0 \quad \text{and} \quad F(x) = x + o(x) \quad \text{as} \quad x \to 0, \tag{4}$$

a condition that simultaneously covers the uniform distribution on $[0,1]$ and the exponential distribution with mean one — our two leading cases.

Next, we consider the power-weighted analog to the MST,

$$L^{\alpha}_{\text{MST}}(G_n) = \sum_e \xi_e^{\alpha} \mathbb{I}\big(e \in \text{MST}(G_n)\big),$$

and we introduce a new sequence

$$Y_\alpha(G_n) = \sum_e (d_n \xi_e)^{\alpha} \mathbb{I}\big(e \in \text{MST}(G_n) \text{ and } R(G_n) \in e\big), \tag{5}$$

where $R(G)$ denotes an element of the vertex set $v(G)$ that is chosen independently according to the uniform distribution. While $Y_\alpha(G_n)$ may not seem natural at first, we will see shortly that its expectation determines the expectation of L^{α}_{MST}; moreover, there are major technical benefits to working with $Y_\alpha(G_n)$. In particular, $Y_\alpha(G_n)$ satisfies a limit law that requires nothing more of the graph sequence $\{G_n\}$ than those features that one needs for the definition of $Y_n(G_n)$ and the statement of the limit. As an easy consequence of the general theory of local weak convergence and an exact calculation on a special infinite tree, one obtains the asymptotic behavior of $\mathbb{E}[L^{\alpha}_{\text{MST}}(G_n)]$.

Theorem 1.2. *If G_n, $n = 1, 2, \ldots$ is a sequence of connected graphs that are nearly regular in the sense of (3), then for any $0 < \alpha < \infty$, one has*

$$\mathbb{E}[L^{\alpha}_{\text{MST}}(G_n)] \sim \Gamma(1+\alpha)\,\zeta(2+\alpha)\,|v(G_n)|\,d_n^{-\alpha} \quad \textit{as} \quad n \to \infty, \tag{6}$$

if and only if the sequence

$$\{\,Y_\alpha(G_n) : n = 1, 2, \ldots\} \quad \textit{is uniformly integrable.}$$

Thus, one finds that even a crude qualitative measure of the good behavior of the sequence $\{Y_\alpha(G_n) : n = 1, 2, \ldots\}$ is enough to guarantee the regular asymptotic behavior of $\mathbb{E}[L^{\alpha}_{\text{MST}}(G_n)]$. Moreover, the good behavior of $\{Y_\alpha(G_n)\}$ turns out to be necessary, so one finds a strong hint that this sequence may be more fundamental to the theory of the MST than first impressions might suggest.

As a quick illustration of this last result, we should note that if one takes $\alpha = 1$ and takes G_n to be Q_n, the n-cube, then it implies

$$\mathbb{E}[L_{\text{MST}}(Q_n)] \sim \frac{2^n}{n}\zeta(3) \quad \text{as} \quad n \to \infty, \tag{7}$$

a limit which was found by Penrose (1998) by different means. A more novel consequence of the limit (6) comes from taking $\alpha = 1/2$ and $\alpha = 3/2$ to find that the limit (7) is nicely sandwiched between

$$\mathbb{E}\big[L^{1/2}_{\text{MST}}(Q_n)\big] \sim \frac{1}{2}\zeta(5/2)2^n\sqrt{\frac{\pi}{n}} \quad \text{and} \quad \mathbb{E}\big[L^{3/2}_{\text{MST}}(Q_n)\big] \sim \frac{3}{4}\zeta(7/2)\sqrt{\pi}n^{-3/2}2^n.$$

Despite the large swath of ground that Theorem 1.2 covers, one should not lose sight of the fact that it really is a simple corollary of more general result from the theory of local weak convergence that has its roots in Aldous (1992) and Aldous (2001). In particular, local weak convergence to the PWIT is a fundamental part of those papers, and the modest generalization PWIT Limit Theorem developed here in Theorem 4.2 is best viewed as part of a longer term effort to make the techniques introduced in Aldous (1992) and Aldous (2001) more easily accessible and more readily applied.

2 Exact Calculations for Finite Graphs

The program begins with the derivation of an exact formula for the expectation $\mathbb{E}[L_{\mathrm{MST}}(G)]$ under the uniform model for the edge lengths. Here the first step is to derive a relationship between the random variable $L_{\mathrm{MST}}(G)$ and an integral of another random variable that measures the connectedness of G when one just uses edges length not greater than $0 \le p \le 1$. Versions of this relationship go back at least to Avram and Bertsimas (1992), and in some way or another it has had a role in most recent investigations of the MST, including the central limit theorem of Janson (1999) and the general graph MST results of Beveridge, Frieze, and McDiarmid (1998) and Frieze, Ruzinkó, and Thoma (2000).

2.1 Length of the MST as an Integral

For any finite graph G and any subset A of the edge set $e(G)$, we write $k(G, A)$ for the number of connected components of the graph with vertex set $v(G)$ and edge set A. If each edge $e \in G$ is assigned length ξ_e, then we also write

$$e_t(G) = \{e \in e(G) : \xi_e \le t\},$$

and we let

$$N_{\mathrm{MST}}(G,t) = \sum_{e \in \mathrm{MST}(G)} \mathbb{I}(\xi_e \le t),$$

so $N_{\mathrm{MST}}(G,t)$ denotes the number of edges of the MST of G that are elements of $e_t(G)$. Now, if G is a connected graph, then by counting the number of elements of $e_t(G)$ in each connected component of $(G, e_t(G))$ one finds

$$N_{\mathrm{MST}}(G,t) + k(G, e_t(G)) = n,$$

so we can simply compute

$$\begin{aligned} L_{\mathrm{MST}}(G) &= \sum_{e \in G} \xi_e \mathbb{I}(\, e \in \mathrm{MST}(G)\,) = \sum_{e \in G} \int_0^1 \mathbb{I}(\, t < \xi_e,\, e \in \mathrm{MST}(G)\,)\, dt \\ &= \int_0^1 \sum_{e \in G} (\, 1 - \mathbb{I}(\xi_e \le t,\, e \in \mathrm{MST}(G)\,)\,)\, dt \\ &= \int_0^1 (n - 1 - N_{\mathrm{MST}}(G,t))\, dt = \int_0^1 \{k(G, e_t(G)) - 1\}\, dt. \end{aligned}$$

In other words, for any connected graph we have the rather pleasing random variable representation

$$1 + L_{\mathrm{MST}}(G) = \int_0^1 k(G, e_t(G))\, dt. \tag{8}$$

Thus, our main task is to understand the expectation of $k(G, e_t(G))$, and this provides a natural roll for the Tutte polynomials.

2.2 The Tutte Polynomial

To define the Tutte polynomial, one needs to go outside the familiar class of simple graphs and to consider graphs that may have loops or parallel edges. Given such a graph G, the Tutte polynomial $T(G; x, y)$ is then defined by a set of four devilishly simple rules:

1. If G has no edges, then $T(G; x, y) = 1$.
2. If e is an edge of G that is neither a loop nor an isthmus, then

   $$T(G; x, y) = T(G'_e; x, y) + T(G''_e; x, y),$$

   where G'_e is the graph G with the edge e *deleted* and G''_e is the graph G with the edge e *contracted.*
3. If e is an isthmus, then $T(G; x, y) = xT(G'_e; x, y)$.
4. If e is a loop, then $T(G; x, y) = yT(G''_e; x, y)$.

To confirm the understanding of these rules, one might want to check that they imply that the Tutte polynomial of K_2 is just x; indeed, by successive applications of Rule 3 one finds that the Tutte polynomial of any tree with n vertices is just the monomial x^{n-1}.

The rules are more amusing when one needs to use contractions, and here the basic exercise is to show that the Tutte polynomial of K_3 is $x + x^2 + y$. Finally, one might want to check that the Tutte polynomial of a *bow tie* (made by two copies of K_3 joined at a vertex) is just $(x + x^2 + y)^2$. The last exercise naturally suggests a general principle for finding the Tutte polynomial for the graph built by joining two arbitrary graphs at a single vertex; one can then recapture Rule 3 as a special case of the general principle.

Much of the usefulness of the Tutte polynomial comes from its relation to the rank function $r(\cdot)$ that associates to each $A \subset e(G)$ the integer $r(A)$ given by

$$r(A) = |v(G)| - k(G, A),$$

where, as before, $k(G, A)$ is the number of connected components of the graph with vertex set $v(G)$ and edge set A. The rank function provides a measure of the extent to which the graph $(v(G), A)$ is connected, and it permits one to express the Tutte polynomial as a large — but informative — sum:

$$T(G; x, y) = \sum_{A \subset e(G)} (x-1)^{r(G)-r(A)} (y-1)^{|A|-r(A)}, \tag{9}$$

where $r(G)$ is shorthand for the more pedantic $r(e(G))$.

One immediate consequence of this formula is that it shows the Tutte polynomial does not depend on the order in which one deletes the edges of G in the recursive definition of $T(G;x,y)$, a fact that may not seem particularly evident from the rules themselves. To return the favor, the defining rules make it evident the coefficients of $T(G;x,y)$ are nonnegative, while this is not so easily seen from the sum.

One obvious consequence of the sum formula (9) is that

$$T(G;2,2) = 2^m \quad \text{where} \quad m = |e(G)|, \tag{10}$$

and a natural use of this triviality is to provide a quick feasibility check on a candidate Tutte polynomial. In fact, the evaluations of the Tutte polynomial at special choices of x and y provide a rich buffet of combinatorial interpretations (cf. Welsh (1999)), and in principle each such evaluation can be used as a check. In practice, the evaluation (10) is the easiest to use; it catches many blunders and offers many hints.

2.3 Connection to the Probability Model

Any sum over all of the subsets of $e(G)$ can be interpreted as an expectation

$$\sum_{A\subset e(G)} p^{|A|}(1-p)^{m-|A|} f(A) \tag{11}$$

for an appropriate choice of f, and when one recalls that

$$r(A) = |v(G)| - k(G,A) = n - k(G,A) \quad \text{and} \quad r(G) = n-1$$

for a connected graph G, then the sum formula (9) is simply

$$T(G;x,y) = \frac{1}{(x-1)(y-1)^n} \sum_{A\subset e(G)} (y-1)^{|A|}\big((x-1)(y-1)\big)^{k(G,A)},$$

which obviously may be written in expectation form as

$$\frac{y^m}{(x-1)(y-1)^n} \sum_{A\subset e(G)} \left(\frac{y-1}{y}\right)^{|A|} \left(\frac{1}{y}\right)^{m-|A|} \big((x-1)(y-1)\big)^{k(G,A)}, \tag{12}$$

provided that we set $m = |e(G)|$ and make the natural identifications

$$p = \frac{y-1}{y} \quad \text{and} \quad 1-p = \frac{1}{y}. \tag{13}$$

This kind of reinterpretation of the Tutte polynomial is bread-and-butter to the theory of the correlated percolation model (cf. Fortuin and Kasteleyn (1972)), and this specific form of the Tutte polynomial has also been useful in the study of the computational complexity of the Tutte polynomial (cf. Welsh (1999) and especially Lemma 1 of Alon, Frieze, Welsh (1994)).

On the other hand, the application of this formula to the problem of calculating the minimal spanning tree for uniformly distributed edges seems to be novel, though admittedly easy and natural. We first note that the first factors under the sum provide the probability under the uniform model that one has $\xi_e \le p$ for exactly those edges in the set A. If one then takes

$$A = e_p(G) \equiv \{e : e \in e(G),\, \xi_e \le p\}$$

then one can write the moment generating function

$$\varphi(t) \equiv \mathbb{E}\big[\exp\big(tk(G, e_p(G))\big)\big]$$

in terms of $T(G;x,y)$ as

$$\varphi(t) = p^{n-1}(1-p)^{m-n+1}\, e^t\, T\bigg(G; 1 + e^t \frac{1-p}{p}, \frac{1}{1-p}\bigg), \tag{14}$$

and this formula gives us a natural way to calculate the expectation of $k\big(G, e_p(G)\big)$. Specifically, if we retain the abbreviations (13), we have

$$\varphi'(t) = \varphi(t)\bigg\{1 + e^t \frac{1-p}{p} \frac{T_x(G;x,y)}{T(G;x,y)}\bigg\},$$

so, when we let $t = 0$, we find for $x = 1/p$ and $y = 1/(1-p)$ that

$$\mathbb{E}[k(G, e_p(G))] = 1 + \frac{1-p}{p} \frac{T_x(G;x,y)}{T(G;x,y)}. \tag{15}$$

Finally, when we expand the abbreviations for x and y and recall the representation (8) for $L_{\mathrm{MST}}(G)$ in as an integral of $k(G, e_p(G))$, we find

$$\mathbb{E}[L_{\mathrm{MST}}(G)] = \int_0^1 \frac{(1-p)}{p} \frac{T_x\big(G; 1/p, 1/(1-p)\big)}{T\big(G; 1/p, 1/(1-p)\big)}\, dp, \tag{16}$$

just as we needed to complete the proof of Theorem 1.1.

2.4 Illustrations and Applications

There are some natural and easy checks one can make to familiarize the formula (16). If we recall that for $G = K_2$ we have $T(G;x,y) = x$, then the integral (16) easily works out to be $1/2$, just as it should. More generally, if G is a tree with n vertices, then $T(G;x,y) = x^{n-1}$ and the integral work out to be $(n-1)/2$, and again this is obviously the correct value of $\mathbb{E}[L_{\mathrm{MST}}(G)]$.

It is perhaps more informative to note that the form of the integrand as a logarithmic derivative is quite natural. If G and H are two graphs that share a common vertex, then the graph $G \cup H$ has Tutte polynomial $T(G;x,y)T(H;x,y)$ so the formula (16) recaptures the obvious fact that in this case one also has

$$\mathbb{E}[L_{\mathrm{MST}}(G \cup H)] = \mathbb{E}[L_{\mathrm{MST}}(G)] + \mathbb{E}[L_{\mathrm{MST}}(H)].$$

For the complete graph on three vertices we have already seen that one has $T(K_3) = x + x^2 + y$, and for this polynomial the integral (16) yields $3/4$, and

n	$\mathbb{E}[L_{\rm MST}(K_n)]$	Numerical Value	Forward Difference
2	1/2	0.50000	0.250000
3	3/4	0.75000	0.135714
4	31/35	0.88571	0.080735
5	893/924	0.96645	0.051864
6	278/273	1.01832	0.035400
7	30739/29172	1.05372	0.025342
8	199462271/184848378	1.07906	0.018843
9	126510063932/115228853025	1.09790	———

Table 2: The exact expected values of the MST of K_n for $n = 2$ to $n = 9$ under the model of independent $U[0,1]$ edge lengths.

yet again one can check independently that $\mathbb{E}[L_{\rm MST}(K_3)] = 3/4$. Nevertheless, for K_4 the situation is much more interesting. Hand computations become tedious, but they still suffice for one to show

$$T(K_4; x, y) = 2x + 2y + 3x^2 + 3y^2 + 4xy + x^3 + y^3.$$

When this polynomial is used in the integral formula (16), one then finds

$$\mathbb{E}[L_{\rm MST}(K_4)] = \frac{31}{35},$$

and now we are on new ground. This appears to be the first time $\mathbb{E}[L_{\rm MST}(K_4)]$ has been computed, and one may be hard pressed to provide an independent calculation that not pass through some integral like that provided by our basic representation (16).

Naturally one can go further, but beyond $n = 4$ it would be masochistic not to use symbolic calculation to determine the Tutte polynomials and to perform the required integrations. In fact, a table of the Tutte polynomials $T(K_n; x, y)$ for the values $n = 2, 3, ..., 8$ is included in Gessel and Sagan (1996), and with help from Maple this table has been extended by Gessel (personal communication) to include all values up to $n = 15$. For convenience of display, we us just the first nine of these polynomials in the construction of Table 2.

The numerical evaluations in the table and their successive differences suggests two compelling conjectures; it seems inevitable that $\mathbb{E}[L_{\rm MST}(K_n)]$ is monotone increasing and concave. This evidence is new and not fully digested, so it is possible that these conjectures will follow from our basic formula (16) and the known properties of the Tutte polynomial for K_n. On the other hand, if such an approach is not successful, the conjectures may prove to be difficult. After all, the analogous monotonicity conjecture for the assignment problem (cf. Steele (1997), p. 94) has resisted all attempts for more than fifteen years.

A final feature of Table 1 worth noting is that the rate of convergence is perhaps slower than one might guess. By the result of Frieze (1985) mentioned in the introduction, we know that $\mathbb{E}[L_{\rm MST}(K_n)]$ converges to $\zeta(3) = 1.202\cdots$, and one might hope that the behavior of $\mathbb{E}[L_{\rm MST}(K_n)]$ would parallel that of the partial sums of $\zeta(3)$ given by $s_n = 1 + 1/2^3 + \cdots + 1/n^3$. Sadly, s_n reaches 1.19 when $n = 6$ and reaches 1.20 when $n = 16$, while $\mathbb{E}[L_{\rm MST}(K_n)]$ lags far behind. By analogy with the Parisi conjecture for the assignment problem (cf. Parisi (1998) and Aldous and

Steele (2002)), one suspects that under the exponential model the corresponding expected values $\mathbb{E}[L_{\text{MST}}(K_n)]$ will indeed be closer to s_n. Nevertheless, such an exploration will have to wait for another day.

3 An Exact Calculation for an Infinite Tree

We now take up a second exact calculation, but this time it will be for a special infinite graph. To explain why this graph deserves to be singled out requires some background on the theory of the Poisson weighted infinite tree and the attending theory of local weak convergence. This background is developed more fully in Aldous (2001) and Aldous and Steele (2002), so the next two subsections recall just the most essential facts.

3.1 An Infinite Tree of Special Significance

The Poisson weighted infinite tree — or the PWIT — is a simple object. Nevertheless, it provides one with a direct and effective understanding of many of the problems of combinatorial optimization for large graphs with edge lengths that are given by independent random variables.

Formally, a PWIT is a rooted tree that one defines recursively. One starts with a single vertex r called the *root*, and one gives the root a countably infinite number of children. The set of these children is called *generation one*, and the edges from the root to the children are then labeled by the realizations of a Poisson process on $[0,\infty)$ that has constant intensity $\mu > 0$. That is, each edge from the root is assigned a unique element of the set

$$\mathcal{P}(\mu) = \{\xi_k : k = 1, 2, ...\} \quad \text{where} \quad \xi_k = Y_1 + Y_2 + \cdots + Y_k$$

and the random variables $\{Y_j : j = 1, 2, ...\}$ are independent and

$$\mathbb{P}(Y_j > x) = \exp(-\mu x) \quad \text{for all } j = 1, 2, ... \text{ and } x \in [0,\infty).$$

After generation k has been defined, one defines generation $k+1$ by taking each element of generation k and applying the same construction that we applied to the root to get the first generation. At each stage the Poisson process that is used to label the edges is taken to be independent of all of the other Poisson processes that have been introduced. This construction is then continued until there is a well defined generation for each of the natural numbers $k = 1, 2, ...$

A tree $\mathcal{T}$ that is produced by this construction is said to be a PWIT with intensity $\mu > 0$, and, as shorthand, we will write

$$\mathcal{T} \stackrel{d}{=} \text{PWIT}(\mu),$$

whenever $\mathcal{T}$ has the same distribution as the PWIT we have just constructed. Shortly, we will be more precise about the metric space in which one understands this distributional equality to take place.

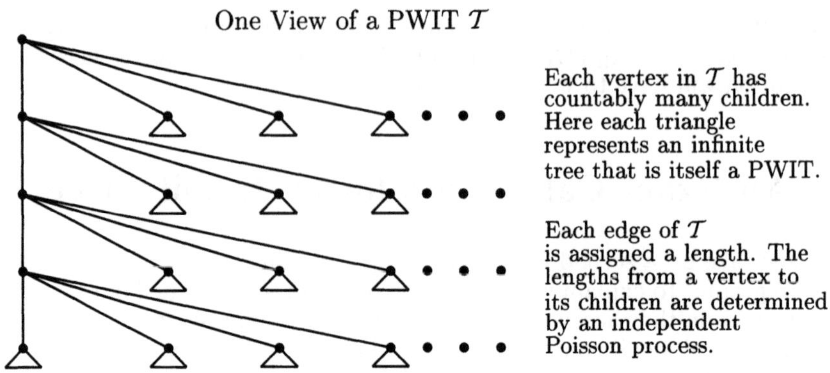

Figure 1: The PWIT is arguably the most fundamental limit object in the theory of randomly weighted graphs. It is the local weak limit of many different sequences, and it offers a unified approach to limit theorems for matching, spanning trees, and many other problems of combinatorial optimization.

3.2 Components of the PWIT

If G is any graph with a real number associated to each edge of G, then G is called a weighted graph, and the numbers on the edges are called the edge lengths. Given such a graph, we let $G(x)$ denote the graph that one obtains when all of the edges of length x or greater are delete, and if G is a rooted graph we also let $G_\star(s)$ denote the component of $G(s)$ that contains the root. If $\mathcal{T}$ is a PWIT with intensity μ and root r, then we may again view $\mathcal{T}_\star(s)$ as a rooted graph with root r, and this graph turns out to be and old friend. It is nothing more than a Poisson Galton-Watson tree.

More precisely, if $PGW(s)$ denotes the distribution of the random tree determined Galton-Watson branching process with a single progenitor and an offspring distribution that is Poisson with mean s, then we have

$$\mathcal{T} \stackrel{d}{=} \text{PWIT}(\mu) \quad \Rightarrow \quad \mathcal{T}_\star(s) \stackrel{d}{=} \text{PGW}(s\mu).$$

Many pleasing computations may be based on this simple observation.

In particular, we will need a qualitative understanding of the size of $\mathcal{T}_\star(s)$ when $\mu = 1$, but everything we need has been known for 120 years or more. Specifically, the probability $p = p(s)$ that a $PGW(s)$ branching process is finite (the so-called *extinction* probability) is one if $0 \leq s \leq 1$ and for $s > 1$ the value of p is given by the unique root in $(0, 1)$ of the equation.

$$p = \exp(-s(1-p)).$$

Lagrange-Bürmann inversion provides an explicit formula

$$p(s) = \frac{1}{s} \sum_{k=1}^{\infty} k^{-k} \frac{(se^s)^k}{k!} \quad \text{for } s > 1,$$

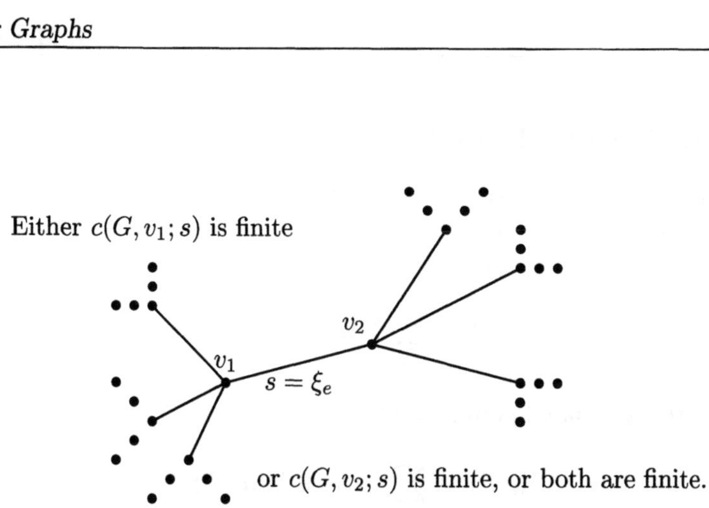

Figure 2: The edge $e = (v_1, v_2)$ of G is in the minimal spanning forest if and only if at least one of the trees $c(G, v_1; s)$ and $c(G, v_2; s)$ is finite when $s = \xi_e$.

but, despite its charm, this sum does not always provide the best way to understand $p(s)$, or the complementary probability $q(s) = 1 - p(s)$. Here we will rely more directly on the fact that $q(s)$ is the unique strictly positive solution of

$$1 - q(s) = \exp(-sq(s)) \quad \text{when } s > 1, \tag{17}$$

and the obvious inverse relationship

$$s(q) = -\frac{\log(1-q)}{q}. \tag{18}$$

that gives us the value of s for which we have we have probability $0 < q < 1$ that the total PGW(s) population is infinite.

3.3 Minimal Spanning Forests

The *minimal spanning forest* of an infinite graph G that has all distinct edge lengths is the subgraph MSF(G) of G with the same vertex set as G and with an edge set that contains each edge $e = (v_1, v_2)$ of G for which

(1) $c(G, v_1; s)$ and $c(G, v_2; s)$ are disjoint, and

(2) $c(G, v_1; s)$ and $c(G, v_2; s)$ are *not both infinite*

when s is taken to be the length of the edge $e = (v_1, v_2) \in G$. An illustration of this definition is given in Figure 2 from Aldous and Steele (2002).

The real utility of this definition can only be brought out by the PWIT Limit Theorem (Theorems 4.2), but a good exercise with the definition is first to show that each component of MSF(G) must be infinite and then to argue that MSF(G) is indeed free of cycles.

3.4 Zeta Meets a PWIT

Let $\mathcal{T}$ be a PWIT with intensity $\mu = 1$, and let r denote its root. If $\mathrm{MSF}(\mathcal{T})$ is the minimal spanning forest of $\mathcal{T}$, then by a natural extension of our earlier notation we denote the sum of the edges incident to the root by

$$Y(\mathcal{T}) = \sum_{e:r\in e} \xi_e \mathbb{I}(e \in \mathrm{MSF}(\mathcal{T})),$$

and denote the associated power sum by

$$Y_\alpha(\mathcal{T}) = \sum_{e:r\in e} \xi_e^\alpha \mathbb{I}(e \in \mathrm{MSF}(\mathcal{T})).$$

The next lemma exploits the method of Lemma 4 of Aldous and Steele (2002) to obtain a slightly more general result. Although the innovation is minor, there do seem to be genuine benefits to having the parameter α at one's disposal. At a minimum, the joint presence of the gamma and zeta functions is amusing.

Lemma 3.1. *The sum of the αth powers of edges of the PWIT that are incident to the root has expectation*

$$\mathbb{E}[Y_\alpha(\mathcal{T})] = 2\Gamma(1+\alpha)\zeta(2+\alpha) \quad \text{for } \alpha \in (-1,\infty), \tag{19}$$

and by analytic continuation the same formula holds for all complex α for which the left-hand side is well defined; consequently, one has the Mellin integral representation

$$\mathbb{E}[Y_\alpha(\mathcal{T})] = \frac{2}{1+\alpha}\int_0^\infty \frac{x^{\alpha+1}}{e^x - 1}\,dx \quad \text{for all} \quad \mathrm{Re}\,\alpha > 0. \tag{20}$$

Proof: If one conditions a Poisson process $\mathcal{P}$ on $[0,\infty)$ to have a point at s, then $\mathcal{P}\setminus\{s\}$ is again a Poisson process, so, if we condition on the event that there is an edge e of length s incident to the root, then the subtrees obtained by cutting that edge are again independent PGW(s) trees.

Now, since the probability that at least one of these is finite is equal to $1 - q^2(s)$, we see that this is also the probability that the edge e is in the minimal spanning forest of T and we have

$$\mathbb{E}[Y_\alpha(\mathcal{T})] = \int_0^\infty s^\alpha(1 - q^2(s))\,ds. \tag{21}$$

To compute the integral, we apply integration-by-parts, the implicit formula for (17) for $q(s)$, and the fact that $q(s)$ vanishes on $[0,1]$ to find

$$\begin{aligned}(1+\alpha)\mathbb{E}[Y_\alpha(\mathcal{T})] &= 2\int_0^\infty s^{1+\alpha} q(s)q'(s)\,ds = 2\int_1^\infty s^{1+\alpha} q(s)q'(s)\,ds \\ &= 2\int_0^1 \frac{\log^2(1-q)}{q^\alpha}\,dq.\end{aligned}$$

A good table would now suffice, but it is as easy to substitute $u = -\log(1-q)$ to find

$$(1+\alpha)\mathbb{E}[Y_\alpha(\mathcal{T})] = 2\int_0^\infty u^{1+\alpha}\frac{e^{-u}}{1-e^{-u}}\,du = 2\int_0^\infty u^{1+\alpha}\sum_{k=1}^\infty e^{-ku}\,du$$
$$= 2\sum_{k=1}^\infty \frac{1}{k^{2+\alpha}}\Gamma(2+\alpha) = 2\Gamma(\alpha+2)\zeta(\alpha+2).$$

Since one has $(1+\alpha)\Gamma(1+\alpha) = \Gamma(2+\alpha)$, the proof of formula (19) is complete. Finally, the analytic continuation of the identity (19) follows from the general principles of function theory, and the validity of the Mellin integral representation (20) is embedded in our calculations. Alternatively, one can note that the representation (20) follows from formula (19) and the well known formula

$$\Gamma(z)\zeta(z) = \int_0^\infty \frac{x^{z-1}}{e^x - 1}\,dx \qquad \mathrm{Re}\, z > 1,$$

which one can prove by expanding $(e^x - 1)^{-1}$ as a geometric series. □

4 Local Weak Convergence Theory

We now need to recall (and to modestly extend) some basic facts from the theory of local weak convergence. The main results in this section are the PWIT Convergence Theorem (Theorem 4.2) and the MST Convergence Theorem (Theorem 4.4). The first of these is implicit in Aldous (1992) and Aldous (2001), and, although Theorem 4.2 is nominally more general than the results that were needed in Aldous (1992) and Aldous (2001), no essentially new ideas are needed. Finally, the MST Convergence Theorem is a direct import from Aldous and Steele (2002).

The real benefit of the present development of the PWIT Limit Theorem is that it is reasonably self-contained. Thus, with very little overhead, one gains direct access to the single most important fact about the PWIT.

4.1 A Poisson Convergence Lemma

We begin with a lemma that is surely part of classic folklore, but the snappy proof via Rényi's characterization of the Poisson process appears to be new. At a minimum, this proof draws the straightest possible line between the hypotheses on F and the required Poisson limit.

Lemma 4.1. *Let F denote a distribution function such that*

$$F(0) = 0 \quad \textit{and} \quad F(x) = \mu x + o(x) \quad \textit{as} \quad x \to 0. \tag{22}$$

If the random variables of the triangular array $\{\, \xi_{i,n} : 1 \le i \le d_n\}$ are independent within each row and if one has

$$P\big(\xi_{i,n} \le x\big) = F(x/d_n) \quad \textit{for all } 1 \le i \le d_n,$$

where $d_n \to \infty$ *as* $n \to \infty$, *then one has the weak convergence*

$$S_n = \{\, \xi_{i,n} : 1 \le i \le d_n \,\} \xrightarrow{d} \mathcal{P}(\mu) \quad \text{as } n \to \infty$$

in the sense of point processes.

Proof: By Rényi's characterization of the Poisson process (cf. Rényi (1967) or Kingman (1993), pp. 34–37), it suffices to show that for each union of disjoint intervals $A = (a_1, b_1] \cup (a_2, b_2] \cup \cdots \cup (a_n, b_n]$ one has

$$P(\,|S_n \cap A| = 0\,) \to e^{-\lambda(A)\mu}, \tag{23}$$

where $\lambda(A)$ denotes the Lebesgue measure of A. By our hypothesis on F and the independence of the $\{\xi_{i,n} : 1 \le i \le d_n\}$, we have

$$\begin{aligned} P(\,|S_n \cap A| = 0\,) &= \Big(1 - \sum_{i=1}^{k} \{F(b_i/d_n) - F(a_i/d_n)\}\Big)^{d_n} \\ &= \big(1 - \mu\lambda(A)/d_n + o(\mu\lambda(A)/d_n)\big)^{d_n}, \end{aligned}$$

so the limit (23) follows instantly. □

There is a sense in which Rényi's criterion is modestly magical; it provides us with independence of a different sort than we assume at the beginning. Also one should note that the only sly aspect of Rényi's Theorem is the requirement that one deal with all A that can be written as finite unions of disjoint intervals; in fact, Moran (1967) shows by example that one cannot get by with less. Finally, there is one small technical point; we have used Rényi's *characterization* of the Poisson process to provide *convergence criterion* for of a sequence of processes. Naturally, one only needs to apply the usual subsequence argument to pass from the characterization to the convergence criterion.

4.2 Local Weak Convergence Defined

We now need to extend the classical notion of weak convergence for point processes to a larger domain that is more directly connected with the convergence of weighted graphs. The treatment given here follows the exposition of Aldous and Steele (2002) which was designed in part to systematize the basic constructions used in Aldous (1992) and Aldous (2001).

To begin, we consider a graph G with a vertex set $v(G)$ that may be finite or countable. We further suppose there is a function ℓ from the edge set $e(G)$ to $(0, \infty]$, and we call $\ell(e)$ the length of the edge e. We then use ℓ to define a metric on $v(G)$ by taking the distance from u to v as the infimum over all paths between u and v of the sum of the lengths of the edges in the path. Naturally, the distance from any vertex v to itself is taken to be zero.

Now, if G is a connected graph with a countable or infinite vertex set and if ℓ is an edge length function that makes G locally finite in the sense that for each vertex v and each real $\rho < \infty$ the number of vertices within distance ρ from v is finite, then we call G a *geometric graph.* Also, when there is a distinguished vertex v, we say that G is a rooted geometric graph with *root* v, and to save space, we denote the set of geometric graphs by $\mathcal{G}$ and the set of rooted geometric graphs by $\mathcal{G}_\star$.

The key issue is to say what one means for a sequence $\{G_n\}$ of elements of $\mathcal{G}_\star$ to converge to a G in $\mathcal{G}_\star$. The driving idea is that for large n, the rooted geometric graph G_n should look very much like G in a neighborhood of the root of G that is as large as we like.

Formally, we take $\rho > 0$ and let $N_\rho(G)$ denote the graph whose vertex set $V_\rho(G)$ is the set of vertices of G that are at a distance of at most ρ from the root of G and whose edge set consists of just those edges of G that have both vertices in $V_\rho(G)$. One again views $N_\rho(G)$ as an element of $\mathcal{G}_\star$ with edge length function and root given by those of G. Also, $\rho > 0$ is called a *continuity point* of G if no vertex of G is exactly at a distance ρ from the root of G.

Now, at last, we say that G_n *converges* to G_∞ in $\mathcal{G}_\star$ provided that for each continuity point ρ of G_∞ there is an $n_0 = n_0(\rho, G_\infty)$ such that for all $n \geq n_0$ there exists a isomorphism[2] $\gamma_{n,\rho}$ from the rooted geometric graph $N_\rho(G_\infty)$ to the rooted geometric graph $N_\rho(G_n)$ such that for each edge e of $N_\rho(G_\infty)$ the length of $\gamma_{n,\rho}(e)$ converges to the length of e as $n \to \infty$.

This definition determines a topology that makes $\mathcal{G}_\star$ into a complete separable metric space. As a consequence, and it gives us access to the usual tools of weak convergence theory. Here, if $\{X_n\}$ is a sequence of $\mathcal{G}_\star$-valued random variables and X is a $\mathcal{G}_\star$-valued random variable we write

$$X_n \xrightarrow{d} X \quad \text{to mean that} \quad \mathbb{E}[f(X_n)] \to \mathbb{E}[f(X)]$$

for each bounded continuous function $f : \mathcal{G}_\star \to \mathbb{R}$. This is just plain vanilla weak convergence $\mathcal{G}_s tar$-valued random variables, but to emphasize the special attention that is paid to the neighborhood of the root we also say that we have the *local weak convergence* of X_n to X.

From examples one finds that local weak convergence is a perfectly natural notion, despite the fact that it takes a while to make precise. In fact, the only subtle feature about local weak convergence is the way in which it force one to focus so myopically on the neighborhoods of the root.

4.3 The PWIT Limit Theorem

We now have the background in place to prove the theorem that makes us interested in PWIT; it shows that the PWIT is arises as the limit of a very natural sequence of geometric graphs. As noted earlier, this particular version of the PWIT limit theorem is intended to make the PWIT limit ideas from Aldous (1992) and Aldous (2001) more explicit, more accessible, and modestly more general

Theorem 4.2 (The PWIT Limit Theorem). *Let G_n, $n = 1, 2, \ldots$, denote a sequence of graphs such that the vertex set $v(G_n)$ has cardinality n for each $n = 1, 2, ..$ and such that the maximum and minimum degrees satisfy the degree conditions*

$$\Delta(G_n) \sim d_n \quad \textit{and} \quad \delta(G_n) \sim d_n \quad \textit{as} \quad d_n \to \infty. \tag{24}$$

[2]Graphs G and G' are isomorphic provided that is a bijection $\phi : v(G) \to v(G')$ such that $(\phi(u), \phi(v)) \in e(G')$ if and only if $(u, v) \in e(G)$.

Also, let F denote a distribution function that satisfies the conditions (22) and associate to each $e \in v(G_n)$ an independent edge length ξ_e with distribution

$$P(\xi_e \leq x) = F(x/d_n) \quad \textit{for all } x \geq 0.$$

Next, independently choose an element of $v(G_n)$ according to the uniform distribution, and let $\mathbf{G}_n$ denote the rooted geometric graph produced by this construction. One then has

$$\mathbf{G}_n \xrightarrow{d} \mathrm{PWIT}(\mu).$$

As one often does in the theory of weak convergence, we prove this limit theorem by passage to an equivalent characterization theorem. Specifically, one first argues (in a step that we leave as an exercise) that the sequence of $\mathcal{G}_\star$-valued random variables $\{\mathbf{G}_n\}$ is tight. Then we consider an arbitrary subsequence, say $\{n_k : k = 1, 2, ...\}$, and we note by tightness that there must exist a further subsequence $\{m_k : k = 1, 2, ...\}$ and a $\mathcal{G}_\star$-valued random variable $\mathbf{G}$ such

$$\mathbf{G}_{m_k} \xrightarrow{d} \mathbf{G} \quad \text{as} \quad n \to \infty.$$

Next, we observe by Skorohod's theorem (cf. Dudley (1989), pp. 325–327) that one can assume without loss of generality that

$$\mathbf{G}_{m_k} \to \mathbf{G} \quad \text{almost surely as} \quad n \to \infty,$$

and now all we have to do is to prove that $\mathbf{G}$ is actually a PWIT.

From the definition of the topology of local weak convergence, we know automatically that $\mathbf{G}$ is connected, so $\mathbf{G}$ will be a tree provided that we show that it has no cycles. This will follow from the next lemma.

For the statement of the lemma, we note that a *path-plus-cycle* is a graph that can be written as a path plus one additional edge that makes a cycle by joining two vertices on the path. Also, to anticipate the application of the lemma, one should recall that the root of $\mathbf{G}_n$ is randomly chosen uniformly from the vertex set $v(G_n)$, so the lemma immediately implies that with probability one the limit graph $\mathbf{G}$ has no cycles in any ρ neighborhood of its root.

Lemma 4.3. *Let $S(n, \rho)$ the set of all vertices $v \in v(\mathbf{G}_n)$ for which there exists a path-plus-cycle $H = H(v) \subset \mathbf{G}_n$ such that*

$$v \in H \quad \textit{and} \quad \sum_{e \in e(H)} \xi_e \leq \rho.$$

One then has

$$|S(n, \rho)|/n \xrightarrow{p} 0 \quad \textit{as} \quad n \to \infty.$$

Proof: We first note that the number of path-plus-cycle subgraphs of $\mathbf{G}_n$ with k vertices cannot be larger than

$$n \cdot \Delta(G_n)^{k-1} \binom{k}{2} < \frac{1}{2} n k^2 \Delta(G_n)^{k-1}.$$

Also, by our hypothesis on F, we know there is an x_0 such that $F(x) \leq 2\mu x$ for all $0 \leq x \leq x_0$, and from this bound, integration by parts, and induction one finds a corresponding bound for the k-fold convolution is given by

$$F^{(k)}(x) \leq (2\mu)^k \frac{x^k}{k!} \quad \text{for all } 0 \leq x \leq x_0.$$

Thus, for any k edges $e_1, e_2, ..., e_k$ of $\mathbf{G}_n$ we find from the distributional assumption $\mathbb{P}(\xi_e \leq x) = F(x/d_n)$ that

$$P(\xi_{e_1} + \xi_{e_2} + \cdots + \xi_{e_k} \leq \rho) \leq (2\mu)^k \frac{\rho^k}{d_n^k k!} \quad \text{for all } 0 \leq \rho/d_n \leq x_0.$$

The expected number vertices of $\mathbf{G}_n$ that are contained in path-plus-cycle subgraph $\mathbf{G}_n$ with k vertices and total length bounded by ρ is therefore no larger than

$$(2\mu)^k nk^2 \frac{\Delta(G_n)^{k-1}\rho^k}{d_n^k k!} \quad \text{provided that} \quad 0 \leq \rho \leq x_0 d_n.$$

Now, since $\Delta(G_n) \sim d_n$, we may chose a constant $C = C(\mu, \rho)$ such that this bound is not larger than $nC^k/d_n k!$, and, thus, one finds

$$\mathbb{E}(|S(n,\rho)|) \leq e^C n/d_n, \quad \text{for all } n \text{ such that } d_n \geq \rho/x_0.$$

Since we assume $d_n \to \infty$, this bound is more than one needs to complete the proof of the lemma. □

Now that we know $\mathbf{G}$ is a tree, the proof of Theorem 4.2 will be complete provided that we confirm that the edge lengths from each vertex to its children are given by the realization of an independent Poisson process. For the root of $\mathbf{G}$ this is quite easy. When we look at the edges incident to the root of $\mathbf{G}_n$ for large n, we see by Lemma 4.1 that the lengths of these edges are approximately the points of a Poisson process, and consequently the lengths of the edges incident to the root of $\mathbf{G}$ must exactly follow a Poisson process.

Now consider a fixed ρ and an n so large that the probability that $\mathbf{G}_n$ contains a cycle in the ρ-neighborhood of the root of $\mathbf{G}_n$ is small. We know that the distances to the children of the root approximately follow the initial segment of a Poisson process, and now we consider the second generation. Let c be a fixed child of the root r, and consider the set S of edges incident to c. The set of edge lengths $\{\xi_e : e \in S \text{ and } e \neq (r,c)\}$ again satisfy the assumptions of Lemma 4.1, so the distribution of the lengths of the descendants of c will again follow a Poisson process as closely as we like. This argument shows that the first two generations of $\mathbf{G}$ are consistent with the construction of the PWIT. There is no change to the argument as one moves from the kth to the $k+1$st generations, so one finds that $\mathbf{G}$ is indeed a PWIT and the proof of Theorem 4.2 is complete.

4.4 Convergence of MSTs

We now need a general result from Aldous and Steele (2002) that tells us that the local weak convergence of a sequence of randomly rooted graphs automatically gives us the local weak convergence of their associated MSTs.

Theorem 4.4 (MST Convergence Theorem). *Let G_∞ denote a $\mathcal{G}_\star$-valued random variable such that with probability one G_∞ has infinitely many vertices and no two of the edges of G have the same length. Further, let $\{G_n : n = 1,2,...\}$ denote a sequence of $\mathcal{G}_\star$-valued random variables such that for each n the distribution of G_n is given by the standard construction and such that for each n the vertex set of G_n has cardinality n with probability one. If*

$$G_n \xrightarrow{d} G_\infty \quad \text{as } n \to \infty, \tag{25}$$

then one has the joint weak convergence in $\mathcal{G}_\star \times \mathcal{G}_\star$,

$$\big(G_n, \mathrm{MST}(G_n)\big) \xrightarrow{d} \big(G_\infty, \mathrm{MSF}(G_\infty)\big). \tag{26}$$

Further, if N_n denotes the degree of the root of $MST(G_n)$ and N denotes the degree of the root of $\mathrm{MSF}(G_\infty)$

$$N_n \xrightarrow{d} N \quad \text{and} \quad \mathbb{E}[N_n] \to \mathbb{E}[N] = 2, \tag{27}$$

and, if L_n denotes the sum of lengths of the edges incident to the root of $MST(G_n)$ and L denotes the corresponding quantities for $\mathrm{MSF}(G_\infty)$, *then*

$$L_n \xrightarrow{d} L. \tag{28}$$

4.5 Closing the Loops

Theorems 1.2 is now a remarkably easy corollary of the PWIT Limit Theorem, the MST Convergence Theorem, and the exact PWIT calculation developed in Section 3. The first step is simply to make the link between $L^\alpha_{\mathrm{MST}}(\mathbf{G}_n)$ and $Y_\alpha(\mathbf{G}_n)$ more explicit.
Here it is useful to let $\mathbb{E}_R[f(\mathbf{G}_n)]$ denote the conditional expectation of $f(\mathbf{G}_n)$ given $\{\xi_e : e \in v(\mathbf{G}_n)\}$; in other words, we just average $f(\mathbf{G}_n)$ over the possible values of the random root R. We now just compute

$$\begin{aligned}
L^\alpha_{\mathrm{MST}}(\mathbf{G}_n) &= d_n^{-\alpha} \sum_e (d_n\xi_e)^\alpha \mathbb{I}\big(e \in \mathrm{MST}(\mathbf{G}_n)\big) \\
&= d_n^{-\alpha}\frac{1}{2}\sum_v \sum_e (d_n\xi_e)^\alpha \mathbb{I}\big(e \in \mathrm{MST}(\mathbf{G}_n)\big) \\
&= \frac{1}{2} d_n^{-\alpha}\,|v(\mathbf{G}_n)|\,\mathbb{E}_R\Big[\sum_e (d_n\xi_e)^\alpha \mathbb{I}\big(e \in \mathrm{MST}(\mathbf{G}_n) \text{ and } R(\mathbf{G}_n) \in e\big)\Big] \\
&= \frac{1}{2} d_n^{-\alpha}\,|v(\mathbf{G}_n)|\,\mathbb{E}_R[Y_\alpha(\mathbf{G}_n)].
\end{aligned}$$

Finally, if one takes expectations in this representation one finds the basic identity

$$\mathbb{E}[L^\alpha_{\mathrm{MST}}(\mathbf{G}_n)] = \frac{1}{2} d_n^{-\alpha}\,|v(\mathbf{G}_n)|\,\mathbb{E}[Y_\alpha(\mathbf{G}_n)]. \tag{29}$$

Now, by the PWIT limit theorem we already know that $\mathbf{G}_n \xrightarrow{d} \mathcal{T}$, and by the MST Convergence Theorem this automatically entails $\text{MST}(\mathbf{G}_n) \to \text{MSF}(\mathcal{T})$. From the defining topology of local weak convergence, we then have

$$\sum_{e \in e(\mathbf{G}_n)} \xi_e^\alpha \mathbb{I}(e \in \text{MST}(\mathbf{G}_n) \text{ and } R \in e) \xrightarrow{d} \sum_{e \in e(\mathcal{T})} \xi_e^\alpha \mathbb{I}(e \in \text{MSF}(\mathcal{T}) \text{ and } R \in e),$$

or in other words

$$Y_\alpha(\mathbf{G}_n) \xrightarrow{d} Y_\alpha(\mathcal{T}). \tag{30}$$

Now, if $\{Y(G_n) : n = 1, 2, ..\}$ uniformly integrable, we can take expectations in the limit (30) and apply our earlier calculation of $\mathbb{E}[Y_\alpha(\mathcal{T})]$ to find

$$\mathbb{E}[Y_\alpha(\mathbf{G}_n)] \to 2\Gamma(1+\alpha)\zeta(2+\alpha),$$

but by the introductory identity (29), this is equivalent to

$$\mathbb{E}[L^\alpha_{\text{MST}}(\mathbf{G}_n)] \sim d_n^{-\alpha} \, |v(\mathbf{G}_n)| \, \Gamma(1+\alpha)\zeta(2+\alpha), \tag{31}$$

so the direct part of Theorem 1.2 is complete.

The converse now comes almost for free. One first notes that we may reverse the path from the limit (31) to the convergence of the expectations $\mathbb{E}[Y_\alpha(\mathbf{G}_n)]$, so when one pairs this fact with the limit (30), the loop is closed by applying the following simple lemma.

Lemma 4.5. *If a sequence of nonnegative random variables X_n, $n = 1, 2, ...$ converges to X in distribution, then one has*

$$\mathbb{E}[X_n] \to \mathbb{E}[X] \quad \textit{if and only if} \quad \{X_n : n = 1, 2, ...\} \quad \textit{is uniformly integrable.}$$

Proof: If we assume that the sequence $\{X_n : n = 1, 2, ...\}$ is uniformly integrable then convergence in probability implies X_n convergence in L^1 and this certainly implies that one has the convergence of the expectations. For the converse, we first note that by the Skorohod embedding theorem (Dudley (1989), pp. 325–327), there is no loss of generality if we assume that X_n converges almost surely to X. In this case, the nonnegativity and convergence of the expectations implies that X_n converges to X in L^1 by Sheffé's lemma (Williams (1991), p. 55). Since L^1 convergence is stronger than uniform integrability of $\{X_n : n = 1, 2, ...\}$, the proof of the lemma is complete. □

4.6 An Illustrative Example

Theorem 1.2 tells us that the limit behavior of $E[L^\alpha_{\text{MST}}(G_n)]$ is determined once one shows the uniform integrability of the sequence $\{Y_\alpha(G_n)\}$. The systematic treatment of this question will be left for another time, but one should note that this reasonably crude and qualitative property of $\{Y_\alpha(G_n)\}$ often follows from known results. Nevertheless, there are certainly many situations where uniform integrability fails or where the proof of uniform integrability can be subtle.

The example we consider here is illustrated in Figure 3, and it has also been used for illustration in Beveridge, Frieze, and McDiarmid (1998) and Frieze, Ruzinkó, and Thoma (2000). If one takes $G_n = C(m_n, K_n^{-1})$, then one can see just from Frieze's $\zeta(3)$ theorem that under the model of uniformly distributed costs that one has

$$E[L_{\mathrm{MST}}(G_n)] \sim m_n(\zeta(3)+1) \quad \text{as} \quad n \to \infty$$

for *any* choice of the sequence $2 \le m_n < \infty$. On the other hand, uniform integrability of $\{Y(G_n) : n = 1, 2, ...\}$ would imply that

$$E[L_{\mathrm{MST}}(G_n)] \sim m_n \zeta(3),$$

so in this case we certainly know $\{Y(G_n) : n = 1, 2, ...\}$ is not uniformly integrable. Nevertheless, one might want to check this directly, and, in fact, a moment's thought about the impact of a random root is all one needs to show

$$\limsup_{n\to\infty} E[Y(G_n)\mathbb{I}(Y(G_n) \ge t)] \ge 1 \quad \text{for all} \quad 0 \le t < \infty.$$

The situation is more interesting in case one takes $0 < \alpha < 1$, in which case one may now show that

$$\lim_{t\to\infty} \limsup_{n\to\infty} E[Y_\alpha(G_n)\mathbb{I}(Y_\alpha(G_n) \ge t)] = 0,$$

so the sequence $\{Y_\alpha(G_n)\}$ is uniform integrable. As a consequence, one obtains a positive result for $C(m_n, K_n^{-1})$ that asserts

$$E[L^{\alpha}_{\mathrm{MST}}(C(m_n, K_n^{-1}))] \sim \Gamma(\alpha+1)\zeta(\alpha+2)m_n n^{1-\alpha},$$

a fact which is perhaps more amusing when made more concrete. If one takes $m_n = \lfloor n^\alpha \rfloor$ and $\alpha = 0.99$, then $\Gamma(1.99) = 0.995\cdots$ and $\zeta(2.99) = 1.204\cdots$, so one finds

$$E[L^{0.99}_{\mathrm{MST}}(C(n, K_n^{-1}))] \sim cn \quad \text{where} \quad c = \Gamma(1.99)\zeta(2.99) = 1.1990\cdots.$$

5 Concluding Observations

As noted earlier, this is a report on work-in-progress and there are many loose ends that time and diligence may suffice to resolve. Perhaps the most compelling questions that have been left open concern the monotonicity and concavity of $E[L_{\mathrm{MST}}(K_n)]$ under the uniform model. Next on the list would be the possible analog of Parisi's conjecture and an exploration of the relationship of $E[L_{\mathrm{MST}}(K_n)]$ to $s_n = 1 + 1/2^3 + \cdots + 1/(n-1)^3$ under the exponential model.

More generally, the exact formula (2) for $E[L_{\mathrm{MST}}(G)]$ provides one with considerable motivation to work out detailed representations for the Tutte polynomials for those graphs that are of most interest in probability theory. One also suspects that interesting consequences may flow from the interpretation of formula (2) in the light of general results for the Tutte polynomial. Specifically, one might speculate that results like the Negami Splitting Formula (Negami (1987)) could lead to interesting inferences. Finally, the appearance of the logarithmic derivative of the Tutte polynomial in formula (2) suggests that this rational function may have informative properties beyond those it inherits from the Tutte polynomial.

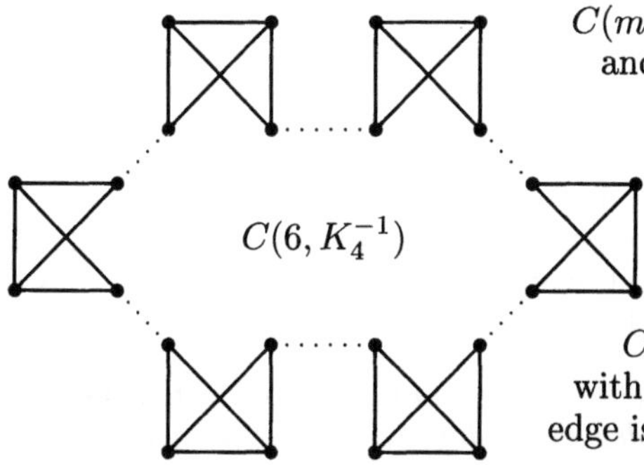

Figure 3: The graph $C(6, K_4^{-1})$ is built out of 6 copies of K_4 that have had one edge removed. These altered graphs — called K_4^{-1}s — are then chained together in a cycle. The result is a regular graph with degree 3 and $6 \cdot 4 = 24$ vertices; in grocer's terms one has a cubic graph with two dozen vertices.

The questions that are left open from the second part of the report are less well formed. Certainly, the question of uniform integrability of $\{Y_\alpha(G_n)\}$ deserves more systematic thought. Right now the easiest paths to uniform integrability freeload on the efforts of the more direct approaches to $E[L_{\text{MST}}(G)]$, especially the recent arguments of Frieze, Ruzinkó, and Thoma (2000) that exploit the lovely bound of Karger (1999) on the number of approximately minimal cuts. Nevertheless, as the easy example of Subsection 4.6 suggests, the sequence $\{Y_\alpha(G_n)\}$ does have an independent character. One suspects that in time the direct investigation of its uniform integrability will lead to arguments that do not poach on other approaches.

Finally, the second part of the report suggests several overarching questions from the theory of local weak convergence. One major line of investigation that surely deserves a sustained effort is the extension of the MST Convergence Theorem. There are many other classes of subgraphs from combinatorial optimization for which one expects an analogous result, and the class of confirmed examples is growing. Nevertheless, the final form of this theory is nowhere in sight.

ACKNOWLEDGEMENTS: It is a pleasure to thank David Aldous for continued coaching in the theory of local weak convergence and to thank Ira Gessel for providing an electronic extension to $n = 15$ of the table of Tutte polynomials for K_n given in Gessel and Sagan (1996).

References

[1] D.J. Aldous (1992) *Asymptotics in the random assignment problem*, Probab. Th. Rel. Fields, 93, 507–534.

[2] D.J. Aldous (2001) *The $\zeta(2)$ limit in the random assignment problem*, Random Structures Algorithms, 18, 381–418.

[3] D.J. Aldous and J.M. Steele (2002) *The Objective Method and the Theory of Local Weak Convergence* in *Discrete and Combinatorial Probability* (ed. H. Kesten), Springer-Verlag, New York. [in press]

[4] N. Alon, A. Frieze, and D. Welsh (1994) *Polynomial time randomized approximation schemes for the Tutte polynomial of dense graphs*, pp. 24–35 in the Proceedings of the 35th Annual Symposium on the Foundations of Computer Science (S. Goldwasser, ed.), IEEE Computer Society Press, 1994.

[5] F. Avram and D. Bertsimas (1992) *The minimum spanning tree constant in geometric probability and under the independent model: a unified approach,* Annals of Applied Probability 2, 113–130.

[6] A. Beveridge, A. Frieze, C. McDiarmid (1998) *Minimum length spanning trees in regular graphs*, Combinatorica 18, 311-333.

[7] R.M. Dudley (1989) *Real Analysis and Probability*, Wadworth Publishing, Pacific Grove CA.

[8] A.M. Frieze (1985) *On the value of a random minimum spanning tree problem*, Discrete Appl. Math. 10, 47–56.

[9] A.M. Frieze and C.J.H. McDiarmid (1989) *On random minimum lenght spanning trees*, Combinatorica, 9, 363–374.

[10] A. M. Frieze, M. Ruszinkó, and L. Thoma (2000) *A note on random minimum lenght spanning trees*, Electronic Journal of Combinatorics 7, Research Paper 5, (5 pp.)

[11] I.M. Gessel and B.E. Sagan (1996) *The Tutte polynomial of a graph, depth-first search, and simpicial complex partitions*, Electronic Journal of Combinatorics 3, Reseach Paper 9, (36 pp.)

[12] T.H. Harris (1989) *The Theory of Branching Processes*, Dover Publications, New York.

[13] S. Janson (1995) *The minimal spanning tree in a complete graph and a functional limit theorem for trees in a random graph*, Random Structures and Algorithms 7, 337–355.

[14] D. Karger (1999) *A randomized fully polynomial time approximation scheme for the all-terminal network reliablity problem*, SIAM Journal of Computing 29, 492–514.

[15] J.F.C. Kingman (1993) *Poisson Processes*, Oxford Universtiy Press, New York, 1993.

[16] P.A.P. Moran (1967) *A non-Markovian quasi-Poisson process*, Studia Scientiarum Mathematicarum Hungarica 2, 425–429.

[17] S. Negami (1987) *Polynomial Invariants of Graphs*, Transactions of the American Mathematical Society 299, 601–622.

[18] G. Parisi (1998) *A conjecture on radom bipartite matching*, ArXiv Condmat 9801176.

[19] M. Penrose (1998) *Random minimum spanning tree and percolation on the n-cube*, Random Structures and Algorithms 12, 63–82.

[20] A. Rényi, (1967) *Remarks on the Poisson Process*, Studia Scientiarum Mathematicarum Hungarica 2, 119–123.

[21] J.M. Steele (1987) *On Frieze's $\zeta(3)$ limit for lengths of minimal spanning trees*, Discrete Applied Mathematics, 18 (1987), 99–103.

[22] J.M. Steele (1997) *Probability Theory and Combinatorial Optimization*, NSF-CBMS Volume 69. Society for Industrial and Applied Mathematics, Philadelphia.

[23] D. Welsh (1999) *The Tutte Polynomial*, Random Structures and Algorithms 15, 210-228.

[24] D. Williams (1991) *Probability with Martingales*, Cambridge University Press, Cambridge.

J. Michael Steele

Department of Statistics, Wharton School
University of Pennsylvania
Philadelphia, PA 19104
steele@wharton.upenn.edu

PART III

Analysis of Algorithms and Trees

Trends in Mathematics, © 2002 Birkhäuser Verlag Basel/Switzerland

Generalized Pattern Matching Statistics

Jérémie Bourdon, Brigitte Vallée

ABSTRACT: *In pattern matching algorithms, a characteristic parameter is the number of occurrences of a given pattern in a random text of length n generated by a source. We consider here a generalization of the pattern matching problem in two ways. First, we deal with a generalized notion of pattern that encompasses classical patterns as well as "hidden patterns". Second, we consider a quite general probabilistic model of sources that may possess a high degree of correlations. Such sources are built with dynamical systems and are called dynamical sources. We determine the mean and the variance of the number of occurrences in this generalized pattern matching problem, and establish a property of concentration of distribution. These results are obtained via combinatorics, formal language techniques, and methods of analytic combinatorics based on generating operators and generating functions. The generating operators come from the dynamical system framework and generate themselves generating functions. The motivation to study this problem comes from an attempt at finding a reliable threshold for intrusion detections, from textual data processing applications, and from molecular biology.*

1 Introduction

Various pattern matching problems. String matching is the basic pattern matching problem. Here, a string w is a sequence of symbols $w = w_1 w_2 \dots w_s$ (of length s), and one searches for occurrences of w (as a block of consecutive symbols) in a text T. However, there are several useful generalizations of this basic problem:

Set of patterns. In the classical string matching problem, the pattern w should appear exactly (and consecutively) in the text, while, in the approximate case, a few mismatches are considered acceptable. The *approximate string matching* is then expressed as matching against a *set $\mathcal{L}$ of words* that contains all the valid approximations of the initial string.

Sequence of patterns. If we are interested in occurrences of the pattern w as a subsequence of the text T (now, the symbols no longer need to be consecutive), the problem is quite different, and it is called the subsequence matching problem. If the lengths of the gaps between successive symbols are not bounded, this matching problem becomes that of finding a *subsequence of symbols*.

Hidden pattern problem. If some of the gap lengths are bounded, while some others are not, one has the hidden pattern problem. A typical hidden pattern may look like $\mathtt{ab\#_2r\#ac\#a\#d\#_4a\#br\#a}$. Here, the notation $\#_i$ abbreviates a sequence of "don't-care-symbols" where the subscript denotes a strict upper bound on the length of the associated gap, $\#$ abbreviates $\#_\infty$ and $\#_1$ is omitted.

Here, we present a *common framework* for all the previously discussed types of pattern matching problems. We consider a generalized pattern which consists in a sequence $\mathcal{L}$ of languages, i.e., $\mathcal{L} := (\mathcal{L}_1, \mathcal{L}_2, \dots, \mathcal{L}_r)$. Each language $\mathcal{L}_i$ itself represents the i-th set of patterns to be used; it may be of infinite cardinality, but it is supposed to be "nondense" (in a formal sense that will be precised in the sequel). This generalized pattern then models a succession of r tasks, the i-th task corresponding to language $\mathcal{L}_i$. Between each task, any event (namely, any word of $\mathcal{A}^\star$) may occur. The corresponding pattern matching problem encompasses all the problems that we have described previously, as well as most of the classical problems of pattern matching, as described in [7] for instance. Surprisingly enough and to the best of our knowledge, there are no studies in the literature that address the question at this level of generality.

Motivations. We cite the introduction of [8] devoted to hidden patterns : "This general problem arises in two domains: intrusion detection and molecular biology. In the area of computer security, the *intrusion detection* [1, 12] searches in an audit file (the text) for certain patterns (known also as signatures) representing suspicious activities that might be indicative of an intrusion by an outsider, or misuse of the system by an insider. The key to this approach is to recognize that these patterns are *subsequences* because an intrusion signature specification requires the possibility of a variable number of events between successive events of the signature.

Molecular biology provides another important source of applications [16, 22, 23]. As a rule, there, one searches for subsequences, not strings. Examples are in abundance: split genes where exons are interrupted by introns, starting and stopping signal in genes, etc.... In general, for gene searching, the generalized pattern matching is the right approach for finding meaningful information.

We wish to study the number of occurrences of the generalized pattern in a random text of length n produced by a source of symbols. In all of the contexts mentioned above, it is of obvious interest to discern what constitutes a meaningful observation of pattern occurrences from what is merely a statistically unavoidable phenomenon ("noise"). This is precisely the problem addressed here. An immediate consequence of our results is the possibility to set *thresholds* at which appearance of a generalized pattern starts being meaningful".

Probabilistic model. In information theory contexts, data items are (infinite) words that are produced by a common mechanism, called a source. While real-life sources are often complex objects, pattern matching analyses only deal with quite idealized sources, such as memoryless sources or Markov chains. We use here a general framework of sources related to dynamical systems theory which goes beyond the cases of memoryless and Markov sources [19]. This model can describe non-Markovian processes, where the dependency on past history is unbounded, and as such, they attain a high level of generality. A probabilistic dynamical source is defined by two objects: a symbolic mechanism and a density. The mechanism is related to symbolic dynamics and associates an infinite word $M(x)$ to a real number x of the $[0,1]$ interval. It can be viewed as a generalization of numeration systems. Once the mechanism has been fixed, the density f on the $[0,1]$ interval can vary. This induces then different probabilistic behaviors for source words.

In dynamical systems theory, an important tool is the *density transformer*; here, we use it in a non classical way, and, since we show that it can easily generate objects that are essential in the analysis, we give it the rôle of a "generating operator".

Results. To the best of our knowledge, all the results that are already obtained in this area deal with classical sources –memoryless sources, Markov chains–. For instance, the number of string occurrences in a random text has been intensively studied over the last two decades, with significant progress in this area being reported [2, 9, 10, 13, 14, 15, 23]. Guibas and Odlyzko [9, 10] have revealed the fundamental rôle played by autocorrelation vectors and their associated polynomials. Régnier and Szpankowski [14, 15] established that the number of occurrences of a string is asymptotically normal under a diversity of models that include Markov chains. Nicodème, Salvy, and Flajolet [13] showed generally that the number of places in a random text at which a motif (i.e., a general regular expression pattern) terminates is asymptotically normally distributed.

We consider here the general problem –a general pattern in a general source–. Let $\Omega_n(\mathcal{L})$ be the number of occurrences of a given nondense generalized pattern $\mathcal{L} = (\mathcal{L}_1, \mathcal{L}_2, \ldots, \mathcal{L}_r)$ in a random text of length n generated by a dynamical source. Here, the number r of components of the generalized pattern $\mathcal{L}$ plays a fundamental rôle, since it measures the degree of freedom of the pattern. We study the expectation and the variance of the random variable $\Omega_n(\mathcal{L})$, and show the following estimates

$$\mathbf{E}\,[\Omega_n(\mathcal{L})] \sim \pi(\mathcal{L})\,\frac{n^r}{r!} \qquad \mathbf{Var}[\Omega_n(\mathcal{L})] \sim \sigma^2(\mathcal{L})\,n^{2r-1}$$

where $\pi(\mathcal{L})$ is the weight of $\mathcal{L}$. Here, the variance coefficient $\sigma^2(\mathcal{L})$ depends on two kinds of correlations. The first kind of is due to the source, and disappears when the source is memoryless. The second kind of correlations is due to the structure of the pattern, and it may exist even in the memoryless case.

When there exists only one degree of freedom (i.e., $r = 1$), the mean and the variance become of linear growth. This situation arises as soon as one considers only one set of patterns: for instance, the basic string matching, the approximate string matching, the totally constrained hidden pattern, etc.... For $r = 1$, it is possible to deal with finite–state models and de Bruijn graphs (as it is announced in [8] for memoryless sources) and obtain a central limit law.

Methodology. The first step we approach the probabilistic analysis is through a formal description of situations of interest by means of regular languages. Basically, as in [8], such a description of contexts of one or two occurrences gives access to expectation and variance, respectively. In previous works, based on the "generating function methodology", as in the main books of the area [11, 18, 17], one operates a systematic translation into generating functions. Due to correlations of the source, such a direct approach is no longer possible. Instead, we perform what we call a "dynamical analysis" and we first operate a systematic translation into *generating operators*. Now, there are many instances of this methodology, that can be applied in two main areas: text algorithms as in [3, 6, 19], or arithmetical algorithms as in [20, 21]. Here, the structure of the implied generating operators at the pole $z = 1$ provides the necessary asymptotic information. Finally, we come back to (classical) generating functions, and obtain the asymptotic behavior of the main parameters.

Plan of the paper. In Section 2, we describe the general pattern matching problem. We introduce the languages and the related generating functions that intervene in the analysis of the first two moments of the characteristic parameter, namely the number of occurrences of a given generalized pattern. Section 3 is devoted to the probabilistic model. Here, we define dynamical sources and introduce the generating operators that are a basic ingredient associated to the correlated sources considered. In Section 4, we come back to the average-case analysis of the characteristic parameter and prove our results. Finally, Section 5 presents important examples of pattern matching problems that fit in our general framework.

2 Collections of words and generating functions

We fix an alphabet $\mathcal{A}$, either finite or denumerable. The set of all possible texts is $\mathcal{A}^\star$, and a text of length n is an element $T = t_1 t_2 \cdots t_n$ of $\mathcal{A}^n$. We distinguish two notions: a language that is a set of words, and a collection (of words) that is a multi-set of words: in a collection, the same word may occur several times.

2.1. Generalized patterns. We give a higher degree of generalization to the pattern matching problem and consider *a sequence of sets of patterns.*

A generalized pattern is then specified by a finite sequence of languages $\mathcal{L} := (\mathcal{L}_1, \mathcal{L}_2, \ldots, \mathcal{L}_r)$. Each language $\mathcal{L}_i$ represents the i-th set of patterns to be used; It is supposed to be *nondense.* A precise definition of this notion will be given in 3.4. When all the components $\mathcal{L}_i$ are nondense, the generalized pattern itself will be said to be nondense. We say that $\mathcal{L}$ *occurs* in the text T if the text T contains as a subsequence a sequence $\ell = (\ell_1, \ell_2, \ldots, \ell_r)$ of $\mathcal{L}$. In this case, T is of the form

$$T = w_0 \ell_1 w_1 \ell_2 \ldots w_i \ell_i w_{i+1} \ldots w_r \ell_r w_{r+1} \qquad \text{with} \qquad w_i \in \mathcal{A}^\star \qquad \text{and} \qquad \ell_i \in \mathcal{L}_i.$$

The set of all valid *occurrences* of generalized pattern $\mathcal{L}$ is then the collection $\rho(\mathcal{L})$,

$$\rho(\mathcal{L}) = \mathcal{A}^\star \times \mathcal{L}_1 \times \mathcal{A}^\star \times \mathcal{L}_2 \times \cdots \times \mathcal{A}^\star \times \mathcal{L}_r \times \mathcal{A}^\star. \tag{1}$$

This operation ρ transforms a finite sequence of languages into a collection of words. This operation is called the completion, and $\rho(\mathcal{L})$ is the completion of the sequence $\mathcal{L}$.

An occurrence, that is an element of $\rho(\mathcal{L})$, defines a position I: this is the sequence of the r disjoint intervals $(I_1, I_2, \ldots I_r)$, where interval $I_j := [a_j, b_j]$ represents the exact portion of text T where the word ℓ_j occurs. Then, for instance, $a_1 = |w_0| + 1, b_1 = a_1 + |\ell_1|$, etc .... We denote by $\mathcal{P}_n(\mathcal{L})$ the set of all valid positions relative to $\mathcal{L}$ satisfying $b_r \le n$. The number $\Omega_n(\mathcal{L})$ of occurrences of pattern $\mathcal{L}$ with size n is then a sum of characteristic variables

$$\Omega_n(\mathcal{L}) = \sum_{I \in \mathcal{P}_n(\mathcal{L})} X_I, \qquad \text{with} \qquad X_I(T) := [\![\mathcal{L} \text{ occurs at position } I \text{ in } T]\!], \tag{2}$$

where $[\![B]\!] = 1$ if the property B holds, and $[\![B]\!] = 0$ otherwise (Iverson's notation).

2.2. An important particular case: hidden patterns. A hidden pattern matching problem is specified by a pair $(\mathcal{W}, \mathcal{D})$: the pattern $\mathcal{W} = w_1 \cdots w_s$ is a word of length s; the constraint $\mathcal{D} = (d_1, \ldots, d_{s-1})$ is an element of $\bar{\mathbb{N}}^{s-1}$. An s-tuple $I = (i_1, i_2, \ldots, i_s)$, $(1 \le i_1 < i_2 < \cdots < i_s)$, satisfies the constraint $\mathcal{D}$ if each gap $i_{j+1} - i_j$ is at most d_j, in which case it defines a *position.* An occurrence of pattern $\mathcal{W}$ subject to the constraint $\mathcal{D}$ is a pair (I, T) formed with a position $I = (i_1, i_2, \ldots, i_s)$ and a text $T = t_1 t_2 \cdots t_n$ for which $t_{i_j} = w_j$ $(1 \le j \le s)$.

The case $\mathcal{D} = (\infty, \ldots, \infty)$ models the *unconstrained problem*: in this case, the i-th language $\mathcal{L}_i$ reduces to the symbol $\{w_i\}$; at the other extreme of the spectrum, there lies the case where all d_j are finite, which we name the *constrained problem*: in this case, there exists only one finite-length language (i.e., $r = 1$) formed by all valid words that begin with the first symbol of $\mathcal{W}$, that end with the last symbol of $\mathcal{W}$, and fulfill all the constraints of $\mathcal{D}$. In the general case, the subset $\mathcal{U}$ of indices j for which d_j is unbounded ($d_j = \infty$) has cardinality $r - 1$. It separates the pattern $(\mathcal{W}, \mathcal{D})$ into r independent sub-patterns that are called the blocks and are denoted by $(\mathcal{W}_1, \mathcal{D}_1), (\mathcal{W}_2, \mathcal{D}_2), \ldots, (\mathcal{W}_r, \mathcal{D}_r)$. Each block $(\mathcal{W}_k, \mathcal{D}_k)$ gives rise to a finite–length language $\mathcal{L}_k$ defined as the language of all valid words that begin with the first symbol of $\mathcal{W}_k$, that end with the last symbol of $\mathcal{W}_k$, and fulfill all the constraints of $\mathcal{D}_k$.

2.3. Probabilistic model and generating functions. As regards the probabilistic model, we consider a *random source* that emits symbols of the text from the fixed alphabet $\mathcal{A}$. For a given length n, a random *text*, denoted by T_n, is drawn according to the induced probability on $\mathcal{A}^n$, and, for any word w of length n, we denote by p_w the probability that the source emits a prefix equal to w.

We associate to any collection $\mathcal{M}$ its generating function $M(z)$, where the complex variable z marks the length of the word w

$$M(z) = \sum_{w \in \mathcal{M}} p_w \, z^{|w|} = \sum_{n \ge 0} p(\mathcal{M}_n) \, z^n.$$

The last expression involves the probability of the collection $\mathcal{M}_n$ of words of $\mathcal{M}$ of length n. If the series $M(1)$ and $M'(1)$ converge, they define two objects, the total weight $p(\mathcal{M})$, and the average–length $N(\mathcal{M})$,

$$p(\mathcal{M}) := \sum_{w \in \mathcal{M}} p_w = M(1), \qquad N(\mathcal{M}) := \frac{\sum_{w \in \mathcal{M}} |w| p_w}{\sum_{w \in \mathcal{M}} p_w} = \frac{M'(1)}{M(1)}. \tag{3}$$

For a generalized pattern $\mathcal{L}$, the weight and the average length are defined by

$$\pi(\mathcal{L}) = \prod_{i=1}^{r} p(\mathcal{L}_i), \qquad N(\mathcal{M}) := \sum_{i=1}^{r} N(\mathcal{L}_i), \tag{4}$$

two quantities that surface throughout the analysis.

2.4. Generating function of the mean number of occurrences. Under this randomness model, all the quantities X_I defined in (2) become random variables whose expectations satisfy, for all allowable $I \in \mathcal{P}_n(\mathcal{L})$,

$$\mathbf{E}\,[X_I] = \sum_{T \in \mathcal{A}^n} p(T) \; [\![\mathcal{L} \text{ occurs in } T \text{ at position } I]\!].$$

Then, the quantity $\Omega_n(\mathcal{L})$ is itself a sum of correlated random variables, and the expectation $\mathbf{E}\,[\Omega_n(\mathcal{L})]$ of the number of occurrences of the generalized pattern $\mathcal{L}$

in a text of length n, equal to

$$\mathbf{E}\left[\Omega_n(\mathcal{L})\right] = \sum_{I \in \mathcal{P}_n(\mathcal{L})} \mathbf{E}\left[X_I\right] = \sum_{w \in \rho(\mathcal{L})_n} p_w = p\left(\rho(\mathcal{L})_n\right),$$

is exactly the probability of the sub-collection $\rho(\mathcal{L})_n$ defined in (1). Finally, the generating function of the expectations coincides exactly with the generating function $L(z)$ of the collection $\rho(\mathcal{L})$,

$$L(z) := \sum_{w \in \rho(\mathcal{L})} p_w \, z^{|w|} = \sum_{n \geq 1} \mathbf{E}\left[\Omega_n(\mathcal{L})\right] \, z^n.$$

2.5. Generating functions that intervene in the second moment analysis. The second moment of the variable $\Omega_n(\mathcal{L})$ equals

$$\mathbf{E}\left[\Omega_n^2(\mathcal{L})\right] = \sum_{I,J \in \mathcal{P}_n(\mathcal{L})} \mathbf{E}\left[X_I X_J\right].$$

As previously, the generating function $\sum_n \mathbf{E}\left[\Omega_n^2(\mathcal{L})\right] z^n$ coincides with the generating function $L^{[2]}(z)$ of the collection $\mathcal{L}^{[2]}$ of pairs of $\mathcal{L}$–occurrences.

We now describe the collection $\mathcal{L}^{[2]}$. The aggregate $\alpha(I,J)$ between two positions I and J is the sequence of system of intervals obtained by merging together all intersecting intervals of I and J. The number $\beta(I,J)$ of intervals of $\alpha(I,J)$ plays a fundamental rôle here, since it measures the *degree of freedom* of pairs. As an example, suppose that $I = ([2,6],[10,13])$, $J = ([5,11],[12,14])$ are two valid positions for a generalized pattern. Then $\alpha(I,J) = [2,14]$ and $\beta(I,J) = 1$.

Next, we group the pairs (I,J) according to the value of $\beta(I,J)$ and consider the sub–collections $\mathcal{L}_p^{[2]}$ formed by the pairs of $\mathcal{L}$–occurrences that arise at a pair (I,J) of positions for which $\beta(I,J)$ equals $2r-p$. However, it is sufficient to deal with the two sub–collections $\mathcal{L}_0^{[2]}$ and $\mathcal{L}_1^{[2]}$ relative to cases $p=0$ and $p=1$ since they give rise to the two main terms in the asymptotics of the second moment.

The first collection $\mathcal{L}_0^{[2]}$ (relative to non-intersecting positions) is just obtained from a completion of a generalized shuffle. Consider two sequences of languages $\mathcal{M}$ and $\mathcal{N}$. The shuffle of $\mathcal{M}$ and $\mathcal{N}$, denoted by $\mathcal{M} \amalg \mathcal{N}$, is a set of sequences obtained by shuffling the two sequences; if the sequence $\mathcal{M}$ has i components, and $\mathcal{N}$ has j components, the set $\mathcal{M} \amalg \mathcal{N}$ contains $\binom{i+j}{i}$ sequences with $i+j$ components. For instance, for a generalized pattern $\mathcal{L}$, the shuffle $\mathcal{L} \amalg \mathcal{L}$ contains $\binom{2r}{r}$ sequences, and

$$\mathcal{L}_0^{[2]} := \rho(\mathcal{L} \amalg \mathcal{L}). \tag{5}$$

The second collection $\mathcal{L}_1^{[2]}$ is relative to positions I and J for which only one pair (I_i, J_j) intersects. It is obtained from the operations of completion and generalized shuffle, but it also uses another operation that we denote by $\uparrow$ that transforms a pair $(\mathcal{B},\mathcal{C})$ of languages into a collection $\mathcal{B} \uparrow \mathcal{C}$. This collection gathers all the words that can be obtained by overlaps of words of $\mathcal{B}$ with words of $\mathcal{C}$. It is itself the union of three collections

$$\mathcal{B} \uparrow \mathcal{C} := <\mathcal{B},\mathcal{C}> \bigcup <\mathcal{C},\mathcal{B}> \bigcup (\mathcal{B} \cap \mathcal{C}) \tag{6}$$

where $< \mathcal{B}, \mathcal{C} >:= \{w = \beta u \gamma \mid \beta, \gamma, u \in \mathcal{A}^\star, u \neq \varepsilon, (\beta, \gamma) \neq (\varepsilon, \varepsilon), \beta u \in \mathcal{B}, u\gamma \in \mathcal{C}\}$.

Then, for each pair (i, j) of indices with $1 \leq i, j \leq r$, the collection $\mathcal{L}_{[i,j]}$ is obtained by mixing $\mathcal{L}_i \uparrow \mathcal{L}_j$ with two beginning sequences $b_i(\mathcal{L}), b_j(\mathcal{L})$ and two ending sequences $e_i(\mathcal{L}), e_j(\mathcal{L})$ of sequence $\mathcal{L}$,

$$\mathcal{L}_{[i,j]} := (b_i(\mathcal{L}) \amalg b_j(\mathcal{L}), \mathcal{L}_i \uparrow \mathcal{L}_j, e_i(\mathcal{L}) \amalg e_j(\mathcal{L})), \tag{7}$$

$$\text{with } b_i(\mathcal{L}) := (\mathcal{L}_1, \ldots, \mathcal{L}_{i-1}), \qquad e_j(\mathcal{L}) := (\mathcal{L}_{j+1}, \ldots, \mathcal{L}_r), \ .$$

The collection $\mathcal{L}_{[i,j]}$ is then the union of $t(i, j)$ sequences of languages, with

$$t(i,j) := \binom{i+j-2}{i-1}\binom{2r-i-j}{r-i}. \tag{8}$$

Note that each sequence has the same components, but with a different order. Finally, the collection $\mathcal{L}_1^{[2]}$ is obtained with the completion operator,

$$\mathcal{L}_1^{[2]} := \sum_{1 \leq i,j \leq r} \rho(\mathcal{L}_{[i,j]}). \tag{9}$$

2.6. Main differences with the memoryless case. In the memoryless case, we can obtain quite easily a direct translation of collections defined in (1, 5, 9) into generating functions. Moreover, it is possible to work directly with centered variables $Y_I := X_I - \mathbf{E}(X_I)$.

Here, this approach does not work (or at least we did not succeed in making it work). Due to correlations of the sources, we no longer use a direct translation in generating functions. However, in the framework of dynamical sources, it is possible to deal with "generating operators".

3 Dynamical sources and generating operators

We now present a quite general model of source –dynamical sources– that are associated to dynamical systems. Then, probabilities are "generated" by some generating operators, and the main generating functions to be studied can be generated themselves by operators. More precisely, we prove that an operator $\mathbf{L}(z)$ can be associated to each generalized pattern $\mathcal{L}$ from which the generating function $L(z)$ is easily deduced. Furthermore, unions and Cartesian products of sets translate into sums and compositions of the associated operators.

Moreover, such dynamical sources encompass and generalize the two classical models of sources, namely, the memoryless sources and Markovian sources. We refer to [19] for more details.

3.1. Dynamical sources. *A dynamical system $\mathcal{S}$ is defined by four elements:*

(a) an alphabet $\mathcal{A}$, either finite or denumerable,

(b) a topological partition of $\mathcal{I} :=]0,1[$ with disjoint open intervals $\mathcal{I}_a, a \in \mathcal{A}$,

(c) an encoding mapping σ which is constant and equal to a on each $\mathcal{I}_a$,

(d) a shift mapping T whose restriction to $\mathcal{I}_a$ is a bijection of class $\mathcal{C}^1$ from $\mathcal{I}_a$ to $\mathcal{I}$. The local inverse of T restricted to $\mathcal{I}_a$ is denoted by h_a.

Such a dynamical system can be viewed as a "dynamical source". since it produces infinite words of $\mathcal{A}^\infty$. With an input x of $\mathcal{I}$, it outputs the word $M(x)$ formed with the sequence of symbols $\sigma T^j(x)$, i.e., $M(x) := (\sigma x, \sigma T x, \sigma T^2 x, \dots)$.

3.2. Generating operators. The mappings $h_w := h_{m_1} \circ h_{m_2} \circ \cdots \circ h_{m_k}$ relative to prefix words $w := m_1 \dots m_k$ are then the inverse branches of T^k. All the infinite words that begin with the same prefix w correspond to real numbers x that belong to the same interval $\mathcal{I}_w =]h_w(0), h_w(1)[$. Then, if the unit interval is endowed with some density f, the probability p_w that a word begins with prefix w is the measure of the interval $\mathcal{I}_w$. Such a probability p_w is now easily generated by the operator $\mathbf{G}_{[w]}$, defined as

$$\mathbf{G}_{[w]}[f](t) = |h'_w(t)| \; f \circ h_w(t),$$

since one has

$$p_w = |\int_{h_w(0)}^{h_w(1)} f(t)dt| = \int_0^1 |h'_w(t)| f \circ h_w(t) dt = \int_0^1 \mathbf{G}_{[w]}[f](t)dt. \tag{10}$$

The following composition property holds:

$$\text{for any prefixes } \; w, w', \qquad \text{one has:} \qquad \mathbf{G}_{[w.w']} = \mathbf{G}_{[w']} \circ \mathbf{G}_{[w]}, \tag{11}$$

and "replaces" the relation $p_{w.w'} = p_w p_{w'}$ which is no longer true when the source has some memory.

The generating operator $\mathbf{B}(z)$ of a collection $\mathcal{B}$ of $\mathcal{A}^\star$ is then defined by

$$\mathbf{B}(z) := \sum_{w \in \mathcal{B}} z^{|w|} \; \mathbf{G}_{[w]}.$$

where the complex variable z marks the length of the word w. If the operator $\mathbf{B}(z)$ is well-defined at $z = 1$, the operator $\mathbf{B} := \mathbf{B}(1)$ is called the normalized operator of collection $\mathcal{B}$. For instance, the operator

$$\mathbf{G} := \sum_{a \in \mathcal{A}} \mathbf{G}_{[a]}, \tag{12}$$

is the normalized operator of the alphabet $\mathcal{A}$ and it plays a fundamental rôle in the sequel. It is the density transformer of the dynamical system in the sense that if X is a random variable with density f, the density of TX is $\mathbf{G}[f]$.

Then, from (11), unions and Cartesian products of languages translates into sums and compositions of the associated operators. For instance, the operator associated to the set $\mathcal{A}^\star$ is the quasi-inverse

$$(I - z\mathbf{G})^{-1} := \sum_{i > 0} z^i \mathbf{G}^i.$$

If $\mathcal{L}$ is a generalized pattern, i.e., a sequence of languages $\mathcal{L} := (\mathcal{L}_1, \mathcal{L}_2, \ldots, \mathcal{L}_r)$, the generating operator $\mathbf{L}(z)$ relative to collection $\rho(\mathcal{L})$ defined in (1) satisfies

$$\mathbf{L}(z) = (I - z\mathbf{G})^{-1} \circ \mathbf{L}_r(z) \circ (I - z\mathbf{G})^{-1} \circ \cdots \circ \mathbf{L}_1(z) \circ (I - z\mathbf{G})^{-1}, \tag{13}$$

and involves $r+1$ occurrences of the quasi-inverse $(I - z\mathbf{G})^{-1}$ "mixed" with the generating operators $\mathbf{L}_i(z)$ of languages $\mathcal{L}_i$.

Then, equation (10) provides a relation between the generating operator $\mathbf{B}(z)$ and the generating function $B(z)$ relative to the same collection $\mathcal{B}$, that is

$$B(z) := \sum_{w \in \mathcal{B}} z^{|w|} p_w = \sum_{w \in \mathcal{B}} z^{|w|} \int_0^1 \mathbf{G}_{[w]}[f](t)dt = \int_0^1 \mathbf{B}(z)[f](t)dt. \tag{14}$$

3.3. Some nice dynamical sources. There exist dynamical sources for which "nice" properties of shift T can be transferred into "nice" properties of their density transformers. Under some "natural" properties of shift T, there exists a spectral decomposition of the density transformer $\mathbf{G}$ defined in (12) that induces a spectral decomposition of the quasi–inverse $(I - z\mathbf{G})^{-1}$. This property will be fundamental in the analysis. More precisely, we give the following definition, perhaps a little bit informal :

Definition. *A dynamical system is said to be decomposable if the density transformer* $\mathbf{G}$ *satisfies the following: It acts on a convenient Banach space* $\mathcal{F}$ *on which it is quasi–compact, and it has positivity properties that entail the existence of dominant spectral objects.*

Then, there exist a unique dominant eigenvalue λ positive and a dominant eigenfunction denoted by φ. Under the normalization condition $\int_0^1 \varphi(t)dt = 1$, this last object is unique too. Then, quasi–compacity entails the existence of a spectral gap between the dominant eigenvalue and the remainder of the spectrum, that separates the operator $\mathbf{G}$ in two parts $\mathbf{G} = \lambda\mathbf{P} + \mathbf{N}$, where $\mathbf{P}$ is the projection of $\mathbf{G}$ onto the dominant eigenspace, and $\mathbf{N}$ is relative to the remainder of the spectrum. Since the operator $\mathbf{G}$ is a density transformer, its dominant eigenvalue satisfies $\lambda = 1$ and φ is also the (unique) stationary density. One then splits the quasi-inverse $(I - z\mathbf{G})^{-1}$ into two parts,

$$(I - z\mathbf{G})^{-1} = \frac{1}{1-z}\mathbf{P} + \mathbf{R}(z), \tag{15}$$

with $\mathbf{R}(z) := (I - z\mathbf{N})^{-1} - \mathbf{P} = \sum_{k \geq 0} z^k(\mathbf{G}^k - \mathbf{P}), \qquad \mathbf{P}[f](t) = \varphi(t) \int_0^1 f(x)dx.$

The first term of (15) defines an operator which has a pole at $z = 1$. Due to the existence of the spectral gap, the operator $\mathbf{N}$ has a spectral radius μ less than 1, and then the second term of (15) defines an operator $\mathbf{R}(z)$ that is analytic in $|z| < (1/\mu)$. At $z = 1$, the operator $\mathbf{R} := \mathbf{R}(1)$ describes the correlations of the source.

A decomposable dynamical source is proven to be ergodic and mixing with exponential rate. The main class of decomposable dynamical sources is provided by the analytic expanding sources, as defined in [19]; however, there are some other instances, as systems described in [5].

In the remainder of the paper, we deal with a decomposable dynamical source and the initial density will be always the stationary density φ. For any function g of $\mathcal{F}$, one has:

$$\mathbf{R}[\varphi] = 0, \qquad \int_0^1 \mathbf{P}[g](t)dt = \int_0^1 g(t)dt, \qquad \int_0^1 \mathbf{R}[g](t)dt = 0. \tag{16}$$

3.4. Nondense collections and correlations. We are now ready to give the precise definition of a nondense collection.

Definition. *The collection $\mathcal{B}$ is said to be nondense if the associated generating operator $\mathbf{B}(z) : \mathcal{F} \to \mathcal{F}$ is analytic on the disk $|z| < (1/\rho)$ for some $\rho < 1$. A generalized pattern $\mathcal{L}$ whose all components $\mathcal{L}_i$ are nondense is said to be nondense.*

Denote by $\mathcal{B}_n$ the subcollection of $\mathcal{B}$ formed with words of length n, and by $\mathbf{B}_n$ its normalized generating operator. The collection $\mathcal{B}$ is nondense if and only if there exists $\rho < 1$ for which one has $||\mathbf{B}_n||_{\mathcal{F}} = O(\rho^n)$. Remark that, if $\mathcal{B}$ is nondense, one has $p(\mathcal{B}_n) = O(\rho^n)$. Then the generating function $B(z)$ is analytic at $z = 1$, so that both quantities $p(\mathcal{B})$ and $N(\mathcal{B})$ defined in (3) are well-defined.

Here are some instances of nondense languages: finite languages, finite–length languages. For memoryless sources, languages where a symbol is forbidden, etc...

For a nondense collection $\mathcal{B}$, the normalized generating operator $\mathbf{B}$ satisfies

$$\int_0^1 \mathbf{P} \circ \mathbf{B} \circ \mathbf{P}[g](t) = p(\mathcal{B}) \left(\int_0^1 g(t)dt \right). \tag{17}$$

In the analysis, two kinds of correlations may occur between two nondense collections $\mathcal{B}, \mathcal{C}$, according to the relative position of $\mathcal{B}$ and $\mathcal{C}$. If $\mathcal{B}$ and $\mathcal{C}$ do not overlap, there are two cases: $\mathcal{B}$ before $\mathcal{C}$, or $\mathcal{C}$ before $\mathcal{B}$. In these cases, the correlation coefficients are $c(\mathcal{B}, \mathcal{C})$ or $c(\mathcal{C}, \mathcal{B})$ where $c(\mathcal{B}, \mathcal{C})$ is defined as

$$p(\mathcal{B})p(\mathcal{C})c(\mathcal{B}, \mathcal{C}) := \sum_{k \geq 0} \left[p(\mathcal{B} \times \mathcal{A}^k \times \mathcal{C}) - p(\mathcal{B})p(\mathcal{C}) \right] = \int_0^1 \mathbf{C} \circ \mathbf{R} \circ \mathbf{B}[\varphi](t). \tag{18}$$

If $\mathcal{B}$ and $\mathcal{C}$ overlap, the correlation coefficient

$$d(\mathcal{B}, \mathcal{C}) := \frac{p(\mathcal{B} \uparrow \mathcal{C})}{p(\mathcal{B})p(\mathcal{C})} \tag{19}$$

involves the nondense collection $\mathcal{B} \uparrow \mathcal{C}$ of words w defined in (6), so that coefficient $d(\mathcal{B}, \mathcal{C})$ is well-defined. Finally, the total correlation coefficient $m(\mathcal{B}, \mathcal{C})$ between $\mathcal{B}$ and $\mathcal{C}$ gathers the three possible cases

$$m(\mathcal{B}, \mathcal{C}) = c(\mathcal{B}, \mathcal{C}) + c(\mathcal{C}, \mathcal{B}) + d(\mathcal{B}, \mathcal{C}), \tag{20}$$

so that

$$p(\mathcal{B})p(\mathcal{C})m(\mathcal{B}, \mathcal{C}) = p(\mathcal{B} \uparrow \mathcal{C}) + \sum_{k > 0} \left[p(\mathcal{B} \times \mathcal{A}^k \times \mathcal{C}) + p(\mathcal{C} \times \mathcal{A}^k \times \mathcal{B}) - 2p(\mathcal{B})p(\mathcal{C}) \right].$$

4 Analysis of the number of pattern occurrences

We are now able to come back to the analysis of the characteristic parameter and we prove a similar asymptotic behavior for the number of occurrences as in the classical case [8]. We first study the mean, then the second moment. We conclude with an expression of the variance, and a proof of the concentration of distributions.

The main steps of our analysis which can be called a "dynamical analysis" are as follows:

(a) We first describe the generating operators relative to each collection, namely $\mathbf{L}(z)$ relative to collection $\rho(\mathcal{L})$ for the mean, and $\mathbf{L}_0^{[2]}, \mathbf{L}_1^{[2]}$ relative to collections $\mathcal{L}_0^{[2]}, \mathcal{L}_1^{[2]}$ for the second moment. Each generating operator will contain some occurrences of the quasi-inverse $(I - z\mathbf{G})^{-1}$, and some occurrences of generating operators $\mathbf{L}_i(z)$.

(b) We then decompose the three main operators with the help of (15). We will consider only the two transformations which give rise to the main asymptotic terms: The transformation called AllP, where we replace all the occurrences of the quasi-inverse by the first term $(1-z)^{-1}\mathbf{P}$, and the transformation AllP-1R, where we replace all occurrences except one of the quasi inverse $(I - z\mathbf{G})^{-1}$ by the first term $(1 - z)^{-1}\mathbf{P}$, the last occurrence being replaced by the operator $\mathbf{R}(z)$. Then, formulae (16,17) are useful in the computations of constants involved.

(c) We thus come back to generating functions thanks to (14).

(d) We finally extract the asymptotic behavior of coefficients of generating functions.

4.1. Mean number of occurrences. We state our first result:

Theorem 1. *Consider a decomposable dynamical source endowed with its stationary density φ and a generalized nondense pattern $\mathcal{L} = (\mathcal{L}_1, \mathcal{L}_2, \ldots, \mathcal{L}_r)$. The expectation $\mathbf{E}[\Omega_n(\mathcal{L})]$ of the number of occurrences of the generalized pattern $\mathcal{L}$ in a text of length n satisfies*

$$\mathbf{E}[\Omega_n(\mathcal{L})] = \binom{n+r}{r}\pi(\mathcal{L}) + \binom{n+r-1}{r-1}\pi(\mathcal{L})\ [C(\mathcal{L}) - N(\mathcal{L})]\ + O(n^{r-2}),$$

where $\pi(\mathcal{L})$ is the total weight and $N(\mathcal{L})$ is the average length. The coefficient $C(\mathcal{L})$ equals the sum of the correlations $c(\mathcal{L}_{i-1}, \mathcal{L}_i)$ between two consecutive languages, where the correlation coefficient $c(\mathcal{B}, \mathcal{C})$ is defined in (18).

Proof. The collection relative to the mean is $\rho(\mathcal{L})$ defined in (1), and the generating function is $\mathbf{L}(z)$ defined in (13). When using the transformation AllP in $\mathbf{L}(z)$, we obtain an operator $\mathbf{M}_1(z)$ which has a pole of order $r+1$ at $z = 1$,

$$\mathbf{M}_1(z) = \left(\frac{1}{1-z}\right)^{r+1} \mathbf{P} \circ \mathbf{L}_r(z) \circ \mathbf{P} \circ \cdots \circ \mathbf{P} \circ \mathbf{L}_1(z) \circ \mathbf{P}.$$

Near $z = 1$, each operator $\mathbf{L}_i(z)$ is analytic and admits the expansion $\mathbf{L}_i(z) = \mathbf{L}_i + (z-1)\mathbf{L}_i'(1) + O(z-1)^2$, so that the main term of the expansion is

$$\left(\frac{1}{1-z}\right)^{r+1} \mathbf{P} \circ \mathbf{L}_r \circ \mathbf{P} \circ \cdots \circ \mathbf{P} \circ \mathbf{L}_1 \circ \mathbf{P} \tag{21}$$

while the second main term is obtained as a sum of r terms, each of them obtained by replacing the operator $\mathbf{L}_i(z)$ by its derivative $\mathbf{L}_i'(1)$ at $z = 1$. The corresponding generating function $M_1(z)$ satisfies near $z = 1$

$$M_1(z) = \left(\frac{1}{1-z}\right)^{r+1} \pi(\mathcal{L}) - \left(\frac{1}{1-z}\right)^{r} \pi(\mathcal{L})N(\mathcal{L}) + O\left(\frac{1}{1-z}\right)^{r-1}. \tag{22}$$

The apparition of the weight $\pi(\mathcal{L})$ and the average length $N(\mathcal{L})$ is due to formulae (16,17), together with definitions (4).

When using now AllP-1R in $\mathbf{L}(z)$, we obtain an operator $\mathbf{M}_2(z)$ which has a pole of order r at $z = 1$. This is a sum of $r+1$ terms, each of the term containing an occurrence of the operator $\mathbf{R}(z)$ between two generating operators of consecutive languages $\mathcal{L}_{i-1}, \mathcal{L}_i$. The relative generating function $M_2(z)$ has also a pole of order r at $z = 1$ and satisfies near $z = 1$

$$M_2(z) = \left(\frac{1}{1-z}\right)^{r} \pi(\mathcal{L}) \sum_{i=2}^{r} c(\mathcal{L}_{i-1}, \mathcal{L}_i) + O\left(\frac{1}{1-z}\right)^{r-1}.$$

Here, the correlation number $c(\mathcal{B}, \mathcal{C})$ between $\mathcal{B}$ and $\mathcal{C}$ is defined in (18). □

4.2. The second moment. We prove here that the analysis of the second moment is very similar to the analysis of the expectation, even if it is more intricate.

Theorem 2. *The second moment* $\mathbf{E}\left[\Omega_n^2(\mathcal{L})\right]$ *of the number of occurrences of the nondense generalized pattern* $\mathcal{L} = (\mathcal{L}_1, \mathcal{L}_2, \ldots, \mathcal{L}_r)$ *in a text of length* n *produced by a decomposable dynamical source endowed with stationary density* φ *satisfies*

$$\mathbf{E}\left[\Omega_n^2(\mathcal{L})\right] = \binom{n+2r}{2r}\binom{2r}{r}\pi^2(\mathcal{L}) + \binom{n+2r-1}{2r-1}\pi^2(\mathcal{L})\widetilde{C}(\mathcal{L}) + O(n^{r-2}),$$

$$\text{with } \widetilde{C}(\mathcal{L}) := 2\binom{2r-1}{r-1}C(\mathcal{L}) - 2\binom{2r}{r}N(\mathcal{L}) + \sum_{1\le i,j\le r} t(i,j)\, m(\mathcal{L}_i, \mathcal{L}_j).$$

Here, $\pi(\mathcal{L})$ *is the total weight,* $N(\mathcal{L})$ *is the average length,* $C(\mathcal{L})$ *is the coefficient of Theorem 1 and the correlation–coefficient* $m(\mathcal{B}, \mathcal{C})$ *is defined in (20). The "choice" coefficients* $t(i,j)$ *are defined in (8).*

Proof. We deal now with collections $\mathcal{L}_0^{[2]}, \mathcal{L}_1^{[2]}$ defined in (5, 9), and the generating functions $\mathbf{L}_0^{[2]}, \mathbf{L}_1^{[2]}$ relative to these collections.

We begin with $\mathbf{L}_0^{[2]}$. When using the transformation AllP in $\mathbf{L}_0^{[2]}$, we obtain an operator $\mathbf{M}_3(z)$ which has a pole of order $2r+1$ at $z = 1$. The corresponding generating function $M_3(z)$ has also a pole of order $2r+1$ at $z=1$ and satisfies near $z=1$

$$M_3(z) = \binom{2r}{r}\left(\frac{1}{1-z}\right)^{2r+1}\pi^2(\mathcal{L}) - 2\binom{2r}{r}\left(\frac{1}{1-z}\right)^{2r}\pi^2(\mathcal{L})N(\mathcal{L}) + O\left(\frac{1}{1-z}\right)^{2r-1}.$$

Here, the weight $\pi(\mathcal{L})$ and the average length $N(\mathcal{L})$ occur thanks to the same arguments as in Theorem 1.

When using AllP-1R in $\mathbf{L}_0^{[2]}$, we obtain an operator $\mathbf{M}_4(z)$ which has a pole of order $2r$ at $z=1$. This is a sum of terms, where each term contains an occurrence of the operator $\mathbf{R}(z)$ between two generating operators of languages $\mathcal{L}_i, \mathcal{L}_j$. Now, there are two cases, according as the two languages $\mathcal{L}_i, \mathcal{L}_j$ "come from" the same occurrence or not. In the second case, these languages are not necessary consecutive, and all pairs (i,j) may intervene. More precisely, each pair (i,j), for $1 \le i,j \le r$, intervenes in a number of terms equal to $2t(i,j)$ –here, the "choice coefficient" $t(i,j)$ is defined in (8)–. In the first case, the pair (i,j) is necessary of the form $(i-1,i)$, with $2 \le i \le r$, and such a pair occurs in exactly $2\binom{2r-1}{r-1}$ terms. Finally, the associated generating function $M_4(z)$ has also a pole of order $2r$ at $z=1$ and satisfies

$$M_4(z) = \left(\frac{1}{1-z}\right)^{2r}\pi^2(\mathcal{L})C_2(\mathcal{L}) + O\left(\frac{1}{1-z}\right)^{2r-1},$$

$$\text{where } C_2(\mathcal{L}) := 2\binom{2r-1}{r-1}C(\mathcal{L}) + \sum_{1\le i,j\le r} t(i,j)\,[c(\mathcal{L}_i,\mathcal{L}_j) + c(\mathcal{L}_j,\mathcal{L}_i)]$$

involves the coefficient $C(\mathcal{L})$ that intervenes in Theorem 1 together with supplementary correlations $c(\mathcal{B},\mathcal{C})$ defined in (18).

The second generating operator $\mathbf{L}_1^{[2]}(z)$ is the sum of generating operators $\mathbf{L}_{[i,j]}$ relative to collections $\rho(\mathcal{L}_{[i,j]})$ defined in (7,9). Each operator $\mathbf{L}_{[i,j]}$ is itself the sum of $t(i,j)$ terms. If, we use AllP in each term, we obtain an operator $\mathbf{M}_5(z)$ which has a pole of order $2r$ at $z=1$. Near $z=1$, the associated generating function

$$M_5(z) = \left(\frac{1}{1-z}\right)^{2r}\pi^2(\mathcal{L})\sum_{1\le i,j\le r} t(i,j)\,d(\mathcal{L}_i,\mathcal{L}_j) + O\left(\frac{1}{1-z}\right)^{2r-1},$$

involves correlation–coefficients $d(\mathcal{B},\mathcal{C})$ defined in (19). □

4.3. Concentration of distributions for the number of occurrences. In the computation of the variance, the two main terms of order n^{2r} in $\mathbf{E}[\Omega_n^2]$ and in $\mathbf{E}[\Omega_n]^2$ cancel, and thus, the standard deviation is of an order, $O(n^{r-1/2})$, that is smaller than the mean, $O(n^r)$. This fact entails concentration of distribution, via a well-known argument based on Chebyshev's inequalities. In summary:

Theorem 3. *The mean and the variance of the number of occurrences $\Omega_n(\mathcal{L})$ relative to a nondense generalized pattern $\mathcal{L} = (\mathcal{L}_1, \mathcal{L}_2, \ldots, \mathcal{L}_r)$ in a text of length*

n produced by a decomposable dynamical source endowed with stationary density φ satisfy

$$\mathbf{E}\left[\Omega_n(\mathcal{L})\right] = \frac{\pi(\mathcal{L})}{r!}\, n^r \left(1 + O(\frac{1}{n})\right), \qquad \mathbf{Var}\left[\Omega_n(\mathcal{L})\right] = \sigma^2(\mathcal{L})\, n^{2r-1} \left(1 + O(\frac{1}{n})\right),$$

where the variance coefficient

$$\sigma^2(\mathcal{L}) = \pi^2(\mathcal{L}) \left[\frac{r - 2N(\mathcal{L})}{r!(r-1)!} + \frac{m(\mathcal{L})}{(2r-1)!}\right]$$

involves the weight $\pi(\mathcal{L})$, the average length $N(\mathcal{L})$ and the total correlation–coefficient

$$m(\mathcal{L}) := \sum_{1 \le i,j \le r} \binom{i+j-2}{i-1} \binom{2r-i-j}{r-i} m(\mathcal{L}_i, \mathcal{L}_j)$$

that deals with the correlation–coefficients $m(\mathcal{B}, \mathcal{C})$, defined as

$$p(\mathcal{B})p(\mathcal{C})m(\mathcal{B},\mathcal{C}) := p(\mathcal{B} \uparrow \mathcal{C}) + \sum_{k \ge 0} \left[p(\mathcal{B} \times \mathcal{A}^k \times \mathcal{C}) + p(\mathcal{C} \times \mathcal{A}^k \times \mathcal{B}) - 2p(\mathcal{B})p(\mathcal{C})\right].$$

Consequently, the distribution of the random variable $\Omega_n(\mathcal{L})$ converges in probability:

$$\text{for any } \epsilon > 0, \qquad \lim_{n \to \infty} \Pr\left\{\left|\frac{\Omega_n(\mathcal{L})}{\mathbf{E}\left[\Omega_n(\mathcal{L})\right]} - 1\right| < \epsilon\right\} = 1.$$

4.4. A Gaussian law ? In the memoryless case, we can adapt the method used in [8], based on the study of higher centered moments. We then obtain a limit Gaussian law.

However, in the case of a general decomposable source, we did not succeed in analysing directly centered moments. We can obtain several main terms of each moment $\mathbf{E}\left[\Omega_n^k\right]$ of order k, but, in order to compute the main term of the moment of order k of the centered variable $X_n := \Omega_n - \mathbf{E}\left[\Omega_n\right]$, we would have to get a very precise expansion of each $\mathbf{E}\left[\Omega_n^k\right]$. We do not succeed to obtain these expansions. However, we conjecture that the asymptotic Gaussian law holds, even for a general decomposable dynamical source.

5 Examples of various pattern matching problems

For memoryless sources, all the correlation–coefficients $c(\mathcal{L}_i, \mathcal{L}_j)$ disappear. We now study some particular cases where the coefficients $d(\mathcal{L}_i, \mathcal{L}_j)$ admit some expressions that are related to classical objects that appear in pattern matching problems, such as correlation polynomials, for instance, and we find again some classical results that are described in the book [18].

5.1. Basic pattern matching. Here, we consider a pattern α that is a finite string of length s. This fits in our general framework with $r = 1$ and $\mathcal{L} = \{\alpha\}$. The set $\mathcal{L} \uparrow \mathcal{L}$ is here a collection that contains all the words of the form $\beta u \gamma$ where u satisfies $\beta u = u\gamma = \alpha$. Classically, as in [18], the set of all such possible suffixes γ is called the autocorrelation set and is denoted by $\mathcal{D}_\alpha$. Then $\mathcal{L} \uparrow \mathcal{L}$ coincides with the collection of the words $\alpha\gamma$, where the suffix γ belongs to $\mathcal{D}_\alpha$. For $\gamma = \varepsilon$, the word has multiplicity 1, while, for $\gamma \neq \varepsilon$, the word has multiplicity 2. Then, the autocorrelation-coefficient $d(\alpha, \alpha)$ is related to the auto-correlation polynomial $D_\alpha(z)$ via the relation $p^2(\alpha)d(\alpha,\alpha) = 2D_\alpha(1) - p(\alpha)$. The following theorem holds:

Theorem 4. *Let α be a single pattern of length s. The number Ω_n of occurrences of α in a text of length n produced by a decomposable dynamical source satisfies*

$$\mathbf{E}\,[\Omega_n] = [n - s + 1]\,p(\alpha) + O(\frac{1}{n}), \quad \mathbf{Var}\,[\Omega_n] = n\,p^2(\alpha)\,[1 - 2s + m(\alpha)] + O(1),$$

$$\textit{with} \quad m(\alpha) := \frac{2}{p^2(\alpha)} \left(\sum_{k \geq 0} [p(\alpha \times \mathcal{A}^k \times \alpha) - p^2(\alpha)] + \sum_{\gamma \in \mathcal{D}_\alpha} p(\alpha\gamma) \right) - \frac{1}{p(\alpha)}.$$

5.2. Matching a set of patterns. Approximate pattern matching as well as multiple pattern matching are very similar to the previous problem. In these cases, the generalized pattern $\mathcal{L}$ still satisfies $r = 1$ and there is a unique finite-length language that we denote by $\mathcal{L}$. The pattern–correlation coefficient is related to a correlation polynomial $A_{(\alpha,\beta)}(z)$ that extends the autocorrelation polynomial for two words α and β of $\mathcal{L}$. Totally constrained hidden patterns, as defined in paragraph 2.2, are also a particular case of the general setting where $r = 1$.

5.3. The gaussian law for a single regular language. In this case, our analysis, described in [4], deals with a generating matrix operator that is a mixing between generating operators relative to the source and the automaton A relative to the regular language $\mathcal{L}$. We associate to $\mathcal{L}$ subsets $\mathcal{B}^{[i,j]}$ of the alphabet $\mathcal{A}$ formed with all the symbols a that allow automaton transitions from the i-th state to the j-th state,

$$\mathcal{B}^{[i,j]} := \{a \in \mathcal{A} \mid A(a,i) = j\}.$$

We denote by $\mathbf{B}^{[i,j]}$ the normalized operator relative to language $\mathcal{B}^{[i,j]}$, and, we mark with variable u all the operators that lead to an accepting state. Finally, we consider the matrix operator $\mathbf{B}_u$ whose general coefficient is

$$\mathbf{B}_u^{[i,j]} := \begin{cases} \mathbf{B}^{[i,j]} & \text{if } j \text{ is not accepting,} \\ u\,\mathbf{B}^{[i,j]} & \text{if } j \text{ is accepting.} \end{cases}$$

This mixed operator $\mathbf{B}_u$, that depends on parameter u, possesses dominant spectral properties when u is near the real axis. We then obtain the following result:

Theorem 5. *The number of occurrences of a regular language in a random text of length n produced by a decomposable dynamical source, once normalized by the mean and standard deviation, obeys in the asymptotic limit a Gaussian law.*

5.4. Examples of problems with $r > 1$. We recall that a hidden pattern $(\mathcal{W}, D)$ gives rise to a generalized pattern with r components where $r+1$ equals the number of infinite constraints. Then, as soon as the hidden pattern contains an infinite constraint, one has $r > 1$.

In the totally unconstrained hidden pattern problem, one considers a pattern $\alpha := \alpha_1 \dots \alpha_s$ of $\mathcal{A}^s$ and we are interested in counting all the occurrences of α as a subsequence (commonly called *hidden word*) of a text T. The associated generalized pattern corresponds exactly to $r = s$ and $\mathcal{L}_i := \{\alpha_i\}$. The pattern–correlation coefficient $d(\alpha_i, \alpha_j)$ between two symbols α_i and α_j satisfies

$$d(\alpha_i, \alpha_j) = \frac{[\![\alpha_i = \alpha_j]\!]}{p_{\alpha_i}},$$

and the constant $N(\mathcal{L})$ equals the length s of the pattern.

The problem of searching for a sequence of words (i.e. $\mathcal{L}_i := \{w_i\}$) is a slight generalization of this problem. In this case, the correlation coefficient d is expressed with the correlation polynomial $A_{(w_i,w_j)}(z)$ between two words w_i and w_j, and the constant $N(\mathcal{L})$ is the sum $N(\mathcal{L}) = \sum_i |w_i|$.

Acknowledgments. This work was partially supported by the Future and Emerging Technologies programme of the EU under contract number IST-1999-14186 (ALCOM-FT).

References

[1] A. Apostolico and M. Atallah, Compact Recognizers of Episode Sequences, Submitted to *Information and Computation.*

[2] E. Bender and F. Kochman, The Distribution of Subword Counts is Usually Normal, *European Journal of Combinatorics*, 14, 265-275, 1993.

[3] J. Bourdon, Size and Path-length of Patricia Tries: Dynamical Sources Context, *Random Structures and Algorithms*, Vol 19 (3–4), 289–315, 2001.

[4] J. Bourdon, B. Vallée, Dynamical Sources in Information Theory : Motif Statistics, not published.

[5] J. Bourdon, B. Daireaux, B. Vallée, Dynamical Analysis of α-Euclidean Algorithms, submitted.

[6] J. Clément, P. Flajolet, and B. Vallée, Dynamical Sources in Information Theory: A General Analysis of Trie Structures, *Algorithmica*, 29, 307–369, 2001.

[7] M. Crochemore and W. Rytter, *Text Algorithms*, Oxford University Press, New York, 1994.

[8] Ph Flajolet, Y. Guivarch, W. Szpankowski and B. Vallée, Hidden Pattern Statistics,*Proc. of ICALP'2001*, LNCS 2076, 152–165, 2001.

[9] L. Guibas and A. M. Odlyzko, Periods in Strings, *J. Combinatorial Theory Ser. A*, 30, 19–43, 1981.

[10] L. Guibas and A. M. Odlyzko, String Overlaps, Pattern Matching and Nontransitive Games, *J. Combinatorial Theory Ser. A*, 30, 183–208, 1981.

[11] D. E. Knuth, *The Art of Computer Programming, Fundamental Algorithms*, Vol. 1, Third Edition, Addison-Wesley, Reading, MA, 1997.

[12] S. Kumar and E.H. Spafford, A Pattern-Matching Model for Intrusion Detection, *Proceedings of the National Computer Security Conference*, 11–21, 1994.

[13] P. Nicodème, B. Salvy, and P. Flajolet, Motif Statistics, *European Symposium on Algorithms, Lecture Notes in Computer Science*, No. 1643, 194–211, 1999.

[14] M. Régnier and W. Szpankowski, On the Approximate Pattern Occurrences in a Text, *Proc. Compression and Complexity of SEQUENCE'97*, IEEE Computer Society, 253–264, Positano, 1997.

[15] M. Régnier and W. Szpankowski, On Pattern Frequency Occurrences in a Markovian Sequence, *Algorithmica*, 22, 631–649, 1998.

[16] I. Rigoutsos, A. Floratos, L. Parida, Y. Gao and D. Platt, The Emergence of Pattern Discovery Techniques in Computational Biology, *Metabolic Engineering*, 2, 159–177, 2000.

[17] R. Sedgewick and P. Flajolet, *An Introduction to the Analysis of Algorithms*, Addison-Wesley, Reading, MA, 1995.

[18] W. Szpankowski, *Average Case Analysis of Algorithms on Sequences*, John Wiley & Sons, New York, 2001.

[19] B. Vallée, Dynamical Sources in Information Theory: Fundamental Intervals and Word Prefixes, *Algorithmica*, 29, 262–306, 2001.

[20] B. Vallée, Dynamical Analysis of a Class of Euclidean Algorithms, to appear in *Theoretical Computer Science* (2002), also available in Les cahiers du GREYC 2000.

[21] B. Vallée, Digits and Continuants in Euclidean Algorithms. Ergodic Versus Tauberian Theorems, JTNB 12, 531–570, 2000.

[22] A. Vanet, L. Marsan, and M.-F. Sagot, Promoter sequences and algorithmical methods for identifying them, *Res. Microbiol.*, 150, 779-799, 1999.

[23] M. Waterman, *Introduction to Computational Biology*, Chapman and Hall, London, 1995.

Jérémie BOURDON, **Brigitte** VALLÉE

GREYC–UMR 6072,
Université de Caen,
F-14032 Caen (France)
e-mail: {jeremie.bourdon, brigitte.vallee}@info.unicaen.fr

Trends in Mathematics, © 2002 Birkhäuser Verlag Basel/Switzerland

A Note on Random Suffix Search Trees

Luc Devroye and Ralph Neininger

ABSTRACT: *A random suffix search tree is a binary search tree constructed for the suffixes $X_i = 0.B_iB_{i+1}B_{i+2}\ldots$ of a sequence $B_1, B_2, \ldots$ of independent identically distributed random b-ary digits B_j. Let D_n denote the depth of the node for X_n in this tree when B_1 is uniform on $\mathbb{Z}_b$. We show that for any value of $b > 1$, $\mathbb{E}\, D_n = 2\log n + O(\log^2 \log n)$, just as for the random binary search tree. We also show that $D_n / \mathbb{E}\, D_n \to 1$ in probability.*

1 Introduction

Current research in data structures and algorithms is focused on the efficient processing of large bodies of text (encyclopedia, search engines) and strings of data (DNA strings, encrypted bit strings). For storing the data such that string searching is facilitated, various data structures have been proposed. The most popular among these are the suffix tries and suffix trees (Weiner, 1973; McCreight, 1976), and suffix arrays (Manber and Myers, 1990). Related intermediate structures such as the suffix cactus (Karkkainen, 1995) have been proposed as well. Apostolico (1985), Crochemore and Rytter (1994) and Stephen (1994) cover most aspects of these data structures, including their applications and efficient construction algorithms (Ukkonen 1995, Weiner 1973, Giegerich and Kurtz, 1997, and Kosaraju, 1994). If the data are thought of as strings $B_1, B_2, \ldots$ of symbols taking values in an alphabet $\mathbb{Z}_b = \{0, 1, \ldots, b-1\}$ for fixed finite b, then the suffix trie is an ordinary b-ary trie for the strings $X_i = (B_i, B_{i+1}, \ldots)$, $1 \le i \le n$. The suffix tree is a compacted suffix trie. The suffix array is an array of lexicographically ordered strings X_i on which binary search can be performed. Additional information on suffix trees is given in Farach (1997), Farach and Muthukrishnan (1996, 1997), Giancarlo (1993, 1995), Giegerich and Kurtz (1995), Gusfield (1997), Sahinalp and Vishkin (1994), Szpankowski (1993). The suffix search tree we are studying in this paper is the search tree obtained for $X_1, \ldots, X_n$, where again lexicographical ordering is used. Care must be taken to store with each node the position in the text, so that the storage comprises nothing but pointers to the text. Suffix search trees permit dynamic operations, including the deletion, insertion, and alteration of parts of the string. Suffix arrays on the other hand are clearly only suited for off-line applications.

The analysis of random tries has a long history (see Szpankowski, 2001, for references). Random suffix tries were studied by Jacquet, Rais and Szpankowski (1995) and Devroye, Szpankowski and Rais (1992). The main model used in these studies is the independent model: the B_i's are independent and identically distributed. Markovian dependence has also been considered. If $p_j = \mathbb{P}\{B_1 = j\}$, $0 \le j < b$, then it is known that the expected depth of a typical node in an n-node suffix trie is close in probability to $(1/\mathcal{E})\log n$, where $\mathcal{E} = \sum_j p_j \log(1/p_j)$ is the entropy of B_1. The height is in probability close to $(b/\xi)\log n$, where $\xi = \log(1/\sum_j p_j^b)$. If ξ or $\mathcal{E}$ are small, then the performance of these structures deteriorates to the point that perhaps more classical structures such as the binary search tree are preferable.

In this paper, we prove that for first order asymptotics, random suffix search trees behave roughly as random binary search trees. If D_n is the depth of X_n, then

$$\mathbb{E}\, D_n = 2 \log n + O(\log^2 \log n)$$

and $D_n / \log n \to 2$ in probability, just as for the random binary search tree constructed as if the X_i's were independent identically distributed strings (Knuth, 1973, and Mahmoud, 1992, have references and accounts). We prove this for $b = 2$ and $p_0 = p_1 = 1/2$. The generalization to $b > 2$ is straightforward as long as B_1 is uniform on $\mathbb{Z}_b$.

The second application area of our analysis is related directly to random binary search trees. We may consider the X_i's as real numbers on $[0, 1]$ by considering the b-ary expansions

$$X_i = 0.B_i B_{i+1} \ldots, \quad 1 \leq i \leq n \,.$$

In that case, we note that $X_{i+1} = \{bX_i\} := (bX_i) \bmod 1$. If we start with X_1 uniform on $[0, 1]$, then every X_i is uniform on $[0, 1]$, but there is some dependence in the sequence $X_1, X_2, \ldots$. The sequence generated by applying the map $X_{i+1} = \{bX_i\}$ resembles the way in which linear congruential sequences are generated on a computer, as an approximation of random number sequences. In fact, all major numerical packages in use today use linear congruential sequences of the form $x_{n+1} = (bx_n + a) \bmod M$, where a, b, x_n, x_{n+1}, M are integers. The sequence x_n / M is then used as an approximation of a truly random sequence. Thus, our study reveals what happens when we replace i.i.d. random variables with the multiplicative sequence. It is reassuring to note that the first order behavior of binary search trees is identical to that for the independent sequence.

The study of the behavior of random binary search trees for dependent sequences in general is quite interesting. For the sequence $X_n = (nU) \bmod 1$, with U uniform on $[0, 1]$, a detailed study by Devroye and Goudjil (1998) shows that the height of the tree is in probability $\Theta(\log n \log \log n)$. The behavior of less dependent sequences $X_n = (n^\alpha U) \bmod 1$, $\alpha > 1$, is largely unknown. The present paper shows of course that $X_n = (2^n U) \bmod 1$ is sufficiently independent to ensure behavior as for an i.i.d. sequence. Antos and Devroye (2000) looked at the sequence $X_n = \sum_{i=1}^n Y_i$, where the Y_i's are i.i.d. random variables and showed that the height is in probability $\Theta(\sqrt{n})$. Cartesian trees (Devroye 1994) provide yet another model of dependence with heights of the order $\Theta(\sqrt{n})$.

This extended abstract is organized as follows. In section 2 we introduce a perturbed version of the random suffix search tree on which we will draw back troughout our analysis. Section 3 provides a rough bound for the mean of the height of the random suffix search tree, which will be used later in the analysis of $\mathbb{E}\, D_n$. In the following two sections we present a key lemma on which our expansion of the mean and a weak law of large numbers for D_n is based, and give a detailed proof for $\mathbb{E}\, D_n = 2 \log n + O(\log^2 \log n)$. From section 6 on we approach the tree from a different path, the spacings formed by $X_1, \ldots, X_n$ on $[0, 1]$. First we show a limit law for the scaled length of a randomly chosen spacing, convergence of all moments and a related limit law when the spacings are chosen with probability according to their length. These results could also be used to find the dominant term in the expansion of $\mathbb{E}\, D_n$. We will derive asymptotic information on the size of the subtree rooted at X_j for a large range of j. In the last section we state some lemmas which were used in the analysis. Complete proofs can be found in Devroye and Neininger (2002).

2 Notation and perturbed tree

Denote the uniform distribution on $[0,1]$ by $U[0,1]$ and the Bernoulli(p) distribution by $Be[p]$. We have given a $U[0,1]$ distributed random variable X_1 and define $X_k := T(X_{k-1})$ for $k \geq 2$, with the map $T : [0,1] \to [0,1], x \mapsto \{2x\} = 2x \mod 1$. In the binary representation $X_1 = 0.B_1B_2\ldots$, the B_k are independent $Be[1/2]$ bits. Then we have

$$X_k = 0.B_kB_{k+1}B_{k+2}\ldots$$

for all $k \geq 1$. For $m \geq 1$ we introduce the corresponding perturbed random variates

$$Y_k^{\langle m\rangle} := 0.B_kB_{k+1}\ldots B_{k+m-1}B_1^{(k)}B_2^{(k)}\ldots, \quad k = 1,\ldots,n,$$

where $\{B_j^{(k)} : k, j \geq 1\}$ is a family of independent $Be[1/2]$ distributed bits, independent of X_1. Then we have for all $k \geq 1$,

$$|X_k - Y_k^{\langle m\rangle}| \leq \frac{1}{2^m}.$$

and $Y_i^{\langle m\rangle}, Y_j^{\langle m\rangle}$ are independent if $|i-j| \geq m$.

Since we will switch in our analysis between the random suffix search tree built from $X_1,\ldots,X_n$ and its perturbed counterpart generated by $Y_1^{\langle m\rangle},\ldots,Y_n^{\langle m\rangle}$ we have to control the probability that they coincide. We denote by $\|\lfloor x\rfloor\| := 2\lfloor x/2\rfloor$ the largest even integer not exceeding x. For a vector $(a_1,\ldots,a_n)$ of distinct real numbers, let $\pi(a_1,\ldots,a_n)$ be the permutation given by the vector.

Lemma 2.1. *If* $m := 18\|\lfloor\log_2 n\rfloor\|$, *then for all* $n \geq 16$,

$$\mathbb{P}\left(\pi(X_1,\ldots X_n) \neq \pi(Y_1^{\langle m\rangle},\ldots,Y_n^{\langle m\rangle})\right) \leq \frac{8}{n^2}.$$

The perturbed tree and the original tree are thus identical with high probability. In the perturbed tree, note that $Y_i^{\langle m\rangle}$ and $Y_j^{\langle m\rangle}$ are independent whenever $|i-j| \geq m$. Unfortunately, it is not true that random binary search trees constructed on the basis of identically distributed m-dependent sequences behave as those for i.i.d. sequences, even when m is as small as 1. For example, the depth of a typical node and the height may increase by a factor of m when m is small and positive.

3 A rough bound for the height

We will need a rough upper bound for the mean of the height of the random suffix search tree.

Lemma 3.1. *Let a binary search tree $\mathcal{T}$ be built up from distinct numbers $x_1, \ldots, x_n$ and denote its height by H. We assume that the set of indices $\{1, \ldots, n\}$ is decomposed into k nonempty subsets $\mathcal{I}_1, \ldots, \mathcal{I}_k$ of cardinalities $|\mathcal{I}_j| = n_j$. Assume that $\mathcal{I}_j$ consists of the indices $n(j,1) < \cdots < n(j, n_j)$ and denote the height of the binary search tree $\mathcal{T}_j$ built up from $x_{n(j,1)}, \ldots, x_{n(j,n_j)}$ by H_j for $j = 1, \ldots, k$. Then we have*

$$H \leq k - 1 + \sum_{j=1}^{k} H_j.$$

This can be turned into a rough estimate for the height using the fact the mean of the height is known to be of the order $\log n$ for the binary search tree in the random permutation model, where each permutation of the keys inserted is equally likely (Devroye 1987). Lemma 3.2 below is valid for our model, but also for any random binary search tree constructed on the basis of $U[0,1]$ random variables that are m-dependent, with $m = O(\log n)$.

Lemma 3.2. *Let H_n denote the height of the random suffix search tree with n nodes. Then $\mathbb{E}\, H_n = O(\log^2 n)$.*

4 A key lemma

We introduce the events $A_j = \{X_j \text{ is ancestor of } X_n \text{ in the tree}\}$. Then we have the representations

$$D_n = \sum_{j=1}^{n-1} \mathbf{1}_{A_j}, \quad \mathbb{E}\, D_n = \sum_{j=1}^{n-1} \mathbb{P}(A_j).$$

We use the notation $\alpha, \beta \rhd \gamma_1, \ldots, \gamma_n$, if there does not exist k with $1 \leq k \leq n$ for which $\alpha < \gamma_k < \beta$ or $\beta < \gamma_k < \alpha$, i.e., α, β are neighbors in $\{\gamma_1, \ldots, \gamma_n\}$. Note that $A_j = \{X_j, X_n \rhd X_1, \ldots, X_{j-1}\}$. We use $A_j^{\langle m \rangle}$ for the corresponding event involving the $Y_k^{\langle m \rangle}$: $A_j^{\langle m \rangle} = \{Y_j^{\langle m \rangle}, Y_n^{\langle m \rangle} \rhd Y_1^{\langle m \rangle}, \ldots, Y_{j-1}^{\langle m \rangle}\}$. Thoughout we abbreviate $m = 18\|\lfloor \log_2 n \rfloor\|$.

Our key lemma consists of an analysis of the depth of the n-th inserted node X_n conditioned on its location. For $x \in [0,1]$ and $1 \leq i \leq n-1$, define

$$p_i(x) := \mathbb{P}\Big(Y_i^{\langle m \rangle}, x \rhd Y_1^{\langle m \rangle}, \ldots, Y_{i-1}^{\langle m \rangle}\Big).$$

We use the following *bad set*:

$$B_n(\xi) := \bigcup_{k=1}^{m} \{x \in [0,1] : |x - T^k(x)| < \xi\}, \quad \xi > 0,$$

where T is the map $T(x) := \{2x\}$ and T^k its k-th iteration, see Figure 1.

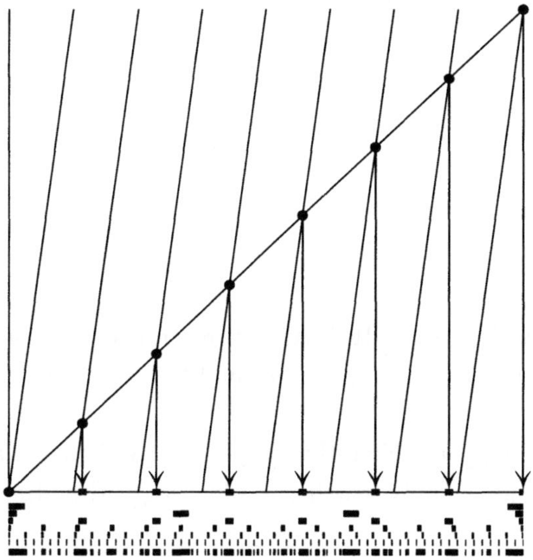

Figure 1: *The last line shows the bad set $B_n(\xi)$ for $m = 6$ and $\xi = 3/50$. The six lines above show the sets $\{|x - T^k(x)| \le \xi\}$ for $k = 1, \dots, 6$. In the square, for the case $k = 3$, it is shown how these sets emerge.*

Lemma 4.1. *For all n sufficiently large, all $x \in [0,1]$, and $1 \le i < n$, we have*

$$p_i(x) = \mathbf{1}_{[m^2/i, 1-m^2/i]}(x) \left(\frac{2}{i} + R_1(n,i) + \mathbf{1}_{B_n(2m^2/\sqrt{i})}(x) R_2(n,i) \right) + \left(1 - \mathbf{1}_{[m^2/i, 1-m^2/i]}(x)\right) R_3(n,i),$$

where for appropriate constants $C_1, C_2, C_3 > 0$,

$$|R_1(n,i)| \le C_1 \frac{\log^6 n}{i^{3/2}},$$
$$|R_2(n,i)| \le C_2 \frac{\log^3 n}{i},$$
$$|R_3(n,i)| \le C_3 \frac{\log n}{i}.$$

5 Analysis of the depth

Based on Lemma 4.1, we obtain an expansion for the mean of the depth D_n as well as a weak law of large numbers. For a random binary search tree based on i.i.d. random variables, it is well-known that $\mathbb{E}\, D_n = 2 \log n + O(1)$, where D_n is the depth of the n-th node (see, e.g., Knuth 1973 or the references in Mahmoud 1992).

Theorem 5.1. *The depth D_n of the n-th node inserted into a random suffix search tree satisfies*

$$\mathbb{E}\, D_n = 2\log n + O(\log^2 \log n).$$

Proof: We define the events $A_j = \{X_j \text{ is ancestor of } X_n \text{ in the tree}\}$ and the representations

$$D_n = \sum_{j=1}^{n-1} \mathbf{1}_{A_j}, \quad \mathbb{E}\, D_n = \sum_{j=1}^{n-1} \mathbb{P}(A_j).$$

For the estimate of $\mathbb{P}(A_j)$ we distinguish three ranges for the index j, namely $1 \le j \le \lceil \log_2^{12} n \rceil$, $\lceil \log_2^{12} n \rceil < j \le n - m$, and $n - m < j < n$, where we choose $m = 18\|\lfloor \log_2 n \rfloor\|$.

<u>The range $1 \le j \le \lceil \log_2^{12} n \rceil$:</u> Note that $\sum_{j=1}^{\lceil \log_2^{12} n \rceil} \mathbf{1}_{A_j}$ is bounded from above by the height of the random suffix search tree with $\lceil \log_2^{12} n \rceil$ nodes. Thus, by Lemma 3.2, we obtain

$$\sum_{j=1}^{\lceil \log_2^{12} n \rceil} \mathbb{P}(A_j) \le \mathbb{E}\, H_{\lceil \log_2^{12} n \rceil} = O(\log_2^2 \log_2^{12} n)) = O(\log^2 \log n).$$

<u>The range $\lceil \log_2^{12} n \rceil < j \le n - m$:</u> We start, using Lemma 2.1, with the representation

$$\begin{aligned} \mathbb{P}(A_j) &= \mathbb{P}(X_j, X_n \triangleright X_1, \dots, X_{j-1}) \\ &= \mathbb{P}(Y_j^{\langle m \rangle}, Y_n^{\langle m \rangle} \triangleright Y_1^{\langle m \rangle}, \dots, Y_{j-1}^{\langle m \rangle}) + O(1/n^2) \\ &= \mathbb{P}(A_j^{\langle m \rangle}) + O(1/n^2). \end{aligned}$$

Note that $Y_n^{\langle m \rangle}$ is independent of $Y_1^{\langle m \rangle}, \dots, Y_j^{\langle m \rangle}$, since $j \le n - m$. Thus for the calculation of $\mathbb{P}(A_j^{\langle m \rangle})$ we may condition on $Y_n^{\langle m \rangle}$. With the notation of Lemma 4.1 and using the fact that $Y_n^{\langle m \rangle}$ is $U[0,1]$ distributed this yields for all $1 \le j \le n - m$,

$$\mathbb{P}(A_j^{\langle m \rangle}) = \mathbb{E}\,[p_j(Y_n^{\langle m \rangle})] = \frac{2}{j} + R_{n,j}, \quad |R_{n,j}| \le C \frac{\log^6 n}{j^{3/2}},$$

for some constant $C > 0$. When summing note that

$$\sum_{j=\lceil \log_2^{12} n \rceil}^{\infty} \frac{\log^6 n}{j^{3/2}} \le \log^6 n \int_{\lceil \log_2^{12} n \rceil - 1}^{\infty} \frac{1}{x^{3/2}}\, dx = O(1).$$

We obtain

$$\begin{aligned} \sum_{j=\lceil \log_2^{12} n \rceil}^{n-m} \mathbb{P}(A_j) &= \sum_{j=\lceil \log_2^{12} n \rceil}^{n-m} \left(\frac{2}{j} + R_{n,j} + O\Big(\frac{1}{n^2}\Big) \right) \\ &= 2\log n + O(\log \log n). \end{aligned}$$

Hence, this range gives the main contribution.

The range $n-m<j<n-1$: With $q:=\lfloor j/m\rfloor-1$ we have

$$\begin{aligned}\mathbb{P}(A_j) &= \mathbb{P}(X_j, X_n \rhd X_1,\dots,X_{j-1})\\ &\le \mathbb{P}(X_j, X_n \rhd X_{j-m},\dots,X_{j-qm})\\ &= \mathbb{P}(Y_j^{\langle m\rangle}, Y_n^{\langle m\rangle} \rhd Y_{j-m}^{\langle m\rangle},\dots,Y_{j-qm}^{\langle m\rangle}) + O(1/n^2).\end{aligned}$$

We have, using Lemma 9.2, for n sufficiently large,

$$\begin{aligned}&\mathbb{P}(Y_j^{\langle m\rangle}, Y_n^{\langle m\rangle} \rhd Y_{j-m}^{\langle m\rangle},\dots,Y_{j-qm}^{\langle m\rangle})\\ &\le \mathbb{P}\Big(\{|Y_j^{\langle m\rangle}-Y_n^{\langle m\rangle}|\ge m^2/j\}\cap\{Y_j^{\langle m\rangle}, Y_n^{\langle m\rangle} \rhd Y_{j-m}^{\langle m\rangle},\dots,Y_{j-qm}^{\langle m\rangle}\}\Big)\\ &\qquad+\mathbb{P}(|Y_j^{\langle m\rangle}-Y_n^{\langle m\rangle}|<m^2/j)\\ &\le \left(1-\frac{m^2}{j}\right)^{j/m-2}+8\frac{m^2}{j}\\ &\le 4\exp(-m)+8\frac{m^2}{j}\\ &\le O\left(\frac{1}{n^{18}}\right)+8\frac{m^2}{j}.\end{aligned}$$

The summation yields

$$\sum_{j=n-m}^{n-1}\mathbb{P}(A_j)=O(1),$$

so that the third range makes an asymptotically negligible contribution. Collecting the estimates of the three ranges, we obtain the assertion. ■

Theorem 5.2. *We have $D_n/\mathbb{E}\,D_n\to 1$ in probability as $n\to\infty$.*

6 Weak convergence of a random spacing

The lengths of the spacings formed by $X_1,\dots,X_n$ on $[0,1]$ are denoted by $S_j^n := X_{(j+1)}-X_{(j)}$ for $j=1,\dots,n-1$ and $S_0^n := X_{(1)}$, $S_n^n := 1-X_{(n)}$, where $X_{(j)}$ denotes the j-th order statistic of $X_1,\dots,X_n$. In this section we provide a limit law for the rescaled length of a spacing chosen uniformly from $S_0^n,\dots,S_n^n$, where by uniform we mean that we choose one of the indices $j=0,\dots,n$ uniformly at random. Later we will choose an index by into which spacing an $U[0,1]$ random variable, independent of X_1, falls.

Lemma 6.1. *We have*

$$nS_{I_n}^n \xrightarrow{\mathcal{L}} E,\quad (n\to\infty),$$

where E is exp(1)-distributed, i.e., has Lebesgue-density e^{-x} on $[0,\infty)$ and I_n is uniformly distributed on $\{0,\dots,n\}$ and independent of X_1.

This can be reduced to the following result on the spacings between fractional parts of lacunary sequences due to Rudnick and Zaharescu (2002). A *lacunary sequence* is a sequence $(a_j)_{j\geq 1}$ of integers such that we have $\liminf_{j\to\infty} a_{j+1}/a_j > 1$. The primary example is $a_j = 2^j$. Now, for an $\alpha \in \mathbb{R}$ we define $S_j^n(\alpha)$ for $j = 0, \dots, n$ as the spacings between the fractional parts of αa_j, $j = 1, \dots, n$, in the unit interval $[0,1]$. More precisely, for $\vartheta_j^n := \{\alpha a_j\}$ we define $S_j^n(\alpha) := \vartheta_{(j+1)} - \vartheta_{(j)}$ for $j = 1, \dots, n-1$ as well as $S_0^n(\alpha) := \vartheta_{(1)}$ and $S_n^n(\alpha) := 1 - \vartheta_{(n)}$. Then Rudnick and Zaharescu (2002) prove:

Theorem 6.2. *Let (a_j) be a lacunary sequence. Then we have for almost all $\alpha \in \mathbb{R}$ and all $0 \leq a < b$,*

$$\lim_{n\to\infty} \frac{1}{n+1} \#\{0 \leq j \leq n : nS_j^n(\alpha) \in [a,b]\} = \int_a^b e^{-x}\,dx.$$

For background, see also Kurlberg and Rudnick (1999, Appendix A). This can directly be turned into a proof of Lemma 6.1.

7 Uniform integrability

In this section we show that the convergence in Corollary 6.1 holds for all moments.

Lemma 7.1. *For all fixed $p > 0$*

$$\sup_{n\in\mathbb{N}} \mathbb{E}\,(nS_{I_n}^n)^p < \infty,$$

where the random index I_n is unif$\{0,\dots,n\}$ distributed and independent of X_1.

The limit law of Theorem 6.1 together with the uniform integrability of Lemma 7.1 implies convergence of all moments (Billingsley 1979, Theorem 25.12). Thus we have

$$\lim_{n\to\infty} \mathbb{E}\,(nS_{I_n}^n)^\ell = \int_0^\infty x^\ell e^{-x}\,dx = \ell!, \quad \ell = 0,1,2,\dots. \tag{1}$$

We turn to the analysis of the rescaled length of a spacing chosen according to into which spacing an indepedendent $U[0,1]$ random variable falls. For this we define the conditional distribution of the index J_n chosen by

$$\mathbb{P}(J_n = k \mid S_0^n, \dots, S_n^n) = S_k^n, \quad k = 0, \dots, n.$$

Then we have the following limit law:

Lemma 7.2. *We have*

$$nS_{J_n}^n \xrightarrow{\mathcal{L}} G_2, \quad (n \to \infty),$$

where G_2 is Gamma(2)-distributed, i.e., has Lebesgue density xe^{-x} on $[0,\infty)$.

8 Applications of spacings

The analysis of the random spacings generated by $X_1, \ldots, X_n$ can be used for the asymptotic analysis of parameters of the random suffix search tree. The leading order term of $\mathbb{E}\, D_n$ can be rediscovered using (1) with $\ell = 2$. This provides an alternative path to that followed in Theorem 5.1. The limit law for the size $N_{n,j}$ of the subtree rooted at X_j can be found for a large range of values j. This result is rooted in the lemmas of section 7.

Theorem 8.1. *The size $N_{n,j}$ of the subtree of the random suffix search tree of size n rooted at X_j satisfies for $j = j(n)$ with $j = o(n/\log^2 n)$ and $j/\log^5 n \to \infty$,*

$$\mathbb{E}\, N_{n,j} \sim \frac{2n}{j}, \quad \frac{j}{n} N_{n,j} \xrightarrow{\mathcal{L}} G_2,$$

as $n \to \infty$, where G_2 denotes the Gamma(2)-distribution.

It can be shown that in the case $j \sim \alpha n$ with $\alpha \in (0,1)$ the size $N_{n,j}$ tends in distribution to the negative binomial distibution with parameters $(2, \alpha)$, given by its generating function $s \mapsto (\alpha/(1-(1-\alpha)s))^2$.

9 Appendix

Lemma 9.1. *Let I be an interval in $[0,1]$ of length $|I|$. Then for all $1 \le i \le -\log_2 |I|$ we have*

$$\mathbb{P}(X_1, X_{1+i} \in I) \le \frac{|I|}{2^i}.$$

Lemma 9.2. *For all integer $1 \le i < j$, $t \ge 1$ and real $\varepsilon > 0$,*

$$\mathbb{P}(|X_i - X_j| \le \varepsilon) \le 2\varepsilon, \quad \mathbb{P}(|Y_i^{\langle t \rangle} - Y_j^{\langle t \rangle}| \le \varepsilon) \le 8\varepsilon.$$

Lemma 9.3. *For all integer $1 \le i < j$, $t \ge 1$ and real $\varepsilon > 0$, and U being $U[0,1]$ distributed and independent of $X_1, Y_i^{\langle t \rangle}, Y_j^{\langle t \rangle}$ we have*

$$\mathbb{P}(X_i, X_j \in [U, U+\varepsilon]) \le 2\varepsilon^2, \quad \mathbb{P}(Y_i^{\langle t \rangle}, Y_j^{\langle t \rangle} \in [U, U+\varepsilon]) \le 8\varepsilon^2.$$

Lemma 9.4. *For any Borel set $A \subseteq [0,1]$, real $\varepsilon, \delta > 0$, integer $i \ge 0$, and U being $U[0,1]$ distributed we have*

$$\mathbb{P}(\lambda(T^{-i}((U, U+\varepsilon)) \cap A) \ge \delta) \le \frac{\varepsilon \lambda(A)}{\delta},$$

where $\lambda(\,\cdot\,)$ denotes Lebesgue measure.

Lemma 9.5. *For all $n \ge 1$, $a \in [0,1)$, and $\Delta \in (0, 1/\sqrt{n})$ with $a + \Delta \le 1$, we have*

$$\mathbb{P}\Big(Y_1^{\langle m \rangle}, \ldots, Y_{L/2}^{\langle m \rangle} \notin [a, a+\Delta]\Big) \le 1 - \frac{L\Delta}{4} + \frac{2L}{n},$$

where $L = \|\lfloor \log_2 n \rfloor\|$ and $m = 18L$.

References

[1] Antos, A. and Devroye, L. (2000) Rawa Trees. *Mathematics and Computer Science (Versailles, 2000)*, 3–15, Birkhäuser, Basel.

[2] Apostolico, A. (1985) The myriad virtues of suffix trees. *Combinatorial Algorithms on Words*, 85–96, Springer-Verlag.

[3] Billingsley, P. (1979) *Probability and Measure.* John Wiley, New York-Chichester-Brisbane.

[4] Chung, K. L. and Erdös, P. (1952) On the application of the Borel-Cantelli lemma. *Trans. Amer. Math. Soc.* **72**, 179–186.

[5] Crochemore, M. and Rytter, W. (1994) *Text Algorithms.* Oxford University Press, New York,

[6] Devroye, L. (1986) A note on the height of binary search trees. *Journal of the ACM* **33**, 489–498.

[7] Devroye, L. (1987) Branching processes in the analysis of the heights of trees. *Acta Inform.* **24**, 277–298.

[8] Devroye, L. (1994) On random Cartesian trees. *Random Structures Algorithms* **5**, 305–327.

[9] Devroye, L. and Goudjil, A. (1998) A study of random Weyl trees. *Random Structures Algorithms* **12**, 271–295.

[10] Devroye, L. and Neininger, R. (2002) Random suffix search trees. Technical Report, McGill University.

[11] Devroye, L., Szpankowski, W. and Rais, B. (1992) A note on the height of suffix trees. *SIAM Journal on Computing* **21**, 48–53.

[12] Farach, M. (1997) Optimal suffix tree construction with large alphabets. *IEEE Symp. Found. Computer Science*, 137—143.

[13] Farach, M. and Muthukrishnan, S. (1996) Optimal logarithmic time randomized suffix tree construction. *Proc. 23rd ICALP*, 550–561.

[14] Farach, M. and Muthukrishnan, S. (1997) An optimal, logarithmic time, randomized parallel suffix tree contruction algorithm. *Algorithmica* **19**, 331–353.

[15] Giancarlo, R. (1993) The suffix tree of a square matrix, with applications. *Proc. of the Fourth Annual ACM-SIAM Symposium on Discrete Algorithms*, 402–411.

[16] Giancarlo, R. (1995) A generalization of the suffix tree to square matrices, with applications. *SIAM Journal on Computing*, 520–562.

[17] Giegerich, R. and Kurtz, S. (1995) A comparison of imperative and purely functional suffix tree constructions. *Science of Computer Programming* **25**, 187–218.

[18] Giegerich, R. and Kurtz, S. (1997) From Ukkonen to McCreight and Weiner: a unifying view of linear-time suffix tree construction. *Algorithmica* **19**, 331–353.

[19] Gusfield, D. (1997) *Algorithms on Strings, Trees, and Sequences. Computer Science and Computational Biology.* Cambridge University Press, Cambridge.

[20] Jacquet, P., Rais, B. and Szpankowski, B. (1995) Compact suffix trees resemble PATRICIA tries: limiting distribution of depth. Technical Report RR-1995, Department of Computer Science, Purdue University.

[21] Karkkainen, J. (1995) Suffix cactus : a cross between suffix tree and suffix array. *Combinatorial Pattern Matching, Proc. 6th Symposium on Combinatorial Pattern Matching, CPM 95* **937**, 191–204.

[22] Knuth, D. E. (1973) *The Art of Computer Programming, Vol. 1: Fundamental Algorithms.* Addison-Wesley, Reading, Mass., 2nd Ed.

[23] Knuth, D. E. (1973) *The Art of Computer Programming, Vol. 3: Sorting and Searching.* Addison-Wesley, Reading, Mass.

[24] Kosaraju, S. (1994) Real-time pattern matching and quasi-real-time construction of suffix trees. *Proc. of the 26th Ann. ACM Symp. on Theory of Computing*, 310–316, ACM.

[25] Kurlberg, P. and Rudnick, Z. (1999) The distribution of spacings between quadratic residues. *Duke Jour. of Math.* **100**, 211–242.

[26] Mahmoud, H. M. (1992) *Evolution of Random Search Trees.* John Wiley, New York.

[27] Manber, U. and Myers, G. (1990) Suffix arrays: a new method for on-line string searches. *Proceedings of the First Annual ACM-SIAM Symposium on Discrete Algorithms*, 319–327. SIAM, Philadelphia.

[28] McCreight, E. M. (1976) A space-economical suffix tree construction algorithm. *Journal of the ACM* **23**, 262–272.

[29] Rudnick, Z. and Zaharescu, A. (2002) The distribution of spacings between fractional parts of lacunary sequences. *Forum Math.*, to appear.

[30] Sahinalp, S. C. and Vishkin, U. (1994) Symmetry breaking for suffix tree construction. *Proc. 26th Symp. on Theory of Computing*, 300–309.

[31] Stephen, G. A. (1994) *String Searching Algorithms.* World Scientific, Singapore.

[32] Szpankowski, W. (1993) A Generalized Suffix Tree and its (Un)Expected Asymptotic Behaviors. *SIAM Journal on Computing* **22**, 1176–1198.

[33] Szpankowski, W. (2001) *Average-Case Analysis of Algorithms on Sequences.* John Wiley, New York.

[34] Ukkonen, E. (1995) On-line construction of suffix trees. *Algorithmica* **14**, 249–260.

[35] Weiner, P. (1973) Linear pattern matching algorithms. *Proceedings 14th Annual Symposium on Switching and Automata Theory*, 1–11. IEEE Press, New York.

Luc Devroye and Ralph Neininger

School of Computer Science
McGill University
3480 University Street
Montreal, H3A 2K6
Canada
{luc, neiningr}@cs.mcgill.ca

Trends in Mathematics, © 2002 Birkhäuser Verlag Basel/Switzerland

On the Profile of Random Forests

Bernhard Gittenberger

ABSTRACT: *An approach via generating functions is used to derive multivariate asymptotic distributions for the number of nodes in strata of random forests. For a certain range for the strata numbers we obtain a weak limit theorem to Brownian motion as well. Moreover, a moment convergence theorem for the width of random forests is derived.*

1 Introduction

We consider the set $F(n, N)$ of random forests consisting of n vertices and N rooted trees which can be viewed as realizations of Galton-Watson branching processes with N initial particles and conditioned to have total progeny n. Such forests consist of simply generated trees according to Meir and Moon [20] and therefore they can easily be described by generating functions: Let $b(z) = \sum_{n\geq 0} b_{n,N} z^n$ denote the generating function for those forests. Then we have $b(z) = a(z)^N$ with $a(z) = z\varphi(a(z))$. Here $a(z)$ is the generating function for a single tree and $\varphi(t) = \sum_{n\geq 0} \varphi_n t^n$ is the generating function of an arbitrary sequence $(\varphi_k)_{k\geq 0}$ of nonnegative numbers with $\varphi_0 > 0$. In this setting $b_{n,N}$ can be viewed as the number of forests in $F(n, N)$, weighted according to the probability on $F(n, N)$, i.e., to each forest F is assigned a weight

$$\omega(F) = \prod_{k\geq 0} \varphi_k^{n_k(F)}$$

where $n_k(F)$ is the number of nodes with out-degree k. The φ_k are related to the offspring distribution ξ via $\mathbf{P}\{\xi = k\} = \tau^k \varphi_k / \varphi(\tau)$, with a positive number τ within the circle of convergence of $\varphi(t)$. This means that the probability that the realization CGW of a conditioned Galton-Watson process as described above (offspring ξ, N initial particles, and conditioned to total progeny n) equals a given forest $f \in F(n, N)$ is proportional to the weight of f, precisely, we have

$$\mathbf{P}\{\mathrm{CGW} = f\} = \omega(f) \Big/ \sum_{f\in F(n,N)} \omega(f).$$

Without loss of generality we may assume $\mathbf{E}\xi = 1$ which equivalently means that τ satisfies $\tau\varphi'(\tau) = \varphi(\tau)$. Then the variance of ξ can also be expressed in terms of $\varphi(t)$ and is given by

$$\sigma^2 = \frac{\tau^2 \varphi''(\tau)}{\varphi(\tau)}. \tag{1}$$

The height of a vertex x is defined by the number of edges comprising the unique path which connects x with the root of the tree containing x. We are interested in the profile of random forests, thus we define $L_{n,N}(k)$ to be the number of vertices at height k in a random forest in $F(n,N)$. First, let us mention that the average height of a random forest in $F(n,N)$ is proportional to $\sqrt{n}$ as $n\to\infty$ and $N=O(\sqrt{n})$, see [21] and [23] for special cases and [24] for general simply generated forests. Thus the most interesting range is $k=O(\sqrt{n})$. Pavlov [22] derived distributional results for various ranges of n,N,k for labeled trees. Different tree classes are treated in [4, 10] and results for other ranges can be found in [26]. For a survey of results on random forests we refer the reader to [25]. Theorems 5 and 6 in [22] give a formula for the limiting distribution as integral with respect to a two-dimensional probability distribution with explicit Fourier transform for the ranges $k/\sqrt{n}\to\alpha>0$ and $N=o(\sqrt{n})$ and $N\sim\sqrt{n}$ (cf. [19] for the random tree analogue). These theorems have been generalized by Pitman [27] who related the profile of simply generated random forests in the above mentioned range for n,N,k to stochastic differential equations and obtained a weak limit theorem

$$\left(\frac{2}{\sigma\sqrt{n}}L_{n,N}\left(\frac{2\kappa\sqrt{n}}{\sigma}\right),\kappa\geq 0\right)\xrightarrow{d}(X_{\alpha,\kappa},\kappa\geq 0) \tag{2}$$

if $2N/\sigma\sqrt{n}\to\alpha$, where $X_{\alpha,\kappa}$ can be characterized by a stochastic differential equation: Let β denote a Brownian motion and set

$$u(X)=\inf\left\{v:\int_0^v X_s\,ds=1\right\}.$$

Then Pitman [27] showed that for each $\alpha>0$ there exists a unique strong solution of the Itô SDE

$$X_0=\alpha,\quad dX_\kappa=\delta_\kappa(X)\,dv+2\sqrt{X\kappa}\,d\beta_\kappa;\qquad \kappa\in[0,u(X)),\quad X_\kappa\equiv 0 \text{ for } \kappa\geq u(X)$$

with

$$\delta_\kappa(X)=4-X_\kappa^2\left(1-\int_0^\kappa X_s\,ds\right)^{-1}.$$

This process can be identified as total local time of a Brownian bridge B of length one conditioned to have total local time α at level 0 (see [27]),

$$X_{\alpha,v}\stackrel{d}{=}(\ell_v(B)|\ell_0(B)=\alpha), \tag{3}$$

which coincides with a Brownian excursion local time if $\alpha=0$ (cf. the analogous results for random trees, see [6] for the combinatorial setting and [27] for the stochastic calculus setting).

In this paper we are interested in the behavior of $L_{n,N}(k)$ in low strata of random forests. Starting point is the following central limit theorem (see [22, 10, 4]):

Theorem 1.1. *Let $n\to\infty$, $N=O(\sqrt{n})$, and $k=o(N)$. Then*

$$\mathbf{P}\left\{\frac{L_{n,N}(k)-N}{\sigma\sqrt{Nk}}\leq x\right\}\to\frac{1}{2\pi}\int_{-\infty}^x e^{-u^2/2}\,du$$

for any fixed x.

In order to simplify the proofs in the following, let us define $L_{n,N}(t)$ also for noninteger t by linear interpolation:

$$L_n(t) = (\lfloor t \rfloor + 1 - t)L_n(\lfloor t \rfloor) + (t - \lfloor t \rfloor)L_n(\lfloor t \rfloor + 1), \quad t \geq 0.$$

This does not change the limit of the finite-dimensional distributions and simplifies the proof of tightness significantly, since we are dealing with continuous functions.

Theorem 1.1 suggests the convergence to a Gaussian limiting process. In fact we will show the following theorem.

Theorem 1.2. *Let $\varphi(t)$ be a generating function associated to a family of simply generated trees. Assume that $\varphi(t)$ has a positive or infinite radius of convergence R and $\zeta = \gcd\{i|\varphi_i > 0\} = 1$. Suppose that the equation $t\varphi'(t) = \varphi(t)$ has a minimal positive solution $\tau < R$ and that σ^2 defined by (1) is finite. Furthermore, let (c_n) be an arbitrary sequence satisfying $c_n \to \infty$ and $c_n = o(N)$. Moreover, assume $N = O(\sqrt{n})$. Then*

$$\left(\frac{1}{\sigma\sqrt{Nc_n}}\left(L_{n,N}(tc_n) - N\right), t \geq 0\right) \xrightarrow{d} W_t$$

where W_t is a standard Brownian motion.

The proof of this theorem is done by first deriving a limit theorem for the finite-dimensional distributions which is done in the next section. This will be established by describing the joint distribution by means of a suitable generating function (see [9] or [14] for a general background) and then determining an asymptotic formula (and thus the limiting distribution) by complex contour integration. Afterwards we have to prove tightness which is done in Section 3. Section 4 is devoted to higher strata of random forests, i.e., the case $c_n/\sqrt{n} \to \eta > 0$. The limiting process for this case has been completely characterized by Pitman [27] (see (3)), however, using the combinatorial scheme of Section 2 we can give more explicit expressions for the finite-dimensional distributions in terms of integral transforms for the characteristic functions. Moreover, due to a tight bound derived in Section 3, it is also possible to derive a moment convergence theorem for the node numbers at this range as well as for the width of random forests, which complements the weak limit theorem of Pitman [27] (cf. also [3] and [7] for the corresponding results for trees). In fact, we will show

Theorem 1.3. *Set $M_\alpha := \sup_{v\geq 0} X_{\alpha,v}$ and $w_{n,N} := \sup_k 2L_{n,N}(k)/\sigma\sqrt{n}$. If $n, N \to \infty$ such that $2N/\sigma\sqrt{n} \to \alpha > 0$, then we have for every $d > 0$*

$$\mathbf{E}w_{n,N}^d \to \mathbf{E}M_\alpha^d \text{ and } \mathbf{E}\left(\frac{2}{\sigma\sqrt{n}}L_{n,N}\left(\frac{2\kappa\sqrt{n}}{\sigma}\right)\right)^d \to \mathbf{E}X_{\alpha,\kappa}^d$$

2 The finite-dimensional distributions

We have to compute the joint distribution of $L_{n,N}(k_1), \ldots, L_{n,N}(k_d)$. This can be done by determining the quotient

$$\mathbf{P}\left\{L_n(k_1) = m_1, \ldots, L_n(k_d) = m_d\right\} = \frac{b_{k_1,m_1,k_2,m_2,\ldots,k_d,m_d,n,N}}{b_{n,N}},$$

where $b_{k_1,m_1,k_2,m_2,\dots,k_d,m_d,n,N}$ is the (weighted) number of forests in $F(n,N)$ with m_i nodes in stratum k_i for $i=1,\dots,d$. Therefore define first the generating function (see [6] for a more detailed description)

$$\sum_{m_1,\dots,m_d,n\geq 0} a_{k_1 m_1 k_2 m_2 \cdots k_d m_d n} u_1^{m_1}\cdots u_d^{m_d} z^n$$
$$= y_{k_1}\left(z, u_1 y_{k_2-k_1}\left(z,\dots y_{k_d-k_{d-1}}\left(z,u_d a(z)\right)\dots\right),\right.$$

where $a_{k_1 m_1 k_2 m_2 \cdots k_d m_d n}$ is the number of single trees with the above property and

$$y_0(z,u)=u, \qquad y_{i+1}(z,u)=z\varphi(y_i(z,u)), \quad i\geq 0.$$

Forests consisting of N trees can now be described by the Nth power of this function and thus the characteristic function of the joint distribution of $\frac{1}{\sigma\sqrt{Nc_n}}L_{n,N}(k_1)$, $\dots, \frac{1}{\sigma\sqrt{Nc_n}}L_{n,N}(k_d)$ is given by the coefficient

$$\phi_{k_1,\cdots,k_d,n,N}(t_1,\dots,t_d)$$
$$= \frac{1}{b_{n,N}}[z^n] y_{k_1}\left(z, e^{it_1/\sigma\sqrt{Nc_n}} y_{k_2-k_1}\left(z,\dots y_{k_d-k_{d-1}}\left(z, e^{it_d/\sigma\sqrt{Nc_n}} a(z)\right)\dots\right)^N\right. \tag{4}$$

where $[z^n]f(z)$ denotes the coefficient of z^n in the power series of $f(z)$.

In order to extract the desired coefficient we will need some lemmas. First we need the tree function and related functions (see [20] or [11]).

Lemma 2.1. *Let $z_0=1/\varphi'(\tau)$ be the point on the circle of convergence of $a(z)$ which lies on the positive real axis. Set $\alpha(z)=z\varphi'(a(z))$ and $\beta(z)=z\varphi''(a(z))$ and assume* $\arg(z-z_0)\neq 0$. *Then the following local expansions hold:*

$$a(z)=\tau-\frac{\tau\sqrt{2}}{\sigma}\sqrt{1-\frac{z}{z_0}}+O\left(\left|1-\frac{z}{z_0}\right|\right) \quad \text{as } z\to z_0$$

$$\alpha(z)=1-\sigma\sqrt{2}\sqrt{1-\frac{z}{z_0}}+O\left(\left|1-\frac{z}{z_0}\right|\right) \quad \text{as } z\to z_0$$

$$\beta(z)=\frac{\sigma^2}{\tau}+O\left(\sqrt{1-\frac{z}{z_0}}\right) \quad \text{as } z\to z_0.$$

The previous two lemmas immediately imply

$$b_{n,N}=\frac{N\tau^N}{\sigma z_0^n\sqrt{2\pi n^3}}\left(\exp\left(-\frac{N^2}{2n\sigma^2}\right)+O\left(\frac{1}{\sqrt{n}}\right)\right) \tag{5}$$

We will need an expansion of the bivariate generating function $y_k(z,u)$ as well (see [15], cf. also [6, Lemmas 2.1 and 3.1]).

Lemma 2.2. *Set $w=u-a(z)$, If $w\to 0$ and $z-z_0\to 0$ in such a way that* $\arg(z-z_0)\neq 0$ *and $|1-\sqrt{z-z_0}|\leq 1+O\left(n^{-1/2}\right)$, then $y_k(z,u)$ admits the local representation*

$$y_k(z,u)=a(z)+\frac{\alpha^k(z)w}{1-\frac{\beta(z)}{2\alpha(z)}\frac{1-\alpha^k(z)}{1-\alpha(z)}w+O\left(\left|\frac{1-\alpha^{2k}(z)}{1-\alpha^2(z)}\right||w|^2\right)} \tag{6}$$

uniformly for $k=O(1/|w|)$.

With the help of these asymptotic expansions we can prove the convergence of the finite-dimensional distributions to a Gaussian limiting distribution now.

Theorem 2.3. *Let $n \to \infty$, $N = O(\sqrt{n})$ and $c_n \to \infty$ such that $c_n = o(N)$. Moreover, set*

$$X_\kappa := \frac{L_{n,N}(\kappa c_n) - N}{\sigma\sqrt{N c_n}}.$$

Then the joint distribution of $X_{\kappa_1}, \dots, X_{\kappa_d}$ converges to a centered Gaussian distribution with covariance $\mathbf{Cov}(X_s, X_t) = \min(s,t)$.

Proof: We have to show that the characteristic function of the centered joint distribution of $\frac{1}{\sigma\sqrt{Nc_n}}L_n(k_1), \dots, \frac{1}{\sigma\sqrt{Nc_n}}L_n(k_d)$ for $k_j = \lfloor \kappa_j c_n \rfloor$, $j = 1, \dots, d$ satisfies

$$\lim_{n\to\infty} \exp\left(-i\frac{\sqrt{N}}{\sigma\sqrt{c_n}}\sum_{j=1}^{d} t_j\right)\phi_{k_1,\cdots,k_d,n,N}(t_1,\dots,t_d)$$
$$= \exp\left(-\sum_{j=1}^{d}\frac{\kappa_j t_j^2}{2} - \sum_{j,\ell=1;j<\ell}^{d}\kappa_j t_j t_\ell\right) \qquad (7)$$

Therefore we apply Cauchy's integral formula on (4) with the integration contour $\Gamma = \Gamma_1 \cup \Gamma_2 \cup \Gamma_3 \cup \Gamma_4$ where

$$\Gamma_1 = \left\{ z = z_0\left(1 + \frac{x}{n}\right) \middle| \Re x \le 0 \text{ and } |x| = 1 \right\}$$
$$\Gamma_2 = \left\{ z = z_0\left(1 + \frac{x}{n}\right) \middle| \Im x = 1 \text{ and } 0 \le \Re x \le n^{1/3} \right\}, \qquad \Gamma_3 = \overline{\Gamma}_2$$
$$\Gamma_4 = \left\{ z : |z| = z_0\left|1 + \frac{\log^2 n + i}{n}\right| \text{ and } \arg\left(1 + \frac{\log^2 n + i}{n}\right) \le |\arg(z)| \le \pi \right\}. \qquad (8)$$

Let us first study the contribution of $\gamma = \Gamma_1 \cup \Gamma_2 \cup \Gamma_3$ which will turn out to be the main term. For notational convenience, let us abbreviate the second term in (6) by

$$R_k := R_k(z,u) = \frac{\alpha^k w}{1 - \frac{\beta}{\alpha}\frac{1-\alpha^k}{1-\alpha}w + O\left(\left|\frac{1-\alpha^{2k}}{1-\alpha^2}\right||w|^2\right)} \qquad (9)$$

and let us omit the function arguments z, u and so forth whenever there is no ambiguity. Furthermore, set $u_j = e^{it_j/\sigma\sqrt{Nc_n}}$ and $w_j = (u_j - 1)a$. Since on γ the equation $|1 - \sqrt{-x/n}| = 1 + O\left(n^{-2/3}\right)$ is valid, the assumptions of Lemma 2.2 are fulfilled. Thus we have on γ

$$y_{k_1}^N = y_{k_1}\left(z, u_1 y_{k_2-k_1}\left(z, \dots y_{k_d-k_{d-1}}\left(z, u_d a(z)\right)\dots\right)^N\right.$$
$$= a^N\left(1 + \frac{\alpha^{k_1}(w_1 + u_1 R_{k_2-k_1})/a}{1 - \frac{\beta}{\alpha}\frac{1-\alpha^{k_1}}{1-\alpha}(w_1 + u_1 R_{k_2-k_1}) + O\left(\frac{1}{N}\right)}\right)^N \qquad (10)$$

Here $R_{k_2-k_1} = R_{k_2-k_1}\left(z, u_2 y_{k_2-k_1}\left(z, u_3 y_{k_3-k_2}(z, \dots y_{k_d-k_{d-1}}\left(z, u_d a(z)\right)\dots\right)\right)$. Expanding the second factor and using the asymptotic relations

$$u_j = 1 + \frac{it_j}{\sigma\sqrt{Nc_n}} + O\left(\frac{1}{Nc_n}\right) \tag{11}$$

$$w_j = \tau\left(\frac{it_j}{\sigma\sqrt{Nc_n}} - \frac{t_j^2}{2Nc_n\sigma^2}\right) + O\left(\frac{1}{N^{3/2}c_n^{3/2}}\right) \tag{12}$$

$$\alpha^k = 1 + O\left(\frac{k\sqrt{|x|}}{\sqrt{n}}\right) = 1 + O\left(\frac{c_n\sqrt{|x|}}{\sqrt{n}}\right) \quad \text{for } k = O(c_n) \tag{13}$$

as well as those in Lemma 2.1 yield

$$\begin{aligned} y_{k_1}^N &= a^N \exp\left(N\alpha^{k_1}\frac{w_1}{a} + N\frac{\alpha^{k_1}u_1 R_{k_2-k_1}}{a}\right. \\ &\quad \left. + N\left(\alpha^{k_1-1}\beta\frac{1-\alpha^{k_1}}{1-\alpha} - \frac{\alpha^{2k_1}}{2a^2}\right)\left(w_1^2 + 2u_1 w_1 R_{k_2-k_1} + u_1^2 R_{k_2-k_1}^2\right) + O\left(k_1 N w_1^3\right)\right) \\ &= a^N \exp\left(it_1\sqrt{\frac{N}{c_n\sigma^2}} - \frac{t_1^2}{2c_n\sigma^2} + N\left(1 + \frac{it_1}{\sigma\sqrt{Nc_n}}\right)\frac{R_{k_2-k_1}}{\tau}\right. \\ &\quad + \left(\frac{k_1\sigma^2}{2} - \frac{1}{2}\right)\left(-\frac{t_1^2}{c_n\sigma^2} + 2it_1\sqrt{\frac{N}{c_n\sigma^2}}\frac{R_{k_2-k_1}}{\tau} + N\left(1 + \frac{it_1}{\sigma\sqrt{Nc_n}}\right)\frac{R_{k_2-k_1}^2}{\tau^2}\right) \\ &\quad \left. + O\left(\frac{k_1}{\sqrt{Nc_n^3}} + \frac{k_1\sqrt{N}}{\sqrt{nc_n}}\sqrt{|x|}\right)\right) \end{aligned} \tag{14}$$

Now observe that for $\ell \le d$ we have by (9) and (10) as well as the asymptotic expansions (11)–(13)

$$\begin{aligned} \frac{R_{k_\ell-k_{\ell-1}}}{\tau} &= \frac{\alpha^{k_\ell-k_{\ell-1}}(w_\ell + u_{\ell+1}R_{k_{\ell+1}-k_\ell})/\tau}{1 - \frac{\beta}{\alpha}\frac{1-\alpha^{k_\ell-k_{\ell-1}}}{1-\alpha}\left(w_\ell + u_\ell R_{k_{\ell+1}-k_\ell}\right) + O\left(\frac{k}{Nc_n}\right)} \\ &= \left(\frac{it_\ell}{\sigma\sqrt{Nc_n}} - \frac{t_\ell^2}{2Nc_n\sigma^2} + \left(1 + \frac{it_\ell}{\sigma\sqrt{Nc_n}}\right)\frac{R_{k_{\ell+1}-k_\ell}}{\tau} + O\left(\frac{1}{\sqrt{N^3c_n^3}} + \sqrt{\frac{|x|}{n}}\right)\right) \\ &\quad \times\left(1 + \frac{(k_\ell - k_{\ell-1})\sigma^2}{2}\left(\frac{it_2}{\sigma\sqrt{Nc_n}} + \frac{R_{k_{\ell+1}-k_\ell}}{\tau}\right) + O\left(\frac{1}{N}\right)\right) \end{aligned}$$

and $R_{k_\ell-k_{\ell-1}} \equiv 0$ for $\ell > d$. Thus $R_{k_\ell-k_{\ell-1}} = O\left(1/\sqrt{Nc_n}\right)$ and in particular

$$R_{k_d-k_{d-1}} = \left(\frac{it_d\tau}{\sigma\sqrt{Nc_n}} - \frac{t_d^2\tau}{2Nc_n\sigma^2}\right)\left(1 + \frac{it_d(k_d - k_{d-1})\sigma^2}{2\sigma\sqrt{Nc_n}}\right) + O\left(\frac{1}{\sqrt{N^3c_n}} + \sqrt{\frac{|x|}{n}}\right).$$

Plugging the expressions for $R_{k_2-k_1}$ into (14) yields

$$y_{k_1}^N = a^N \exp\left(it_1\sqrt{\frac{N}{c_n\sigma^2}} - \frac{t_1^2}{2c_n\sigma^2} + it_2\sqrt{\frac{N}{c_n\sigma^2}} - \frac{t_2^2}{2c_n\sigma^2} - \frac{t_1t_2}{\sigma^2 c_n} \right.$$
$$+ N\left(1 + \frac{it_1}{\sigma\sqrt{Nc_n}} + \frac{it_2}{\sigma\sqrt{Nc_n}}\right)\frac{R_{k_3-k_2}}{\tau} + N\frac{(k_2-k_1)\sigma^2}{2}\left(\frac{it_2}{\sigma\sqrt{Nc_n}} + \frac{R_{k_3-k_2}}{\tau}\right)^2$$
$$+ \left(\frac{k_1\sigma^2-1}{2}\right)\left(-\frac{t_1^2}{c_n\sigma^2} - \frac{2t_1t_2}{c_n\sigma^2} + \sqrt{\frac{N}{c_n\sigma^2}}\frac{2it_1R_{k_3-k_2}}{\tau}\right.$$
$$\left.\left. + N\left(\frac{it_2}{\sigma\sqrt{Nc_n}}\frac{R_{k_3-k_2}}{\tau}\right)^2\right) + O\left(\frac{1}{\sqrt{Nc_n}} + \sqrt{\frac{Nc_n}{n}}\sqrt{|x|}\right)$$

and then by substituting $R_{k_3-k_2}, R_{k_4-k_3}, \ldots$, step by step we arrive at

$$y_{k_1}^N = a^N \exp\left(i(t_1+\cdots+t_d)\sqrt{\frac{N}{c_n\sigma^2}} - \frac{k_1t_1^2+\cdots+k_dt_d^2}{2c_n} - \sum_{j_1=1}^{d-1}\sum_{j_2=j_1+1}^{d}\frac{k_{j_1}t_{j_1}t_{j_2}}{c_n} \right.$$
$$\left. + O\left(\frac{1}{\sqrt{Nc_n}} + \sqrt{\frac{Nc_n}{n}}\sqrt{|x|}\right)\right) \tag{15}$$

If we substitute $z = z_0(1+x/n)$ on γ, we get

$$\frac{[z^n]y_{k_1}^N}{b_{n,N}} = \frac{1}{2\pi i b_{n,N} z_0^n n}\left(\int_\gamma y_{k_1}^N e^{-x}\,dx\left(1+O\left(\frac{1}{n^{1/3}}\right)\right)\right) + O\left(\frac{1}{b_{n,N}}\int_{\Gamma_4} y_{k_1}^N\frac{dz}{z^{n+1}}\right). \tag{16}$$

Moreover, observe that for any $M > 0$ we have

$$\frac{1}{2\pi i}\int_{\gamma'} e^{-\lambda\sqrt{-x}-x}\,dx = \frac{\lambda}{2\sqrt{\pi}}e^{-\lambda^2/4} + O\left(e^{-M}\right)$$

for $\gamma' = \{x: \ |x| = 1, \Re x \le 0\} \cup \{x : 0 \le \Re x \le M, \Im x = \pm 1\}$, as can be easily seen by substituting $u^2 = x$. Thus, since $\sqrt{Nc_n/n} = o(1)$, the error terms in (15) are negligibly small, and hence the first term in (16) in conjunction with (15) yields (7).

So let us estimate the second term in (16). By Taylor's theorem we have

$$y_{k_d-k_{d-1}}(z, ua(z))^N = a^N\left(1 + \alpha(z)^{k_d-k_{d-1}}(u_d-1)a(z) + O\left((u_d-1)^2\right)\right)^N.$$

Since we required $\zeta = 1$ (see Theorem 1.2), we get

$$\max_{z\in\Gamma_4}|\alpha(z)| = |\alpha(\tilde z)| \quad \text{and} \quad \max_{z\in\Gamma_4}|a(z)| = |a(\tilde z)|,$$

where $\tilde z \in \gamma \cap \Gamma_4$. There the local expansions of Lemma 2.1 are still valid and hence $|\alpha| < 1$ and $|a| < \tau$. Consequently, with $u = e^{it_d/\sigma\sqrt{Nc_n}}$ we get

$$\left|y_{k_d-k_{d-1}}(z, ua(z))^N\right| = \tau^N \exp\left(\sqrt{\frac{N}{c_n}} + O\left(\frac{1}{c_n^2}\right)\right).$$

Inserting this into $y_{k_\ell - k_{\ell-1}}$, $\ell = 2, \ldots, d-1$, and y_{k_1} and arguing as above, we get the same estimate for $y_{k_1}^N$. Finally, using $|z|^{-n} \sim z_0^{-n} \exp\left(-n^{1/3}\right)$ implies the existence of some positive constant C such that

$$\int_{\Gamma_4} \left|y_{k_1}^N\right| \frac{|dz|}{|z^{n+1}|} = O\left(\exp\left(-n^{1/3} + Cn^{1/4}\right)\right)$$

which is exponentielly small compared to the integral over γ and the proof is complete. ◇

3 Tightness

In order to complete the proof of Theorem 1.2 we have to show that the sequence of random variables $L_{n,N}(c_n t)/\sigma\sqrt{Nc_n}$, $t \geq 0$, is tight in $\mathrm{C}[0,\infty)$. By [18, Theorem 4.10] it suffices to establish tightness in $\mathrm{C}[0,T]$. Thus by [1, Theorem 12.3] we only have to show that $L_{n,N}(0)$ is tight, which is obviously true, and that there exist constants $\alpha > 1$, $\beta \geq 0$, and $C > 0$ such that

$$\mathbf{P}\left\{|L_{n,N}(\rho c_n) - L_{n,N}((\rho+\theta)c_n)| \geq \varepsilon\sigma\sqrt{Nc_n}\right\} \leq C\frac{\theta^\alpha}{\varepsilon^\beta} \tag{17}$$

This inequality follows from the following theorem.

Theorem 3.1. *There exists a constant $C > 0$ such that for all $r, h \geq 0$ and for $N = O\left(\sqrt{n}\right)$ the following inequality holds:*

$$\mathbf{E}\left|L_{n,N}(r+h) - L_{n,N}(r)\right|^4 \leq CN^2h^2 \tag{18}$$

In order to show this inequality we will investigate a more general situation. First, observe that the left-hand side can be represented by the coefficient of a proper generating function. In fact we have

$$\mathbf{E}\left(L_{n,N}(r) - L_{n,N}(r+h)\right)^4 = \frac{1}{b_{n,N}}[z^n]H_{rh}^{(4)}(z),$$

where

$$\begin{aligned} H_{rh}^{(4)}(z) &= \left(u\frac{\partial}{\partial u}\right)^4 y_r(z, uy_h(z, u^{-1}a(z)))^N\bigg|_{u=1} \\ &= \left[\left(\frac{\partial}{\partial u} + 7\frac{\partial^2}{\partial u^2} + 6\frac{\partial^3}{\partial u^3} + \frac{\partial^4}{\partial u^4}\right) y_r(z, uy_h(z, u^{-1}a(z)))^N\right]_{u=1}. \end{aligned} \tag{19}$$

Since $b_{n,N} \sim (N\tau^N/\sqrt{2\pi\sigma^2}) z_0^{-n} n^{-3/2} \exp(-N^2/2n\sigma^2)$ (see (5)), (18) is valid if

$$[z^n]H_{rh}^{(4)}(z) = O\left(\frac{N^3\tau^N h^2}{z_0^n n^{3/2}}\right) \tag{20}$$

holds uniformly for $r, h \geq 0$. We will estimate this coefficient by analyzing the function $H_{rh}(z)$ and using Flajolet and Odlyzko's [13] transfer lemma:

Lemma 3.2. *Let $F(z)$ be analytic in Δ defined by*

$$\Delta = \{z : |z| < z_0 + \eta,\ |\arg(z - z_0)| > \vartheta\},$$

where z_0 and η are positive real numbers and $0 < \vartheta < \pi/2$. Furthermore suppose that there exists a real number β such that

$$F(z) = O\left((1 - z/z_0)^{-\beta}\right) \qquad (z \in \Delta).$$

Then

$$[z^n]F(z) = O\left(z_0^{-n} n^{\beta-1}\right).$$

Set $Y_{rh}(z,u) = y_r(z, u y_h(z, u^{-1}a(z)))$. We analyze the derivatives of $Y_{rh}(z,u)$ with respect to u in the next lemma.

Lemma 3.3. *Let Δ be the domain defined in Lemma 3.2. Then there exists a finite index set I and functions $\alpha_{i\ell rh}(z)$ such that for all $\ell > 0$*

$$\frac{\partial^\ell}{\partial u^\ell} Y_{rh}(z,1) = \sum_{i \in I} \alpha_{i\ell rh}(z), \tag{21}$$

where the functions $\alpha_{ilrh}(z)$ satisfy for $z \in \Delta$

$$\alpha_{i\ell rh}(z) = O\left(|\alpha^r| \left| \frac{1-\alpha^r}{1-\alpha} \right|^{\mu_1} \left| \frac{1-\alpha^h}{1-\alpha} \right|^{\mu_2} |1-\alpha|^{\mu_3} \right)$$

for some nonnegative integers μ_1, μ_2, μ_3 with $\mu_1 + \mu_2 - \mu_3 \le \ell - 1$.

Proof: First compute the first few derivatives of $\frac{\partial^\ell}{\partial u^\ell} y_r(z, a(z))$,

$$\frac{\partial y_r}{\partial u}(z, a(z)) = \alpha^r, \quad \frac{\partial^2 y_r}{\partial u^2}(z, a(z)) = \frac{\beta}{\alpha}\alpha^r \frac{1-\alpha^r}{1-\alpha},$$
$$\frac{\partial^3 y_r}{\partial u^3}(z, a(z)) = \frac{\tilde\beta}{\alpha}\alpha^r \frac{1-\alpha^{2r}}{1-\alpha^2} + 3\frac{\beta^2}{\alpha}\alpha^r \frac{(1-\alpha^r)(1-\alpha^{r-1})}{(1-\alpha)(1-\alpha^2)},$$

where $\beta = z\varphi''(a(z))$ and $\tilde\beta = z\varphi'''(a(z))$. Noticing that Faà di Bruno's formula (see e.g. [5]) gives

$$\frac{\partial^\ell y_r}{\partial u^\ell}(z,1) = \sum_{\sum_{i=1}^{\ell-1} i k_i = \ell} \frac{\ell! z \varphi^{(k_1+\cdots+k_{\ell-1})}(a(z))}{k_1! \cdots k_{\ell-1}!} \prod_{j=1}^{\ell-1} \left(\frac{1}{j!} \frac{\partial^j y_{r-1}}{\partial u^j} \right)^{k_j} + \alpha(z) \frac{\partial^\ell y_{r-1}}{\partial u^\ell}(z,1).$$

and that hence $\frac{\partial^\ell}{\partial u^\ell} y_r(z,1)$ is the solution of an inhomogeneous first order linear recurrence, the estimate

$$\frac{\partial^\ell}{\partial u^\ell} y_r(z,1) = O\left(|\alpha^r| \left| \frac{1-\alpha^r}{1-\alpha} \right|^{l-1} \right) \tag{22}$$

is now easily proved by induction.

Now, setting $D = \partial/\partial u$ and employing again Faà di Bruno's formula, this time to Y_{rh}, yields

$$\begin{aligned}\frac{\partial^\ell Y_{rh}}{\partial u^\ell}(z,1) &= \sum_{\sum ik_i=\ell} \frac{\ell!}{k_1!\cdots k_\ell!} D^{k_1+\cdots+k_\ell} y_r(z,a(z)) \prod_{j=1}^{\ell} \left(\frac{1}{j!} D^j \left(u y_h\left(z, \frac{a(z)}{u}\right)\right)\right)^{k_j} \\ &= \sum_{\sum ik_i=\ell} \frac{\ell!}{k_1!\cdots k_\ell!} D^{k_1+\cdots+k_\ell} y_r(z,a(z)) \prod_{i=1}^{\ell} (A_i - A_{i-1})^{k_i}\end{aligned}$$

where

$$A_i = (-1)^i \sum_{\sum jm_j=i} \frac{i!}{m_1!\cdots m_i!} D^{m_1+\cdots+m_i} y_h(z,1).$$

Since

$$D u y_h(z, a(z)/u)\big|_{u=1} = 1-\alpha^h$$

we get by (22)

$$\begin{aligned}&D^{k_1+\cdots+k_\ell} y_r(z,1) \left(D u y_h(z,a(z)/u)\big|_{u=1}\right)^{k_1} \prod_{j=2}^{\ell} \left(\frac{1}{j!}\, D^j u y_h(z,a(z)/u)\big|_{u=1}\right)^{k_j} \\ &\quad = O\left(|\alpha^r| \left|\frac{1-\alpha^r}{1-\alpha}\right|^{k_1+\cdots+k_\ell-1} |1-\alpha|^{k_1} \left|\frac{1-\alpha^h}{1-\alpha}\right|^{k_1+\sum(i-1)k_i}\right).\end{aligned}$$

Note that we omitted a factor α^h coming from y_h. This is justified since $|\alpha|<1$ in Δ. So we could also neglect the factor α^r but this one is needed in the sequel. If we set $\mu_1 = \sum_i k_i - 1$, $\mu_2 = k_1 + \sum_i (i-1)k_i$, and $\mu_3 = k_1$, then obviously $\mu_1 + \mu_2 - \mu_3 = \ell - 1$ which yields (21) and completes the proof. ◇

For the tightness inequality (18) we need the derivatives of Y_{rh}^N. These are investigated in the next lemma.

Lemma 3.4. *There exist bounded functions $\beta_{\ell,M}$ on Δ such that*

$$\left(u\frac{\partial}{\partial u}\right)^M Y_{rh}^N = \sum_{\ell=1}^{M} \beta_{\ell,M}(z) N^\ell a(z)^{n-\ell} \prod_i \left(\frac{\partial^i Y_{rh}}{\partial u^i}\right)^{c_i}$$

where the c_i satisfy $\sum_i (i-1)c_i \le M-\ell$.

Proof: Faà di Bruno's formula yields

$$(uD)^M Y_{rh}^N = \sum_{\sum ik_i=M} \frac{M!N!}{k_1!\cdots k_M!(N-\sum_i k_i)!} Y_{rh}^{N-\sum_i k_i} \prod_{j=1}^{\ell} \left(\frac{1}{j!}\frac{\partial^j Y_{rh}}{\partial u^j}\right)^{k_j}$$

and because of $Y_{rh}(z,1) = a(z)$ and $\sum_i (i-1)k_i = M-\ell$ we are done. ◇

Now we are able to prove Theorem 3.1:

Proof of Theorem 3.1. By Lemmas 3.3 and 3.4, all terms of $H^{(4)}_{rh(z)}$ are bounded by functions of the form

$$\begin{aligned} &N^{\ell}|\alpha^{r}||a|^{N}\left|\frac{1-\alpha^{r}}{1-\alpha}\right|^{\mu_1}\left|\frac{1-\alpha^{h}}{1-\alpha}\right|^{\mu_2}|1-\alpha|^{\mu_3} \\ &= O\left(N^{\ell}\tau^{N}|\alpha^{r}||1-\alpha^{r}|^{\mu_1}|1-\alpha^{h}|^{\mu_2-d}h^{d}|1-\alpha|^{\mu_3-\mu_1-\mu_2+d}\right), \end{aligned} \tag{23}$$

where $\mu_3-\mu_1-\mu_2 \geq 2d-\ell$ and $d=2$. Thus by Lemma 3.2 and the fact that $N = O(\sqrt{n})$ we get

$$[z^n]H_{rh}(z) = O\left(\frac{\tau^N N^{\ell} h^2}{z_0^n n^{1+(\ell-2)/2}}\right) = O\left(\frac{\tau^N N^3 h^2}{z_0^n n^{3/2}}\right)$$

as desired. ◇

4 The profile in the range $c_n/\sqrt{n} \to \eta > 0$ and the width of random forests

Equations (2) and (3) characterize the distributions of $L_{n,N}(k)$ in the range $k \approx \sqrt{n}$ by a limiting process given implicitly by a stochastic differential equation and by conditioning a well known process. The same ideas as in Section 2 allow us to make the distributions more explicit, leading to a representation in terms of an integral transform for the characteristic functions of the finite-dimensional distributions. Starting again with (6) we get as above

$$y_{k_1}^N = a^N\left(1+\frac{\alpha^{k_1}(w_1+u_1R_{k_2-k_1})/a}{1-\frac{\beta}{\alpha}\frac{1-\alpha^{k_1}}{1-\alpha}(w_1+u_1R_{k_2-k_1})+O\left(\frac{1}{N}\right)}\right)^N.$$

Insert the asymptotic approximations for $u_j = e^{2it_j/\sigma\sqrt{n}}$, w_j, β,

$$\begin{aligned} a^N &= \tau^N \exp\left(-\frac{N\sqrt{-2x}}{\sigma\sqrt{n}} + O\left(\frac{N}{n}\right)\right), \\ \alpha^k &= \exp\left(-2\kappa\sqrt{-2x} + O\left(\frac{|x|}{\sqrt{n}}\right)\right), \end{aligned}$$

for $k = 2\kappa\sqrt{n}/\sigma$ we finally arrive at

$$\begin{aligned} y_{k_1}^N =&\tau^N \exp\left(\frac{N\sqrt{-x}\exp\left(-\kappa_1\sqrt{-2x}\right)\left(2it_1/\sigma\sqrt{n}+R_{k_2-k_1}/\tau\right)}{\sqrt{-x}\exp\left(\kappa_1\sqrt{-2x}\right)-(it_1\sqrt{2}+(\sigma\sqrt{n}/\tau\sqrt{2})R_{k_2-k_1}\sinh\left(\kappa_1\sqrt{-2x}\right)}\right. \\ &\left. + -\frac{N\sqrt{-2x}}{\sigma\sqrt{n}} + O\left(\frac{N}{n}\right)\right) \end{aligned}$$

Error estimation for Γ_4 works similar as in Section 2 and therefore we get the following theorem.

Theorem 4.1. *Assume $2N/\sigma\sqrt{n} \to \alpha > 0$. Furthermore, let $k_j = 2\kappa_j\sqrt{n}/\sigma$. Then the characteristic function of the joint distribution of $\frac{2}{\sigma\sqrt{n}}L_{n,N}(k_1), \dots, \frac{2}{\sigma\sqrt{n}}L_{n,N}(k_d)$, satisfies*

$$\tilde{\phi}_{k_1,\cdots,k_d,n,N}(t_1,\dots,t_d) = \frac{\sqrt{2}}{i\alpha\sqrt{\pi}}\int_{\gamma'} \exp\Big(-x - \alpha\sqrt{-x/2}$$
$$+ \Psi_{\kappa_1}(x, it_1 + \Psi_{\kappa_2-\kappa_1},(\dots \Psi_{\kappa_{p-1}-\kappa_{p-2}}(x, it_{p-1} + \Psi_{\kappa_d-\kappa_{d-1}}(x, it_d))\cdots)\Big)\, dx$$

with

$$\Psi_\kappa(x,t) = \frac{\alpha t\sqrt{-x}e^{-\kappa\sqrt{-2x}}}{\sqrt{-x}e^{\kappa\sqrt{-2x}} - t\sqrt{2}\sinh\left(\kappa\sqrt{-2x}\right)}.$$

Now we turn to the width: The structure of the functions in the previous section allows us to prove an even tighter bound for the moments of $L_{n,N}(r+h) - L_{n,N}(r)$ with the help of the following lemma (cf. [7] and [16, Lemma 3.5]).

Lemma 4.2. *Let $f(z)$ and $g(z)$ be analytic functions in Δ which satisfy*

$$|f(z)| \le \exp\left(-C\sqrt{\left|1 - \frac{z}{z_0}\right|}\right), \quad z \in \Delta,$$
$$g(z) = 1 - D\sqrt{1 - \frac{z}{z_0}} + O\left(1 - \frac{z}{z_0}\right), \quad z \in \Delta,$$

for some positive constants C, D. Then for any fixed ℓ there exists a constant $C' > 0$ such that

$$[z^n]\frac{f(z)^r}{(1-g(z))^\ell} = O\left(e^{-C'r/\sqrt{n}}n^{(\ell-2)/2}\right)$$

uniformly for all $r, n \ge 0$.

Theorem 4.3. *For every fixed positive integer d there exist constants c_1, c_2 such that for every $r, h > 0$*

$$\mathbf{E}\left|L_{n,N}(r) - L_{n,N}(r+h)\right|^{2d} \le c_1 e^{-c_2 r/\sqrt{n}} h^d n^{d/2}. \tag{24}$$

The constants c_1 and c_2 are independent of n and N, provided that $N = O(\sqrt{n})$

Proof: Since

$$\left(u\frac{\partial}{\partial u}\right)^{2d} = \sum_{k=1}^{2d} s_{2d,k}u^k\left(\frac{\partial}{\partial u}\right)^k,$$

where $s_{n,k}$ are the Stirling numbers of the second kind, we can apply Lemmas 3.3 and 3.4 directly to $H_{rh}^{(2d)}$ and get (23). Keep in mind that $\alpha(z)$ admits a representation like $g(z)$ in Lemma 4.2 due to Lemma 2.1 and thus there exists a constant $C > 0$ such that in Δ the inequality $|\alpha(z)| \le \exp\left(-C\sqrt{|1 - z/z_0|}\right)$ holds. Hence we obtain

$$\mathbf{E}\left|L_{n,N}(r)-L_{n,N}(r+h)\right|^{2d} = \frac{1}{b_{n,N}}[z^n]\left(u\frac{\partial}{\partial u}\right)^{2d} Y_{rh}(z,1)$$
$$= O\left(\frac{1}{b_{n,N}} e^{-c_2 r/\sqrt{n}} \frac{N\tau^N h^d n^{(d-3)/2}}{z_0^n}\right)$$

and this immediately implies (24). ◇

By [8, Theorem 1], this property in conjunction with the fact that there exists a $t \geq 0$ (in fact we can choose $t=0$) such that $|\sup_n \mathbf{E}(L_{n,N}(t)/\sqrt{n})^k| < \infty$ for all $k \geq 0$ and for $2N/\sigma\sqrt{n} \in [\alpha - f_n, \alpha + f_n]$ with $\alpha > 0$, $f_n \to 0$ implies that the sequence $(L_{n,N}(\cdot)/\sqrt{n})_{n>0}$ is polynomially convergent in the sense of Drmota and Marckert [8], i.e.,

$$\mathbf{E}F\left(\left(\frac{2}{\sigma\sqrt{n}}L_{n,N}\left(\frac{2\sqrt{n}\kappa}{\sigma}\right)_{\kappa\geq 0}\right)\right) \to \mathbf{E}F((X_{\alpha,\kappa})_{\kappa\geq 0}) \qquad n\to\infty$$

for every functional F satisfying $|F(f)| \leq C(1+\|f\|_\infty)^k$ for some constants $C, k > 0$. If we choose, in particular, $F(f) = \|f\|_\infty^d$ and $F(f) = f$, respectively, then Theorem 1.3 is proved.

Remark 1. *Note that Drmota and Marckert [8] studied the concept of polynomial convergence only for processes with compact support. But since the height of random forests is asymptotically a.s. bounded by $c\sqrt{n}$, it suffices to study truncated processes with arbitrarily large but compact support and then argue in the same way as in [7, Lemma 5]. Thus [8, Theorem 1] is applicable in this case as well.*

Remark 2. *Note that the connection between random forests and conditioned Brownian bridge can be used to compute functionals of the latter one. Recently, this has been done for the conditioned Brownian bridge area, see [12, 17, 2]. Thus this approach can help us to derive expressions for the moments of $X_{\alpha,v}$. This will be done in a forthcoming paper with G. Louchard.*

References

[1] P. Billingsley, (1968) *Convergence of Probability Measures*, John Wiley & Sons, New York.

[2] P. Chassaing and G. Louchard, (2001) *Reflected Brownian bridge area conditioned on its local time at the origin*, submitted, `http://www.ulb.ac.be/di/mcs/louchard/`.

[3] P. Chassaing and J.-F. Marckert, (2001) *Parking functions, empirical processes, and the width of rooted labeled trees*, Electronic J. Combinatorics 8(1), R14.

[4] I. A. Cheplyukova, (1997) *Limit distributions for the number of vertices in layers of a random forest*, Diskret. Mat. 9, No. 4, 150–157.

[5] L. Comtet, (1974) *Advanced Combinatorics*, Reidel, Dordrecht, Netherlands.

[6] M. Drmota and B. Gittenberger, (1997) *On the profile of random trees*, Random Structures and Algorithms 10, 421–451.

[7] M. Drmota and B. Gittenberger, (1999) *The width of Galton-Watson trees*, submitted, see `http://www.geometrie.tuwien.ac.at/drmota/width.ps.gz`

[8] M. Drmota and J. F. Marckert, (2001) *Reinforced weak convergence of stochastic processes*, submitted, see `http://www.geometrie.tuwien.ac.at/drmota/reinf.ps.gz`.

[9] M. Drmota and M. Soria, (1995) *Marking in combinatorial constructions: generating functions and limiting distributions*, Theoretical Comp. Science 144, 67–99.

[10] I. A. Egorova, (1997) *The distribution of vertices in strata of plane planted forests*, in Probabilistic methods in discrete mathematics (Petrozavodsk, 1996), 179–188, VSP, Utrecht.

[11] P. Flajolet and A. M. Odlyzko, (1989) *Random mapping statistics*, in Advances in Cryptology (1990), J.-J. Quisquater and J. Vandewalle, Eds., vol. 434 of Lecture Notes in Computer Science, Springer Verlag, pp. 329-354. Proceedings of Eurocrypt '89, Houtalen, Belgium.

[12] P. Flajolet, P. Poblete, and A. Viola, (1998) *On the analysis of linear probing hashing*, Algorithmica 22, 490–515.

[13] P. Flajolet and A. M. Odlyzko, (1990) *Singularity analysis of generating functions*, SIAM J. on Discrete Math. 3, No. 2, 216–240.

[14] P. Flajolet and J. S. Vitter, (1990) *Average-Case Analysis of Algorithms and Data Structures*, in *Handbook of Theoretical Computer Science,* J. van Leeuwen, Ed., vol. A: Algorithms and Complexity. North Holland, ch. 9, pp. 431–524.

[15] B. Gittenberger, (1998) *Convergence of branching processes to the local time of a Bessel process*, Random Structures and Algorithms 13, 423–438.

[16] B. Gittenberger, (1999) *On the contour of random trees*, SIAM J. Discrete Math. 12, No. 4, 434–458.

[17] S. Janson, (2001) Asymptotic distributions for the cost of linear probing hashing, Random Structures and Algorithms, to appear, `http://www.math.uu.se/svante/papers/`.

[18] I. Karatzas and S. E. Shreve, (1988) *Brownian Motion and Stochastic Calculus,* Springer, New York.

[19] V. F. Kolchin, (1986) *Random Mappings,* Optimization Software, New York.

[20] A. Meir and J. W. Moon, (1978) *On the altitude of nodes in random trees*, Canadian Journal of Mathematics 30, 997–1015.

[21] Yu. L. Pavlov, (1983) *Limit distributions of the height of a random forest*, Theory Probab. Appl. 28, 471–480.

[22] Yu. L. Pavlov, (1988) *Distributions of the number of vertices in strata of a random forest*, Theory Probab. Appl. 33, 96–104.

[23] Yu. L. Pavlov, (1994) *Limit distributions of the height of a random forest consisting of plane rooted trees*, Discr. Math. Appl. 4, 73–88, (translated from Diskret. Mat. 6, 137–154, 1994).

[24] Yu. L. Pavlov, (1994) *Asymptotic behavior of the height of a random forest,* (Russian), A collection of papers, No. 1, (Russian), 4–17, 139, Ross. Akad. Nauk, Karel. Nauchn. Tsentr, Petrozavodsk, 1994.

[25] Yu. L. Pavlov, (1997) *Random forests, in Probabilistic methods in discrete mathematics (Petrozavodsk, 1996)*, 11–18, VSP, Utrecht.

[26] Yu. L. Pavlov and I. A. Cheplyukova, (1999) *Limit distributions of the number of vertices in strata of a simply generated forest*, Discr. Math. Appl. 9, 137–154, (translated from Diskret. Mat. 11, 97–112, 1999).

[27] J. Pitman, (1999) *The SDE solved by local times of a Brownian excursion or bridge derived from the height profile of a random tree*, Ann. Probab. 27, 261–283.

Bernhard Gittenberger

Department of Geometry
Technische Universität Wien
Wiedner Hauptstraße 8-10/113
A-1040 Wien, Austria
Bernhard.Gittenberger@geometrie.tuwien.ac.at

Trends in Mathematics, © 2002 Birkhäuser Verlag Basel/Switzerland

On the Number of Heaps and the Cost of Heap Construction

Hsien-Kuei Hwang, Jean-Marc Steyaert

ABSTRACT: *Heaps constitute a well-known data structure allowing the implementation of an efficient $O(n \log n)$ sorting algorithm as well as the design of fast priority queues. Although heaps have been known for long, their combinatorial properties are still partially worked out: exact summation formulae have been stated, but most of the asymptotic behaviors are still unknown. In this paper, we present a number of general (not restricting to special subsequences) asymptotic results that give insight on the difficulties encountered in the asymptotic study of the number of heaps of a given size and of the cost of heap construction. In particular, we exhibit the influence of arithmetic functions in the apparently chaotic behavior of these quantities and study their extremal and average properties. It is also shown that the distribution function of the cost of heap construction using Floyd's algorithm and other variants is asymptotically normal.*

1 Introduction

Heap is an elementary data structure often used in applications concerned with priority queues and partial (as well as total) ordering. It first appeared in Williams' Heapsort algorithm [29], which happened to be the first *in-place* $O(n \log n)$ sorting algorithm. Besides its original applications to sorting, heap has wide applications in algorithm design, see Aho et al. [1]. It serves as the prototype, both conceptually and in actual implementations, of many complex data structures in computational geometry (see Preparata and Shamos [25]) and in graph-theoretical problems (see Noltemeier [21], Mehlhorn and Tsakalidis [20]).

A (max) *heap* is an array with elements a_j, $1 \le j \le n$, satisfying the *path-monotonic property*: $a_j \le a_{\lfloor j/2 \rfloor}$ for $j = 2, 3, \ldots, n$, where $\lfloor x \rfloor$ denotes the integral part of x. It can be viewed as a binary tree in which the value of each element is not smaller than that of its children.

The construction of a heap from an arbitrary set of n keys can be performed efficiently in linear time using Floyd's algorithm (see [18, §5.2.3]), but it is easy to see that the exact number of operations used varies according to the nature of the permutation on the input keys. Likewise, the number of heaps of a given size n is not a steadily growing function of n, but proves to have a chaotic rate of growth. Several authors have proposed partial answers to these problems. Knuth states the basic recurrence relation for the number of heaps, and gives an explicit solution as a product; he also shows that the expected cost used by Floyd's algorithm to construct a heap is linear; Doberkat [5] derives the probability generating

[1]Most results of this paper appeared as Rapport de Recherche, LIX/RR/93/07, Ecole polytechnique, 1993; only quite recently did we revise the paper; a full version of this extended abstract is available via the link `http://algo.stat.sinica.edu.tw`. This work was partially supported by the ESPRIT Basic Research Action No. 7141 (ALCOM II) and by the AQSI research action of the French Ministry for Research and Space.

functions for the number of exchanges and comparisons used by Floyd's algorithm, he also finds the asymptotics of the first two moments; Sprugnoli [27] examines in more detail the average case behavior of Floyd's heap-construction algorithm [18] for general values of the heap size. The method of Doberkat is via probability generating functions and that of Sprugnoli relies largely on refining the results on special sequences.

We propose a new approach to the asymptotic study of the number of heaps and the cost of heap construction. It is based on explicit decomposition of the solution to the basic recurrence characteristic of heap structures. The decomposition is in terms of some digital sums (sums expressible using the binary digits of n in base 2). From this canonical representation, elementary asymptotic methods then give the asymptotic behavior of the quantity in question.

This paper is organized as follows. We state the basic recurrence together with some notations in the next section. The general solution, in terms of digital sums, of this recurrence is presented in Section 3. Then we apply this result to determine an asymptotic expression for the number of heaps and the cost (mean and variance) of Floyd's heap-construction algorithm. Our results on the mean and the variance improve upon that of Doberkat [5] and that of Sprugnoli [27]. We also propose means of deriving extremal and typical behaviors of digital sums of the form $\sum_{0\le j\le \lfloor log_2 n\rfloor} \psi(\{n/2^j\})$, which appear naturally in our analysis. In the final section, we establish the asymptotic normality of the cost used by Floyd's algorithm under the uniform permutation model.

2 The basic divide-and-conquer recurrence

A characteristic property of a heap, when viewing as a binary tree, is that at least one of the two subtrees of the root node is complete (i.e., it contains $2^k - 1$ elements for some positive integer k), and then the size of the other subtree is at least half and at most twice this number. Furthermore, this property recursively applies to each node. Given a heap $\mathcal{H}$ of size n with left and right subtrees $\mathcal{L}$ and $\mathcal{R}$, an additive cost function φ on heaps is defined by a relation

$$\varphi[\mathcal{H}] = \tau[\mathcal{H}] + \varphi[\mathcal{L}] + \varphi[\mathcal{R}], \tag{1}$$

for some cost function τ. Summing over all heaps of size n, which obviously decompose into subheaps of smaller sizes, we thus define a function f that satisfies the same recurrence on sizes. Since at least one of $\mathcal{L}$ or $\mathcal{R}$ is complete, the relation (1) can be written into a more precise form as follows. For positive integer k and a given sequence $\{t_n\}_{n\ge 1}$

$$\begin{cases} f_{2^k+j} = t_{2^k+j} + \begin{cases} f_{2^{k-1}-1} + f_{2^{k-1}+j}, & \text{if } 0 \le j < 2^{k-1}, \\ f_{2^k-1} + f_j, & \text{if } 2^{k-1} \le j < 2^k, \end{cases} \\ f_0 = 0, f_1 = t_1, \end{cases}$$

which is referred to as the *(additive) heap recurrence* (see [27], [13, Ch. 3]), where t_n denotes the sum function for τ and is usually referred to as the "toll function". Observe that we can incorporate the two cases into one by writing

$$f_n = f_{2^{\lfloor \log_2 2n/3\rfloor}-1} + f_{n-2^{\lfloor \log_2 2n/3\rfloor}} + t_n \qquad (n \ge 2), \tag{2}$$

where $2^{\lfloor \log_2 2n/3 \rfloor}$ represents the unique power of two lying between $n/2$ and $2n/3$. Here and in the following, $\mathcal{H}_n$ denotes the total number of ways to rearrange the integers $\{1, 2, ..., n\}$ into a heap. It is obvious that $\mathcal{H}_n$ satisfies the *multiplicative heap recurrence:*

$$\mathcal{H}_{2^k+j} = \begin{cases} \binom{2^k+j-1}{2^{k-1}-1} \mathcal{H}_{2^{k-1}-1} \mathcal{H}_{2^{k-1}+j}, & \text{if } 0 \leq j < 2^{k-1}, \\ \binom{2^k+j-1}{2^k-1} \mathcal{H}_{2^k-1} \mathcal{H}_j, & \text{if } 2^{k-1} \leq j < 2^k, \end{cases}$$

since the root has to contain the largest key and the remaining elements can be assigned independently to the left and right subheaps. The sequence

$$\begin{aligned} \{\mathcal{H}_n\}_{n\geq 2} &= 1, 2, 3, 8, 20, 80, 210, 896, 3360, 19200, 79200, 506880, 2745600, \\ &\quad 21964800, 108108000, 820019200, 5227622400, 48881664000... \end{aligned}$$

increases very fast and is the A056971 entry in Sloane's On-Line Encyclopedia of Integer Sequences (`http://www.research.att.com/~njas/sequences`). It is more convenient to work with $h_n = \log(n!/\mathcal{H}_n)$ since h_n satisfies (2) with $t_n = \log n$.

Knuth [18, p. 154] expresses the number of heaps of size n via the product of the sizes of all its subtrees. Denoting by s_i, $i = 1, 2, \ldots, n$, these sizes, the number $\mathcal{H}_n$ satisfies

$$\mathcal{H}_n = \frac{n!}{\prod_{1\leq i \leq n} s_i}.$$

The s_i's are generally of the form $2^k - 1$, for some positive integer k, except for the nodes lying on the *special path* whose ranks (in hierarchical order) are of the form $2^j(1 + \{n/2^j\})$, where $\{x\}$ denotes the fractional part of x.

If we consider the (backward) difference of f_n: $\varphi_n = \nabla f_n = f_n - f_{n-1}$, then we obtain,

$$\varphi_{2^k+j} = \tau_{2^k+j} + \begin{cases} \varphi_{2^{k-1}+j}, & 0 \leq j < 2^{k-1}, \\ \varphi_j, & 2^{k-1} \leq j < 2^k, \end{cases} \qquad (k \geq 1),$$

with the initial condition $\varphi_0 = 0$, where $\tau_n = \nabla t_n$. Equivalently, if we write $n = (1b_{L-1} \ldots b_0)_2$ in base 2, then this recurrence can be re-written as

$$\varphi_n = \varphi_{(1b_{L-1}\ldots b_0)_2} = \tau_{(1b_{L-1}\ldots b_0)_2} + \varphi_{(1b_{L-2}\ldots b_0)_2} = \sum_{0\leq j \leq L} \tau_{(1b_{j-1}\ldots b_0)_2}.$$

Before solving (2), we note that there is another very similar type of recurrences [10]

$$\phi_{2^k+j} = \tau_{2^k+j} + \begin{cases} \phi_{2^{k-1}} + \phi_{2^{k-1}+j}, & \text{if } 0 \leq j \leq 2^{k-1}; \\ \phi_{2^k} + \phi_j, & \text{if } 2^{k-1} \leq j \leq 2^k, \end{cases} \tag{3}$$

which occurs as the solution of the following equation

$$\phi_n = \tau_n + \min_{1\leq j \leq \lfloor n/2 \rfloor} (\phi_j + \phi_{n-j}) \qquad (n \geq 2),$$

(with ϕ_1 given) when the sequence $\{\tau_n\}_{n\geq 0}$ is strictly concave, namely, $\tau_{n+2} - 2\tau_{n+1} + \tau_n < 0$ for all $n \geq 0$. The same recurrence also appeared in the analysis of queue-mergesort and other problems; see [4].

On the other hand, when the sequence $\{\tau_n\}_{n\geq 0}$ is strictly convex, the solution ϕ_n satisfies [10]

$$\phi_n = \phi_{\lfloor n/2 \rfloor} + \phi_{\lceil n/2 \rceil} + \tau_n \qquad (n \geq 2). \tag{4}$$

This type of recurrences occurs very often in the analysis of algorithms and was systematically treated by an analytic approach by Flajolet and Golin [6]; see also [15].

Recurrences of the types (2) and (4) are typical representatives of the divide-and-conquer recurrences that one naturally encounters from its designing principle

$$A_n = A_{x(n)} + A_{y(n)} + B_n \qquad (n \geq n_0 > 1),$$

with some initial conditions, where B_n is some given sequence (called "toll function"), A_n is certain cost measure related to the algorithm in question, $x(n)$ and $y(n)$ usually satisfy $1 \leq x(n), y(n) \leq n-1$, their values depending upon the dividing strategy.

Besides (2) and (4), let us mention another recurrence studied by Panny and Prodinger [22] for the analysis of bottom-up mergesort algorithm

$$A_n = A_{x(n)} + A_{n-x(n)} + B_n \qquad (n \geq 2), \tag{5}$$

with $x(n) = 2^{\lceil \log_2 n \rceil - 1}$, where $\lceil y \rceil$ denotes the least integer $\geq y$.

A typical phenomenon of the behaviors of divide-and-conquer recurrences is that they often involve certain periodic fluctuations depending on the instance size. Intuitively, this is due to the fact that in each dividing step, the sizes of the two subproblems are not necessarily identical. Consequently, the cumulative (recursive) effect renders the global behavior less smooth, or even chaotic. Such a fluctuating behavior relies sensitively upon the dividing strategy. Roughly, the more "balanced" the two sequences $x(n)$ and $y(n)$ are, the more "smooth" the resulting oscillating behavior is. For example, the recurrence (4) should be expected to involve oscillating functions that are more smooth than those of (2) and (5). A detailed comparison of the two recurrences (5) and (4) can be found in [22] and in [4].

Notations. Throughout this paper, n is a positive integer, and $n = (b_L b_{L-1} ... b_0)_2$, the binary representation of n, where $L = \lfloor \log_2 n \rfloor$ and $b_L = 1$. Denote by $n_j = (1b_{j-1}...b_0)_2$ for $j = 1, 2, \ldots, L$ with $n_0 = 1$. All limits are taken to be $n \to \infty$.

3 Explicit formula

To solve the heap recurrence (2) explicitly, we observe that when $n = 2^{k+1} - 1$, the recurrence is essentially linear: $f_{2^{k+1}-1} = t_{2^{k+1}-1} + 2f_{2^k-1}$, which can be solved easily by iteration. From this, we can find the solution for the sequences $\{2^k\}$, $\{2^k + 2^{k-1} - 1\}, \ldots$ But this process does not lead readily to a general solution; see Sprugnoli [27]. Hence, we begin with another approach, which extends a classical counting argument for Young tableaux of a given shape (see [18, §§5.1.4, 5.2.3]).

Lemma 1. *For $n \geq 1$, the solution f_n of the heap recurrence (2) is given by*

$$f_n = \sum_{1\leq j\leq L} \left(\left\lfloor \frac{n}{2^{j-1}} \right\rfloor - \left\lfloor \frac{n}{2^j} \right\rfloor - 1 \right) t_{2^j-1} + \sum_{0\leq j\leq L} t_{n_j}, \tag{6}$$

for any given sequence $\{t_n\}_{n\geq 1}$.

Proof. We call, as in [18], the nodes lying on the path from 1 to n *special nodes.* Let Σ_1 and Σ_2 denote the first sum and the second sum on the right-hand side of (6), respectively. Then Σ_1 counts the weights of nonspecial nodes under the cost function t_n, and Σ_2 counts similarly the weights of special nodes. Nonspecial nodes are always perfectly balanced, so their sizes are of the form $2^j - 1$. Now any node of rank $\beta = (c_r...c_0)_2$ lying in the range $\lfloor n/2^j \rfloor < \beta < \lfloor n/2^{j-1} \rfloor$ has as subtree size $s_\beta = \sum_{0\leq i<j} 2^i = 2^j - 1$, since[1] s_β is the number of positive integers $\leq n$ whose binary representation is of the form $(c_r...c_0*)_2$, where $*$ is any 0-1 string. Hence, the number of nonspecial subtrees of size $2^j - 1$ is

$$\left\lfloor \frac{n}{2^{j-1}} \right\rfloor - \left\lfloor \frac{n}{2^j} \right\rfloor - 1.$$

On the other hand, the number of positive integers $\leq n$ whose binary representation has the form $(b_L...b_j*)$ for $|*| = 0, 1, ..., j-1$ is $\sum_{0\leq i<j} 2^i + (b_{j-1}\dots b_0)_2 + 1 = (1b_{j-1}\dots b_0)_2$. Hence, the special subtree sizes are $n, n_{L-1}, ..., n_0 = n_j$. This completes the proof. ∎

An interesting consequence of Lemma 1 is the following necessary and sufficient condition for the asymptotic linearity of f_n.

Lemma 2. *Assume that f_n satisfies the heap recurrence (2). Then f_n is asymptotically linear: $f_n \sim cn$ iff*

$$t_n = o(n), \quad \text{and} \quad \sum_{j\geq 1} t_{2^j-1} 2^{-j} < \infty. \tag{7}$$

The constant c is given by $c = \sum_{j\geq 1} t_{2^j-1} 2^{-j}$.

This result says that without loss of generality, we can, under the hypotheses of Lemma 2, consider only the special sequence $\{f_{2^k-1}\}_k$, as far as the dominant asymptotics is concerned.

Proof. Assume that (7) holds. By (6) and $t_n = o(n)$, we deduce that

$$f_n = n \sum_{1\leq j\leq L} t_{2^j-1} 2^{-j} + o(n). \tag{8}$$

Since the sum $\sum_{1\leq j\leq L} t_{2^j-1} 2^{-j}$ converges (by assumption), we obtain $f_n \sim cn$. Note that the expression (8) holds as long as $t_n = o(n)$.

On the other hand, assume that $f_n = c_0 n + o(n)$ for some constant c_0. Then by definition (2)

$$\begin{aligned} t_n &= f_n - f_{2^{\lfloor \log_2 2n/3 \rfloor}-1} - f_{n-2^{\lfloor \log_2 2n/3 \rfloor}} \\ &= c_0 n - c_0 2^{\lfloor \log_2 2n/3 \rfloor} - c_0 \left(n - 2^{\lfloor \log_2 2n/3 \rfloor} \right) + o(n) \\ &= o(n). \end{aligned}$$

[1] By the basic property of heap: if a node β has rank $(c_r...c_0)_2$, then its left and right children have respective ranks $(c_r...c_0 0)_2$ and $(c_r...c_0 1)_2$.

This implies, by (8) and $f_n \sim c_0 n$, that $c_0 = c = \sum_{j\geq 1} t_{2^j-1}2^{-j} < \infty$. This proves the necessity part and the lemma. ■

4 The number of heaps

In this section, we derive an asymptotic expression for the number of heaps $\mathcal{H}_n$ of size n. To this end, we apply Lemma 1 to the sequence $h_n = \log(n!/\mathcal{H}_n)$. The extremal and average order of the arithmetic functions involved are also discussed.

4.1 The asymptotic behavior of $\mathcal{H}_n$

First of all, since h_n satisfies (2) with $t_n = \log n$, Lemma 2 gives,

$$h_n \sim n \sum_{j\geq 1} \frac{\log(2^j-1)}{2^j} = n\left(2\log 2 + \sum_{j\geq 1} 2^{-j}\log(1-2^{-j})\right) = 0.945755\ldots n.$$

Let $\alpha = 2\log 2 + \sum_{j\geq 1} 2^{-j}\log(1-2^{-j})$. In order to obtain an asymptotically equivalent approximation for $\mathcal{H}_n$, we need to determine h_n up to the constant term.

Theorem 1. *The number of heaps of size n satisfies the asymptotic expression*

$$\mathcal{H}_n = 2Q\sqrt{2\pi}\,\Pi(\log_2 n)R(n)\, n^{n+3/2}\, e^{-(\alpha+1)n}\left(1+\frac{1}{2^{L+1}}+\frac{1}{12n}+O\left(n^{-2}\right)\right),$$

where $Q=\prod_{j\geq 1}(1-2^{-j})=0.288788\ldots$, $R(n)=\prod_{j\geq 1}[(1-2^{-j-1})/(1-2^{-j})]^{\{n/2^j\}}$ *and*

$$\Pi(\log_2 n) = 2^{2^{\{\log_2 n\}}-\{\log_2 n\}} \prod_{0\leq j\leq L} \frac{2^{\{n/2^j\}}}{1+\{n/2^j\}}. \tag{9}$$

Proof. (Sketch) Apply (6) with $t_n = \log n$ and simplify; see [17] for details. ■

Note that this is not an usual asymptotic formula since

$$\frac{1}{\log n}\log\frac{\mathcal{H}_n}{n^{n+3/2}e^{-(\alpha+1)n}}$$

varies between 0 and $-\log_4(10/9)$ (see (10) below).

4.2 Extremal orders of $R(n)$ and $\Pi(u)$

Taking logarithms and making use of the inequalities $0 \leq \{n/2^j\} \leq 1-2^{-j}$, we have, for all $n \geq 1$,

$$1 \leq R(n) \leq \exp\left(-\sum_{j\geq 1} 2^{-j}\log(1-2^{-j})\right) = 1.553544\ldots$$

Proposition 1. *For all $n \geq 1$, the inequalities*

$$0 < c_1 n^{-\eta/\log 2} \leq \Pi(\log_2 n) \leq 2 \tag{10}$$

hold, where $\eta = \log(10/9)/2$ and c_1 is some constant.

Proof. Observe first that by (9) and the elementary inequalities

$$e \log 2 \leq 2^{2^x - x} \leq 2 \qquad (0 \leq x \leq 1),$$
$$\frac{e}{2} \log 2 \leq \frac{2^x}{1+x} \leq 1 \qquad (0 \leq x \leq 1),$$

we obtain the upper bound $\Pi(\log_2 n) \leq 2$ and the lower bounds

$$\Pi(\log_2 n) \geq e(\log 2) \left(\frac{e}{2} \log 2\right)^L \geq e(\log 2) n^{-1+1/\log 2 + \log\log 2/\log 2}.$$

Such a lower bound obtained by considering only the maximum of each term (as a continuous function) is, however, too crude and unattainable.
To prove the better lower bound in (10), we consider the function

$$\varpi(n) := \sum_{1 \leq j \leq L} \psi\left(\{n/2^j\}\right), \quad \text{where} \quad \psi(x) := \log(1+x) - x\log 2,$$

so that

$$\Pi(\log_2 n) = 2^{2^{\{\log_2 n\}} - \{\log_2 n\}} e^{-\varpi(n)}.$$

Note that the function $\psi(x)$ attains the maximum value at $(1-\log 2)/\log 2 \approx 0.44$. Also $\varpi(n)$ satisfies the recurrence

$$\varpi(n) = \varpi(n - 2^L) + \sum_{\lfloor \log_2(n-2^L) \rfloor < j \leq L} \psi\left(\frac{n-2^L}{2^j}\right).$$

We prove that

$$\limsup_{n\to\infty} \frac{\varpi(n)}{L} = \eta = \frac{1}{2}(\psi(1/3) + \psi(2/3)) = \frac{1}{2}\log\frac{10}{9}. \tag{11}$$

The proof consists mainly of two parts: First, by induction, it is straightforward to show that if

$$n = (1010\cdots101)_2 = \frac{4^{L/2+1} - 1}{3} \qquad (L \text{ even}), \tag{12}$$

then

$$\frac{L}{2}\eta \leq \varpi(n) \leq \frac{L}{2}\eta + \eta,$$

so that

$$\limsup_{n\to\infty} \frac{\varpi(n)}{L} \geq \eta. \tag{13}$$

Second, if n is not of the form (12), then n can be written as

$$n = (1010\cdots 101\underbrace{0\cdots 0}_{\lambda}1*)_2, \tag{14}$$

where $\lambda \geq 0$, $\lambda \neq 1$ and $*$ is a binary sequence. We prove again by induction and numerical bounds for $\psi(x)$ that

$$\varpi(n) \leq (L+1)\eta \qquad (n \geq 1), \tag{15}$$

which together with (13) proves (11); see [17] for details.

Remarks. The proof does not reveal why repeating the pattern 10 yields the best possible upper bound for $\varpi(n)/L$. Roughly, this is because the average of $\psi(1/3)$ and $\psi(2/3)$ is somehow the most balanced choice. Here is another rough explanation (by calculations). Write first $x = \{n/2^L\}$. Then $\varpi(n) = \sum_{0\leq j<L} \psi(\{2^j x\})$. The function $b - \psi(x)$ is positive for $0 \leq x \leq 0.27$ and $0.62 \leq x < 1$. So if we assume that (15) holds, then, by induction, we are left with the interval $[0.27, 0.62]$ for x. Then $2b - \psi(x) - \psi(\{2x\}) > 0$ for $1/3 < x \leq 0.62$, and we are left with the smaller interval $[0.27, 1/3]$. Continuing this way, we see that $1/3$ is a limiting point in some sense, which means that n is of the form (12).

4.3 Average order

Although the function $\varpi(n)$ oscillates between $O(1)$ and $O(\log n)$, its sum function has a much smoother behavior. We determine its average order, namely $n^{-1}\sum_{1\leq m\leq n} \varpi(m)$.

Lemma 3. *Let $\varphi(x)$ be any real, continuous function on $[0,1]$ and differentiable in $(0,1)$. Define $\phi(m) = \sum_{0\leq j\leq \log_2 m} \varphi(\{m/2^j\})$. Then*

$$\frac{1}{n}\sum_{1\leq m\leq n} \phi(m) = \left(\int_0^1 \varphi(x)\,dx\right)\log_2 n + O(1).$$

In other words, the lemma says that the average order of the function $\phi(m)$ is equal to $\log_2 n$ times the mean value of the function φ in $[0,1]$.

Proof. (Sketch) Let $\Phi(n) := \sum_{1\leq m\leq n} \phi(m)$. Then

$$\Phi(n) = \int_1^{n+1} \phi(x)\,dx - \int_1^{n+1} (\phi(x) - \phi(\lfloor x\rfloor))\,dx + O(1).$$

The first integral on the right-hand side is easily seen to be

$$\int_1^{n+1} \phi(x)\,dx = \left(\int_0^1 \varphi(x)\,dx\right) n\log_2 n + O(n).$$

The second integral is $O(n)$ by the differentiability of $\varphi(x)$. ■

If we take $\varphi(x) = x$, we obtain, in view of the relation $\nu(n) = \sum_{0\le j\le L}\{n/2^j\} + n/2^L$, where $\nu(n) = \sum_{0\le j\le L} b_j$,

$$\frac{1}{n}\sum_{1\le m\le n}\nu(m) = \frac{1}{2}\log_2 n + O(1);$$

and in general

$$\frac{1}{n}\sum_{1\le m\le n}\nu_q(m) = \frac{1}{q+1}\log_2 n + O(1) \qquad (q>0),$$

where $\nu_q(n) := \sum_{0\le j\le L}\{n/2^j\}^q$. The implied constant in the O-term depends on q. Other examples are given in the next section.

5 The cost of constructing heaps

In this section, we again apply the solution (6) of the heap recurrence to refine previous analyses on the mean and the variance of the cost used by Floyd's algorithm to construct a heap from a random sequence. Moreover, the asymptotic normality of the cost is also established. Our methods apply equally to other variants of Floyd's algorithm.

Let us first recall the algorithm. It is divide-and-conquer in nature. Briefly, to construct a heap of size n, construct the left and the right subheaps, respectively, by the same procedure and then find the proper place in one of the two subheaps for the root element (by comparisons and key-exchanges). We assume, throughout this section, that *a uniform probability measure is assumed on the set of permutations of size n, $n \ge 1$.* Knuth [18, p. 155] showed that, given a random permutation of size n, Floyd's algorithm preserves randomness in each recursive step. Thus given a random permutation of n elements, the output obtained by Floyd's algorithm is a random heap of size n.

Thus, with the notations of (1), the probability generating function $P_n(z)$ for some cost measure of the algorithm like the number of exchanges running over all heaps of size n satisfies (see (2))

$$\begin{aligned} P_0(z) &= 1 \\ P_n(z) &= P_{2^{\lfloor \log_2 2n/3\rfloor}-1}(z)P_{n-2^{\lfloor \log_2 2n/3\rfloor}}(z)Q_n(z) \qquad (n\ge 1), \end{aligned} \tag{16}$$

where $Q_n(z)$ is the probability generating function for the cost of placing the root element of a heap of size n into one of its two subheaps. Note that $Q_n(z)$ is always a polynomial for all finite n. From (6), we have the explicit representation

$$P_n(z) = Q_n(z)\prod_{1\le j\le L} Q_{2^j-1}(z)^{\lfloor n/2^{j-1}\rfloor - \lfloor n/2^j\rfloor - 1}Q_{n_j}(z) \qquad (n\ge 1).$$

Our analysis below applies to all versions of heap-construction algorithm using divide-and-conquer paradigm such that the randomness is preserved in each "conquering" step, the randomness is preserved.

For concreteness, let ξ_n denote the number of exchanges used by Floyd's algorithm to convert a random permutation into a heap. We derive precise asymptotic expansions for the mean and variance of ξ_n and show that ξ_n is asymptotically normally distributed with convergence rate of order $n^{-1/2}\log n$.

5.1 The mean of ξ_n

>From [18, §5.2.3], the mean number of exchanges $E(\xi_n)$ satisfies the heap recurrence (2) with $t_n = n^{-1}\sum_{1\le j\le n}\lfloor \log_2 j\rfloor = L + (L+2)/n - 2^{L+1}/n$, the average height of a node in a random heap of size n (see [18, p. 155]). Applying formulae (6), we get the following result that is more general than that of Sprugnoli [27]; see also [13, Ch. 3].

Theorem 2. *The expected number of exchanges $E(\xi_n)$ used by Floyd's heap construction algorithm satisfies*

$$E(\xi_n) = \mu n + \varpi_1(n) + \varpi_2(n) + O\left(\frac{\log n}{n}\right) \qquad (n\to\infty), \tag{17}$$

where $\mu = -2 + \sum_{j\ge1} j(2^j-1)^{-1} = 0.744033...$, $\varpi_1(n)$ is defined by

$$\varpi_1(n) = -\sum_{1\le j\le L}\frac{1+\{n/2^j\}^2}{1+\{n/2^j\}},$$

and $\varpi_2(n) = O(1)$ is given by

$$\varpi_2(n) = -1 - \frac{n}{2^L} - \sum_{j\ge1}\frac{j}{2^j-1} + \sum_{j\ge1}\frac{j+2}{2^j(1+\{n/2^j\})} + \sum_{j\ge1}\left\{\frac{n}{2^j}\right\}\frac{j2^j-2^j+1}{(2^j-1)(2^{j+1}-1)}.$$

Furthermore, $\varpi_1(n)$ satisfies

$$\liminf_{n\to\infty}\frac{\varpi_1(n)}{L} = -1, \quad \text{and} \quad \limsup_{n\to\infty}\frac{\varpi_1(n)}{L} = -\frac{17}{20}.$$

By Lemma 4, the average order of the arithmetic function $\varpi_1(n)$ is $(3/2 - 2\log 2)\log_2 n + O(1)$.

Proof. (Sketch) With the exact formula (6), simple manipulations using Maple lead to (17). The inequalities for $\varpi_1(n)$ are derived by similar arguments as in Proposition 1 (both approaches apply). ■

In particular, if $n = (2^{(m+1)d}-1)/(2^d-1)$, $d\ge 2$, then

$$E(\xi_n) = \mu n + K_d L + O(1), \quad \text{where } K_d = -\frac{1}{d(2^d-1)}\sum_{0\le j<d}\frac{(2^d-1)^2+4^j}{2^d-1+2^j}.$$

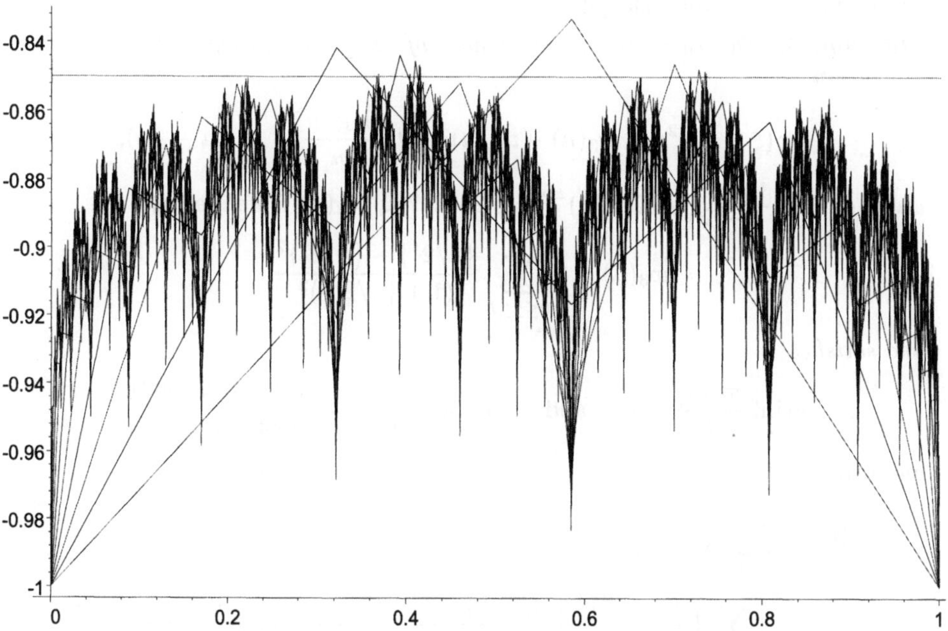

Figure 1: *The function $(E(\xi_n) - \mu n)/L$ plotte dagainst $\{\log_2 n\}$ for n from 2 to 1024; the upper b ound as predicte d by Thorem 2 is also shown.*

5.2 The variance of ξ_n

We now consider the variance of ξ_n. Since the variance of a sum of independent random variables is equal to the sum of the variance of individual random variable, the variance of ξ_n satisfies (2) with

$$\begin{aligned} t_n &= \frac{1}{n}\sum_{1\le j\le n} \lfloor \log_2 j\rfloor^2 - \left(\frac{1}{n}\sum_{1\le j\le n}\lfloor \log_2 j\rfloor\right)^2 \\ &= 6\frac{2^L}{n} - \frac{4^{L+1}}{n^2} - \frac{L^2}{n} + \frac{2^{L+2}L}{n^2} - 4\frac{L}{n} - \frac{6}{n} + \frac{2^{L+3}}{n^2} - \frac{L^2}{n^2} - 4\frac{L}{n^2} - \frac{4}{n^2}, \end{aligned}$$

by partial summation. With the help of Maple, we obtain the following result, improving that of Doberkat [5].

Theorem 3. *The variance of the number of exchanges satisfies the asymptotic expression*

$$\mathrm{Var}\,(\xi_n) = \sigma^2 n + \varpi_3(n) + \varpi_4(n) + O\left(\frac{\log^2 n}{n}\right) \qquad (n\to\infty),$$

where $\sigma^2 = 2 - \sum_{j\ge1} j^2(2^j-1)^{-2} = 0.261217...$, $\varpi_3(n)$ *is defined by*

$$\varpi_3(n) = 2\sum_{0\le j\le L} \frac{\{n/2^j\} - \{n/2^j\}^2}{(1+\{n/2^j\})^2},$$

and satisfies

$$\liminf_{n\to\infty} \frac{\varpi_3(n)}{L} = 0, \quad \text{and} \quad \limsup_{n\to\infty} \frac{\varpi_3(n)}{L} = \frac{40137372396071}{188961240258000},$$

and $\varpi_4(n) = O(1)$*:*

$$\begin{aligned} \varpi_4(n) &= \sum_{j\ge1} \frac{j^2 2^j}{(2^j-1)^2} \\ &\quad + \sum_{j\ge1} \left\{\frac{n}{2^j}\right\} \frac{2^j(j^2+4j+2) - 4^{j+1}(2j+1) - 2\cdot 8^j(j^2-2j-1)}{(2^j-1)^2(2^{j+1}-1)^2}. \end{aligned}$$

The average order of $\varpi_3(n)$ is $(6\log 2 - 4)\log_2 n + O(1)$.

Note that, unlike the two functions $\varpi(n)/L$ and $\varpi_1(n)/L$ whose maximum values are achieved when $n = (1010\cdots101)_2$, the maximum of ϖ_3/L is (asymptotically) attained when n is of the form

$$n = (10100\,10100\cdots 101001)_2,$$

so that the ratio 40137372396071/188961240258000 is equal to

$$\frac{1}{5}\left(\psi_3(5/31) + \psi_3(10/31) + \psi_3(20/31) + \psi_3(9/31) + \psi_3(18/31)\right),$$

where $\psi_3(x) = 2(x-x^2)/(1+x)^2$. The main difference is that the underlying function $\psi_3(x)$ attains its maximum value 1/4 at $x = 1/3$ and not between 1/3 and 1/2. The proof follows the same arguments used in the proof of Proposition 1; we omit the proof here since it is very involved and uninteresting.

5.3 Asymptotic normality

Theorem 4. *The distribution functions of the random variables* $(\xi_n - \mu n)/(\sigma\sqrt{n})$ *converge to the standard normal distribution:*

$$\sup_{-\infty<x<\infty}\left|P\left(\frac{\xi_n - \mu n}{\sigma\sqrt{n}} < x\right) - \frac{1}{\sqrt{2\pi}}\int_{-\infty}^{x} e^{-t^2/2}\,dt\right| = O\left(\frac{\log n}{\sqrt{n}}\right), \tag{18}$$

where the two constants μ *and* σ^2 *are defined in Theorems 2 and 3, respectively.*

This result is a special case of the following lemma refined from that of Haigh [11], where we explicitly characterize the convergence rate to the normal law by a direct approach based upon characteristic functions and the Berry-Esseen smoothing inequality (see [23]).

Lemma 4. *Let* $\{X_n\}_n$ *be a sequence of random variables taking only non-negative integral values with mean* μ_n *and variance* σ_n^2. *Suppose that the probability generating function* $P_n(z)$ *of* Ω_n *can be decomposed as* $P_n(z) = \prod_{1\le j\le k_n} P_{nj}(z)$, *for some sequence* $\{k_n\}_n$, *where the* $P_{nj}(z)$ *are polynomials such that (i) each* $P_{nj}(z)$ *is itself a probability generating function of some random variable, say,* X_{nj} $(1 \le j \le k_n)$; *and (ii)*

$$\frac{M_n}{\sigma_n} \to 0, \tag{19}$$

where $M_n = \max_{1\le j\le k_n} \deg P_{nj}(z)$. *Then the distribution of* X_n *is asymptotically Gaussian:*

$$\sup_{-\infty<x<\infty}\left|P\left(\frac{X_n - \mu_n}{\sigma_n} < x\right) - \frac{1}{\sqrt{2\pi}}\int_{-\infty}^{x} e^{-t^2/2}\,dt\right| = O\left(\frac{M_n}{\sigma_n}\right). \tag{20}$$

Proof. (Sketch) By the expansion

$$\log P_{nj}(e^{it/\sigma_n}) = \frac{E(X_{nj})}{\sigma_n}\,it - \frac{\operatorname{Var}(X_{nj})}{2\sigma_n^2}\,t^2 + O\left(\frac{E(|X_{nj} - \mu_{nj}|)}{\sigma_n^3}\,t^3\right),$$

for $1 \le j \le k_n$, $|t| \le T_n$, where $T_n := \varepsilon\sigma_n/M_n$, we obtain

$$\log\varphi_n(t) = -\frac{t^2}{2} + O\left(\frac{M_n}{\sigma_n}\,|t|^3\right) \qquad (|t| \le T_n),$$

where $\varphi_n(t) := E(e^{(X_n-\mu_n)it/\sigma_n})$, since

$$\frac{1}{\sigma_n^3}\sum_{1\le j\le k_n} E(|X_{nj} - E(X_{nj})|^3) \le \frac{2\max_{1\le j\le k_n}|X_{nj}|}{\sigma_n} = \frac{2M_n}{\sigma_n}.$$

Using $|e^w - 1| \le |w|e^{|w|}$ and the Berry-Esseen smoothing inequality [23, p. 109]

$$\sup_{-\infty<x<\infty}\left|P\left(\frac{X_n - \mu_n}{\sigma_n} < x\right) - \frac{1}{\sqrt{2\pi}}\int_{-\infty}^{x} e^{-t^2/2}\,dt\right|$$

$$= O\left(T_n^{-1} + \int_{-T_n}^{T_n} \left|\frac{\varphi_n(t) - e^{-t^2/2}}{t}\right| dt\right),$$

we deduce (20). See [17] for details. ∎

This simple lemma can be used to derive the asymptotic normality (with convergence rate) for many combinatorial quantities; see [14].

Proof of Theorem 4. By (16) and Lemma 5, we need only check that the degree of the polynomials $Q_n(z)$ in (16) is small compared to the standard deviation of ξ_n, which is $O(\sqrt{n})$. The special form of $Q_n(z)$ given in [5] has no importance here. Since $Q_n(z)$ is the probability generating function for inserting the root into one of the two subheaps that are of height $O(\log n)$, it is obvious that the degree of $Q_n(z)$ is $O(\log n)$. From this, an application of Lemma 5 yields

$$\sup_{-\infty<x<\infty} \left| P\left(\frac{\xi_n - E(\xi_n)}{\sqrt{\mathrm{Var}(\xi_n)}} < x\right) - \frac{1}{\sqrt{2\pi}} \int_{-\infty}^{x} e^{-t^2/2}\, dt \right| = O\left(\frac{M_n}{\sigma_n}\right).$$

In view of Theorems 2 and 3, the above equation is asymptotically equivalent to (18). ∎

Our approach also applies to the number of comparisons used by Floyd's algorithm and similar results as Theorems 2–4 hold.

Here is a more general rule.

The cost (the number of key-exchanges or the number of comparisons) used to construct a heap from a random permutation of n elements by algorithms using divide-and-conquer paradigm such that the randomness is preserved in each "conquer" step is asymptotically normal in the sense of convergence in distribution.

It is merely a rule since a formal statement of a precise version would be too heavy. In particular, this rule applies to the heap construction algorithms in [3, 19, 12, 28], the basic ideas of improvement being more or less due to Floyd. Our approach also gives in most cases more precise quantitative results in the form of (18).

In connection with this, it should be pointed out that the original on-line algorithm proposed by Williams [29] to construct a heap is not linear in the worst case and that the randomness is not guaranteed (see [24]) in each step. The average case analysis of its behavior is more difficult; see [2, 8, 12]. Likewise, a precise analysis of the expected behavior of heapsort is very involved since successive deletions destroy the initial random character; see for example Schaffer and Sedgewick [26].

Acknowledgements

The first author thanks Tsung-Hsi Tsai for helpful suggestions.

References

[1] A. V. Aho, J. E. Hopcroft and J. D. Ullman, *The Design and Analysis of Computer Algorithms*, Addison-Wesley, MA, 1974.

[2] B. Bollabás and I. Simon, Repeated random insertion into a priority queue structure, *Journal of Algorithms*, **6** (1985) 466–477.

[3] S. Carlsson, Average-case results on heapsort, *BIT*, **27** (1987) 2–17.

[4] W.-M. Chen, H.-K. Hwang and G.-H. Chen, The cost distribution of queue-mergesort, optimal mergesorts, and power-of-two rules, *Journal of Algorithms*, **30** (1999), 423–448.

[5] E. E. Doberkat, An average case analysis of Floyd's algorithm to construct heaps, *Information and Control*, **61** (1984) 114–131.

[6] P. Flajolet and M. Golin, Mellin transforms and asymptotics: the mergesort recurrence, *Acta Informatica*, **31** (1994) 673–696.

[7] R. W. Floyd, Algorithm 245, tree 3, *Communications of the ACM*, **7** (1964) 701.

[8] A. Frieze, On the random construction of heaps, *Information Processing Letters*, **27** (1988) 103–109.

[9] G. H. Gonnet and J. I. Munro, Heaps on heaps, *SIAM Journal on Computing*, **15** (1986) 964–971.

[10] J. M. Hammersley and G. R. Grimmett, Maximal solutions of the generalised subadditive inequality, in Chapter 4 of "*Stochastic Geometry*", Ed. by E. F. Harding and D. G. Kendall, John Wiley and Sons, 1974.

[11] J. Haigh, A neat way to prove asymptotic normality, *Biometrika*, **58** (1971) 677–678.

[12] R. Hayward and C. McDiarmid, Average case analysis of heap building by repeated insertion, *Journal of Algorithms*, **12** (1991) 126–153.

[13] H.-K. Hwang, On the solutions of some divide-and-conquer recurrences, Rapport de Recherche, LIX/RR/93/01, Ecole Polytechnique, 1993.

[14] H.-K. Hwang, On convergence rates in the central limit theorems for combinatorial structures, *European Journal of Combinatorics*, **19** (1998) 329–343.

[15] H.-K. Hwang, Asymptotic expansions of the mergesort recurrences, *Acta Informatica*, **35** (1998), 911–919.

[16] H.-K. Hwang, Limit theorems for mergesort, *Random Structures Algorithms*, **8** (1996), 319–336.

[17] H.-K. Hwang and J.-M. Steyaert, On the number of heaps and the cost of heap construction, full version available via the link `http://algo.stat.sinica.edu.tw`.

[18] D. E. Knuth, *The Art of Computer Programming, Volume III—Sorting and Searching*, Addison-Wesley, Second Edition, 1998.

[19] C. J. H. McDiarmid and B. A. Reed, Building heaps fast, *Journal of Algorithms*, **10** (1989) 352–365.

[20] K. Mehlhorn and A. Tsakalidis, Data structures, Chapter 6 of *Handbook of Theoretical Computer Science*, pp. 301–341, Edited by J. van Leeuwen, Elsevier Science Publishers, Amsterdam, 1990.

[21] H. Noltemeier, Dynamic partial orders and generalised heaps, *Computing (Supplement)*, **7** (1990) 125–139.

[22] W. Panny and H. Prodinger, Bottom up mergesort, a detailed analysis, *Algorithmica*, **14** (1995) 340–354.

[23] V. V. Petrov, *Sums of Independent Random Variables*, Translated from the Russian by A. A. Brown, Springer-Verlag, Berlin-Heidelberg, New-York, 1975.

[24] T. Porter and I. Simon, Random insertion into a priority queue structure, *IEEE Transactions on Software Engineerings*, **SE-1** (1975) 292–298.

[25] F. P. Preparata and M. I. Shamos, *Computational Geometry*, Springer-Verlag, Berlin, 1985.

[26] R. Schaffer and R. Sedgewick, The analysis of heapsort, *Journal of Algorithms* **15** (1993) 76–100.

[27] R. Sprugnoli, Recurrence relations on heaps, *Algorithmica*, **15** (1996), 467–480.

[28] I. Wegener, Bottom-up-Heapsort, a new variant of HEAPSORT beating, on an average, Quicksort (if n is not very small), *Theoretical Computer Science*, **118** (1993) 81–98.

[29] J. W. J. Williams, Algorithm 232: Heapsort, *Communications of the ACM*, **7** (1964) 347–348.

Hsien-Kuei Hwang
Institute of Statistical Science
Academia Sinica
Taipei 115
TAIWAN
hkhwang@stat.sinica.edu.tw

Jean-Marc Steyaert
Laboratoire d'Informatique
École Polytechnique
91128 Palaiseau
FRANCE
steyaert@lix.polytechnique.fr

Trends in Mathematics, © 2002 Birkhäuser Verlag Basel/Switzerland

A Combinatorial Problem Arising in Information Theory: Precise Minimax Redundancy for Markov Sources

Philippe Jacquet and Wojciech Szpankowski

ABSTRACT: *Redundancy of a code is defined as the excess of the code length over the optimal code length. When the source of information is unknown, then one wants to design the best code for the worst source (within the class of sources that are being considered). This is called the minimax redundancy. It can come in two flavors: either on average or the worst case. The latter is known as the maximal minimax redundancy, and it is studied in this paper for Markovian sources. Surprisingly, this problem led us to an interesting combinatorial problem on directed graphs that we shall solve using analytic tools. To be more precise, we need to count the number of Eulerian cycles in a directed multi–graph. The maximal minimax redundancy turns out to be a sum over such Eulerian paths. In particular, we shall prove that the maximal minimax redundancy for Markov sources of order r is asymptotically equal to $\frac{1}{2}m^r(m-1)\log n + \log A_m + O(1/n)$, where n is the length of source sequences, m is the size of the alphabet and A_m is an explicit constant that depends on m.**

1 Introduction

We start with a quick introduction of the *redundancy problem* of source coding (better known as data compression). A code $C_n : \mathcal{A}^n \to \{0,1\}^*$ is defined as a mapping from the set $\mathcal{A}^n$ of all source sequences $x_1^n = (x_1, \ldots, x_n)$ of length n over the finite alphabet $\mathcal{A}$ to the set $\{0,1\}^*$ of all binary sequences (i.e., block-to-variable code). For a probabilistic source model, we let $P(x_1^n)$ be the probability of the message x_1^n and $L(C_n, x_1^n)$ be the code length for x_1^n.
It is known that the entropy $H_n(P) = -\sum_{x_1^n} P(x_1^n)\log P(x_1^n)$ is the absolute lower bound on the *expected* code length, where $\log := \log_2$ throughout the paper will denote the binary logarithm. Hence $-\log P(x_1^n)$ can be viewed as the "ideal" code length and therefore one may ask by how much the code length $L(C_n, x_1^n)$ exceeds the ideal code length, either for individual sequences or on average. The *pointwise redundancy* is

$$R_n(C_n, P; x_1^n) = L(C_n, x_1^n) + \log P(x_1^n),$$

while the *average redundancy* $\bar{R}_n(C_n, P)$ and the *maximal* redundancy $R_n^*(C_n, P)$ are defined, respectively, as

$$\begin{aligned}\bar{R}_n(C_n, P) &= \mathbf{E}_P[R_n(C_n, P; X_1^n)] = \mathbf{E}_P[L(C_n, X_1^n)] - H_n(P),\\ R_n^*(C_n, P) &= \max_{x_1^n}[R_n(C_n, P; x_1^n)],\end{aligned}$$

*This work was supported by the NSF Grants CCR-9804760 and CCR-0208709, and contract 1419991431A from sponsors of CERIAS at Purdue.

where the underlying probability measure P represents a particular source model and $\mathbf{E}$ denotes the expectation.

In practice, however, the probability distribution (i.e., source) P is unknown, hence one is looking for optimal codes for sources with unknown probabilities. In fact, for unknown probabilities, the redundancy rate can be also viewed as the penalty paid for estimating the underlying probability measure. The *redundancy-rate problem* consists in determining for a class of sources $\mathcal{S}$ the rate of growth of the minimax quantities either on average

$$\bar{R}_n(\mathcal{S}) = \min_{C_n} \max_{P\in\mathcal{S}} [\bar{R}_n(C_n, P)], \tag{1}$$

or in the worst case

$$R_n^*(\mathcal{S}) = \min_{C_n\in C} \max_{P\in\mathcal{S}} [R_n^*(C_n, P)], \tag{2}$$

where C_n denotes prefix codes (*i.e.*, satisfying Kraft's inequality).

In this paper we deal with the maximal *minimax redundancy* $R_n^*(\mathcal{S})$ defined by (2) for Markov sources. Precise asymptotics of $R_n^*(\mathcal{S})$ for memoryless sources are known (cf. [6, 11, 23]). But there is lack of similar results for Markovian sources $\mathcal{M}_r$ of order r. Rissanen [18] obtained the first two terms of a related quantity called the *regret function*, while Atteson [1] derived similar asymptotics for the *average* minimax redundancy of $\mathcal{M}_r$. In this paper, we focus on analyzing Markov sources of order one, $\mathcal{M}_1$, and build a novel analytic framework upon which we obtain asymptotics of $R_n^*(\mathcal{M}_r)$. Here, we present our preliminary results. Interestingly enough, in the course of deriving our findings we encounter some combinatorial problems on directed multigraphs that are quintessential to the solution of our problem. Among others, we must enumerate Eulerian paths in a multigraph whose multiplicity is expressed by a given matrix satisfying the so called conservation flow property. In fact, this quantity turns out to be the number of types for Markov sources (cf. [28]).

2 Main Results

We first review some known results about the *maximal* minimax redundancy $R_n^*(\mathcal{S})$ for a class of sources $\mathcal{S}$ defined above in (2). In 1987 Shtarkov [21] established the following bound

$$\log\left(\sum_{x_1^n} \sup_{P\in\mathcal{S}} P(x_1^n)\right) \le R_n^*(\mathcal{S}) \le \log\left(\sum_{x_1^n} \sup_{P\in\mathcal{S}} P(x_1^n)\right) + 1.$$

Recently, Drmota and Szpankowski [11] replaced these bounds by an exact formula, namely

$$R_n^*(\mathcal{S}) = \log\left(\sum_{x_1^n} \sup_{P\in\mathcal{S}} P(x_1^n)\right) + R^{GS}(Q^*), \tag{3}$$

where $R^{GS}(Q^*)$ is the maximal redundancy of the generalized Shannon code (i.e., a code which assigns $\lceil \log 1/P(x_1^n)\rceil$ for some source sequences x_1^n and $\lfloor \log 1/P(x_1^n)\rfloor$

for remaining source sequences) designed for the maximal likelihood distribution

$$Q^*(x_1^n) = \frac{\sup_P P(x_1^n)}{\sum_{x_1^n} \sup_P P(x_1^n)}. \tag{4}$$

In $R_n^{GS}(Q^*)$ the distribution Q^* is assumed to be known.
Formula (3) suggests that $R_n^*(\mathcal{S}) = \log D_n(\mathcal{S}) + R_n^{GS}(Q^*)$ where

$$D_n := D_n(\mathcal{S}) = \sum_{x_1^n} \sup_{P\in\mathcal{S}} P(x_1^n). \tag{5}$$

The above decomposition of $R_n^*(\mathcal{S})$ is quite interesting. The first part $d_n(\mathcal{S}) := \log D_n(\mathcal{S})$ can be proved to be a nondecreasing function of n that depends *only* on the underlying class $\mathcal{S}$ of probability distributions, while the second (bounded) term contains a coding component and may be a fluctuating function of n.

In general, our goal is to estimate asymptotically both terms of $R_n^*(\mathcal{S})$ for a class of memoryless sources, Markov sources, mixing sources, and other non-parameterized class of sources. We aim to develop precise results of practical consequence using a combination of tools from average case analysis of algorithms, information theory, and combinatorics (cf. [25]).

Let us first consider, as a warm-up, a memoryless sources $\mathcal{M}_0$ over an m-ary alphabet. This is well-studied problem (cf. [2, 18, 23]). Indeed, it is easy to see that for the memoryless case (5) becomes

$$D_n(\mathcal{M}_0) = \sum_{k_1+\cdots+k_m=n} \binom{n}{k_1,\ldots,k_m} \left(\frac{k_1}{n}\right)^{k_1} \cdots \left(\frac{k_m}{n}\right)^{k_m}, \tag{6}$$

where k_i is the number of elements in x_1^n containing symbol $i \in \mathcal{A}$. Indeed, we have

$$\sup_{p_1,\ldots,p_m} p_1^{k_1}\cdots p_m^{k_m} = \left(\frac{k_1}{n}\right)^{k_1} \cdots \left(\frac{k_m}{n}\right)^{k_m}$$

and $\binom{n}{k_1,\ldots,k_m}$ is equal to the number of strings x_1^n having k_i symbols $i \in \mathcal{A}$.

In [23] we argued that such a sum can be analyzed through the so-called *tree generating function.* Let us define

$$B(z) = \sum_{k=0}^{\infty} \frac{k^k}{k!} z^k = \frac{1}{1-T(z)}, \tag{7}$$

where $T(z)$ satisfies $T(z) = ze^{T(z)}$ and also $T(z) = \sum_{k=1}^{\infty} \frac{k^{k-1}}{k!} z^k$. Let us now define another tree-like generating function, namely: $D(z) = \sum_{k=0}^{\infty} \frac{k^k}{k!} D_k(\mathcal{S})$. Then the convolution formula for generating functions (cf. [25]) immediately implies

$$D(z) = (B(z))^m$$

which further leads to

$$D_n = \frac{n!}{n^n} [z^n] (B(z))^m,$$

where $[z^n]f(z)$ is the standard notation for the coefficient of $f(z)$ at z^n. To extract an asymptotic expansion from the above one must know the singular expansion of $B(z)$. But a minor modification of [7] gives

$$B(z)=\frac{1}{\sqrt{2(1-ez)}}+\frac{1}{3}-\frac{\sqrt{2}}{24}\sqrt{(1-ez)}+\frac{4}{135}(1-ez)-\frac{23\sqrt{2}}{1728}(1-ez)^{3/2}+O((1-ez)^2)\,.$$

Then an application of the Flajolet and Odlyzko *singularity analysis* [12] yields for $d_m(\mathcal{M}_0)=\log D_n(\mathcal{M}_0)$

$$\begin{aligned} d_n(\mathcal{M}_0) &= \frac{m-1}{2}\log\left(\frac{n}{2}\right)+\log\left(\frac{\sqrt{\pi}}{\Gamma(\frac{m}{2})}\right)+\frac{\Gamma(\frac{m}{2})m}{3\Gamma(\frac{m}{2}-\frac{1}{2})}\cdot\frac{\sqrt{2}}{\sqrt{n}} \\ &+\left(\frac{3+m(m-2)(2m+1)}{36}-\frac{\Gamma^2(\frac{m}{2})m^2}{9\Gamma^2(\frac{m}{2}-\frac{1}{2})}\right)\cdot\frac{1}{n}+O\left(\frac{1}{n^{3/2}}\right)\end{aligned}$$

for large n. The first two terms were known before (cf. [22, 26]) *but not* the constant term of $R_n^*(\mathcal{M}_0)$ which involves $R_n^{GS}(Q^*)$. In [11] it was proved that

$$R_n^{GS}(Q^*)=-\frac{\ln\frac{1}{m-1}\ln m}{\ln m}+o(1).$$

In general, the term $o(1)$ can not be improved. Putting everything together we obtain

$$R_n^*(\mathcal{M}_0)=\frac{m-1}{2}\log\left(\frac{n}{2}\right)-\frac{\ln\frac{1}{m-1}\ln m}{\ln m}+\log\left(\frac{\sqrt{\pi}}{\Gamma(\frac{m}{2})}\right)+o(1).$$

Now, we turn our attention to the main topic of this paper, namely, maximal minimax redundancy of Markov sources. Markov sources $\mathcal{M}_1$ (in general $\mathcal{M}_r$ for Markov of order r) still present a challenge even if we know that $R_n^*(\mathcal{M}_r)\sim\frac{K}{2}\log n$ (cf. [2, 18, 26]) where $K=m(m-1)$ is the number of degrees of freedom (or more precisely, the dimension of the parameter space). There are some results (cf. [18]) concerning the second term of the asymptotic expansion for the minimax *regret* function [18] and the *average* minimax redundancy [1], but not for the maximal redundancy. We should point out that our methodology is quite different from the others since we apply throughout analytic combinatorics and algorithmics while previous attempts were pure probabilistic. We aim at deriving a full asymptotic expansion, however, in this conference version we present only the first two terms and an error term.

We start with a precise formulation of the problem. We concentrate on the non-fluctuating part $D_n(\mathcal{M}_1)$. After some calculation, we arrive at

$$D_n(\mathcal{M}_1)=\sum_{[k]}N_{[k]}\left(\frac{k_{11}}{k_1}\right)^{k_{11}}\cdots\left(\frac{k_{m,m}}{k_m}\right)^{k_{m,m}},$$

where $k_i=\sum_{j=1}^m k_{ij}$, the matrix† $[k]=\{k_{ij}\}_{i,j=1}^m$ is an integer matrix whose (i,j)-th coefficients satisfy the condition $\sum_{1\le i,j\le m}k_{ij}=n-1$. In the above, k_{ij}

†To simply notation, we shall write $[a]$ for a matrix $\{a_{i,j}\}_{i,j=1}^m$.

denotes the number of pairs (i, j) in x_1^n, that is the number of times symbol j follows symbol i. The quantity $N_{[k]}$ is the number of string x_1^n generated over $\mathcal{A}$ having k_{ij} positions in x_1^n where j follows i. It is known under the name *frequency count* (cf. [3]), but in fact it is the number of types for a Markov chain. We call the matrix $[k]$ the *pair occurrence* (PO) matrix for x_1^n.

It turns out that an estimation of $N_{[k]}$ leads to an interesting combinatorial problem, namely, enumeration of Eulerian paths in a directed multigraph with multiplicity matrix $[k]$. To find an analytic expression for this expression, we will go through the concept of *cyclic strings* in which the last symbol is followed by the first symbol. Cyclic strings have their PO matrices $[k]$ satisfying the following two conditions:

$$\sum_{1\le i,j\le m} k_{ij} = n, \tag{8}$$

$$\sum_{j=1}^{m} k_{ij} = \sum_{j=1}^{m} k_{ji}, \qquad \forall\, i. \tag{9}$$

Our main result proved in the next section can be formulated as follows.

Theorem 2.1. *Let $\mathcal{M}_1$ be a class of Markov sources over a finite alphabet $\mathcal{A}$ of size m. Then the non-fluctuating term $D_n(\mathcal{M}_1)$ of the maximal minimax redundancy attains the following asymptotics as $n \to \infty$*

$$D_n(\mathcal{M}_1) = \left(\frac{n}{2\pi}\right)^{m(m-1)/2} A_m \times \left(1 + O\left(\frac{1}{n}\right)\right) \tag{10}$$

with

$$A_m = \int_{\mathcal{K}(1)} mF_m(y_{ij}) \prod_i \frac{\sqrt{\sum_j y_{ij}}}{\prod_j \sqrt{y_{ij}}} d[y_{ij}]$$

where $\mathcal{K}(1) = \{y_{ij} : \sum_{ij} y_{ij} = 1\}$ and $F_m(\cdot)$ is a polynomial expression of degree $m-1$ defined in Theorem 3.4 of the next section.

In particular, for a binary alphabet ($m = 2$) we found that $A_2 = 16 \cdot G$ where $G = \sum_i \frac{(-1)^i}{(2i+1)^2} \approx 0.915965594$ is the Catalan constant.

Next, we extend Theorem 2.1 to Markov sources of order r.

Theorem 2.2. *Let $\mathcal{M}_r$ be a class of Markov sources of order r over a finite alphabet $\mathcal{A}$ of size m. Then the non-fluctuating term $D_n(\mathcal{M}_r)$ of the maximal minimax redundancy attains the following asymptotics as $n \to \infty$*

$$D_n(\mathcal{M}_r) = \left(\frac{n}{2\pi}\right)^{m^r(m-1)/2} A_m^r \times \left(1 + O\left(\frac{1}{n}\right)\right) \tag{11}$$

where A_m^r is defined in Theorem 3.9 of the next section.

3 Analysis and Proofs

In this section we prove our main result, including the enumeration of Eulerian paths in a directed multigraph with a given multiplicity. We observe that the main novelty of our approach lies in analytical treatment of certain sums over matrices satisfying the conservation flow property.

3.1 Combinatorics on strings

A cyclic string is a string in which the first symbol follows the last one. If X is a cyclic string we call $k_{ij}(X)$ the number of positions in X where symbol j follows symbol i. It is clear that in cyclic strings we have one pair occurrence more than in linear strings, that is, $\sum_{ij} k_{ij} = n$, where n is the length of the cyclic string. We call the matrix $[k(X)] = \{k_{ij(X)}\}_{i,j=1}^{m}$ the *pair occurrence* (PO) matrix of X. The PO matrix obviously satisfies the *conservation flow property* (CFP) defined in (9). Let $\mathcal{F}_*$ be the set of all the integer matrices which satisfy the CFP. For a given n, we let $\mathcal{F}_n$ be a subset of $\mathcal{F}_*$ consisting of matrices $[k]$ such that $\sum_{ij} k_{ij} = n$.

The key parameters needed to enumerate cyclic strings with a given matrix $[k] \in \mathcal{F}_n$ are the so called *frequency counts* defined as follows:

- The number $N_{[k]}$ (i.e., frequency count) of cyclic strings for which $[k]$ is the PO matrix;
- The number $N_{[k]}^{a}$ of cyclic strings starting with a symbol a and having $[k]$ as the PO matrix;
- The number $N_{[k]}^{b,a}$ of cyclic strings starting with a pair of symbols ba and for which $[k]$ is the PO matrix.

Notice that the frequency count $N_{[k]}^{b,a}$ is of most interest to linear strings since it gives the number of strings starting with symbol a and ending with symbol b as a function of the PO matrix $[k]$. Indeed, we know that one occurrence of the pair (b, a) has to be removed to make it a linear string.

Let us start with some notation. Throughout, we shall use the following quantity:

$$B_{[k]} = \prod_i \frac{(k_i)!}{\prod_j (k_{i,j})!} = \binom{k_1}{k_{11} \cdots k_{1m}} \cdots \binom{k_m}{k_{m1} \cdots k_{mm}} \tag{12}$$

where, we recall, $k_i = \sum_j k_{ij}$. Let also $[z] = \{z_{ij}\}_{ij=1}^{m}$ be a complex $m \times m$ matrix and $[k]$ an integer matrix. In the sequel, we write $[z]^{[k]} = \prod_{ij} z_{ij}^{k_{ij}}$. Let also $g_{[k]}$ be a sequence of scalars indexed by matrix $[k]$, and

$$g([z]) = \sum_{[k]} g_{[k]} [z]^{[k]}$$

be its generating function. We denote

$$\mathcal{F}g(z) = \sum_{[k] \in \mathcal{F}_*} g_{[k]} [z]^{[k]}.$$

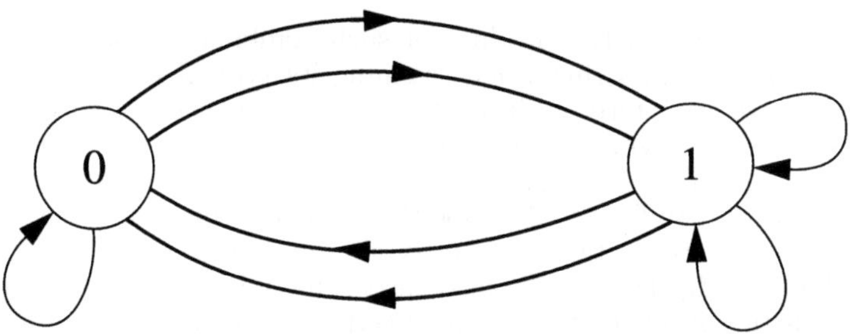

Figure 1: A directed multigraph for a binary alphabet $\mathcal{A} = \{0,1\}$ with $[k] = k_{00} = 1$, $k_{01} = 2$, $k_{10} = 2$ and $k_{11} = 2$.

For example,

$$B([z]) = \sum_{[k]} B_{[k]}[z]^{[k]} = \prod_{a\in\mathcal{A}} (1 - \sum_{b\in\mathcal{A}} z_{a,b})^{-1}.$$

We will also write $B_{\mathcal{A}}([z]) := \mathcal{F}B([z])$.

We have the following theorems:‡

Theorem 3.1. *For $n \geq 1$ and $[k] \in \mathcal{F}_n$ the frequency count $N^a_{[k]}$ is the coefficient of $[z]^{[k]}$ of $\frac{B([z])}{B_{\mathcal{A}-\{a\}}([z])}$, that is,*

$$N^a_{[k]} = [z]^{[k]} \frac{B([z])}{B_{\mathcal{A}-\{a\}}([z])}, \tag{13}$$

where $B_{\mathcal{A}-\{a\}}([z])$ is the generating function of $B_{[k]}$ over $\mathcal{A} - \{a\}$ satisfying the conservation flow property.

Proof The proof proceeds via the enumeration of Euler cycles (paths) in a directed multigraph over m vertices. In such a graph vertices are labeled by symbols from the alphabet $\mathcal{A}$ with the edge multiplicity given by the matrix $[k]$: there are k_{ij} edges from vertex i to j. The number of Eulerian paths starting from vertex 1 in a such multigraph is equal to $N^1_{[k]}$. It is illustrated in Figure 1 for $\mathcal{A} = \{0,1\}$ where the matrix $[k]$ is

$$[k] = \begin{bmatrix} 1 & 2 \\ 2 & 2 \end{bmatrix}.$$

Let us now define a "combination" as a set of m arbitrary permutations, one permutation per vertex, corresponding to a combination of the edges that depart from the vertex. In other words, for a given vertex i there are

$$\binom{k_i}{k_{i1} \cdots k_{im}}$$

‡Recently, it was pointed to us by Dr. Marcelo Weinberger, HPL, that $N_{[k]}$ was studied before by Whittle [28] (cf. [3]). Our formula is a generating function version of Whittle's formula.

combinations, and $B_{[k]}$ is the product of such combinations as shown in (12). Observe that for a given string when scanning its symbol we trace an Eulerian path. However, we are interested in an "inverse" problem: given an initial symbol a and a matrix $[k]$ satisfying the flow property (with non zero weights on symbol a), does a combination corresponds to a string x_1^n, that is, does it trace an Eulerian path. Observe that it suffices to describe the string obtained by successive symbol visits by the combination, starting from symbol a: the next symbol is the vertex visited by the first edge from a, the next-next symbol is the symbol visited by the first edge departing from the the previous vertex (if different of a, otherwise it is the symbol visted by the second edge departing from a), and so on. So that, the next symbol to symbol i on its kth occurence is the symbol visited by the kth edge departing from i in the multigraph combination. The answer to the above question is very likely negative, since the process may prematurely end at a symbol $c \in \mathcal{A}$ exhausting all edges departing from symbol c but not using all the other edges of the multigraph (i.e., the length of the traced string is shorter than n). Let $[k_1]$ be the matrix composed of the remaining non-visited edges of the multigraph (the matrix $[k]-[k_1]$ has been exhausted by the trace). Notice that matrix $[k_1]$ satisfies the flow property but the row and column corresponding to symbol a contain only zeros. Notice also that these edges can be taken in any combination, it will not prevent the visiting process to end the same way on symbol a, provided that the combination on $[k_1]$ are always concatenated after the combination over $[k]-[k_1]$ that have been exhausted by the trace.

Given that $[k]$ and $[k']$ are members of $\mathcal{F}_*$, let $N^a_{[k],[k']}$ be the number of combinations on the multigraph (*i.e.* matrix) $[k]$, starting on symbol a, that leaves an unexplored set of edges which corresponds to matrix $[k']$. Notice that $k'_a = 0$. We have $N^a_{[k],[0]} = N^a_{[k]}$, but also the following

$$N^a_{[k],[k']} = N^a_{[k]-[k']} \times B_{[k']}.$$

Summing over all matrices $[k']$ we get all possible combinations on $[k]$, that is, $\sum_{[k']} N^a_{[k],[k']} = B_{[k]}$. This leads to the following

$$B_{[k]} = \sum_{[k'],k'_a=0} N^a_{[k]-[k']} \times B_{[k']}.$$

Summing now over all $[z]^{[k]}$ such that $k_a \neq 0$ it yields

$$\sum_{[k]\in\mathcal{F}_*,k_a\neq 0} B_{[k]}[z]^{[k]} = (\sum_{[k]} N^a_{[k]}[z]^{[k]}) \times \left(\sum_{[k]\in\mathcal{F}_*,k_a=0} B_{[k]}[z]^{[k]} \right)$$

Denoting $N^a([z]) = \sum_{[k]} N^a_{[k]}[z]^{[k]}$, we finally arrive at

$$B_{\mathcal{A}}([z]) - B_{\mathcal{A}-\{a\}}([z]) = N^a([z]) B_{\mathcal{A}-\{a\}}([z]).$$

Since $\mathcal{F}_*$ is an additive semi-group, we observe that for $[k] \in \mathcal{F}_n$ we also have

$$[[z]^{[k]}]\frac{B_{\mathcal{A}}([z])}{B_{\mathcal{A}-\{a\}}([z])} = [[z]^{[k]}]\frac{B([z])}{B_{\mathcal{A}-\{a\}}([z])},$$

which completes the proof. ∎

The frequency count $N_{[k]}$ can be computed as

$$N_{[k]} = [[z]^{[k]}]B([z]) \sum_{a \in \mathcal{A}} (B_{\mathcal{A}-\{a\}}([z]))^{-1}.$$

In the next theorem we compute $N_{[k]}^{a,b}$.

Theorem 3.2. *For $n \geq 1$ and $[k] \in \mathcal{F}_n$, the frequency count $N_{[k]}^{b,a}$ is the coefficient of $[z]^{[k]}$ in $\frac{B([z]) z_{b,a}}{B_{\mathcal{A}-\{b\}}([z])}$.*

Proof The proof proceeds the same way as in the previous theorem except that we have to consider combinations such that the first edge departing from symbol b is always (b,a) (*i.e.* the first occurrence of symbol b is always followed by symbol a), and we let $B_{[k]}^{ba}$ be the number of such combinations. Observe that $B_{[k]}^{ba} = B_{[k]} \frac{k_{ba}}{k_b} = B_{[k]-[\delta_{ba}]}$, where $[\delta_{ba}]$ is the matrix with all zero coefficients except the ba-th coefficient which is set to one. Let $[k] \in \mathcal{F}_*$, using the convolution we find

$$B_{[k]}^{ba} = \sum_{[k'], k'_b = 0} N_{[k]-[k']}^{ba} \times B_{[k']}.$$

Computing the generating function we arrive at

$$\sum_{[k] \in \mathcal{F}_*, k_{ba} \neq 0} B_{[k]}^{ba} [z]^{[k]} = \left(\sum_{[k]} N_{[k]}^{ba} [z]^{[k]} \right) \times \left(\sum_{[k] \in \mathcal{F}_*, k_b = 0} B_{[k]} [z]^{[k]} \right).$$

In other words, $\sum_{[k] \in \mathcal{F}_*, k_{ba} \neq 0} B_{[k]}^{ba} [z]^{[k]} = N^{ba}([z]) B_{\mathcal{A}-\{a\}}([z])$. Using the fact that

$$\sum_{[k] \in \mathcal{F}_*, k_{ba} \neq 0} B_{[k]}^{ba} [z]^{[k]} = \mathcal{F} B^{ba}([z]),$$

with

$$\begin{aligned} B^{ba}([z]) &= \sum_{[k], k_{ba} > 0} B_{[k]}^{ba} [z]^{[k]} = \sum_{[k], k_{ba} > 0} B_{[k]-[\delta_{ba}]} [z]^{[k]} \\ &= B([z]) z_{ba} \end{aligned}$$

we complete the proof. ∎

The next result is important. It provides asymptotics of $N_{[k]}^{b,a}$. We were not able to verify whether such asymptotics appeared before in literature. The closest result we found was a bound suggested by Boza [4]. We recall that $N_{[k]}^{ba}$ is the number of types of the underlying Markov chain.

Theorem 3.3. *For a PO matrix* $[k]$ *such that* $k_{ba} > 0$, *we have the following asymptotic estimate*

$$N_{[k]}^{b,a} \sim \frac{k_{ba}}{k_b B_{\mathcal{A}-\{b\}}([k^*])} B_{[k]} \tag{14}$$

where $[k^*] > 0$ *is the matrix whose the* ij*-th coefficient is* k_{ij}/k_i, *that is,* $[k^*] = \{k_{ij}/k_i\}_{i,j=1}^{m}$.

Proof First, we derive an asymptotic estimate of $B_{[k]}$. We should point out that the method we propose is not very satisfying since the result is not explicit but for our purposes it suffices. (For more explicit results the Stirling formula performs much better.)
We start with the identity:

$$B_{[k]} = \left(\frac{1}{2\mathbf{i}\pi}\right)^{m^2} \prod_{i,j} \oint \frac{dz_{ij}}{z_{ij}} \times \frac{B([z])}{[z]^{[k]}}. \tag{15}$$

Recall that $B([z]) = \sum_{[k]} B_{[k]}[z]^{[k]} = \prod_i (1 - \sum_j z_{ij})^{-1}$. We make the change of variable $z_{ij} = \frac{k_{ij}}{k_i} e^{-\mathbf{i}t_{ij}/k_{ij}}$, where as before $k_i = \sum_j k_{ij}$. We have

$$\begin{aligned} 1 - \sum_j z_{ij} &= \sum_j \frac{k_{ij}}{k_i}(1 - e^{-\mathbf{i}t_{ij}/k_{ij}}) \\ &= \frac{\mathbf{i}}{k_i}\left(\sum_j t_{ij}\right)\left(1 + O\left(\frac{1}{n}\right)\right), \qquad k_{ij} = O(n). \end{aligned}$$

Thus

$$\begin{aligned} B_{[k]} &= (1 + O(n^{-1}))\left(\frac{1}{2\mathbf{i}\pi}\right)^{m^2} \prod_i \frac{\prod_j k_{ij}^{k_{ij}-1}}{k_i^{k_i-1}} \prod_{i,j} \int_{-\pi k_{ij}}^{\pi k_{ij}} dt_{ij} \prod_i \left(\frac{1}{\sum_j t_{ij}}\right) \exp(\mathbf{i}\sum_j t_{ij}) \\ &= (1 + O(n^{-1}))\left(\frac{1}{2\mathbf{i}\pi}\right)^{m^2} \prod_i \frac{\prod_j k_{ij}^{k_{ij}-1}}{k_i^{k_i-1}} \prod_{i,j} \int_{-\infty}^{+\infty} dt_{ij} \prod_i \left(\frac{1}{\sum_j t_{ij}}\right) \exp(\mathbf{i}\sum_j t_{ij}) \end{aligned}$$

We now directly apply this method it to the estimate of $N_{[k]}^{ba}$. Since

$$N_{[k]}^{ba} = \left(\frac{1}{2\mathbf{i}\pi}\right)^{m^2} \prod_{i,j} \oint \frac{dz_{ij}}{z_{ij}} \times \frac{B([z])z_{ba}}{B_{\mathcal{A}-\{b\}}([z])[z]^{[k]}}$$

Therefore, using the same change of variable $z_{ij} = \frac{k_{ij}}{k_i}\exp(-\mathbf{i}t_{ij})$, we obtain

$$\begin{aligned} N_{[k]}^{ba} = &\left(1 + O\left(\frac{1}{n}\right)\right)\left(\frac{1}{2\mathbf{i}\pi}\right)^{m^2} \prod_i \frac{\prod_j k_{ij}^{k_{ij}-1}}{k_i^{k_i-1}} \prod_{i,j} \int_{-\infty}^{+\infty} dt_{ij} \\ &\prod_i \frac{1}{\sum_j t_{ij}} \exp(\mathbf{i}\sum_j t_{ij}) \frac{z_{ba}}{k_b B_{\mathcal{A}-\{b\}}([z])} \end{aligned}$$

with $z_{ba} = \frac{k_{ba}}{k_b}\exp(-\mathrm{i}t_{ba}/k_{ba}) = k^*_{ba}(1+O(1/n))$ and $[z] = [k^*]+O(1/n))$. We denote by $[k^*]$ the matrix whose ij-th coefficient is k_{ij}/k_i. Since the function $B_{\mathcal{A}-\{b\}}([z])$ is defined and bounded in a neighborhood of $[k^*]$, we have $B_{\mathcal{A}-\{b\}}([z]) = B_{\mathcal{A}-\{b\}}([k^*])(1+O(\frac{1}{n}))$. Therefore

$$
\begin{aligned}
N^{ba}_{[k]} &= \left(1+O\left(\frac{1}{n}\right)\right)\left(\frac{1}{2\mathrm{i}\pi}\right)^{m^2}\prod_i \frac{\prod_j k_{ij}^{k_{ij}-1}}{k_i^{k_i-1}} \\
&\quad \prod_{i,j}\int_{-\infty}^{+\infty} dt_{ij}\prod_i\left(\frac{1}{\sum_j t_{ij}}\exp\left(\sum_j t_{ij}\right)\frac{k^*_{ba}}{B_{\mathcal{A}-\{b\}}([k^*])}\right. \\
&= \left(1+O\left(\frac{1}{n}\right)\right)B_{[k]}\frac{k_{ba}}{k_b B_{\mathcal{A}-\{b\}}([k^*])}.
\end{aligned}
$$

This completes the proof. ∎

3.2 Asymptotics of redundancy

We first restate our main result in a more precise form.

Theorem 3.4. *When $n\to\infty$ the following holds*

$$D_n = (n/2\pi)^{(m^2-m)/2}A_m.\left(1+O(n^{-1})\right)$$

with

$$A_m = \int_{\mathcal{K}(1)} mF_m([y])\prod_i \frac{\sqrt{\sum_j y_{ij}}}{\prod_j \sqrt{y_{ij}}}d[y_{ij}] \tag{16}$$

where $F_m([y]) = \sum_b(\det_{bb}(1-[y^*]))^{-1}$, and $[y^*]$ is the matrix whose ij-th coefficient is $y_{ij}/\sum_{j'} y_{ij'}$, and $\det_{ij}([a])$ is the determinant of the matrix obtained by omitting the ith row and the jth column from the matrix $[a]$.

In order to establish this result, we need to estimate certain sums over matrices $[k]\in\mathcal{F}_n$, We start with the following lemma.

Lemma 3.5. *Let $g([z]) = \sum_{[k]} g_{[k]}[z]^{[k]}$ be the generating function of a complex matrix $[g]$. Then*

$$\mathcal{F}g([z]) := \sum_{n\geq 0}\sum_{[k]\in\mathcal{F}_n} g_{[k]}[z]^{[k]} = \left(\frac{1}{2\mathrm{i}\pi}\right)^m \oint\frac{dx_1}{x_1}\cdots\oint\frac{dx_m}{x_m}g([z\frac{x_i}{x_j}]) \tag{17}$$

with the convention that the ij-th coefficient of $[z\frac{x_i}{x_j}]$ is $z_{ij}\frac{x_i}{x_j}$. In other words, $[z\frac{x_i}{x_j}] = \Delta^{-1}(x)[z]\Delta(x)$ where $\Delta(x)$ is the diagonal matrix with coefficient $x_1,\ldots,$ x_m. By change of variable $x_i = \exp(\mathrm{i}\theta_i)$ we also have

$$\mathcal{F}g([z]) = \frac{-i}{(2\pi)^m}\int_{-\pi}^{\pi}d\theta_1\cdots\int_{-\pi}^{\pi}d\theta_m g([z\exp((\theta_i-\theta_j)i)]$$

where $[z\exp(\theta_i-\theta_j)] = \exp(-\Delta(\theta))[z]\exp(\Delta(\theta))$.

Proof. We show (17). We proceed as follows

$$g([z\frac{x_i}{x_j}]) = \sum_{[k]} g_{[k]}[z]^{[k]} \prod_{i=1}^{m} x_i^{\sum_j k_{ij} - \sum_i k_{ij}} \tag{18}$$

Therefore $\mathcal{F}g([z]) = \left[x_1^0 \cdots x_m^0\right] g([z\frac{x_i}{x_j}])$ and the result follows from the Cauchy integration (17). ∎

Corollary 3.6. *We have* $\mathcal{F}B([z]) = (\det(\mathbf{I} - [z]))^{-1}$, *where* **I** *is the identity* $m \times m$ *matrix.*

Proof: For completeness we give a proof that is due to Whittle [28] who showed that

$$\left(\frac{1}{2\mathrm{i}\pi}\right)^m \oint dx_1 \cdots \oint dx_m \prod_i \left(\sum_j a_{ij}x_j\right)^{-1} = (\det([a]))^{-1} \tag{19}$$

provided that $[a]$ is not singular. Indeed, one makes the linear change of variables $y_i = \sum_j a_{ij}x_j$ to obtain

$$\begin{aligned}\left(\frac{1}{2\mathrm{i}\pi}\right)^m \oint dx_1 \cdots \oint dx_m \prod_i \left(\sum_j a_{ij}x_j\right)^{-1} &= (\det([a]))^{-1} \left(\frac{1}{2\mathrm{i}\pi}\right)^m \oint \frac{dy_1}{y_1} \cdots \oint \frac{dy_m}{y_m} \\ &= (\det([a]))^{-1}\end{aligned}$$

which completes the proof. ∎

Remark: From the above corollary one concludes that $B_{\mathcal{A}-\{a\}}([z]) = (\det_{aa}(\mathbf{I} - [z]))^{-1}$, where $\det_{ij}([a])$ is the (i,j)th coefficient of the covariant matrix of $[a]$.

We will also need a continuous version of Lemma 3.5. Let $\mathcal{K}(x)$ the hyper-polygon (simplex) of matrices $[y_{ij}]$ with non-negative *real* coefficients that satisfies the conservation flow property and such that $\sum_{ij} y_{ij} = x$. Recall that $\mathcal{F}_n$ is the set of non-negative integer matrices $[k]$ that belongs to $\mathcal{K}(n)$. Let $a(x)$ the area (hyper-volume) of $\mathcal{K}(x)$.

Lemma 3.7. *Let* $g([x])$ *be a function of real matrices* $[x]$. *Let* $G([t])$ *be the Laplace transform of* g, *that is,*

$$G([t]) = \int g([x]) \exp\left(-\sum_{ij} t_{ij}x_{ij}\right) d[x],$$

and let

$$\tilde{G}([t]) = \int_0^\infty dy \int_{\mathcal{K}(y)} g([x]) \exp(-\sum_{ij} t_{ij}x_{ij}) d[x].$$

We have

$$\tilde{G}([t]) = (\frac{1}{2\mathrm{i}\pi})^m \int_{-\mathrm{i}\infty}^{+\mathrm{i}\infty} d\theta_1 \cdots \int_{-\mathrm{i}\infty}^{+\mathrm{i}\infty} d\theta_m G([t + \theta_i - \theta_j]) \tag{20}$$

where $[t + \theta_i - \theta_j]$ *is a matrix whose the* ij*-th coefficient is* $t_{ij} + \theta_i - \theta_j$.

This lemma allow us to prove the following result that we shall use in the asymptotic evaluation of D_n.

Lemma 3.8. *We have*

$$\frac{|\mathcal{F}_n|}{a(n)} = 1 + O(1/n),$$

that is, the density of $\mathcal{F}_n$ *in* $\mathcal{K}(n)$ *tends to* 1 *for large* n.

Proof We will show that $|\mathcal{F}_n| = a(n)(1 + O(\frac{1}{n}))$. First, we give an estimate of $a(n)$. In Lemma 3.7 we set $g([x]) = 1$ to find that $\int a(x)e^{-tx}dx$ is $\tilde{G}(t[1])$ where $[1]$ is the matrix with all coefficients equal to 1. Since

$$G([t]) = \int \exp(-\sum_{ij} t_{ij}x_{ij})d[x] = \prod_{ij} \frac{1}{t_{ij}},$$

it turns out that

$$a(n) = \frac{1}{(2\mathrm{i}\pi)^{m+1}} \int dt \int_{-\mathrm{i}\infty}^{+\mathrm{i}\infty} d\theta_1 \cdots \int_{-\mathrm{i}\infty}^{+\mathrm{i}\infty} d\theta_m e^{nt} \prod_{ij} \frac{1}{t + \theta_i - \theta_j}$$

where the integration path in t is parallel to the imaginary axis with non-negative real parts. With the change of variable $(t', \theta_1', \ldots, \theta_m') = \frac{1}{n}(t, \theta_1, \ldots, \theta_m)$ we obtain

$$a(n) = \frac{n^{m^2-m-1}}{(2\mathrm{i}\pi)^{m+1}} \int dt' \int_{-\mathrm{i}\infty}^{+\mathrm{i}\infty} d\theta_1' \cdots \int_{-\mathrm{i}\infty}^{+\mathrm{i}\infty} d\theta_m' e^{t'} \prod_{ij} \frac{1}{t' + \theta_i' - \theta_j'}.$$

Now we turn to $|\mathcal{F}_n|$. Let $F(z) = \sum_n |\mathcal{F}_n| z^n$. First we look for an expression for $F(z)$. Let $G([z]) = \sum_{[k]} [z]^{[k]}$. We have $G([z]) = \prod_{ij}(1 - z_{ij})^{-1}$, and $F(z) = \mathcal{F}G(z[1])$. By Lemma 3.5

$$\mathcal{F}G([z]) = \left(\frac{1}{2\mathrm{i}\pi}\right)^m \int_{-\mathrm{i}\pi}^{+\mathrm{i}\pi} d\theta_1 \cdots \int_{-\mathrm{i}\pi}^{+\mathrm{i}\pi} d\theta_m \prod_{ij}(1 - z_{ij}\exp(\theta_j - \theta_i))^{-1},$$

and therefore

$$F(z) = \left(\frac{1}{2\mathrm{i}\pi}\right)^m \int_{-\mathrm{i}\pi}^{+\mathrm{i}\pi} d\theta_1 \cdots \int_{-\mathrm{i}\pi}^{+\mathrm{i}\pi} d\theta_m \prod_{ij}(1 - z\exp(\theta_j - \theta_i))^{-1}.$$

Observe also that

$$\begin{aligned} |\mathcal{F}_n| &= \frac{1}{2\mathrm{i}\pi} \oint \frac{dz}{z^{n+1}} F(z) \\ &= \left(\frac{1}{2\mathrm{i}\pi}\right)^{m+1} \oint \frac{dz}{z} \int_{-\mathrm{i}\pi}^{+\mathrm{i}\pi} d\theta_1 \cdots \int_{-\mathrm{i}\pi}^{+\mathrm{i}\pi} d\theta_m \prod_{ij}(1 - z\exp(\theta_j - \theta_i))^{-1} \frac{1}{z^n}. \end{aligned}$$

With the change of variable $z = e^{-t}$ we find

$$|\mathcal{F}_n| = \left(\frac{1}{2\mathbf{i}\pi}\right)^{m+1} \int dt \int_{-\mathbf{i}\pi}^{+\mathbf{i}\pi} d\theta_1 \cdots \int_{-\mathbf{i}\pi}^{+\mathbf{i}\pi} d\theta_m \prod_{ij} (1 - \exp(-t + \theta_j - \theta_i))^{-1} e^{nt}.$$

Let $(t', \theta'_1, \ldots, \theta'_m) = \frac{1}{n}(t, \theta_1, \ldots, \theta_m)$, then $1 - \exp(-t - \theta_i + \theta_j) = \frac{1}{n}(t' + \theta'_i - \theta'_j)\left(1 + O\left(\frac{1}{n}\right)\right)$, and finally we arrive at

$$\begin{aligned}
|\mathcal{F}_n| &= \frac{n^{m^2-m-1}}{(2\mathbf{i}\pi)^{m+1}} \int dt' \int_{-\mathbf{i}n\pi}^{+\mathbf{i}n\pi} d\theta'_1 \cdots \int_{-\mathbf{i}n\pi}^{+\mathbf{i}n\pi} d\theta'_m \Big(\prod_{ij} \frac{1}{t' + \theta'_i - \theta'_j}\Big) e^{t'} \left(1 + O\left(\frac{1}{n}\right)\right) \\
&= \frac{n^{m^2-m-1}}{(2\mathbf{i}\pi)^{m+1}} \int dt' \int_{-\mathbf{i}\infty}^{+\mathbf{i}\infty} d\theta'_1 \cdots \int_{-\mathbf{i}\infty}^{+\mathbf{i}\infty} d\theta'_m \Big(\prod_{ij} \frac{1}{t' + \theta'_i - \theta'_j}\Big) e^{t'} \left(1 + O\left(\frac{1}{n}\right)\right) \\
&= a(n) \left(1 + O\left(\frac{1}{n}\right)\right).
\end{aligned}$$

This completes the proof. ■

Now we are ready to prove Theorem 3.4 which is our main result. To simplify our analysis we first handle the partial redundancy D_n^a restricted to all strings starting with a symbol a. It is clear that $D_n = mD_n^a$. We have

$$\begin{aligned}
D_n^a &= \sum_b \sum_{[k] \in \mathcal{F}_n, k_{ba} > 0} \frac{N_{[k]}^{ba}}{B_{[k]}} B_{[k]}([k] - [\delta_{ba}])^{[k]-[\delta_{ba}]} (k_b - 1)^{-k_b+1} \prod_{i \neq b} (k_i)^{-k_i} \\
&= \sum_b \sum_{[k] \in \mathcal{F}_n, k_{ba} > 0} \frac{k_{ba}}{k_a} \det{}_{bb}(\mathbf{I} - [k^*]) B_{[k]}([k] - [\delta_{ba}])^{[k]-[\delta_{ba}]} (k_b - 1)^{k_b - 1} \\
&\times \prod_{i \neq b} (k_i)^{-k_i} (1 + O(n^{-1})).
\end{aligned}$$

Using Stirling's formula we obtain for $[k] \in \mathcal{F}_n$ and $k_{ij} = \Theta(n)$

$$\frac{k_{ba}}{k_a} B_{[k]}([k] - [\delta_{ba}])^{[k]-[\delta_{ba}]} (k_b - 1)^{-k_b+1} \prod_{i \neq b} (k_i)^{-k_i} = \prod_i \frac{\sqrt{2\pi k_i}}{\prod_j \sqrt{2\pi k_{ij}}} (1 + O(1/n)),$$

and this leads to

$$D_n^a = (1 + O(1/n)) \sum_{[k] \in \mathcal{F}_n} F_m\left(\frac{1}{n}[k]\right) \prod_i \frac{\sqrt{2\pi k_i}}{\prod_j \sqrt{2\pi k_{ij}}}.$$

But $F_m([x]) = \sum_a \det_{aa}(\mathbf{I} - [x^*])$ where $[x^*]$ is the matrix whose (i,j) coefficient is x_{ij}/x_i, with $x_i = \sum_{j'} x_{ij'}$.

Using the Euler–Maclaurin summation formula, we finally arrive at the following function of

$$D_n = \left(1 + O\left(\frac{1}{n}\right)\right) \frac{|\mathcal{F}_n|}{a(n)} \int_{\mathcal{K}(n)} F_m([y]) \prod_i \frac{\sqrt{2\pi \sum_j y_{ij}}}{\prod_j \sqrt{2\pi y_{ij}}} d[y]. \qquad (21)$$

Via trivial change of variable $[y'] = \frac{1}{n}[y]$, and since $F_m(\frac{1}{n}[y]) = F_m([y])$, we find

$$\int_{\mathcal{K}(n)} F_m([y]) \prod_i \frac{\sqrt{2\pi \sum_j y_{ij}}}{\prod_j \sqrt{2\pi y_{ij}}} d[y] = \left(\frac{n}{2\pi}\right)^{(m-1)m/2} \int_{\mathcal{K}(1)} F_m([y']) \prod_i \frac{\sqrt{\sum_j y'_{ij}}}{\prod_j \sqrt{y'_{ij}}} d[y'] \tag{22}$$

Since $|\mathcal{F}_n|/a(n) = 1 + O(1/n)$, we obtain the final result of our theorem, that is $D_n = \left(1 + O\left(\frac{1}{n}\right)\right)\left(\frac{n}{2\pi}\right)^{(m-1)m/2} A_m$ for large n. ∎

Let us now estimate the constant A_m for $m = 2$. We have

$$A_2 = 2\int_{\mathcal{K}(1)} (\det_{11}(\mathbf{I} - [y^*]) + \det_{22}(\mathbf{I} - [y^*])) \frac{\sqrt{y_1}}{\sqrt{y_{11}}\sqrt{y_{12}}} \frac{\sqrt{y_2}}{\sqrt{y_{21}}\sqrt{y_{22}}} dy_{11}dy_{12}dy_{21}dy_{22} \tag{23}$$

Since $\det_{11}(\mathbf{I} - [y^*]) = \frac{y_{21}}{y_2}$ and $\det_{22}(\mathbf{I} - [y^*])$ obtained by symmetry, and since the condition $[y] \in \mathcal{K}(1)$ means $y_1 + y_2 = 1$ and $y_{12} = y_{21}$ we arrive at

$$A_2 = 4\int_{y_{11}+2y_{12}+y_{22}=1} \frac{1}{\sqrt{y_{11}}\sqrt{y_1}\sqrt{y_{22}}\sqrt{y_2}} dy_{11}dy_{12}dy_{22}$$

which further yields

$$\begin{aligned} A_2 &= 4\int_0^1 \frac{dx}{\sqrt{(1-x)x}} \int_0^{\min\{x,1-x\}} \frac{dy}{\sqrt{(1-x-y)(x-y)}} \\ &= 8\int_0^{1/2} \frac{\log(1-2x) - \log(1-2\sqrt{(1-x)x})}{\sqrt{(1-x)x}} dx \\ &= 16\int_0^{\pi/4} \log\left(\frac{\cos(2\theta)}{1-\sin(2\theta)}\right) d\theta \\ &= 16 \cdot G \end{aligned}$$

with the change of variable $x = \sin^2(\theta)$, where G is the Catalan constant.

3.3 Redundancy of Markov Sources of Higher Order

In this section we show that the maximal redundancy of a Markov source of order r can be derived in a similar manner as for $r = 1$.

We define the PO matrix $[k]$ as an $m^r \times m$ matrix whose coefficient $k_{w,j}$th ($w \in \mathcal{A}^r$) is the number of times the string w is followed by symbol j in the string x_1^n. Then

$$\sup_{P \in \mathcal{M}_r} P(x_1^n) = \prod_{w,j} \left(\frac{k_{w,j}}{k_w}\right)^{k_{w,j}} \tag{24}$$

with the convention that $k_w = \sum_j k_{w,j}$.

Using the approach developed in the previous sections, we arrive at the following generalization.

Theorem 3.9. *The maximum redundancy $D_n(\mathcal{M}_r)$ of Markov sources of order r attains the following asymptotics*

$$D_n(\mathcal{M}_r) = \left(\frac{n}{2\pi}\right)^{(m-1)m^r/2} A_m^r \times \left(1 + O\left(\frac{1}{n}\right)\right)$$

with

$$A_m^r = \int_{\mathcal{K}_r(1)} m^r F_m^r([y]) \prod_w \frac{\sqrt{y_w}}{\prod_j \sqrt{y_{w,j}}},$$

where $\mathcal{K}_r(1)$ is the convex set of $m^r \times m$ matrices $[y]$ with non-negative coefficients such that $\sum_{w,j} y_{w,j} = 1$. The function $F_m^r([y]) = \sum_w \det_{ww}(\mathbf{I} - [y^*]_r)$, where $[y]_r$ is the $m^r \times m^r$ matrix whose (w, w') coefficient is equal to $y_{w,a}$ if there exist a in $\mathcal{A}$ such that w' is suffix of wa, otherwise the (w, w') coefficient is equal to 0.

Here, we present only a sketch of the proof. The main combinatorial results that we need are as follows:

$$B_r([z]) = \prod_w (1 - \sum_j z_{w,j})^{-1}, \tag{25}$$

$$\mathcal{F}B_r([z]) = (\det(\mathbf{I} - [z]_r))^{-1}. \tag{26}$$

Furthermore, $N_{[k]}^{w,w'}$ is the $[z]^{[k]}$ coefficient in

$$B_r([z])\det_{w,w}(\mathbf{I} - [z]_r) \prod_{i=1}^{r} z_{(ww')_i^{i+r-1},(ww')_{i+r}^{i+r}} \tag{27}$$

where $X_i^j = X_i X_{i+1} \dots X_j$ $(i \le j)$. The rest follows the footsteps of our previous discussion.

References

[1] K. Atteson, The Asymptotic Redundancy of Bayes Rules for Markov Chains, *IEEE Trans. on Information Theory*, 45, 2104–2109, 1999.

[2] A. Barron, J. Rissanen, and B. Yu, The Minimum Description Length Principle in Coding and Modeling, *IEEE Trans. Information Theory*, 44, 2743-2760, 1998.

[3] P. Billingsley, Statistical Methods in Markov Chains, *Ann. Math. Statistics*, 32, 12–40, 1961.

[4] L. Boza, Asymptotically Optimal Tests for Finite Markov Chains, *Ann. Math. Statistics*, 42, 1992-2007, 1971.

[5] L. Campbell, A Coding Theorem and Rényi's Entropy, *Information and Control*, 8, 423–429, 1965.

[6] B. Clarke and A. Barron, Information-theoretic Asymptotics of Bayes Methods, *IEEE Trans. Information Theory*, 36, 453–471, 1990.

[7] R. Corless, G. Gonnet, D. Hare, D. Jeffrey and D. Knuth, On the Lambert W Function, *Adv. Computational Mathematics*, 5, 329–359, 1996.

[8] T. Cover and J.A. Thomas, *Elements of Information Theory*, John Wiley & Sons, New York 1991.

[9] I. Csiszár and P. Shields, Redundancy Rates for Renewal and Other Processes, *IEEE Trans. Information Theory*, 42, 2065–2072, 1996.

[10] A. Dembo and I Kontoyiannis, Critical Behvaior in Lossy Coding, *IEEE Trans. Inform. Theory*, March 2001.

[11] M. Drmota and W. Szpankowski, Generalized Shannon Code Minimizes the Maximal Redundancy, *Proc. LATIN 2002*, Cancun, Mexico, 2002.

[12] P. Flajolet and A. Odlyzko, Singularity Analysis of Generating Functions, *SIAM J. Disc. Methods*, 3, 216-240, 1990.

[13] P. Flajolet and W. Szpankowski, Analytic Variations on Redundancy Rates of Renewal Processes, *2000 International Symposium on Information Theory*, pp. 499, Sorento, Italy, June 2000; also INRIA RR No. 3553, 1998. 1995.

[14] I. Kontoyiannis, Pointwise Redundancy in Lossy Data Compression and Universal Lossy Data sCcompression, *IEEE Trans. Inform. Theory*, 46, 136-152, 2000.

[15] G. Louchard and W. Szpankowski, On the Average Redundancy Rate of the Lempel-Ziv Code, *IEEE Trans. Information Theory*, 43, 2–8, 1997.

[16] J. Rissanen, Complexity of Strings in the Class of Markov Sources, *IEEE Trans. Information Theory*, 30, 526–532, 1984.

[17] J. Rissanen, Universal Coding, Information, Prediction, and Estimation, *IEEE Trans. Information Theory*, 30, 629–636, 1984.

[18] J. Rissanen, Fisher Information and Stochastic Complexity, *IEEE Trans. Information Theory*, 42, 40–47, 1996.

[19] S. Savari, Redundancy of the Lempel-Ziv Incremental Parsing Rule, *IEEE Trans. Information Theory*, 43, 9–21, 1997.

[20] P. Shields, Universal Redundancy Rates Do Not Exist, *IEEE Trans. Information Theory*, 39, 520-524, 1993.

[21] Y. Shtarkov, Universal Sequential Coding of Single Messages, *Problems of Information Transmission*, 23, 175–186, 1987.

[22] Y. Shtarkov, T. Tjalkens and F.M. Willems, Multi-alphabet Universal Coding of Memoryless Sources, *Problems of Information Transmission*, 31, 114-127, 1995.

[23] W. Szpankowski, On Asymptotics of Certain Recurrences Arising in Universal Coding, *Problems of Information Transmission*, 34, 55-61, 1998.

[24] W. Szpankowski, Asymptotic Redundancy of Huffman (and Other) Block Codes, *IEEE Trans. Information Theory*, 46, 2434-2443, 2000.

[25] W. Szpankowski, *Average Case Analysis of Algorithms on Sequences*, Wiley, New York, 2001.

[26] Q. Xie, A. Barron, Minimax Redundancy for the Class of Memoryless Sources, *IEEE Trans. Information Theory*, 43, 647-657, 1997.

[27] Q. Xie, A. Barron, Asymptotic Minimax Regret for Data Compression, Gambling, and Prediction, *IEEE Trans. Information Theory*, 46, 431-445, 2000.

[28] P. Whittle, Some Distribution and Moment Formulæ for Markov Chain, *J. Roy. Stat. Soc.*, Ser. B., 17, 235–242, 1955.

[29] A. J. Wyner, The Redundancy and Distribution of the Phrase Lengths of the Fixed-Database Lempel-Ziv Algorithm, *IEEE Trans. Information Theory*, 43, 1439–1465, 1997.

Philippe Jacquet

INRIA
Rocquencourt
78153 Le Chesnay Cedex
France
`jacquet@blagny.inria.fr`

Wojciech Szpankowski

Department of Computer Science
Purdue University
W. Lafayette, IN 47907
U.S.A.
`spa@cs.purdue.edu`

Trends in Mathematics, © 2002 Birkhäuser Verlag Basel/Switzerland

Analysis of Quickfind with Small Subfiles

Conrado Martínez, Daniel Panario and Alfredo Viola

ABSTRACT: *In this paper we investigate variants of the well-known Hoare's Quickfind algorithm for the selection of the j-th element out of n when recursion stops for subfiles whose size is below a predefined threshold and a simpler algorithm is run instead. We provide estimates for the combined number of passes, comparisons and exchanges under three policies for the small subfiles: insertion sort and two variants of selection sort, but the analysis could be easily adapted for alternative policies. We obtain the average cost for each of these variants and compare them with the costs of the standard variant which does not use cutoff. We also give the best explicit cutoff bound for each of the variants.*

1 Introduction

Hoare's quickfind [2] selects the j-th element (equivalently, the element of *rank* j in ascending order, the j-th order statistic) out of an array of n elements by picking an element from the array —the pivot— and rearranging the array so that elements smaller than the pivot are to its left and elements larger than the pivot are to its right. If the pivot has been brought to position $k = j$ then it is the sought element; otherwise, if $j < k$ then the procedure is recursively applied to the subarray to the left of the pivot, and if $j > k$ the process continues in the right subarray. A similar principle is used in the celebrated quicksort algorithm [3], also by Hoare; once the pivot is brought into place by the partitioning of the array, the subarrays to its left and right are recursively sorted.

Quickfind performs well in practice, its average cost being linear. Knuth [6] has shown that the average number of comparisons $C_{n,j}$ needed to locate the j-th element out of n is

$$C_{n,j} = 2(n + 3 + (n+1)H_n - (n+3-j)H_{n+1-j} - (j+2)H_j),$$

where $H_n = \sum_{1\le i\le n} 1/i = \log n + \mathcal{O}(1)$ denotes the n-th harmonic number. Clearly, $C_{n,j}$ is $\Theta(n)$ for any j, $1 \le j \le n$. More recently, Hwang and Tsai [4] have shown that the limiting distribution of the number of comparisons made by this algorithm when given a random permutation of n elements for finding the j-th smallest element (with $j = o(n)$), is the Dickman function. They also give the limit distribution of the number of exchanges.

In order to make uneven partitions leading to $\Theta(n^2)$ worst-case performance more unlikely and to reduce the average cost, one could use the median of a small sample of s elements as the pivot of each recursive stage [1, 5, 8].

Besides this major, but rather specific, optimization of the algorithm, other general optimization techniques should be used in a carefully engineered implemen-

*The first author was supported by project DGES PB98-0926 (AEDRI) of the Spanish Ministry for Education and Science and the Future and Emergent Technologies programme of the EU under contract IST-1999-14186 (ALCOM-FT). The second author was supported by NSERC of Canada (grant # 238757). The third author was supported by the CSIC of Uruguay (project *Desarrollo de métodos matemáticos para analizar algoritmos y su aplicación al análisis de algoritmos criptográficos*).

tation of quickfind. For instance, recursion removal and loop unwrapping; the former is straightforward, because quickfind has tail recursion. Another general optimization technique for divide-and-conquer algorithms is *recursion cutoff*: for sufficiently small inputs, we switch from the recursive divide-and-conquer algorithm to a simpler algorithm. For instance, it is well known that switching from quicksort to insertion sort when subfiles have $n_0 = 9$ elements improves the overall performance of the algorithm. Cutoff values in the range 6 to 15 do about as well, but $n_0 = 9$ is the optimal choice for most implementations. Actually, it is typically better to ignore small subfiles and perform a single pass of insertion sort over the entire file [9, 10].

In this paper, we tackle the analysis of quickfind with recursion cutoff. Although computing the optimal cutoff value can be readily done using dynamic programming, solving the recurrences for "small" values of the parameters, this approach does not guarantee success if the cost function is not increasing with n and provides very limited information about the modified variant. Besides that, on this paper we want also to concentrate on the methodological and analytical aspects of the problem. Our approach, although a bit more complex and difficult to apply, yields much more information about the problem at hand.

We consider using insertion sort and two variants of selection sort for small subfiles. Actually, in the case of selection sort we consider two variants which stop as soon as the desired element is found (unlike insertion sort, which must sort the whole subfile in order to locate the j-th element). The first one, which we call selection sort locates first the minimum, then the second minimum, and so on. The other variant, which we call optimized selection sort looks for the j-th element, starting from the most suitable end: so if $j \leq n/2$, it behaves just as the first variant, but if $j > n/2$ then it first locates the maximum, then the second maximum, etc, until the $n + 1 - j$-th maximum is found. We perform our analysis trying to keep it as general as possible, in order to facilitate a similar analysis using other algorithms for small subfiles. This analysis could also be generalized for the median-of-three variant.

The original motivation of our analysis was to determine the optimal choices of cutoff values n_0 for each combination of small subfiles algorithm and pivot selection strategy. But in the course of this investigation, we have shifted our original motivation to more ambitious goals, namely, the study of general techniques for the analysis of recursive algorithms with cutoff.

The paper is organized as follows. The methodology used in this paper is based on solving recurrences via generating functions and related differential equations. The general approach for analyzing quickfind with small subfiles is presented in Section 2. The required functions when we use insertion sort and the two variants of selection sort for small subfiles are given in Section 3. In the last section, we discuss the benefits of the different small subfiles policies and compute explicit cutoff bounds for each of them.

2 General analysis

In this section, we study the standard quickfind (a random pivot is chosen from the elements in the current subfile) when we use a different policy for the small subfiles. An interesting feature of our approach is to provide a common presentation for all the costs involved. In the next sections we particularize these results to a certain model of costs and consider three different algorithms for small subfiles, namely insertion and two variants of selection sort.

Our methodology is based upon the translation of recurrence relations into functional equations over corresponding generating functions. This is typical in this area of research (see for instance [5]). The analysis of quickfind and its variants leads systematically to differential equations. In many instances, we can solve them exactly or we can provide suitable asymptotic estimates of the solution near the dominant singularities.

Conditioning the expected value of the cost of selecting the j-th out of n on the event that the selected pivot is the k-th element and summing up for all k, one can easily verify that for all $1 \le j \le n$ the following recurrence relation for the expected cost $C_{n,j}$ of selecting the j-th out of n holds:

$$C_{n,j} = \begin{cases} t_{n,j} + \sum_{k=1}^{j-1} \pi_{n,k} C_{n-k,j-k} + \sum_{k=j+1}^{n} \pi_{n,k} C_{k-1,j} & \text{if } n > n_0, \\ b_{n,j} & \text{if } n \le n_0, \end{cases} \tag{1}$$

where $\pi_{n,k}$ is the probability that the chosen pivot is the k-th element, $t_{n,j}$ is the average cost of a single pass of the algorithm (for instance, the combined cost of the selection of the pivot, the partitioning of the subfile and eventually any associated bookkeeping), and $b_{n,j}$ is the same average cost but for the small subfiles algorithm to find the j-th out of n. Since recursion cutoff affects the lower order terms of the performance of the algorithm, not its main linear order term, it only makes sense to consider the expected *total cost of the algorithm*, not just the expected number of comparisons. Therefore, we need to work with a measure of cost that combines the cost of comparisons, exchanges, passes, etc.

Since the pivot is chosen at random (equivalently, the array contains a random permutation) we have $\pi_{n,k} = 1/n$ for all $1 \le k \le n$. We take $C_{n,j} = 0$ for all values n, j such that either $j < 1$ or $j > n$. Now, we consider two cases: $n_0 \ge j$ and $n_0 < j$.

If $j \le n_0$, we multiply both sides of (1) by nz^{n-1} and sum over all $n > n_0$ to obtain

$$\sum_{n=n_0+1}^{\infty} nC_{n,j} z^{n-1} = \sum_{n=n_0+1}^{\infty} nt_{n,j} z^{n-1} + \sum_{n=n_0+1}^{\infty} \sum_{k=1}^{j-1} C_{n-k,j-k} z^{n-1} + \sum_{n=n_0+1}^{\infty} \sum_{k=j+1}^{n} C_{k-1,j} z^{n-1}.$$

Using the definition $C_j(z) = \sum_{n\geq 0} C_{n,j} z^n$, the equation above implies:

$$\begin{aligned} C_j'(z) &= \sum_{n\geq 1} nC_{n,j} z^{n-1} = \sum_{n=j}^{n_0} nC_{n,j} z^{n-1} + \sum_{n=n_0+1}^{\infty} nC_{n,j} z^{n-1} \\ &= \sum_{n=j}^{n_0} nb_{n,j} z^{n-1} + \sum_{n=n_0+1}^{\infty} nt_{n,j} z^{n-1} \\ &\quad + \sum_{k=1}^{j-1} \sum_{n=n_0+1}^{\infty} C_{n-k,j-k} z^{n-1} + \sum_{k=j+1}^{\infty} \sum_{n=\max(n_0+1,k)}^{\infty} C_{k-1,j} z^{n-1}. \end{aligned}$$

Rearranging the third and fourth sums in the previous equation we have

$$\begin{aligned} C_j'(z) &= \sum_{n=j}^{n_0} nb_{n,j} z^{n-1} + \sum_{n=n_0+1}^{\infty} nt_{n,j} z^{n-1} + \sum_{k=1}^{j-1} z^{k-1} \sum_{n=j}^{\infty} C_{n-k,j-k} z^{n-k} \\ &\quad - \sum_{k=1}^{j-1} \sum_{n=j}^{n_0} C_{n-k,j-k} z^{n-1} + \sum_{k=j+1}^{\infty} C_{k-1,j} \frac{z^{\max(n_0+1,k)-1}}{1-z}. \end{aligned}$$

Writing the third summand in terms of $C_{j-k}(z)$ and simplifying the fourth term, we obtain

$$\begin{aligned} C_j'(z) &= \sum_{n=j}^{n_0} nb_{n,j} z^{n-1} + \sum_{n=n_0+1}^{\infty} nt_{n,j} z^{n-1} + \sum_{k=1}^{j-1} z^{k-1} C_{j-k}(z) \\ &\quad - \sum_{k=1}^{j-1} \sum_{n=j}^{n_0} b_{n-k,j-k} z^{n-1} + \sum_{k=j+1}^{n_0+1} C_{k-1,j} \frac{z^{n_0}}{1-z} + \sum_{k=n_0+2}^{\infty} C_{k-1,j} \frac{z^{k-1}}{1-z}. \end{aligned}$$

Simplifying the last two terms, we have

$$\begin{aligned} C_j'(z) &= \sum_{n=j}^{n_0} nb_{n,j} z^{n-1} + \sum_{n=n_0+1}^{\infty} nt_{n,j} z^{n-1} + \sum_{k=1}^{j-1} z^{k-1} C_{j-k}(z) \\ &\quad - \sum_{k=1}^{j-1} \sum_{n=j}^{n_0} b_{n-k,j-k} z^{n-1} + \frac{z^{n_0}}{1-z} \sum_{k=j}^{n_0} b_{k,j} + \frac{1}{1-z} \left(C_j(z) - \sum_{k=j}^{n_0} C_{k,j} z^k \right). \end{aligned}$$

Finally, we obtain

$$\begin{aligned} C_j'(z) &= \sum_{n=j}^{n_0} nb_{n,j} z^{n-1} + \sum_{n=n_0+1}^{\infty} nt_{n,j} z^{n-1} + \sum_{k=1}^{j-1} z^{k-1} C_{j-k}(z) \\ &\quad - \sum_{k=1}^{j-1} \sum_{n=j}^{n_0} b_{n-k,j-k} z^{n-1} + \frac{z^{n_0}}{1-z} \sum_{k=j}^{n_0} b_{k,j} + \frac{C_j(z)}{1-z} - \frac{1}{1-z} \sum_{k=j}^{n_0} b_{k,j} z^k. \end{aligned}$$

Therefore, for all $j \le n_0$, we have

$$C_j'(z) = \sum_{k=1}^{j-1} z^{k-1} C_{j-k}(z) + \frac{C_j(z)}{1-z} + f_j^{\text{low}}(z), \tag{2}$$

where

$$f_j^{\text{low}}(z) = \sum_{n=j}^{n_0} n b_{n,j} z^{n-1} + \sum_{n=n_0+1}^{\infty} n t_{n,j} z^{n-1} - \sum_{k=1}^{j-1} \sum_{n=j}^{n_0} b_{n-k,j-k} z^{n-1}$$
$$+ \frac{z^{n_0}}{1-z} \sum_{k=j}^{n_0} b_{k,j} - \frac{1}{1-z} \sum_{k=j}^{n_0} b_{k,j} z^k.$$

Similarly, for $j > n_0$, we multiply both sides of (1) by nz^{n-1} and sum for all $n \ge j$:

$$\begin{aligned}
C_j'(z) &= \sum_{n=j}^{\infty} n t_{n,j} z^{n-1} + \sum_{n=j}^{\infty} \sum_{k=1}^{j-1} C_{n-k,j-k} z^{n-1} + \sum_{n=j}^{\infty} \sum_{k=j+1}^{n} C_{k-1,j} z^{n-1} \\
&= \sum_{n=j}^{\infty} n t_{n,j} z^{n-1} + \sum_{k=1}^{j-1} z^{k-1} \sum_{n=j}^{\infty} C_{n-k,j-k} z^{n-k} + \sum_{k=j+1}^{\infty} \sum_{n=k}^{\infty} C_{k-1,j} z^{n-1} \\
&= \sum_{n=j}^{\infty} n t_{n,j} z^{n-1} + \sum_{k=1}^{j-1} C_{j-k}(z) z^{k-1} + \sum_{k=j+1}^{\infty} C_{k-1,j} \frac{z^{k-1}}{1-z} \\
&= \sum_{n=j}^{\infty} n t_{n,j} z^{n-1} + \sum_{k=1}^{j-1} C_{j-k}(z) z^{k-1} + \frac{C_j(z)}{1-z}.
\end{aligned}$$

Therefore all $j > n_0$, we have

$$C_j'(z) = \sum_{k=1}^{j-1} z^{k-1} C_{j-k}(z) + \frac{C_j(z)}{1-z} + f_j^{\text{high}}(z), \tag{3}$$

where $f_j^{\text{high}}(z) = \sum_{n=j}^{\infty} n t_{n,j} z^{n-1}$. If we define $g_j(z)$ as follows

$$g_j(z) = \begin{cases} f_j^{\text{low}}(z) & \text{if } j \le n_0, \\ f_j^{\text{high}}(z) & \text{if } j > n_0, \end{cases} \tag{4}$$

then, (2) and (3) imply that for all $j \ge 1$,

$$C_j'(z) = \sum_{k=1}^{j-1} z^{k-1} C_{j-k}(z) + \frac{C_j(z)}{1-z} + g_j(z). \tag{5}$$

Let $C(z,u) = \sum_{j \ge 1} C_j(z) u^j$ and $g(z,u) = \sum_{j \ge 1} g_j(z) u^j$. Multiplying both sides of (5) by u^j and summing up for all $j \geq 1$ entails that

$$\begin{aligned}\frac{\partial C(z,u)}{\partial z} &= g(z,u) + \frac{C(z,u)}{1-z} + \sum_{j=1}^{\infty}\sum_{k=1}^{j-1} C_{j-k}(z) z^{k-1} u^j \\ &= g(z,u) + \frac{C(z,u)}{1-z} + \sum_{k=1}^{\infty} u^k z^{k-1} \sum_{j=k+1}^{\infty} C_{j-k}(z) u^{j-k} \\ &= g(z,u) + \frac{C(z,u)}{1-z} + \frac{u}{1-uz} C(z,u),\end{aligned} \tag{6}$$

with the initial condition $C(0,u) = 0$. The homogeneous solution of the equation above is $1/((1-z)(1-uz))$. Using it, one can easily obtain the general solution of the partial differential equation (6) as follows

$$C(z,u) = \frac{1}{(1-z)(1-uz)} \left(\int (1-z)(1-uz) g(z,u) dz + K \right), \tag{7}$$

where K is a constant which should be determined according to the initial condition $C(0,u) = 0$. Now, if we introduce

$$\begin{aligned}T(z,u) &= \sum_{n=1}^{\infty} z^n \sum_{j=1}^{n} t_{n,j} u^j, \\ T_{n_0}(z,u) &= \sum_{n=1}^{n_0} z^n \sum_{j=1}^{n} t_{n,j} u^j, \quad B_{n_0}(z,u) = \sum_{n=1}^{n_0} z^n \sum_{j=1}^{n} b_{n,j} u^j,\end{aligned}$$

we obtain the following expression for $g(z,u)$ in terms of the toll generating function $T(z,u)$ and the polynomials $T_{n_0}(z,u)$ and $B_{n_0}(z,u)$:

$$\begin{aligned}g(z,u) = &\frac{\partial}{\partial z}\left(B_{n_0}(z,u) + T(z,u) - T_{n_0}(z,u)\right) + \frac{u(uz)^{n_0}}{1-uz} B_{n_0-1}(1/u,u) \\ &- \frac{u}{1-uz} B_{n_0-1}(z,u) - \frac{1}{1-z}\left(B_{n_0-1}(z,u) - z^{n_0} B_{n_0-1}(1,u)\right).\end{aligned}$$

As a consequence, the solution of the differential equation (6) is

$$
\begin{aligned}
C(z,u) = {} & \frac{\int_0^z (1-y)(1-uy)\frac{\partial}{\partial y}T(y,u)\,dy}{(1-z)(1-uz)} + B_{n_0}(z,u) - T_{n_0}(z,u) \\
& + \frac{\int_0^z (1+u-2uy)\left(y^{n_0}\sum_{j=1}^{n_0} b_{n_0,j}u^j - T_{n_0}(y,u)\right)dy}{(1-z)(1-uz)} \\
& + \frac{B_{n_0-1}(1,u)\frac{z^{n_0+1}}{n_0+2}\left((1-uz)+\frac{1}{n_0+1}\right)}{(1-z)(1-uz)} \\
& + \frac{B_{n_0-1}(1/u,u)\frac{(uz)^{n_0+1}}{n_0+2}\left((1-z)+\frac{1}{n_0+1}\right)}{(1-z)(1-uz)} \\
= {} & \frac{\int_0^z (1-y)(1-uy)\frac{\partial}{\partial y}T(y,u)\,dy}{(1-z)(1-uz)} + C_{n_0}(z,u) \\
= {} & C_T(z,u) + C_{n_0}(z,u),
\end{aligned}
$$

where $C_{n_0}(z,u)$ represents the part of $C(z,u)$ that depends on n_0 and

$$
C_T(z,u) = \frac{\int_0^z (1-y)(1-uy)\frac{\partial}{\partial y}T(y,u)\,dy}{(1-z)(1-uz)},
$$

a part which is independent of the cutoff value and the policy for small subfiles. This is a rather general phenomenon which is not particular to quickfind. Such a decomposition appears whenever we analyze a recursive algorithm with recursion cutoff and it is not a mere mathematical device.

Assuming $t_{n,j} = \alpha n + \beta + \gamma/(n-1)$ (see Section 3), it is almost straightforward to get the explicit formula

$$
\begin{aligned}
C_T(z,u) = {} & \frac{(-3\alpha+\alpha u-\beta u+\beta)}{(1-u)^2}\left(\frac{1}{1-z}-\frac{u}{1-uz}\right)\log\frac{1}{1-uz} \\
& + \frac{(3\alpha u-\alpha-\beta u+\beta)u}{(1-u)^2}\left(\frac{1}{1-z}-\frac{u}{1-uz}\right)\log\frac{1}{1-z} \\
& + \frac{(\alpha-\beta)u}{(1-u)(1-z)} + \frac{2\alpha u}{(1-u)(1-z)^2} - \frac{(\alpha-\beta)u}{(1-u)(1-uz)} \\
& - \frac{2\alpha u}{(1-u)(1-uz)^2} + \left(\frac{\gamma u(1+u-2uz)}{6(1-u)} - \frac{\gamma(1-u)}{6(1-z)}\right)\log\frac{1}{1-uz} \\
& - \left(\frac{\gamma(1+u-2uz)}{6(1-u)} - \frac{\gamma(1-u)}{6(1-uz)}\right)\log\frac{1}{1-z} \\
& - \frac{\gamma(1+33u)}{36(1-u)(1-uz)} + \frac{\gamma(1+u-8uz)}{36} + \frac{\gamma u(u+33)}{36(1-u)(1-z)},
\end{aligned}
$$

and then to extract coefficients.

On the other hand, if $n > n_0$ then we have $C_{n_0}(z,u) = p(z,u)/((1-z)(1-uz))$, with

$$
p(z,u) = \sum_{k=2}^{n_0+2}\sum_{i=1}^{n_0} p_{k,i}z^k u^i
$$

a polynomial in z and u that has neither $(1-z)$ nor $(1-uz)$ as a factor. Since $[z^n u^j](1-z)^{-1}(1-uz)^{-1} = [\![j \le n]\!]$ (with $[\![P]\!] = 1$ if the predicate P is true, and 0 otherwise), then

$$[z^n u^j]C_{n_0}(z,u) = \sum_{k=2}^{n_0+2} \sum_{i=1}^{n_0} p_{k,i}\, [\![j - i \le n - k]\!].$$

Therefore, if $n_0 + 1 \le j \le n - n_0$ and $n \ge 2n_0 + 1$, then for $2 \le k \le n_0 + 2$ and $1 \le i \le n_0$, we have $j - i \le n - k$, and so we find

$$r_{n_0} := [z^n u^j]C_{n_0}(z,u) = p(1,1), \qquad n_0 < j \le n - n_0, \quad n \ge 2n_0 + 1.$$

If we do the calculations, we obtain for $n_0 + 1 \le j \le n - n_0$ and $n \ge 2n_0 + 1$:

$$\begin{aligned} r_{n_0} &= \frac{\sum_{i=1}^{n_0}\sum_{k=i}^{n_0} b_{k,i}}{(n_0+1)(n_0+2)} + \frac{\sum_{k=1}^{n_0-1}\sum_{i=1}^{n_0-k} b_{i+k-1,i}}{(n_0+1)(n_0+2)} \\ &+ \frac{\sum_{i=1}^{n_0} b_{n_0,i}}{(n_0+1)(n_0+2)} - \frac{(n_0-1)(17n_0^2+32n_0+12)}{18(n_0+2)(n_0+1)n_0}\gamma \\ &+ \left(6H_{n_0+2} - 2n_0 - 10 + \frac{2}{n_0+2}\right)\alpha \\ &+ \left(4 - 2H_{n_0+2} - \frac{2}{n_0+2}\right)\beta. \end{aligned} \tag{8}$$

Equation (8) covers most of the interesting values of j, in particular $j = n/2$. The fact that $[z^n u^j]C_{n_0}(z,u)$ does not depend on j for a wide range of values of j and given a sufficiently large value of n, has an easy intuitive explanation: after the $\Theta(\log n)$ passes needed to reduce the size of the subfile from n to $m \le n_0$ the rank of the sought element relative to the subfile will be any of $1, \ldots, m$ with identical probability, no matter what the initial value of j is, apart from the extreme cases $j \le n_0$ or $j > n - n_0$.

3 Toll and cutoff functions

Up to now, we have investigated the average total cost of quickfind under a general framework for which no particular values of the toll function $t_{n,j}$ nor the cutoff function $b_{n,j}$ are given.

Comparisons	$n-1$
Passes	1
Exchanges (basic)	$n/6 + 1/3 + 2/(3(n-1))$

Table 3: Toll functions.

In Table 3 we find the values of $t_{n,j}$ for three important measures: comparisons, exchanges and passes. None of these depends on j and for our subsequent development we shall assume that

$$t_{n,j} = \alpha n + \beta + \gamma/(n-1).$$

	Comparisons	Exchanges	Data movements
Insertion	$n(n-1)/4$		$(n+4)(n-1)/4$
Selection	$jn - j(j+1)/2$	j	

Table 4: Cutoff functions ($j \leq \lceil n/2 \rceil$).

On the other hand, Table 4 lists the average number of comparisons, exchanges, and single data movements made by insertion sort and selection sort. The analysis of these basic algorithms can be found elsewhere; see for instance [7, 9, 11, 12].

Recall that even though we use the term "selection sort", the algorithm that we consider proceeds as selection sort but it terminates as soon as the j-th minimum in the subfile is located (and brought to its correct position). For the optimized selection sort which either looks for the j-th minimum or the $n+1-j$-th maximum, depending on whether $j \leq n/2$ or $j > n/2$, we have that the number of comparisons is

$$n \cdot \min(j, n+1-j) - \frac{1}{2}\min(j, n+1-j)\big(1 + \min(j, n+1-j)\big),$$

and the number of exchanges is $\min(j, n+1-j)$.

Guided by these algorithms, but in order to facilitate the analysis of quickfind with alternative small subfiles algorithms, we shall work with cutoff functions of the form

$$b_{n,j} = K_1 n^2 + K_2 n + K_3 j^2 + K_4 j + K_5 jn + K_6 + \\ K_7 m^2 + K_8 m + K_9 mn + K_{10} j H_n + K_{11} m H_n + K_{12} H_n,$$

where $m \equiv \min(j, n+1-j)$ and the inclusion of the three last terms is required, for instance, in the analysis of quickfind with heapselect (standard or optimized) for small subfiles. In this extended abstract, we will set $K_{10} = K_{11} = K_{12} = 0$, though.

Last but not least, we conclude this section with the simplified model of costs that we will use in the sequel. In this model, the unit of measurement is a single comparison between any two elements. We denote by ξ the cost of an exchange of elements, and by ϕ the cost of a pass (this cost includes the selection of the pivot and the cost of additional bookkeeping). Typically, $\xi \approx 3$ and $\phi \approx 2\xi \approx 6$; these values agree with the experimental values given for usual implementations and also with the values obtained from more delicate and precise cost models [7, 9, 11]. We also assume that a single data movement has cost $\xi/3$; in a real setting it would be a bit more, since exchanges use some intermediate register rather than three data movements.

Under these assumptions and measuring the total cost as the sum of the costs of comparisons, exchanges and passes, we have

$$\alpha = 1 + \xi/6, \quad \beta = \xi/3 + \phi - 1, \quad \text{and} \quad \gamma = 2\xi/3.$$

The bookkeeping costs in the outermost loop of insertion and the selection sort variants are assimilable to "passes", but we shall introduce suitable constants ϕ_{ins} and ϕ_{sel} to account for these costs (usually, ϕ_{ins} and ϕ_{sel} are $\approx 3/2$).

In the case of insertion sort we have $n-1$ iterations through the main loop and hence

$$K_1 = \frac{1}{4}(1+\xi/3), K_2 = \frac{1}{4}(-1+\xi) + \phi_{\text{ins}}, K_6 = -\xi/3 - \phi_{\text{ins}},$$
$$K_3 = K_4 = K_5 = K_7 = K_8 = K_9 = 0.$$

In the case of selection sort, we perform the outermost loop j times and we have thus

$$K_3 = -\frac{1}{2}, K_4 = -\frac{1}{2} + \xi + \phi_{\text{sel}}, K_5 = 1,$$
$$K_1 = K_2 = K_6 = K_7 = K_8 = K_9 = 0.$$

Finally, if we use the optimized variant of selection sort that scans the array from the most convenient end, we have

$$K_7 = -\frac{1}{2}, K_8 = -\frac{1}{2} + \xi + \phi_{\text{sel}}, K_9 = 1,$$
$$K_1 = K_2 = K_3 = K_4 = K_5 = K_6 = 0.$$

4 Optimal cutoff bounds

We are now ready to tie the general framework of Section 2 with the particular toll and cutoff functions of Section 3, to provide the optimal values of n_0. We have used MAPLE to instantiate (8) with the proper values of the toll and cutoff functions of Section 3, and to help with the tedious calculations*.

Since r_{n_0} is the contribution originated by the term that depends on n_0, r_0 should be 0, since in this case we have the basic quickfind algorithm. When r_{n_0} is negative, then cutting off the algorithm (when we have files of size n_0 or less) improves its efficiency. Therefore, the optimal value of n_0 is the one that minimizes r_{n_0}.

In these calculations we have considered $\xi = 3, \phi = 6, \phi_{\text{ins}} = 3/2$, and $\phi_{\text{sel}} = 3/2$. For the case of insertion sort, we obtain

n_0	1	2	3	4	5	6
r_{n_0}	-2.5000	-5.0000	-6.5166	-7.3277	-7.5182	-7.1277

n_0	7	8	9	10	11	12
r_{n_0}	-6.1787	-4.6850	-2.6562	-0.9902	2.9817	6.5823

Table 5: Values of r_{n_0} for $1 \leq n_0 \leq 12$, for insertion sort.

From the table we can see that the optimal value of n_0 is 5, while quickfind with cutoff has better performance than the standard algorithm only up to $n_0 = 10$. For the case of selection sort, we obtain

From this table we can see that the optimal value of n_0 is 6, and that switching to selection sort improves the performance of the standard algorithm only up to $n_0 = 11$.

*The MAPLE program and C code for dynamic programming computations is available from request from the contact author (`conrado@lsi.upc.es`).

n_0	1	2	3	4	5	6
r_{n_0}	-1.0000	-2.8333	-4.1166	-4.9944	-5.4944	-5.6277

n_0	7	8	9	10	11	12
r_{n_0}	-5.4009	-4.8183	-3.8835	-2.5990	-0.9669	1.0108

Table 6: Values of r_{n_0} for $1 \leq n_0 \leq 12$, for selection sort.

Finally, Table 7 shows the first values of r_{n_0} when the optimized selection sort is used. The optimal value is $n_0 = 11$ and $r_{n_0} < 0$ for $n_0 \leq 22$ (not shown in the table). Also, comparing this last table with the two previous ones, it is clear (and not surprising) that this variant outperforms the other two variants.

n_0	1	2	3	4	5	6	7
r_{n_0}	-1.0000	-3.5833	-5.5667	-7.4278	-8.9468	-10.3063	-11.3731

n_0	8	9	10	11	12	13	14
r_{n_0}	-12.2628	-12.8835	-13.3187	-13.4990	-13.4891	-13.2336	-12.7848

Table 7: Values of r_{n_0} for $1 \leq n_0 \leq 14$, for optimized selection sort.

The variation of r_{n_0} as a function of the performance constants is small for reasonable ranges of variation of ξ, ϕ, ϕ_{ins} and ϕ_{sel}, which indicates that setting n_0 in a range around the optimal choice provides a tuned but portable implementation with good performance. For example, if we consider $\xi = 1.5$ and $\phi = 4.5$, then the optimal value of n_0 is 5 for insertion sort and selection sort, and 9 for optimized selection, while if we consider $\xi = 2.5$ and $\phi = 7.5$ then the optimal values of n_0 are respectively 6, 7 and 12.

On the other hand, experimental determination of the performance constants (or even better, of the constants α, β, γ and K_i) together with (8) would yield a precise estimation of the optimal cutoff value n_0 for any particular implementation and environment under which the algorithm is run.

References

[1] Anderson, D., and Brown, R. Combinatorial aspects of C.A.R. Hoare's Find algorithm. *Australasian Journal of Combinatorics 5* (1992), 109–119.

[2] Hoare, C. Find (Algorithm 65). *Communications of the ACM 4* (1961), 321–322.

[3] Hoare, C. Quicksort. *Computer Journal 5* (1962), 10–15.

[4] Hwang, H.-K., and Tsai, T.-H. Quickselect and Dickman function. *Combinatorics, Probability and Computing* (2002). To appear.

[5] Kirschenhofer, P., Prodinger, H., and Martínez, C. Analysis of Hoare's Find algorithm with median-of-three partition. *Random Structures & Algorithms 10* (1997), 143–156.

[6] Knuth, D. Mathematical analysis of algorithms. In *Information Processing '71, Proc. of the 1971 IFIP Congress* (Amsterdam, 1972), North-Holland, pp. 19–27.

[7] KNUTH, D. *The Art of Computer Programming: Sorting and Searching*, 2nd ed., vol. 3. Addison-Wesley, Reading, Mass., 1998.

[8] MARTÍNEZ, C., AND ROURA, S. Optimal sampling strategies in Quicksort and quickselect. *SIAM Journal on Computing 31*, 3 (2001), 683–705.

[9] SEDGEWICK, R. The analysis of Quicksort programs. *Acta Informatica 7* (1976), 327–355.

[10] SEDGEWICK, R. Implementing Quicksort programs. *Communications of the ACM 21* (1978), 847–856.

[11] SEDGEWICK, R. *Quicksort.* Garland, New York, 1978.

[12] SEDGEWICK, R., AND FLAJOLET, P. *An Introduction to the Analysis of Algorithms.* Addison-Wesley, Reading, Mass., 1996.

Conrado Martínez
Departament de Llenguatges i Sistemes Informàtics.
Universitat Politècnica de Catalunya.
E-08034, Barcelona, Spain.
conrado@lsi.upc.es.

Daniel Panario
School of Mathematics and Statistics.
Carleton University.
K1S 5B6, Ottawa, Canada.
daniel@math.carleton.ca.

Alfredo Viola
Instituto de Computación.
Universidad de la República.
Casilla de Correo 16120,
Distrito 6, Montevideo, Uruguay.
viola@fing.edu.uy.

Trends in Mathematics, © 2002 Birkhäuser Verlag Basel/Switzerland

Distribution of the Size of Simplified or Reduced Trees

Michel Nguyễn Thế

ABSTRACT: *Previous works by Casas, Fernández-Camacho and Steyaert have determined the mean and variance of the size of simplified or reduced trees of m-ary trees. Using the implicit function and the Quasi Powers theorems, we show that the limit distribution of that size is Gaussian.*

1 Introduction

Many analyses, but mostly limited to the average case, have been done which concern algorithms in symbolic computation: symbolic differentiation [10] and more general rewriting rules [4], unification [1].

This paper deals with the analysis of a number of simplifying rewriting rules, as used in computer algebra systems. Instances of such rewriting rules commonly encountered in mathematics are for example $x^2 = x$ in Boolean rings, $p \cdot x = 0$ and $x^p = x$ in fields of characteristic p, $x \cdot 0 = 0$, $x \cdot 1 = 1$, $x - x = 0$ over the reals, and so on.

We will consider two types of simplifying rewriting rules.

For the first type, we consider expressions generated by a finite algebra starting from a finite number $q \geq 2$ of constants, and an idempotent p-ary law $\odot$ $(p \geq 2)$: given any expression f, the expression $\odot(f, \cdots, f)$ (p times) is equivalent to f. The operation of rewriting an expression to its shortest equivalent expression will be called *simplification*. For the binary case, we will use the infix notation $f \odot f$ instead of $\odot(f, f)$.

For the second type, we start from a finite number $c_0' = c_0 + 1$ of constants ($c_0 \geq 1$), among which one is distinguished and denoted by e, and c_m m-ary laws $\odot_i$ $(m \geq 2)$ such that for any $i = 1, 2, \cdots, m$ and any expression f, the expression $\odot_i(f, \cdots, f)$ (m times) can be rewritten to e. We will call this operation *reduction*.

An issue at stake consists in knowing the gain in space brought by these two types of operation. The study of that problem was initiated in [2] and further developed in [7] and [3], where the average size and the variance of the size of simplified and reduced trees were computed and were proved to be linear in the size of the initial tree, using the Darboux-Pólya method. Progress has been made with singularity analysis and the Quasi Powers Theorem developed by Hwang [11]. In this paper, we will show that these new tools allow us to establish that the limit distribution of the size of simplified and reduced trees is Gaussian.

This paper is quite reminiscent of [5, 6], where Drmota studied systems of functional equations leading to Gaussian limits by a repeated use of Weierstrass preparation theorem to reduce the problem to dimension 1. His framework turned out to be quite adapted to rewriting systems. One must nevertheless notice that we deal here with an infinity of different rewriting rules, like $f \odot f \to f$, with f of arbitrary size, and hence that we cannot directly apply Drmota's work here. That is why we will directly state our problem in only one functional equation and apply Weierstrass preparation theorem just once. The core difficulty of our problem consists in showing that all hypotheses are satisfied in order to apply the theorem.

2 Simplification of trees

2.1 Algorithm and generating functions

Let us start with studying a few examples of simplification as defined in the introduction: $a \odot b$ cannot be simplified, $a \odot a$ simplifies to a, $(a \odot b) \odot (a \odot b)$ simplifies to $a \odot b$, $(a \odot a) \odot (a \odot b)$ simplifies to $a \odot (a \odot b)$, $(a \odot b) \odot (b \odot a)$ cannot be simplified, and $(a \odot a) \odot (a \odot a)$ can be simplified to a through the chains of rewriting rules:

$$(a \odot a) \odot (a \odot a) \to a \odot (a \odot a) \to a \odot a \to a,$$

$$(a \odot a) \odot (a \odot a) \to (a \odot a) \odot a \to a \odot a \to a.$$

It should be clear that every expression admits one and only one shortest equivalent expression, and that the chain of simplifying rules used does not matter. Nevertheless, a good strategy to simplify an expression consists in using the procedure *simplify* defined below, depending on a procedure *equal* testing the equality of trees. The procedure *simplify* recursively simplifies trees in prefix order, and clearly transforms any expression tree into its shortest equivalent.

```
function simplify(tree:Q):Q;
local t: array of p Q; eq: boolean; i: integer;
    if tree.degree = 0 then simplify:=tree;
            else for i:=1 to p do
                    t[i]:=simplify(tree.child[i]);
                od;
                i:=1; eq:=true;
                while i≤p-1 and eq do
                    if equal(tree.child[i],tree.child[i+1]) then i:=i+1;
                    else eq:=false; fi od
                if eq then simplify:=t[1];
                    else simplify:=⊙(t[1],···,t[p]) fi fi.
```

Program for simplification of expressions with an idempotent law.

Tree-variables consist of three fields: *degree* (0 for leaves, p for internal nodes), *info* (operand e_i (for $i = 1, 2, \cdots, q$), or operator $\odot$), *child* array of p subtrees.

There corresponds to this procedure only one chain of rewriting rules leading to the shortest equivalent expression. For example, the expression

$$\big((a \odot a) \odot (b \odot b)\big) \odot (a \odot b)$$

simplifies to $a \odot b$ according to the chain of rules

$$\big((a \odot a) \odot (b \odot b)\big) \odot (a \odot b) \to \big(a \odot (b \odot b)\big) \odot (a \odot b) \to (a \odot b) \odot (a \odot b) \to a \odot b.$$

Such a procedure is called a bottom-up algorithm, because it operates firstly on leaves, then on small subtrees and so on climbing up to the root.

The family of $\mathcal{Q}$ of p-ary trees with q possible symbols at external nodes is defined by the formal series

$$\mathcal{Q} = \sum_{i=1}^{q} \boxed{e_i} + \odot(\mathcal{Q}, \cdots, \mathcal{Q}).$$

The generating function Q of the class $\mathcal{Q}$ can be written as

$$Q(z) = q + zQ^p(z). \tag{1}$$

Let ρ be the dominant singularity of Q and $\tau = Q(\rho)$. Of course ρ's value could be recovered from the closed form $\frac{1}{pn+1}\binom{pn+1}{n}q^{(p-1)n+1}$ of the coefficients of Q, that can be computed for example through Lagrange inversion theorem applied to $zQ(z^p)$, but for reasons to be apparent below, we will perform the analysis with generating functions and singularity analysis.

As long as the function Q can be defined implicitly by (1), it remains analytic ($|z| < \rho$); it is no longer analytic when the functional equation defining Q ceases to be regular, that is when the derivative of the functional equation (1) with respect to Q becomes null: this corresponds to the moment when z becomes equal to ρ. Then it is often said that the dominant singularity corresponds to a failure of the implicit function theorem [8].

Hence ρ and τ are solutions of the system

$$\tau = q + \rho\tau^p, \ 1 = p\rho\tau^{p-1},$$

from which we deduce

$$\rho = \frac{(p-1)^{p-1}}{p^p q^{p-1}}. \tag{2}$$

Let an *irreducible* tree be a tree that cannot be further simplified. Let $\mathcal{I}$ the class of irreducible trees recursively defined by

$$\begin{aligned}\mathcal{I} &= \sum_{i=1}^{q} \boxed{e_i} + \{\odot(t_1, \cdots, t_p) \mid \exists\, i, j \text{ such that } t_i \neq t_j\} \\ &= \sum_{i=1}^{q} \boxed{e_i} + \odot(\mathcal{I}, \cdots, I) - \{\odot(t, \cdots, t) \mid t \in I\},\end{aligned}$$

of generating function $I(z) = \sum_{t \in \mathcal{I}} z^{|t|} = \sum_{n \geq 0} I_n z^n$, which is solution of equation $I(z) = q + zI(z)^p - zI(z^p)$. It is easy to see that the number of irreducible trees is small:

Lemma 2.1. *Let ρ_I be the dominant singularity of I. Then $\rho < \rho_I < 1$; hence the number of irreducible trees of size n is asymptotically exponentially small, compared to the total number of trees of that size.*

Proof. It is easy to see — e.g. by using a proof by induction — that, for each tree of size n, there exist at least two irreducible trees having the same structure. As

$(p-1)^{p-1}p^{-p}$ is the singularity of the generating function of p-ary trees without symbols at its nodes (formula (2) remains valid for $q = 1$), we have

$$\rho_I < \frac{(p-1)^{p-1}}{p^p} < 1.$$

We deduce that $\rho_I^p < \rho_I$, and that the radius of convergence of $z \mapsto I(z^p)$ is strictly greater than the radius of convergence of I. Hence the failure of the application of the implicit function theorem on $(z, y) \mapsto y - q - zy^p - zI(z^p)$ gives $\rho < \rho_I$. □

For each $u \in \mathcal{Q}$, let $s(u)$ be the irreducible tree to which u simplifies. Define for each irreducible t the generating function $Q_t(z) = \sum_{s(u)=t} z^{|u|}$, and introduce the bivariate generating function $X(z, y) = \sum_{t \in \mathcal{Q}} y^{|s(t)|} z^{|t|}$. Then the following lemma gives a functional equation for X:

Lemma 2.2.

$$X(z, y) = q + yzX(z, y)^p - z \sum_{t \in \mathcal{I}} \left(y^{p|t|+1} - y^{|t|} \right) Q_t^p(z).$$

Proof. We have the equations

$$\mathcal{Q}_e = \boxed{e} + \odot(\underbrace{\mathcal{Q}_e, \cdots, \mathcal{Q}_e}_{p}), \; \mathcal{Q}_t = \odot(\mathcal{Q}_{t_1}, \cdots, \mathcal{Q}_{t_p}) + \odot(\underbrace{\mathcal{Q}_t, \cdots, \mathcal{Q}_t}_{p}),$$

where $e \in \{e_1, \cdots, e_q\}$ and $t \in \mathcal{I}$ with $t = \odot(t_1, \cdots, t_p)$.

Summing the equalities $y^{|t|}Q_t(z) = yz \prod_{i=1}^{p} y^{|t_i|} Q_{t_i}(z) + zy^{|t|} Q_t(z)$, we obtain the desired expression for $X(z, y)$. □

2.2 An algebraic singularity

This section is devoted to show that the singularity $z = \rho(y)$ of the bivariate generating function satisfies certain properties, so that $\rho(1)/\rho(u)$ resembles a probability generating function. These properties are summarized in the following proposition:

Proposition 2.3. *There exist real numbers $z_1 > \rho$, $y_1 > 1$, and analytic functions A, B and C such that, in the domain*

$$\mathcal{D} = \{(z, y) \mid |z| < z_1, \; |y| < y_1\},$$

X admits the representation

$$X(z, y) = A(z, y) + B(z, y)\sqrt{C(z, y)}.$$

This proposition states first that for every y, the singularity z is of the algebraic type, and second that the function $y \mapsto \rho(y)$ is analytic near 1.

To prove the proposition, it suffices to show (as will be done in Lemma 2.5 below) that the function $(z, y) \mapsto \sum_{t \in \mathcal{I}} \left(y^{p|t|+1} - y^{|t|} \right) Q_t^p(z)$ is analytic for (z, y) beyond $(\rho, 1)$. In order to do so, we will use the fact that the functions Q_t are small enough:

Lemma 2.4. *There exist* $z_0 > \rho$, γ, $c > 0$, *such that* $\gamma^p < \rho$ *and, for all* $t \in \mathcal{I}$, $Q_t(z_0) < c\gamma^{|t|}$.

Proof. Due to lack of space, we cannot give a proof of that crucial lemma, already fully proved in [7]. We can only say that straightforward computations are enough for the binary case, but that more efforts are needed for $p \geq 3$. The proof is by induction, using the formula

$$Q_t(z_0) - z_0 \prod_{i=1}^{p} Q_{t_i}(z_0) = z_0 Q_t(z_0)$$

and closely examining the function $x \mapsto x - z_0 x^p$. □

Lemma 2.5. *Let* $d(z,y) = \sum_{t \in \mathcal{I}} \left(y^{p|t|+1} - y^{|t|}\right) Q_t^p(z)$ *be the function such that*

$$X(z,y) = q + yzX(z,y)^p - zd(z,y).$$

There exists $z_0 > \rho$ *and* $y_0 > 1$ *such that* $d(z,y)$ *is holomorphic in the two variables* z *and* y *in the domain* $|z| < z_0$, $|y| < y_0$.

Proof. Define z_0 and γ as in Lemma 2.4. Since we have $1 < \rho\gamma^{-p} < \rho_I \gamma^{-p}$, we can find a real number y_0 in the interval $]1, \rho_I^{1/p}\gamma^{-1}[$ and, for $|z| < z_0$, $|y| < y_0$:

$$\sum_{t \in \mathcal{I}} y^{p|t|+1} Q_t^p(z) \leq \sum_{t \in \mathcal{I}} y_0^{p|t|+1} Q_t^p(z) \leq \sum_{n \geq 0} I_n y_0^{pn+1} c^p \gamma^{pn}$$

and

$$\sum_{t \in \mathcal{I}} y^{|t|} Q_t^p(z) \leq \sum_{t \in \mathcal{I}} y_0^{|t|} Q_t^p(z) \leq \sum_{n \geq 0} I_n y_0^n c^p \gamma^{pn},$$

from which we deduce that the numerical series

$$\sum_{t \in \mathcal{I}} y^{p|t|+1} Q_t^p(z) \text{ and } \sum_{t \in \mathcal{I}} y^{|t|} Q_t^2(z)$$

converge. Since they have nonnegative coefficients in $z^n y^k$, they are normally convergent, and hence holomorphic in z and y. As a consequence, their difference $d(z,y)$ is also holomorphic in z and y for $|z| < z_0$, $|y| < y_0$. □

We are now fully equipped to show that the bivariate generating function X admits an algebraic singularity, and prove Proposition 2.3.

Proof of Proposition 2.3. Define $F(z,y,x) = x - q - yzx^p + zd(z,y)$, which satisfies $F(z,y,X(z,y)) = 0$ for all z, y in the domain of convergence of X. As long as d is analytic and $\frac{\partial F}{\partial x}$ is non zero, the implicit function theorem asserts that X is analytic in z and y (it is a direct consequence of the residues theorem which enables to write $X(z,y) = \displaystyle\int_{\mathcal{C}} \frac{F_x(z,y,u)}{F(z,y,u)} u \, du$ for an appropriate curve $\mathcal{C}$).

Let z_0 and y_0 be real numbers as in Lemma 2.5. Let us show that for all y such that $|y| < y_0$, $z \mapsto X(z,y)$ admits a singularity $z = \rho(y)$ that is implicitly defined by an analytic function and is of square root type in a vicinity of $y = 1$.

For all y, the system of equations

$$\begin{cases} F(z,y,x) = 0 \\ \frac{\partial F}{\partial x} = 0 \end{cases}$$

defines a singularity $\rho(y)$ depending on y. Hence this system implicitly defines a function $z = \rho(y)$. Let us reduce it to a single equation. The system can be written

$$\begin{cases} x - q - yzx^p + zd(z,y) = 0 & (2.2.a) \\ 1 - pyzx^{p-1} = 0. & (2.2.b) \end{cases}$$

(2.2.b) rewrites into $x^{p-1} = \frac{1}{pyz}$, and replacing x^p by $\frac{x}{pyz}$ in (2.2.a) gives a linear equation in x which solves to $x = \frac{q - zd}{1 - \frac{1}{p}}$. Injecting this expression in (2.2.b) gives

$$1 - pyz\left(\frac{q - zd}{-1 + \frac{1}{p}}\right)^{p-1} = 0,$$

or $C(z,y) = 0$, with

$$C(z,y) = -\frac{1}{p}\left(1 - \frac{1}{p}\right)^{p-1} + yz\big(q - zd(z,y)\big)^{p-1}.$$

According to Lemma 2.5 about $d(z,y)$, C is analytic on a domain of the form $\{\ (z,y) \mid |z| < z_0,\ |y| < y_0\}$, with $z_0 > \rho$ and $y_0 > 1$. Moreover, it satisfies $C_z(\rho,1) \neq 0$. As a consequence, the singularity $z = \rho(y)$ of $z \mapsto X(z,y)$ is implicitly defined in a neighborhood of $y = 1$ by $C(\rho(y), y) = 0$.
As the function implicitly defining X satisfies

$$F_{x^2}(\rho, 1, X(\rho,1)) = (p-1)p\rho X(\rho,1)^{p-2} \neq 0,$$

$\rho(1)$ is a square root singularity, and it remains true for $\rho(y)$ if y stays in the vicinity of 1. Hence, there exist real numbers z_1, y_1, and analytic functions A and B such that $\rho < z_1$, $1 < y_1$, and X admits the representation

$$X(z,y) = A(z,y) + B(z,y)\sqrt{C(z,y)}.$$

□

2.3 Limit law

Before using the Quasi Powers theorem, we need to establish a variability condition, contained in the following lemma.

Lemma 2.6. *The expectation μ_n and the variance v_n of the size of the simplified trees are linear in the size of the initial trees.*

Proof. Singularity analysis provides us with the asymptotics

$$\mu_n = \frac{[z^n]X_y(z,1)}{[z^n]Q(z)} = \mu n + O(1) \text{ with } \mu > 0,$$

and

$$v_n = \frac{[z^n]\left(X_{y^2}(z,1) + X_y(z,1)\right)}{[z^n]Q(z)} - \mu_n^2 = \overline{v}n + O(1).$$

The fact that $\overline{v} > 0$ is not obvious and requires a proof, that has been overlooked in literature so far, apart from numerical computations in particular cases. Let us suppose that $\overline{v} = 0$ (it of course cannot be negative). Then the variance would be bounded by a constant V. Define K such that $VK^{-2} < 1/2$. Then if Y_n is the random variable equal to the size of the simplified tree of a tree of size n, Chebyshev inequality writes

$$\forall n, \ \Pr\left[|Y_n - \mu_n| \geq K\right] \leq \frac{v_n}{K^2} < \frac{1}{2},$$

from which we deduce

$$\lim_{n\to\infty} \Pr\left[|Y_n - \mu_n| < K\right] = \lim_{n\to\infty} \frac{[z^n]\sum_{t\in\mathcal{I},\ \mu_n - K<|t|<\mu_n+K} Q_t(z)}{[z^n]Q(z)} \geq \frac{1}{2}. \tag{3}$$

But from Lemma 2.4, there exist $c > 0$ and $z_0 > \rho$ such that $Q_t(z_0)$ is uniformly bounded by c. As Q_t has nonnegative coefficients for all t, it is easy to see that, for all n and t, $[z^n]Q_t < cz_0^{-n}$, and

$$[z^n] \sum_{\substack{t\in\mathcal{I} \\ \mu_n - K<|t|<\mu_n+K}} Q_t(z) \leq 2Kcz_0^{-n}.$$

As on the other hand we have $[z^n]Q(z) \simeq C_Q\rho^n/\sqrt{\pi n^3}$ for some constant C_Q, it entails that

$$\overline{\lim_{n\to\infty}} \frac{[z^n]\sum_{t\in\mathcal{I},\ \mu_n - K<|t|<\mu_n+K} Q_t(z)}{[z^n]Q(z)} = 0,$$

which contradicts (3). Hence $\overline{v} = 0$ is impossible. □

It is now possible to find the limit distribution of the size of simplified trees.

Theorem 2.7. *The size of simplified p-ary trees has a Gaussian limit distribution, with mean μ_n and variance v_n satisfying*

$$\mu = \lim_{n\to\infty} \frac{\mu_n}{n} = 1 - \rho\frac{p-1}{q}d_y(\rho,1)$$

and

$$\overline{v} = \lim_{n\to\infty} \frac{v_n}{n} = \left[\rho\frac{p-1}{q} - \frac{(p-1)(p-2)}{q^2}\rho^2\right] d_y(\rho,1) - 3(p-1)^2\frac{\rho^2}{q^2}d_y^2(\rho,1)$$
$$+2\rho^2\frac{p-1}{q}d_{yz}(\rho,1) - 2\rho^3\left(\frac{p-1}{q}\right)^2 d_y(\rho,1)d_{yz}(\rho,1) - (p-1)\frac{\rho}{q}d_{y^2}(\rho,1).$$

Proof. According to Lemma 2.3, in a domain of the form

$$\mathcal{D} = \{(z,y) \mid |z| < z_1,\ |y| < y_1\}$$

for some $z_1 > \rho$ and $y > 1$, X admits the representation

$$X(z,y) = A(z,y) + B(z,y)\sqrt{C(z,y)}$$

with A, B and C analytic,

$$B(\rho,1) \neq 0$$

and

$$C(z,y) = -\frac{1}{p}\left(1-\frac{1}{p}\right)^{p-1} + yz\big(q - zd(z,y)\big)^{p-1}.$$

As $d(z,1) \equiv 0$, it is easy to see that $\zeta = \rho$ is a simple root of $\zeta \mapsto C(\zeta,1)$. Hence X satisfies the algebraic-logarithmic conditions of theorem 9.8 of [9] p.68-69, which states a Gaussian limit for any random variable Y_n with probability generating function

$$\frac{[z^n]X(z,y)}{[z^n]X(z,1)},$$

given the variability condition

$$\lim_{n\to\infty} \frac{\mathrm{Var}(Y_n)}{n} > 0,$$

that we have just proved in the lemma above.

We know from the same theorem that $\mu = \lim \frac{1}{n}E[Y_n]$ and $\overline{v} = \lim \frac{1}{n}\mathrm{Var}(Y_n)$ are given by the expressions

$$\mu = -\frac{\rho_y(1)}{\rho(1)},$$

$$\overline{v} = -\frac{\rho_{y^2}(1) + \rho_y(1)}{\rho(1)} + \left(\frac{\rho_y(1)}{\rho(1)}\right)^2.$$

To compute these values, one only has to solve the system of equations

$$C_y(\rho,1) = 0,\ C_{y^2}(\rho,1) = 0,$$

and notice that $d_z(\rho,1) = d_{z^2}(\rho,1) = 0$, which is simpler than using the partial derivatives $X_y(z,1)$ and $X_{y^2}(z,1)$ and the asymptotic equalities of the former lemma. □

Numerical computations for $p = q = 2$ give the values $\mu = 0.8196...$ and $\overline{v} = 0.2166...$. Heuristically, we also obtained that μ and $\overline{v}$ are respectively increasing and decreasing when p and q are increasing.

3 Nilpotent m-ary laws

3.1 Algorithm and generating functions

We consider now an algorithm of reduction slightly different from simplification. We work with c_m different operators of arity m, which are nilpotent instead of being idempotent: there exists a particular operand e such that, if a tree has identical subtrees, it is reduced to e.

Examples of reduction, as defined in the introduction, are, for $c_m = 1$ and a nilpotent operator $\odot$: $a \odot b$ cannot be reduced, $a \odot a$ reduces to e, $(a \odot b) \odot (a \odot b)$ reduces to e, $(a \odot a) \odot (a \odot b)$ reduces to $e \odot (a \odot b)$, $(a \odot b) \odot (b \odot a)$ cannot be reduced, and $(a \odot a) \odot (a \odot a)$ can be reduced to e through the chains of rewriting rules:

$$(a \odot a) \odot (a \odot a) \to e \odot (a \odot a) \to e \odot e \to e,$$

$$(a \odot a) \odot (a \odot a) \to (a \odot a) \odot e \to e \odot e \to e.$$

We define a procedure *reduce* quite similarly as the procedure *simplify*: just replace occurrences of *simplify* by *reduce* in the algorithm, and define the result to be equal to e when one finds identical subtrees instead of the common subtree itself.

As an example, the expression $\big((a \odot a) \odot (b \odot b)\big) \odot (a \odot b)$ reduces to $e \odot (a \odot b)$ according to the chain of rules

$$\big((a\odot a)\odot(b\odot b)\big)\odot(a\odot b) \to \big(e\odot(b\odot b)\big)\odot(a\odot b) \to (e\odot e)\odot(a\odot b) \to e\odot(a\odot b).$$

One can notice that the reduction of the initial expression leads to a tree with more nodes than the tree obtained through simplification.

We consider the class $\mathcal{R}$ of m-ary trees whose external nodes can be labeled with c_0 constants a_i or with the specific symbol e, and whose internal nodes can be labeled with an m-ary operator among m different operators. We will denote the total number of possible labels for the external nodes with $c_0' = c_0 + 1$. Hence the generating function enumerating the objects of class $\mathcal{R}$ by their size satisfies

$$R(z) = \sum_{t \in \mathcal{R}} z^{|t|} = c_0' + c_m z R^m(z).$$

Let ρ_R be the dominant singularity of R, and $\| t \|$ denote the total number of nodes of the tree t. We also introduce the generating function

$$M(z) = \sum_{t \in \mathcal{R}} z^{\|t\|} = zR(z^m).$$

It is obvious that M admits m singularities on its circle of convergence, of modulus $\rho_M = \rho_R^{1/m}$.

We define again a notion of *irreducible* tree, that is a tree which cannot be reduced. The class of irreducible trees is isomorphic to the former one defined in the case of simplification, as long as $c_0' = q$ and $m = p$, so there cannot be any confusion, and we still denote it $\mathcal{I}$. We introduce two subclasses of $\mathcal{I}$, namely $\mathcal{I}^-$ and $\mathcal{I}^+$, defined by:

$$\mathcal{I}^- = \left\{ t \in \mathcal{I} \;\middle|\; |t|_e \le \frac{\|t\|}{2} \right\}; \; \mathcal{I}^+ = \left\{ t \in \mathcal{I} \;\middle|\; |t|_e > \frac{\|t\|}{2} \right\}.$$

The generating function enumerating the objects of class $\mathcal{I}$ by their size satisfies $I(z) = \sum_{t \in \mathcal{I}} z^{|t|} = c_0' + c_m z I^m(z) - c_m z I^m(z)$. Let ρ_I be its dominant singularity. We introduce the series $J(z) = \sum_{t \in \mathcal{I}} z^{\|t\|} = zI(z^m)$, of radius of convergence ρ_J.

With obvious notations, let also I^-, I^+, J^-, J^+ be the generating functions of objects in classes $\mathcal{I}^-$ and $\mathcal{I}^+$, respectively by size and by total number of nodes. Their respective radii of convergence will be denoted ρ_{I^-}, ρ_{I^+}, ρ_{J^-}, ρ_{J^+}. Once again we have $\rho_I = \rho_J^m$, $\rho_{I^-} = \rho_{J^-}^m$ and $\rho_{I^+} = \rho_{J^+}^m$.

It is easy to prove, as in Lemma 2.1, that $\rho_R < \rho_I < 1$. Besides, we also have $\rho_I \le \rho_{I^-} < 1$ and $\rho_I \le \rho_{I^+} < 1$.

For each $u \in \mathcal{R}$, let $s(u)$ be the irreducible tree to which u simplifies. Define for all irreducible t the generating function $R_t(z) = \sum_{r(u)=t} z^{|u|}$, and introduce the bivariate generating function $H(z,y) = \sum_{t\in\mathcal{R}} y^{|r(t)|} z^{|t|}$. H is given by the following functional equation:

Lemma 3.1.

$$H(z,y) = c_0' + c_m y z H(z,y)^m - c_m z \sum_{t\in\mathcal{I}} \left(y^{m|t|+1} - 1\right) R_t^m(z).$$

Proof. We have:

$$\forall a \in \{a_1, \cdots, a_{c_0}\},\ \mathcal{R}_a = \boxed{a},$$

$$\mathcal{R}_e = \boxed{e} + \sum_{i=1}^{c_m} \sum_{t\in\mathcal{I}} \odot_i(\underbrace{\mathcal{R}_t, \cdots, \mathcal{R}_t}_{m}),$$

$$\forall t = \odot_i(t_1, \cdots, t_m) \in \mathcal{I},\ \mathcal{R}_t = \odot_i(\mathcal{R}_{t_1}, \cdots, \mathcal{R}_{t_m}),$$

which in terms of generating functions translate in:

$$\forall a \in \{a_1, \cdots, a_{c_0}\},\ R_a(z) = 1,$$

$$R_e(z) = 1 + c_m z \sum_{t\in\mathcal{I}} R_t(z)^m,$$

$$\forall t = \odot_i(t_1, \cdots, t_m) \in \mathcal{I},\ R_t(z) = z \prod_{j=1}^{m} R_{t_j}(z).$$

Summing these equalities, we obtain the desired expression for $H(z,y)$. □

3.2 An algebraic singularity

To get a representation of the bivariate generating function H, we have to show that the function $(z,y) \mapsto \sum_{t\in\mathcal{I}} \left(y^{m|t|+1} - 1\right) R_t^m(z)$ is analytic far enough (Lemma 3.4). In order to do so, we will once again begin with providing upper bounds for $R_t(z_0)$ for some z_0. The m-ary case will prove more difficult than in the precedent section. That is why we will separate the binary case (Lemma 3.2) from the m-ary general one (Lemma 3.3).

In the binary case, we can find an upper bound for R_t that is similar to the one we found for Q_t:

Lemma 3.2. *There exist $z_0 > \rho_R$, $c > 0$ and $\gamma > 0$ such that $\gamma^2 < \rho_I$ and for all $t \in \mathcal{I}$, $R_t(z_0) \le c\gamma^{|t|}$.*

Proof. We will admit the following results proved in [7]: first, we have $M_e(\rho_M) < 1$; second, there exists $\varepsilon > 0$ such that for all $t \in \mathcal{I}$, $M_t(z)$ is convergent on the disk $\{z \in \mathbb{C} \mid |z| \geq \rho_M + \varepsilon\}$. From these two results we deduce that there exists $\varepsilon > 0$ such that $M_e(\rho_M + \varepsilon)$ is convergent and $M_e(\rho_M + \varepsilon) < 1$. In terms of R_e it can be rewritten:

$$(\rho_M + \varepsilon)R_e((\rho_M + \varepsilon)^2) < 1,$$

and there exists $\varepsilon' > 0$ such that

$$(\rho_R + \varepsilon')R_e(\rho_R + \varepsilon') < \sqrt{\rho_R + \varepsilon'}.$$

As $z \mapsto \sqrt{z}R_e(z)$ is increasing, we have

$$\forall z \leq \rho_R + \varepsilon',\ zR_e(z) < \sqrt{z}.$$

Let us choose $z_0 \leq \rho_R + \varepsilon'$ such that $\rho_R < z_0 < \rho_I$, and define $\gamma = \sqrt{z_0}$. Then $z_0 R_e(z_0) < \gamma$ with $\gamma^2 < \rho_I$. Define $c = R_e(z_0)$ and use the equalities $R_a(z) = 1$ for any label $a \neq e$, and $R_t = zR_uR_v$ for any $t = u \odot v \in \mathcal{I}$. We have

$$R_a(z_0) < R_e(z_0) \leq c\gamma^0,$$

and by induction on the size of t, one can prove

$$\forall t \in \mathcal{I},\ R_t(z_0) \leq c\gamma^{|t|}.$$

Indeed, if $R_u \leq c\gamma^{|u|}$ and $R_v \leq c\gamma^{|v|}$ with $u \neq v$, then if $t = u \odot v$ we have

$$R_t(z_0) = z_0 R_u(z_0)R_v(z_0) \leq z_0 c^2 \gamma^{|u|}\gamma^{|v|} \leq c(z_0 c)\gamma^{|t|-1} \leq c\gamma^{|t|}.$$

□

The m-ary case proves a little more difficult in so far as we have to use the classes $\mathcal{I}^-$ and $\mathcal{I}^+$.

Lemma 3.3. *There exists $\varepsilon' > 0$ such that, if $c = \rho_I^{1/2m}(\rho_R + \varepsilon')^{-1/m}$:*

a) if $c_0' = 2$ or $c_0' < m - 2$: $R_t(\rho_R + \varepsilon') < c\rho_I^{\frac{|t|}{2}}\ \forall t \in \mathcal{I} \setminus e$.

b) if $c_0' \neq 2$ and $c_0' \geq m - 2$: $\begin{cases} R_t(\rho_R + \varepsilon') < c\rho_{I^+}^{|t|/2}\ \forall t \in \mathcal{I}^+ \setminus e \\ R_t(\rho_R + \varepsilon') < c\rho_{I^-}^{|t|/2}\ \forall t \in \mathcal{I}^- \setminus e. \end{cases}$

Proof. A lemma from [7] states that there exists $\varepsilon > 0$ such that:

a) if $c_0' = 2$ or $c_0' < m - 2$: $M_t(\rho_M + \varepsilon) < \rho_J^{\frac{\|t\|}{2}}\ \forall t \in \mathcal{I} \setminus e$.

b) if $c_0' \neq 2$ and $c_0' \geq m - 2$: $\begin{cases} M_t(\rho_M + \varepsilon) < \rho_{J^+}^{\|t\|/2}\ \forall t \in \mathcal{I}^+ \setminus e \\ M_t(\rho_M + \varepsilon) < \rho_{J^-}^{\|t\|/2}\ \forall t \in \mathcal{I}^- \setminus e. \end{cases}$

These results are proved by induction, using elementary methods like in 2.4, but with more intermediate steps and different cases. Why do we have to split $\mathcal{I}$ in two? As can be seen from formula $M_t(z) = z^{\|t\|-|t|_e}\big(M_e(z)\big)^{|t|_e}$ for $t \in \mathcal{I}$,

majorizing $M_t(z)$ is more difficult when $\| t \| - |t|_e$ is small. This is why the case $|t|_e > \| t \|/2$ has to be studied more closely.
Let us now choose ε such as above. Define ε' such that $\rho_R + \varepsilon' = (\rho_M + \varepsilon)^m$. We have
a) if $c_0' = 2$ or $c_0' < m - 2$, then, for all $t \in \mathcal{I} \setminus e$,

$$M_t(\rho_M + \varepsilon) < \rho_J^{\frac{\|t\|}{2}} \iff (\rho_M + \varepsilon) R_t((\rho_M + \varepsilon)^m) < \rho_J^{\frac{\|t\|}{2}}$$
$$\iff (\rho_R + \varepsilon')^{1/m} R_t(\rho_R + \varepsilon') < \rho_J^{\frac{m|t|+1}{2}} \iff R_t(\rho_R + \varepsilon') < (\rho_R + \varepsilon')^{-\frac{1}{m}} \rho_I^{\frac{2}{m}} \rho_I^{\frac{|t|}{2}}.$$

b) if $c_0' \neq 2$ and $c_0' \geq m - 2$, similar computations provide the result. □

Lemma 3.4. *Let $g(z,y) = \sum_{t \in \mathcal{I}} \left(y^{m|t|+1} - 1\right) R_t^m(z)$ be the function such that $H(z,y) = c_0' + c_m yz H(z,y)^m - c_m z g(z,y)$. There exist $z_0 > \rho_R$ and $y_0 > 1$ such that $g(z,y)$ is holomorphic in the two variables z and y in the domain $|z| < z_0$, $|y| < y_0$.*

Proof. Case $m = 2$. — Define z_0 and γ as in Lemma 3.2. We have $\gamma^2 < \rho_I$, hence $1 < \rho_I \gamma^{-2}$. Let $y_0 \in \left]1, \sqrt{\rho_I \gamma^{-2}}\right[$. Recall that there exists $c > 0$ such that, for all $t \in \mathcal{I}$, $R_t(z_0) < c\gamma^{|t|}$. Hence for $|z| < z_0$, $|y| < y_0$, the following two relations are satisfied:

$$\sum_{t \in \mathcal{I}} \left(y^{2|t|+1}\right) R_t^2(z) \leq \sum_{t \in \mathcal{I}} \left(y_0^{2|t|+1}\right) R_t^2(z_0) \leq \sum_{n \geq 0} I_n \left(y_0^{2n+1}\right) c^2 \gamma^{2n},$$

$$\sum_{t \in \mathcal{I}} R_t^2(z) \leq \sum_{t \in \mathcal{I}} R_t^2(z_0) \leq \sum_{n \geq 0} I_n c^2 \gamma^{2n}.$$

From $y_0^2 \gamma^2 < \rho_I$ and $\gamma^2 < \rho_I$, we deduce that the numerical series

$$\sum_{t \in \mathcal{I}} y^{2|t|+1} R_t^2(z) \qquad \text{and} \qquad \sum_{t \in \mathcal{I}} R_t^2(z)$$

converge. Since the two series have nonnegative coefficients in $z^n y^k$, they are normally convergent, and hence holomorphic in z and y. Their difference $g(z,y)$ is consequently also holomorphic in z and y.

Case $m \geq 3$. — If we are in case a) with $c_0' = 2$ or $c_0' < m - 2$, then Lemma 3.3 gives $\varepsilon' > 0$ such that $R_t(\rho_R + \varepsilon') < c\rho_I^{|t|/2}$ for all $t \in \mathcal{I} \setminus e$. As $1 < \rho_I (\rho_I^{-1/2})^m$ (because $m \geq 3$), similar computations allow us to establish that there exists $y_0 \in \left]1, \sqrt[m]{\rho_I (\rho_I^{-1/2})^m}\right[$ such that the series $g(z,y)$ is holomorphic in the domain $|z| < \rho_R + \varepsilon'$, $|y| < y_0$.
If we are in case b) with $c_0' \neq 2$ and $c_0' \geq m-2$, we can prove, still from Lemma 3.3, that there exists $\varepsilon' > 0$ such that

$$\sum_{t \in \mathcal{I}^+} y^{m|t|+1} R_t^m(z) \qquad \text{and} \qquad \sum_{t \in \mathcal{I}^-} y^{m|t|+1} R_t^m(z)$$

are holomorphic respectively in the domains

$$\{z \mid |z| < \rho_R + \varepsilon'\} \times \{y \mid |y| < \rho_{I+}^{1/m}(\rho_{I+}^{-1/2})\}$$

and

$$\{z \mid |z| < \rho_R + \varepsilon'\} \times \{y \mid |y| < \rho_{I+}^{1/m}(\rho_{I-}^{-1/2})\}.$$

Hence if we define $z_0 = \rho_R + \varepsilon'$ and $y_0 = \inf(\rho_{I+}^{1/m}(\rho_{I+}^{-1/2}), \rho_{I-}^{1/m}(\rho_{I-}^{-1/2}))$, then $y_0 > 1$ (because $\rho_{I+} < 1$, $\rho_{I-} < 1$ and $m > 2$), and $g(z, y)$ is holomorphic in the domain $|z| < z_0$, $|y| < y_0$. □

As was the case in section 2, we can prove that H follows an algebraic schema:

Proposition 3.5. *There exist real numbers $z_1 > \rho$, $y_1 > 1$, and analytic functions A, B and C such that, in the domain*

$$\mathcal{D} = \{(z, y) \mid |z| < z_1, \ |y| < y_1\},$$

H admits the representation $H(z, y) = A(z, y) + B(z, y)\sqrt{C(z, y)}$.

Proof. We proceed as in the proof of Proposition 2.3, by considering the function $(z, y, x) \mapsto x - c_0' - c_m yzx^p + c_m zg(z, y)$. □

3.3 Limit law

We are now able to determine the limit distribution of the size of reduced trees.

Theorem 3.6. *The size of reduced m-ary trees has a Gaussian limit distribution, and the mean μ_n and standard deviation σ_n are asymptotically linear in n, with*

$$\lim \frac{\mu_n}{n} = \mu \text{ and } \lim \frac{\sigma_n}{n} = \overline{v},$$

where, with $q = \frac{c_0'}{c_m}$, $\mu = 1 - \rho\frac{m-1}{q} g_y(\rho, 1)$ and

$$\overline{v} = \left[\rho\frac{m-1}{q} - \frac{(m-1)(m-2)}{q^2}\rho^2\right] g_y(\rho, 1) - 3(m-1)^2\frac{\rho^2}{q^2}g_y^2(\rho, 1)$$
$$+ 2\rho^2\frac{m-1}{q}g_{yz}(\rho, 1) - 2\rho^3\left(\frac{m-1}{q}\right)^2 g_y(\rho, 1)g_{yz}(\rho, 1) - (m-1)\frac{\rho}{q}g_{y^2}(\rho, 1).$$

Proof. The proof is the same as in Theorem 2.7. □

Numerical computations for $c_0' = c_m = 2$ provide the values $\mu = 0.8162...$ and $\overline{v} = 0.2469....$ The average reduction ratio is not much smaller than the average simplification ratio.

4 Conclusion and open questions

We proved that the size of simplified and reduced trees are asymptotically Gaussian, whereas only the mean and variance were known before. Our framework also allows to assert that the rate of convergence to the Gaussian limit is in $O(1/\sqrt{n})$ thanks to the Berry-Essén inequality. A natural development would be to extend this study to a larger family of trees, and to mix different types of operators.

A more interesting challenge would be to study the distributions of the costs of the algorithms of simplification and reduction. Their means are known to be linear in n, but their variances have not been computed yet. Their distributions are all the more so still unknown.

References

[1] Luc Albert, Rafael Casas, François Fages, *Average-Case Analysis of Unification Algorithms*, Theoretical Computer Science **113**(1) (1993), 3-34.

[2] Rafael Casas, Jean-Marc Steyaert, *Bottom-Up Recursion in Trees*, in Proceedings of CAAP'86, Lecture Notes in Computer Science **214** (1986), 172-182.

[3] Rafael Casas, María-Inés Fernández-Camacho, Jean-Marc Steyaert, *Algebraic Simplification in Computer Algebra: an Analysis of Bottom-Up Algorithms*, Theoretical Computer Science **74** (1990), 273-298.

[4] Christine Choppy, Stéphane Kaplan, Michèle Soria, *Complexity Analysis of Term Rewriting Systems*, Theoretical Computer Science **67** (1989), 261-282.

[5] Michael Drmota, *Asymptotic distributions and a multivariate Darboux method in enumeration problems*, Journal of Combinatorial Theory, Series A **67** (1994), 169-184.

[6] Michael Drmota, *Systems of Functional Equations*, Random Structures and Algorithms **10** (1997), 103-124.

[7] María-Inés Fernández-Camacho, *Analisis Medio de Algoritmos de reducción sobre árboles*, Tesis, Universidad Complutense de Madrid (1988).

[8] Philippe Flajolet, Robert Sedgewick, *The Average Case Analysis of Algorithms: Complex Asymptotics and Generating Functions*, Research Report n^o**2026** (1993), INRIA. 100 pages.

[9] Philippe Flajolet, Robert Sedgewick, *The Average Case Analysis of Algorithms: Multivariate Asymptotics and Limit Distributions*, Research Report n^o**3162** (1997), INRIA. 123 pages.

[10] Philippe Flajolet, Jean-Marc Steyaert, *A Complexity Calculus for Recursive Tree Algorithms*, Mathematical Systems Theory **19** (1987), 301-331.

[11] Hsien-Kuei Hwang, *Théorèmes limites pour les structures combinatoires et les fonctions arithmétiques*, thèse, École Polytechnique (1994).

Michel Nguyễn Thế

Laboratoire d'Informatique de l'X
École Polytechnique
91128 PALAISEAU CEDEX
mnguyen@lix.polytechnique.fr

Trends in Mathematics, © 2002 Birkhäuser Verlag Basel/Switzerland

Digits and Beyond

Helmut Prodinger

ABSTRACT: *This is a survey about digits from a personal point of view. Counting the occurrences of digits (and, more generally, subblocks) is discussed in the context of various positional number systems. The methods to achieve this are Delange's elementary method and Flajolet's idea to use the Mellin–Perron summation (or integral) formula.*

Then we move to problems from Theoretical Computer Science (register function of binary trees, number of exchanges in Baxter's odd–questions) are also discussed. An open problem from physicists Yekutieli and Mandelbrot can also be treated in that fashion.

Furthermore, we consider representations of numbers where some digits are forbidden. As a representative example, we discuss the Cantor distribution and its moments, asymptotically analyzed by Mellin transforms. Other problems in this context lead to sums involving Bernoulli numbers which can be attacked by analytic Depoissonization.

Very briefly we mention carry propagation, mergesort parameters and jump interpolation search trees.

1 Introduction

This is a survey about digits from a personal point of view, i. e., I will talk about digit expansions but stress those things that are in one way or another related to my own research.

We are interested in the binary expansion, and, more generally in base q–expansion, with an integer $q \geq 2$, and digits $\{0, \ldots, q-1\}$. Delange [3] has analyzed the sum–of–digits function $S(n)$ (sometimes, especially for $q = 2$, written as $\nu(n)$), and found in particular the following result for the mean value

$$\frac{1}{m} \sum_{0 \leq n < m} S(n) = \frac{q-1}{2} \log_q m + F(\log_q m),$$

with a periodic function $F(x)$ of period 1, which is continuous but nowhere differentiable, and with explicitly known Fourier coefficients. Perhaps even more important than his result, is his *method* which is based on finding an explicit formula for the individual digits, in terms of floor functions, rewriting them as integrals over piecewise constant functions. Forming the mean value just means the union of the intervals of integration, which makes things easier. Then one pulls out the main (logarithmic) term and finds that what remains is a periodic function of $\{\log_q m\}$ (the fractional part of $\log_q m$) which can be brought into some standard form from which one also gets the Fourier coefficients.

In Section 2 we consider such a Delange type analysis for the system with base q but digits $\{d, d+1, \ldots, d+q-1\}$, where $d = 0$ means the standard case. The restriction for the parameter d is that 0 must be in the set of digits. We are counting how often a given digit i occurs in the representation of a natural number, more precisely, we consider the mean value of this counting function. This is based on the paper [26] which contains slightly more general results, namely not only counting individual digits i but rather a given (contiguous) subblock w. So, as an example, one knows how often "on average" the subblock 123014 occurs in the base 5 representation of n.

Delange's technique is elementary and nice, however, in more complicated situations, it can become quite messy. Flajolet came up with an alternative method [10]: It is based on the Mellin–Perron summation formula, which allows to write the counting function of interest as a contour integral in the complex plane. Shifting the line of integration produces then the terms in the explicit formula. As it often happens in this type of problems, one gets exact formulæ. The fact that the error term disappears completely is based on

$$0 = \int_{-\frac{1}{4}-i\infty}^{-\frac{1}{4}+i\infty} \zeta(s) n^s \frac{ds}{s(s+1)}$$

or similar results. In most instances, this technique is somewhat superior to Delange's elementary approach. As an additional reading we recommend Marko Riedel's master's thesis [43].

In Section 3, we describe the technique for Delange's result ($q = 2$), and also for a related number system called the *Gray code*, which is obtained from the binary representation by a series of reflections. Interestingly enough, the sections of Knuth's *The Art of Computer Programming, Vol. 4*, dealing with Gray codes, are already available at the web, see Knuth's homepage http://Sunburn.Stanford.EDU/knuth/.

In Section 4, we discuss a symmetric number system with base q (an even integer) and digits $\{-q/2, \ldots, q/2\}$. This redundant system is made unique by some conditions which in the instance $q = 2$ simply say that no two adjacent nonzero digits can occur. This instance $q = 2$ was rediscovered many times; perhaps the earliest reference is [42]. Some historical remarks can be found in [40]. We discuss again the problem of counting digits and subblocks in such a *symmetric signed system*.

Digits occur however in counting problems related to the *Analysis of Algorithms*. In Section 6 we deal with the *Register function* of binary trees (in the literature often appearing as Horton–Strahler numbers; here are some (random) references: [35, 30, 5, 4, 49, 33, 34]).

The question about the average value of the register function was solved independently by Flajolet–Raoult–Vuillemin [14] and Kemp [25], compare also [32]. The quantity $v_2(n)$, defined to be k if $n = 2^k(2i+1)$, plays a crucial role here. It is related to the sum-of-digits function $S(n)$, since $\sum_{k\le n} v_2(k) = n - S(n)$. Consequently one can, after some manipulations, invoke Delange's result. This was the route taken by Flajolet–Raoult–Vuillemin. Another approach (Kemp) is, after approximating the explicit quantities using real analysis, to consider Mellin transforms, which leads to Dirichlet series of the type $\sum_{n\ge 1} v_2(n)/n^s$. Perhaps the method that is most advanced [12], also uses Mellin transforms, but *on the level of generating functions*, in order to find the local behaviour of a generating function closely related to the desired average values. Once this is established, *singularity analysis of generating functions* (transfer theorems), developed in [11], provides the asymptotics of the, say, average value of the register function of binary trees of size n.

Section 7 is in a sense the companion of Section 6. The analysis of Batcher's odd–even method (see Knuth [29]) was originally left open and eventually solved by Sedgewick [44]. What was the binary expansion and the quantity $v_2(n)$ for the register problems is now the *Gray code* representation and the function $\vartheta(k)$ which is 1 for $k = 2^m(4i+1)$, and -1 for $k = 2^m(4i+3)$. Flajolet and Ramshaw [13] use a result analogous to Delange's but for the Gray code representation (also included in Section 3), and proceed along the lines of [14]. The more advanced technique (singularity analysis of generating functions) is also described here and was worked out a long time ago for the proceedings of a French summer school in Ile de Ré [38].

The next Section 8 deals with an open problem which is related to the register function of binary trees. It was posed by Yekutieli and Mandelbrot [49] and solved in [39]. The register function can be extended by attaching a value to each internal node in the binary tree (which is the register function of the subtree having that node as the root), the value at the root being the register function of the tree. The question was to count the number of internal nodes having register function exactly 1 less than the register function of the entire tree. It was observed empirically that the expected value of this parameter is asymptotically a periodic function of $\log_4 n$ if all trees of size n (n internal nodes) are considered to be equally likely. Here, the following arithmetic function $\psi(m)$ plays a role: $\psi(m) = -k$ if $m = 2^i(4k+1)$ and $\psi(m) = k+1$ if $m = 2^i(4k+3)$. Otherwise, the same comments about the methodology as for the previous sections apply here.

The Section 9 deals with sets of numbers which have representations avoiding digits from a subset of *forbidden digits.* The most common case is $q = 3$ with forbidden digit 1. If one considers numbers of the form $\sum_{k\ge1} a_k 3^{-k}$ where $a_k \in \{0,2\}$, one obtains the classical Cantor set. One question is about the moments of the Cantor distribution (strings of the form $\sum_{k\ge1} a_k 3^{-k}$ are produced at random, where "random" means that each digit might be 0 or 2, with probability 1/2). Grabner and I could satisfactorily solve this problem, which originated in [31], using the method of Mellin transforms. Another problem, suggested in [23], was eventually solved by A. Knopfmacher and myself [27]: If one draws n random numbers, according to the Cantor distribution, what is the expected value of their *minimum*? Using exponential generating functions to solve the relevant difference equations, one can relate it to "the hardest asymptotic nut" from [28]. These days, one has alternative methods to attack the sums coming out in the analysis à la "asymptotic nut"; these are techniques known as Rice's method [15] which are again intimately related to Mellin transforms, and (analytic) Depoissonization, which was floating around for years in the literature, an early example being perhaps [45]. Recently it has been *polished* and systematized by Jacquet and Szpankowski [24].

The last sections are very brief and mention *carry propagation* in positional number systems, mergesort, and jump interpolation search trees.

2 A Delange type analysis

We describe Delange's technique by considering the number system with base q and digits $\{d, d+1, \dots, d+q-1\}$ where $1-q \le d \le 0$. We are going to study $B_i(n)$, the number of occurrences of digit $i \ne 0$ in the representation of n. This is not a real restriction since the number of occurrences of digit 0 can be obtained

as the difference of the length of the representation and the number of the other digits. A full treatment, even for subblocks instead of just digits, is in [26].
First, one needs an explicit formula: If n is written as $\dots a_2a_1a_0$ in the (q,d)–system, then $a_k = i$ iff

$$\left\lfloor \frac{n}{q^{k+1}} + 1 - \frac{i}{q} - \frac{d}{q(q-1)} + \frac{1}{q} \right\rfloor - \left\lfloor \frac{n}{q^{k+1}} + 1 - \frac{i}{q} - \frac{d}{q(q-1)} \right\rfloor = 1.$$

Consequently,

$$B_i(n) = \sum_{k\ge 1} \left(\left\lfloor \frac{n}{q^{k}} + 1 - \frac{i}{q} - \frac{d}{q(q-1)} + \frac{1}{q} \right\rfloor - \left\lfloor \frac{n}{q^{k}} + 1 - \frac{i}{q} - \frac{d}{q(q-1)} \right\rfloor \right).$$

Delange's simple but extremely useful observation is that one can write the floor functions as integrals. We are considering an average of the function $B_i(n)$:

$$\begin{aligned}\frac{1}{m}\sum_{n=0}^{m-1} B_i(n) &= \frac{1}{m}\sum_{n=0}^{m-1}\sum_{k\ge 1}\left(\left\lfloor \tfrac{n}{q^k} + 1 - \tfrac{i}{q} - \tfrac{d}{q(q-1)} + \tfrac{1}{q} \right\rfloor - \left\lfloor \tfrac{n}{q^k} + 1 - \tfrac{i}{q} - \tfrac{d}{q(q-1)} \right\rfloor \right)\\ &= \frac{1}{m}\int_0^m \sum_{k\ge 1}\left(\left\lfloor \tfrac{t}{q^k} + 1 - \tfrac{i}{q} - \tfrac{d}{q(q-1)} + \tfrac{1}{q} \right\rfloor - \left\lfloor \tfrac{t}{q^k} + 1 - \tfrac{i}{q} - \tfrac{d}{q(q-1)} \right\rfloor \right) dt.\end{aligned}$$

It can be readily checked that nonzero contributions are only possible for $k \le l+1$, with $l = \lfloor \log_q m - \log_q(1+\frac{d}{q-1})\rfloor$. After a simple change of variable in the integrals, it turns out that one should study functions like

$$g_\beta(x) = \int_0^x \left(\left\lfloor u + \beta + \tfrac{1}{q} \right\rfloor - \lfloor u+\beta \rfloor - \tfrac{1}{q} \right) du, \qquad h_\beta(x) = \sum_{k\ge 0} q^{-k} g_\beta(xq^k)$$

and

$$H_\beta(x) = \frac{1-\gamma-\{-\gamma+x\}}{q} + q^{1-\gamma-\{-\gamma+x\}} h_\beta(q^{\{-\gamma+x\}-1+\gamma}),$$

with $\{x\} = x - \lfloor x \rfloor$ (the fractional part of x) and $\gamma = \log_q(1+\frac{d}{q-1})$.
Eventually one finds that

$$\frac{1}{m}\sum_{n=0}^{m-1} B_i(n) = \frac{\log_q m}{q} + H_\beta(\log_q m),$$

where β depends on the digit i in a simple way: $\beta = 1 - \frac{i}{q} - \frac{d}{q(q-1)} - \frac{1}{q}$.
Since the function $H_\beta(x)$ is periodic with period 1, it is natural to study its *Fourier series* expansion. The result is: $H_\beta(x) = \sum_{k\in\mathbb{Z}} h_k e^{2\pi i k x}$, with

$$\begin{aligned} h_0 &= \log_q \frac{\Gamma(1-\{\beta+\frac{1}{q}\})}{\Gamma(1-\{\beta\})} - \frac{1}{q}\Big(\frac{1}{2} + \frac{1}{\log q}\Big),\\ h_k &= \frac{\zeta(\chi_k, 1-\{\beta+\frac{1}{q}\}) - \zeta(\chi_k, 1-\{\beta\})}{\log q\cdot \chi_k(1+\chi_k)}, \qquad k\ne 0,\end{aligned}$$

with $\chi_k = 2\pi ik/\log q$ and the Hurwitz' zeta function $\zeta(z,a)$.
The computation of the Fourier coefficients is relatively straight forward; one has to use results like

$$\int_1^\infty \frac{\lfloor u\rfloor\,du}{u^{s+1}} = \sum_{k\geq 1} k\int_k^{k+1}\frac{du}{u^{s+1}} = \frac{1}{s}\sum_{k\geq 1} k\Big(\frac{1}{k^s}-\frac{1}{(k+1)^s}\Big) = \frac{1}{s}\sum_{k\geq 1}\frac{1}{k^s} = \frac{1}{s}\zeta(s).$$

The computations in the case of counting subblocks are more complicated, and in the end an additional error term of order $1/m$ occurs.

3 The Mellin–Perron summation formula

This section is based on [10]. Let $f(x)$ be a function defined over $[0,+\infty)$. Its Mellin transform $f^*(s) = \mathcal{M}[f(x);s]$ is defined by $f^*(s) = \int_0^\infty f(x)x^{s-1}dx$. By linearity and the rescaling property we have

$$F(x) = \sum_k \lambda_k f(\mu_k x) \implies F^*(s) = \Big(\sum_k \lambda_k \mu_k^{-s}\Big) f^*(s). \tag{1}$$

The condition is for s to belong to a 'fundamental strip' defined by the property that the integral giving $f^*(s)$ and the sum $\sum_k \lambda_k\mu_k^{-s}$ are both absolutely convergent. Similar to the Laplace transform there is an inversion theorem (cf. [7]). When applied to (1), it provides

$$\sum_k \lambda_k f(\mu_k x) = \frac{1}{2\pi i}\int_{c-i\infty}^{c+i\infty}\Big(\sum_k \lambda_k\mu_k^{-s}\Big) f^*(s)x^{-s}ds, \tag{2}$$

with c in the fundamental strip.
Formula (2) could be called Mellin's summation formula. It is especially useful when the integral can be computed by *residues*, and in that case each residue contributes a term in an asymptotic expansion of $F(x)$.
This formula lends itself to various number theoretic applications, most notably proofs of the prime number theorem. Introduce the step function $H_0(x)$ defined by

$$H_0(x) = \begin{cases} 1 & \text{if } x\in[0,1], \\ 0 & \text{if } x>1, \end{cases}$$

together with the functions $H_m(x) = (1-x)^m H_0(x)$. In the interesting case where $\mu_k \equiv k$, we obtain from (2), formulæ of the Perron type that provide integral representations for the iterated summations of arithmetic functions in terms of their Dirichlet generating function.
Let $c>0$ lie in the half–plane of absolute convergence of $\sum_k \lambda_k k^{-s}$. Then for any $m\geq 1$, we have

$$\frac{1}{m!}\sum_{1\leq k<n}\lambda_k\Big(1-\frac{k}{n}\Big)^m = \frac{1}{2\pi i}\int_{c-i\infty}^{c+i\infty}\Big(\sum_{k\geq 1}\frac{\lambda_k}{k^s}\Big)n^s\frac{ds}{s(s+1)\cdots(s+m)}. \tag{3}$$

For $m = 0$,

$$\sum_{1\le k<n} \lambda_k + \frac{\lambda_n}{2} = \frac{1}{2\pi i}\int_{c-i\infty}^{c+i\infty} \Big(\sum_{k\ge 1}\frac{\lambda_k}{k^s}\Big) n^s \frac{ds}{s}.$$

Formula (3) is obtained from (2) by setting $x \equiv n^{-1}$, $f(x) \equiv H_m(x)$, and observing that $H_m^*(s) = m!(s(s+1)\cdots(s+m))^{-1}$. For $m = 0$ the formula has to be modified slightly by taking a principal value for the sum, since $H_0(x)$ is discontinuous at $x = 1$. For instance, if we use $\lambda_k \equiv 1$ and $m = 1$, we get

$$\frac{n-1}{2} = \frac{1}{2\pi i}\int_{2-i\infty}^{2+i\infty} \zeta(s) n^s \frac{ds}{s(s+1)}.$$

Shifting the line of integration to the left and taking residues into account we obtain

$$0 = \int_{-\frac14-i\infty}^{-\frac14+i\infty} \zeta(s) n^s \frac{ds}{s(s+1)}. \tag{4}$$

Identity (4) is the basis for the existence of several *exact* rather than plainly *asymptotic* summation formulæ. We use this Mellin–Perron technique to derive an alternative proof of Delange's theorem: The sum–of–digits function $S(n)$ satisfies

$$S(n) = \frac{1}{2} n\log_2 n + nF_0(\log_2 n),$$

where $F_0(u)$ can be represented by the Fourier series $F_0(u) = \sum_{k\in\mathbb{Z}} f_k e^{2\pi i k u}$ and

$$f_0 = \frac{\log_2 \pi}{2} - \frac{1}{2\log 2} - \frac{3}{4},$$
$$f_k = -\frac{1}{\log 2}\frac{\zeta(\chi_k)}{\chi_k(\chi_k+1)} \quad \text{for} \quad \chi_k = \frac{2\pi i k}{\log 2}, \quad k \ne 0.$$

Let $v_2(k)$ be the exponent of 2 in the prime decomposition of k and $\nu(k)$ be the number of 1-digits in the binary representation of k. We have $\nu(k) - \nu(k-1) = 1 - v_2(k)$, so that $S(n)$ resembles a double summation of $v_2(k)$. Furthermore, it is well-known (and also rederived in Section 6) that

$$\sum_{k\ge 1} \frac{v_2(k)}{k^s} = \frac{\zeta(s)}{2^s - 1}.$$

Thus, from (3), with $\lambda_k = v_2(k)$ and $m = 1$, we get the basic integral representation

$$S(n) = \frac{n(n-1)}{2} - \frac{n}{2\pi i}\int_{2-i\infty}^{2+i\infty} \frac{\zeta(s)}{2^s-1} n^s \frac{ds}{s(s+1)}.$$

The integrand has a simple pole at $s = 1$ (originationg from the ζ–function), a double pole at $s = 0$ and simple poles at $s = \chi_k$. Shifting the line of integration to $\Re(s) = -\frac14$ and taking residues into account we get

$$S(n) = \frac{1}{2} n\log_2 n + nF_0(\log_2 n) - nR(n),$$

where the Fourier series (akin to F_0) $\frac{1}{\log 2}\sum_{k\in\mathbb{Z}}\frac{\zeta(\chi_k)}{\chi_k(\chi_k+1)}n^{\chi_k}$ occurs as the sum of residues of the integrand at the imaginary poles $s=\chi_k$. The remainder term is

$$R(n)=\frac{1}{2\pi i}\int_{-\frac{1}{4}-i\infty}^{-\frac{1}{4}+i\infty}\frac{\zeta(s)}{2^s-1}n^s\frac{ds}{s(s+1)}, \tag{5}$$

so that there only remains to prove that $R(n)\equiv 0$ when n is an integer. The integral converges since $|\zeta(-\frac{1}{4}+it)|\ll |t|^{3/4}$ (cf. [48]). Using the expansion $\frac{1}{2^s-1}=-1-2^s-2^{2s}-2^{3s}-\cdots$ in (5), which is legitimate since now $\Re(s)<0$, we find that $R(n)$ is a sum of terms of the form

$$\frac{1}{2\pi i}\int_{-\frac{1}{4}-i\infty}^{-\frac{1}{4}+i\infty}\zeta(s)(2^k n)^s\frac{ds}{s(s+1)},$$

and each of these terms is 0 by virtue of (4).

It is clear from the discussion above that an exact formula for a sum–of–digits function is obtained each time a similar Dirichlet generating function can be introduced. Let us illustrate this point by the integral representation for the sum–of–digits function associated to *Gray code* representations. The Gray code representation of the integers starts like 0, 1, 11, 10, 110, 111, 101, 100, 1100, 1101,...; its characteristic is that the representations of n and $n+1$ differ in exactly one binary position and it is constructed in a simple manner by reflections based on powers of two (for a definition, see, e. g., [13]). Let $\gamma(k)$ be the number of 1-digits in the Gray code representation of k, and $\delta_k=\gamma(k)-\gamma(k-1)$. It is easy to see that $\delta_{2k}=\delta_k$, and the pattern for odd values is $\delta_{2k+1}=(-1)^k$. Thus the Dirichlet generating function $\delta(s)$ of $\{\delta_k\}$ is

$$\delta(s)=\frac{2^sL(s)}{2^s-1}\quad\text{with}\quad L(s)=\sum_{k\geq 0}\frac{(-1)^k}{(2k+1)^s}.$$

Thus by (2.4), the summatory function $G(n)=\sum_{k<n}\gamma(k)$ admits the integral representation

$$G(n)=\frac{n}{2\pi i}\int_{2-i\infty}^{2+i\infty}\frac{2^sL(s)}{2^s-1}n^s\frac{ds}{s(s+1)}.$$

The summatory function $G(n)$ of sum–of–digits function of Gray–code satisfies

$$G(n)=\frac{1}{2}n\log_2 n+nF_1(\log_2 n),$$

where $F_1(x)$ is representable by the Fourier series

$$F_1(x)=2\log_2\Gamma\Big(\frac{1}{4}\Big)-\frac{3}{2}-\log_2\pi+\frac{1}{\log 2}\sum_{k\in\mathbb{Z}\setminus\{0\}}\frac{L(\chi_k)}{\chi_k(\chi_k+1)}e^{2k\pi ix}.$$

4 Symmetric signed digit expansions

Recently, Heuberger and I [22] have considered a symmetric system with an *even* base q and digits $-\frac{q}{2},\ldots,\frac{q}{2}$. Such a system is a priori redundant because of the existence of both $\pm\frac{q}{2}$ but made unique by the condition that $n = \sum_{j\geq 0} \varepsilon_j q^j$, where $|\varepsilon_j| \leq q/2$ and $|\varepsilon_j| = q/2$ implies that $0 \leq \operatorname{sign}(\varepsilon_j)\varepsilon_{j+1} < q/2$. (Equivalent conditions were discussed in [22].) We call this expansion the *symmetric signed digit expansion* of n and denote it by $(\ldots\varepsilon_2(n)\varepsilon_1(n)\varepsilon_0(n))$. For $q=2$, this system was already considered by Reitwiesner in a computer science context [42]. Here is an algorithm to compute it from right to left (Algorithm 1):

Algorithm 1 Computation of the symmetric signed digit expansion of n

Input: $n > 0$, $q \geq 2$ integers.
Output: The representation ε.

```
    ε ← (); m ← n
while m > 0 do
    a ← (m mod q)
    if a > q/2 or (a = q/2 and {m/q²} ≥ 1/2) then
        a ← a − q
    end if
    m ← (m − a)/q; ε ← ε & a
end while
```

We were able to derive an *explicit formula* for the digit ε_r, which for $q=2$ is

$$\varepsilon_r = \left\lfloor \frac{n}{2^{r+2}} + \frac{5}{6} \right\rfloor - \left\lfloor \frac{n}{2^{r+2}} + \frac{4}{6} \right\rfloor - \left\lfloor \frac{n}{2^{r+2}} + \frac{2}{6} \right\rfloor + \left\lfloor \frac{n}{2^{r+2}} + \frac{1}{6} \right\rfloor ,$$

and was already obtained in [40]. For $q=6$ it reads

$$\begin{aligned}\varepsilon_r = &+ \left\lfloor y+\tfrac{248}{252}\right\rfloor + \left\lfloor y+\tfrac{241}{252}\right\rfloor + \left\lfloor y+\tfrac{234}{252}\right\rfloor - 5\left\lfloor y+\tfrac{228}{252}\right\rfloor + \left\lfloor y+\tfrac{221}{252}\right\rfloor + \left\lfloor y+\tfrac{214}{252}\right\rfloor \\ &+ \left\lfloor y+\tfrac{206}{252}\right\rfloor + \left\lfloor y+\tfrac{199}{252}\right\rfloor + \left\lfloor y+\tfrac{192}{252}\right\rfloor - 5\left\lfloor y+\tfrac{186}{252}\right\rfloor + \left\lfloor y+\tfrac{179}{252}\right\rfloor + \left\lfloor y+\tfrac{172}{252}\right\rfloor \\ &+ \left\lfloor y+\tfrac{164}{252}\right\rfloor + \left\lfloor y+\tfrac{157}{252}\right\rfloor + \left\lfloor y+\tfrac{150}{252}\right\rfloor - 5\left\lfloor y+\tfrac{144}{252}\right\rfloor + \left\lfloor y+\tfrac{137}{252}\right\rfloor + \left\lfloor y+\tfrac{130}{252}\right\rfloor \\ &+ \left\lfloor y+\tfrac{122}{252}\right\rfloor + \left\lfloor y+\tfrac{115}{252}\right\rfloor - 5\left\lfloor y+\tfrac{108}{252}\right\rfloor + \left\lfloor y+\tfrac{102}{252}\right\rfloor + \left\lfloor y+\tfrac{95}{252}\right\rfloor + \left\lfloor y+\tfrac{88}{252}\right\rfloor \\ &+ \left\lfloor y+\tfrac{80}{252}\right\rfloor + \left\lfloor y+\tfrac{73}{252}\right\rfloor - 5\left\lfloor y+\tfrac{66}{252}\right\rfloor + \left\lfloor y+\tfrac{60}{252}\right\rfloor + \left\lfloor y+\tfrac{53}{252}\right\rfloor + \left\lfloor y+\tfrac{46}{252}\right\rfloor \\ &+ \left\lfloor y+\tfrac{38}{252}\right\rfloor + \left\lfloor y+\tfrac{31}{252}\right\rfloor - 5\left\lfloor y+\tfrac{24}{252}\right\rfloor + \left\lfloor y+\tfrac{18}{252}\right\rfloor + \left\lfloor y+\tfrac{11}{252}\right\rfloor + \left\lfloor y+\tfrac{4}{252}\right\rfloor ,\end{aligned}$$

where $y = n/6^{r+2}$. In general, it has q^2 terms.

One sees the pattern $0,1,2,3,-2,-1$ appearing 3 times followed by $0,1,2,-3,-2,-1$, also 3 times. The digits $0,1,2,3,-2,-1$ are not symmetric around 0, and $0,1,2,-3,-2,-1$ are not, either. However, in combination, both coming with "probability" $1/2$, the distribution of the digits becomes symmetric.

In a recent paper with Grabner and Heuberger [16] we are addressing the subblock counting problem in symmetric signed digit expansions. To announce our principal findings, we need some notation. If a block $\mathbf{b} = (b_s, \ldots, b_0)$ is given, we denote its *value* by $\mathsf{value}(\mathbf{b}) = \sum_\ell b_\ell q^\ell$.

We also use Iverson's notation, popularized in [19]: $[P]$ is defined to be 1 if condition P is true, and 0 otherwise. With this notation we can count the number of subblock occurrences of $\mathbf{b}$ in (the symmetric signed digit expansions of) n via

$$\sum_{k\ge 0}[(\varepsilon_{k+r-1}(n),\dots,\varepsilon_k(n))=\mathbf{b}].$$

We only consider *admissible* blocks $\mathbf{b}$: these blocks represent the number $\mathsf{value}(\mathbf{b})$ in the symmetric signed digit expansion. For interest we note that there are $\frac{2+q}{1+q}q^r-\frac{1}{1+q}(-1)^r$ admissible blocks of length r; this was implicitly proved in [22]. We are studying the quantity

$$S_{\mathbf{b}}(N)=\sum_{n<N}\sum_{k\ge 0}[(\varepsilon_{k+r-1}(n),\dots,\varepsilon_k(n))=\mathbf{b}].$$

This is our main result:

Let $q\ge 2$ be an even integer and $r\ge 1$. For an admissible block $\mathbf{b}=(b_{r-1},\dots,b_0)$ with $|b_{r-1}|<\frac{q}{2}$ and $\mathbf{b}\ne 0^r$ the number of occurrences of the block $\mathbf{b}$ in the symmetric signed digit expansions of the positive integers less than N satisfies

$$S_{\mathbf{b}}(N)=\frac{Q(b_0)}{q^r(q+1)}N\log_q N+h_0(\mathbf{b})N+NH_{\mathbf{b}}(\log_q N)+o(N),$$

where

$$\begin{aligned}
Q(\eta)&=q+\begin{cases}2 & \text{for } \eta=0,\\ 0 & \text{for } \eta=\pm\frac{q}{2},\\ 1 & \text{else,}\end{cases}\\
H_{\mathbf{b}}(x)&=\sum_{k\in\mathbb{Z}\setminus\{0\}}h_k(\mathbf{b})e^{2k\pi i x},\\
h_k(\mathbf{b})&=\frac{\log q}{2k\pi i(\log q+2k\pi i)}\Bigg(\zeta\left(\tfrac{2k\pi i}{\log q},[\mathsf{value}(\mathbf{b})<0]+q^{-r}\,\mathsf{value}(\mathbf{b})+\tfrac{R_{\min}(b_0)}{q^r(q+1)}\right)\\
&\quad-\zeta\left(\tfrac{2k\pi i}{\log q},[\mathsf{value}(\mathbf{b})<0]+q^{-r}\,\mathsf{value}(\mathbf{b})+\tfrac{R_{\max}(b_0)}{q^r(q+1)}\right)\Bigg)\quad\text{for } k\ne 0,\\
h_0(\mathbf{b})&=\log_q\Gamma\left([\mathsf{value}(\mathbf{b})<0]+q^{-r}\,\mathsf{value}(\mathbf{b})+\frac{R_{\min}(b_0)}{q^r(q+1)}\right)\\
&\quad-\log_q\Gamma\left([\mathsf{value}(\mathbf{b})<0]+q^{-r}\,\mathsf{value}(\mathbf{b})+\frac{R_{\max}(b_0)}{q^r(q+1)}\right)\\
&\quad-\frac{Q(b_0)}{q^r(q+1)}\left(r+\frac12+\frac{1}{\log q}-\frac{1}{q+1}\right)+\frac{1}{q^{r-1}(q+1)},\\
R_{\min}(\eta)&=-\frac{q}{2}-\left[(\eta-1)\bmod q\ge\frac{q}{2}\right],\qquad R_{\max}(\eta)=\frac{q}{2}+\left[\eta\bmod q<\frac{q}{2}\right].
\end{aligned}$$

The function $H_{\mathbf{b}}(x)$ is a periodic continuous function of period 1 and mean 0. As usual $\zeta(s,x)$ denotes the Hurwitz ζ–function.

The case of blocks $\mathbf{b}$ with most significant digit $b_{r-1}=\pm\frac{q}{2}$ can also be obtained from that, by a trivial argument taking differences.

The instance $r=1$ (counting digits) was discussed in [22], although without mentioning the periodic fluctuations in explicit form. Thuswaldner [47] has used

Dirichlet series and the Mellin–Perron summation formula to exhibit this fluctuating behaviour in the case $q = 2$ and $r = 1$.

The analysis is based on the Mellin–Perron formula, but with a slight twist:

$$\sum_{n<N}(N-n)a_n = \frac{1}{2\pi i}\int_{c-i\infty}^{c+i\infty}\sum_{n\geq 1}\frac{a_n}{(n-\alpha)^s}(N-\alpha)^{s+1}\frac{ds}{s(s+1)},$$

where $0 \leq \alpha < 1$ and c is in the half-plane of absolute convergence of the Dirichlet series. This slightly more general situation $0 \leq \alpha < 1$ was discussed in [17]. Without this version with the parameter α, one could still proceed successfully, as in [47], but that would be considerably more cumbersome and less elegant.

5 Representations of natural numbers as sums of 3 squares

This brief account is based on [46, 36, 43]. It is an old result that n is not representable as a sum of 3 squares iff $n = 4^i(8k+7)$, which means that in base-four representation, the last two nonzero digits are either 13 or 33. Let $Q(m)$ be the counting function of the numbers that are representable as a sum of 3 squares:

$$Q(m) = \sum_{0<n\leq m}[n \text{ representable as a sum of 3 squares}] = m - \sum_{0<n\leq m}\kappa(n),$$

with $\kappa(n) = 1$ if $n = 4^i(8k+7)$, $\kappa(n) = 0$ otherwise. It is a good idea to separate the main term and write $Q(m) = \frac{5}{6}m + \Delta(m)$. Osbaldestin and Shiu provide a Delange type analysis of the following formula:

$$\frac{1}{N}\sum_{0\leq n<N}\Delta(n) = \frac{3}{8}\log_4 N + F(\log_4 N) + \frac{1}{8}[N \text{ is odd}],$$

with $F(x) = \sum_{k\in\mathbb{Z}} f_k e^{2\pi i k x}$ and

$$f_0 = -\frac{31}{48} - \frac{3}{8\log 4} - \frac{1}{\log 4}\Big(\zeta'(0,\tfrac{7}{16}) + \zeta'(0,\tfrac{15}{16})\Big),$$
$$f_k = -\frac{1}{\log 4}\frac{1}{\chi_k(1+\chi_k)}\Big(\zeta(\chi_k,\tfrac{7}{16}) + \zeta(\chi_k,\tfrac{15}{16})\Big), \quad k \neq 0, \quad \chi_k = \frac{2\pi i k}{\log 4}.$$

It was announced already in [10] that a proof of this result using the Mellin–Perron summation formula would be possible, and Marko Riedel has worked out the details in his Master's thesis [43].

6 The average number of registers to evaluate a binary tree of size n

The function "number of registers" *reg* is given recursively by $\mathrm{Reg}(\square) = 0$ and

Reg 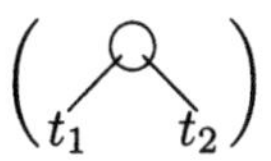

$$= \begin{cases} \max\{\mathrm{Reg}(t_1), \mathrm{Reg}(t_2)\} & \text{if } \mathrm{Reg}(t_1) \neq \mathrm{Reg}(t_2), \\ 1 + \mathrm{Reg}(t_1) & \text{otherwise.} \end{cases}$$

The problem is the evaluation of R_n, the average value of the function *reg*, based on the set of binary trees of size n. This was first studied in [14, 25]; for a tutorial see [38]. We have

$$R_n = \frac{1}{b_n}[z^n]E(z), \qquad \text{with} \qquad E(z) = \sum_{n \geq 0} z^n \sum_{t \text{ of size } n} \mathrm{Reg}(t)$$

and $b_n = \frac{1}{n+1}\binom{2n}{n}$ being the number of binary trees with n nodes. One can derive the following for the function $E(z)$, using the substitution $z = u/(1+u)^2$:

$$E(z) = \frac{1-u^2}{u} \sum_{p \geq 1} \frac{u^{2^p}}{1-u^{2^p}} = \frac{1-u^2}{u} \sum_{p \geq 1} \sum_{\lambda \geq 1} u^{\lambda 2^p} = \frac{1-u^2}{u} \sum_{n \geq 2} v_2(n) u^n,$$

where $v_2(n)$ is the exponent of 2 in the prime number decomposition of n. If we write in a unique way $n = 2^m(1+2i)$, $(m \geq 1,\ i \geq 0)$, we have $v_2(n) = \sum_{p=1}^{m} 1 = m$. The quantity $v_2(n)$ (which we have seen already in Section 3) is often called "dyadic valuation," for obvious reasons. One can get the coefficients, e. g., using the Lagrange inversion formula:

$$[z^n]E(z) = \sum_{k \geq 1} v_2(n) \left[\binom{2n}{n-k} - 2\binom{2n}{n-k-1} + \binom{2n}{n-k-2} \right].$$

One strategy to proceed is the following (that was the method of Flajolet, Raoult, and Vuillemin): One application of partial summation brings, via $\nu(k) = k - \sum_{j \leq k} v_2(j)$, the sum of digits function in, and a second application of partial summation makes the summatory function of the sum–of–digits function appear, which, according to Delange's analysis, is completely known! That means on the other side that the second difference of the binomial coefficients becomes eventually a fourth difference, but that causes no problems. The final result is this: The average number R_n of registers to evaluate a binary tree of size n satisfies

$$R_n = \log_4 n + D(\log_4 n) + o(1),$$

where $D(x)$ is a continuous periodic function of period 1; this function can be expanded in a Fourier series as $D(x) = \sum_{k \in \mathbb{Z}} d_k e^{2k\pi i x}$, with

$$d_0 = \frac{1}{2} - \frac{\gamma}{2\log 2} - \frac{1}{\log 2} + \log_2 \pi$$

and

$$d_k = \frac{1}{\log 2}\zeta(\chi_k)\Gamma\Big(\frac{\chi_k}{2}\Big)(\chi_k - 1), \quad k \neq 0, \quad \chi_k = \frac{2\pi i k}{\log 2}.$$

Another strategy, used by Kemp, is as follows: Approximation of the binomial coefficients leads eventually to the study of series of the type

$$f(x) := \sum_{k\ge 1} v_2(k) e^{-k^2x^2};$$

the Mellin transform then brings in Dirichlet series $g(s) := \sum_{k\ge 1} v_2(k)k^{-s}$; but this one is easily evalutated, using $v_2(2k) = 1 + v_2(k)$ and $v_2(2k+1) = 0$:

$$g(s) = \sum_{k\ge 1} \big(1 + v_2(k)\big)(2k)^{-s} = 2^{-s}\big(\zeta(s) + g(s)\big),$$

from which one gets $g(s) = \zeta(s)/(2^s - 1)$.

A third strategy is more on the lines of [12] and based on singularity analysis of generating functions: One must study $E(z)$ for $z \to \frac{1}{4}$, i. e., for $u \to 1$, and sets $u = e^{-t}$ and considers $E(z)$ for $t \to 0$. The second factor of $E(z)$, which is denoted by $V(t)$, is the challenging one. The Mellin transform of $V(t)$ is denoted by $V^*(s)$:

$$V^*(s) = \int_0^\infty x^{s-1} \sum_{k\ge 1} v_2(k) e^{-kx} dx = \sum_{k\ge 1} v_2(k) k^{-s} \int_0^\infty x^{s-1} e^{-x} dx.$$

This gives $V^*(s) = \frac{\Gamma(s)\zeta(s)}{2^s - 1}$. The Mellin inversion formula then leads to

$$V(t) = \frac{1}{2\pi i}\int_{2-i\infty}^{2+i\infty} V^*(s) t^{-s} ds,$$

and we shift the line of integration to the left, taking the residues of the integrand $V^*(s)t^{-s}$ into account. Using some classical expansions we get

$$V(t) \sim \frac{1}{t} + \frac{1}{2}\log_2 t - \frac{1}{2}\log_2 2\pi + \frac{1}{4} + \frac{\gamma}{2\log 2} + \frac{1}{\log 2}\sum_{k\neq 0}\Gamma(\chi_k)\zeta(\chi_k)t^{-\chi_k} + \dots$$

which gives the local expansion for the function $E(z)$; the rest is then singularity analysis of generating functions [11].

7 Odd–even merge

In the analysis of a sorting algorithm (Batcher's odd–even merge [28]), one has to consider the quantity

$$B_n = \sum_{k>1} \frac{\binom{2n}{n-k}}{\binom{2n}{n}} \Big(2F(k) + k\Big),$$

representing the number of exchanges in the odd–even merge of $2n$ elements. In this equation, $F(k) = \sum_{0 \le j < k} f(j)$ and $f(j)$ is the number of digits "1" in the Gray code representation of j. The numbers $\vartheta(j) := f(j) - f(j-1)$ equal ± 1: if k is of the form $k = 2^m(4i+1)$, $\vartheta(k) = 1$; if k is of the form $k = 2^m(4i+3)$, $\vartheta(k) = -1$. We might recognize a great similarity between $\vartheta(k)$ and $v_2(k)$ from Section 6.

Indeed, there are again basically 3 lines of attack: The Flajolet–Ramshaw approach [13] uses partial summation and uses an explicit result for the summatory function of the sum of digits function in the Gray code representation, as derived in Section 3; this is analogous to [14]. The next one, analogous to [25], was used by Sedgewick [44]: Approximating binomial coefficients, one has to study series à la

$$f(x) := \sum_{k \ge 1} \vartheta(k) e^{-k^2 x^2},$$

or, after using Mellin transforms, Dirichlet series like $g(s) := \sum_{k \ge 1} \vartheta(k) k^{-s}$, but this one is again easily evaluated:

$$g(s) = \sum_{m,i \ge 0} \big(2^m(4i+1)\big)^{-s} - \sum_{m,i \ge 0} \big(2^m(4i+3)\big)^{-s} = \frac{1}{2^s(2^s-1)} \Big[\zeta(s, \tfrac{1}{4}) - \zeta(s, \tfrac{3}{4})\Big],$$

where the Hurwitz $\zeta(s,a)$-function is defined by $\zeta(s,a) = \sum_{n \ge 0} (n+a)^{-s}$ for $\Re(s) > 1$. Recall that the function from Section 3 can be expressed as $L(s) = 4^{-s}\big[\zeta(s, \tfrac{1}{4}) - \zeta(s, \tfrac{3}{4})\big]$.

The third approach, using singularity analysis of generating functions, is presented in the tutorial [38]: One has to study the function

$$\frac{u(1+u)}{(1-u)^3} \sum_{i \ge 1} \vartheta(i) u^i$$

near $u = 1$. Eventually one finds:

The average number of exchanges in the odd–even merge of $2n$ elements satisfies

$$B_n \sim \frac{1}{4} n \log_2 n + nB\big(\log_4 n\big),$$

where $B(x)$ is a continuous periodic function of period 1; this function can be expanded as a Fourier series $B(x) = \sum_{k \in \mathbb{Z}} b_k e^{2k\pi i x}$, with

$$b_0 = -\frac{1}{2\log 2} - \frac{\gamma}{4 \log 2} - \frac{3}{4} + 2\log_2 \Gamma\Big(\frac{1}{4}\Big) - \log_2 \pi$$

and for $k \ne 0$

$$b_k = \frac{1}{\log 2} \zeta(\chi_k, \tfrac{1}{4}) \frac{\Gamma(\chi_k/2)}{1 + \chi_k}.$$

8 A problem of Yekutieli and Mandelbrot

The following problem was left open (and attacked empirically) in [49] and later solved in [39].

If we have an extended binary tree, we label the leaves with 0, and, recursively, if the left subtree of a node is labeled with a and the right subtree with b, we label the node with $\max\{a, b\}$ if $a \neq b$ and with $a + 1$ otherwise. The value attached to the root is called the *register function* of the tree t. The value attached to a particular node is the register function of the subtree having this node as its root, as already discussed earlier in this paper.

The authors in [49] consider the *bifurcation ratio (at the root)*. It is meant to be the *number of maximal subtrees (which is not the same as the number of internal nodes (!)) having register function exactly 1 less than the register function of the entire tree*. It was observed empirically that the expected value of this parameter is asymptotically a periodic function of $\log_4 n$ if all trees of size n (n internal nodes) are considered to be equally likely. Here, we want to settle this problem by explicitly describing the periodic function in terms of the Fourier coefficients. In principle, a full asymptotic expansion could be given, but the computation of the lower order term becomes more and more complicated.

Now, let $w_{p,k,n}$ be the number of binary trees with n nodes, register function p, and Yekutieli–Mandelbrot–parameter k, and let

$$W_p(z, y) = \sum_{n,k\geq 0} w_{p,k,n}\, y^k z^n$$

be its bivariate generating function.

To find the expected values, we have to work with $T_p(z) = \frac{\partial}{\partial y} W_p(z, y)|_{y=1}$ and $T(z) = \sum_{p\geq 1} T_p(z)$. The coefficient of z^n in $T(z)$, divided by $\frac{1}{n+1}\binom{2n}{n}$, is the expected value sought by Yekutieli and Mandelbrot. One finds for $p \geq 1$

$$W_p(z, y) = zy^2 R_{p-1}^2(z) + 2zyW_p(z, y)R_{p-1}(z) + 2zW_p(z, y)\big(B(z) - S_{p-1}(z)\big),$$

with $R_p(z)$ being the generating function of binary trees and register function $= p$, and $S_p(z) = \sum_{j\geq p} R_j(z)$. Therefore

$$T_p(z) = 2zR_{p-1}^2(z) + 2zR_p(z)R_{p-1}(z) + 2zT_p(z)\big(B(z) - S_p(z)\big),$$

and eventually (again with $z = u/(1 + u)^2$)

$$T(z) = 2u + 2\frac{1-u^2}{u} \sum_{p\geq 1} \frac{u^{3\cdot 2^{p-1}}}{(1 + u^{2^p})^2(1 - u^{2^p})}.$$

One has to study the series

$$\begin{aligned}\sigma = \sum_{p\geq 1} \frac{u^{3\cdot 2^{p-1}}}{(1 + u^{2^p})^2(1 - u^{2^p})} &= \sum_{p\geq 0, k\geq 1} u^{2^p(2k+1)}(-1)^{k-1}\left\lfloor \tfrac{k+1}{2}\right\rfloor \\ &= -\sum_{p,k\geq 0} ku^{2^p(4k+1)} + \sum_{p,k\geq 0} (k + 1)u^{2^p(4k+3)}.\end{aligned}$$

Now, one performs a Mellin transform analysis in order to find the local behaviour

$$T(z) \sim 3 - \left(\frac{4C}{(\log 2)\pi} + 5\right)\sqrt{1 - 4z} - \frac{8}{\log 2}\sum_{k\neq 0} \Gamma(\chi_k)\beta(\chi_k - 1)(1 - 4z)^{(1-\chi_k)/2}.$$

Here, $\beta(s) = \zeta(s, \frac{1}{4}) - \zeta(s, \frac{3}{4})$, which is the function that appeared already in the odd–even merge problem, $\chi_k = 2\pi ik/\log 2$, and $\mathcal{C} = \sum_{k\geq 0}(-1)^k/(2k+1)^2 = 0.9159655942\ldots$ is Catalan's constant. Consequently we get this result: The average value of the Yekutieli–Mandelbrot parameter, if all binary trees of size n are considered to be equally likely, is given by

$$\frac{2\mathcal{C}}{(\log 2)\pi} + \frac{5}{2} + \delta(\log_4 n) + \mathcal{O}\Big(\frac{1}{n}\Big).$$

The periodic function $\delta(x)$ has mean value 0 and admits the following representation as a Fourier series,

$$\delta(x) = -\frac{2}{\log 2}\sum_{k\neq 0}(\chi_k - 1)\Gamma\Big(\frac{\chi_k}{2}\Big)\beta(\chi_k - 1)e^{2k\pi ix}.$$

One could also perform an analysis along the lines of Kemp resp. Sedgewick [25, 44]; it would be based on the exact formula

$$[z^n]T(z) = \frac{2}{n+1}\binom{2n}{n} + 2\sum_{m\geq 1}\psi(m)\left[\binom{2n}{n+1-m} - 2\binom{2n}{n-m} + \binom{2n}{n-1-m}\right]$$

with $\psi(m) = \begin{cases} -k & \text{if } m = 2^i(4k+1) \quad \text{for some } i \text{ and } k, \\ k+1 & \text{if } m = 2^i(4k+3) \quad \text{for some } i \text{ and } k. \end{cases}$

9 Missing digits

Interesting phenomena occur if several digits are forbidden. We describe here the instance of $q = 3$ and digits $\{0, 2\}$, i. e., the digit 1 is forbidden. The set of strings $0.a_1a_2\ldots$, with $a_i \in \{0, 2\}$ is the classical *Cantor set*. Thus considering $2\sum_{i\geq 1} X_i 3^{-i}$, where the X_i are independent and identically distributed with probability distribution $\mathbb{P}\{X = 0\} = \mathbb{P}\{X = 1\} = 1/2$, we get a probability distribution on the interval $[0, 1]$ which is conveniently called *Cantor distribution*. In [18], the moments of a slightly more general distribution were investigated, motivated by an earlier paper of Lad and Taylor [31]; we sketch the procedure. Denote a_n the nth moment of the Cantor distribution; it is not hard to see that

$$a_n = \frac{1}{2(3^n - 1)}\sum_{i=0}^{n-1}\binom{n}{i}2^{n-i}a_i, \quad n \geq 1, \quad a_0 = 1.$$

If one rewrites it as $(2 - 3^{-n})a_n = \sum_{i=0}^{n}\binom{n}{i}2^{n-i}a_i$ and introduces the exponential generating function $A(z) = \sum_{n\geq 0} a_n z^n/n!$, it translates into

$$2A(z) - 2 - A(\tfrac{z}{3}) + 1 = e^{2z/3}A(\tfrac{z}{3}) - 1,$$

or

$$A(z) = \frac{1 + e^{2z/3}}{2}A(\tfrac{z}{3}) = \prod_{k>1}\frac{1 + e^{2z/3^k}}{2};$$

the last step was by iteration of the functional equation. Slightly more useful than this infinite product is the *Poisson transformed* generating function $B(z) = e^{-z}A(z)$, since by a process called *Depoissonization* one finds that $a_n \sim B(n)$. There is a well written survey paper on the subject by Jacquet and Szpankowski [24]. We get

$$B(z) = \frac{1+e^{-2z/3}}{2}B(\tfrac{z}{3}) = \prod_{k\geq 1}\frac{1+e^{-2z/3^k}}{2}.$$

To find the asymptotic behaviour of $B(z)$ for large z (and thus $B(n)$ and thus a_n) we compute the *Mellin transform* $B^*(s)$ of $B(z)$, viz.

$$B^*(s) = \frac{1}{2}3^sB^*(s) + \frac{1}{2}\int_0^\infty B(\tfrac{z}{3})e^{-2z/3}z^{s-1}dz,$$

or

$$B^*(s) = \frac{1}{2-3^s}\int_0^\infty \prod_{k\geq 2}\frac{1+e^{-2z/3^k}}{2}\,e^{-2z/3}z^{s-1}dz.$$

The standard reference for the Mellin transform in the context of asymptotic enumeration is [9]. The Mellin inversion formula gives

$$B(z) = \frac{1}{2\pi i}\int_{c-i\infty}^{c+i\infty} B^*(s)z^{-s}ds,$$

where the constant c might be chosen to be $-1/2$. Now one shifts the line of integration to the right and collects negative residues. They come from the solutions of $2-3^s = 0$, i. e., $s = \log_3 2 + 2\pi ik/\log 3$, for $k \in \mathbb{Z}$. If one calls them f_k, then $B(n) \sim \sum_{k\in\mathbb{Z}} f_k n^{-\log_3 2 - 2\pi ik/\log 3}$, which is of the form $n^{-\log_3 2}\delta(\log_3 n)$, with a periodic function $\delta(x)$ of period 1. Typically, the amplitudes of such periodic functions are quite small, so the most interesting term is obtained for $k=0$. The negative residue at $s = \log_3 2$ is

$$\frac{1}{2\log 3}\int_0^\infty \prod_{k\geq 2}\frac{1+e^{-2z/3^k}}{2}\,e^{-2z/3}z^{\log_3 2-1}dz = 0.734\ldots,$$

and so $a_n \sim B(n) \simeq f_0 n^{-\log_3 2} \approx 0.734\,n^{-0.631}$.

Another interesting question was triggered by Hosking [23] and solved in [27]: Assume that we draw n random numbers (independently) according to the Cantor distribution. What is the expected value of the minimum of them? If one would draw from the interval $[0,1]$ according to the uniform distribution, then it is fairly easy to see that it is $1/(n+1)$. If we call that minimum a_n, we get

$$a_n = \frac{1}{3}2^{-n}\sum_{k=1}^{n}\binom{n}{k}a_k + \frac{1}{3}2^{-n}(2+a_n), \quad a_0 := 0.$$

This recursion is obtained by observing that the minimum is obtained from one of the random strings starting with a zero, provided there is one. In the rare event that all strings start with 2, this first digit contributes 2/3 (additively), and the minimum is sought from all the n participants. Rearranging the recursion and introducing the exponential generating function $A(z) := \sum_{k\geq 0} a_k z^k/k!$, one finds

$$A(2z) = \frac{1}{3}(1+e^z)A(z) + \frac{2}{3}(e^z-1).$$

Now we introduce, following Knuth [28], the function $\hat{A}(z) := A(z)/(e^z-1) = \sum_{k\geq 0} \hat{a}_k z^k/k!$ and get

$$\hat{A}(2z) = \frac{1}{3}\hat{A}(z) + \frac{2}{3}\frac{1}{e^z+1},$$

from which one finds

$$\hat{a}_n = -\frac{2}{3}\frac{B_{n+1}}{n+1}\frac{2^{n+1}-1}{2^n-\frac{1}{3}},$$

with Bernoulli numbers B_n. Hence one gets the explicit solution

$$a_n = -\frac{2}{3}\sum_{k=0}^{n-1}\binom{n}{k}\frac{B_{k+1}}{k+1}\frac{2^{k+1}-1}{2^k-\frac{1}{3}}.$$

In order to study the asympotic behaviour of that quantity one can use Rice's method [15], which is related to Mellin transforms, and eventually write

$$a_n = -\frac{2}{3(n+1)}\cdot\frac{1}{2\pi i}\int_{\mathcal{C}}\frac{(-1)^n n!}{z(z-1)\dots(z-n)}\big(-z\zeta(1-z)\big)\frac{2^z-1}{2^{z-1}-\frac{1}{3}}dz,$$

where the contour $\mathcal{C}$ encloses the poles $1,2,\dots,n-1$ and no others. Deforming the contour of integration reduces the problem to the computation of residues; the dominant ones are the solutions of $2^{z-1}-\frac{1}{3}=0$, i. e., $1-\log_2 3 + 2k\pi i/\log 2$, for all $k \in \mathbb{Z}$. Eventually one finds, apart from a periodic function as above, that

$$a_n \simeq n^{\log_2 3}\frac{2}{3\log 2}\Gamma(\log_2 3)\zeta(\log_2 3).$$

Another type of moments are *Cantor's singular moments*: Consider the unique non–decreasing function on $[0,1]$ such that, if $x = 2\sum_{j\geq 1}\varepsilon_j 3^{-j}$ with $\varepsilon_j \in \{0,1\}$, then $F(x) = \sum_{j\geq 1}\varepsilon_j 2^{-j}$. It was a question in the problem section of the American Mathematical Monthly [6] to compute the moments; the following solution came out:

$$J_n = \int_0^1 \big(F(x)\big)^n dx = \frac{2}{3(n+1)}\sum_{j=0}^{n}\binom{n+1}{j}\frac{B_j}{3\cdot 2^{j-1}-1} \quad \text{for } n\geq 1 \text{ and } J_0 = 1;$$

the case $n=-1$ was left open and treated by me in [41]: Since one can show that $J_{-1} = \sum_{n\geq 0} J_n$, and, again by Rice's method,

$$J_n = \frac{2}{3}\cdot\frac{1}{2\pi i}\int_{-\frac{1}{2}-i\infty}^{-\frac{1}{2}+i\infty}\frac{\Gamma(n+1)\Gamma(1-s)}{\Gamma(n+2-s)}\frac{\zeta(1-s)}{3\cdot 2^{s-1}-1}ds,$$

one finds

$$\sum_{n=0}^{N} J_n = J_0 + \frac{2}{3}\cdot\frac{1}{2\pi i}\int_{-\frac12-i\infty}^{-\frac12+i\infty}\frac{\Gamma(N+2)\Gamma(1-s)}{\Gamma(N+2-s)s}\frac{\zeta(1-s)}{3\cdot 2^{s-1}-1}\,ds$$
$$-\frac{2}{3}\cdot\frac{1}{2\pi i}\int_{-\frac12-i\infty}^{-\frac12+i\infty}\frac{\Gamma(2)\Gamma(1-s)}{\Gamma(2-s)s}\frac{\zeta(1-s)}{3\cdot 2^{s-1}-1}\,ds.$$

Now we perform the limit $N\to\infty$ and get eventually (the computations are not displayed here):

$$\sum_{n\ge 0} J_n = \frac{4}{3}+\frac{2}{3}\sum_{k,m\ge1}3^{-k}\cdot\frac{1}{2\pi i}\int_{\frac32-i\infty}^{\frac32+i\infty}\frac{1}{s(s-1)}\Big(\frac{2^k}{m}\Big)^s ds$$
$$= -\frac{1}{3}+\frac{2}{3}\sum_{k\ge1}\Big(\frac{2}{3}\Big)^k H_{2^k} = 3.36465\,07281\,00925\,16083\,89349\,6289\ldots,$$

with harmonic numbers $H_n = \sum_{1\le k\le n}\frac{1}{k}$.

More general results of this type are currently worked out in collaboration with F. Bassino [1].

10 Von Neumann's addition algorithm

This brief description is based on a recent paper with Heuberger [21].

Knuth [29] has analyzed von Neumann's addition algorithm: Assume that two integers are given in q–ary notation, say $(\ldots y_2y_1y_0)_q$ and $(\ldots y_2y_1y_0)_q$; then the integer $(\ldots s_2s_1s_0)_q$ with $s_i=(x_i+y_i) \mod q$ is formed, as well as $(\ldots c_2c_1c_0)_q$ (the carries), where $c_{i+1}=[x_i+y_i\ge q]$. The process is iterated by adding $(\ldots s_2s_1s_0)_q$ with $s_i=(x_i+y_i) \mod q$ and $(\ldots c_2c_1c_0)_q$ until the string of carries contains only zeros. Knuth studied the average number of iterations, assuming two random integers with n digits. The result is $\sim \log_q n$; it turns out that the longest subsequence of the form $\ldots i(q-1)(q-1)\ldots(q-1)j\ldots$ with $i\ne q-1$ and $j\ge q$ in $\big(\ldots(x_2+y_2)(x_1+y_1)(x_0+y_0)\big)_q$ is responsible for the number of iterations. We extended Knuth's results to other positional number systems, namely for the basis q and the set of q digits $\{d,d+1,\ldots,d+q-1\}$. Note carefully that carries might now be ±1 and that the sequence of sums might be oscillating, being smaller or larger than the true value of the sum of the two integers. This is in sharp contrast to the traditional q–ary system, where the sums are monotonically increasing until the algorithm stops. Thus it is perhaps natural that the description of subsequences being responsible for the number of iterations is significantly more complicated. Here is an example for $q=5$, $d=-1$, and $\bar{1}=-1$:

The generating function where the coefficient of z^n counts the number of pairs of integers of length $\le n$, such that $\le k+2$ iterations are necessary, is given by

$$G^{\le k}(z)=\frac{s_0(z)+(z/q)^k r_1(z)+(z/q^2)^k r_2(z)+(z^2/q^3)^k r_3(z)}{(1-z)s_0(z)+(z/q)^k s_1(z)+(z/q^2)^k s_2(z)+(z^2/q^3)^k s_3(z)},$$

$$
\begin{array}{rcr}
(12\bar{1}\bar{1}\bar{1}13)_{(5,-1)} = (\dots x_2x_1x_0)_{(5,-1)} & = & 21108 \\
(22\bar{1}0\bar{1}23)_{(5,-1)} = (\dots y_2y_1y_0)_{(5,-1)} & = & 36863 \\
\hline
(3\bar{1}3\bar{1}331)_{(5,-1)} = (\dots z_2z_1z_0)_{(5,-1)} & = & 45591 \\
(1\bar{1}0\bar{1}010)_{(5,-1)} = (\dots c_2c_1c_0)_{(5,-1)} & = & 12380 \\
\hline
(\bar{1}3333\bar{1}1)_{(5,-1)} = (\dots z_2z_1z_0)_{(5,-1)} & = & -3929 \\
(1\bar{1}0\bar{1}0100)_{(5,-1)} = (\dots c_2c_1c_0)_{(5,-1)} & = & 61900 \\
\hline
(13323\bar{1}\bar{1}1)_{(5,-1)} = (\dots z_2z_1z_0)_{(5,-1)} & = & 135971 \\
(\bar{1}0001000)_{(5,-1)} = (\dots c_2c_1c_0)_{(5,-1)} & = & -78000 \\
\hline
(0332\bar{1}\bar{1}\bar{1}1)_{(5,-1)} = (\dots z_2z_1z_0)_{(5,-1)} & = & 57346 \\
(00010000)_{(5,-1)} = (\dots c_2c_1c_0)_{(5,-1)} & = & 625 \\
\hline
(0333\bar{1}\bar{1}\bar{1}1)_{(5,-1)} = (\dots z_2z_1z_0)_{(5,-1)} & = & 57971 \\
(00000000)_{(5,-1)} = (\dots c_2c_1c_0)_{(5,-1)} & = & 0
\end{array}
$$

where $s_0(z) = -2q^4(q-1)(q^2 - z(1-d))(q^2 - z(q+d))$. The terms $r_1(z)$, $r_2(z)$, $r_3(z)$, $s_1(z)$, $s_2(z)$, $s_3(z)$ are polynomials in z, q, d which are independent of k. From this one can derive, essentially by bootstrapping, approximations, and Mellin transforms, that the expected number t_n of carry propagations satisfies

$$t_n = \log_q n + \log_q \delta + \frac{\gamma}{\log q} + \frac{1}{2} + \psi(\log_q n + \log_q \delta) + O\left(\frac{\log^4 n}{n}\right),$$

where $\delta = \frac{(q^3+(2d-2)q^2+(2d-1)(d-1)q-(d-1)d)(q-1)(q+1)}{2(q^2-q-d)(q^2+d-1)}$ and $\psi(x)$ is again a periodic function.

For the instance of the symmetric signed digit expansion, one can also perform such an analysis, but this is more involved, and we refer to the original paper.

11 Mergesort

Because of space restrictions, we cannot describe the algorithm and/or any details. Basically, there is a top–down version where one needs to study functions like

$$(k+1)n - 2(2^{k+1}-1) + 2\sum_{0\le j\le k} \frac{2^j}{\lfloor \frac{n}{2^{j+1}} \rfloor + 2} + 2\sum_{0\le j\le k} \frac{(1-b_j)(2^j - (b_{j-1}\dots b_0)_2)}{(\lfloor \frac{n}{2^{j+1}} \rfloor + 1)(\lfloor \frac{n}{2^{j+1}} \rfloor + 2)},$$

where $(b_k \dots b_0)_2$ is the binary representation of n. This can be efficiently done by a method introduced by Flajolet and Golin [8]. This worked well because such quantities were obtained by so–called divide–and–conquer recursions.

The bottom–up version that I studied with Panny [37] led to even more involved expressions, and no divide–and–conquer recursions were available, so we had to resort to some Delange type analysis. More recently, Hwang and his coworkers [2] developed techniques to deal with quite general versions of Mergesort, including the top–down and bottom–up variants; further mergesort papers of Hwang's can be found on his homepage http://algo.stat.sinica.edu.tw/.

12 Jump interpolation search trees

Güntzer and Paul have used symmetric signed digit representations for the classical case $q=2$ to construct a data structure that they called *jump interpolation search trees* [20]. All numbers with representations of length $\le n$ are in the tree, and the number 0, its length being zero; it serves as the root. Node y is a child of node x, if the least significant nonzero digit in y is replaced by 0, resulting in x. Thus, the depth of a node is the number of nonzero digits. This construction can be *verbatim* translated to the case of general even q. The computations are analogous to the ones in [40]. The admissible words allow the representation

$$\varepsilon+\left(\tfrac{q}{2}+L+\tfrac{-q}{2}+\bar{L}\right)\left(L\tfrac{q}{2}+0\tfrac{q}{2}+\bar{L}\tfrac{-q}{2}+0\tfrac{-q}{2}+L+\bar{L}+0\right)^{*}$$

with $L=\{1,\dots,\frac{q}{2}-1\}$ and $\bar{L}=\{-\frac{q}{2}+1,\dots,-1\}$,
To mark the length of the representations by z and the depth by u, we replace L and $\bar{L}$ by $(\frac{q}{2}-1)zu$, $\pm\frac{q}{2}$ by zu, and 0 by z, and of course f^* by $1/(1-f)$. The coefficient of z^n then refers to words of length n; since we are interested in words of length $\le n$, we divide the result by $1-z$ and obtain

$$\frac{1+(-1+2u)z-u(uq-2u+2)z^2}{(1-z)(1+(2u-1-uq)z-u(uq-2u+2)z^2)}. \tag{6}$$

From this, we get first, by setting $u=1$, that the number of words of length $\le n$ is given by

$$\frac{q^2}{(q+1)(q-1)}q^n+\frac{q-2}{2(q-1)}-\frac{q}{2(q+1)}(-1)^n. \tag{7}$$

Differentiating (6) with respect to u, and then setting $u=1$ leads to $\frac{zq(1-z+(q-2)z^2)}{(1-z)(1+z)^2(1-zq)^2}$, and the coefficient of z^n in it is given by

$$\left[\tfrac{q(q^2-2)}{(q-1)(q+1)^2}n+\tfrac{q(q^2-q+2)}{(q-1)^2(q+1)^3}\right]q^n+\frac{q(q-2)}{4(1-q)^2}-\left[\tfrac{q^2}{2(q+1)^2}n+\tfrac{(q^2+3q+6)q}{4(q+1)^3}\right](-1)^n. \tag{8}$$

Taking the quotient of (8) and (7) gives us then that the average depth of a random node in the tree is

$$\frac{q^2-2}{q(q+1)}n-\frac{q^2-q+2}{q(q+1)^2(q-1)}+o(1);$$

for $q=2$, we get again the old result $n/3$ from [20].

References

[1] F. Bassino and H. Prodinger. (q,δ)-numeration systems with missing digits. *In preparation*, 2002.

[2] W. M. Chen, H. K. Hwang, and G. H. Chen. The cost distribution of queue-mergesort, optimal mergesorts, and power-of-2 rules. *J. Algorithms*, 30:423–448, 1999.

[3] H. Delange. Sur la fonction sommatoire de la fonction somme des chiffres. *Enseignement Mathématique*, 21:31–47, 1975.

[4] L. Devroye and P. Kruszewski. A note on the Horton–Strahler number for random trees. *Inform. Process. Lett.*, 56:95–99, 1995.

[5] L. Devroye and P. Kruszewski. On the Horton–Strahler number for random tries. *RAIRO Inform. Théor. Appl.*, 30:443–456, 1996.

[6] H. G. Diamond and B. Reznick. Problems and solutions. *American Mathematical Monthly*, 106:175–176, 1999.

[7] G. Doetsch. *Handbuch der Laplace-Transformation, 3 volumes.* Birkhäuser Verlag, Basel, 1972.

[8] P. Flajolet and M. Golin. Mellin transforms and asymptotics. The mergesort recurrence. *Acta Inform.*, 31:673–696, 1994.

[9] P. Flajolet, X. Gourdon, and P. Dumas. Mellin transforms and asymptotics: Harmonic sums. *Theoretical Computer Science*, 144:3–58, 1995.

[10] P. Flajolet, P. Grabner, P. Kirschenhofer, H. Prodinger, and R. Tichy. Mellin transforms and asymptotics: Digital sums. *Theoretical Computer Science*, 123:291–314, 1994.

[11] P. Flajolet and A. Odlyzko. Singularity analysis of generating functions. *SIAM Journal on Discrete Mathematics*, 3:216–240, 1990.

[12] P. Flajolet and H. Prodinger. Register allocation for unary–binary trees. *SIAM J. Comput.*, 15:629–640, 1986.

[13] P. Flajolet and L. Ramshaw. A note on Gray code and odd–even merge. *SIAM J. Comput.*, 9(1):142–158, 1980.

[14] P. Flajolet, J.-C. Raoult, and J. Vuillemin. The number of registers required for evaluating arithmetic expressions. *Theoretical Computer Science*, 9:99–125, 1979.

[15] P. Flajolet and R. Sedgewick. Mellin transforms and asymptotics: Finite differences and Rice's integrals. *Theoretical Computer Science*, 144:101–124, 1995.

[16] P. Grabner, C. Heuberger, and H. Prodinger. Subblock occurrences in signed digit representations. *Submitted.*

[17] P. Grabner, P. Kirschenhofer, and H. Prodinger. The sum–of–digits function for complex bases. *J. London Math. Soc.*, 57:20–40, 1998.

[18] P. Grabner and H. Prodinger. Asymptotic analysis of the moments of the Cantor distribution. *Statistics and Probability Letters*, 26:243–248, 1996.

[19] R. L. Graham, D. E. Knuth, and O. Patashnik. *Concrete mathematics. A foundation for computer science.* Addison-Wesley, second edition, 1994.

[20] U. Güntzer and M. Paul. Jump interpolation search trees and symmetric binary numbers. *Information Processing Letters*, 26:193–204, 1987/88.

[21] C. Heuberger and H. Prodinger. Carry propagation in signed digit representations. *submitted*, 2001.

[22] C. Heuberger and H. Prodinger. On minimal expansions in redundant number systems: Algorithms and quantitative analysis. *Computing*, 66:377–393, 2001.

[23] J. R. M. Hosking. Moments of order statistics of the Cantor distribution. *Statistics and Probability Letters*, 19:161–165, 1994.

[24] P. Jacquet and W. Szpankowski. Analytical de-Poissonization and its applications. *Theoretical Computer Science*, 201:1–62, 1998.

[25] R. Kemp. The average number of registers needed to evaluate a binary tree optimally. *Acta Inform.*, 11:363–372, 1978/79.

[26] P. Kirschenhofer and H. Prodinger. Subblock occurrences in positional number systems and Gray code representation. *Journal of Information and Optimization Sciences*, 5:29–42, 1984.

[27] A. Knopfmacher and H. Prodinger. Exact and asymptotic formulæ for average values of order statistics of the Cantor distribution. *Statistics and Probability Letters*, 27:189–194, 1996.

[28] D. E. Knuth. *The Art of Computer Programming*, volume 3: Sorting and Searching. Addison-Wesley, 1973. Second edition, 1998.

[29] D. E. Knuth. The average time for carry propagation. *Indagationes Mathematicæ*, 40:238–242, 1978.

[30] P. Kruszewski. A note on the Horton–Strahler number for random binary search trees. *Inform. Process. Lett.*, 69:47–51, 1999.

[31] F. R. Lad and W. F. C. Taylor. The moments of the Cantor distribution. *Statistics and Probability Letters*, 13:307–310, 1992.

[32] A. Meir, J. W. Moon, and J. R. Pounder. On the order of random channel networks. *SIAM J. Algebraic Discrete Methods*, 1:25–33, 1980.

[33] J. W. Moon. An extension of Horton's law of stream numbers. *Math. Colloq. Univ. Cape Town*, 12:1–15 (1980), 1978/79.

[34] J. W. Moon. On Horton's law for random channel networks. *Ann. Discrete Math.*, 8:117–121, 1980. Combinatorics 79 (Proc. Colloq., Univ. Montréal, Montreal, Que., 1979), Part I.

[35] M. Nebel. On the Horton–Strahler number for combinatorial tries. *Theor. Inform. Appl.*, 34:279–296, 2000.

[36] A. H. Osbaldestin and P. Shiu. A correlated digital sum problem associated with sums of three squares. *Bull. London Math. Soc.*, 21:369–374, 1989.

[37] W. Panny and H. Prodinger. Bottom-up mergesort—a detailed analysis. *Algorithmica*, 14:340–354, 1995.

[38] H. Prodinger. Some analytic techniques for the investigation of the asymptotic behaviour of tree parameters. *EATCS Bulletin*, 47:180–199, 1992.

[39] H. Prodinger. On a problem of Yekutieli and Mandelbrot about the bifurcation ratio of binary trees. *Theoretical Computer Science*, 181:181–194, 1997.

[40] H. Prodinger. On binary representations of integers with digits −1, 0, 1. *Integers*, pages A8, 14 pp. (electronic), 2000.

[41] H. Prodinger. On Cantor's singular moments. *Southwest J. Pure Appl. Math.*, 2000(1):27–29 (electronic), 2000.

[42] G. Reitwiesner. Binary arithmetic. *Vol. 1 of Advances in Computers, Academic Press*, pages 231–308, 1960.

[43] M. Riedel. Applications of the Mellin–Perron formula in number theory. Technical report, University of Toronto, 1996.

[44] R. Sedgewick. Data movement in odd–even merging. *SIAM J. Comput.*, 7:239–272, 1978.

[45] L. A. Shepp and S. P. Lloyd. Ordered cycle lengths in a random permutation. *Trans. Amer. Math. Soc.*, 121:340–357, 1966.

[46] P. Shiu. Counting sums of three squares. *Bull. London Math. Soc.*, 20:203–208, 1988.

[47] J. Thuswaldner. Summatory functions of digital sums occurring in Cryptography. *Periodica Math. Hungarica*, 38:111–130, 1999.

[48] E. T. Whittaker and G. N. Watson. *A Course of Modern Analysis.* Cambridge University Press, fourth edition, 1927. Reprinted 1973.

[49] I. Yekutieli and B. Mandelbrot. Horton–Strahler ordering of random binary trees. *J. Phys. A*, 27:285–293, 1994.

Helmut Prodinger

The John Knopfmacher Centre for Applicable Analysis and Number Theory
School of Mathematics, University of the Witwatersrand
P. O. Wits, 2050 Johannesburg, South Africa
helmut@staff.ms.wits.ac.za
http://www.wits.ac.za/helmut/index.htm

PART IV

Branching Processes and Trees

Trends in Mathematics, © 2002 Birkhäuser Verlag Basel/Switzerland

Growth Rate and Ergodicity Conditions for a Class of Random Trees

Guy Fayolle, Maxim Krikun

ABSTRACT: *The main substance of the paper concerns growth rate and classification (ergodicity, transience) of a family of random trees. In the basic model, new edges appear according to a Poisson process of parameter λ and leaves can be deleted at a rate μ. The main results lay the stress on the famous number e. A complete classification of the process is given in terms of the intensity factor $\rho = \lambda/\mu$: it is ergodic if $\rho \leq e^{-1}$, and transient if $\rho > e^{-1}$. There is a phase transition phenomenon: the usual region of null recurrence (in the parameter space) here does not exist. This fact is rare for countable Markov chains with exponentially distributed jumps. A theorem, much of ergodic type, is derived for the height of the tree at time t, which in the transient case is shown to grow linearly as $t \to \infty$, at a rate explicitly computed.*

1 Introduction and model description

So far, very few results seem to exist for random trees as soon as insertions and deletions are simultaneously permitted (see e.g. [9]). We shall study one of the simplest models in this class, which offers both interesting and non trivial properties. Broadly speaking, one might think of a vertex as being a node of a network (e.g. the Internet) or of some general data structure. This paper is a self-contained continuation of [8].

Let $G = \{G(t), t \geq 0\}$ be a continuous time Markov chain with state space the set of finite directed trees rooted at some fixed vertex v_0. Throughout the study, the *distance* between two vertices is the number of edges in the path joining them, and the *height* $h(v)$ of a vertex v is the distance from the root. The set of vertices having the same height k form the k-th *level* of the tree, the root v_0 being at level 0. Hence the height of G is a stochastic process $\{H_G(t), t \geq 0\}$, where

$$H_G(t) \stackrel{\text{def}}{=} \max_{v \in G(t)} h(v).$$

Wherever the meaning is clear from the context, the subscript G will be omitted, and we simply write H. The *indegree* of a vertex v is the number of edges starting at v and a vertex with indegree 0 is a *leaf*. Finally, we will also need the classical notion of *subtree* with root v, which goes without saying.

At time $t = 0$, $G(0)$ consists of the single vertex v_0. Then at time $t > 0$, the evolution rules on G mimic those coming in ordinary birth and death processes, and they are quite natural. Indeed the transitions are of two kinds

- **Adjunction.** At each vertex v, a new edge having its origin at v can be appended to the tree at the epochs of a Poisson process with parameter $\lambda > 0$. In this case, the *indegree* of v is increased by one and the new edge produces a new leaf.

- **Deletion.** From its birth, a leaf (but the root) can be deleted at a rate μ. In other words, a vertex *as long as it has no descendant* has an exponentially distributed lifetime with parameter $\mu \geq 0$.

1.1 Organization of the paper and main results

Section 2 is devoted to the birth and death model described above, with $\lambda, \mu > 0$. An exact and complete classification of G is given. Indeed, necessary and sufficient conditions are derived for the process to be ergodic ($\mu \geq \lambda e$) or transient ($\mu < \lambda e$). A phase transition phenomenon is enlightened, which corresponds precisely to the absence of a null recurrence region. The main outcome of section 3 is a general theorem for $H_G(t)$, much of ergodic type. It shows in the transient case that $H_G(t)$ has a linear gows rate δ, as $t \to \infty$. In the particular case $\mu = 0$ (pure birth-process), one rediscovers the known result $\delta = \lambda e$ (see [3, 10]).

2 Classification of G(t) in the birth and death case

The random tree G evolves according to the rules given in the introduction, and the first important question is to find exact conditions for this process to be recurrent or transient. The main results are stated in Theorem 2.2.

For convenience, we define the *lifetime* τ_v of an arbitrary vertex v, which measures the length of the time interval between the birth and the death of v (for consistency $\tau_v = \infty$ if v is never erased).

Lemma 2.1. *All vertices, but the root, have the same lifetime distribution* $p(t)$, *which satisfies the following system (S)*

$$\beta(t) = \mu \exp\Big\{-\lambda \int_0^t (1 - p(x))dx\Big\}, \tag{1}$$

$$\beta(t) = \frac{dp(t)}{dt} + \int_0^t \beta(t - y)dp(y), \tag{2}$$

with the initial condition $p(0) = 0$.

Sketch of proof : Let v be a particular vertex of $G(t)$ and consider the related random subtree with root v. Its evolution does not depend on anything below v, as long as v exists. Therefore all these subtrees are identically distributed and, accordingly, their vertices have the same lifetime distribution. To capture more precisely the evolution of the process, we introduce two important random variables associated with each vertex v:

- t_v, the *proper time* of v, such that v appears at $t_v = 0$;
- $X(t_v)$, the number of direct descendants of v (i.e. who are located at a distance 1 from v).

At rate λ, a vertex v produces descendants whose lifetimes are independent, with the common distribution $p(t)$. As soon as $X(t_v) = 0$, v can die at rate μ, in which case the process of production stops. It is actually useful to extend $X(t_v)$ for all $t_v \geq 0$ by deciding that, instead of deleting v, a *μ-event* occurs without stopping the production of descendants. With this convention, the number of descendants of the root vertex v_0 evolves as $X(t)$, for all $t \geq 0$.

Let τ_v denote the random epoch of the first μ-event, which is distributed according to $p(t_v)$.Clearly the process X is regenerative with respect to the μ-events. Thus that the random variables $X(t_v)$ and $X(\tau_v + t_v)$ have the same distribution.

For any fixed t_v, we write down a sum of conditional probabilities, expressing the fact that v had exactly k descendants, who all have died in $[0, t_v]$, their birth-times being independent and uniformly spread over $[0, t_v])$. This yields equation (1), putting $\beta(t) \stackrel{\text{def}}{=} \mu \mathsf{P}\{X(t_v) = 0\}$.

As for equation (2), we note that the process X is regenerative with respect to the μ-events, and hence the random variables $X(t_v)$ and $X(\tau_v + t_v)$ have the same distribution. ■

Theorem 2.2.

(A) The Markov chain G is ergodic if, and only if,

$$\rho \stackrel{\text{def}}{=} \frac{\lambda}{\mu} \leq \frac{1}{e}. \tag{3}$$

(B) When the system is ergodic, the mean lifetime $m \stackrel{\text{def}}{=} \mathsf{E}(\tau_v)$ is given by

$$m = \frac{r}{\lambda},$$

where $r \leq 1$ denotes the smallest root of the equation

$$re^{-r} = \rho \tag{4}$$

and represents the mean number of descendants of an arbitrary vertex at steady state.

(C) When $\rho > \dfrac{1}{e}$, then the system is transient. In this case,

$$\lim_{t \to \infty} p(t) \stackrel{\text{def}}{=} \ell < 1.$$

As a rule, x being the positve root of $xe^x = \rho^{-1}$, we have for any ρ

$$x \leq \ell \leq \min\left(1, \frac{1}{\rho}\right) \quad \text{and} \quad \lim_{\rho \to \infty} \rho\ell = 1.$$

The proof of the theorem is spread over the next two subsections.

2.1 Ergodicity

Relying on the standard theory of Markov chains with countable state space (see [4, vol. I]), we claim the system ergodic if, and only if, $m < \infty$. As a matter of fact, the μ-events are regeneration points for the process $X(t)$, which is precisely the number of descendants of the *root* v_0. Hence when $\mathsf{E}(\tau_v) < \infty$ (i.e. $\beta(\infty) > 0$), the event $\{X(t) = 0\}$ has a positive probability, so that G is ergodic. Conversely, if $\mathsf{E}(\tau_v) = \infty$ then $X(t)$ is transient and so is G.

For an arbitrary positive function f, denote by f^* its ordinary Laplace transform

$$f^*(s) \stackrel{\text{def}}{=} \int_0^\infty e^{-st} f(t)dt, \quad \Re(s) \geq 0.$$

Later on we will also need the associated inversion formula (see e.g. [5])

$$f(t) = \frac{1}{2i\pi} \int_{\sigma-i\infty}^{\sigma+i\infty} e^{st} f^*(s)ds, \quad \Re(\sigma) > 0. \tag{5}$$

To show the necessity of condition (3), suppose G is ergodic. In this case, by (1), the quantity $\lim_{t\to\infty} \beta(t)$ does exist and, applying relations of Abelian type (see e.g. [5]) in equations (1) and (2), we obtain easily

$$\rho = \lambda m e^{-\lambda m}.$$

As for the sufficiency of (3), we have to get a deeper insight into system (S). There two main steps.

(a) Although (S) reduces to a second order nonlinear integro-differential equation, this does not help much. What is more useful is that all derivatives $p^{(n)}(0)$, $\beta^{(n)}(0)$, taken at the the origin in the complex t-plane, can be recursively computed for all n. This can be checked at once, rewriting (1) in the differential form

$$\frac{d\beta(t)}{dt} + \lambda(1 - p(t))\beta(t) = 0. \tag{6}$$

Noticing the derivatives $p^{(n)}(0)$—resp. $\beta^{(n)}(0)$—have alternate signs when n varies, it is direct to verify that β and p are analytic functions around the origin, and that their respective power series have a non-zero radius of convergence. The first singularities of p and β are on the negative real axis, but not easy to locate precisely. Thus (S) has a solution, which is unique, remarking also that uniqueness is a mere consequence of the Lipschitz character of $dp(t)/dt$ with respect to β in the Volterra integral equation (2) (see e.g. [1]). En passant, it is worth noting that the solution in the whole complex plane—which is not really needed for our purpose—could be obtained by analytic continuation directly on system (S).

(b) When (3) holds, the next stage consists in exhibiting a *non-defective* probabilistic solution $p(t)$ [necessarily unique by step (a)], with a finite mean $m < \infty$. This is more intricate and is achieves by means of a converging iterative scheme.

Consider the system

$$\begin{cases} \beta_0(t) & = \mu, \quad t \geq 0, \\ \beta_k(t) & = \dfrac{dp_k(t)}{dt} + \displaystyle\int_0^t \beta_k(t-y)dp_k(y), \\ \beta_{k+1}(t) & = \mu \exp\Big\{-\lambda \displaystyle\int_0^t (1-p_k(y))dy\Big\}, \\ p_k(0) & = 0, \; \forall k \geq 0. \end{cases} \tag{7}$$

The second equation in (7) is equivalent to

$$sp_k^*(s) = \frac{\beta_k^*(s)}{1+\beta_k^*(s)}, \tag{8}$$

allowing to derive p_k from β_k by means of (5) (see also [4] for various inversion formulas in the real plane). Then computational algorithm is reasonably simple:

1. $p_0(t) = 1 - e^{-\mu t}$.
2. Compute $\beta_1(t) = \mu \exp\big[-\rho(1-e^{-\mu t})\big]$.
3. Compute $p_1(t)$, then $\beta_2(t), p_2(t)$, etc.

At each step, the successive p_k's are non-defective probability distributions, with finite means denoted by m_k. Indeed, one has to check first that the right-hand side of (8) is the Laplace transform of a positive measure, since a priori it does not correspond to a *completely monotone* function, according to the classical definition of [4]. The scheme (7) enjoys two nice properties.

(1) It is monotone decreasing: the positive sequences $\{p_k(t), \beta_k(t), k \geq 0\}$ are uniformly bounded and non-increasing for each fixed t. Consequently,

$$p(t) = \lim_{k\to\infty} \searrow p_k(t) \quad \text{and} \quad \beta(t) = \lim_{k\to\infty} \searrow \beta_k(t)$$

form the unique solutions of (S).

(2) Letting $r_k \stackrel{\text{def}}{=} \lambda m_k$ and combining the two main equations of (7), we get

$$r_{k+1} = \rho e^{r_k}, \; \forall k \geq 0, \quad \text{with } r_0 = \rho.$$

When $\rho \leq e^{-1}$, the r_k's form an increasing sequence of positive real numbers, with a finite positive limit r satisfying equation (4). Since $1-p_k(t)$ is also an increasing sequence of positive functions, the theorem of Beppo Levi ensures the equality

$$\int_0^\infty (1-p(t))dt = \lim_{k\to\infty} \int_0^\infty (1-p_k(t))dt = \lim_{k\to\infty} m_k = \frac{r}{\lambda}. \tag{9}$$

It is worth to point out that (7) is equivalent to the construction of a sequence of trees $\{G_k, k \geq 0\}$, such that, for any finite k, G_k is ergodic and has a height not greater than k.

This completes the sketch of proof of points (A) and (B) of Theorem 3.

Remarks One could have considered the scheme

$$\begin{cases} \gamma_0(t) &= \mu e^{-\lambda t}, \quad t \geq 0, \\ \gamma_k(t) &= \dfrac{dq_k(t)}{dt} + \displaystyle\int_0^t \gamma_k(t-y)dq_k(y), \\ \gamma_{k+1}(t) &= \mu \exp\Big\{-\lambda \displaystyle\int_0^t (1-q_k(y))dy\Big\}, \\ q_k(0) &= 0, \ \forall k \geq 0, \end{cases} \tag{10}$$

which differs from (7) only by its first equation, but this difference is crucial and corresponds to a fictitious function $q_{-1}(t) = 0, \ \forall t \geq 0$.

Actually, the scheme (10) produces a sequence of trees $\{L_k, k \geq 0\}$, with the property that the leaves of L_k at level k never die, and the following is true:

- the q_k's form an *increasing* sequence of *defective* distributions;
- for all $k \geq 0$, the tail distribution of q_k dominates a defective exponential distribution with Laplace transform of the form $\dfrac{a_k b_k}{b_k + s}$;
- under condition (3), we have

  $$\lim_{k\to\infty} a_k = 1, \quad \lim_{k\to\infty} b_k = \frac{\lambda}{r}$$

  and q_k converges in L_1 to the proper distribution p.

Actually, the scheme (10 is useful to analyze the transient regime, and it plays a decisive role in the forthcoming lemma 2.3.

2.2 Transience

It turns out that the classification of the process for $\rho > e^{-1}$ can be obtained rather straightforwardly from analytic arguments.
Recalling that $\ell = \lim_{t\to\infty} p(t)$, we define

$$\varepsilon(t) \stackrel{\text{def}}{=} \lambda \int_0^t (\ell - p(x))dx. \tag{11}$$

A direct computation yields

$$\beta^*(s) = \mu \int_0^\infty \exp\big[-\big(\varepsilon(t) + (\lambda(1-\ell) + s)t\big)\big]dt,$$

together with the functional equation

$$\frac{s^2 \varepsilon^*(s)}{\lambda} = \frac{\ell + (\ell - 1)\beta^*(s)}{1 + \beta^*(s)}. \tag{12}$$

When $\rho > e^{-1}$, one can show that necessarily $\ell < 1$, which is tantamount to saying the system is transient.

The last point concerns an exact computation of ℓ. This is a difficult task, involving a forest of technicalities. In fact one can hardly expect more than approximate formulas. Hereafter, we pick out some salient results (both formal and concrete) yielding some interesting bounds for ℓ.

2.2.1 Formal approach

Using the definition (11), it appears that the right-hand side member of (12) can be analytically continued to the region $\Re(s) < -\lambda(1-\ell)$. Thus an analysis of singularities becomes theoretically possible, which should hopefully allow to compute ℓ.

Owing to the inversion formula (5), we can rewrite (12) in the functional form

$$\frac{1}{2i\pi}\int_{\sigma-i\infty}^{\sigma+i\infty} e^{st}\beta^*(s)ds = \mu \exp\left[\frac{-\lambda}{2i\pi}\int_{\sigma-i\infty}^{\sigma+i\infty} \frac{e^{st}ds}{s^2\big(1+\beta^*(s)\big)}\right], \quad \Re(\sigma) > 0. \quad (13)$$

Arguing by analytic continuation in (13), it is possible to prove, that $\beta^*(s)$ is a meromorphic function with real negative poles. Hence, $\beta(t)$ can be represented by the Dirichlet series

$$\beta(t) = C\exp\left[\frac{-\lambda t}{1+\beta^*(0)}\right] + \sum_{i\geq 0} u_i e^{-\sigma_i t}, \quad (14)$$

where C is a constant, the σ_i's form a sequence of positive increasing numbers satisfying

$$\sigma_i > \frac{\lambda}{1+\beta^*(0)}, \quad \forall i \geq 0,$$

and the u_i's are *ad hoc* residues. In the ergodic case $\beta^*(0) = \infty$ and the first term in (14) reduces to the constant C. Then, $\varepsilon(t)$ could be obtained by formal inversion of $\beta^*(s)$. Alas, the computation becomes formidable and we did not get an exact tractable form (if any at all !) for ℓ, since this is equivalent to compute $u_i, \sigma_i, i \geq 0$.

2.2.2 Bounds and tail distribution

Beforehand, it is worth quoting some simple facts. First, the value of ℓ does solely depend on ρ, as can be seen by scaling in system (1–2). Secondly, combining (1) and (2) leads to the inequality

$$\beta(t) \leq \mu - \lambda p(t),$$

whence immediately

$$\ell \leq \min\Big(1, \frac{1}{\rho}\Big), \quad \forall \rho < \infty. \quad (15)$$

The iterative scheme (10) is convergent *for all* ρ, but the distributions $q_k(t)$, $k \geq 0$, are *defective*, their limit being proper if and only if $\rho \leq e^{-1}$. When $\rho > e^{-1}$, the limiting function $p(t)$ remains defective and

$$\lim_{t\to\infty} p(t) = \lim_{k\to\infty}\lim_{t\to\infty} q_k(t) = \ell < 1.$$

Showing by induction that $q_k(t)$, for large t, dominates an exponential distribution, we can derive bounds on ℓ. The tail-ordering stated in the forthcoming lemma gives, quite pleasantly, the *exact* value $\ell = 1$, when $\rho \leq e^{-1}$.

Lemma 2.3.

$$q_k(t) \geq a_k(1 - e^{-b_k t}) + o(e^{-b_k t}), \quad \forall\, k \geq 0, \tag{16}$$

where the sequence (a_k, b_k) *satisfies the recursive scheme*

$$\begin{cases} a_{k+1}b_{k+1} = \mu \exp\Bigl(\dfrac{-\lambda a_k}{b_k}\Bigr), \\ b_{k+1}(1 - a_{k+1}) = \lambda(1 - a_k), \end{cases} \tag{17}$$

with $a_0 = \mu/(\lambda+\mu)$ *and* $b_0 = \lambda + \mu$.
Setting $a \stackrel{def}{=} \lim_{k\to\infty} a_k$ *and* $b \stackrel{def}{=} \lim_{k\to\infty} b_k$ *in (17), one has the limits*

$$\begin{cases} a = 1,\ b = \dfrac{\lambda}{r},\ \text{if} \quad \rho \leq e^{-1}, \\ a = x,\ b = \lambda,\ \text{if} \quad \rho \geq e^{-1}, \end{cases} \tag{18}$$

where $x \leq 1$ *is the root of the equation*

$$xe^x = \frac{1}{\rho}. \tag{19}$$

Subsidiary comments The method of schemes to analyze nonlinear operators in a probabilistic context is extremely powerful (see e.g. [2] for problems related to systems in thermodynamical limit), and in some sense deeply related to the construction of Lyapounov functions. Up to sharp technicalities, the schemes (7) and (10) can be exploited to derive precise information about the speed of convergence as $t \to \infty$, for any ρ, $0 < \rho < \infty$, and when pushing exact computations slightly farther, one perceives underlying relationships with intricate continued fractions. Finally, we note that the question of transience could be studied from a large deviation point of view, by considering $\varepsilon(t)$ as the member of a family indexed by the parameter $(\rho - e^{-1})$.

3 An ergodic theorem for $H(t)$

The key result of this section is a limit theorem for the height of the tree, which indeed is of special interest when the system is transient.

For every integer k and all $t > 0$, define the quantities

$$\begin{cases} X_k(t) \stackrel{\text{def}}{=} \#\{v \in G(t) : h(v) = k\}, \\ Y_k(t) \stackrel{\text{def}}{=} \sum_{j=k}^{\infty} X_j(t) \mathbb{1}_{\{t \le \tau\}}. \end{cases}$$

Thus $X_k(t)$ represents the number of vertices at level k in the whole tree at time t.

Let

$$b(s,c) \stackrel{\text{def}}{=} \frac{s}{c} + \log\left[\frac{\lambda(1 - sp^*(s))}{s}\right]. \tag{20}$$

Theorem 3.1. *With probability* 1,

$$\lim_{t\to\infty} \frac{H(t)}{t} = \delta,$$

where δ *is the unique positive solution of the system*

$$b(s,\delta) = \frac{\partial b(s,\delta)}{\partial s} = 0.$$

In the ergodic case $\delta = 0$.

The proof needs three intermediate lemmas.

Lemma 3.2. *Define the events*

$$A_c = \left\{\liminf_{t\to\infty} \frac{H(t)}{t} \ge c\right\}, \qquad B_c = \left\{\limsup_{t\to\infty} \frac{H(t)}{t} \le c\right\}.$$

Then $\mathsf{P}\{A_c\} = \mathbb{1}_{\{A_c\}}$ *and* $\mathsf{P}\{B_c\} = \mathbb{1}_{\{B_c\}}$. *In other words,* A_c *and* B_c *satisfy a zero-one law and can only be trivial events (i.e. sure or impossible).*

Lemma 3.3.

(i) If, for some integer n *and real number* $c > 0$, $\mathsf{E}[Y_n(n/c)] > 1$, *then*

$$\mathsf{P}\{A_c\} = 1.$$

(ii) If, for some n *and real number* $c > 0$, $\sum_{k=0}^{\infty} \mathsf{E}[X_{kn}(kn/c)] < \infty$, *then*

$$\mathsf{P}\{B_c\} = 1.$$

Sketch of proof : The result follows from the relations

$$\mathsf{P}\{A_c \mid G(t_0) = G_0\} \ge \mathsf{P}\{A_c\},$$
$$\mathsf{P}\{B_c \mid G(t_0) = G_0\} \le \mathsf{P}\{B_c\}.$$

■

Lemma 3.4. *With $b(s,c)$ defined in (20), the two following transforms hols:*

$$\mathsf{E}[Y_n(n/c)] = \frac{1}{2i\pi}\int_{\sigma-i\infty}^{\sigma+i\infty} \frac{e^{nb(s,c)}\left[1-sp^*(s)\right]}{s-\lambda(1-sp^*(s))}ds, \tag{21}$$

in the region $\mathcal{U} \stackrel{def}{=} \{\sigma > 0,\ \sigma > \lambda(1-\sigma p^*(\sigma))\}$;

$$\sum_{k=0}^{\infty} \mathsf{E}[X_{kn}(kn/c)] = \frac{1}{2i\pi}\int_{\sigma-i\infty}^{\sigma+i\infty} \frac{ds}{s}\left[1-e^{nb(s,c)}\right]^{-1}, \tag{22}$$

in the region $\mathcal{V} \stackrel{def}{=} \{\sigma > 0,\ \sigma > \lambda(1-\sigma p^*(\sigma))e^{\sigma/c}\}$.

The proof of Theorem 3 follows from the above lemma, remarking that when the system is ergodic, $\lim_{s\to 0} b(s,c) = \log\lambda m = \log r \le 0$, which yields $\delta = 0$ as might be expected. ■

As a by-product, we state the following corollary, of which the almost sure convergence part has been derived in [3, 10] through different and less terse methods.

Corollary 3.5. *In the pure birth case $\mu = 0$, almost surely and in L_1,*

$$\lim_{t\to\infty}\frac{H(t}{t} = \lambda e. \tag{23}$$

Proof : Since here

$$\mathsf{E}X_n(t) = \frac{(\lambda t)^n}{n!},$$

a direct use of criteria *(i)* and *(ii)* of lemma 3.3 yields the first part of (23). On the other hand, the function $\mathsf{E}H(t)$ is superadditive, namely

$$\mathsf{E}H(s+t) \ge \mathsf{E}H(s) + \mathsf{E}H(t),$$

so that, by a variant of Kingman's theorem (see [7]), the limit $\displaystyle\lim_{t\to\infty}\frac{\mathsf{E}H(t)}{t}$ does exist. Consequently, the convergence in L_1 in (23) will follow if one can show

$$\mathsf{E}H(t) \le At, \quad \forall t > 0,$$

for some positive finite constant A. Using the obvious inequality

$$\mathsf{P}\{H(t) \ge k\} = \mathsf{P}\{X_k(t) > 0\} \le \mathsf{E}X_k(t),$$

we can write

$$\mathsf{E}H(t) = \sum_{k=1}^{\infty}\mathsf{P}\{H(t)\ge k\} \le \sum_{k=1}^{m} 1 + \sum_{k=m+1}^{\infty}\frac{(\lambda t)^k}{k!}.$$

Then, taking $m = \lceil \lambda e t\rceil$ and using Stirling's formula, we obtain

$$\mathsf{E}H(t) \le \lambda e t + \frac{\sqrt{\lambda e t}}{\sqrt{2\pi}(e-1)}. \tag{24}$$

■

Acknowledgements The authors thank V.A. Malyshev for bringing the problem to their attention and Th. Deneux for useful numerical experiments. They also want to mention a joint study in progress with J.-M Lasgouttes, which will includes stationary laws and multiclass models.

References

[1] H. Cartan (1977) *Cours de calcul différentiel,* Hermann, Collection Méthodes.

[2] F. Delcoigne and G. Fayolle (1999) *Thermodynamical limit and propagation of chaos in polling systems,* Markov Processes and Related Fields, 5 (1), pp. 89–124.

[3] L. Devroye (1987) *Branching processes in the analysis of the height of trees,* Acta Informatica, 24, pp. 277–298.

[4] W. Feller (1971) *An Introduction to Probability Theory and its Applications,* Vol. I and II, Wiley.

[5] B. A Fuchs and V. I. Levin (1961) *Functions of a Complex Variable,* Vol. II, Pergamon Press.

[6] I.S. Gradshteyn and I.M Ryzhik (1980) *Table of Integrals, Series, and Products,* Academic Press, corrected and enlarged edition.

[7] O. Kallenberg (2001) *Foundations of Modern Probability,* Springer, Probability and its Applications.

[8] M. Krikun (2000) *Height of a random tree,* Markov Processes and Related Fields, 6 (2), pp. 135–146.

[9] H. M. Mahmoud (1992) *Evolution of Random Search Trees,* Wiley Intersciences. Series.

[10] B. Pittel (1994) *Note on the heights of random recursive trees and random m-ary search trees,* Random Struct. Algorithms, 5, pp. 337–347.

Guy Fayolle

INRIA Rocquencourt
Domaine de Voluceau BP 105
78153 Le Chesnay, France
Guy.Fayolle@inria.fr

Maxim Krikun

Laboratory of Large Random Systems
Faculty of Mathematics and Mechanics
Moscow State University
119899, Moscow, Russia
krikun@lbss.math.msu.su

Trends in Mathematics, © 2002 Birkhäuser Verlag Basel/Switzerland

Ideals in a Forest, One-Way Infinite Binary Trees and the Contraction Method

Svante Janson

ABSTRACT: *The analysis of an algorithm by Koda and Ruskey for listing ideals in a forest poset leads to a study of random binary trees and their limits as infinite random binary trees. The corresponding finite and infinite random forests are studied too. The infinite random binary trees and forests studied here have exactly one infinite path; they can be defined using suitable size-biazed Galton–Watson processs. Limit theorems are proved using a version of the contraction method.*

1 Introduction

The vertices of a rooted forest may be regarded as a poset in a natural way, with the roots being the minimal elements. Consider the family of all ideals (or down-sets) of this poset. If the forest consists of trees $T_1,\dots,T_k$, then the ideals are the sets of the form $V_1 \cup \cdots \cup V_k$, where each V_i is either empty or the vertex set of a rooted subtree of T_i.

Koda and Ruskey [11] described two algorithms for listing the ideals of a forest poset in a Gray code manner, i.e. such that consecutive ideals differ by exactly one element. (For background and applications, see [11]. For actual implementations, see Knuth [10].) We are here concerned only with their first algorithm, `Algorithm P` in [11]. Since the algorithm operates on ordered forests, we assume from now on that all forests and trees are rooted and ordered.

As noted in [11], the running time per ideal of `Algorithm P`, i.e. the total running time divided by the number of ideals listed, is not bounded. However, it is conjectured in [11, Section 6] that the expected running time per ideal for a randomly selected rooted tree on n vertices is bounded as $n \to \infty$.

In the present paper, we study random ordered rooted trees, and verify the conjecture of [11] in this case. (As pointed out by a referee, the conjecture in [11] is really stated for random rooted trees; the algorithm operates on ordered trees, but the probability distribution depends on whether the ordering is imposed before or after the random selection. Presumably, the result holds for random rooted trees and other families of simply generated trees too.) Moreover, we show that both the expectation and the distribution of the running time per ideal converges as $n \to \infty$ (without further normalization).

The proofs use a version of the contraction method, which earlier has been used to study many other algorithms, see e.g. [16, 17, 18, 19]. The present application includes some novel features, however, which we find at least as interesting as the results themselves. Thus, although the paper exclusively studies `Algorithm P`, it should mainly be seen as an example illustrating a method that we hope may be useful for the study of other algorithms as well.

In the proofs we find it convenient to transfer the problem to an equivalent one for random binary trees, see Section 3. Note that we consider random binary trees with the uniform distribution over all binary trees of a given size (sometimes called Catalan trees), in contrast to the binary search trees that appear in connection with other applications of the contraction method (in particular, `Quicksort`). The distributions are quite different, with the uniform binary trees studied here tending to be much more unbalanced and stringy, which leads to new phenomena.

In the present case (unlike the case of binary search trees), there is a natural limit of the random binary tree as its size tends to infinity; this is a non-trivial infinite random binary tree. Similarly, there is an infinite random forest that is the limit of the random ordered rooted forest. We study these infinite trees and forests in Sections 5 and 6 and show that the cost per ideal can be defined (a.s.) for these infinite objects too, in such a way that its distribution is the limiting distribution of the cost per ideal for finite random forests. This enables us to deduce some properties of the limiting distribution. For example, we show that the distribution is continuous (Theorem 5.10). It is, however, an open problem whether it is absolutely continuous.

The infinite random forests and binary trees studied here have exactly one infinite branch. They can be defined using a size-biased Galton–Watson branching process, see Section 5. We include some further comments on the structure of these infinite objects in Sections 5 and 6.

Acknowledgement. I thank Donald Knuth for suggesting this problem.

2 Preliminaries

We let $|F|$ denote the number of vertices in a forest, or tree, F. If T is a tree, we let T^* be the forest obtained by deleting the root, letting the children of the old root be the new roots. Thus $|T^*| = |T| - 1$.

Denote the number of ideals of a forest F by $N(F)$. If F consists of the trees $T_1, \dots, T_m$, then

$$N(F) = \prod_{i=1}^{m} N(T_i). \tag{1}$$

In particular, $N(\emptyset) = 1$. Moreover, it is easily seen that if T is a tree, then

$$N(T) = 1 + N(T^*). \tag{2}$$

Note that (1) and (2) together determine N recursively. It is easily seen by induction, or directly, considering the ideals consisting of paths from the root and the empty ideal, that

$$N(F) \geq |F| + 1. \tag{3}$$

We let $W(F)$ denote a measure of the running time (work) of `Algorithm P` on a forest F. Of course, the actual running time depends on details in the implementation, but we make a precise definition as follows, using the descriptions in the proof of Lemma 3.1 in [11]:

If F is empty, we let $W(F) = 0$.

If F consists of a single tree T, then `Algorithm P` lists first $\emptyset$ and then all ideals of T^*, in the order given by `Algorithm P` on T^*, with the root of T added to each. The work required by the algorithm on T is therefore the same as for T^*, but with one extra unit for each ideal of T^* (because of the added root) and two extra units for the additional ideal. Hence

$$W(T) = W(T^*) + N(T^*) + 2 = W(T^*) + N(T) + 1. \tag{4}$$

If F consists of several trees $T_1, \dots, T_k$, $k \geq 2$, let $F' = F \setminus T_1 = T_2 \cup \cdots \cup T_k$. Then `Algorithm P` first lists all ideals of T_1, ignoring F', then acts in F' as if running on F', then lists all ideals of T_1 in reverse order with the first nonempty ideal of F' added to each, then acts in F' again, then lists the ideals of T_1 in order with the second nonempty ideal of F' added to each, and so on. Hence the ideals of T_1 are run through $N(F')$ times (in alternating directions) with a work $W(T_1)$ each time, while the remaining steps together are equivalent to running the algorithm on F', which requires $W(F')$. Hence

$$W(F) = N(F')W(T_1) + W(F'). \tag{5}$$

This completes our (recursive) definition of W. (This definition of W by (4) and (5) was given, in an equivalent form, by Knuth [personal communication].)

Remark 2.1. There is some arbitrariness in the definition; in particular, (4) might be modified to $W(T) = W(T^*) + aN(T^*) + b$ for some other positive constants a and b. This would not cause any important differences to the results of this paper (although numerical values will differ); we can assume that $a = 1$ by dividing W by a, and a value of b different from 2 would cause only routine changes below.

Note that $N(F)$ and $W(F)$ vary wildly among forests of the same size. The extreme cases are, as is easily verified by induction:

(i) n isolated roots; $N = 2^n$, $W = 3(2^n - 1)$.

(ii) n vertices in a path; $N = n + 1$, $W = \binom{n+1}{2} + 2n = (n^2 + 5n)/2$.

We therefore study $Q(F) := W(F)/N(F)$, the work per ideal. Note that the path in (ii) shows that Q is unbounded (on trees as well as on forests). ($N(F)$ and $N(T)$ are studied in [8, 20], but we do not use the results there.)

There are C_n ordered forests with n vertices, where

$$C_n = \frac{1}{n+1}\binom{2n}{n} = \frac{(2n)!}{n!\,(n+1)!} \tag{6}$$

is the n:th Catalan number, and thus there are C_{n-1} ordered trees with n vertices [9, 2.3.4.4].

Let F_n denote a random ordered rooted forest with n vertices, uniformly selected among the C_n possibilities; let similarly T_n be a uniformly selected random ordered rooted tree with n vertices. We can now state the main results of the paper, proved in Section 4.

Theorem 2.2. *There exists a positive random variable* $\mathbf{Q}$ *with finite mean such that, as* $n \to \infty$, $Q(F_n) \xrightarrow{\mathrm{d}} \mathbf{Q}$ *and* $\mathbb{E}\, Q(F_n) \to \mathbb{E}\, \mathbf{Q}$.

Corollary 2.3. *As* $n \to \infty$, *with* $\mathbf{Q}$ *as in Theorem 2.2,* $Q(T_n) \xrightarrow{\mathrm{d}} \mathbf{Q} + 1$ *and* $\mathbb{E}\, Q(T_n) \to \mathbb{E}\, \mathbf{Q} + 1$.

Proof: By (2) and (4),

$$Q(T_n) = \frac{W(T_n)}{N(T_n)} = \frac{W(T_n^*) + N(T_n^*) + 2}{N(T_n^*) + 1} = Q(T_n^*) + 1 + \frac{1 - Q(T_n^*)}{N(T_n^*) + 1}.$$

Since T_n^* is distributed as F_{n-1}, and $N(T_n^*) \geq n$ by (3), the results follow from Theorem 2.2.

We have no explicit description of the limit distribution $\mathcal{L}(\mathbf{Q})$, but it is characterized by a fixed point equation. This fixed point equation is more complicated than in many other similar cases, so we postpone it to Section 5, see Theorems 5.7 and 5.8. In Sections 5 and 6 we further show that $\mathbf{Q}$ may be interpreted as an extension of Q to random infinite forests.

3 Binary trees

We find it convenient to consider binary trees instead of forests, using the well-known correspondence in [9, Section 2.3.2], which can be defined recursively as follows: If $F = \emptyset$, then $B(F) = \emptyset$. If F is a forest consisting of trees $T_1, \dots, T_k$, $k \geq 1$, then $B(F)$ is the binary tree with a root, a left subtree $B(T_1^*)$ and a right subtree $B(T_2 \cup \dots \cup T_k)$. Note that $|B(F)| = |F|$.

We define N, W and Q for binary trees by this correspondence, setting $N(B(F)) = N(F)$ and so on.

It is easily seen that (1), (2), (4), (5) translate as follows: If B is a nonempty binary tree with left and right subtrees L and R, then

$$N(B) = \big(N(L) + 1\big)N(R) \tag{7}$$

$$W(B) = N(R)\big(W(L) + N(L) + 2\big) + W(R). \tag{8}$$

Together with $N(\emptyset) = 1$ and $W(\emptyset) = 0$, (7) and (8) define N and W directly on binary trees by recursion. Taking the quotient, we further obtain

$$\begin{aligned} Q(B) &= \frac{W(L) + N(L) + 2}{N(L) + 1} + \frac{W(R)}{N(R)(N(L) + 1)} \\ &= Q(L) + 1 + \frac{1 - Q(L) + Q(R)}{N(L) + 1}. \end{aligned} \tag{9}$$

Let B_n denote a (uniformly selected) random binary tree with n vertices ($n \geq 0$). Then Theorem 2.2 is equivalent to the following.

Theorem 3.1. *As $n \to \infty$, $Q(B_n) \xrightarrow{\mathrm{d}} \mathbf{Q}$ and $\mathbb{E}\, Q(B_n) \to \mathbb{E}\,\mathbf{Q}$, with $\mathbf{Q}$ as in Theorem 2.2.*

Let, for $n \geq 1$, L_n and R_n denote the left and right subtrees of B_n. Note that

$$|L_n| + |R_n| = n - 1. \tag{10}$$

Let $(p_{n,k})_{k=0}^{n-1}$ be the probability distribution of the size of the left (or, by symmetry, the right) subtree of B_n, i.e.

$$p_{n,k} := \mathbb{P}(|L_n| = k) = \mathbb{P}(|R_n| = k). \tag{11}$$

By (10),

$$p_{n,k} = \mathbb{P}(|L_n| = k) = \mathbb{P}(|R_n| = n-1-k) = p_{n,n-1-k}. \tag{12}$$

There are C_n binary trees with n vertices, where again C_n is the Catalan number (6). Hence, the number of binary trees with k vertices in the left subtree and $n-1-k$ in the right is $C_k C_{n-1-k}$, and

$$p_{n,k} = \frac{C_k C_{n-1-k}}{C_n}.$$

Stirling's formula easily yields

$$C_n = \frac{1}{n+1}\binom{2n}{n} = \pi^{-1/2} n^{-3/2} 2^{2n}\bigl(1+O(n^{-1})\bigr).$$

Hence, uniformly for $0 \le k \le n/2$,

$$\begin{aligned} p_{n,k} &= C_k (n-1-k)^{-3/2} n^{3/2} 2^{2(n-1-k)-2n}\bigl(1+O(n^{-1})\bigr) \\ &= C_k 4^{-k-1}\Bigl(1+O\Bigl(\frac{k+1}{n}\Bigr)\Bigr) \\ &= \pi_k\Bigl(1+O\Bigl(\frac{k+1}{n}\Bigr)\Bigr), \end{aligned}$$

where

$$\pi_k = \lim_{n\to\infty} p_{n,k} = C_k 4^{-k-1}, \qquad k \ge 0.$$

By the generating function for Catalan numbers

$$B(z) := \sum_{k=0}^{\infty} C_k z^k = \frac{1-\sqrt{1-4z}}{2z}, \tag{13}$$

see e.g. [9, (2.3.4.4-13)], we have

$$\sum_{k=0}^{\infty} \pi_k = \sum_{k=0}^{\infty} C_k 4^{-k-1} = \frac{1}{4} B\Bigl(\frac{1}{4}\Bigr) = \frac{1}{2}. \tag{14}$$

Hence $(\pi_k)_0^\infty$ is not a probability distribution. This reflects the symmetry of the left and right sides; roughly speaking, for n large, with probability 1/2 $|L_n|$ is small, and with probability 1/2 $|R_n|$ is small and then $|L_n| \approx n$. In particular, a large random binary tree is extremely unbalanced. We state this more precisely.

Lemma 3.2. *For each $\varepsilon > 0$ there exists M such that, for every n,*

$$\mathbb{P}(|L_n| < M) > \tfrac{1}{2} - \varepsilon \tag{15}$$

$$\mathbb{P}(|L_n| \ge n - M) > \tfrac{1}{2} - \varepsilon \tag{16}$$

$$\mathbb{P}(M \le |L_n| < n - M) < 2\varepsilon. \tag{17}$$

Proof: Choose M_1 such that $\sum_{k=M_1}^{\infty} \pi_k < \varepsilon$. Then, as $n \to \infty$,

$$\mathbb{P}(|L_n| < M_1) \to \sum_{k=0}^{M_1-1} \pi_k > \tfrac{1}{2} - \varepsilon,$$

so (15) holds with $M = M_1$ for sufficiently large n, say $n \geq n_0$. Taking $M := \max(M_1, n_0)$, (15) holds for all n. Furthermore, (16) holds by (12), while (17) is an immediate consequence of (15) and (16).

We can modify (π_k) to make it into a probability distribution in two ways, both of which will be used below. First, we can allow the value $+\infty$, giving it the probability 1/2 because of (14). We let ξ^* be a random variable with values in $N^* := \{0, 1, \ldots, \infty\}$ having this distribution, i.e.

$$\mathbb{P}(\xi^* = k) = \begin{cases} \pi_k, & 0 \leq k < \infty, \\ \frac{1}{2}, & k = \infty. \end{cases}$$

Alternatively, we can renormalize (π_k) and consider the probability distribution $(2\pi_k)_{0 \leq k < \infty}$. We let ξ be a random variable with this distribution, i.e.

$$\mathbb{P}(\xi = k) = 2\pi_k = 2^{-2k-1} C_k, \qquad k \geq 0. \tag{18}$$

Note that ξ can be defined as ξ^* conditioned on $\xi^* < \infty$.

With this notation, the following lemmas are immediate consequences of the results above.

Lemma 3.3. *Let $n \to \infty$. Then $|L_n| \xrightarrow{\mathrm{d}} \xi^*$, as random variables in N^*.* □

Lemma 3.4. *Let $n \to \infty$. Then $|L_n|$, conditioned on $|L_n| < n/2$, converges in distribution to ξ.* □

Of course, the same results hold for $|R_n|$.

Remark 3.5. These results for the uniform random binary tree studied here should be compared with the corresponding results for random binary search trees, which have a different distribution and for which $|L_n|$ is uniformly distributed on $\{0, \ldots, n-1\}$. The stronger imbalance in our case is a source of phenomena quite different from the binary search tree case.

Finally we record a simple but important observation. We let $\widetilde{B}_n$ be another (uniform) random binary tree, independent of $\{B_k\}_{k=0}^{\infty}$.

Lemma 3.6. *Let $0 \leq k < n$. The conditional joint distribution of L_n and R_n given $|L_n| = k$ equals the distribution of $(B_k, \widetilde{B}_{n-1-k})$.* □

In other words, conditioned on the sizes of the subtrees L_n and R_n, they are two independent random binary trees.

4 Proof of the limit theorems

We begin with a preliminary estimate, which verifies the conjecture that the expected running time is bounded.

Lemma 4.1. $\sup_{n\geq 0} \mathbb{E}\, Q(B_n) < \infty$.

Proof: Define, for $n \geq 0$,

$$a_n := \mathbb{E}\, Q(B_n) + 1,$$
$$b_n := \mathbb{E}\, \frac{1}{N(B_n)+1} \leq \frac{1}{2}.$$

From (9), conditioning on $|L_n|$ and using Lemma 3.6,

$$\begin{aligned} a_n &\leq \mathbb{E}\Big(Q(L_n) + 1 + (1 + Q(R_n))\frac{1}{N(L_n)+1}\Big) + 1 \\ &= \sum_{l=0}^{n-1} p_{n,l} a_l + \sum_{r=0}^{n-1} p_{n,r} a_r b_{n-1-r} + 1 \\ &= \sum_{l=0}^{n-1} p_{n,l} a_l (1 + b_{n-1-l}) + 1 \\ &\leq \tfrac{3}{2} \sum_{l=0}^{n-1} p_{n,l} a_l + 1. \end{aligned} \tag{19}$$

Let $a_n^* := \max_{0\leq k\leq n} a_k$. By (19), for any $M \geq 0$,

$$\begin{aligned} a_n &\leq \tfrac{3}{2} a_M^* \,\mathbb{P}(|L_n| \leq M) + \tfrac{3}{2} a_{n-1}^* \,\mathbb{P}(|L_n| > M) + 1 \\ &\leq 1 + \tfrac{3}{2} a_M^* + \tfrac{3}{2} a_{n-1}^* \big(1 - \mathbb{P}(|L_n| \leq M)\big). \end{aligned}$$

We choose M as in Lemma 3.2 with $\varepsilon = 0.1$. Then $\mathbb{P}(|L_n| \leq M) > 0.4$ and thus, for all $n \geq 1$,

$$a_n \leq 1 + \tfrac{3}{2} a_M^* + \tfrac{3}{2} \cdot 0.6 a_{n-1}^* = 0.9 a_{n-1}^* + 1 + \tfrac{3}{2} a_M^*.$$

An easy induction yields

$$a_n \leq 10(1 + \tfrac{3}{2} a_M^*), \qquad n \geq 0.$$

The lemma follows because $\mathbb{E}\, Q(B_n) \leq a_n$.

We prove Theorem 3.1, and thus Theorem 2.2 and Corollary 2.3, using the Mallows metric d_1 for probability distributions with finite expectations. (This metric is also known under many other names, such as the Dudley, Fortet-Mourier, Kantorovich or Wasserstein distance.) It has several equivalent definitions, see e.g. [15]; for us the following is convenient.

If f is a real (or complex) function on $\mathbb{R}$, let

$$\|f\|_{\mathrm{Lip}} := \sup_{x\neq y} \frac{|f(x) - f(y)|}{|x-y|}.$$

If μ and ν are probability measures on $\mathbb{R}$ with finite expectations, then

$$d_1(\mu,\nu) := \sup\Big\{ \Big| \int f\, d\mu - \int f\, d\nu \Big| : \|f\|_{\mathrm{Lip}} \leq 1 \Big\}. \tag{20}$$

In other words, we take the supremum in (20) over all functions f satisfying the Lipschitz condition $|f(x)-f(y)| \le |x-y|$. (It does not matter whether we consider real or complex functions.)

If X and Y are random variables with finite expectations, we will for simplicity write $d_1(X,Y)$ for the d_1 distance between their distributions. Thus

$$d_1(X,Y) := \sup\{|\mathbb{E} f(X) - \mathbb{E} f(Y)| : \|f\|_{\mathrm{Lip}} \le 1\}. \tag{21}$$

It is easily seen that $d_1(X_n, X) \to 0$ implies $X_n \xrightarrow{\mathrm{d}} X$ and $\mathbb{E} X_n \to \mathbb{E} X$. (Take $f(x) = |t|^{-1}e^{itx}$, $t \neq 0$, and $f(x) = x$.) We will show that $d_1(Q(B_n), \mathbf{Q}) \to 0$ for some random variable $\mathbf{Q}$; this thus proves Theorem 3.1. (Indeed, Theorem 3.1 is equivalent to $d_1(Q(B_n), \mathbf{Q}) \to 0$, using the fact that $d_1(X_n, X) \to 0$ is equivalent to $X_1 \xrightarrow{\mathrm{d}} X$ and $\mathbb{E}|X_n| \to \mathbb{E}|X|$ for any random variables with finite expectations.)

Remark 4.2. That $\mathbb{E} Q(B_n)$ converges could also be shown directly using a simplified version of the proof below, taking $f(x) = x$.

Note first that replacing f by $f - f(0)$ does not change $\mathbb{E} f(X) - \mathbb{E} f(Y)$. Hence we may in (21) further impose $f(0) = 0$. Since then $|f(x)| = |f(x) - f(0)| \le |x|$, we have the bound

$$d_1(X,Y) \le \mathbb{E}|X| + \mathbb{E}|Y|. \tag{22}$$

We now consider $Q(B_n)$. For notational convenience we write $X_n = Q(B_n)$, $Y_n = (N(B_n)+1)^{-1}$ and $\widetilde{X}_n = Q(\widetilde{B}_n)$. Thus $\widetilde{X}_n$ has the same distribution as X_n but is independent of all X_k and Y_k. Note that, by (3),

$$Y_n \le \frac{1}{n+2}.$$

We further define

$$\delta_N := \sup\{d_1(X_n, X_m) : n, m \ge N\}. \tag{23}$$

Let $c_1 := \sup_n \mathbb{E} X_n$, which is finite by Lemma 4.1. By (22), $\delta_N \le 2c_1 < \infty$.

Fix a function f with $\|f\|_{\mathrm{Lip}} \le 1$ and $f(0) = 0$. By (9) and Lemma 3.6, conditioning on $|L_n|$, for $n \ge 1$,

$$\begin{aligned}
\mathbb{E} f(X_n) = \mathbb{E} f(Q(B_n)) &= \mathbb{E} f\Big(Q(L_n) + 1 + \frac{1 - Q(L_n) + Q(R_n)}{N(L_n)+1}\Big) \\
&= \sum_k p_{n,k}\, \mathbb{E} f\Big(Q(B_k) + 1 + \frac{1 - Q(B_k) + Q(\widetilde{B}_{n-1-k})}{N(B_k)+1}\Big) \\
&= \sum_k p_{n,k}\, \mathbb{E} f\big(X_k + 1 + Y_k(1 - X_k + \widetilde{X}_{n-1-k})\big) \\
&= \sum_k p_{n,k}\, \mathbb{E} f(U_{n,k})
\end{aligned} \tag{24}$$

where

$$U_{n,k} := X_k + 1 + Y_k(1 - X_k + \widetilde{X}_{n-1-k}). \tag{25}$$

We have, for $0 \le k \le n$,

$$|\mathbb{E} f(U_{n,k})| \le \mathbb{E}|U_{n,k}| \le \mathbb{E}(X_k + 2 + \widetilde{X}_{n-1-k}) \le c_2 := 2c_1 + 2 \tag{26}$$

and

$$\begin{aligned} |\mathbb{E} f(U_{n,k}) - \mathbb{E} f(X_k+1)| &\le \mathbb{E}|f(U_{n,k}) - f(X_k+1)| \\ &\le \mathbb{E}|U_{n,k} - X_k - 1| = \mathbb{E}|Y_k(1 - X_k + \widetilde{X}_{n-1-k})| \\ &\le \mathbb{E}\frac{1}{k+2}|1 - X_k + \widetilde{X}_{n-1-k}| \le \frac{c_2}{k+2}. \end{aligned} \tag{27}$$

Let $\varepsilon > 0$ and let M be as in Lemma 3.2. By (26),

$$\left|\sum_{k=0}^{M} p_{n,k}\,\mathbb{E} f(U_{n,k}) - \sum_{k=0}^{M} \pi_k\,\mathbb{E} f(U_{n,k})\right| \le \sum_{k=0}^{M} |p_{n,k} - \pi_k| c_2 \tag{28}$$

and

$$\begin{aligned} \left|\sum_{k=M+1}^{n-M-2} p_{n,k}\,\mathbb{E} f(U_{n,k})\right| &\le c_2 \sum_{k=M+1}^{n-M-2} p_{n,k} = c_2\,\mathbb{P}(M < |L_n| < n-1-M) \\ &< 2c_2\varepsilon. \end{aligned} \tag{29}$$

Furthermore,

$$\sum_{k=n-1-M}^{n-1} p_{n,k}\,\mathbb{E} f(U_{n,k}) = \sum_{j=0}^{M} p_{n,n-j-1}\,\mathbb{E} f(U_{n,n-1-j}) = \sum_{j=0}^{M} p_{n,j}\,\mathbb{E} f(U_{n,n-1-j})$$

and thus by (26) and (27)

$$\begin{aligned} &\left|\sum_{k=n-1-M}^{n-1} p_{n,k}\,\mathbb{E} f(U_{n,k}) - \sum_{j=0}^{M} \pi_j\,\mathbb{E} f(X_{n-1-j}+1)\right| \\ &\le \sum_{j=0}^{M} |p_{n,j} - \pi_j|\,|\mathbb{E} f(U_{n,n-1-j})| + \sum_{j=0}^{M} \pi_j |\mathbb{E} f(U_{n,n-1-j}) - \mathbb{E} f(X_{n-1-j}+1)| \\ &\le c_2 \sum_{j=0}^{M} |p_{n,j} - \pi_j| + \frac{c_2}{n-M} \sum_{j=0}^{M} \pi_j. \end{aligned} \tag{30}$$

We define R_n by

$$\mathbb{E} f(X_n) = \sum_{k=0}^{M} \pi_k\,\mathbb{E} f(U_{n,k}) + \sum_{k=0}^{M} \pi_k\,\mathbb{E} f(X_{n-1-k}+1) + R_n \tag{31}$$

and obtain by (24) and (28)–(30), for $n \ge 2M$,

$$|R_n| \le 2c_2 \sum_{k=0}^{M} |p_{n,k} - \pi_k| + 2c_2\varepsilon + \frac{c_2}{n-M}. \tag{32}$$

Let N_0 be so large that $N_0 > 2M$, $N_0 > M + 1/\varepsilon$ and $|p_{n,k} - \pi_k| < \varepsilon/(M+1)$ when $n \geq N_0$ for $k \leq M$. Then (32) yields

$$|R_n| \leq 5c_2\varepsilon, \qquad n \geq N_0. \tag{33}$$

Now suppose $N \geq N_0$ and $m, n \geq N$. Using (31) and (33) we have

$$\begin{aligned} &|\mathbb{E}\, f(X_n) - \mathbb{E}\, f(X_m)| \\ &\leq \sum_{k=0}^{M} \pi_k |\mathbb{E}\, f(U_{n,k}) - \mathbb{E}\, f(U_{m,k})| \\ &\qquad + \sum_{k=0}^{M} \pi_k |\mathbb{E}\, f(X_{n-1-k}+1) - \mathbb{E}\, f(X_{m-1-k}+1)| + 10c_2\varepsilon. \end{aligned} \tag{34}$$

Since $f_1(x) := f(x+1)$ is a function with $\|f_1\|_{\mathrm{Lip}} = \|f\|_{\mathrm{Lip}} \leq 1$,

$$|\mathbb{E}\, f(X_{n-1-k}+1) - \mathbb{E}\, f(X_{m-1-k}+1)| \leq d_1(X_{n-1-k}, X_{m-1-k}). \tag{35}$$

Similarly, for any given X_k and Y_k, the function

$$g(x) := f\big(X_k + 1 + Y_k(1 - X_k + x)\big)$$

has Lipschitz norm $\|g\|_{\mathrm{Lip}} \leq Y_k$, and thus by (25)

$$\begin{aligned} |\mathbb{E}\big(f(U_{n,k}) - f(U_{m,k}) \mid X_k, Y_k\big)| &= |\mathbb{E}\, g(\widetilde{X}_{n-1-k}) - \mathbb{E}\, g(\widetilde{X}_{m-1-k})| \\ &\leq \|g\|_{\mathrm{Lip}} d_1(\widetilde{X}_{n-1-k}, \widetilde{X}_{m-1-k}) \leq Y_k d_1(X_{n-1-k}, X_{m-1-k}). \end{aligned}$$

Using the crude bound $Y_k \leq 1/2$ and taking the expectation we have

$$\begin{aligned} |\mathbb{E}\, f(U_{n,k}) - \mathbb{E}\, f(U_{m,k})| &\leq \mathbb{E}\,|\mathbb{E}\big(f(U_{n,k}) - f(U_{m,k}) \mid X_k, Y_k\big)| \\ &\leq \tfrac{1}{2} d_1(X_{n-1-k}, X_{m-1-k}). \end{aligned} \tag{36}$$

Consequently, by (34), (36) and (23), for $m, n \geq N \geq N_0$,

$$\begin{aligned} |\mathbb{E}\, f(X_n) - \mathbb{E}\, f(X_m)| &\leq \sum_{k=0}^{M} \pi_k \tfrac{1}{2} \delta_{N-1-k} + \sum_{k=0}^{M} \pi_k \delta_{N-1-k} + 10c_2\varepsilon \\ &\leq \tfrac{3}{2} \delta_{N-1-M} \sum_{k=0}^{M} \pi_k + 10c_2\varepsilon \\ &\leq \tfrac{3}{4} \delta_{N-1-M} + 10c_2\varepsilon. \end{aligned}$$

Taking the supremum over all f with $\|f\|_{\mathrm{Lip}} \leq 1$ and $f(0) = 0$ we find, for $n, m \geq N \geq N_0$,

$$d_1(X_n, X_m) \leq \tfrac{3}{4} \delta_{N-1-M} + 10c_2\varepsilon$$

and thus

$$\delta_N \leq \tfrac{3}{4} \delta_{N-1-M} + 10c_2\varepsilon, \qquad N \geq N_0.$$

Letting $N \to \infty$ we obtain

$$\limsup_{N\to\infty} \delta_N \le \tfrac{3}{4} \limsup_{N\to\infty} \delta_N + 10c_2\varepsilon$$

and thus, since $\limsup_{N\to\infty} \delta_N \le 2c_1 < \infty$,

$$\limsup_{N\to\infty} \delta_N \le 40c_2\varepsilon.$$

Finally, letting $\varepsilon \to 0$, we obtain $\delta_N \to 0$ as $N \to \infty$.

By the definition (23), this shows that $(X_n)_n$, or rather the corresponding sequence of distributions, is a Cauchy sequence in the d_1 metric. It is easily seen that the space of all probability measures on $\mathbb{R}$ with finite expectation is complete with the metric d_1 [15]. Hence there exists a limit distribution, and thus a random variable $\mathbf{Q}$ such that $d_1(X_n, \mathbf{Q}) \to 0$, which completes the proof. □

Remark 4.3. The proof above shows that the distributions of $(X_n)_n$ form a Cauchy sequence, and thus converge to some limit. The limit will in the next section be characterized by fixed point equations. An anonymous referee has pointed out that, alternatively, it is possible to first define the limit distribution by the fixed point equation in Theorem 5.7 and then use arguments similar to the proof above to show that $d_1(X_n, \mathbf{Q}) \to 0$. This is the usual procedure in applications of the contraction method; it has some advantages in the current setting too but also some disadvantages, and we do not find the differences decisive. Anyone interested in extending the present results should consider both versions of the method.

5 More on binary trees

We begin with some more or less well-known (folk-lore?) observations on random binary trees.

Define a random binary tree B_* with random size by the following construction: Flip a fair coin. If it comes up tails, let B_* be empty, otherwise begin with a root. In the latter case, flip the coin again twice and add a left child of the root if the first flip results in heads and a right child if the second flip results in heads. Continue in this way, flipping the coin twice for every new vertex, as long as new vertices are added. (Equivalently, do site percolation on the complete infinite binary tree by flipping a fair coin for each vertex and removing the vertices that get tails, and let B_* be the component of the root, if any.)

We can regard B_* as the family tree of a Galton–Watson branching process with offspring distribution $\mathrm{Bi}(2, 1/2)$ (and children labelled as left or right), starting with $\mathrm{Bi}(1, 1/2)$ individuals. We thus call B_* the *Galton–Watson binary tree.* Since this Galton–Watson process is critical, it a.s. dies out, and thus B_* is finite.

The probability that B_* equals a given binary tree with n vertices is 2^{-2n-1}, since the vertices have to be chosen by n specified coin flips coming up heads, while $n+1$ other coin flips have to yield tails. Hence

$$\mathbb{P}(|B_*| = n) = C_n 2^{-2n-1} = 2\pi_n. \tag{37}$$

In other words, $|B_*|$ has the same distribution as ξ defined in (18). Moreover, the conditional distribution of B_* given $|B_*| = n$ is uniform, and thus equals the distribution of B_n. This yields yet another possibility of defining B_*: select its size by (37) and then select uniformly a binary tree with this size. Equivalently, if ξ is independent of $(B_n)_{n=0}^{\infty}$, we can take $B_* = B_\xi$.

Lemmas 3.4 and 3.6 imply the following:

Lemma 5.1. *Let $n \to \infty$. Then L_n, conditioned on $|L_n| < n/2$, converges in distribution to B_*.* □

Consequently, a large random binary tree has one branch at the root distributed (asymptotically) as B_*, while the other is large. We may continue recursively with the large branch, which suggests the following construction.

Define a (noncomplete) *infinite* random binary tree B_∞ as follows. Begin with the root and create an infinite path from it by randomly adding, an infinite number of times, either a left or a right child to the last added vertex. Finally, add independent copies of B_* at the free sites of the vertices in the path, i.e. as left or right subtrees depending on which side is not already occupied by the remainder of the infinite path. Note that B_∞ has exactly one infinite path from the root; we call this path the *trunk*.

It is easily seen that B_∞ can be defined by the following modification of the branching process above creating B_*. Consider a Galton–Watson process with two types of individuals, *mortals* and *immortals*. Let a mortal have $\mathrm{Bi}(2, 1/2)$ children, all mortal, and let an immortal have exactly one immortal child and $\mathrm{Bi}(1, 1/2)$ mortal children. Moreover, label each child as left or right, at random but ensuring that two siblings get different labels. The resulting family tree is B_∞.

It is now easy to see that Lemma 5.1 implies the following description of the asymptotic shape of large random binary trees. Let, for any tree B, $B^{(M)}$ be the first M levels of B, i.e. the tree with all branches pruned at height M.

Lemma 5.2. *As $n \to \infty$, $B_n \xrightarrow{\mathrm{d}} B_\infty$ in the sense of finite-dimensional distributions, i.e., $B_n^{(M)} \xrightarrow{\mathrm{d}} B_\infty^{(M)}$ for every finite M.* □

Remark 5.3. If we regard the space $\mathcal{B}$ of all finite or infinite binary trees as a subset of the power set of the vertex set of the complete infinite binary tree, with the natural product space topology on the power set, $\mathcal{B}$ is a metrizable compact space. A metric can be defined e.g. by $d(B, B') = 1/(k+1)$ if B and B' differ in the k-th level but not below it. The conclusion of Lemma 5.2 is equivalent to $B_n \xrightarrow{\mathrm{d}} B_\infty$ in this compact metric space $\mathcal{B}$.

Remark 5.4. The construction of B_∞ is a special case of the following general construction of the *size-biased* Galton–Watson process (regarded as a family tree); see e.g. [1] and [13]. Starting from a Galton–Watson process with an offspring distribution μ having finite, positive mean, the size-biased process can be obtained by considering a branching process with two types: mortals with an offspring distribution μ and all children mortals, and immortals with the size-biased offspring distribution $\widehat{\mu}$ and exactly one immortal child (in a random position among its siblings). The process starts with a single immortal. In the critical case studied here (and in the subcritical case), the size-biased process is the same as the *Q-process* studied in [3, Section I.14]. It is shown there that this process arises as the limit (in the sense of finite-dimensional distributions) as $t \to \infty$ of the original process conditioned on extinction occuring after time t (see also [5]). Informally (for critical and subcritical processes), it is the process conditioned on living forever. Similarly, it is easily shown that for a critical Galton–Watson process with finite offspring variance, the size-biased process is the limit as $n \to \infty$ of the process conditioned on the total progeny being n [7, 1]. In the case of random binary trees, this conditioning yields B_n, and we recover Lemma 5.2.

Having proved that both the trees B_n and the functional $Q(B_n)$ defined on them converge in distribution, it is natural to try to interpret the limit in Theorem 3.1 as $Q(B_\infty)$ for an extension of Q to infinite trees. Unfortunately, we cannot define this extension by continuity on the space $\mathcal{B}$ in Remark 5.3. Indeed, it is easily seen that for any infinite binary tree b, there is a sequence b_n of finite binary trees such that $b_n \to b$ in $\mathcal{B}$ but $Q(b_n) \to \infty$; for example, construct b_n by pruning b at height n and adding a sufficiently large complete binary tree at one of the cuts. (We leave the verification to the reader.) Hence, Q has no continuous extension to $\mathcal{B}$.

However, we can extend Q in the following, somewhat weaker, way. We let $N(B) = \infty$ for any infinite tree B. We further let L_∞ and R_∞ denote the left and right subtree of the root of B_∞. Note that exactly one of L_∞ and R_∞ is finite.

Theorem 5.5. *There exists an extension of Q to infinite binary trees such that $\mathbb{E}\,|Q(B_\infty^{(M)}) - Q(B_\infty)| \to 0$ as $M \to \infty$. This extension satisfies a.s. the equation*

$$Q(B_\infty) = Q(L_\infty) + 1 + \frac{1}{N(L_\infty)+1}\big(1 - Q(L_\infty) + Q(R_\infty)\big). \tag{38}$$

Moreover, the limit random variable $\mathbf{Q}$ *in Theorems 2.2 and 3.1 can be taken as $Q(B_\infty)$, i.e. $Q(B_n) \xrightarrow{\mathrm{d}} Q(B_\infty)$ as $n \to \infty$.*

We do not know whether $Q(B_\infty^{(M)}) \to Q(B_\infty)$ a.s. as $M \to \infty$.

We begin with a lemma on truncations of finite trees.

Lemma 5.6. *Let, for $M \geq 1$,*

$$\delta^{(M)} := \sup_n \mathbb{E}\,|Q(B_n^{(M)}) - Q(B_n)|.$$

Then $\delta^{(M)} \to 0$ as $M \to \infty$.

Proof: Note first that $\delta^{(M)} < \infty$ by Lemma 4.1 because $Q(B_n^{(M)})$ attains only a finite number of values for each M.

For any n and $M \geq 1$, the left and right subtrees of $B_n^{(M+1)}$ are $L_n^{(M)}$ and $R_n^{(M)}$, and thus (9) yields

$$\begin{aligned} Q(B_n^{(M+1)}) - Q(B_n) &= \frac{N(L_n^{(M)})}{N(L_n^{(M)})+1}\big(Q(L_n^{(M)}) - Q(L_n)\big) \\ &+ \frac{1}{N(L_n^{(M)})+1}\big(Q(R_n^{(M)}) - Q(R_n)\big) \\ &+ \Big(\frac{1}{N(L_n^{(M)})+1} - \frac{1}{N(L_n)+1}\Big)\big(1 - Q(L_n) + Q(R_n)\big). \end{aligned}$$

Since either $L_n^{(M)} = L_n$ or $|L_n| > |L_n^{(M)}| > M$, this implies

$$|Q(B_n^{(M+1)}) - Q(B_n)| \le |Q(L_n^{(M)}) - Q(L_n)| + \tfrac{1}{2}|Q(R_n^{(M)}) - Q(R_n)| + \frac{1}{M}|1 - Q(L_n) + Q(R_n)|. \tag{39}$$

For any $k \ge 0$, by Lemma 3.6,

$$\mathbb{E}\big(|Q(L_n^{(M)}) - Q(L_n)| \;\big|\; |L_n| = k\big) = \mathbb{E}\,|Q(B_k^{(M)}) - Q(B_k)| \le \delta^{(M)}.$$

Moreover, if $|L_n| = k \le M$, then $L_n^{(M)} = L_n$. Hence,

$$\mathbb{E}\,|Q(L_n^{(M)}) - Q(L_n)| \le \delta^{(M)}\,\mathbb{P}(|L_n| > M).$$

The same estimate holds for $\mathbb{E}\,|Q(R_n^{(M)}) - Q(R_n)|$. We thus obtain from (39), again letting $c_1 := \sup_n \mathbb{E}\,Q(B_n) < \infty$, see Lemma 4.1,

$$\mathbb{E}\,|Q(B_n^{(M+1)}) - Q(B_n)| \le \tfrac{3}{2}\delta^{(M)}\,\mathbb{P}(|L_n| > M) + \frac{1 + 2c_1}{M}. \tag{40}$$

Let M_0 be as in Lemma 3.2 with $\varepsilon = 0.1$. Then, for every $M \ge M_0$, we have $\mathbb{P}(|L_n| > M) \le \mathbb{P}(|L_n| > M_0) < 0.6$, and thus by (40)

$$\mathbb{E}\,|Q(B_n^{(M+1)}) - Q(B_n)| \le 0.9\delta^{(M)} + \frac{c_2}{M}.$$

for every n and thus

$$\delta^{(M+1)} \le 0.9\delta^{(M)} + \frac{c_2}{M}, \qquad M \ge M_0. \tag{41}$$

It follows by induction that $\delta^{(M)} \le \delta^{(M_0)} + 10c_2$, $M \ge M_0$, and thus $\delta := \limsup_{M\to\infty} \delta^{(M)} < \infty$. Furthermore, (41) implies $\delta \le 0.9\delta$, and consequently $\delta = 0$.

Proof: [Proof of Theorem 5.5] For any $M, N \ge 1$, it follows from Lemma 5.2 that $Q(B_n^{(M)}) - Q(B_n^{(N)}) \xrightarrow{\text{d}} Q(B_\infty^{(M)}) - Q(B_\infty^{(N)})$ as $n \to \infty$, and thus by Fatou's lemma

$$\mathbb{E}\,|Q(B_\infty^{(M)}) - Q(B_\infty^{(N)})| \le \liminf_{n\to\infty} \mathbb{E}\,|Q(B_n^{(M)}) - Q(B_n^{(N)})| \le \delta^{(M)} + \delta^{(N)}.$$

It follows from Lemma 5.6 that $B_\infty^{(M)}$, $M \ge 1$, is a Cauchy sequence in L^1, and thus this sequence converges to a limit, which can be written $Q(B_\infty)$.

This proves the first assertion and the third follows from this and Lemmas 5.2 and 5.6 by a standard 3ε argument, see e.g. [4, Theorem 4.2].

For (38), we observe again that by (9)

$$Q(B_\infty^{(M+1)}) = Q(L_\infty^{(M)}) + 1 + \frac{1 - Q(L_\infty^{(M)}) + Q(R_\infty^{(M)})}{N(L_\infty^{(M)}) + 1}. \tag{42}$$

As $M \to \infty$, the left hand side converges to $Q(B_\infty)$ in L^1 and thus in probability by the first part of the theorem. Similarly, conditioned on $|L_\infty| = \infty$, $Q(L_\infty^{(M)}) \to Q(L_\infty)$ in L^1 and thus in probability, since the conditional distribution of L_∞ given that it is infinite equals the distribution of B_∞. On the other hand, conditioned on $|L_\infty| < \infty$, obviously $Q(L_\infty^{(M)}) \to Q(L_\infty)$ a.s.. Combining the two cases, $Q(L_\infty^{(M)}) \xrightarrow{\text{p}} Q(L_\infty)$. Similarly $Q(R_\infty^{(M)}) \xrightarrow{\text{p}} Q(R_\infty)$, while $N(L_\infty^{(M)}) \xrightarrow{\text{p}} N(L_\infty) \leq \infty$ is evident. Letting $M \to \infty$ in (42) thus yields (38).

We can develop (38) further. First, L_∞ is infinite with probability 1/2. In this case, $N(L_\infty) = \infty$ and (38) reduces to

$$Q(B_\infty) = Q(L_\infty) + 1. \tag{43}$$

Moreover, the conditional distribution of L_∞ given $|L_\infty| = \infty$ equals the unconditional distribution of B_∞.

The other possibility is L_∞ finite; in this case R_∞ is infinite and its (conditional) distribution equals the unconditional distribution of B_∞, while L_∞ has the same distribution as B_*. We rewrite (38) as

$$Q(B_\infty) = \beta(L_\infty) + \alpha(L_\infty)Q(R_\infty) \tag{44}$$

where, for a finite binary tree B,

$$\alpha(B) := \frac{1}{N(B)+1}$$
$$\beta(B) := \frac{N(B)Q(B)+N(B)+2}{N(B)+1} = \frac{W(B)+N(B)+2}{N(B)+1}.$$

We can combine (43) and (44) into the following fixed point equation.

Theorem 5.7. *The limit random variable $\mathbf{Q} = Q(B_\infty)$ in Theorems 2.2 and 3.1 satifies the fixed point equation $\mathbf{Q} \overset{\text{d}}{=} A\mathbf{Q} + B$, where (A, B) is independent of $\mathbf{Q}$ and has the distribution given by*

$$(A,B) \overset{\text{d}}{=} \begin{cases} (1,1), & \eta = 0, \\ \big(\alpha(B_*), \beta(B_*)\big), & \eta = 1, \end{cases}$$

where $\eta \sim \mathrm{Bi}(1, 1/2)$ and B_* are independent. □

We can obtain a slightly simpler fixed point equation if we follow the leftmost branch of B_∞ until we find a vertex v with a finite left subtree, i.e. until the infinite path makes its first right turn. (In the branching process construction above, we continue until the left child is mortal.) Let $\zeta \geq 0$ be the height of v, and denote its left and right subtrees by L and R. Then, ζ, L and R are independent; ζ has a geometric distribution $\mathrm{Ge}(1/2)$; L is finite and $L \overset{\text{d}}{=} B_*$; and R is infinite and $R \overset{\text{d}}{=} B_\infty$. Applying (43) ζ times followed by (44), we find

$$Q(B_\infty) = \beta(L) + \alpha(L)Q(R) + \zeta.$$

This yields the following alternative fixed point equation.

Theorem 5.8. *The limit random variable* $\mathbf{Q}$ *in Theorems 2.2 and 3.1 satifies the fixed point equation* $\mathbf{Q} \overset{\mathrm{d}}{=} A'\mathbf{Q} + B'$, *where* (A', B') *is independent of* $\mathbf{Q}$ *and has the distribution given by*

$$(A', B') \overset{\mathrm{d}}{=} (\alpha(B_*), \beta(B_*) + \zeta),$$

where $\zeta \sim \mathrm{Ge}(1/2)$ *and* B_* *are independent.* □

Corollary 5.9. *The limit of* $\mathbb{E}\, Q(F_n)$ *and* $\mathbb{E}\, Q(B_n)$ *is given by*

$$\begin{aligned}\mathbb{E}\,\mathbf{Q} = \mathbb{E}\, Q(B_\infty) &= \frac{\mathbb{E}\, B'}{1 - \mathbb{E}\, A'} = \frac{\mathbb{E}\,\beta(B_*) + 1}{1 - \mathbb{E}\,\alpha(B_*)} \\ &= \frac{(2 - \mathbb{E}\,\alpha(B_*))\,\mathbb{E}\, Q(B_*) + 1}{1 - \mathbb{E}\,\alpha(B_*)} = \mathbb{E}\, Q(B_*) + \frac{\mathbb{E}\, Q(B_*) + 1}{1 - \mathbb{E}\,\alpha(B_*)}.\end{aligned}$$

Proof: Taking expectations in Theorem 5.8 we find $\mathbb{E}\,\mathbf{Q} = \mathbb{E}\, A' \,\mathbb{E}\,\mathbf{Q} + \mathbb{E}\, B'$, which yields the second inequality, and the third follows by the definitions of A' and B', since $\mathbb{E}\,\zeta = 1$. (Theorem 5.7 leads to the same result.)

Next, we argue as above for the finite random tree B_* too. In this case, the tree is empty and $Q(B_*) = 0$ with probability 1/2, and otherwise $Q(B_*) = \beta(L_*) + \alpha(L_*)Q(R_*)$, where L_* and R_* are independent with the same distribution as B_*. This can be written, in analogy with Theorem 5.7, $Q(B_*) \overset{\mathrm{d}}{=} A_0 Q(B_*) + B_0$, where (A_0, B_0) is independent of $Q(B_*)$ and has the distribution given by

$$(A_0, B_0) \overset{\mathrm{d}}{=} \begin{cases} (0,0), & \eta = 0, \\ (\alpha(B_*), \beta(B_*)), & \eta = 1, \end{cases}$$

where $\eta \sim \mathrm{Bi}(1, 1/2)$ as above is independent of B_*. Taking expectations we find

$$\mathbb{E}\, Q(B_*) = \frac{\mathbb{E}\, B_0}{1 - \mathbb{E}\, A_0} = \frac{\mathbb{E}\,\beta(B_*)}{2 - \mathbb{E}\,\alpha(B_*)}$$

or $\mathbb{E}\,\beta(B_*) = (2 - \mathbb{E}\,\alpha(B_*))\,\mathbb{E}\, Q(B_*)$, and the result follows.

Note that the variables A, B, A' and B' are discrete and take only rational values; for example, A' takes the values $\{1/k\}_{k=2}^{\infty}$, while B and B' are unbounded. (We do not know whether the range B and B' is the set of all nonnegative rational numbers.) Since B_* and the auxiliary variables η and ζ only take countably many values, with explicitly given probabilities, the distributions of these variables, and in particular their expectations, can in principle be determined numerically with arbitrary accuracy. In practice, the slow convergence of $\mathbb{P}(|B_*| > n)$ to zero together with the exponential growth of the number of trees of a given size may make it difficult to attain high precision.

We have found, using `Maple`, the estimates $\mathbb{E}\,\alpha(B_*) = \mathbb{E}(1 + N(B_*))^{-1}) \doteq 0.318$, $\mathbb{E}\,\beta(B_*) \doteq 2.9$, and $\mathbb{E}\, Q(B_*) \doteq 1.7$, which yields $\mathbb{E}\,\mathbf{Q} \doteq 5.7$; we have no sharp rigorous error bounds, however, so these values should not be taken as absolute truths.

The fixed point equations imply further some qualitative properties of $\mathbf{Q}$.

Theorem 5.10. *The limit random variable* $\mathbf{Q}$ *has a continuous distribution with support* $[3, \infty)$.

Remark 5.11. Although A and B (and A' and B') are discrete, $\mathbf{Q}$ is continuous. Indeed, this is very general, and the proof below uses only $A' \neq 0$ a.s.. However, we have not been able to resolve whether $\mathbf{Q}$ is absolutely continuous, although it seems very plausible. Note that singular distributions may occur in this type of fixed point equations. For example, $A = 1/3$ and $B \sim \mathrm{Bi}(1, 1/2)$ yields the Cantor measure (up to a scale factor).

Proof: Let $p(x) := \mathbb{P}(\mathbf{Q} = x)$ and suppose that $p(x) > 0$ for some x. Let $p_0 := \sup_x p(x) > 0$. It is easily seen that this supremum is attained, since $\sum_x p(x) \leq 1$, so we can choose x with $p(x) = p_0$. By Theorem 5.8,

$$p_0 = \mathbb{P}(A'\mathbf{Q} + B' = x) = \mathbb{E}\big(\mathbb{P}(\mathbf{Q} = (x - B')/A')\big) = \mathbb{E}\,p\big((x - B')/A'\big).$$

Since $p(y) \leq p_0$, this is possible only if $p\big((x - B')/A'\big) = p_0$ for all values of A' and B', but this implies that $p(y) = p_0$ for infinitely many values of y, which contradicts $\sum_y p(y) \leq 1$. Hence $p(x) = 0$ for every x, i.e., the distribution of $\mathbf{Q}$ is continuous.

Next, it is easily shown by (7) and (8) and induction that for any finite binary tree B,

$$W(B) \geq 2N(B) - 2.$$

Consequently,

$$B' \geq \beta(B_*) \geq \frac{3N(B_*)}{N(B_*) + 1}.$$

Hence, for any $\varepsilon > 0$, again using Theorem 5.8,

$$\begin{aligned} \mathbb{P}(\mathbf{Q} < 3 - \varepsilon) &= \mathbb{P}(A'\mathbf{Q} + B' < 3 - \varepsilon) \leq \mathbb{P}\Big(\frac{\mathbf{Q} + 3N(B_*)}{N(B_*) + 1} < 3 - \varepsilon\Big) \\ &= \mathbb{P}\big(\mathbf{Q} < 3 - (N(B_*) + 1)\varepsilon\big) \leq \mathbb{P}(\mathbf{Q} < 3 - 2\varepsilon). \end{aligned}$$

Evidently, this implies $\mathbb{P}(\mathbf{Q} < 3 - \varepsilon) = 0$ for every $\varepsilon > 0$, and thus $\mathbf{Q} \geq 3$ a.s..

Conversely, let E be the support of the distribution of $\mathbf{Q}$; by definition, E is closed. It follows from the fixed point equation that if $x \in E$ and $\mathbb{P}\big((A', B') = (a, b)\big) > 0$, then $ax + b \in E$. In particular, taking $B_* = \emptyset$ which yields $\alpha(B_*) = 1/2$ and $\beta(B_*) = 3/2$, we find

$$x \in E \implies (x + 3)/2 + n \in E \quad \text{for every integer } n \geq 0. \tag{45}$$

Starting with any $x \in E$, taking $n = 0$ and iterating (45), we find in the limit $3 \in E$. Taking $x = 3$ in (45), we find $3 + n \in E$ for every $n \geq 0$. Finally, again taking $n = 0$ in (45), we find by induction on k, that E contains every dyadic rational $3 + m2^{-k}$ with $m, k \geq 0$. Since E is closed, $E \supseteq [3, \infty)$.

Remark 5.12. Although B_* is finite, its size has infinite expectation. Indeed, for every critical branching process, the expected size of each generation is the same, in this case $1/2$; this follows also from the fact that each of the 2^k possible vertices at height k appears with probability 2^{-k-1}.

In B_∞, there is at height k one immortal and on the average $1/2$ mortal in each of the k finite branches descending from the k immortals closer to the root. Hence the expected number of vertices at height k is $k/2 + 1$ and, by symmetry, each of the 2^k possible vertices appears with probability $(k+2)2^{-k-1}$. This illustrates that the infinite tree B_∞ is sparse and stringy.

As a further illustration, consider the intersection of two independent copies of B_∞; the expected size is $\sum_{k=0}^{\infty} 2^k(k+2)^2 2^{-2k-2} = 11/2$. (This can also be seen by considering the two independent two-type branching processes generating the trees as a single branching process with 4 types representing the common vertices and the pairs of types there. We leave the details as an exercise.) Hence, two independent random large binary trees have on the average close to 5.5 vertices in common.

For the finite trees B_n, and more generally for any conditioned Galton–Watson trees with finite offspring variance, it is known that the bulk of the vertices have heights of the order $\sqrt{n}$; see e.g. [1], [2] and [14] for much more detailed results.

6 Back to the forest

The results on binary trees in Section 5 can be translated to results on forests by the correspondence discussed in Section 3, which extends to infinite forests and binary trees. Note that the number of trees in a forest equals the number of vertices in the rightmost branch of the corresponding binary tree. Again, we begin with some simple, more or less well-known observations.

We let F_* be the (finite) random forest corresponding to the random binary tree B_*. The construction of B_* in Section 5 shows that the number of vertices in the rightmost branch has the geometric distribution Ge(1/2). Consequently, the number of trees in F_* is Ge(1/2). Similarly, the number of children of any vertex is Ge(1/2), and all these numbers are independent. Consequently, F_* is a Galton–Watson forest obtained from a Galton–Watson process with Ge(1/2) initial individuals (roots) and offspring distribution Ge(1/2). Note that this, too, is a critical Galton–Watson process.

Equivalently, if T_* is the Galton–Watson tree with offspring distribution Ge(1/2), then T_* equals F_* with all components joined to a common added root; conversely, $F_* = T_*^*$.

It follows immediately that $|F_*| = |B_*| \overset{\mathrm{d}}{=} \xi$ and $|T_*| = |F_*| + 1 \overset{\mathrm{d}}{=} \xi + 1$, that F_* conditioned on $|F_*| = n$ has the distribution of F_n, and that T_* conditioned on $|T_*| = n$ has the distribution of T_n.

Similarly, let F_∞ be the random infinite forest corresponding to B_∞, and let T_∞ be the random infinite tree obtained by adding a root to F_∞; thus $F_\infty = T_\infty^*$. We can decompose the rightmost branch of B_∞ into the part belonging to the infinite path, which has $1 + \mathrm{Ge}(1/2)$ vertices, and the part after it, which is independent of the first part and has the same distribution as the rightmost branch in B_*, i.e. it has Ge(1/2) vertices. Hence, if $\widehat{\zeta}$ is the total number of vertices in the rightmost branch of B_∞, then $\widehat{\zeta} = 1 + \zeta + \zeta'$, where ζ and ζ' are independent and Ge(1/2). (In the equivalent branching process construction, the rightmost branch has $1+\zeta$ immortal and ζ' mortal vertices.) It follows that $\widehat{\zeta}$, which also is the number of components in F_∞ and the degree of the root in T_∞, has a shifted negative binomial distribution,

$$\mathbb{P}(\widehat{\zeta} = k) = k2^{-k-1}, \qquad k = 1, 2, \ldots; \tag{46}$$

this is the size-biased distribution $\widehat{\mathrm{Ge}}(1/2)$.

Using the branching process construction of B_∞, exposing first the rightmost branch, then the rightmost branches in the left subtrees sprouting from it, and so on, it is now easily seen that T_∞ is the tree produced by the size-biased Galton–Watson process defined in Remark 5.4 with the offspring distribution Ge(1/2) for the mortals, and thus $\widehat{\mathrm{Ge}}(1/2)$ for the immortals. F_∞ is obtained by chopping off the root of T_∞, or by starting with $\widehat{\mathrm{Ge}}(1/2)$ individuals (roots), one of them immortal.

Note that F_∞ and T_∞ are locally finite and have exactly one infinite path (the immortals). The equation $\widehat{\zeta} = \zeta + 1 + \zeta'$ above shows that F_∞ and T_∞ also can be constructed by starting with an infinite path (the trunk) and adding to each vertex in it a Ge(1/2) number of branches to each side, each branch being an independent copy of T_*; for F_∞ we further add a Ge(1/2) number of copies of T_* on each side of the infinite component as separate components. (All random choices should be independent.)

It is easily seen that Lemma 5.2 implies the corresponding statements for forests and trees. (This is another instance of the general result given in Remark 5.4.) Note, however, that the truncation $F_\infty^{(M)}$ does not correspond to the truncation $B_\infty^{(M)}$; it corresponds to $B_\infty^{[M]}$, where we let $B^{[M]}$ denote the binary tree B with each branch truncated after M steps to the left. (Note that $B_\infty^{[M]}$ a.s. is a finite tree.) We give a formal statement.

Lemma 6.1. *As $n \to \infty$, $F_n \xrightarrow{\mathrm{d}} F_\infty$ and $T_n \xrightarrow{\mathrm{d}} T_\infty$ in the sense of finite-dimensional distributions, in the sense $F_n^{(M)} \xrightarrow{\mathrm{d}} F_\infty^{(M)}$ and $T_n^{(M)} \xrightarrow{\mathrm{d}} T_\infty^{(M)}$ for every finite M.*

Proof: Fix $M \geq 0$. Lemma 5.2 implies that for each fixed finite binary tree b, $\mathbb{P}(B_n^{[M]} = b) \to \mathbb{P}(B_\infty^{[M]} = b)$, and thus $B_n^{[M]} \xrightarrow{\mathrm{d}} B_\infty^{[M]}$.

Using the correspondence between forests and binary trees, we now define Q for infinite forests too; thus $Q(F_\infty) = Q(B_\infty)$, and the limit $\mathbf{Q}$ in Theorem 2.2 can be taken as $Q(F_\infty)$. The following theorem shows that Q can be defined (a.s.) directly on infinite forests without our use of binary forests as a convenient technical tool.

Theorem 6.2. *There exists an extension of Q to infinite forests such that we have $\mathbb{E}\,|Q(F_\infty^{(M)}) - Q(F_\infty)| \to 0$ as $M \to \infty$. We have $Q(F_n) \xrightarrow{\mathrm{d}} Q(F_\infty)$ as $n \to \infty$. Furthermore, $Q(F_\infty) = Q(B_\infty)$ when B_∞ corresponds to F_∞.*

Proof: It remains only to prove that $\mathbb{E}\,|Q(F_\infty^{(M)}) - Q(F_\infty)| \to 0$, or equivalently, transfering to binary trees again, that $\mathbb{E}\,|Q(B_\infty^{[M]}) - Q(B_\infty)| \to 0$.

Let, for $M \geq 1$,

$$\delta^{[M]} := \sup_n \mathbb{E}\,|Q(B_n^{[M]}) - Q(B_n)|.$$

The proof of Lemma 5.6 shows with minor modifications that $\delta^{[M]} \to 0$; note that $B_n^{[M+1]}$ has the subtrees $L_n^{[M]}$ and $R_n^{[M+1]}$, but this causes no significant problem. Moreover, we now need a preliminary step to ensure that $\delta^{[M]} < \infty$; this is easily done using induction, since (9) implies $Q(B) < Q(L) + 2 + \frac{1}{2}Q(R)$, and we omit the details.

For every M we have as $n \to \infty$, see the proof of Lemma 6.1, $B_n^{[M]} \xrightarrow{d} B_\infty^{[M]}$. Moreover, since $B^{(M)}$ is a truncation of $B^{[M]}$, we have joint convergence of $(B_n^{[M]}, B_n^{(M)})$ to $(B_\infty^{[M]}, B_\infty^{(M)})$, and consequently

$$Q(B_n^{[M]}) - Q(B_n^{(M)}) \xrightarrow{d} Q(B_\infty^{[M]}) - Q(B_\infty^{(M)}).$$

Since $\mathbb{E}\,|Q(B_n^{[M]}) - Q(B_n^{(M)})| \le \delta^{[M]} + \delta^{(M)}$ for each n, Fatou's lemma yields

$$\mathbb{E}\,|Q(B_\infty^{[M]}) - Q(B_\infty^{(M)})| \le \delta^{[M]} + \delta^{(M)},$$

which tends to 0 as $M \to \infty$ by Lemma 5.6 and the claim above. Finally, the triangle inequality and Theorem 5.5 yields

$$\mathbb{E}\,|Q(F_\infty^{(M)}) - Q(F_\infty)| = \mathbb{E}\,|Q(B_\infty^{[M]}) - Q(B_\infty)| \to 0, \qquad M \to \infty. \qquad \square$$

Remark 6.3. We saw above that the number of components of the infinite random forest F_∞ has the shifted negative binomial distribution in (46); hence, by Lemma 6.1, the number of components of the random forest F_n has asymptotically this distribution. It is easy to find the exact distribution for finite n as follows. The generating function for ordered trees is $zB(z)$, with $B(z)$ given in (13), and thus the generating function for ordered forests with k components is $z^k B(z)^k$. It follows as an easy exercise, using e.g. [6, (5.70)], that, with $n^{\underline{k}}$ denoting the falling factorial,

$$\mathbb{P}(F_n \text{ has } k \text{ components}) = \frac{\frac{k}{2n-k}\binom{2n-k}{n}}{C_n} = k\frac{(n+1)^{\underline{k+1}}}{(2n)^{\underline{k+1}}}.$$

This evidently converges to $k2^{-k-1}$ as $n \to \infty$, as asserted above.

Remark 6.4. There is a well-known correspondence between (random) trees and (random) walks on the non-negative integers by means of the depth first walk, see e.g. [1]. In this context, several nice results are known for the random trees studied here.

The random tree T_n corresponds to a simple random walk of length $2n$ conditioned on returning to 0 at the end but not before (sometimes called Dyck paths). The random tree T_* corresponds to a simple random walk stopped at its first return to 0 [12]. For the infinite tree T_∞, the depth first walk only captures the structure on one side of the infinite trunk; the other side is described by a depth first walk running in the opposite direction. The two depth first walks are independent, and each is a biased random walk which is a discrete version of the three-dimensional Bessel process, see Le Gall [12].

References

[1] D. Aldous (1991), *The continuum random tree II: an overview.* In Proc. Durham Symp. Stoch. Analysis 1990, Cambridge Univ. Press, 23–70.

[2] D. Aldous (1993), *The continuum random tree III.* Ann. Probab. 21, no. 1, 248–289.

[3] K.B. Athreya & P.E. Ney (1972), *Branching Processes.* Springer-Verlag, Berlin, 1972.

[4] P. Billingsley (1968), *Convergence of Probability Measures.* Wiley, New York.

[5] T.E. Harris (1951), *Some mathematical models for branching processes.* In Second Berkeley Symposium on Mathematical Statistics and Probability, Univ. California Press, Berkeley, 305–328.

[6] R.L. Graham, D.E. Knuth & O. Patashnik (1994), *Concrete Mathematics.* 2nd ed., Addison–Wesley, Reading, Mass.

[7] D.P. Kennedy (1975), *The Galton–Watson process conditioned on the total progeny.* J. Appl. Probab. 12, 800–806.

[8] M. Klazar (1997), *Twelve countings with rooted plane trees.* European J. Combin. 18, no. 2, 195–210. *Addendum:* European J. Combin. 18, no. 6, 739–740.

[9] D.E. Knuth (1997), *The Art of Computer Programming. Vol. 1: Fundamental algorithms.* 3rd ed., Addison–Wesley, Reading, Mass.

[10] D.E. Knuth, `KODA-RUSKEY`. CWEB program. Available from `http://Sunburn.Stanford.EDU/~knuth/programs/koda-ruskey.w`

[11] Y. Koda & F. Ruskey (1993), *A Gray code for the ideals of a forest poset.* J. Algorithms 15, no. 2, 324–340.

[12] J.-F. Le Gall (1986), *Une approche élémentaire des théorèmes de décomposition de Williams.* In Séminaire de Probabilités, XX, 1984/85, Lecture Notes in Math., 1204, Springer, Berlin, 447–464.

[13] R. Lyons, R. Pemantle & Y. Peres (1995), *Conceptual proofs of* $L \log L$ *criteria for mean behavior of branching processes.* Ann. Probab. 23, no. 3, 1125–1138.

[14] J.-F. Marckert & A. Mokkadem (2001), *The depth first processes of Galton–Watson trees converge to the same Brownian excursion.* To appear in Ann. Probab.

[15] S.T. Rachev (1991), *Probability metrics and the stability of stochastic models.* Wiley, Chichester, U.K.

[16] S.T. Rachev & L. Rüschendorf (1995), *Probability metrics and recursive algorithms,* Adv. Appl. Probab. 27, 770–799.

[17] U. Rösler(1991), *A limit theorem for "Quicksort".* RAIRO Inform. Théor. Appl. 25, no. 1, 85–100.

[18] U. Rösler(1992), *A fixed point theorem for distributions.* Stochastic Process. Appl. 42, no. 2, 195–214.

[19] U. Rösler & L. Rüschendorf (2001), *The contraction method for recursive algorithms.* Algorithmica 29, 3-33.

[20] F. Ruskey (1981), *Listing and counting subtrees of a tree.* SIAM J. Comput. 10, no. 1, 141–150.

Svante Janson

Department of Mathematics
Uppsala University
PO Box 480
S-751 06 Uppsala, Sweden
svante.janson@math.uu.se
http://www.math.uu.se/~svante/

Trends in Mathematics, © 2002 Birkhäuser Verlag Basel/Switzerland

On Random Walks in Random Environment on Trees and Their Relationship with Multiplicative Chaos

Mikhail Menshikov and Dimitri Petritis

ABSTRACT: *The purpose of this paper is to report on recent results concerning random walks in a random environment on monochromatic and coloured trees and their relationship with multiplicative chaos. The proofs are omitted since they are extensively given elsewhere [12]. It is worth noticing that for the random walk on monochromatic tree the results we give were previously known [11]; we provide however a totally new proof, based solely on multiplicative chaos results, that allows to relax some stringent conditions on independence properties of the random transition probabilities. For the random walk on a coloured tree the results are new; the classification of the asymptotic behaviour of the random walk allows to obtain some hints for the classification of the yet unsolved corresponding multiplicative chaos problem.*

1 Notation

Let d be a fixed non-negative integer. We consider the rooted regular tree of order d, *i.e.* a connected graph without loops with a denumerable set of vertices $\mathbb{V}$ and a denumerable set of non oriented edges $\mathbb{A}(\mathbb{V})$. There is a distinguished vertex called the root that has degree d; all other vertices have degree $d+1$. Vertices are completely determined by giving their genealogical history from their common ancestor, the root; hence they are bijectively indexed by the set of sequences of arbitrary length over an alphabet of d letters. We use the same symbol for the indexing set so that $\mathbb{V} = \cup_{n=0}^{\infty}\mathbb{V}_n$ with $\mathbb{V}_0 = \{\emptyset\}$ and $\mathbb{V}_n = \{v = (v_1, \ldots, v_n) : v_i \in \{1, \ldots, d\}, i = 1 \ldots n\}$ for $n \geq 1$. For every $v \in \mathbb{V}$, we denote $|v|$ the length of the path from v to the root *i.e.* the number of edges encountered. For $v \in \mathbb{V}$ and $k \leq |v|$ we denote by $v|_k$ the truncation of the sequence v to its k first elements, *i.e.* if $v = (v_1 \ldots v_n) \in \mathbb{V}_n$ and $k \leq n$, then $v|_k = (v_1 \ldots v_k) \in \mathbb{V}_k$; the symbol $v|_k$ must not be confused therefore with v_k, representing the letter appearing at the k-th position of the sequence. For $0 \leq k < \ell \leq |v|$ we denote $v|_\ell^k$ the subsequence of length $\ell - k$ defined by $v|_\ell^k = (v_{k+1}, \ldots, v_\ell)$. If $u \in \mathbb{V}$, we write $u \leq v$ if $|u| \leq |v|$ and $v = (u_1, \ldots u_{|u|}, v_{|u|+1}, \ldots v_{|v|})$ *i.e.* if u is the initial sequence of v; we write $u < v$ when $u \leq v$ and $|u| < |v|$. Similarly for every sequence u and any letter $\ell \in \{1 \ldots d\}$, the sequence $u\ell$ will have length $|u|+1$ and last letter ℓ.

Edges are unordered pairs $\langle u, v \rangle$ of adjacent vertices u and v. Since every vertex has an unique ancestor, every edge is uniquely defined by its most remote vertex. Hence, every vertex $v \in \mathbb{V}\setminus\{\emptyset\}$ defines an edge $a(v) = \langle\, v|_{|v|-1}, v\,\rangle$. Edges are thus also indexed by the set $\mathbb{V}$, more precisely by $\overset{\circ}{\mathbb{V}} = \mathbb{V} \setminus \{\emptyset\}$ and we denote $a(v)$ the edge defined by v; therefore $\mathbb{A}(\mathbb{V}) \simeq \overset{\circ}{\mathbb{V}}$.

If $u, v \in \mathbb{V}$ and $u < v$, we denote $[u; v]$ the (unique) path from u to v *i.e.* the collection of edges $(a_1, a_2, \ldots)$ with $a_j \equiv a(v|_{|u|+j})$, for $j = 1, \ldots, |v|$. For every $u \in \mathbb{V}$, the symbol $[u; u]$ denotes an empty set of edges. If u and v are not comparable vertices, *i.e.* neither $u \leq v$ nor $v \leq u$ holds, although there is a canonical way to define the path $[u; v]$, this definition is not necessary in the present paper and hence omitted. We write simply $[v]$ to denote the path joining the root to v, namely $[\emptyset; v]$.

At every edge a we assign a number $\xi_a \in [0, \infty[$ in some specific manner. This specification differs from model to model and since various models are considered here, we don't wish to be more explicit about these variables at the present level. Mind however that the numbers $(\xi_a)_{a \in \mathbb{A}(\mathbb{V})}$ are random variables neither necessarily independent nor necessarily equi-distributed. For the time being, we only assume that we dispose of a specific collection $(\xi_a)_{a \in \mathbb{A}}$, called the *edge-environment.*

2 Multiplicative chaos

Let $(\mathbb{V}, (\xi_a)_{a \in \mathbb{A}(\mathbb{V})})$ be a given tree and a given edge environment. For $u, v \in \mathbb{V}$, with $u < v$ we denote

$$\xi[u; v] = \prod_{a \in [u;v]} \xi_a$$

the product of environment values encountered on the path of edges from u to v; the symbol $\xi[v]$ is defined to mean $\xi[\emptyset; v]$ and $\xi[v; v]$ — as a product over an empty set — is consistently defined to be 1. It is not necessary for the purpose of the present article to define the value of $\xi[u; v]$ when u and v are not comparable.

For every $u \in \mathbb{V}$, we consider the process $Y_n(u)_{n \in \mathbb{N}}$ defined by $Y_0(u) = 1$ and

$$Y_n(u) = \sum_{v \in \mathbb{V}_{n+|u|}; v > u} \xi[u; v] = \sum_{v \in \mathbb{V}_{n+|u|}; v > u} \prod_{a \in [u;v]} \xi_a,$$

for $n \geq 1$. This process is known as the *multiplicative chaos process.* Notice that even when $(\xi_a)_{a \in \mathbb{A}}$ is a family of independent random variables, the random variables $\xi[u; v]$ are not independent for v scanning the set $\mathbb{V}_{n+|u|}$. Hence the asymptotic behaviour of $Y_n(u)$ when $n \to \infty$ is far from trivial and it is studied for several particular cases of dependences of the family (ξ_a) in an extensive literature; see for instance [7, 5, 6, 3, 8, 9, 14, 2, 10].

The study of the asymptotics of the process (Y_n) is done by various techniques:

1. If the limit $\lim_{n \to \infty} Y_n(u) \stackrel{\mathrm{d}}{=} Y(u)$ exists in distribution for all $u \in \mathbb{V}$ then it must verify the functional equation

$$Y(u) \stackrel{\mathrm{d}}{=} \sum_{w \in \mathbb{V}_{|u|+1}; w > u} \xi[u; w] \, Y(w). \tag{1}$$

   The process $(Y_n(u))_n$ and the corresponding functional equation (1) are thoroughly studied in the literature for some particular choices of dependencies of the family (ξ_a).

2. A second technique of study of the asymptotics is by martingale analysis. If for any fixed $u \in \mathbb{V}$, $(\mathcal{F}_n^{(u)})$ denotes the natural filtration $\mathcal{F}_k^{(u)} = \sigma(\xi_{a(uv_1\ldots v_k)}, v_i \in \{1,\ldots,d\}, i = 1,\ldots,k\}$ for $k \in \mathbb{N}$, we have

$$\mathbb{E}(Y_n(u)|\mathcal{F}_{n-1}^{(u)}) = \sum_{v_1,\ldots,v_n \in \{1,\ldots,d\}} \xi_{a(uv_1)} \cdots \xi_{a(uv_1\ldots v_{n-1})} \mathbb{E}(\xi_{a(uv_1v_2\ldots v_n)}|\mathcal{F}_{n-1}^{(u)})$$

and in the special case where the distribution of $\xi_{a(uv_1v_2\ldots v_n)}$ depends solely on v_n and the random variables are independent for different generations, the previous formula simplifies into

$$\mathbb{E}(Y_n(u)|\mathcal{F}_{n-1}^{(u)}) = Y_{n-1}(u) \sum_{v_n=1}^{d} \mathbb{E}(\xi_{a(uv_1v_2\ldots v_n)}).$$

Although the process (Y_n) is thoroughly studied, the closely related process

$$Z_n(u) = \sum_{k=0}^{n} Y_k(u) \quad \text{for} \quad n \geq 0$$

does not seem — to the best of our knowledge — to have attracted much attention. However, if we are interested in connections between multiplicative chaos and random walks in random environment on a tree, it is this latter process that naturally appears in both subjects.

3 Nearest neighbours random walk on a tree in an inhomogeneous environment

To every vertex $u = (u_1,\ldots,u_{|u|}) \in \mathring{\mathbb{V}}$ are assigned $d+1$ numbers $(p_{u,0}, p_{u,1},\ldots, p_{u,d})$ with $p_{u,0} > 0$, $p_{u,i} \geq 0$ $\forall i = 1,\ldots,d$ and $\sum_{i=0}^{d} p_{u,i} = 1$. To $u \in \mathbb{V}_0 = \{\emptyset\}$ are assigned only d numbers $(p_{\emptyset,1},\ldots p_{\emptyset,d})$ with $p_{\emptyset,i} \geq 0$ $\forall i = 1,\ldots,d$ and $\sum_{i=1}^{d} p_{\emptyset,1} = 1$. These numbers will be random variables with some specific dependence properties that will be defined later. These numbers stand for transition probabilities of a reversible Markov chain $(X_n)_{n\in\mathbb{N}}$ on the tree verifying for $|u| \geq 1$

$$P_{u,v} = \mathbb{P}(X_{n+1} = v|X_n = u) = \begin{cases} p_{u,0} & \text{if } v = u|_{|u|-1} \\ p_{u,v_{|v|}} & \text{if } u = v|_{|v|-1} \\ 0 & \text{otherwise.} \end{cases}$$

For $u = (\emptyset)$ we have the slightly modified transition probabilities

$$P_{\emptyset,v} = \mathbb{P}(X_{n+1} = v|X_n = (\emptyset)) = \begin{cases} p_{\emptyset,v_1} & \text{if } v \in \mathbb{V}_1 \\ 0 & \text{othrewise.} \end{cases}$$

For $u \in \mathbb{V}$ with $|u| \geq 2$ we consider the edge $a(u) = \langle\ u|_{|u|-1}, u\ \rangle$ and attach to this edge the variable

$$\xi_{a(u)} = \frac{p_{u|_{|u|-1},u_{|u|}}}{p_{u|_{|u|-1},0}} \in [0,\infty[.$$

For $u \in \mathbb{V}_1$ we attach $\xi_{a(u)} = p_{\emptyset,u_1}$. One can easily check the validity of the following

Lemma 3.1. *For every $v \in \mathbb{V}$ define the variable*

$$\pi[v] = \begin{cases} \pi[\emptyset]\, \xi[v] \frac{1}{p_{v,0}} & \text{if } v \in \overset{\circ}{\mathbb{V}} \\ \pi[\emptyset] & \text{if } v = (\emptyset), \end{cases}$$

with $\pi[\emptyset]$ an arbitrary constant. Then $\pi[v]$ verifies the stationarity condition

$$\sum_{v \in \mathbb{V}} \pi[v] P_{v,v'} = \pi[v'], \quad \forall v' \in \mathbb{V}.$$

To avoid technical difficulties, we assume that

$$\mathbb{E}((p_{v,0})^{-1}) < \infty.$$

Then, apart the factor $\frac{1}{p_{v,0}}$, the expression for the invariant measure $\pi[v]$ involves the product $\xi[v]$ of variables along the edges of the path form $\emptyset$ to v as was the case in the expression of multiplicative chaos.

4 Models covered by the present formalism and main results

We present below a unified treatment of both the multiplicative chaos process and the random walk problem stating in the same theorem the asymptotic behaviour of the limiting chaos process and of the random walk. Several models fit the present formalism; by making appropriate identifications of random variables, the random walk in random environment on $\mathbb{N}$ or the problem of random strings in a random environment can be rephrased in the present language.

4.1 Random walk in a random environment on a regular tree

At every vertex $v \in \overset{\circ}{\mathbb{V}}$ is assigned a $(d+1)$-dimensional random vector with positive components $(p_{v,0}, \ldots, p_{v,d})$ verifying $\sum_{j=0}^{d} p_{v,j} = 1$. For the vertex $v = \emptyset$, the corresponding random vector is d-dimensional and its components verify $\sum_{j=1}^{d} p_{\emptyset,j} = 1$. These random vectors are independent for different v's and, for $v \in \overset{\circ}{\mathbb{V}}$ they have the same distribution. Let $\boldsymbol{\eta} = (\eta_1, \ldots, \eta_d)$, be a vector of non-negative random variables η_i, $i = 1, \ldots, d$, having the same distribution with $p_{v,i}/p_{v,0}$, for $v \in \overset{\circ}{\mathbb{V}}$, with not necessarily independent nor identically distributed components. We assume the law of the random vector is explicitly known with $\mathbb{E}\eta_i < \infty$ and $\mathbb{E}\eta_i \log^+ \eta_i < \infty$, $\forall i = 1, \ldots, d$. Moreover, to avoid technicalities we assume that although the support of the random variables η_i extends up to 0, their law has no atom at 0.

To the edge $a(v)$, having most remote vertex $v \in \overset{\circ}{\mathbb{V}}$, we assign the random variable $\xi_{a(v)}$ having the same distribution as $\eta_{v_{|v|}}$; the variables $\xi_{a(v)}$ and $\xi_{a(v')}$ are independent if $v|_{|v|-1} \neq v'|_{|v'|-1}$. Notice that if the components of the random vector η are not independent, the variables $\xi_{a(v)}$ and $\xi_{a(v')}$ with $|v| = |v'|$ and $v|_{|v|-1} = v'|_{|v'|-1}$ are not independent either.
The results are expressed in terms of the functions

$$f(x) = \mathbb{E}\,(\sum_{i=1}^{d} \eta_i^x), x \in \mathbb{R}^+ \quad \text{and} \quad g(x) = \log f(x),$$

and of the parameter $\lambda = \inf_{x\in[0,1]} f(x)$.

Theorem 4.1. *Let $\lambda = \inf_{x\in[0,1]} f(x)$ and $x_0 \in [0,1]$ be such that $f(x_0) = \lambda$. Then*

1. *If $\lambda < 1$, then almost surely the random walk is positive recurrent and $Z_\infty < \infty$.*
2. *If $\lambda > 1$, then almost surely the random walk is transient, $Y_\infty = \infty$, and $Z_\infty = \infty$.*
3. *If $\lambda = 1$ and moreover $f'(1) < 0$, then almost surely $0 < Y_\infty < \infty$, $Z_\infty = \infty$, and the random walk is null-recurrent.*

This theorem is already formulated, with some stringent conditions on the random variables, in [11]. In [12] a totally new proof of this result is provided, based on the multiplicative chaos results of [8].

4.2 Random walk in a random environment on a coloured tree

This problem is reminiscent of the problem on random strings in a random environement, studied in [4], where non reversible Markov chains on the tree $\mathbb{V}$ are considered and general conditions for transience/null recurrence/ergodicity are given in terms of Lyapuonov exponent of a product of matrices. To describe the problem of random strings in a random environment, we distinguish the d children of every vertex by assigning a colour index, chosen without replacement from the set $\{1,\ldots,d\}$, to each child. The root is assigned an arbitrary colour $\alpha \in \{1,\ldots,d\}$. Consequently, every edge $a(v)$ with $v \in \overset{\circ}{\mathbb{V}}$ is assigned the bicolour $(ij) \in \{1,\ldots,d\}^2$, where $i = v_{|v|-1}$ and $j = v_{|v|}$.
Passing to the edge-indexed ratio of outwards over inwards probabilities, the model can be rephrased to fit the present formalism. Let

$$\eta = \begin{pmatrix} \eta_{11} & \cdots & \eta_{1d} \\ \vdots & & \\ \eta_{d1} & \cdots & \eta_{dd} \end{pmatrix}$$

be a matrix of non-negative random elements of known joint distribution. The matrix elements are not necessarily independent.

Theorem 4.2. *Let* $m(x) = \begin{pmatrix} \mathbb{E}(\eta_{11}^x) & \cdots & \mathbb{E}(\eta_{1d}^x) \\ \vdots & & \\ \mathbb{E}(\eta_{d1}^x) & \ldots & \mathbb{E}(\eta_{dd}^x) \end{pmatrix}$ *for* $x \in [0,1]$. *Assume that the matrix* $m(x)$ *is regular i.e. there exists some integer* N *such that that for every* $x \in [0,1]$, $(m(x)^N)_{ij} > 0 \quad \forall i,j$. *Denote by* $\rho(x)$ *the largest eigenvalue of* $m(x)$ *for* $x \in [0,1]$ *and* $\lambda = \inf_{x \in [0,1]} \rho(x)$.

1. *If* $\lambda < 1$ *the random walk is almost surely positive recurrent and* $Z_\infty < \infty$ *almost surely*
2. *If* $\lambda > 1$ *the random walk is almost surely transient and* $Y_\infty = \infty$ *almost surely.*

5 Some open problems on multiplicative chaos and further development

We demonstrated the close relationship between results on multiplicative chaos and reversible Markov chains. In particular, the most difficult part for the Markov chain problem, namely the critical case $\lambda = 1$ becomes an immediate consequence of the theorem on the existence of non trivial solutions of the functional equation and the uniform integrability of the corresponding martingale, once the conditions for the existence of non trivial solutions are known. This analogy can even be extended on more general settings to include the case of random trees and of general distributions for the environment that correspond to situations much more general than the one considered in [11]. Actually, what plays an important *rôle* is the theorem (1) of [5] but this theorem is properly generalised by Liu [9] to include random number of variables *d*. Therefore, the treatment of random walks in general random environment on random trees becomes accessible by virtue of the results of Liu on multiplicative chaos.

We got conditions under which the chaos processes Y_n and Z_n tend to ∞ or remain finite according to the values of the parametre λ. The precise study of this classification gives rise to a multiplicative chaos functional equation of the type

$$Y^{(\alpha)} \stackrel{\mathrm{d}}{=} \sum_{\beta} \eta_{\alpha\beta} Y'^{(\alpha\beta)}$$

for which the conditions of existence of non trivial solutions are not known. In view of the results on the random walk problems it is expected that the classifying parametre in this problem is the largest eigenvalue of the matrix of moments $m(x)$. This problem is actually under investigation. The above mentioned intuition is confirmed by some preliminary results, by the partial results of [1] and by physical intuition. As a matter of fact the random walk in a random environment can also be viewed as a physical system of spins in a quenched disorder. In the random string problem the quenching is quite stringent so that the Lyapunov's exponent appear. On the contrary, the random walk in random environment on the coloured tree behaves very much like a self-averaging problem.

Other random walk models on more general trees (multiplexed coloured trees) can also be introduced that involve matrix valued multiplicative chaos [13]. Again, classification of the random walk problem can be used as a hint for the classification of the chaos process.

References

[1] F. Ben Nasr. Mesures aléatoires de Mandelbrot associées à des substitutions. *C. Rendus Acad. Sciences (Paris)*, 304:253–258, 1987.

[2] Brigitte Chauvin. Martingales produits et lignes d'arrêt pour un processus de branchement brownien. *C. R. Acad. Sci. Paris Sér. I Math.*, 306(12):499–502, 1988.

[3] P. Collet and F. Koukiou. Large deviations for multiplicative chaos. *Commun. Math. Phys.*, 147:329–342, 1992.

[4] F. Comets, M. Menshikov, and S. Popov. Lyapunov functions for random walks and strings in random environment. *Ann. Probab.*, 26:1433–1445, 1998.

[5] R. Durrett and Th. Liggett. Fixed points of smoothing transformation. *Z. Wahrscheinlichkeitstheorie verw. Gebiete*, 64:275–301, 1983.

[6] Y. Guivarc'h. Sur une extension de la notion semi-stable. *Ann. Inst. H. Poicaré*, 26:261–285, 1990.

[7] J.-P. Kahane and J. Peyrière. Sur certaines martingales de Benoît Mandelbrot. *Adv. Math.*, 22:131–145, 1976.

[8] Q. Liu. Sur une équation fonctionnelle et ses applications: une extension du théorème de Kesten-Stigun concernant les processus de branchement. *Ann. Appl. Probab.*, 29:353–373, 1997.

[9] Q. Liu. Fixed points of a generalised smoothing transformation and applications to the branching random walk. *Ann. Appl. Probab.*, 30:85–112, 1998.

[10] Quansheng Liu and Alain Rouault. Limit theorems for Mandelbrot's multiplicative cascades. *Ann. Appl. Probab.*, 10(1):218–239, 2000.

[11] R. Lyons and R. Pemantle. Random walk in a random environment and first passage percolation on trees. *Ann. Probab.*, 20:125–136, 1991.

[12] M. Menshikov and D. Petritis. Random walks in random environment on trees and multiplicative chaos, preprint 2001 submitted for publication, eprint arXiv:math.PR/0112103.

[13] M. Menshikov, D. Petritis, and S. Popov. Matrix multiplicative chaos and bindeweeds, in preparation 2002.

[14] E. Waymire and S. Williams. A cascade decomposition theory with applications to Markov and exchangeable cascades. *Trans. Amer. Math. Soc.*, 348:585–632, 1996.

Mikhail Menshikov

Department of Mathematical Sciences
University of Durham
South Road
Durham DH1 3LE
United Kingdom
Mikhail.Menshikov@durham.ac.uk

Dimitri Petritis

IRMAR
Université de Rennes I
35042 Rennes Cedex
France
Dimitri.Petritis@univ-rennes1.fr

Trends in Mathematics, © 2002 Birkhäuser Verlag Basel/Switzerland

Note on Exact and Asymptotic Distributions of the Parameters of the Loop-Erased Random Walk on the Complete Graph

Boris Pittel

ABSTRACT: *We study the loop-erased random walk algorithm for generating a random spanning tree of the complete graph on n vertices. The number of moves is shown to be distributed as $n-2$ plus $G_{1/n}$, a Geometric with expectation n. The lengths of the paths (branches) that are added to a subtree are jointly distributed as the consecutive waiting times for heads in a sequence of time-biased, but independent, coin flips. As a corollary, the subtree size is shown to grow, with high probability, at the rate $(rn)^{1/2}$, r being the number of branches added. The lengths of the largest path and the largest loop are shown to scale with $n^{1/2}$ and $(n \log n)^{1/2}$; the limiting distributions are obtained as well.*

Introduction.

Broder [2] and Aldous [1] independently found a random-walk algorithm for generating a uniformly random spanning tree of an undirected (connected) graph. The walk starts at an arbitrary vertex and, at each move, chooses the next vertex uniformly at random among all the neighbors of the current vertex. (Call this walk simple.)The spanning tree is formed by the first edges leading away from the vertices already visited. Wilson [6] discovered a different random-walk algorithm for generating a uniform spanning (rooted) tree. Here is how it works. Pick an arbitrary vertex different from root, and start the simple random walk. If at some moment the walk returns to a vertex v previously visited, then the loop is erased, and the walk continues from v. So what remains of the walk trajectory is always a path, possibly consisting of the starting vertex only. With probability 1, eventually this path hits the root. The path that includes the root becomes a branch (subtree) of the future tree. Recursively, given a current subtree, one picks an arbitrary external vertex and generates the loop-erased random walk until it hits the subtree, and thus gives birth to the next branch. As mentioned in [6], it had been known (Burton and Pemantle [3], Pemantle [5]) that the path from a vertex to the root of a uniform spanning tree is the loop-erased random walk. Still it was Wilson who suggested to use this walk as a macrostep of a provably efficient algorithm to generate a uniform spanning (rooted) tree.

Our goal was to find the asymptotic distributions of the leading parameters of Wilson's algorithm when applied to the complete graph K_n. To our surprise, it turned out possible to obtain exact distributions in some cases. We proved that the number of moves (microsteps) in the walk is distributed as $n-2+G_{1/n}$, where

*Supported by the NSF grant DMS-0104104

G_p denotes the Geometric with parameter p. (Thus the expected number of steps is $2(n-1)$, the fact already established in [6].) More generally, once the subtree reaches size k, $(1 \le k \le n-1)$, the number of remaining microsteps is distributed as $n-k-1+G_{k/n}$. The lengths of the paths (macrosteps) are shown to be jointly distributed as the consecutive waiting times till appearance of Heads in a sequence of $n-1$ independent coin flips, with probability of Heads in j-th flip being equal $(j+1)/n$. This connection shows immediately that the number of macrosteps is, in the limit, Gaussian, with mean and variance $n/2$ and $n/6$ respectively. The subtree size is shown with high probability (whp) to grow at the rate $(rn)^{1/2}$, r being the number of macrosteps so far. And, for $r = o(n^{1/3})$, the subtree sizes up to the r-th macrostep are sharply approximated by the sequence $\{(2n\sum_{j=1}^{s} Z_j)^{1/2}\}_{s\le r}$, where $Z_1, Z_2, \dots$ are independent Exponentials with parameter 1. Finally, using the factorial moments method, we prove that the longest path and the number of loops, each scaled by $n^{1/2}$, and the longest loop scaled by $(n\log n)^{1/2}$, all have the nondegenerate limiting distributions.

To conclude, we mention that Marchal [4] has obtained some determinantal formulas for the joint distribution of the consecutive subtrees for the loop-erased walk on a general graph.

1 Distribution of running time and branches lengths

Let X_n be the total number of branches (paths) added to a current subtree, and let $\mathbf{L} = \{L_j : 1 \le j \le X_n\}$ be the lengths of (the number of edges in) the successive paths (macrosteps). Let Y_n be the total number of moves (microsteps) of the random walk. And let G_x denote the geometrically distributed random variable (Geometric) with parameter $x \in (0,1]$,

$$\mathrm{P}(G_x = j) = x(1-x)^j,\ j \ge 0 \Longrightarrow \mathrm{E}(G_x) = \frac{1}{x}.$$

Theorem 1. *(a) X_n and Y_n are independent. (b) Y_n and $n-2+G_{1/n}$ are equidistributed,*

$$Y_n \stackrel{\mathcal{D}}{\equiv} n-2+G_{1/n}, \tag{1}$$

in short. Let $\boldsymbol{\varepsilon} = (\varepsilon_2, \dots, \varepsilon_n)$ be a sequence of independent Bernoulli variables, $P(\varepsilon_j = 1) = j/n$; let $\varepsilon_1 = 1$. Let $t_1 = \min\{j \ge 1 : \varepsilon_{1+j} = 1\}$, and recursively, if $\tau_s := 1 + \sum_{r=1}^{s} t_r < n$, then $t_{s+1} = \min\{r > 1 : \varepsilon_{\tau_s + r} = 1\}$, and denote $\mathbf{t} = \{t_1, t_2, \dots\}$. Then (c)

$$\mathbf{L} \stackrel{\mathcal{D}}{\equiv} \mathbf{t}; \tag{2}$$

in words, in distribution the sequence of branches lengths is the same as the sequence of time intervals between the success events $\{\varepsilon_j = 1\}$ for the Bernoulli sequence $\boldsymbol{\varepsilon}$. In particular,

$$X_n \stackrel{\mathcal{D}}{\equiv} \big|\{r > 1 : \varepsilon_r = 1\}\big|. \tag{3}$$

Note. At first glance, the part (b) is counterintuitive, and may even appear blatantly wrong. After all, by its definition, Y_n equals the number of moves until the walk hits the vertex 1, which is G_{n-1} in distribution, plus a genuinely random number of the remaining moves that assumes the value $n-2$ with a probability strictly less than 1. However disconcerting, this observation does not rule out validity of the statement, since these two Geometrics (n^{-1}) are different, interdependent, parameters of the walk.

Proof of Theorem 1. Let $\mathcal{L}_s = 1+\sum_{r=1}^{s} L_r$, so that $\mathcal{L}_s$ is the size of the subtree after s branches (paths) have been attached. Clearly $\{\mathcal{L}_s\}$ is a Markov chain. And so is $\{t_s\}$. So, to prove (c) ((2), (3)), it suffices to show that

$$\mathrm{P}(L_{s+1} = j|\mathcal{L}_s = k) = \mathrm{P}(\varepsilon_{k+j} = 1) \cdot \prod_{r=k+1}^{k+j-1} \mathrm{P}(\varepsilon_r = 0). \tag{4}$$

Let k be a generic size of the current subtree (core), i.e. a subtree already grown by adding several paths. (At the start $k = 1$.) Let $P_{nk}(i, j)$ denote the probability that it takes i moves to hit the k-core, and that the path which gets attached to the k-core has length j. Clearly, $P_{nk}(i,j) = 0$ if $i < j$. Let $j = 1$. Then

$$P_{nk}(i,1) = \begin{cases} \frac{k}{n}, & \text{if } i = 1, \\ \frac{1}{n} \cdot \frac{k}{n} \cdot \left(\frac{n-k}{n}\right)^{i-2}, & \text{if } i \geq 2. \end{cases} \tag{5}$$

(For $i \geq 2$, the walk first has to spend $i-2$ moves outside the core, then to loop back on the starting point, and finally to hit the core on the next move.) Let $j \geq 2$. The corresponding paths are partitioned into two groups, first and second type. A first-type walk moves to a new outside vertex, then spends the next $i-1$ moves never going back to the starting point in such a way that it hits the core for the first time on the last of those $i-1$ moves. Conditioned on this event, we have the loop-erased random walk on the set of $n-1$ vertices. A second-type walk spends some $r \geq 0$ moves outside the core, hits the starting point on the next move, then moves immediately to a different outside vertex, and spends the next $i-r-2$ moves outside the core, without ever going back to the starting point, and hits the core on the last of those $i-r-2$ moves. Therefore

$$\begin{aligned} P_{nk}(i,j) = {} & \frac{n-k-1}{n} \cdot \left(\frac{n-1}{n}\right)^{i-1} \cdot P_{n-1,k}(i-1,j-1) \\ & + \sum_{r\geq 0} \left(\frac{n-k}{n}\right)^{r} \cdot \frac{1}{n} \cdot \frac{n-k-1}{n} \cdot \left(\frac{n-1}{n}\right)^{i-r-2} \cdot P_{n-1,k}(i-r-2,j-1). \end{aligned} \tag{6}$$

The remaining argument is purely algebraic. To handle the recurrence (6) and the initial condition (5), let us introduce

$$\mathrm{P}_{nk}(z;j) = \sum_{i\geq 0} z^i P_{nk}(i,j),$$

which is the expected value of $z^{\#\text{of moves}}$ times the indicator of the event "added path has length"j. From (4) it follows that

$$\begin{aligned} \mathrm{P}_{nk}(z;1) &= z\frac{k}{n} + \sum_{i\geq 2} z^i \left(\frac{n-k}{n}\right)^{i-2} \cdot \frac{1}{n} \cdot \frac{k}{n} \\ &= z\frac{k}{n} + z^2 \cdot \frac{k}{n^2} \cdot \frac{1}{1 - z\frac{n-k}{n}} \\ &= z\frac{k}{n} \cdot \frac{1 - z\frac{n-k-1}{n}}{1 - z\frac{n-k}{n}}. \end{aligned} \tag{7}$$

Further, using (6), for $j \geq 2$ we have:

$$\begin{aligned} \mathrm{P}_{nk}(z;j) &= z\frac{n-k-1}{n} \sum_i \left(z\frac{n-1}{n}\right)^{i-1} P_{n-1,k}(i-1, j-1) \\ &+ \sum_{r\geq 0} \left(z\frac{n-1}{n}\right)^{i-r-2} P_{n-1,k}(i-r-2, j-1) \cdot z^{r+2} \left(\frac{n-k}{n}\right)^r \cdot \frac{1}{n} \cdot \frac{n-k-1}{n} \\ &= \frac{n-k-1}{n} \cdot \mathrm{P}_{n-1,k}\left(z\frac{n-1}{n}; j-1\right) \cdot \left(z + \frac{z^2/n}{1 - z\frac{n-k}{n}}\right) \\ &= z\frac{n-k-1}{n} \cdot \frac{1 - z\frac{n-k-1}{n}}{1 - z\frac{n-k}{n}} \cdot \mathrm{P}_{n-1,k}\left(z\frac{n-1}{n}; j-1\right). \end{aligned} \tag{8}$$

Introduce

$$z_\nu = z \prod_{j=1}^{\nu} (n-j)/(n-j+1) = z\frac{n-\nu}{n-\nu+1}.$$

Applying (8) repeatedly and using (7) at the end, we obtain

$$\begin{aligned} \mathrm{P}_{nk}(z;j) &= \prod_{\nu=0}^{j-2} z_\nu \frac{n-k-1-\nu}{n-\nu} \cdot \frac{1 - z_\nu \frac{n-k-1-\nu}{n-\nu}}{1 - z_\nu \frac{n-k-\nu}{n-\nu}} \cdot \mathrm{P}_{n-j-1,k}(z_{j-1}; 1) \\ &= k\left(\frac{z}{n}\right)^j \cdot (n-k-1)_{j-1} \cdot \frac{1 - z\frac{n-j-k}{n}}{1 - z\frac{n-k}{n}}. \end{aligned} \tag{9}$$

Extracting the coefficient by z^i, we get:

$$P_{nk}(i,j) = \begin{cases} \frac{k}{n^j}(n-k-1)_{j-1}, & \text{if } i = j, \\ \frac{k}{n^j}(n-k-1)_{j-1} \cdot \frac{j}{n}\left(\frac{n-k}{n}\right)^{i-j-1}, & \text{if } i > j. \end{cases} \tag{10}$$

The rest is short. Summing over $i \geq j$, we get:

$$P_{nk}(j) := \Pr(\text{pathlength} = j) = \frac{k+j}{n^j}(n-k-1)_{j-1}. \tag{11}$$

And of course

$$P_{nk}(i) := \Pr(\text{number of moves} = i) = \frac{k}{n}\left(\frac{n-k}{n}\right)^{i-1}, \tag{12}$$

that is, the number of moves is equal, in distribution, to $G_{k/n}$. The relation (11) can be rewritten as

$$P_{nk}(j) = \frac{k+j}{n} \cdot \prod_{r=1}^{j-1} \left(1 - \frac{k+r}{n}\right), \tag{13}$$

which is equivalent to (4). Let Y_n be the total number of moves. We want to prove that Y_n is independent of X_n, and

$$Y_n \stackrel{\mathcal{D}}{\equiv} G_{1/n} + n - 2. \tag{14}$$

To this end, let us consider a more general case of X_{nk} and Y_{nk}, the number of paths and the number of moves remaining after the moment the subtree has grown to size k. ($X_n = X_{n1}, Y_n = Y_{n1}$.) Introduce the joint (bivariate) generating function

$$g_{nk}(x,y) := \mathrm{E}\big(x^{X_{nk}} y^{Y_{nk}}\big).$$

By (9),

$$\begin{aligned} g_{nk}(x,y) = & \quad x \textstyle\sum_{j>0} \mathrm{P}_{nk}(y;j) g_{n,k+j}(x,y) \\ = & \; x \sum_{j>0} k \left(\frac{y}{n}\right)^j (n-k-1)_{j-1} \frac{1 - y^{\frac{n-j-k}{n}}}{1 - y^{\frac{n-k}{n}}} \cdot g_{n,k+j}(x,y). \end{aligned} \tag{15}$$

Introducing

$$h_{nk}(x,y) := g_{nk}(x,y) \frac{\left(1 - y^{\frac{n-k}{n}}\right) \left(\frac{y}{n}\right)^k}{(n-k-1)!}, \tag{16}$$

we transform (15), after multiplying its both sides by the appropriate factor and denoting $k + j = s$, into

$$\frac{1}{k} h_{nk}(x,y) = x \sum_{s>k} \frac{1}{n-s} h_{ns}(x,y), \tag{17}$$

a much simpler relation. Writing (17) for $k+1$ instead of k and subtracting the result from both sides of (17), we get

$$\frac{1}{k} h_{nk} - \frac{1}{k+1} h_{n,k+1} = \frac{x}{n-k-1} h_{n,k+1},$$

or

$$h_{nk} = \frac{k\big[n-k-1+x(k+1)\big]}{(k+1)(n-k-1)} \cdot h_{n,k+1}, \quad k < n-1. \tag{18}$$

Notice that

$$g_{n,n-1}(x,y) = x \sum_{j>0} \left(\frac{1}{n}\right)^{j-1} \left(1 - \frac{1}{n}\right) \cdot y^j = xy \frac{1 - \frac{1}{n}}{1 - \frac{y}{n}},$$

so that

$$h_{n,n-1}(x,y) = \frac{\left(1-\frac{y}{n}\right)\left(\frac{y}{n}\right)^{n-1}}{0!} \cdot g_{n,n-1}(x,y) = x(n-1)\left(\frac{y}{n}\right)^n. \tag{19}$$

Combining (18) and (19) we have

$$\begin{aligned} h_{nk}(x,y) &= \left(\prod_{j=k+1}^{n-1} \frac{(j-1)(n-j+xj)}{j(n-j)}\right) \cdot h_{n,n-1}(x,y) \\ &= \frac{k}{n(n-k-1)!} \cdot \prod_{j=k+1}^{n} (n-j+xj) \cdot \left(\frac{y}{n}\right)^n. \end{aligned}$$

So, by (16),

$$\begin{aligned} g_{nk}(x,y) &= \frac{\left(\frac{k}{n}\right) \cdot \prod_{j=k+1}^{n} (n-j+xj) \cdot \left(\frac{y}{n}\right)^n}{\left(1-y\frac{n-k}{n}\right) \cdot \left(\frac{y}{n}\right)^k} \\ &= \prod_{j=k+1}^{n} \left(1-\frac{j}{n}+x\frac{j}{n}\right) \cdot \frac{y\frac{k}{n}}{1-y\left(1-\frac{k}{n}\right)} y^{n-k-1}. \end{aligned}$$

The product shape of the last formula shows that X_{nk} and Y_{nk} are independent, and that marginally X_{nk} is distributed as the total number of successes in the sequence $\{\varepsilon_j\}_{k<j\le n}$ (no news here!), while

$$Y_{nk} \stackrel{\mathcal{D}}{\equiv} G_{k/n} + n - k - 1.$$

For $k = 1$ this yields (1).□

The process of building a path leading to a given subtree (of size k) is a Markov process on the state space $\{1, \ldots, n-k\} \cup \{r^*\}$, the union of the set of all possible lengths r of a current path and the absorbing state r^*, hitting which corresponds to hitting the subtree. The transition probabilities are given by

$$p(r'|r) = \begin{cases} \frac{n-k-r}{n}, & r' = r+1, \\ 1/n, & r' \in [1,r], \\ k/n, & r' = r^*. \end{cases} \tag{20}$$

We could have used this Markov chain to obtain a recurrence equation for

$$\mathrm{P}_{nkr}(z;j) = \sum_{i\ge 0} z^i P_{nkr}(i,j),$$

where $P_{nkr}(i,j)$ is the probability analogous to $P_{nk}(i,j)$, with r corresponding to the length r of a starting path; so that $P_{nk}(i,j) = P_{nk1}(i,j)$. Following this route, relatively easily one obtains a formula for $\mathrm{P}_{nkr}(z;j) - \mathrm{P}_{nk,r-1}(z;j)$, but getting the equation (9) from it requires some messy computations. There is definitely a conservation energy law in the play! This Markov chain is quite useful though in dealing with the distribution of cycles erased during the walk.

Theorem 2. *Let c_{nt} denote the expected number of cycles of length t erased in the course of the random walk, $1 \le t \le n-1$. Then*

$$c_{nt} = \frac{(n-2)_{t-1}}{n^{t-1}} + \frac{1}{t} \cdot \frac{(n-2)_t}{n^t}, \tag{21}$$

so that, for every fixed $t \ge 1$,

$$\lim_{n\to\infty} c_{nt} = \frac{t+1}{t}. \tag{22}$$

Proof of Theorem 2. As in the proof of Theorem 1, consider the macrostep that adds a path to a subtree of size k. Using the Markov chain defined above, introduce $c_{nt}(r,k)$, the expected number of cycles in question until the moment of absorption, given that the chain starts at state r, i.e. path of length r, $1 \le r \le n-k$. Dropping the indices n, t, k for simplicity, we have a recurrence

$$\begin{aligned} c(r) &= \tfrac{n-k-r}{n} c_{r+1} + \tfrac{1}{n} \textstyle\sum_{j=1}^{r} \left[\delta_{t,r-j+1} + c(j)\right] \\ &= \qquad \tfrac{n-k-r}{n} + \tfrac{1}{n} \textstyle\sum_{j=1}^{r} \left[\delta_{tj} + c(j)\right]. \end{aligned} \tag{23}$$

Consequently, for $2 \le s \le n-k$,

$$c(s) - c(s-1) = \frac{n-k-r}{n}(c_{s+1} - c_s) + \frac{1}{n}\delta_{ts}, \tag{24}$$

with the first summand on the right missing for $s = n-k$. Given r, for $s \in [r, n-k]$, we multiply both sides of (24) by $\frac{(n-k-r)_{s-r}}{n^{s-r}}$ and, adding the results, obtain

$$c(r) - c(r-1) = \frac{1}{n} \sum_{s=r}^{n-k} \frac{(n-k-r)_{s-r}}{n^{s-r}} \cdot \delta_{ts}.$$

That is,

$$c(r) - c(r-1) = \begin{cases} \dfrac{(n-k-r)_{t-r}}{n^{t-r+1}}, & r \le t \le n-k, \\ 0, & t < r. \end{cases} \tag{25}$$

Using (23) for $r=1$ and (25) for $r=2$, we get two equations for $c(1), c(2)$, which —putting the dropped indices back—yield

$$c_{nt}(1,k) = \frac{1}{k} \cdot \frac{(n-k-1)_{t-1}}{n^{t-1}}.$$

Consequently

$$\begin{aligned} c_{nt} &= \sum_{k=1}^{n-1} \mathrm{P}(\varepsilon_k = 1) c_{nt}(1,k) \\ &= \frac{(n-2)_{t-1}}{n^{t-1}} + \frac{(t-1)!}{n^t} \cdot \sum_{k=2}^{n-1} \binom{n-k-1}{t-1} \\ &= \frac{(n-2)_{t-1}}{n^{t-1}} + \frac{(t-1)!}{n^t} \cdot \binom{n-2}{t}, \end{aligned}$$

which is equivalent to (21). □

Note. That $\lim c_{nt} > 0$ for every t makes it plausible that with high probability (whp) the random walk develops cycles of substantial length. In Section we will show that the lenght of the longest cycle is asymptotic, in probability, to $(n \log n)^{1/2}$.

2 Asymptotic distribution of branches lengths.

First of all, from Theorem 1 (c) and the central limit theorem for Bernoulli random variables, it follows that X_n (the number of paths, macrosteps) is asymptotically Gaussian with mean and variance equal $n/2$ and $n/6$ respectively. The next lemma tells us that whp the subtree grows essentially as $(rn)^{1/2}$, r being the number of steps.

Lemma 3. *Let $\omega(n) \to \infty$ however slowly. Introduce $m_r = [\sqrt{rn/\omega(n)}]$ and $M_r = [\sqrt{rn\omega(n)}]$. Then*

$$P(m_r < \mathcal{L}_r \leq M_r, \forall r \geq 1) \geq 1 - O(\omega^{-1}(n)).$$

Proof of Lemma 3. Using Theorem 1 (c),we have

$$\mathrm{P}(\mathcal{L}_r \leq m_r) = \mathrm{P}(\varepsilon_2 + \cdots + \varepsilon_{m_r} \geq r).$$

So (Chernoff-type bound), for every $u > 0$ we bound

$$\begin{aligned}\mathrm{P}(\mathcal{L}_r \leq m_r) &\leq e^{-ur} \prod_{j=2}^{m_r} \left(1 - \frac{j}{n} + e^u \frac{j}{n}\right) \\ &\leq \exp\left(-ur + (e^u - 1)\frac{m_r(m_r+1)}{2n}\right).\end{aligned}$$

And the best u is given by

$$u = \log \frac{2nr}{m_r(m_r+1)} = \log(2\omega(n)) + o(1).$$

Therefore

$$\mathrm{P}(\mathcal{L}_r \leq m_r) \leq \left(\exp(-u + 1 - e^{-u})\right)^r \leq \omega^{-r}(n),$$

and consequently

$$\mathrm{P}(\exists\, r \geq 1 : \mathcal{L}_r \leq m_r) \leq \sum_{r \geq 1} \omega^r(-n) \leq 2\omega^{-1}(n). \tag{26}$$

Analogously, for $u = \log \frac{M_r(M_r+1)}{2nr}$,

$$\mathrm{P}(\mathcal{L}_r > M_r) \leq \left(\exp(u + 1 - e^u)\right)^r \leq \left(\exp(-\omega(n)/3)\right)^r,$$

so that

$$\mathrm{P}(\exists r \geq 1 : \mathcal{L}_r > M_r) \leq \sum_{r\geq 1} e^{-r\omega(n)/3} < \omega^{-1}(n). \tag{27}$$

(26),(27) together complete the proof. □

Note. Here is an interesting application of the lemma. A starting vertex of the first path is trivially a pendant vertex of the resulting subtree. Given $r > 1$, the probability that the starting vertices of the first r paths and the root itself are pendant vertices of the subtree formed by all these paths is at least

$$1 - \sum_{s=1}^{r-1} \frac{s+1}{m_s} - \mathrm{P}(\exists\, s \in [1, r-1] : \mathcal{L}_s \leq m_s) \;=\; 1 - O(\sqrt{(r^3/n)\omega(n)}) - O(\omega^{-1}(n))$$
$$\to 1,$$

if $r = o((n/\omega(n))^{1/3}$, that is if $r = o(n^{1/3})$, since in Lemma 3 $\omega(n) \to \infty$ however slowly. It can be shown that for $r \gg n^{1/3}$ the number of the starting points of the first r paths that fail to have degree 1 in the subtree formed by those paths is unbounded in probability. Thus $r = n^{1/3}$ is a threshold value for the property "none of the first r paths is an extension of an earlier path".

Let us have a close look at the Markov chain $\{\mathcal{L}_j\}$, for $j \leq r$, $r = o(n^\alpha)$, $\alpha \in (0,1)$ to be chosen later. Let $\mathbf{j} = (j_1, \ldots, j_r)$ denote the sequence of generic values of the paths lengths $L_1, \ldots, L_r$. Then

$$\mathcal{L}_t = j^{(t)} := 1 + j_1 + \cdots + j_t, \quad 1 \leq t \leq r.$$

By Lemma 3, we may and will concentrate on $\mathbf{j}$ such that

$$j^{(r)} \leq (rn\omega(n))^{1/2} \Longrightarrow \frac{j^{(r)}}{n} \ll \frac{\omega^{1/2}(n)}{n^{(1-a)/2}} \to 0.$$

Using Theorem 1 (c), (see also (13)), and setting $j^{(0)} = 1$, we have: for $s \leq r$,

$$\mathrm{P}(\mathcal{L}_t = j^{(t)}, 1 \leq t \leq s) = \prod_{t=1}^{s} \mathrm{P}\big(L_t = j_t | \mathcal{L}_{t-1} = j^{(t-1)}\big)$$
$$= \prod_{t=1}^{s} P_{nj^{(t-1)}}(j_t) = \prod_{t=1}^{s} \frac{j^{(t)}}{n^{j_t}}(n - j^{(t-1)} - 1)_{j_t - 1}$$
$$= \frac{(n-2)_{j^{(s)}-1}}{n^{j^{(s)}-1}} \cdot \prod_{t=1}^{s} \frac{j^{(t)}}{n - j^{(t)}}$$
$$= \exp\left(-\frac{(j^{(s)})^2}{2n} + O\big(sj^{(s)}/n + (j^{(s)})^3/n^2\big)\right) \times \prod_{t=1}^{s}\left(\frac{j^{(t)}}{n^{1/2}} \cdot n^{-1/2}\right). \tag{28}$$

Here

$$\frac{sj^{(s)}}{n} \leq \left(\frac{r^3}{n}\omega(n)\right)^{1/2}, \quad \frac{(j^{(s)})^3}{n^2} \leq \left(\frac{r^3}{n}\omega^3(n)\right)^{1/2},$$

so the remainder term in (28) approaches zero, uniformly for $s \le r$, provided that

$$r = o(n^{1/3}\omega^{-1}(n)). \tag{29}$$

($r = n^{1/3}$ raises its head again!) For $s = 1$, (28) shows that $\mathcal{L}_1/n^{1/2}$ has a limiting density $f(x) = xe^{-x^2/2}$. Introduce

$$f(x'|x) = x' \exp\left(-\frac{(x')^2}{2} + \frac{x^2}{2}\right);$$

as a function of x', this is a density on $[x, \infty)$, and $f(x|0) = f(x)$. Using $f(x'|x)$, we can rewrite (28) as

$$\mathrm{P}(\mathcal{L}_t = j^{(t)}, 1 \le t \le s) = (1 + o(1)) \prod_{t=1}^{s} f\left(\frac{j^{(t)}}{n^{1/2}} \,\Big|\, \frac{j^{(t-1)}}{n^{1/2}}\right) n^{-1/2}, \tag{30}$$

uniformly for $s \le r$, if r meets (29) *and* $j^{(r)} = O((rn\omega(n))^{1/2})$. Thus

Theorem 4. *Let $r = o(n^{1/3}\omega^{-1}(n))$. For $\mathcal{L}_r = O((rn\omega(n))^{1/2})$, the Markov chain $\{\mathcal{L}_s/n^{1/2}\}_{s\le r}$ is asymptotic, in terms of the local probabilities, to the Markov chain $\{\ell_s\}_{s\le r}$, with a state space $[0, \infty)$, the initial state 0, and the one-step transitional density $f(x'|x)$.*

Surprisingly, there is a very simple description of the chain $\{\ell_s\}_{s\ge1}$. Let $Z_1, Z_2, \dots$ be i.i.d. Exponentials, with parameter 1. Then $\{\ell_s\}_{s\ge1}$ has the same distribution as $\{\ell_s^*\}_{s\ge1}$, where

$$\ell_s^* = \left(2\sum_{t=1}^{s} Z_t\right)^{1/2}.$$

Indeed,

$$\mathrm{P}(\ell_1^* \le x) = \mathrm{P}(2Z_1 \le x^2) = 1 - e^{-x^2/2},$$

so that ℓ_1^* has density $xe^{-x^2/2} = f(x)$. Furthermore,

$$\begin{aligned}\mathrm{P}(\ell_s^* \le x' | \ell_{s-1}^* = x) &= \mathrm{P}\big(\sqrt{2Z_s + x^2} \le x'\big)\\ &= \mathrm{P}\left(Z \le \frac{(x')^2 - x^2}{2}\right)\\ &= 1 - \exp\left(-\frac{(x')^2}{2} + \frac{x^2}{2}\right).\end{aligned}$$

Hence the conditional density of ℓ_s^* is indeed $f(x'|x)$.□

So Theorem 4 can be reformulated as

Theorem 4′. *Let $r = o(n^{1/3}\omega^{-1}(n))$. For $\mathcal{L}_r = O((rn\omega(n))^{1/2})$, the Markov chain $\{\mathcal{L}_s/n^{1/2}\}_{s\le r}$ approaches, in terms of the local probabilities, the sequence $\{\ell_s^*\}$, $\ell_s^* = \sqrt{2\sum_{t=1}^{s} Z_s}$, where $Z_1, Z_2, \dots$ are independent copies of the Exponential Z with mean 1.*

3 Longest path length.

Let $L_{\max}$ denote the length of the longest path. Let us show that $L_{\max}/n^{1/2}$ converges, in distribution, to a nondegenerate random variable. From Theorem 1, $L_{\max} \overset{\mathcal{D}}{\equiv} D_{\max}$, where $D_{\max}$ is the largest distance between successes, the moments j for which $\varepsilon_j = 1$ in the Bernoulli sequence $\{\varepsilon_j\}_{1\le j\le n}$. Pick $a > 0$ and introduce $N_n(a)$, the total number of successes that are at distance $\ell := \lceil an^{1/2}\rceil$ at least from the next success. (Every such success j meets the condition

$$\varepsilon_j = 1, \quad \varepsilon_{j+1} = \cdots \varepsilon_{\ell-1} = 0,$$

for some $1 \le j \le n-\ell$; recall that $\varepsilon_n = 1$ with probability 1.) Let us evaluate the binomial moments $B_n^{(k)} := \mathrm{E}\left[\binom{N_n(a)}{k}\right]$. Since $\binom{N_n(a)}{k}$ is the total number of k-long (chronologically ordered) samples of the successes meeting the above condition, we have

$$B_n^{(k)} = \sum_{\substack{1\le j_1,\dots,j_k\le n-\ell\\ j_s+\ell\le j_{s+1},\,1\le s\le k-1}} \Pr\bigl(\forall 1\le s\le k : \varepsilon_{j_s}=1, \varepsilon_{j_s+1}=\cdots\varepsilon_{j_s+\ell-1}=0\bigr). \quad (31)$$

Denote the generic term $P(\mathbf{j})$. Suppose first that $j_1 > 1$. Then

$$P(\mathbf{j}) = \prod_{r=1}^{k}\left[\frac{j_r}{n}\left(1-\frac{j_r+1}{n}\right)\cdots\left(1-\frac{j_r+\ell-1}{n}\right)\right].$$

A standard argument shows that the dominant $\mathbf{j}$ are such that each $j_s = O(n^{1/2})$, and for those $\mathbf{j}$, denoting $x_r = \frac{j_r}{n^{1/2}}$,

$$\begin{aligned} P(\mathbf{j}) &= (1+O(n^{-1/2}))\prod_{r=1}^{k}\frac{x_r}{n^{1/2}}\exp\left(-\frac{(x_r+a)^2-x_r^2}{2}\right)\\ &= (1+O(n^{-1/2}))\left(e^{-a^2/2}\right)^k\prod_{r=1}^{k}\frac{x_r}{n^{1/2}}e^{-ax_r}. \end{aligned}$$

So the contribution of those $\mathbf{j}$ to $B_n^{(k)}$ converges to

$$\left(e^{-a^2/2}\right)^k \int\!\!\cdots\!\!\int_{\substack{x_1,\dots,x_k\ge 0\\ x_s+a\le x_{s+1},\,s\le k-1}} \prod_{r=1}^{k} x_r e^{-ax_r}\,d\mathbf{x} = \left(\frac{e^{-a^2/2}}{a^2}\right)^k \int\!\!\cdots\!\!\int_{\substack{y_1,\dots,y_k\ge 0\\ y_s+1\le y_{s+1},\,s\le k-1}} \prod_{r=1}^{k} y_r e^{-y_r}\,d\mathbf{y}.$$

Notice that ye^{-y} is the density of $Y = Z_1 + Z_2$, with Z_1, Z_2 the independent exponentials, of mean 1. So the last integral equals

$$\frac{1}{k!}\mathrm{P}\bigl(\min_{1\le i\ne j\le k}|Y_i - Y_j| \ge 1\bigr),$$

where $Y_1, \dots, Y_k$ are independent copies of Y. The same derivation shows that the limiting contribution to $B_n^{(k)}$ of the $\mathbf{j}$ with $j_1 = 1$ is

$$\frac{\left(e^{-a^2/2}\right)^k}{(a^2)^{k-1}} \int_{\substack{y_1,\dots,y_{k-1}\ge 0 \\ y_{s-1}+1\le y_s,\, s\le k-1}} \cdots \int \prod_{r=1}^{k-1} y_r e^{-y_r}\, d\mathbf{y},$$

where $y_0 := 0$. And the integral equals

$$\frac{1}{k!}\mathrm{P}\Big(\min_{0\le i\ne j\le k-1} |Y_i - Y_j| \ge 1\Big),$$

where $Y_0 := 0$. Thus we have proved that, for every fixed $k \ge 1$,

$$\lim_{n\to\infty} B_n^{(k)} = B^{(k)}(a),$$

where

$$B^{(k)}(a) = \frac{1}{k!}\left(\frac{e^{-\frac{a^2}{2}}}{a^2}\right)^k \cdot \left[\mathrm{P}\Big(\min_{1\le i\ne j\le k} |Y_i - Y_j| \ge 1\Big) + a^2\mathrm{P}\Big(\min_{0\le i\ne j\le k-1} |Y_i - Y_j| \ge 1\Big)\right]. \tag{32}$$

Since $B^{(k)}(a)$ goes to zero as $k\to\infty$ faster than k^{-m} for every integer m, we infer that there exists a unique probability distribution $\{p_j(a)\}_{j\ge 0}$ with the binomial moments equal $B_k(a)$, that is

$$p_j(a) = \sum_{k\ge j} (-1)^{k-j} B_k(a)\binom{k}{j}, \quad j \ge 0. \tag{33}$$

And, since $B_n^{(k)} \to B^{(k)}(a)$ for every k, we have

$$\lim_{n\to\infty} \mathrm{P}(N_n(a) = k) = p_k(a), \quad k \ge 0.$$

Using the last relation for $k = 0$, and $L_{\max} \overset{\mathcal{D}}{\equiv} D_{\max}$, we arrive at

Theorem 5. *For every $a > 0$*

$$\lim_{n\to\infty} P\big(L_{max} < [an^{1/2}]\big) = \lim_{n\to\infty} P\big(N_n(a) = 0\big) = p_0(a). \tag{34}$$

Notes. 1. We were not able to find a formula for $p_0(a)$ more explicit than the series in (34). Since $1 - e^{-a^2/2}$ is the limiting distribution of $L_1/n^{1/2}$, we have

$$p_0(a) \le 1 - e^{-a^2/2}.$$

The bound follows also from the enveloping property of the series in (34). Using the series, one can show easily that

$$1 - p_0(a) \sim e^{-a^2/2}, \quad a \to \infty.$$

2. In principle, there is an alternative approach to $L_{\max}$. Let $L_{\max}^{(k)}$ denote the length of the longest path added after the subtree reached size k. Then

$$L_{\max}^{(k)} = \max\{j, L_{\max}^{(k+j)}\},$$

where j is the random length of the path added next. Then

$$\mathrm{P}(L_{\max}^{(k)} \le \ell) = \sum_{j \le \ell} P_{nk}(j) \mathrm{P}(L_{\max}^{(k+j)} \le \ell). \tag{35}$$

Assuming that $\mathrm{P}(L_{\max}^{(k)} \le \ell) \sim h(k/n^{1/2}, \ell/n^{1/2})$, and using (11), we transform (on a heuristic level) (35) into an integral equation

$$h(y,x) = \int_0^x (y+z) \exp\left(-\frac{(y+z)^2 - y^2}{2}\right) h(y+z,x)\, dz, \tag{36}$$

or, introducing $g(y,x) = e^{-y^2/2} h(y,x)$,

$$g(y,x) = \int_y^{x+y} u g(u,x)\, du. \tag{37}$$

Does (36-37) have a closed-form solution?

4 Looping back on cycles.

Let C_n denote the total number of loops created and deleted in the course of the algorithm. And let $O_{\max}$ denote the length of the longest loop. By Theorem 2, c_{nt} the expected number of cycles of length t is given by

$$c_{nt} = c'_{nt} + c''_{nt},$$

where

$$c'_{nt} = \frac{(n-2)_{t-1}}{n^{t-1}}, \quad c''_{nt} = \frac{1}{t} \cdot \frac{(n-2)_t}{n^t}.$$

And, see the proof, c'_{nt}, c''_{nt} is the expected number of cycles of length t created during the first macrostep, and during all the subsequent macrosteps respectively. Given $\ell \in [1, n-1]$, let $C_{n\ell}$ denote the total number of cycles of length $\ell + 1$ or more. Then $C_{n\ell} = C'_{n\ell} + C''_{n\ell}$, the summands equal to the number of those cycles formed during the first macrostep, and all other macrosteps. Then

$$\begin{aligned} m_{n\ell} &:= \mathrm{E}(C_{n\ell}) = m'_{n\ell} + m''_{n\ell}, \\ m'_{n\ell} &:= \mathrm{E}(C'_{n\ell}) = \sum_{t \ge \ell} c'_{nt}, \\ m''_{n\ell} &:= \mathrm{E}(C''_{n\ell}) = \sum_{t > \ell} c''_{nt}. \end{aligned}$$

In particular,

$$\begin{aligned} m'_{n1} &:= \sum_{t\geq 1} c'_{nt} \sim \sum_{t=1}^{n} e^{-t^2/(2n)} \sim \sqrt{\frac{\pi n}{2}}, \\ m''_{n1} &:= \sum_{t\geq 1} c''_{nt} \sim \sum_{t=1}^{n} t^{-1} e^{-t^2/(2n)} \sim \frac{\log n}{2}, \end{aligned} \tag{38}$$

so that

$$m''_{n1}/m'_{n1} = O(n^{-1/2}\log n). \tag{39}$$

In fact, an easy argument shows that, for every $\ell = \ell(n)$,

$$m''_{n1}/m'_{n\ell} = O(n^{-1/2}\log n).$$

Our main result is

Lemma 6. *Let $\mu_{n\ell k} = E[(C'_{n\ell})_k]$ denote the k-th factorial moment of $C'_{n\ell}$; in particular $\mu_{n\ell 1} = m'_{n\ell}$. If $\ell = \ell(n)$ is such that* $\liminf m'_{n\ell} > 0$*, then, for every $k \geq 1$,*

$$\frac{\mu_{n\ell k}}{(\mu_{n\ell 1})^k} \to k!. \tag{40}$$

Let us apply this Lemma to C_n and $O_{\max}$. We have $C'_n = C'_{n1}$, $m_{n1} = \Theta(n^{1/2})$ and by (38), $C''_n/m'_{n1} \to 0$ in probability. Then, as $\mu_{n11} = m'_{n1}$, it follows from (40) that, for every $k \geq 1$,

$$\mathrm{E}\left(\frac{C'_n}{m_{n1}}\right)^k \to k!.$$

Since $\{k!\}$ are the moments of the exponential random variable with parameter 1, we have proved

Theorem 7. *For every $x > 0$,*

$$\lim_{n\to\infty} P\left(\frac{C_n}{\sqrt{\frac{\pi n}{2}}} \leq x\right) = 1 - e^{-x}.$$

Next, let $x > 0$, and

$$\ell = \left[\sqrt{n \log\left(\frac{nx^2}{\log n}\right)}\right].$$

Then

$$\begin{aligned} m'_{n\ell} &= \sum_{t>\ell} c'_{nt} \sim \sum_{t>\ell} e^{-t^2/n} \sim n^{1/2} \int_{y\geq \ell/n^{1/2}} e^{-y^2/2}\,dy \\ &\sim n^{1/2} \cdot \left.\frac{e^{-y^2/2}}{y}\right|_{y=\ell/n^{1/2}} \to x^{-1} > 0. \end{aligned}$$

Then, by Lemma 6, for every $k \geq 1$,

$$\lim_{n\to\infty} \mathrm{E}[(C'_{n1})_k] = x^{-k}k!.$$

Now $\{x^{-k}k!\}$ are the factorial moments of the Geometric with parameter

$$p = \frac{1}{1+x^{-1}} = \frac{x}{1+x}.$$

So, for every $k \geq 0$,

$$\mathrm{P}(C'_{n1} = k) \to \mathrm{P}(G_p = k) = p(1-p)^k.$$

Observing that $\mathrm{P}(O_{\max} \leq \ell) = \mathrm{P}(C_{n\ell} = 0)$ and that $\mathrm{P}(C''_{n\ell} = 0) \to 1$, we have proved

Theorem 8. *For every fixed $x > 0$,*

$$\lim_{n\to\infty} P\left(O_{max} \leq \sqrt{n \log\left(\frac{nx^2}{\log n}\right)}\right) = \frac{x}{1+x},$$

and consequently, in probability,

$$\lim_{n\to\infty} \frac{O_{max}}{\sqrt{n\log n}} = 1.$$

Proof of Lemma 6. We use the Markov chain from the proof of Theorem 2. Introduce $f_{n\ell r}(z) = \mathrm{E}_r(z^{C'_{n\ell}})$, the probability generating function of the number of cycles (during the first macrostep) if the starting path contains r vertices; so $f_{n\ell 1}(z) = \mathrm{E}(z^{C'_{n\ell}})$. Then $\mu_{n\ell rk} := f^{(k)}_{n\ell r}(1) = \mathrm{E}_r[(C'_{n\ell})_k]$, and $\mu_{n\ell 11} = \mu_{n\ell 1} = m'_{n\ell}$. For simplicity, we drop the subindices n and ℓ. By the Markov property, for $1 \leq r \leq n-1$,

$$f_r(z) = \frac{1}{n} + \frac{n-1-r}{n} f_{r+1}(z) + \frac{1}{n}\sum_{j=1}^{r-\ell} z f_j(z) + \frac{1}{n}\sum_{j=r-\ell+1}^{r} f_j(z);$$

($f_j(z) := 0$ for $j \leq 0$). So, for every $k \geq 1$,

$$\mu_{rk} = \frac{n-1-r}{n}\mu_{r+1,k} + \frac{1}{n}\sum_{j=1}^{r}\mu_{jk} + \frac{k}{n}\sum_{j=1}^{r-\ell}\mu_{j,k-1}; \tag{41}$$

($\mu_{j\kappa} := 0$ for $j \leq 0$, $\kappa \geq 0$). Using (41) for $r-1$ instead of r and taking the difference, we have

$$\mu_{rk} - \mu_{r-1,k} = \frac{n-1-r}{n}(\mu_{r+1,k} - \mu_{rk}) + \frac{k}{n}\mu_{r-\ell,k-1}. \tag{42}$$

Iterating this relation upward to $r = n-1$,

$$\mu_{rk} - \mu_{r-1,k} = \frac{k}{n}\sum_{j=0}^{n-1-r}\frac{(n-1-r)_j}{n^j}\mu_{r+j-\ell,k-1}. \tag{43}$$

Consider $k = 1$. As $\mu_{r+j-\ell,0} = \mathbb{I}(r + j > \ell)$, and

$$\frac{(n-1-r)_j}{n^j} \le c \exp\left(-\frac{(r+j)^2 - r^2}{2n}\right),$$

for some absolute constant $c > 0$, we get easily:

$$\mu_{r1} - \mu_{r-1,1} \le_b \frac{\exp\left[r^2/(2n) - (\max(r,\ell))^2/(2n)\right]}{\max(r,\ell)}. \tag{44}$$

(Here and below $A \le_b B$ stands for $A = O(B)$, uniformly for all values of parameters involved in A and B.) And, summing the last bounds downward to $r = 2$, we obtain that

$$\mu_{r1} - \mu_{11} \le_b \begin{cases} e^{-\ell^2/(2n)}, & r \le \min(\ell, n^{1/2}), \\ \frac{n}{\ell r} \cdot e^{r^2/(2n) - \ell^2/(2n)}, & n^{1/2} \le r \le \ell, \\ \log n, & r \ge \ell. \end{cases} \tag{45}$$

Inductively, suppose that for some $k \ge 1$

$$\lim_{n\to\infty} \frac{\mu_{1k}}{(\mu_{11})^k} = k!,$$

and that

$$\mu_{rk} - \mu_{1k} \le_b \begin{cases} e^{-\ell^2/(2n)}(\mu_{11}\log n)^{k-1}, & r \le \min(\ell, n^{1/2}), \\ \dfrac{n}{\ell r} \cdot e^{r^2/(2n) - \ell^2/(2n)}(\mu_{11}\log n)^{k-1}, & n^{1/2} \le r \le \ell, \\ \log^k n, & r \ge \ell. \end{cases} \tag{46}$$

For the first two bounds in (46) to be of any use, both should be negligible compared to $\mu_{1k} \sim k!(\mu_{11})^k$. Now the ratio of either of them to $(\mu_{11})^k$ is bounded above by $\mu_{11}^{-1} e^{-\ell^2/(2n)}$, and we recall that

$$\mu_{11} = \sum_{t \ge \ell} \frac{(n-2)_t}{n^t}. \tag{47}$$

Since $\liminf \mu_{11} > 0$, we have $\ell = O(\sqrt{n \log n})$, and from (47) it follows easily that

$$\frac{\frac{(n-2)_\ell}{n^\ell}}{\mu_{11}} \le n^{-\beta}, \quad \forall \beta < 1/2,$$

which implies the same bound for $e^{-\ell^2/(2n)}/\mu_{11}$. Thus,

$$\mu_{rk} = (1 + O(n^{-\beta}))\mu_{11}^k = O(\mu_{11}^k), \quad r \le \ell. \tag{48}$$

Using (43) with k replaced by $k+1$, the condition $\mu_{r+j-\ell,k} = 0$ for $r + j - \ell \le 0$, and (48), we get

$$\begin{aligned} \mu_{r,k+1} - \mu_{r-1,k+1} &\le_b n^{-1}(\mu_{11}\log n)^k \textstyle\sum_{j>\ell-r} e^{r^2/(2n) - (r+j)^2/(2n)} \\ &\le_b (\mu_{11}\log n)^k \cdot \dfrac{e^{r^2/(2n) - (\max(\ell,r))^2/(2n)}}{\max(r,\ell)}. \end{aligned} \tag{49}$$

So, just like the bound (44) led to (45), the bound (49) leads now to

$$\mu_{r,k+1} - \mu_{1,k+1} \leq_b \begin{cases} e^{-\ell^2/(2n)}(\mu_{11}\log n)^k, & r \leq \min(\ell, n^{1/2}), \\ \dfrac{n}{\ell r} \cdot e^{r^2/(2n)-\ell^2/(2n)}(\mu_{11}\log n)^k, & n^{1/2} \leq r \leq \ell, \\ \log^{k+1} n, & r \geq \ell, \end{cases} \tag{50}$$

which is the counterpart of (46) with k replaced by $k+1$. It remains to prove that

$$\lim_{n\to\infty} \frac{\mu_{1,k+1}}{(\mu_{11})^{k+1}} = (k+1)!. \tag{51}$$

To this end we combine (42) for $r = 2$ and (41) for $r = 1$, and solving these two equations for $\mu_{1,k+1}, \mu_{2,k+1}$, we get

$$\mu_{1,k+1} = (k+1)\sum_{j=\ell}^{n-2} \frac{(n-2)_j}{n^j}\mu_{1+j-\ell,k}.$$

Using (48) for the summands with $j \in [\ell, 2\ell-1]$ and the bottom bound in (46) for $j \geq 2\ell$, we obtain

$$\begin{aligned} \mu_{1,k+1} &= (k+1)\mu_{1k}\sum_{j=\ell}^{n-2} \frac{(n-2)_j}{n^j} + R_1 + R_2 \\ &= (k+1)\mu_{1k}\mu_{11} + R_1 + R_2. \end{aligned} \tag{52}$$

Here

$$R_1 \leq_b n^{-\beta}\mu_{11}^k \sum_{j=\ell}^{n-2} \frac{(n-2)_j}{n^j} = n^{-\beta}(\mu_{11})^{k+1}, \tag{53}$$

and

$$\begin{aligned} R_2 &\leq_b \log^k n \sum_{j\geq 2\ell} \frac{(n-2)_j}{n^j} \\ &= (\mu_{11})^{k+1} \cdot \left(\frac{\log n}{\mu_{11}}\right)^k \cdot \frac{\sum_{j\geq 2\ell} e^{-j^2/(2n)}}{\sum_{j>\ell} e^{-j^2/(2n)}}. \end{aligned}$$

If $\ell \leq 0.5\sqrt{n\log n}$, then $\mu_{11} \gg n^{1/4} \gg \log n$, so that $R_2 = o(\mu_{11}^{k+1})$. If $\ell \geq 0.5\sqrt{n\log n}$, then—using $\liminf \mu_{11} > 0$—we have that

$$\begin{aligned} \frac{R_2}{(\mu_{11})^{k+1}} &\leq_b \log n \cdot \frac{e^{-(2\ell)^2/(2n)}}{e^{-\ell^2/(2n)}} \\ &\leq \log n \cdot e^{-(3/8)\log n} \to 0. \end{aligned}$$

So $R_2 = o((\mu_{11})^{k+1})$ in both cases. Combining this estimate with (52), (53), and the inductive hypothesis $\mu_{1k}/\mu_{11}^k \to k!$, we conclude that

$$\lim_{n\to\infty} \frac{\mu_{1,k+1}}{(\mu_{11})^{k+1}} = (k+1)!.$$

Thus the inductive step is verified completely, and this proves the lemma.□

Acknowledgement. It gives me a real pleasure to thank David Wilson for introducing me, few years ago, to the loop-erased random walk. I am very indebted to David for his interest in this study, many insightful comments, and for drawing my attention to the reference [4].

References

[1] D. J. Aldous, *A random walk construction of uniform spanning trees and uniform labelled trees*, SIAM Journal of Discrete Mathematics **3** (1990), 450–465.

[2] A. Broder, *Generating random spanning trees*, Foundations of Computer Sciencer (1989), 442–447.

[3] R. Burton and R. Pemantle, *Local characteristics, entropy and limit theorems for spanning trees and domino tilings via transfer- impedances*, The Annals of Probability **21** (1993), 1329–1371.

[4] P. Marchal, *Loop-erased random walks, spanning trees and Hamiltonian cycles*, Electronic Communications in Probability **5** (2000), 39–50.

[5] R. Pemantle, *Choosing a spanning tree for the integer lattice uniformly*, The Annals of Probability **19** (1991),1559-1574.

[6] D. B. Wilson, *Generating random spanning trees more quickly than the cover time*, Proceedings of the Twenty-Eigth Annual ACM Symposium on the Theory of Computing (1996), ACM, New York, 296–303.

Boris Pittel

Ohio State University, Columbus,
Ohio, USA
Department of Mathematics, The Ohio State University,
231 W 18-th Avenue, Columbus, Ohio, 43210, USA
bgp@math.ohio-state.edu

Trends in Mathematics, © 2002 Birkhäuser Verlag Basel/Switzerland

Convergence Rate for Stable Weighted Branching Processes

Rösler Uwe, Topchii Valentin, Vatutin Vladimir

ABSTRACT: *Let the martingale $W_n = m^{-n} Z_n$, where Z_n is a weighted branching process and $m = \mathbf{E} \sum_j T_j$ is the expected sum of the random factors T_j, converge to a limiting random variable W. We give conditions in terms of the factors under which W belongs to the domain of attraction or to the domain of normal attraction of an α–stable distribution with $1 < \alpha \le 2$. The convergence rate of W_n to W is evaluated in the sense that $W_n - W$ correctly normalized converges to a nondegenerate random variable* *.

One of the basic facts for supercritical Galton-Watson processes Z_n, $n = 0, 1, \ldots$ is the almost everywhere convergence of the martingale $W_n = m^{-n} Z_n$ to a limit W ($m = \mathbf{E}[Z_1 | Z_0 = 1]$) [1, 2]. The random variable W (given $Z_0 = 1$) satisfies the equation

$$W \stackrel{d}{=} \sum_{j=1}^{\hat{Z}} \frac{W^{(j)}}{m} \tag{1}$$

where $W^{(j)}$ are independent copies of W and $\hat{Z}$ is independent of $W^{(j)}$ and has the same distribution as the offspring Z_1 of a particle.
In addition, the convergence rate of W_n to W as $n \to \infty$ is of the order $m^{-\frac{n}{2}}$. More precisely, for finite variance $\sigma^2 \stackrel{def}{=} Var Z_1 < \infty$ the random variable

$$G_n = \sqrt{m^n \frac{m^2 - m}{\sigma^2}} (W_n - W)$$

converges in distribution to a limiting variable G (see [6] or Theorem A in [7]). Here G is a mixture of normal distributions with mean 0 and the random variance W and is described by the characteristic function $\mathbf{E} \exp\{itG\} = \mathbf{E} \exp\left\{-\frac{t^2 W}{2}\right\}$.

To formulate the up-to-date results in the case $\sigma^2 = \infty$ we recall some basic definitions and facts related with stable laws. Let $\varphi_Y(t) \stackrel{def}{=} \mathbf{E} \exp\{itY\}$ denote the characteristic function of a nondegenerate random variable Y. It is known (see [8], Chapter II, Section 6) that Y belongs to the domain of attraction of a stable law of index $1 < \alpha \le 2$ if and only if $\ln \varphi_Y(t)$ has the form

$$\ln \varphi_Y(t) = i\gamma t - |t|^\alpha M(t) B(t), \tag{2}$$

where $M(t)$ is a function which is positive in a vicinity of zero and slowly varying (in the sense of Karamata [8, 13]) as $t \to +0$, $M(t) \sim M(-t)$ as $t \to 0$, and

$$B(t) \stackrel{def}{=} 1 - i\beta \frac{t}{|t|} \tan \frac{\pi \alpha}{2},$$

*Supported by grants RFBR: 02-01-00266, INTAS: 99-01317, 00-0265.

where γ, and $|\beta| \leq 1$.
If the limit $M = \lim_{t \to 0} M(t) \in (0, \infty)$ in (2) exists then Y is said to belong to the domain of normal attraction of an α–stable law.
In both cases for Y_j, $j \in \mathbb{N}$ being iid copies of Y there exists a monotone deterministic sequence b_n such that the sequence $b_n \sum_{j=1}^n (Y_j - \mathbf{E}Y)$ converges, as $n \to \infty$ in distribution to a stable law with characteristic function $\varphi(t) = \exp\{-|t|^\alpha B(t)\}$.
Now we return to the supercritical Galton-Watson processes and take $Y = Z_1$. It is known that in general Z_1 belongs to the domain of attraction of a stable law of index $1 < \alpha \leq 2$ if and only if W belongs to the same domain (see [7] Theorems 1 and 2). If this is the case, the rate of convergence is of order $m^n b_{[m^n]}$, or, more precisely,

$$G_n = (m^\alpha - m)^{\frac{1}{\alpha}} m^n b_{[m^n]} (W_n - W)$$

converges in distribution to a limiting random variable G, where G is a mixture of α–stable distributions with characteristic function $\mathbf{E}\varphi^W(t)$.
In the present note we establish similar results for the weighted branching processes [11, 12].
We briefly recall the definition of weighted branching processes.
Let $V = \bigcup_{n=0}^\infty \mathbb{N}^n$ be the set of finite words $v = (v_1, v_2, \ldots, v_n)$ over the alphabet of natural numbers. The length n of the word v is denoted by $|v|$, and sometimes is called the generation.
By means of V we construct a space of elementary outcomes Ω, where an elementary outcome $\omega \in \Omega$ is a result of assigning to each knot $v \in V$ a vector $\mathbf{t}(v) = (t_1(v), t_2(v), \ldots)$, $t_j(v) \in \mathbb{R}$, and take the natural σ-algebra $\mathcal{F}$ on Ω. Now we define a probability measure $\mathcal{P}$ on $\mathcal{F}$ by assigning to each vertex $v \in V$ random vectors $\mathbf{T}(v)$ with values in $R^{\mathbb{N}}$. The vector $\mathbf{T}(v, \omega) = \mathbf{T}(v) = (T_1(v), T_2(v), \ldots)$, respectively, the coordinates $T_j(v), j \in \mathbb{N}$ are called factors. We use $T = (T_1, T_2, \ldots)$ for the coordinates, suppressing the ω and the v if possible. Notice that we allow arbitrary dependence of the factors $T_j(v)$, $j \in \mathbb{N}$ for fixed knot v. The (random) weight $L(v)$ or length of a knot $v \in V$ is recursively given by $L(\emptyset) = 1$

$$L(vj) = L(v)T_j(v),$$

which is the weight $L(v)$ of the mother v times a random factor $T_j(v)$. We skip the detailed construction of the resulting probability space $(\Omega, \mathcal{F}, \mathcal{P})$, since it can be done by the standard procedure.

Let $Z_0 \stackrel{def}{=} L(\emptyset) = 1$, $Z_n \stackrel{def}{=} \sum_{|v|=n} L(v) = \sum_{|v|=n-1} L(v) \sum_j T_j(v)$ be the total weight of the individuals of the n–th generation and let

$$m \stackrel{def}{=} \mathbf{E}Z_1 = \mathbf{E}\sum_j T_j \neq 0.$$

One can check that $W_n = m^{-n} Z_n$ is a martingale which converges (under mild conditions of Proposition 4 below) to W. Observe, that $W_1 = m^{-1} \sum_j T_j$.

For several models of ordinary (non-weighted) branching processes Z_n the problem of the relationship between the asymptotic behavior of the distribution tail for the offspring size of a single particle Z_1 and that of the appropriate limiting variable

W has been studied in [4, 5, 7, 10]. Here we restrict ourselves to the case of stable distributions with exponent $\alpha \in (1, 2]$ and generalize the mentioned convergence results due to Heyde [7] in several directions.

In the sequel we shall use the representations

$$\ln \varphi_{W_1}(t) = it - |t|^\alpha M(t)B(t), \tag{3}$$

$$\ln \varphi_W(t) = it - |t|^\alpha M_\infty(t)B_\infty(t) \tag{4}$$

with the obvious meaning.

Introduce the notation

$$m^-(\alpha) \stackrel{def}{=} \mathbf{E}\sum_j |T_j|^\alpha 1\!\!1_{T_j<0}, \quad m^+(\alpha) \stackrel{def}{=} \mathbf{E}\sum_j |T_j|^\alpha 1\!\!1_{T_j>0},$$

$$m(\alpha) \stackrel{def}{=} \mathbf{E}\sum_j |T_j|^\alpha, \quad \Delta m(\alpha) \stackrel{def}{=} m^+(\alpha) - m^-(\alpha).$$

Our first main result is the following statement.

Theorem 1. *Let $1 < \alpha \le 2$ and assume that $m \ne 0$, $m(1) < \infty$, $m(\alpha) < |m|^\alpha$ and W_n converges to W in L_1.*

i) Suppose there exist constants $0 < C_1 < C_2 < \infty$ such that

$$C_1 < |T_j| < C_2 \tag{5}$$

for all $T_j \ne 0$.

Then W belongs to the domain of attraction of a stable law of index α if and only if Z_1 belongs to the domain of attraction of a stable law of the same index α. In addition, in this case

$$M_\infty(t) \sim M(t)\frac{|m|^\alpha}{|m|^\alpha - m(\alpha)}$$

as $t \to 0$ and

$$B(t) = 1 - i\beta\frac{t}{|t|}\tan\frac{\pi\alpha}{2}, \tag{6}$$

$$B_\infty(t) = 1 - i\beta\frac{t}{|t|}\frac{|m|^\alpha - m(\alpha)}{|m|^\alpha - \Delta m(\alpha)}\tan\frac{\pi\alpha}{2}. \tag{7}$$

ii) Assume there exists an $\alpha_1 > \alpha$ such that

$$m(\alpha_1) < |m|^{\alpha_1}. \tag{8}$$

Then W belongs to the domain of normal attraction of a stable law of index $\alpha \in (1, 2]$ if and only if Z_1 belongs to the domain of normal attraction of a stable law of the same index α. In addition, in this case

$$M_\infty \stackrel{def}{=} \lim_{t\to 0} M_\infty(t) = \frac{|m|^\alpha}{|m|^\alpha - m(\alpha)}\lim_{t\to 0} M(t) = \frac{|m|^\alpha}{|m|^\alpha - m(\alpha)} M$$

and representations (6) and (7) are valid.

Observe that under the conditions of Theorem 1 inequality (5) implies $m(\alpha) < \infty$ for all $\alpha > 0$ and, as we show below there exists an $\alpha_1 > \alpha$ such that relation (8) holds.

There is a well known connection between the fact that a distribution belongs to the domain of attraction of an $\alpha-$stable distribution and the tail behavior of this distribution ([8], Chapter II, Section 6). Using this connection it is not difficult to establish by direct calculations the validity of the following corollary being an extension and generalization of Theorem 2 in [5] which was stated for branching random walks.

Corollary 2. *Under the basic conditions of Theorem 1 the following representations are equivalent for* $1 < \alpha < 2$*:*

$$\mathbf{P}(W_1 < -x) = \frac{p_1 + o(1)}{x^\alpha} H(x), \qquad \mathbf{P}(W_1 > x) = \frac{p_2 + o(1)}{x^\alpha} H(x)$$

as $x \to \infty$ *with* $p_1, p_2 \geq 0$, $p_1 + p_2 > 0$, *and*

$$\mathbf{P}(W < -x) = \frac{q_1 + o(1)}{x^\alpha} H(x), \qquad \mathbf{P}(W > x) = \frac{q_2 + o(1)}{x^\alpha} H(x)$$

as $x \to \infty$ *with* $q_1, q_2 \geq 0$, $q_1 + q_2 > 0$. *Here* $H(x)$ *is a slowly varying function at* ∞ *if* (5) *is true and, in addition,* $\lim_{x\to\infty} H(x) = H \in (0, \infty)$ *if* (8) *is valid.*

Further, in this case

$$\begin{aligned} q_1 &= |m|^\alpha \frac{p_1 + p_2}{2(|m|^\alpha - m(\alpha))} + |m|^\alpha \frac{p_1 - p_2}{2(|m|^\alpha - \Delta m(\alpha))}, \\ q_2 &= |m|^\alpha \frac{p_1 + p_2}{2(|m|^\alpha - m(\alpha))} - |m|^\alpha \frac{p_1 - p_2}{2(|m|^\alpha - \Delta m(\alpha))}. \end{aligned}$$

Remark. This corollary complements a result of Liu [9] in which the case of nonnegative T_j was investigated and where (in our terms) under the assumption that there exists an χ such that

$$\mathbf{E} \sum_j T_j^\chi = 1, \quad \mathbf{E} \sum_j T_j^\chi \log^+ T_j < \infty, \quad \mathbf{E} \left(\sum_j T_j \right)^\chi < \infty$$

it is shown that $\lim_{x\to\infty} x^\chi \mathbf{P}(W > x)$ exists and is strictly positive.

Observe that under our conditions the equality $\mathbf{E} \sum_j T_j^\chi = 1$ for some $\chi > \alpha$ implies $\mathbf{E} \left(\sum_j T_j \right)^\chi = \infty$. And this is an essential difference with [9]. Indeed, in our paper the assumption that the respective functionals of $\mathbf{T}$ belong to a stable distribution means that there is a large number of $T_j \neq 0$ each of which is bounded from above and is separated from zero, while in [9] the number of $T_j \neq 0$ may be finite or infinite but for a fixed N_0 the distribution tail of $\sum_{j<N_0} T_j$ should be "heavy".

The sequence $W_n = m^{-n} Z_n$ is a martingale with respect to the filtration $\mathcal{F}_n \stackrel{def}{=} \sigma\{T(v),\ |v| < n\}$ (or also $\sigma\{L(v),\ |v| \leq n\}$) for $n \geq 1$. Under Doob's condition $\sup_n \mathbf{E}|W_n| < \infty$ this martingale converges almost everywhere to a random variable W, which might be degenerate.

For $\alpha \geq 0$ we consider

$$Z_n(\alpha) \stackrel{def}{=} \sum_{|v|=n} |L(v)|^\alpha.$$

Recall that $m(\alpha) = \mathbf{E}Z_1(\alpha)$. Let $W_n(\alpha) \stackrel{def}{=} m^{-n}(\alpha)Z_n(\alpha)$.

The next result gives the convergence rate of W_n to W and generalizes Theorem 2 in [7].

Theorem 3. *Let Z_1 belong to the domain of normal attraction of a stable law of index $\alpha \in (1,2]$ with $\lim_{t\to 0} M(t) = M > 0$ in representation* (3). *Suppose that*

1) $m \neq 0,\ m(1) < \infty$ *and* $m(\alpha) < |m|^\alpha$*;*

2) W_n *converges in* L_1 *to* W*;*

3) *there is an* $\alpha_1 > \alpha$ *such that* (8) *is true;*

4) *the condition*

$$\mathbf{E}\sum_j |T_j|^\alpha \ln |T_j|^\alpha < m(\alpha)\ln m(\alpha) \tag{9}$$

is valid.

Then

i) there exists a sequence of constants $\{c_n\}$ such that $c_n^{-1}c_{n+1} \to 1$ as $n \to \infty$ and

$$c_n^{-1}W_n(\alpha) \underset{n\to\infty}{\to} Y \tag{10}$$

in probability to a random variable Y which is strictly positive whenever the process $W_n(\alpha)$ survives;

ii) for any $t \in (-\infty, \infty)$

$$\lim_{n\to\infty} \mathbf{E}\exp\left\{it\left(m^n(\alpha)c_n\right)^{-1/\alpha}(W - W_n)\right\}$$
$$= \mathbf{E}\exp\left\{-|t|^\alpha M \frac{|m|^\alpha}{|m|^\alpha - m(\alpha)} Y\left(1 - i\beta^*\frac{t}{|t|}\tan\frac{\pi\alpha}{2}\right)\right\},$$

where

$$\beta^* \stackrel{def}{=} \begin{cases} 1 & \text{if } m^-(\alpha) = 0, \\ 0 & \text{if } m^-(\alpha) \neq 0. \end{cases}$$

For the Galton-Watson process the factors T_j are 0 or 1 and, therefore $L(v)$ are 0 or 1 as well. Since the Galton-Watson dies out almost everywhere as $m \leq 1$, only the supercritical case $m > 1$ is interesting. This implies our condition $m(\alpha) \equiv m < |m|^\alpha$. The factors are bounded below and above. Our Theorem 1 contains convergence in domains of attractions for the Galton-Watson process.

It is necessary to note that for Galton-Watson processes we always have $\beta^* = 1$ and $m^-(\alpha) = 0$ while the case $m^-(\alpha) > 0$ has not been studied before and gives essentially new results.

One of the key problems of the given paper is to investigate the convergence rate of W_n to W. If all T_j are nonnegative then condition (9) for $\alpha = 1$ is a necessary one for the convergence of W_n to W with $W \neq 0$ by Biggins' criterion (see Theorem

5). Convergence conditions for the case $T_j \in \mathbb{R}$ are investigated in [12]. To get the convergence of $W_n(\alpha)$ to an $W(\alpha)$ with $W(\alpha) \neq 0$ we need (again by Biggins' criterion) condition (9) in both cases.

We say that a random variable X satisfies the fixed point equation with factors $\mathbf{T}/m = (T_1/m, T_2/m, \dots)$ if

$$X \stackrel{d}{=} \sum_j \frac{T_j}{m} X_j, \tag{11}$$

where X_j are independent copies of X and $\mathbf{T}$, X_j, $j \in \mathbb{N}$ are independent. The right-hand side is understood in the sense $\sum_{j=1}^{n} T_j X_j$ converges as $n \to \infty$ in distribution.

Proposition 4. *Assume that* $0 < m(1) < \infty$ *and* $\sup_n \mathbf{E}|W_n| < \infty$. *Then the limiting random variable* W *satisfies the fixed point equation* (11) *with factors* $\mathbf{T}/m$.

As it was mentioned the ordinary Galton-Watson process is a special case of the weighted branching processes with factors in $\{0, 1\}$. In this case $\{W_n\}$ is a positive martingale which converges according to Doob almost everywhere to a random variable W. A theorem by Kesten-Stigum [2] states that $\mathbf{E}W$ is either 1 or 0. The equality $\mathbf{E}W = 1$ is equivalent to the so-called $L \log L$ condition $\mathbf{E}(Z_1 \ln Z_1) < \infty$. (We use $0 \ln 0 = 0$.) In both cases W satisfies the fixed point equation for the factor T.

The branching random walk is a special case of the weighted branching processes with finitely many positive factors (see [3] and [11] for more details). Again $\{W_n\}$ is a positive martingale and converges according to Doob almost everywhere to W. The limit W satisfies the fixed point equation of form (11). The problem of L_1 convergence for branching random walk is settled in the following theorem.

Theorem 5. [3] *$W_n(\alpha)$ converges to some random variable called $W(\alpha)$ in L_1 if and only if*

$$\mathbf{E}W_1(\alpha) \ln W_1(\alpha) < \infty \tag{12}$$

and inequality (9) *is true.*

If only inequality (9) *is satisfied then there exists a sequence of constants* $\{c_n\}$ *such that* $c_n^{-1} c_{n+1} \to 1$ *as* $n \to \infty$ *and*

$$c_n^{-1} W_n(\alpha) \underset{n\to\infty}{\to} Y$$

in probability to a random variable Y, *which is strictly positive whenever the process* $W_n(\alpha)$ *survives.*

Remark. Under condition (12) the second part of Theorem 5 is valid for $c_n \equiv 1$ and $Y = W(\alpha)$. It will be convenient for us to use equation (10) in both cases.

According to Proposition 4 W satisfies the fixed point equation

$$W \stackrel{d}{=} \sum_j \frac{T_j}{m} W^{(j)} \tag{13}$$

where $W^{(j)}$ are independent copies of W and $\mathbf{T} = (T_1, T_2, \ldots)$ is independent of $W^{(j)}$. Observe that equation (1) is a particular case of (13). An equivalent description via characteristic functions is

$$\varphi_W(t) = \mathbf{E}\prod_j \varphi_W\left(t\frac{T_j}{m}\right).$$

The fixed point equation, assuming $\mathbf{E}W = 1$, rewrites

$$W - 1 \stackrel{d}{=} \sum_j \frac{T_j}{m}\left(W^{(j)} - 1\right) + \left(\sum_j \frac{T_j}{m} - 1\right).$$

In terms of characteristic functions

$$\varphi_{W-1}(t) = \varphi_{W_1-1}(t) + \mathbf{E}\exp\{it(W_1 - 1)\}\left(\prod_j \varphi_{W-1}\left(t\frac{T_j}{m}\right) - 1\right). \qquad (14)$$

Proposition 6. *Let* $0 < m(\alpha) < \infty$. *Then*

$$\mathbf{E}\sum_{|v|=n} |L(v)|^\alpha \frac{L(v)}{|L(v)|} = (\Delta m(\alpha))^n$$

and the sequence

$$\xi_n \stackrel{def}{=} (\Delta m(\alpha))^{-n} \sum_{|v|=n} |L(v)|^\alpha \frac{L(v)}{|L(v)|}$$

is a martingale with respect to the sequence of σ*-algebras* $\mathcal{F}_n = \sigma\{L(v), |v| \le n\}$.

The proposition can be checked by induction arguments.

Sketch of the Proof of Theorem 1. In order to avoid complicated notation we always assume $m = 1$ and write Z_n for W_n and Z for W. We use representations (3) and (4).

First we show that if Z belongs to the domain of attraction of an α-stable law $1 < \alpha \le 2$ then Z_1 also possesses this property.

Using (14) for $W = Z$ we have

$$\varphi_{Z_1-1}(t) = \varphi_{Z-1}(t) - E\left(\exp\{it(Z_1 - 1)\}\left(\prod_j \varphi_{Z-1}(tT_j) - 1\right)\right).$$

Hence, setting

$$a \stackrel{def}{=} -|t|^\alpha \sum_j |T_j|^\alpha B_\infty(tT_j) M_\infty(tT_j) \quad b \stackrel{def}{=} -|t|^\alpha M_\infty(t) \sum_j |T_j|^\alpha B_\infty(tT_j),$$

we get

$$\varphi_{Z_1-1}(t) = \varphi_{Z-1}(t) - \mathbf{E}(\exp\{it(Z_1 - 1)\}(e^a - 1)),$$

or

$$\begin{aligned}\varphi_{Z_1-1}(t) &= \varphi_{Z-1}(t)e^{-\mathbf{E}b} + (\varphi_{Z-1}(t)-1)\left(1-e^{-\mathbf{E}b}\right) + \left(2-\mathbf{E}e^{a}-e^{-\mathbf{E}b}\right)\\ &\quad - \mathbf{E}\left[(\exp\{it(Z_1-1)\}-1)(e^{a}-1)\right] \stackrel{def}{=} Q(1)+Q(2)+Q(3)+Q(4).\end{aligned}$$

By standard arguments one can show that

$$Q(1) = \exp\left\{-|t|^{\alpha}(1-m(\alpha))M_{\infty}(t)B(t)\right\},$$

where $B(t)$ is given in (6) and $Q(j) = o(Q(1))$ as $t \to 0$ for $j = 2,3,4$.
Collecting the estimates above we deduce the representation

$$\ln \varphi_{Z_1}(t) = it - |t|^{\alpha}(1-m(\alpha))M_{\infty}(t)B(t)(1+o(1))$$

which proves the direct statement of point $i)$ of Theorem 1.
Now we turn to the converse statement of point $i)$ of Theorem 1: Z_1 in the domain of attraction of an $\alpha-$stable law $1 < \alpha \leq 2$ implies Z in the same domain.
This time we use the following simple identity

$$Z_n - 1 = (Z_{n-1}-1) + \sum_{|v|=n} L(v)(Z_1^{(v)}-1)$$

where the random variables $Z_1^{(v)} = \sum_j T_j(v)$ are independent for different v. Hence, letting $Z_0 = 1$, we obtain

$$\begin{aligned}1-\varphi_{Z_n-1}(t) &= 1-\mathbf{E}e^{it(Z_{n-1}-1)}\prod_{|v|=n-1}\varphi_{Z_1-1}(tL(v))\\ &= \sum_{k=0}^{n-1}\mathbf{E}e^{it(Z_k-1)}\Big(1-\prod_{|v|=k}\varphi_{Z_1-1}(tL(v))\Big).\end{aligned} \tag{15}$$

Denote

$$\begin{aligned}\tilde{a}_k &\stackrel{def}{=} -|t|^{\alpha}\sum_{|v|=k}|L(v)|^{\alpha}M(tL(v))B(tL(v)),\\ \tilde{b}_k &\stackrel{def}{=} -|t|^{\alpha}M(t)\sum_{|v|=k}|L(v)|^{\alpha}B(tL(v)).\end{aligned}$$

In the subsequent arguments we deal with the representation (recall, $m = 1$)

$$\begin{aligned}\varphi_{Z-1}(t) &= \sum_{k=0}^{\infty}\mathbf{E}e^{it(Z_k-1)}\left(1-e^{\tilde{b}_k}\right)\\ &\quad + \sum_{k=0}^{\infty}\mathbf{E}e^{it(W_k-1)}\left(e^{\tilde{b}_k}-e^{\tilde{a}_k}\right) \stackrel{def}{=} Q(5)+Q(6)\end{aligned}$$

which follows from (15).

By standard but a bit lengthy arguments one can demonstrate that

$$\lim_{t\to 0} \frac{Q(5)}{|t|^\alpha M(t)B_\infty(t)} = -\frac{1}{1-m(\alpha)},$$

where $B_\infty(t)$ is given in (7) and $Q(6) = o(Q(5))$ as $t \to 0$. This proves the first equivalence i).
The proof of the second equivalence ii) is similar to the proof of i).

Lemma 7. *Let $1 < \alpha \le 2$ and assume that $m \ne 0$, $m(1) < \infty$, $m(\alpha) < |m|^\alpha$ and W_n converges to W in L_1.*
Then

$$\varphi_{W-W_n}(t) = \mathbf{E} \prod_{|v|=n} \exp\left\{-it\frac{L(v)}{m^n}\right\} \varphi_W\left(t\frac{L(v)}{m^n}\right) = \mathbf{E}\prod_{|v|=n} \varphi_{W-1}\left(t\frac{L(v)}{m^n}\right).$$

The proof of the lemma is straightforward and is omitted.

We demonstrate now Theorem 3, a generalization of Theorem 2 [7].
Proof of Theorem 3. Point i) follows from Theorem 5 for $|T_j|^\alpha$ and the subsequent remark.
Let us prove point ii). Without loss of generality we may take $m = 1$. Thus, $W_n = Z_n$ and $W = Z$.
Let $t \ne 0$ and $\delta > 0$ be fixed. Denote $a_n = (m^n(\alpha)c_n)^{-1/\alpha}$ (recall Theorem 5) and put

$$\begin{aligned}
\hat{a} &\stackrel{def}{=} -|t|^\alpha a_n^\alpha \sum_{|v|=n} |L(v)|^\alpha B_\infty(ta_nL(v))M_\infty(ta_nL(v)) \stackrel{def}{=} -|t|^\alpha \zeta_a,\\
\hat{b} &\stackrel{def}{=} -|t|^\alpha a_n^\alpha M_\infty \sum_{|v|=n} |L(v)|^\alpha B_\infty(ta_nL(v)) \stackrel{def}{=} -|t|^\alpha \zeta_b,\\
D_n &= D_n(\delta) \stackrel{def}{=} \left\{\omega : \sum_{|v|=n} a_n|L(v)| \le \delta^n\right\}.
\end{aligned}$$

With this notation we have

$$\begin{aligned}
\varphi_{a_n(Z-Z_n)}(t) &= \varphi_{Z-Z_n}(ta_n) = \mathbf{E}\prod_{|v|=n} \varphi_{Z-1}(ta_nL(v)) = \mathbf{E}e^{\hat{a}}\\
&= \mathbf{E}e^{\hat{b}} + \mathbf{E}1\!\!1_{D_n^c}\left(e^{\hat{a}} - e^{\hat{b}}\right) + \mathbf{E}1\!\!1_{D_n}\left(e^{\hat{a}} - e^{\hat{b}}\right). \qquad (16)
\end{aligned}$$

The scheme of proving point ii) is as follows. First we show that the limit of the first summand in the right-hand side of(16) exists as $n \to \infty$ and coincides with the limit we are searching for. Then we establish existence of an $\delta \in (0,1)$ such that $\mathbf{P}(D_n^c) \underset{n\to\infty}{\longrightarrow} 0$, proving that the second summand is negligible as $n \to \infty$. Finally, we demonstrate that the difference $(\zeta_a - \zeta_b)1\!\!1_{D_n}$, being dependent on t, also vanishes as $n \to \infty$ (in the sense of weak convergence). The mentioned weak convergence is equivalent to the convergence to zero of the third term in the right-hand side of (16). This will complete the proof of Theorem 3.

We proceed to fulfill the plan.

- $\mathbf{E}e^{\hat{b}} \underset{n\to\infty}{\to} \mathbf{E}\exp\left\{-|t|^\alpha M_\infty Y\left(1-i\beta^*\frac{t}{|t|}\tan\frac{\pi\alpha}{2}\right)\right\}$, where

$$\beta^* \stackrel{def}{=} \begin{cases} 1 & \text{if } m^-(\alpha)=0, \\ 0 & \text{if } m^-(\alpha)\neq 0. \end{cases}$$

Note that $\beta = 1$ in the definition of $B_\infty(t)$ (see (7)) if $m^-(\alpha) = 0$ (see Theorem 2.6.1 in [8]). The possibility $m^+(\alpha) = 0$ is excluded by the assumption $m = 1$ while in the remaining cases β may take any value in $[-1, 1]$ admissible for the stable law in question.

We consider separately the real and imaginary parts of the exponent.

By Theorem 5 (the symbol $\stackrel{d}{\to}$ denotes convergence in distribution)

$$\operatorname{Re}\hat{b} = -|t|^\alpha a_n^\alpha M_\infty \sum_{|v|=n} |L(v)|^\alpha = -|t|^\alpha M_\infty c_n^{-1} W_n(\alpha) \stackrel{d}{\to} -|t|^\alpha M_\infty Y$$

as $n \to \infty$. On the other hand (recall Proposition 6)

$$\begin{aligned}\operatorname{Im}\hat{b} &= |t|^\alpha a_n^\alpha M_\infty \frac{t}{|t|} \sum_{|v|=n} |L(v)|^\alpha \frac{L(v)}{|L(v)|}\beta \frac{1-m(\alpha)}{1-\Delta m(\alpha)} \\ &= |t|^\alpha M_\infty \frac{t}{|t|}\left(\frac{\Delta m(\alpha)}{m(\alpha)}\right)^n c_n^{-1}\xi_n \beta \frac{1-m(\alpha)}{1-\Delta m(\alpha)}.\end{aligned}$$

If $m^-(\alpha) = 0$ then $L(v) \geq 0$ with probability 1, and, therefore, $\beta = 1$, $\Delta m(\alpha) = m(\alpha)$ and $\xi_n = W_n(\alpha)$. Hence, the same as for $\operatorname{Re}\hat{b}$ (see Theorem 5)

$$\operatorname{Im}\hat{b} \stackrel{d}{\to} |t|^\alpha M_\infty Y \frac{t}{|t|}\tan\frac{\pi\alpha}{2}.$$

If $m^-(\alpha) > 0$ then $\Delta m(\alpha)/m(\alpha) < 1$ and in view of $\frac{c_n}{c_{n+1}} \underset{n\to\infty}{\to} 1$ it follows that

$$c_n^{-1}\left(\frac{\Delta m(\alpha)}{m(\alpha)}\right)^n \underset{n\to\infty}{\to} 0.$$

Since ξ_n is a martingale by Proposition 6, we get immediately $\operatorname{Im}\hat{b} \stackrel{d}{\to} 0$ completing the proof of this point.

- $\lim_{n\to\infty} \mathbf{P}(D_n^c) = 0$ for any fixed t and $\delta < 1$ such that

$$\frac{m(\alpha_2)}{m^{\alpha_2/\alpha}(\alpha)} < \delta^{\alpha_2}$$

for some $\alpha_2 \in (\alpha, \alpha_1]$.

First we show that one can find δ meeting the restrictions above. Condition (9) implies $\alpha m'(\alpha) = \mathbf{E}\sum_i |T_j|^\alpha \ln |T_j|^\alpha < m(\alpha)\ln m(\alpha) < 0$ if $m(\alpha) < 1$. Therefore, $m(t)$ is decreasing in t in a right vicinity of α. In addition, taking into account the value of the derivative and its estimate we have

$$\left(m^{1/t}(t)\right)'\Big|_{t=\alpha} = -\frac{1}{\alpha^2} m^{1/\alpha}(\alpha)\left(\ln m(\alpha) - \frac{\mathbf{E}\sum_j |T_j|^\alpha \ln |T_j|^\alpha}{m(\alpha)}\right) < 0,$$

that is, $m^{1/t}(t)$ decreases in t in the same right vicinity of α. This means that there exists $\alpha_2 \in (\alpha, \alpha_1]$ such that $\dfrac{m(\alpha_2)}{m^{\alpha_2/\alpha}(\alpha)} < 1$ establishing the existence of the needed δ.

By the Chebyshev inequality we obtain

$$\mathbf{P}(D_n^c) \le \frac{\mathbf{E}a_n^{\alpha_2}\sum_{|v|=n}|L(v)|^{\alpha_2}}{\delta^{n\alpha_2}} = \frac{m^n(\alpha_2)}{(\delta^{\alpha_2} m^{\alpha_2/\alpha}(\alpha))^n c_n^{\alpha_2}}. \tag{17}$$

Since $c_n^{-1}c_{n+1} \underset{n\to\infty}{\to} 1$, there exists an $\epsilon > 0$ such that

$$\frac{m(\alpha_2)}{\delta^{\alpha_2} c_n^{\alpha_2/n} m^{\alpha_2/\alpha}(\alpha)} < 1 - \epsilon$$

for all sufficiently large n. Combining this estimate with (17) completes the proof of the desired statement.

- $\lim\limits_{n\to\infty} \mathbf{E}\left(e^{\hat{a}} - e^{\hat{b}}\right) 1\!\!1_{D_n^c} = 0$ for any fixed t.

This statement immediately follows from the previous point, since for $\operatorname{Re}\hat{a} < 0$ and $\operatorname{Re}\hat{b} < 0$ the inequality $\left|\mathbf{E}\left(e^{\hat{a}} - e^{\hat{b}}\right) 1\!\!1_{D_n^c}\right| \le 2\mathbf{P}(D_n^c)$ is true.

- $\lim\limits_{n\to\infty} (\zeta_a - \zeta_b)\, 1\!\!1_{D_n} \overset{d}{=} 0$ for any fixed t.

The following inequality is valid on D_n:

$$a_n \sup_{|v|=n} |L(v)| \le a_n \sum_{|v|=n} |L(v)| \le \delta^n.$$

Therefore, $d_n \overset{def}{=} \sup\limits_{|v|=n} |M_\infty(ta_nL(v)) - M_\infty| \underset{n\to\infty}{\to} 0$ on D_n. Note, finally, that

$$a_n^\alpha \sum_{|v|=n} |L(v)|^\alpha 1\!\!1_{D_n} \le a_n^\alpha \sum_{|v|=n} |L(v)|^\alpha \underset{n\to\infty}{\overset{d}{\to}} Y.$$

Hence it follows easily that for some constant $C_3 < \infty$

$$|\zeta_a - \zeta_b| 1\!\!1_{D_n} \le C_3 a_n^\alpha \sum_{|v|=n} |L(v)|^\alpha 1\!\!1_{D_n} d_n \underset{n\to\infty}{\overset{d}{\to}} 0.$$

References

[1] K.B. Athreya and P. Ney, *Branching Processes.* Die Grundlehren der mathematischen Wissenschaften, Band 196, Springer Verlag, New York-Heidelberg, 1972.

[2] S. Asmussen and H. Hering, *Branching Processes.* Progress in Probability and Statistics, Birkhäuser, Boston-Basel-Stuttgart, 1983.

[3] J.D. Biggins and A.E. Kyprianou, *Seneta-Heyde norming in the branching random walk.* Ann. Probab. **25**, No.1 (1997), 337-360.

[4] N.H. Bingham and R.A. Doney, *Asymptotic properties of supercritical branching processes. I: The Galton-Watson process.* Adv. Appl. Probab. **6** (1974), 711-731.

[5] N.H. Bingham and R.A. Doney, *Asymptotic properties of supercritical branching processes. II: Crump–Mode and Jirina processes.* Adv. Appl. Probab. **7** (1975), 66-82.

[6] C.C. Heyde, *A rate of convergence result for the super-critical Galton-Watson process.* J. Appl. Probab. **7** (1970), 451-454.

[7] C.C. Heyde, *Some central limit analogues for super-critical Galton-Watson processes.* J. Appl. Probab. **8** (1971), 52-59.

[8] I.A. Ibragimov, Ju.V. Linnik *Independent and stationary sequences of random variables.* Edited by J.F.C. Kingman Wolters-Noordhoff Series of Monographs and Textbooks on Pure and Applied Mathematics. Groningen, The Netherlands: Wolters-Noordhoff Publishing Company. 443 p. (1971).

[9] Q. Liu, *On generalized multiplicative cascades.* Stochastic Processes and their Applications **86** (2000), 263-286.

[10] A. de Meyer, *On a theorem of Bingham and Doney.* J. Appl. Probab. **19** (1982), 217-220.

[11] U. Rösler, *The weighted branching process.* Dynamics of complex and irregular systems (Bielefeld, 1991), 154-165, Bielefeld Encounters in Mathematics and Physics VIII, World Science Publishing, River Edge, NJ, 1993.

[12] U. Rösler, V.A. Topchii, V.A. Vatutin, *Convergence Conditions for Weighted Branching Processes.* Discrete Mathematics **12**, No 1 (2000), 7-23.

[13] E. Seneta, *Regularly varying functions.* Lect. Notes in Math., v. 508, 1976, Springer-Verlag, New York.

Rösler Uwe

Mathematisches Seminar
der Christian-Albrechts-Universität zu Kiel
Ludewig-Meyn-Str. 4, 24098 Kiel
Germany
roesler@math.uni-kiel.de

Topchii Valentin
Omsk Branch of Sobolev Institute of Mathematics
Pevtcova Street, 13, 644099 Omsk
Russia
topchij@iitam.omsk.net.ru

Vatutin Vladimir
Steklov Mathematical Institute
ul. Gubkina, 8
117966, GSP-1 Moscow
Russia
vatutin@mi.ras.ru

Trends in Mathematics, © 2002 Birkhäuser Verlag Basel/Switzerland

Reduced Branching Processes in Random Environment

Vatutin Vladimir and Dyakonova Elena

ABSTRACT: *Let Z_n be the number of particles at time n in a critical branching process in random environment and $Z_{m,n}$ be the number of particles in this process at time $m \leq n$ which have non-empty offspring at time n. We prove limit theorems for the process $\{Z_{[nt],n},\ t \in (0,1]\}$ conditioned on the event $\{Z_n > 0\}$. Quenched and annealed approaches are considered.* *

1 Introduction

Let $\{Z_n,\ n \in N_0\}$ be a branching process in random environment $\{\pi_n,\ n \in N_0\}$ where

$$\pi_n = \left\{\pi_n^{(0)}, \pi_n^{(1)}, \pi_n^{(2)}, ...\right\},\ \pi_n^{(i)} \geq 0,\ \sum_{i=0}^{\infty} \pi_n^{(i)} = 1,\ n \in N_0 = \{0, 1, 2, ...\},$$

and the tuples π_n are identically distributed and independent. By definition of branching processes in random environment (BPRE) we have

$$Z_0 = 1,\ E\left\{s^{Z_{n+1}} \middle| f_0, f_1, ..., f_n; Z_0, Z_1, ..., Z_n\right\} = (f_n(s))^{Z_n},$$

where

$$f_n(s) = \sum_{i=0}^{\infty} \pi_n^{(i)} s^i.$$

We assume that Z_n is critical, that is $E \ln f_0^{'}(1) = 0$. Such processes were investigated by many authors (see [1] and the relevant bibliography in [8]). One of the important characteristics of a branching process is its genealogical tree. It may be described to a certain extent by the so-called reduced branching process $\{Z_{m,n},\ 0 \leq m \leq n\}$ in which $Z_{m,n}$ is equal to the number of particles in the process $\{Z_k,\ 0 \leq k \leq n\}$ at time $m \leq n$ each of which has a non-empty offspring at time n. Reduced Galton-Watson branching processes were introduced by Fleischmann and Prehn [4], who analyzed the subcritical case. The critical reduced Galton-Watson processes were investigated by Zubkov [10] and Fleischmann and Siegmund-Schultze [5]. The first results for reduced branching processes in random environment were obtained by Borovkov and Vatutin [2] and Fleischmann and Vatutin [6]. In the framework of the annealed approach they studied the case of iid fractional-linear generating functions f_n. The present paper deals with both annealed and quenched approaches.

Let $X_k = \ln f_{k-1}^{'}(1)$ and $\eta_k = f_{k-1}^{''}(1)(f_{k-1}^{'}(1))^{-2}$, $k = 1, 2, ...$ It is known that properties of a critical BPRE essentially depend on the accompanying random

*Supported by grants RFBR: 02-01-00266, 00-15-96136, INTAS 99-01317, and 436RUS113/683/0-1(R).

walk $S_0 = 0,\ S_n = X_1 + ... + X_n$. We assume that $\{S_n\}_{n\geq 0}$ is nonlattice and satisfies the following conditions

$$EX_1 = 0,\ \sigma^2 := EX_1^2 \in (0,\infty), \tag{1}$$

and

$$E\eta_1\left(1 + X_1^+\right) < \infty\,. \tag{2}$$

The symbols E and P are used below for the expectation and probability with respect to the measure over environment, while the symbols E_π and P_π stand for the expectation and probability under fixed environment $\pi = (\pi_1, \pi_2, ..., \pi_n, ...)$. Our first result concerns the survival probability of an ordinary critical BPRE Z_n in the quenched setting. Let

$$\tau(n) = \max\{k \in [0,n] :\ S_j \geq S_k,\ j \neq k\} \tag{3}$$

be the right-most point at which the minimal value of $S_j,\ j = 0, 1, ..., n,$ is attained. According to ([7], Ch. IV, Sec. 20) if condition (1) is valid then $\tau(n)n^{-1} \xrightarrow{d} \tau$ as $n \to \infty$, where τ is a random variable obeying the arcsin law on [0,1].

Theorem 1.1. *If conditions (1) and (2) are valid, then the distributions of the random variables*

$$\zeta_n := e^{-S_{\tau(n)}} P_\pi\left(Z_n > 0\right),\ n = 0, 1, 2, ..., \tag{4}$$

converge, as $n \to \infty$, to the distribution of a random variable $\zeta \in [0,1]$ which is positive with probability 1.

The next statement establishes a Yaglom type conditional limit theorem for the critical BPRE.

Theorem 1.2. *If conditions (1) and (2) are valid then for any $\lambda \in [0,\infty)$*

$$E_\pi\left[\exp\left\{-\lambda \frac{Z_n}{E_\pi\left[Z_n | Z_n > 0\right]}\right\} \,\Big|\, Z_n > 0\right] \xrightarrow{d} \phi_\pi(\lambda)\,,\ n \to \infty,$$

where $\phi_\pi(\lambda)$ is the Laplace transform of a proper nondegenerate random variable which is not concentrated at zero.

Here and below $\xrightarrow{d}$ means convergence in distribution with respect to the measure over environment,

Remark. If the offspring generating functions are fractional-linear then

$$\phi_\pi(\lambda) = \frac{1}{1+\lambda} \tag{5}$$

which gives a result in the spirit of the ordinary Galton-Watson branching processes. However, it is not clear if this is the case in the general situation.

For $T \geq 0$ put

$$\phi_\pi^{(T)}(t,\lambda) := \begin{cases} e^{-\lambda} & \text{if} \quad T > t, \\ \phi_\pi(\lambda) & \text{if} \quad 0 \leq T \leq t. \end{cases}$$

To simplify notation, let us agree to understand the product nt, $t \in (0,1)$, $n = 0,1,2,...$, as $[nt]$. Set

$$\beta_n(t) := \frac{1}{E[Z_{nt,n} \mid Z_n > 0]} = \frac{e^{-S_{nt}}(1 - f_{0,n}(0))}{1 - f_{nt,n}(0)}.$$

Theorem 1.3. *Under conditions (1) and (2) for any fixed* $t \in [0,1]$

$$\phi_{n,\pi}(t,\lambda) := E_\pi(\exp\{-\lambda Z_{nt,n}\beta_n(t)\} \mid Z_n > 0) \xrightarrow{d} \phi_\pi^{(\tau)}(t,\lambda),\ n \to \infty.$$

In the case when the offspring generating functions are fractional-linear one can get a more detailed information about the form of $\phi_\pi^{(\tau)}(t,\lambda)$.

Theorem 1.4. *Let* $0 < t_1 < ... < t_k \le 1$. *If conditions (1) and (2) are valid and all offspring generating functions are fractional-linear, then for any tuple* $\lambda_i \ge 0, i = 1,2,...,k$

$$E_\pi\left(\exp\left\{-\sum_{i=1}^{k}\lambda_i Z_{nt_i,n}\beta_n(t_i)\right\} \Big| Z_n > 0\right) \xrightarrow{d} \prod_{i=1}^{i_\tau - 1} e^{-\lambda_i} \times \frac{1}{1 + \sum_{i=i_\tau}^{k}\lambda_i}$$

as $n \to \infty$, *where* $i_\tau = \max\{i : \tau < t_i\}$.

Now we pass to the limiting behavior of $Z_{nt,n}$ in the annealed situation.

Theorem 1.5. *Let condition (1) be valid and* $E\eta_1(1 + X_1) + Ee^{-X_1} < \infty$. *Then*

$$\left\{\frac{1}{\sigma\sqrt{n}}\ln Z_{nt,n},\ t \in [0,1] \Big| Z_n > 0\right\} \xrightarrow{d} \left\{\inf_{t \le u \le 1} W^+(u),\ t \in [0,1]\right\},$$

where $W^+(t)$, $t \in [0,1]$ *is the Brownian meander.*

Remark. Let $a_n = max\{m < n : Z_{m,n} = 1\}$. The difference $d_n = n - a_n$ is called the distance to the closest mutual ancestor of all individuals existing at time n and the respective particle is called the closest mutual ancestor for the particles of the nth generation. It is known [5] that for ordinary critical Galton-Watson processes d_n is asymptotically uniformly distributed over $[0,n]$. In the random environment setting we have quite different situation.

Indeed, in the framework of the quenched approach it follows from Theorem 1.3 and Lemma 2.3 below that, as $n \to \infty$

$$E_\pi(\exp\{-\lambda Z_{nt,n}\} \mid Z_n > 0) \to e^{-\lambda}$$

on the set $\{\tau(n)n^{-1} > t\}$. On the other hand, if $\{\tau(n)n^{-1} < t\}$ then $Z_{nt,n}$ is large. This means, roughly speaking that $n^{-1}d_n \approx 1 - \tau(n)n^{-1}$ for a *frozen* environment. Thus, the distance to the closest mutual ancestor in the critical BPRE and a random but *frozen* environment is still spreading over the whole interval $[0,n]$. However, since the distribution of $\tau(n)n^{-1}$ converges, as $n \to \infty$, to the arcsin law on $[0,1]$, the distribution of this distance is NOT asymptotically uniformly distributed over $[0,n]$.

For the annealed approach Theorem 1.5 states, roughly speaking, that the number of individuals $Z_{nt,n}$ grows as $\exp\{\sqrt{n}\inf_{t\leq u\leq 1} W^{+}(u)\}$. Recalling that $P(\inf_{t\leq u\leq 1} W^{+}(u) > 0) = 1$ for any $t \in (0,1]$, we conclude that in the annealed setting the closest mutual ancestor is located at the distance $o(n)$ from the origin. In fact, as shown by Borovkov and Vatutin [2] for the fractional linear case, the closest mutual ancestor is located with positive probability even at the origin !

2 Quenched setting

In this section we collect some preliminary results and prove statements for the BPRE in the quenched setting. Let

$$\begin{aligned} f_{k,n}(s) &= f_k(f_{k+1}(...(f_{n-1}(s))...)),\ 0 \leq k \leq n-1,\ f_{n,n}(s) = 1, \\ f_{n,0}(s) &= f_{n-1}(f_{n-2}(...(f_0(s))...)),\ n \geq 1, \end{aligned}$$

$$g_k(s) = \frac{1}{1-f_k(s)} - \frac{1}{f_k'(1)(1-s)},\ 0 \leq s \leq 1.$$

The following lemma plays a crucial role in our subsequent arguments.

Lemma 2.1. *(see [3]). Let $f_k \neq 1, 0 \leq k \leq n-1$. Then for every $0 \leq s < 1$,*

$$\frac{1}{1-f_{0,n}(s)} = \frac{e^{-S_n}}{1-s} + \sum_{k=0}^{n-1} \eta_{k,n}(s)\, e^{-S_k}, \tag{6}$$

where

$$0 \leq \eta_{k,n}(s) = g_k(f_{k+1,n}(s)) \leq \eta_{k+1} := \frac{f_k''(1)}{(f_k'(1))^2}. \tag{7}$$

Now we are ready to show that the asymptotic behavior of the survival probability of Z_n depends essentially on the point of minimum of the accompanying random walk.

Lemma 2.2. *Let conditions (1) and (2) be valid, and let N and M be positive integers such that $N < M \leq n$. Then, on the set $\{\tau(n) = M > N\}$*

$$\frac{e^{S_M}}{1-f_{0,n}(s)} = \frac{e^{S_M - S_{M-N}}}{1-f_{M-N,n}(s)} + \varepsilon_N(M,n,s),$$

where, for any $\varepsilon > 0$

$$\limsup_{N\to\infty} \limsup_{M\to\infty} P\Big(\sup_{s\in[0,1],n\geq M} \varepsilon_N(M,n,s) > \varepsilon \,\Big|\, \tau(n) = M\Big) = 0. \tag{8}$$

Proof. By Lemma 2.1 we have on the set $\{\tau(n) = M > N\}$

$$\begin{aligned} \frac{e^{S_M}}{1-f_{0,n}(s)} &= \frac{e^{S_M}}{1-f_{0,M-N}(f_{M-N,n}(s))} \\ &= \frac{e^{S_M - S_{M-N}}}{1-f_{M-N,n}(s)} + \sum_{k=0}^{M-N-1} \eta_{k,M-N}(f_{M-N,n}(s))\, e^{S_M - S_k}. \end{aligned}$$

Now

$$\varepsilon_N(M,n,s) := \sum_{k=0}^{M-N-1} \eta_{k,M-N}\left(f_{M-N,n}(s)\right)e^{S_M-S_k}$$
$$\leq \sum_{k=0}^{M-N-1} \eta_{k+1}e^{S_M-S_k} \overset{d}{=} \sum_{k=N+1}^{M} \eta_k^* e^{S_k^*}, \tag{9}$$

where η_k^* and S_k^* are defined by $S_0^* = 0,\ S_k^* = S_M - S_{M-k} = X_M + X_{M-1} + \cdots + X_{M-k+1},\ k = 0, 1, ..., M$, and

$$(\eta_1^*, \eta_2^*, ..., \eta_M^*) \overset{d}{=} \{(\eta_M, \eta_{M-1}, ..., \eta_1) \mid S_M - S_k \leq 0,\ k = 1, 2, ..., M-1\}.$$

Following the line of arguments used in the proof of Lemma 4.1 in [3], one can demonstrate that under the conditions of our lemma

$$\overline{\lim_{M\to\infty}}\, E\sum_{k=1}^{M} \eta_k^* e^{S_k^*} < \infty. \tag{10}$$

We know that the probability of the event $\{\tau(n) > N\}$ tends to 1 as $n \to \infty$ for each N. Hence the desired statement follows. $\diamondsuit$

Proof of Theorem 1.1. By the total probability formula we write

$$P(\zeta_n^{-1} \leq x) = P\Big(\frac{e^{S_{\tau(n)}}}{1 - f_{0,n}(0)} \leq x\Big)$$
$$= \sum_{M=0}^{n} P\left(\tau(n) = M\right) P\left(\frac{e^{S_M}}{1 - f_{0,n}(0)} \leq x \,\Big|\, \tau(n) = M\right).$$

Since $\tau(n)\,n^{-1}$ obeys the arcsin law as $n \to \infty$, for any $\delta > 0$ one can select an $\varepsilon \in (0, 1/2)$ such that $P\left(\tau(n)\,n^{-1} \leq \varepsilon,\ \tau(n)\,n^{-1} > 1 - \varepsilon\right) \leq \delta$ for all sufficiently large n. Thus, we may consider only M belonging to the interval $[n\varepsilon,\ n(1-\varepsilon)]$. By Lemma 2.2 on the set $\{\tau(n) = M > N\}$

$$\frac{e^{S_M}}{1 - f_{0,n}(0)} = \frac{e^{S_M}e^{-S_{M-N}}}{1 - f_{M-N,M}\left(f_{M,n}(0)\right)} + \varepsilon_N(M, n, 0).$$

In view of (9) and (10) for any fixed $\varepsilon_1 > 0$ and any $\delta > 0$ one can find sufficiently large N such that $P\left(\varepsilon_N(M,n,0) > \varepsilon_1 \mid \tau(n) = M\right) \leq \delta$ for all $M \in [n\varepsilon,\ n(1-\varepsilon)]$. Therefore, for such N, M, and sufficiently large n

$$P\left(\xi(N, M, n) \leq x - \varepsilon_1 \,\Big|\, \tau(n) = M\right) - \delta \leq P\left(\frac{e^{S_M}}{1 - f_{0,n}(0)} \leq x \mid \tau(n) = M\right)$$
$$\leq P\left(\xi(N, M, n) \leq x \mid \tau(n) = M\right),$$

where

$$\begin{aligned}
\xi(N,M,n) :&= \frac{e^{S_M}e^{-S_{M-N}}}{1-f_{M-N,M}\left(f_{M,n}(0)\right)} \\
&= \frac{1}{1-f_{M,n}(0)} + \sum_{k=M-N}^{M-1} \eta_{k,M}\left(f_{M,n}(0)\right)e^{-(S_k-S_M)} \\
&\stackrel{d}{=} \frac{1}{1-f^{**}_{0,n-M}(0)} + \sum_{k=1}^{N} \eta^{*}_{k,0}\left(f^{**}_{0,n-M}(0)\right)e^{S^*_k} \\
&= e^{-S^{**}_{n-M}} + \sum_{k=0}^{n-M-1} \eta^{**}_{k,n-M}(0)\,e^{-S^{**}_k} + \sum_{k=1}^{N} \eta^{*}_{k,0}\left(f^{**}_{0,n-M}(0)\right)e^{S^*_k} \\
&= \frac{e^{S^*_N}}{1-f^{*}_{N,0}\left(f^{**}_{0,n-M}(0)\right)} =: \xi_1(N,n-M). \qquad (11)
\end{aligned}$$

Here S^*_k is the same as before,

$$S^{**}_0 = 0,\ S^{**}_j = S_{j+M} - S_M = X_{M+1} + \cdots + X_{M+j},\ j = 0,1,...,n-M-1,$$

and S^*_k and S^{**}_j are two independent random walks, the first of them conditioned to stay non-positive for $k = 1,...,M$, while the second one conditioned to stay positive for $j = 0,1,...,n-M-1$,

$$\eta^*_{k,0}(s) = g^*_k\left(f^*_{k-1}\left(...f^*_0(s)\right)\right) \stackrel{d}{=} \eta_{M-k,M}(s),$$

$$\eta^{**}_{k,n-M}(s) = g^{**}_k\left(f^*_{k+1}\left(...f^*_{n-M}(s)\right)\right) \stackrel{d}{=} \eta_{k+M,n}(s),$$

where $f^*_{m,0}(s) = f^*_{m-1}\left(...f^*_0(s)\right)$ and $f^{**}_{0,n}(s) = f^*_0\left(f^*_1\left(...f^*_{n-1}(s)\right)\right)$ are iterations of random generating functions with distributions

$$\left(f^*_m, f^*_{m-1}, ..., f^*_0\right) \stackrel{d}{=} \{f_0, f_1, ..., f_m \mid S_k \le 0,\ k = 0,1,...,m\},$$

$$\left(f^{**}_0, f^{**}_1, ..., f^*_{n-1}\right) \stackrel{d}{=} \{f_0, f_1, ..., f_{n-1} \mid S_j > 0,\ j = 0,1,...,n-1\}.$$

Clearly, $f^{**}_{0,n-M}(0)$, being a monotone sequence, converges almost surely as $n - M \to \infty$ to a limit q^{**}. Let us show that $q^{**} < 1$ with probability 1. Indeed, it follows from Lemma 2.1 that

$$\frac{1}{1-f^{**}_{0,n-M}(0)} \le e^{-S^{**}_{n-M}} + \sum_{j=0}^{n-M-1} \eta^{**}_{j+1}e^{-S^{**}_j}$$

with natural definition of η^{**}_{j+1} (compare with (7)).

By Lemma 4.1 in [3]

$$\lim \sup_{n-M\to\infty} E\left(e^{-S^{**}_{n-M}} + \sum_{j=0}^{n-M-1} \eta^{**}_{j+1}e^{-S^{**}_j}\right) < \infty. \qquad (12)$$

Hence it follows that $P(q^{**} < 1) = 1$.

Now everything is ready to complete the proof of the theorem. As we know, $\eta^*_{k,0}(s)$, $k = 1, ..., N$ are continuous in $s \in [0,1]$, the sequence of random variables $f^{**}_{0,n-M}$ converges to q^{**} as $n - M \to \infty$ with probability 1 and is independent of S^*_k and, finally,

$$\frac{e^{S^*_N}}{1 - f^*_{N,0}(s)} \le \frac{e^{S^*_{N+1}}}{1 - f^*_{N+1,0}(s)}, \; s \in [0,1].$$

These facts and (11) show that the limit

$$\begin{aligned}\lim_{N\to\infty} \lim_{n-M\to\infty} \xi_1(N, n-M) &= \lim_{N\to\infty} \frac{e^{S^*_N}}{1 - f^*_{N,0}(q^{**})} \\ &= \frac{1}{1-q^{**}} + \sum_{k=1}^{\infty} \eta^*_{k,0}(q^{**}) e^{S^*_k} := \xi \qquad (13)\end{aligned}$$

exists $P_{\pi*} \times P_{\pi**}$ almost surely, where $P_{\pi*} \times P_{\pi**}$ is the measure generated by the "double"-sided environment $..., f^*_m, f^*_{m-1}, ..., f^*_0; f^{**}_0, f^{**}_1, ..., f^*_{n-1}, ...$. Observe that ξ is finite with probability 1 by (7) and (10). As a result we get

$$\begin{aligned}&P(\xi \le x - \varepsilon_1) - \delta \\ &\quad \le \lim_{N\to\infty} \liminf_{n\to\infty} \min_{M\in[n\varepsilon,\, n(1-\varepsilon)]} P(\xi(N, n-M) \le x \mid \tau(n) = M) \\ &\quad \le \lim_{N\to\infty} \limsup_{n\to\infty} \max_{M\in[n\varepsilon,\, n(1-\varepsilon)]} P(\xi(N, n-M) \le x \mid \tau(n) = M) \\ &\quad \le P(\xi \le x)\end{aligned}$$

for all $M \in [n\varepsilon, n(1-\varepsilon)]$. Hence,

$$\lim_{n\to\infty} P(\zeta_n^{-1} \le x) = P(\zeta^{-1} \le x)$$

where $\zeta = \xi^{-1}$. ◇

Combining Theorem 1.1 and Lemma 2.1 it is not difficult to check the validity of the following statement.

Lemma 2.3. *Under conditions (1) and (2) for any $t \in (0,1)$ and $\varepsilon \in (0,1)$*

$$\lim_{n\to\infty} P\left(\left|\frac{1 - f_{0,nt}(0)}{1 - f_{0,n}(0)} - 1\right| > \varepsilon \;\middle|\; \tau(n) < nt\right) = 0, \qquad (14)$$

$$\lim_{n\to\infty} P(\beta_n(t) > \varepsilon \mid \tau(n) < nt) = 0, \qquad (15)$$

and

$$\lim_{n\to\infty} P(|\beta_n(t) - 1| > \varepsilon \mid \tau(n) > nt) = 0. \qquad (16)$$

Lemma 2.4. *Under conditions (1) and (2) for any $\lambda \in [0, \infty)$*

$$E_\pi \left[e^{-\lambda Z_n e^{S_{\tau(n)} - S_n}} \middle| Z_n > 0 \right] \xrightarrow{d} \psi_\pi(\lambda), \quad n \to \infty,$$

where $\psi_\pi(\lambda)$ is the Laplace transform of a proper random variable not concentrated at zero.

Proof. Clearly,

$$E_{\pi}\left[e^{-\lambda Z_n e^{S_{\tau(n)}-S_n}}\,\middle|\, Z_n>0\right]=1-\frac{1-f_{0,n}\left(e^{-\lambda e^{S_{\tau(n)}-S_n}}\right)}{1-f_{0,n}(0)}.$$

Thus, in view of Theorem 1.1 it suffices to establish that the limit

$$\xi(\lambda):=\lim_{n\to\infty}e^{S_{\tau(n)}}\left(1-f_{0,n}\left(e^{-\lambda e^{S_{\tau(n)}-S_n}}\right)\right)^{-1}$$

in distribution exists and possesses the needed properties. Similarly to (11) we have on the set $\{\tau(n)=M>N\}$

$$\begin{aligned}\xi(N,M,n,\lambda)&:=\frac{e^{S_M}e^{-S_{M-N}}}{1-f_{M-N,M}\left(f_{M,n}\left(e^{-\lambda e^{S_M-S_n}}\right)\right)}\\&\stackrel{d}{=}\frac{e^{S_N^*}}{1-f_{N,0}^*\left(f_{0,n-M}^{**}\left(e^{-\lambda e^{-S_{n-M}^{**}}}\right)\right)}=:\xi_1(N,n-M,\lambda).\qquad(17)\end{aligned}$$

Observe now that $f_{M,n}\left(e^{-\lambda e^{S_M-S_n}}\right)\stackrel{d}{=}f_{0,n-M}^{**}\left(e^{-\lambda e^{-S_{n-M}^{**}}}\right)$, $n\geq M$, are the Laplace transforms of the distributions of the random variables $e^{-S_{n-M}^{**}}Z_{n-M}^{**}$ (with natural meaning for Z_{n-M}^{**}) which constitute a positive martingale for each fixed sequence $\{S_n^{**}\}$ and, therefore,

$$\lim_{n-M\to\infty}f_{0,n-M}^{**}\left(e^{-\lambda e^{-S_{n-M}^{**}}}\right)=\psi_{\pi^{**}}(\lambda),\ \lambda\in(0,\infty),$$

exists $P_{\pi^{**}}$ almost surely. Now repeating almost literally the proof of the previous theorem we get sequentially that $P\left(\psi_{\pi^{**}}(\lambda)<1\right)=1$ and that the limit

$$\lim_{N\to\infty}\lim_{n-M\to\infty}\xi_1(N,n-M,\lambda)=\frac{1}{1-\psi_{\pi^{**}}(\lambda)}+\sum_{k=1}^{\infty}\eta_{k,0}^*\left(\psi_{\pi^{**}}(\lambda)\right)e^{S_k^*}\qquad(18)$$

exists $P_{\pi^*}\times P_{\pi**}$ a.s. . Hence, for any $\lambda>0$

$$\lim_{n\to\infty}P\left(\frac{e^{S_{\tau(n)}}}{1-f_{0,n}\left(e^{-\lambda e^{S_{\tau(n)}-S_n}}\right)}\leq x\right)=P\left(\xi(\lambda)\leq x\right)=P\left(\zeta^{-1}(\lambda)\leq x\right),\qquad(19)$$

where $\zeta(\lambda)=\xi^{-1}(\lambda)$ is a proper random variable which is positive with probability 1. In fact, we can perform the arguments above for $1-f_{0,n}\left(e^{-\lambda e^{S_{\tau(n)}-S_n}}\right)$ and $1-f_{0,n}(0)$ simultaneously. This remark allows us to combine (19) with Theorem 1.1 to get

$$\lim_{n\to\infty}E_{\pi}\left[e^{-\lambda Z_n e^{S_{\tau(n)}-S_n}}\,\middle|\, Z_n>0\right]\stackrel{d}{=}1-\frac{\zeta(\lambda)}{\zeta}:=\psi_{\pi}(\lambda).\qquad\diamond$$

Lemma 2.4 is the crucial step for proving Theorem 1.2.

Proof of Theorem 1.2. We have

$$\begin{aligned} &E_{\pi}\left[\exp\left\{-\frac{\lambda Z_n}{E_{\pi}[Z_n \mid Z_n>0]}\right\}\Big| Z_n>0\right] \\ &\quad= E_{\pi}\left[\exp\left\{-\lambda Z_n e^{-S_n}\left(1-f_{0,n}(0)\right)\right\} \mid Z_n>0\right] \\ &\quad= E_{\pi}\left[\exp\left\{-\lambda Z_n e^{S_{\tau(n)}-S_n}\zeta_n\right\} \mid Z_n>0\right]. \end{aligned}$$

Hence, following the line of arguments used in the proofs of Theorem 1.1 and Lemma 2.4 and recalling that we the limit in (13) holds almost surely, we obtain

$$E_{\pi}\left[\exp\left\{-\lambda Z_n e^{S_{\tau(n)}-S_n}\zeta_n\right\} \mid Z_n>0\right] \xrightarrow{d} \psi_{\pi}(\lambda\zeta) =: \phi_{\pi}(\lambda). \Diamond$$

Remark. In the fractional linear case representation (7) looks like this

$$\frac{1}{1-f_{0,n}(s)} = \frac{e^{-S_n}}{1-s} + \frac{1}{2}\sum_{k=0}^{n-1}\eta_{k+1}e^{-S_k}. \tag{20}$$

Using this identity it is not difficult to show that for a random variable ξ^{**}

$$\lim_{n-M\to\infty}\frac{1}{1-f^{**}_{0,n-M}(0)} = \xi^{**}$$

and

$$\lim_{n-M\to\infty}\frac{1}{1-f^{**}_{0,n-M}\left(e^{-\lambda e^{-S^{**}_{n-M}}}\right)} = \frac{1}{1-\psi_{\pi^{**}}(\lambda)} = \frac{1}{\lambda}+\xi^{**}$$

$P_{\pi^{**}}$ almost surely. Inserting this into (18) gives for a random variable ξ^{*}

$$\lim_{N\to\infty}\lim_{n-M\to\infty}\frac{1}{1-f^{*}_{N,0}\left(f^{**}_{0,n-M}(0)\right)} = \xi^{**}+\xi^{*} = \zeta^{-1}$$

and

$$\begin{aligned} \lim_{N\to\infty}\lim_{n-M\to\infty}\frac{1}{1-f^{*}_{N,0}\left(f^{**}_{0,n-M}\left(e^{-\lambda e^{-S^{**}_{n-M}}}\right)\right)} &= \frac{1}{\lambda}+\xi^{**}+\xi^{*} \\ &= \frac{1}{\lambda}+\zeta^{-1} = \zeta^{-1}(\lambda) \end{aligned}$$

$P_{\pi^{*}} \times P_{\pi^{**}}$ almost surely. Therefore,

$$\phi_{\pi}(\lambda) = \psi_{\pi}(\lambda\zeta) \stackrel{d}{=} 1-\frac{\zeta(\lambda\zeta)}{\zeta} = \frac{1}{1+\lambda}$$

as claimed in (5).

Proof of Theorem 1.3. Direct calculations show (see [5]) that

$$\begin{aligned} &E_{\pi}(\exp\{-\lambda Z_{nt,n}\beta_n(t)\} \mid Z_n>0) \\ &\quad= 1-\frac{1-f_{0,nt}\left(1-\left(1-e^{-\lambda\beta_n(t)}\right)\left(1-f_{nt,n}(0)\right)\right)}{1-f_{0,n}(0)}. \end{aligned} \tag{21}$$

Clearly,

$$\begin{aligned}\exp\{-\lambda\beta_n(t)\} &\geq E_\pi\left(\exp\{-\lambda Z_{nt,n}\beta_n(t)\}\,|Z_n>0\right)\\ &\geq 1-\frac{e^{S_{nt}}\left(1-e^{-\lambda\beta_n(t)}\right)\left(1-f_{nt,n}(0)\right)}{1-f_{0,n}(0)}\\ &=1-\left(1-e^{-\lambda\beta_n(t)}\right)\left(\beta_n(t)\right)^{-1},\end{aligned}$$

where we used the inequality $1-f(s)\leq f'(1)(1-s)$ being valid for any probability generating function. This and (16)show that for any $\varepsilon>0$

$$P\left(\left|E_\pi\left(\exp\{-\lambda Z_{nt,n}\beta_n(t)\}\,|Z_n>0\right)-\exp\{-\lambda\}\right|>\varepsilon\mid\tau(n)>nt\right)\to 0$$

as $n\to\infty$. On the other hand, (14) and (15) yield

$$\begin{aligned}\left(1-e^{-\lambda\beta_n(t)}\right)\left(1-f_{nt,n}(0)\right) &= \lambda\beta_n(t)\left(1+\delta_n^{(1)}(t)\right)\left(1-f_{nt,n}(0)\right)\\ &=\frac{\lambda}{E_\pi\left[Z_{nt}|Z_{nt}>0\right]}\frac{\left(1-f_{0,n}(0)\right)}{\left(1-f_{0,nt}(0)\right)}\left(1+\delta_n^{(1)}(t)\right)\\ &=1-\exp\left\{-\frac{\lambda}{E_\pi\left[Z_{nt}|Z_{nt}>0\right]}\left(1+\delta_n^{(2)}(t)\right)\right\},\end{aligned}$$

where $\delta_n^{(i)}(t)\xrightarrow{p}0,\ n\to\infty,\ i=1,2$ on the set $\{\tau(n)<nt\}$. Hence, recalling Theorem 1.2 and (21), we get

$$\begin{aligned}&E_\pi(\exp\{-\lambda Z_{nt,n}\beta_n(t)\}\mid Z_n>0)\\ &\quad=1-\frac{1-f_{0,nt}\left(\exp\left\{-\frac{\lambda}{E_\pi\left[Z_{nt}|Z_{nt}>0\right]}\left(1+\delta_n^{(2)}(t)\right)\right\}\right)}{1-f_{0,nt}(0)}\times\\ &\quad\quad\times\frac{1-f_{0,nt}(0)}{1-f_{0,n}(0)}\xrightarrow{d}\phi_\pi(\lambda)\end{aligned}$$

on $\{\tau(n)<nt\}$ as $n\to\infty$, completing the proof. $\diamondsuit$

Using Theorem 1.3 and induction arguments one can check the validity of Theorem 1.4. We omit the details.

3 Annealed setting

The study of reduced BPRE in the annealed setting is more involved and we demonstrate main steps only. The reader can find the detailed proofs in [9].

Lemma 3.1. *Let the conditions of Theorem 1.5 be valid. Then*

$$\left\{\frac{1}{\sigma\sqrt{n}}\ln\left(1-f_{0,nt}(0)\right),\ t\in[0,1]\right\}\xrightarrow{d}\left\{\inf_{0\leq u\leq t}W(u),\ t\in[0,1]\right\}\tag{22}$$

in Skorokhod topology in the space $D[0,1]$, where $W(t)$ is the standard Brownian motion with $W(0)=0$.

Proof. By Theorem 1.1 for any $x < 0$ and any $t \in (0, 1]$

$$\begin{aligned}
&\lim_{n\to\infty} P\left(\frac{1}{\sigma\sqrt{n}} \ln\left(1 - f_{0,nt}(0)\right) \le x\right) \\
&\quad = \lim_{n\to\infty} P\left(\frac{1}{\sigma\sqrt{n}}\left(\min_{0\le k\le nt} S_k + \ln e^{-S_{\tau(nt)}}\left(1 - f_{0,nt}(0)\right)\right) \le x\right) \\
&\quad = P\left(\inf_{0\le u\le t} W(u) \le x\right).
\end{aligned}$$

Using this fact it is not difficult to establish (22).
The next lemma shows that in the annealed setting a critical BPRE survives only if the trajectories of the accompanying random walk S_n behave "nicely".

Lemma 3.2. *Let the conditions of Theorem 1.5 be valid. Then for any* $0 < t_1 < ... < t_r \le 1$ *and* $y_1, ..., y_r \in \mathbf{R}^r$

$$\begin{aligned}
&\lim_{n\to\infty} P\Big(\frac{1}{\sigma\sqrt{n}} \min_{0\le k\le nt_1} S_k > y_1, \; ..., \; \frac{1}{\sigma\sqrt{n}} \min_{0\le k\le nt_r} S_k > y_r \mid Z_n > 0\Big) \\
&\qquad\qquad = P\left(\inf_{t_1\le u\le 1} W(u) > y_1, ..., \inf_{t_r\le u\le 1} W(u) > y_r\right).
\end{aligned}$$

The next important step is the representation

$$Z_{nt,n} = Z_{nt} \times \frac{1}{Z_{nt}} \sum_{i=1}^{Z_{nt}} I\left(Z_{nt}^{(k)}(n) > 0\right),$$

where $Z_{nt}^{(k)}(n)$ is the offspring size at moment n in the population generated by the kth particle among those existed at moment nt and $I(A)$ is the indicator of the event A.

Lemma 3.3. *Let the conditions of Theorem 1.5 be valid. Then for* $x > 0$ *and* $t \in (0, 1]$

$$\lim_{n\to\infty} P\left(\frac{1}{\sigma\sqrt{n}} \ln Z_{nt,n} > x \,|\, Z_n > 0\right) = P\left(\inf_{t\le u\le 1} W^+(u) > x\right).$$

Proof. Given Z_{nt} and $f_{nt}(0), f_{nt+1}(0), ..., f_{n-1}(0)$ we have by the strong law of large numbers on the set $\{Z_{nt} \to \infty\}$:

$$\frac{1}{Z_{nt}} \sum_{i=1}^{Z_{nt}} I\left(Z_{nt}^{(k)}(n) > 0\right) \to 1 - f_{nt.n}(0)$$

P_π-almost surely. On the other hand, one can show that under the conditions of the lemma for any $\varepsilon > 0$ and any $t \in (0, 1]$

$$\lim_{n\to\infty} P\left(\frac{1}{\sigma\sqrt{n}} \left|\ln\left(Z_{nt} e^{-S_{nt}}\right)\right| > \varepsilon \,\Big|\, Z_n > 0\right) = 0.$$

Hence, in view of Lemma 3.2

$$P\Big(\frac{1}{\sigma\sqrt{n}}\ln Z_{nt,n} > x \ \Big|\ Z_n > 0\Big)$$
$$\approx P\left(\frac{1}{\sigma\sqrt{n}}\left(\ln Z_{nt} + \ln\left(1 - f_{nt.n}(0)\right)\right) > x \middle| Z_n > 0\right)$$
$$\approx P\left(\frac{1}{\sigma\sqrt{n}}\left(S_{nt} + \min_{nt\le k\le n}(S_k - S_{nt})\right) > x \middle| Z_n > 0\right)$$
$$= P\left(\frac{1}{\sigma\sqrt{n}}\min_{nt\le k\le n} S_k > x \middle| Z_n > 0\right)$$
$$\to P\left(\inf_{t\le u\le 1} W^+(u) > x\right), n \to \infty,$$

as desired. ◊

Using Lemmas 3.2 and 3.3 one can complete the proof of Theorem 1.5.

References

[1] B.K.Athreya and P.Ney, (1972) *Branching processes*, Springer-Verlag, Berlin.

[2] K.A.Borovkov and V.A.Vatutin, (1997) *Reduced critical branching processes in random environment*, Stoch. Proc. Appl., 71, No.2, 225–240.

[3] J.Geiger and G.Kersting, (2000) *The survival probability of a critical branching process in random environment*, Theory Probab. Appl., 45, No. 3, 607–615.

[4] K.Fleischmann and U.Prehn, (1974) *Ein Grenzfersatz für subkritische Verzweigungsprozesse mit eindlich vielen Typen von Teilchen*, Math. Nachr. 64, 233–241.

[5] K.Fleischmann and R.Siegmund-Schultze, (1977) *The structure of reduced critical Galton–Watson processes*, Math. Nachr. 79, 357–362.

[6] K.Fleischmann and V.A.Vatutin, (1999) *Reduced subcritical branching processes in random environment*, Adv. Appl. Probab., 31, No.1, 88–111.

[7] F.Spitzer, (1964) *Principles of random walk*, Princeton, New Jersey.

[8] V.A.Vatutin and A.M. Zubkov, (1993) *Branching Processes II.*, Journal of Soviet Mathematics, 67, 3407–3485.

[9] V.A.Vatutin, (2002) *Reduced branching processes in a random environment: the critical case*, Theory Probab. Appl., 47, No.1, 21–38 (In Russian.).

[10] A.M.Zubkov, (1975) *Limit distributions of the distance to the closest mutual ancestor*, Theory Probab. Appl., 20, No. 3, 602–612.

Vatutin Vladimir

Steklov Mathematical Institute
Gubkin street, 8

117966, GSP-1 Moscow
vatutin@mi.ras.ru

Dyakonova Elena

Steklov Mathematical Institute
Gubkin street, 8
117966, GSP-1 Moscow
elena@mi.ras.ru

PART V

Applied random combinatorics

Trends in Mathematics, © 2002 Birkhäuser Verlag Basel/Switzerland

A Cooperative Approach to Rényi's Parking Problem on the Circle

Thierry Huillet, Anna Porzio

ABSTRACT: *A cooperative approach to* Rényi*'s parking problem is investigated as a circle covering problem. In this approach, the jamming constant is* $\log 2$.

1 Introduction

Place at random a unit interval of length 1 (a car) in the interval of length $x > 1$ (the street), then place a second one, independently of the first in such a way that overlap is avoided, and repeat the process until the largest gap between cars is less than 1 so that no additional car can be inserted. The final number of cars is $\mathcal{M}(x)$ and using *Laplace* transform methods, *Rényi* [8] proved that $\mathbf{E}\mathcal{M}(x)/x \to_{x\uparrow\infty} .748..$ is the average space filling rate.

In Statistical Mechanics, this problem is the 1−dimensional version of the irreversible hard-sphere model, also called the Random Sequential Adsorption (*RSA*) model which has become important in the physical and biological sciences [4], [6]. Some extensions of the sequential approach (the so-called Cooperative Sequential Adsorption model) have also been investigated recently [2] in the Physics' literature.

In this manuscript, a purely cooperative approach to the *Rényi* parking problem is investigated. In more details, the considered problem is the following one:

Consider the circle of circumference 1 (or the interval $[0, 1]$). Throwing at random points on the circle and appending clockwise cars of length s to each such points, we focus on s−packing configurations that avoid overlap for cars and that do not allow for adjunction of a new car. Pick at random one such s−packing configuration and let $\mathcal{N}(s)$ be the number of its "cars". Using tools designed by *Steutel* [10], we prove $s\mathcal{N}(s) \to_{s\downarrow 0} \log 2 \approx .693..$ (in probability) which is slightly less than *Rényi*'s packing constant.

In our approach, the random subset of the circle arising from random throws of atoms of length s on it, is considered globally: the whole configuration is disregarded if it is not a s-packing configuration; in *Rényi*'s model, only the new car inserted is discarded if overlap occurs with the previously obtained non-overlapping configuration, until the space filling process terminates in a packing configuration.

Whereas *Rényi*'s formulation is more like a Random Sequential Adsorption model, the one discussed here can therefore be seen as a Random Cooperative Adsorption (*RCA*) model.

Perhaps not surprisingly, the *RCA* model yields a slightly less effective space-filling procedure but it could be relevant, in opposition to the sequential *RSA* one, whenever a simultaneous interaction (physical, chemical or biological) is involved; for example, in toxicology, the simultaneous action of some toxic substances is necessary to obtain some (often lethal) biological effect, say on genes. As the input flow of toxic particles is intense, one may wish to consider those particles' configurations saturating the available space, provoking the undesirable effect.

2 Background: the number of random intervals needed to cover the circle and the length of the covered set

2.1 Preliminaries

Consider a circle of unit circumference. Throw at random $n > 1$ points on this circle and let $\mathcal{S}_n := \{X_1, .., X_n\}$ be this set of points (thus, with $X_1, .., X_n$ independent and identically distributed, say *iid,* and uniform). Let $s \in (0, 1)$. Consider the coarse-grained random set of intervals

$$\mathcal{S}_n(s) := \{X_1 + x, .., X_n + x, 0 \leq x < s\} \tag{1}$$

appending clockwise an arc of length s to each starting-point atom of $\mathcal{S}_n$.

Let $P_n(s)$ be the number of connected components of $\mathcal{S}_n(s)$ (which is also the number of gaps), with, by convention, $P_n(s) = 0$ as soon as the circle is covered by $\mathcal{S}_n(s)$. With $X_{1:n} := X_1$, consider the ordered set of points $(X_{m:n}, m = 1, .., n)$, putting $X_1, .., X_n$ in order while turning clockwise on the circle. Let $S_{m,n} = X_{m+1:n} - X_{m:n}$, $m = 1, .., n-1$, be the consecutive spacings, with $S_{n,n} = X_{1:n} - X_{n:n}$, modulo 1, closing the loop. Under our hypothesis, $S_{m,n}$ all share the same distribution, say $S_{m,n} \stackrel{d}{=} S_n$, $m = 1, .., n$, which is given by $\mathbf{P}(S_n > s) = (1-s)^{n-1}$, with $\mathbf{E}S_n = 1/n$.

It is indeed an old result [7] that identically distributed (*id*) spacings $S_{m,n}$, with $\sum_m S_{m,n} = 1$ can be generated as the ratio $S_{m,n} = E_m / \overline{E}_n$, with $\overline{E}_n := \sum_{m=1}^n E_m$ the sum of n *iid* exponential variables with parameter 1 and so S_n has the claimed distribution.

Note also that the same construction starting with a circle of circumference $t > 0$ gives *id* consecutive spacings, say $S_{m,n}(t) \stackrel{d}{=} S_n(t)$, $m = 1, .., n$ with $S_{m,n}(t) \stackrel{d}{=} tS_{m,n}(1) := tS_{m,n}$ and $\sum_m S_{m,n}(t) = t$.

The following *Poisson*-type results of *Steutel* [10], which we recall, deepens this connection and allows to compute many spacings functionals in terms of functionals of *iid* exponential random variables or processes. This may require more or less effort, but in general the computational task is far less than the ones which have been designed in the literature, using combinatorial, geometrical and/or *Laplace-Fourier* theoretic methods (see *e.g.* [1] for an overview).

Theorem 2.1. *([10])*

(i) Let f be any Borel-measurable function for which

$$\int_0^\infty \mathbf{E}\left|f(S_{1,n}(t), .., S_{n,n}(t))\right| t^{n-1} e^{-pt} dt < \infty.$$

Then, with $E_1(p), .., E_n(p)$ n iid exponential random variables with common distribution $\mathbf{P}(E(p) > x) = e^{-px}$,

$$\int_0^\infty \mathbf{E}f(S_{1,n}(t), .., S_{n,n}(t))\, t^{n-1} e^{-pt} dt = \Gamma(n) p^{-n} \mathbf{E}f(E_1(p), .., E_n(p)). \tag{2}$$

(ii) At $p=1$, recalling that $S_{m,n}(t) \overset{d}{=} S_{m,n}t$, with $E_m(1) := E_m$, we get as a consequence of (i):

If f is homogeneous of degree d, i.e. if $f(\gamma s_1,..,\gamma s_n) = \gamma^d f(s_1,..,s_n)$, $\gamma > 0$, and if $\mathbf{E}|f(S_{1,n},..,S_{n,n})| < \infty$ then

$$\mathbf{E}f(S_{1,n},..,S_{n,n}) = \frac{\Gamma(n)}{\Gamma(n+d)}\mathbf{E}f(E_1,..,E_n). \tag{3}$$

Comments:

Part (i) derives from the fact that, with $\overline{E}_m(p) := \sum_{m=1}^n E_m(p)$ and with $\overline{S}_{m,n}(t) := \sum_{m=1}^n S_{m,n}(t)$, $m=1,..,n-1$, the random vector $(\overline{E}_1(p),..,\overline{E}_{n-1}(p))$ given $\overline{E}_n(p) = t$ has the same distribution as $\overline{S}_{1,n}(t),..,\overline{S}_{n-1,n}(t)$.

>From (i), $\Gamma(n)p^{-n}\mathbf{E}f(E_1(p),..,E_n(p))$ interprets as the *Laplace* transform of $\mathbf{E}f(S_{1,n}(t),..,S_{n,n}(t))t^{n-1}$. Inverting this *Laplace* transform and putting $t=1$ yields $\mathbf{E}f(S_{1,n},..,S_{n,n})$.

>From (ii), any homogeneous functional of sample spacings can be computed from the simpler one of *iid* exponential variables. Famous examples of such homogeneous functionals of interest are: $f(s_1,..,s_n) = \left[\sum_{m=1}^n \pi_m s_m\right]^\lambda$, $\pi_m > 0$, $\sum \pi_m = 1$ with degree $d = \lambda$ and $f(s_1,..,s_n) = \left[\sum_{m=1}^n s_m^{q+1}\right]^{\lambda/q}$ with degree $d = \lambda(q+1)/q$. Formula (3) allows (with some computational effort) to evaluate the moment functions of respectively the π-average fragments' size $\sum_{m=1}^n \pi_m S_{m,n}$ and the q-average fragments' size $\left[\sum_{m=1}^n S_{m,n}^{q+1}\right]^{1/q}$ of the circle's partition, *i.e.* the quantities $\mathbf{E}\left[\sum_{m=1}^n \pi_m S_{m,n}\right]^\lambda$ and $\mathbf{E}\left[\sum_{m=1}^n S_{m,n}^{q+1}\right]^{\lambda/q}$ respectively.

These formulae are also very useful in the context of large n asymptotics.

Indeed, putting $p = n$ in (2), with $\delta_n(t-1) := \frac{n^n}{\Gamma(n)}t^{n-1}e^{-nt}$, approaching the *Dirac* delta function at $t=1$, with $\phi_n(t) := \mathbf{E}f(S_{1,n}(t),..,S_{n,n}(t))$ and $\psi_n(n) := \mathbf{E}f(E_1(n),..,E_n(n))$ this equation reads $(\mathcal{E}_n) \equiv \int_0^\infty \phi_n(t)\delta_n(t-1)\,dt = \psi_n(n)$ and one expects

Lemma 2.2. *If $\phi_n(t)$ is bounded and continuous, uniformly in n, or if $\phi_n(t)$ is homogeneous,*

$$\lim_{n\uparrow\infty}\frac{\phi_n(1)}{\psi_n(n)} = 1.$$

Proof: Suppose f is homogeneous of degree d. In this case, $\phi_n(t) := t^d\phi_n(1)$ is homogeneous and $(\mathcal{E}_n)$ reads $\frac{\psi_n(n)}{\phi_n(1)} = \int_0^\infty t^d\delta_n(t-1)\,dt$. From *Stirling*'s formula, for large n

$$\frac{\psi_n(n)}{\phi_n(1)} = \frac{n^n}{n^{n+d}}\frac{\Gamma(n+d)}{\Gamma(n)} \sim n^{-d}(n+d)^d \to_{n\uparrow\infty} 1.$$

If f is not homogeneous, the result holds if $\phi_n(t)$ is a bounded and continuous function of t, uniformly in n . Indeed, putting $h_n(t) := \frac{\phi_n(t)}{\psi_n(n)}$, $(\mathcal{E}_n)$ reads:

$\mathbf{E}h_n(X_n) = 1$ for all n, with $X_n = \frac{1}{n}\sum_{m=1}^{n} E_m \to_{n\uparrow\infty} 1$ (almost surely). Note that X_n has density $\delta_n(t-1)$. Suppose $h_n(1) \to a \neq 1$ (or that no limit exists), then, if h_n is uniformly continuous, $h_n(X_n) \overset{a.s.}{\to} a \neq 1$ (or has no limit) and if h_n is uniformly bounded $\mathbf{E}h_n(X_n) \to a \neq 1$ (or has no limit) which contradicts $\mathbf{E}h_n(X_n) = 1$ for all n.

Note that a monotonicity hypothesis instead of the uniform continuity in n will do as well.

In some cases, if the limits exist, one expects $\lim_{n\uparrow\infty}\phi_n(1) = \lim_{n\uparrow\infty}\psi_n(n)$. The precise setting of this intuition is *Steutel*'s Theorem

Theorem 2.3. *([10]) Suppose the functions $\phi_n(t)$ are uniformly bounded and monotonic in t and that $\lim_{n\uparrow\infty}\int_0^\infty \phi_n(\alpha t)\,\delta_n(t-1)\,dt := \psi(\alpha)$ exists and is continuous in α for all $\alpha > 0$, then $\lim_{n\uparrow\infty}\phi_n(\alpha) = \psi(\alpha)$.*

2.2 Illustrations on circle covering problems

Let us illustrate the power of these results. Let $S_{m:n}$ denote the ordered spacings, $m = 1,..,n$, with $S_{1:n} < .. < S_{n:n}$. Using (2) with respectively $f(s_1,..,s_n) = \prod_{m=1}^{n}\mathbf{1}(s_m > s)$ and $f(s_1,..,s_n) = \prod_{m=1}^{n}\mathbf{1}(s_m \leq s)$, with $x_+ := \max(x,0)$, we recover the well-known results (see [7] for example)

$$\mathbf{P}(S_{1:n} > s) = (1-ns)_+^{n-1} \text{ and } \mathbf{P}(S_{n:n} \leq s) = \sum_{m=0}^{n}(-1)^m\binom{n}{m}(1-ms)_+^{n-1}. \tag{4}$$

>From the above definitions, the following two events coincide:

$$(P_n(s) \geq n-m+1) \equiv (S_{m:n} > s) \text{ or } (P_n(s) = p) \equiv (S_{n-p:n} \leq s,\ S_{n+1-p:n} > s),$$

and the distribution of $P_n(s)$ can be computed.

Indeed, with $f(s_1,..,s_n) = \prod_{m=1}^{n-p}\mathbf{1}(s_m \leq s)\prod_{m=n+1-p}^{n}\mathbf{1}(s_m > s)$ in (2), we recover a result due to [11]

$$\mathbf{P}(P_n(s) = p) = \binom{n}{p}\sum_{m=p}^{n}(-1)^{m-p}\binom{n-p}{m-p}(1-ms)_+^{n-1}. \tag{5}$$

Let then $\mathcal{L}_n(s)$ be the total length of $\mathcal{S}_n(s)$. As there are $n - P_n(s)$ spacings covered by s and $P_n(s)$ gaps each contributing of s to the covered length, it can be expressed as a contribution of two terms

$$\mathcal{L}_n(s) = \sum_{m=1}^{n-P_n(s)} S_{m:n} + sP_n(s). \tag{6}$$

Note also that the vacancy, which is the length of the circumference not covered by any arc is

$$1-\mathcal{L}_n(s) = \sum_{p=1}^{P_n(s)}(S_{n-p+1:n} - s) = \sum_{m=1}^{n}(S_{m,n} - s)_+, \tag{7}$$

summing the gaps' lengths over the gaps.
>From these facts, with $p \in \{1,..,n\}$ and with $l \in (0,1]$, we get

$$\mathbf{P}\left(\mathcal{L}_n\left(s\right) < l\right) = \sum_{p=1}^{n} \mathbf{P}\left(P_n\left(s\right) = p\right) \mathbf{P}\left(\sum_{m=1}^{n-p} S_{m:n} + sp < l \mid P_n\left(s\right) = p\right) = \tag{8}$$

$$\sum_{p=1}^{n} \mathbf{P}\left(\sum_{m=1}^{n-p} S_{m:n} < l - sp,\ S_{n-p:n} \leq s,\ S_{n+1-p:n} > s\right) = \tag{9}$$

$$\sum_{p=1}^{\inf(n,\lceil l/s \rceil)} \mathbf{P}\left(\sum_{m=1}^{n-p} S_{m:n} < l - sp,\ S_{n-p:n} \leq s,\ S_{n+1-p:n} > s\right), \tag{10}$$

which is the full distribution of $\mathcal{L}_n\left(s\right)$. This distribution can be obtained in the considered uniform case. From the second expression of the vacancy in (7), it can also be seen to be [9], [5]

$$\mathbf{P}\left(\mathcal{L}_n\left(s\right) < l\right) = \sum_{m=1}^{n} \sum_{k=0}^{m-1} (-1)^{m+k-1} \binom{n}{m} \binom{n-1}{k} \binom{m-1}{k} (1-l)^k \left(l - ms\right)_+^{n-1-k} \tag{11}$$

and this expression can also be obtained (with some difficulty) from (10), using properties of the *Poisson* process as in (2).
Finally, let $N\left(s\right)$ be the number of random intervals of length s needed to cover the circle, that is $N\left(s\right) := \inf\left(n > 1 : P_n\left(s\right) = 0\right)$. The following four events coincide

$$(i)\ N\left(s\right) > n,\quad (ii)\ S_{n:n} > s,\quad (iii)\ P_n\left(s\right) \geq 1,\quad (iv)\ \mathcal{L}_n\left(s\right) < 1.$$

>From (4), the law of $N\left(s\right)$ is given by the alternated sum $\mathbf{P}\left(N\left(s\right) > n\right) = \mathbf{P}\left(S_{n:n} > s\right)$. From the above equivalence of events, an alternative expression of this awkward probability is obtained while putting $l = 1$ in (10).
Note that the cover probability $\mathbf{P}\left(N\left(s\right) \leq n\right)$ is from (5)

$$\mathbf{P}\left(P_n\left(s\right) = 0\right) = \sum_{m=0}^{n} (-1)^m \binom{n}{m} \left(1 - ms\right)_+^{n-1} = \mathbf{P}\left(S_{n:n} \leq s\right). \tag{12}$$

and that, as required

$$\mathbf{P}\left(P_n\left(s\right) = n\right) = \left(1 - ns\right)_+^{n-1} = \mathbf{P}\left(S_{1:n} > s\right). \tag{13}$$

>From these definitions and preliminary facts, we are in the position to define the packing configurations and to formulate the announced alternative approach to the parking problem.

3 A cooperative approach to Rényi's parking problem

3.1 Setting of the problem

Fix an arc length $s < .5$. Throw independently n atoms uniformly on the circle and consider the set $\mathcal{S}_n(s)$, appending an arc of length s to each atom. Suppose $S_{1:n} > s$. Then, the number of $\mathcal{S}_n(s)$'s connected components, $P_n(s)$, is maximal ($P_n(s) = n$) and there is no overlap between the arcs. Suppose in addition that the largest gap's length, which is $S_{n:n} - s$, is less than s. Then, no additional arc of length s can be added, wherever on the circle, without provoking overlap and the length of the covered set in such configurations is ns.

Definition 3.1. *We call a configuration of $\mathcal{S}_n(s)$ for which $S_{1:n} > s$ and $S_{n:n} < 2s$ a $s-$packing configuration: due to lack of space, no additional arc of length s can be added, wherever on the circle, and the number of connected components is maximal (there is no arcs' overlap).*

Thus, an s-packing configuration of $\mathcal{S}_n(s)$ (*i.e.* with n atoms) occurs with probability $\mathbf{P}(S_{1:n} > s,\ S_{n:n} < 2s)$.
Note that the values of n for which the probability $\mathbf{P}(S_{1:n} > s,\ S_{n:n} < 2s) > 0$ vary in the range $\{n_0(s) := [1/(2s)] + 1, .., n_1(s) := [1/s]\}$ and that the number

$$\mathbf{P}(s) := \sum_{n=n_0(s)}^{n_1(s)} \mathbf{P}(S_{1:n} > s,\ S_{n:n} < 2s) \tag{14}$$

is the probability that some $s-$packing configuration occurs. In all these aspects, the joint law of $(S_{1:n}, S_{n:n})$ is involved.

Definition 3.2. *Pick at random an $s-$packing configuration and let $\mathcal{N}(s)$ be the number of connected components of the output. Call $\mathcal{N}(s)$ the $s-$packing number. By this definition*

$$\mathbf{P}(\mathcal{N}(s) = n) = \mathbf{P}(S_{1:n} > s,\ S_{n:n} < 2s) / \mathbf{P}(s), \tag{15}$$

with $n \in \{n_0(s), .., n_1(s)\}$.

To compute this distribution, it follows that the exact law of $(S_{1:n}, S_{n:n})$ is needed. We supply some of the details. Let $0 < a < b < 1$. Applying *Steutel*'s formula, with $f(s_1, .., s_n) = \prod_{m=1}^{n} \mathbf{1}(a < s_m < b)$ yields

$$\int_0^\infty \mathbf{P}(S_{1:n}(t) > a, S_{n:n}(t) < b)\, t^{n-1} e^{-pt} dt = \Gamma(n)\, p^{-n} \left(e^{-pa} - e^{-pb}\right)^n.$$

Inverting the *Laplace* transform on the right hand-side gives, using the Binomial identity

$$\mathbf{P}(S_{1:n}(t) > a, S_{n:n}(t) < b) = t^{-(n-1)} \sum_{m=0}^{n} (-1)^m \binom{n}{m} (t - (na + m(b-a)))_+^{n-1}.$$

Putting $t=1$ gives $\mathbf{P}(S_{1:n}>a, S_{n:n}<b)$.
Putting $(a=s, b=1)$ and $(a=0, b=s)$ gives the two aspects of (4). This formula was first obtained by [1]. Putting next $a=s$, $b=2s$, we get

$$\mathbf{P}(S_{1:n}>s,\ S_{n:n}<2s)=\sum_{m=0}^{n}(-1)^m\binom{n}{m}(1-s(n+m))_+^{n-1}. \tag{16}$$

3.2 Asymptotic results

We give now asymptotic results on this formulation of the problem. In this case, the following limiting results hold.

1/ Consider the probability $p_{n,l}(s)=\mathbf{P}(\ s<S_{m,n}<2s \text{ for } l \text{ values of } m)$, with $0\leq l<n$ a fixed constant. Then, [3], for all $x>0$

$$p_{n,l}\left(x/n^2\right)\to_{n\uparrow\infty}\frac{x^l}{l!}e^{-x},\ x>0, \tag{17}$$

showing that $p_{n,l}\left(x/n^2\right)$ is asymptotically *Poisson*(x) with mean $x>0$.

Indeed, consider $q_{n,l}(s)=\sum_{k=0}^{l}p_{n,k}(s)$. Then, for each fixed $l<n$, the function $\phi_n(t/x)=q_{n,l}\left(\frac{x}{tn^2}\right)$ is uniformly bounded and monotonic in t (see [10] page 241) and, computing $\psi_n(n)$ in that case

$$\begin{aligned}\int_0^{\infty}\phi_n(t/x)\,\delta_n(t-1)\,dt &= \sum_{k=0}^{l}\binom{n}{k}\left(e^{-\frac{x}{n}}-e^{-\frac{2x}{n}}\right)^k\left(1-e^{-\frac{x}{n}}+e^{-\frac{2x}{n}}\right)^{n-k}\\ &\to_{n\uparrow\infty}\sum_{k=0}^{l}\frac{x^k}{k!}e^{-x}.\end{aligned}$$

Picking the term $k=l$ in this sum and applying *Steutel*'s Theorem 3 yields the asymptotic result for $p_{n,l}\left(x/n^2\right)$.
Note finally that $p_{n,n}(s)=\mathbf{P}(S_{1:n}>s,\ S_{n:n}<2s)$, is not concerned by this asymptotic.

2/ The right asymptotic concerning the joint distribution of $(S_{1:n},\ S_{n:n})$ is easily found to be, using similar arguments

$$\mathbf{P}\left(n^2S_{1:n}>a,\ \frac{1}{n}e^{nS_{n:n}}<b\right)\to_{n\uparrow\infty}e^{-(a+b)}. \tag{18}$$

This result can be found in ([1], page 252).

3/ A related result of interest to our purpose is the following. Consider the functions $\phi_n(t)$ and $\psi_n(n)$ defined above with associated test function $f(s_1,..,s_n)=\prod_{m=1}^{n}\mathbf{1}(s<s_m<2s)$. We have, intuitively,

$$\phi_n(1)=\mathbf{P}(S_{1:n}>s,\ S_{n:n}<2s)\sim_{n\uparrow\infty}\psi_n(n)=\left(e^{-ns}-e^{-2ns}\right)^n=e^{-n^2s}\left(1-e^{-ns}\right)^n$$

Recalling the range of n, set $ns = x$, with $x \in (1/2, 1)$, we obtain the estimate

$$\mathbf{P}\left(nS_{1:n} > x,\ nS_{n:n} < 2x\right) \sim_{n\uparrow\infty} \left[e^{-x}\left(1 - e^{-x}\right)\right]^n . \tag{19}$$

>From lemma 2, this may also be seen to follow from the fact that for all $x > 0$, the function $\phi_n(t/x) = \mathbf{P}\left(nS_{1:n} > x/t,\ nS_{n:n} < 2x/t\right)$ is a uniformly bounded in n and monotonic function on its domain $t > x > 0$.

The probability in (19) tends to 0 exponentially fast with n. To have an estimate of the normalizing constant, as a function of n for large n, for all values of s, we have to integrate the above probability over x in the corresponding range. We find the saddle-point estimate

$$\int_{1/2}^{1} \left[e^{-x}\left(1 - e^{-x}\right)\right]^n dx \sim_{n\uparrow\infty} \left[\sup_{x\in(1/2,1)} \left[e^{-x}\left(1 - e^{-x}\right)\right]\right]^n = 4^{-n}, \tag{20}$$

because the function $e^{-x}\left(1 - e^{-x}\right) > 0$ is maximal at $x = \log 2$, with value $1/4$ there.

This shows, from (15), (16), (19), (20), that, as s tends to 0, $s\mathcal{N}(s)$ has a density at x approximated by

$$\mathbf{P}\left(s\mathcal{N}(s) = x\right) \sim_{s\downarrow 0} \left[4e^{-x}\left(1 - e^{-x}\right)\right]^{x/s}, \tag{21}$$

concentrating at $x = \log 2$. As a result, $s\mathcal{N}(s) \xrightarrow{\mathbf{P}}_{s\downarrow 0} \log 2 \approx .693..$ (with an entropy flavor) and for any $\varepsilon > 0$, however small, with $B_\varepsilon(x) :=]x - \varepsilon, x + \varepsilon[$

$$\lim_{s\downarrow 0} s \log \mathbf{P}\left(s\mathcal{N}(s) \in B_\varepsilon(x)\right) = \sup_{z\in B_\varepsilon(x)} r(z), \tag{22}$$

giving the concave large deviation rate function as

$$r(x) = x \log\left[4e^{-x}\left(1 - e^{-x}\right)\right] \le 0. \tag{23}$$

Remark:

Consider again the sequential approach on the circle. The question here is how to define the event $\mathcal{N}^{\rightarrow}(s) = n$, were $\mathcal{N}^{\rightarrow}(s)$ to be a sequential version of the $s-$packing number. Suppose that, given $n-1$ uniform throws have been performed on the circle, the following event is realized:

$$\left(S_{1:n-1} > s,\ S_{n-1:n-1} > 2s,\ S_{n-2:n-1} < 2s\right).$$

If $S_{1:n-1} > s$, there is no overlap between the $n-1$ cars of size s; if $S_{n-1:n-1} > 2s$, there still is some vacant place where to insert at least one car of size s and if $S_{n-2:n-1} < 2s$, there is a single gap where this additional car can be inserted. Suppose that this is the available configuration given the number of cars is $n-1$. Insert at random a new atom $(n-1 \rightarrow n)$ and append to it a car of length s, clockwise. Then $\mathcal{N}^{\rightarrow}(s) = n$ if this new atom precisely falls in the gap remaining to be filled, at distance larger than s of its right-most neighbor (clockwise) so as not to provoke the fatal overlap and to fill the last available gap. We have not tried to compute nor evaluate the probability that $\mathcal{N}^{\rightarrow}(s) = n$.

References

[1] D.A. Darling, (1953) *On a class of problems related to the random division of an interval.* Ann. Math. Statistics, **24**, 239-253.

[2] J.W. Evans, (1993) *Random and cooperative sequential adsorption.* Rev. Mod. Physics, **65**(4), 1281-1329.

[3] L. Flatto, A.G. Konheim, (1962) *The random division of an interval and the random covering of a circle.* SIAM Review, **4**, 211-222.

[4] P.C. Hemmer, (1989) *The random parking problem.* J. Stat. Phys., **57**(3/4), 865-869.

[5] L. Holst, (1983) *A note on random arcs on the circle.* Probability and Mathematical Statistics. Essays in honour of Carl–Gustav Esseen. Ed. by Allan Gut and Lars Holst, Uppsala, 40-45.

[6] M. Penrose, (2001) *Random parking, sequential adsorption, and the jamming limit.* Comm. Math. Phys., **218**(1), 153-176.

[7] R. Pyke, (1965) *Spacings (With discussion).* J. Roy. Statist. Soc. Ser. B, **27**, 395-449.

[8] A. Rényi, (1958) *On a one-dimensional problem concerning random space filling.* Magyar Tud. Akad. Mat. Kutato Int. Kozl, **3**(1/2), 109-127.

[9] A.F. Siegel, (1978) *Random arcs on the circle.* J. Appl. Prob., **15**, 774-789.

[10] F.W. Steutel, (1967) *Random division of an interval.* Statistica Neerlandica, **21**, 231-244.

[11] W.L. Stevens, (1939) *Solution to a geometrical problem in probability.* Ann. Eugenics, **9**, 315-320.

Thierry Huillet

Laboratoire de Physique Théorique et Modélisation
CNRS-ESA 8089, Université de Cergy-Pontoise
5 mail Gay-Lussac
95031, Neuville sur Oise, FRANCE.
Thierry.Huillet@ptm.u-cergy.fr

Anna Porzio

LAGA, Université Paris XIII, Institut Galilée
93430, Villetaneuse and
Centre de Physique Théorique
CNRS-UMR 7644 at Ecole Polytechnique
91128 Palaiseau, FRANCE.
Anna.Porzio@cpht.polytechnique.fr

Trends in Mathematics, © 2002 Birkhäuser Verlag Basel/Switzerland

On the Noise Sensitivity of Monotone Functions

Elchanan Mossel, Ryan O'Donnell

ABSTRACT: *It is known that for all monotone functions* $f : \{0,1\}^n \to \{0,1\}$, *if* $x \in \{0,1\}^n$ *is chosen uniformly at random and* y *is obtained from* x *by flipping each of the bits of* x *independently with probability* ϵ, *then* $\mathbf{P}[f_n(x) \neq f_n(y)] < c\epsilon\sqrt{n}$, *for some* $c > 0$.

Previously, the best construction of monotone functions satisfying $\mathbf{P}[f_n(x) \neq f_n(y)] \geq \delta$, *where* $0 < \delta < 1/2$, *required* $\epsilon \geq c(\delta)n^{-\alpha}$, *where* $\alpha = 1 - \ln 2/\ln 3 = 0.36907\ldots$, *and* $c(\delta) > 0$. *We improve this result by achieving for every* $0 < \delta < 1/2$, $\mathbf{P}[f_n(x) \neq f_n(y)] \geq \delta$, *with:*

- $\epsilon = c(\delta)n^{-\alpha}$ *for any* $\alpha < 1/2$, *using the recursive majority function with arity* $k = k(\alpha)$;

- $\epsilon = c(\delta)n^{-1/2}\log^t n$ *for* $t = \log_2\sqrt{\pi/2} = .3257\ldots$, *using an explicit recursive majority function with increasing arities; and,*

- $\epsilon = c(\delta)n^{-1/2}$, *non-constructively, following a probabilistic CNF construction due to Talagrand.*

The constructions have implications for learning theory, computational complexity, and neural networks, and they shed some light on the American electoral system.

1 Introduction

1.1 Noise sensitivity and Fourier coefficients

The papers [KKL88, BL90] suggested the importance of the *Fourier expansion* and the *influence of variables* on f for the study of boolean functions. The ideas developed in these papers proved to be extremely fruitful in later work, e.g., [LMN93, FK96, F98, BKS98] and the material in Subsection 2, to name just a few examples.

Let $\Omega_n = \{-1,+1\}^n$ be the Hamming cube endowed with the uniform probability measure $\mathbf{P}$. We look at boolean functions $f : \Omega_n \to \{-1,+1\}$. We are mostly concerned with *monotone* boolean functions. Recall that a function f is *monotone* if for all $x, y \in \Omega_n$ we have $f(x) \leq f(y)$ whenever $x \leq y$ (in the sense $x_i \leq y_i$ for all i).

For $-1 \leq \eta \leq 1$ and $x \in \Omega_n$, define $N_\eta(x)$ to be a random element y of Ω_n which satisfies $\mathbf{E}[y_i x_i] = \eta$ (equivalently, $\mathbf{P}[x_i \neq y_i] = (1-\eta)/2$) independently for all i. It is natural to measure how stable f is to η-noise by the correlation between $f(x)$ and $f(N_\eta(x))$,

$$Z(f,\eta) = \mathbf{E}[f(N_\eta(x))f(x)] = 1 - 2\mathbf{P}[f(N_\eta(x)) \neq f(x)]. \quad (1)$$

If f is stable under the noise operator N_η, then typically $f(x)$ and $f(N_\eta(x))$ should have the same value and therefore $Z(f,\eta)$, the expression in (1), should be close to 1; if f is sensitive to noise, then $Z(f,\eta)$ should be close to 0.

The space Ω_n with the uniform probability measure naturally gives rise to an inner product space on all functions $f : \Omega_n \to \mathbb{R}$:

$$\langle f, g\rangle = \mathbf{E}[fg] = 2^{-n} \sum_{x\in\Omega_n} f(x)g(x).$$

For a set $S \subseteq [n]$, define $u_S(x) = \prod_{i\in S} x_i$. Since $u_S u_{S'} = u_{S\Delta S'}$, where Δ denotes symmetric difference, it follows that $(u_S)_{S\subseteq[n]}$ is an orthonormal basis. We call $\hat{f}(S) = \langle u_S, f\rangle$ the S *Fourier coefficient* of f, and $f = \sum_{S\subseteq[n]} \hat{f}(S)u_S$ the *Fourier expansion* of f.

The basis $(u_S)_{S\subseteq[n]}$ has very nice properties with respect to the noise operator; most notably, for all x and S, $\mathbf{E}[u_S(N_\eta(x))] = \eta^{|S|}u_S(x)$, which implies

$$Z(f,\eta) = \mathbf{E}[f(N_\eta(x))f(x)] = \sum_{S\subseteq[n]} \eta^{|S|}\hat{f}^2(S) \tag{2}$$

(see e.g. [BKS98, BJT99, O02]).

The stability of the function f under noise, $Z(f,\eta)$, is therefore closely related to how much of the ℓ_2 mass of the Fourier coefficients of f lies on coefficients $\hat{f}(S)$ for large sets S.

In addition to the sum in (2), it is common to study several other weighted sums of f's squared Fourier coefficients. By Parseval's identity, $\sum_S \hat{f}^2(S) = 1$. The average sensitivity of f is defined by $I(f) := \sum_S |S|\hat{f}^2(S)$. It is shown in [KKL88] that $I(f) = \sum_{k=1}^n I_k(f)$, where $I_k(f)$ is the probability the value of the function flips, when the k'th bit is flipped. Note that if f is monotone, then $I_k(f) = |\hat{f}(\{k\})|$. Finally, we have the quantity $II(f) := \sum_{k=1}^n I_k^2(f)$, introduced in [BKS98].

1.2 Sensitivity of monotone functions

The parity function, $f = u_{[n]} = \oplus$, is the boolean function most sensitive to noise: $Z(f,\eta) = \eta^n$ is minimal, and $I(f) = n$ is maximal.

It is natural to ask if monotone functions can be as sensitive to noise as non-monotone functions. It is known (see Lemma 6.1 of [FK96]) that the majority function has maximal I among all monotone functions on n inputs. Since its average sensitivity is easily computed to be $\sqrt{2/\pi}\sqrt{n} + o(\sqrt{n})$, we get that for all for all monotone f on n inputs,

$$I(f) \le (\sqrt{2/\pi} + o(1))\sqrt{n}. \tag{3}$$

It remains to determine how small $N_\eta(f)$ can be for monotone functions. A natural goal is to find a monotone function f on n bits such that $Z(f, 1-\delta) \le 1 - \Omega(1)$ for the smallest possible quantity δ. This problem was implicitly posed in [BKS98].

An easy folklore argument (see long version for proof) uses (3) to deduce:

Proposition 1.1. *For all monotone f on n inputs,*

$$Z(f, 1-\delta) \geq (1-\delta)^{(1+o(1))\sqrt{(2/\pi)n}}.$$

Therefore if $Z(f, 1-\delta) \leq 1-\epsilon$, then $\delta \geq \sqrt{\frac{2}{\pi}}\frac{\epsilon}{\sqrt{n}} + o(1/\sqrt{n})$.

In particular, in order to obtain $Z(f, 1-\delta) \leq 1-\Omega(1)$, δ must satisfy $\delta \geq \Omega(n^{-1/2})$. Prior to this work, the best sensitivity with respect to N_η was achieved via the recursive majority of 3 function (folklore, see [BL90, BKS98]). This function satisfies $Z(f, 1-\delta) \leq 1-\Omega(1)$, for $\delta = n^{-\alpha}$, where $\alpha = 1 - \ln 2/\ln 3 = 0.36907\ldots$

1.3 Our results

Recursive majority functions seem to be sensitive to noise. Previous techniques for analyzing recursive majorities had suggested that recursive majority of 5, 7, etc. might be less sensitive than recursive majority of 3. However, this is not the case.

Theorem 1.2. *Let $k = 2r+1$ and let $\mathsf{REC\text{-}MAJ}\text{-}k_\ell$ denote the ℓ level k recursive majority. Let*

$$b = \frac{2r+1}{2^{4r}}\binom{2r}{r}^2, \qquad a = \frac{2r+1}{2^{2r}}\binom{2r}{r}.$$

Then $Z(\mathsf{REC\text{-}MAJ}\text{-}k_\ell, 1-\delta) \leq \epsilon$ for $\ell \geq \Big(\log_a(1/\delta) + \log_{1/b}(1/\epsilon)\Big)(1 + r(\epsilon,\delta))$, where $r(\epsilon,\delta) \to 0$ as $\epsilon \to 0$ and $\delta \to 0$. Hence for every $\alpha < 1/2$, and $0 < \delta < 1$, there exists an odd $k \geq 3$ such that for $n = t^\ell$, $f_n = \mathsf{REC\text{-}MAJ}\text{-}k_\ell : \Omega_n \to \{-1,+1\}$ is a balanced function with

$$Z(f_n, 1-n^{-\alpha}) \leq 1-\delta+o(1).$$

Note that this construction is explicit. Moreover, using k-majority gates, we obtain a read-once, log-depth circuit which implements the function. The proof technique is closely related to techniques in classical branching processes [AN72] (see also [M98]).

By relaxing the bounded degree property, and using instead majority gates of varying fan-in, we obtain an explicit read-once construction of log log-depth which is sensitive to a noise rate of about $n^{-1/2}$, up to a sub-logarithmic correction.

Theorem 1.3. *For every $0 < \delta < 1$, there exists an explicit infinite family of balanced monotone functions $f_n : \Omega_n \to \{-1,+1\}$ with the following property:*

$$Z(f_n, 1-1/M) \leq 1-\delta+o(1),$$

where $M = \sqrt{n}/\Theta(\log^t n)$, and $t = \log_2\sqrt{\pi/2} = .3257\ldots$.

It is interesting to note that the t parameter is optimal for this construction.

Finally, analyzing a probabilistic construction due to Talagrand [T96], we obtain a tight result up to constant factors.

Theorem 1.4. *For every $0 < \delta < 1$, there exists an infinite family of monotone functions $f_n : \Omega_n \to \{-1, +1\}$ with the following property:*

$$Z(f_n, 1 - n^{-1/2}) \leq 1 - \delta + o(1). \tag{4}$$

In this extended abstract, we sketch the proof of a slightly weakened version of Theorem 1.4, i.e., instead of (4) we prove

$$Z(f_n, 1 - n^{-1/2}) \leq 1 - \Omega(1). \tag{5}$$

2 Implications for other problems

2.1 Learning monotone functions

In the field of computational learning theory, one of the most widely studied models is Valiant's Probably Approximately Correct (PAC) model [V84]. In PAC learning, a *concept class* $\mathcal{C}$ is a collection $\cup_{n \geq 1} \mathcal{C}_n$ of boolean functions, where each function (concept) $f \in \mathcal{C}_n$ is a boolean function on n bits. Let $f \in \mathcal{C}_n$ be an unknown *target function*, and let $\mathcal{D}$ be an unknown probability *distribution* on $\{-1, +1\}^n$. A learning algorithm A for $\mathcal{C}$ takes as input an *accuracy parameter* $0 < \epsilon < 1$ and a *confidence parameter* $0 < \delta < 1$. During its execution, A has access to an *example oracle* $\mathsf{EX}(f)$ which, when queried, generates a random labeled example $\langle x, f(x) \rangle$, where x is drawn from distribution $\mathcal{D}$. A's goal is to output a hypothesis h which is a boolean function on n bits, which is "close" to f under distribution $\mathcal{D}$. Specifically, we say that A is a *PAC learning algorithm for* $\mathcal{C}$ if for every $f \in C$ and every ϵ, δ, with probability $1 - \delta$ algorithm A outputs a hypothesis h satisfying $\Pr_{x \leftarrow \mathcal{D}}[f(x) \neq h(x)] \leq \epsilon$. Ideally one likes for A to run in time $\mathrm{poly}(n, s, 1/\epsilon, \log(1/\delta))$, where s is a "size parameter" of the concept class.

An important and well-studied restriction of the PAC model is *uniform PAC learning*, which is simply the case in which $\mathcal{D}$ is the uniform distribution on $\{-1, +1\}^n$. Linial, Mansour, and Nisan [LMN93] introduced a very powerful and general uniform PAC learning algorithm, which has come to be known as the "low degree algorithm" (see Mansour's survey [M94]). The low degree algorithm works for any concept class which has a *Fourier concentration bound.* Specifically, suppose that for every function f in a given concept class, $\sum_{|S| \geq m} \hat{f}^2(S) \leq \epsilon$. Then the low degree algorithm will PAC-learn this class under the uniform distribution in time $\exp(O(m \log(n/m))) \log(1/\delta)$. The algorithm works by drawing many examples for f, and using these to calculate empirical estimates for all Fourier coefficients $\hat{f}(S)$ with $|S| < m$. The hypothesis outputted is simply the sign of the resulting truncated Fourier expansion.

Bshouty and Tamon [BT96] give the fastest known uniform PAC learning algorithm for the concept class of monotone functions. Their algorithm is the low degree algorithm, and they show a Fourier concentration bound for the class of monotone functions with $m = O(\epsilon^{-1}\sqrt{n})$. (It is simple to derive this from (3); Bshouty and Tamon also extend these results to general product distributions on $\{-1, +1\}^n$.) This leads to a learning algorithm running in time $\exp(O(\frac{1}{\epsilon}\sqrt{n} \log(\epsilon\sqrt{n}))) \log(1/\delta)$.

As a tightness result, [BT96] prove via a counting argument that there is a monotone f which does not satisfy $\sum_{|S| \geq m} \hat{f}^2(S) \leq n^{-1/2} \log n$ unless $m = \Omega(n)$.

However this leaves open the question of $\epsilon = \Omega(n^{-1/2}\log n)$. To show that the low degree algorithm for monotone functions cannot be improved, we need to exhibit a monotone f for which $\sum_{|S|\geq\Omega(\epsilon^{-1}\sqrt{n})}\hat{f}^2(S) > \epsilon$. The functions f from Theorem 1.4 satisfy $\sum_{|S|\geq\sqrt{n}}\hat{f}^2(S)\geq\Omega(1)$. Hence the low degree algorithm will have $\Omega(1)$ error unless it goes up to degree $\sqrt{n}$. In fact, our Corollary 7.2 gives us an explicit function f with $\sum_{|S|\geq\tilde{\Omega}(\sqrt{\epsilon n})}\hat{f}^2(S)\geq 1-\epsilon$.

See [BJT99, KOS02] for more on noise sensitivity in the context of computational learning theory.

2.2 Hardness amplification within NP

The central problem in computational complexity theory is whether or not NP = P; i.e., deciding if proving a proposition is harder than verifying the proof of that proposition. In studying this problem, many researchers have considered the slightly weaker question of whether or not every language in NP can be computed by circuits of polynomial size. (See any standard text such as [Pa93, BDG88, DK00] for the definitions of P, NP, circuits, etc.) Let us phrase this question precisely. A language $F\in$ NP gives rise to a family of characteristic functions $\langle f_n\rangle$, where $f_n:\{0,1\}^n\to\{0,1\}$ is defined by $f_n(x)=1$ iff $x\in F$. We often abuse language by saying f_n is a function NP (we always have a particular family of functions in mind). A family of boolean circuits $\langle C_n\rangle$ is said to have polynomial size if there is a finite k such that size$(C_n)\leq O(n^k)$. We say NP has polynomial-sized circuits if for every family of functions $\langle f_n\rangle$ in NP, there is a circuit family $\langle C_n\rangle$ of polynomial size such that $C_{|x|}(x)=f_{|x|}(x)$ for every boolean string x.

Most researchers believe that NP does *not* have polynomial-sized circuits; i.e., NP is *hard* for polynomial-sized circuits. One might then ask *how* hard NP is for polynomial circuits. One way of viewing this to question is to ask on how large a fraction of the inputs in $\{0,1\}^n$ can a polynomial-sized circuit compute a given NP function. We say that f is "$(1-\delta)$-hard for polynomial circuits" is for every family $\langle C_n\rangle$ of polynomial-sized circuits, $\mathbf{P}[f(x)=C_n(x)]\leq 1-\delta$. Note that asserting NP is hard for polynomial circuits is the same as saying that there is a function $f\in$ NP which is $(1-2^{-n})$-hard for polynomial circuits. Also note that no function is $(1-\delta)$-hard for $\delta\geq 1/2$ because either the circuit which always outputs 1 or the circuit that always outputs 0 gets f right on at least half of all inputs. Under the assumption that NP does not have polynomial circuits, it is of interest to know just how hard NP is in this sense.

In [O02], the second author addresses this question. Starting from the assumption that there is a function in NP which is $(1-1/n^{O(1)})$-hard for polynomial circuits, [O02] shows the existence of a function in NP which is $(1/2+n^{-1/2+\delta})$-hard for polynomial circuits (for any small $\delta>0$). The main technical theorem in [O02] is that if f is a balanced function which is $(1-\delta)$-hard for polynomial circuits, and g is a function satisfying $Z(g,1-2\delta)\leq\eta$, then $g\otimes f$ is essentially $(\frac{1}{2}+\frac{1}{2}\sqrt{\eta})$-hard for polynomial circuits.

In order to apply this technical theorem to convert a slightly hard function in NP to a very hard function in NP, it is necessary to ensure that $g\otimes f\in$ NP when $f\in$ NP. Recall that NP is the class of functions f which have easily verified proofs of $f=1$. In order for $g\otimes f$ to have easily verified proofs of $g\otimes f=1$, it suffices for g to be (a) in NP, and (b) monotone. For in this case, we can prove that $g \otimes f = 1$ by proving that some subset of the inputs to g are 1, and each of these is a statement of the form $f = 1$, which has an easily verified proof because $f \in$ NP. Hence to amplify hardness within NP, [O02] needs to find a monotone function in NP such that $Z(g, 1 - 1/n^{O(1)})$ is very small. This exact problem is addressed in the present paper. Take g to be the function from Theorem 7.1 on k inputs. This function is easily seen to be in P, hence in NP. If we pick $k = n^C$ and $\epsilon = n^c/k$ for some constants C and c, then Theorem 7.1 tells us that $Z(g, 1 - 1/\tilde{\Omega}(n^{c/2})) \leq 1/k^{1-c/C}$. Hence if f $(1 - 1/n^{O(1)})$-hard for polynomial circuits, by choosing c and C sufficiently large, we can arrange for $g \otimes f$ — which has input length $kn = n^{C+1}$ — to be $(1/2 + (kn)^{-1/2+\delta})$-hard for any small $\delta > 0$. This is the result of [O02]. Note that Theorem 1.4 is not useful in this context, since the amplifying function g must be in NP, and Talagrand's function is not even explicit.

2.3 Neural networks

In the theory of neural networks (see e.g. [H99] for background), a neuron is modeled as a weighted majority function. For physical and biological reasons, it is expected that such a function would be noise stable. In [BKS98] it is shown that there exists a universal constant C such that for all weighted majority functions M, $Z(M, 1 - \epsilon) \geq 1 - C\epsilon^{1/4}$. Peres [Pe98] has improved this to $1 - C\epsilon^{1/2}$.

If we consider the simplest kind of neural network, in which every variable and every majority output is read only once, we obtain a tree circuit of weighted majority gates. Using a simple exchange of variables, we may assume that all the weights of the majority functions are positive and hence that the network represents a monotone function. Proposition 1.1 implies that the network is insensitive to noise rate of $n^{-\alpha}$ for $\alpha > 1/2$, where n is the number of inputs to the function. Our construction in Theorem 1.2 implies on the other hand that this is tight, i.e., for every $\alpha < 1/2$, there exists a neural network in which every variable and every output is read once, and the network is sensitive to noise rate $n^{-\alpha}$.

2.4 Sensitivity of election schemes

One of the desired properties of election schemes is robustness. Consider the following simple model: There are n voters who have to decide between candidate -1 and candidate 1. Suppose that voter i wants to vote x_i, and that the x_i's are uniformly random and independent. Suppose furthermore that due to confusion and some technical errors, the vote of voter i is recorded as y_i where $\mathbf{P}[x_i = y_i] = 1 - \epsilon$ independently for all i. In this setting it is natural to require that the vote outcome $f(y_1, \ldots, y_n)$ be governed by a symmetric balanced monotone function. Moreover, if we want to minimize the effect of the confusion and errors, we want to maximize $\mathbf{P}[f(x_1, \ldots, x_n) = f(y_1, \ldots, y_n)] = Z(f, 1 - 2\epsilon)$.

Let us compare two election schemes. In the first scheme, f is the simple majority function. Here $Z(f, 1 - 2\epsilon)$ is of order $1 - \epsilon^{1/2}$. In the second scheme, we have a two level majority function; e.g., each state votes by simple majority for an elector, and the majority of the electors' votes chooses the president. Here, if we assume $n^{1/2}$ electors, a calculation as in the proof of Theorem 1.2 shows that $Z(f, 1 - 2\epsilon)$ is of order $\epsilon^{1/4}$. Hence the "electoral college" system is much more sensitive to noise. In fact, Theorem 1.2 suggests that adding more levels of sub-electors (such as voting by county first) increases the sensitivity of the election to noise, up to its maximum possible level for a monotone function.

3 Sensitivity of majorities

3.1 Majority

We denote the majority function on k bits by MAJ_k. Using asymptotic results for random walks, one can prove (cf. [O02]):

Proposition 3.1. *For every* $\eta \in [-1,1]$,

$$|Z(\mathsf{MAJ}_k,\eta) - \frac{2}{\pi}\arcsin(\eta)| \leq O(1/\sqrt{k}).$$

Much more can be said when η is very close to 1, specifically, when $1-\eta$ is small compared to $1/k$. For η close to 1, we prefer to view $Z(f,\eta)$ in terms of the probability that flipping input bits of f flips the output bit. We use the following lemma in the proof of Theorem 1.3.

Lemma 3.2. *Suppose* $k \geq 3$ *and* $\delta \leq 1/k$. *Say we pick a random input to* MAJ_k *— call it* x *— and then construct* y *by flipping each bit of* x *independently with probability* δ. *Then*

$$\mathbf{P}[\mathsf{MAJ}_k(x) \neq \mathsf{MAJ}_k(y)] \geq \sqrt{\frac{2}{\pi}}\sqrt{k}\,\delta\, \exp(-1/3k)\exp(-\delta k).$$

Proof: Clearly,

$$\begin{aligned}&\mathbf{P}[\mathsf{MAJ}_k(x) \neq \mathsf{MAJ}_k(y)]\\ &\qquad \geq \mathbf{P}[\mathsf{MAJ}_k(x) \neq \mathsf{MAJ}_k(y)|\text{exactly one flip}] \times \mathbf{P}[\text{exactly one flip}], \qquad (6)\end{aligned}$$

and $\mathbf{P}[\text{exactly one flip}] = k\delta(1-\delta)^{k-1}$. By elementary calculus, $(1-\delta)^{k-1} \geq \exp(-\delta k)$ for $\delta \leq 1/k$. Therefore,

$$\mathbf{P}[\text{exactly one flip}] = k\delta(1-\delta)^{k-1} \geq k\delta\exp(-\delta k). \qquad (7)$$

The probability that the majority flips given that there is exactly one flipped bit in x, is exactly the probability that the remaining input bits split evenly — i.e.,

$$\begin{aligned}\mathbf{P}[\mathsf{MAJ}_k(x) \neq \mathsf{MAJ}_k(y)|\text{exactly one flip}] &= \binom{k-1}{(k-1)/2}2^{-(k-1)}\\ &\geq \sqrt{\frac{2}{\pi k}}(1-1/4k) \geq \sqrt{\frac{2}{\pi k}}\exp(-1/3k), \qquad (8)\end{aligned}$$

where the first inequality follows by Stirling's formula and the second since $1-1/4k \leq \exp(-1/3k)$ for $k \geq 3$. Combining (6), (7) and (8) we obtain the required result. □

3.2 Recursive majority

We begin with a formal definition of the recursive majority function.

Definition 3.3. *For $f : \Omega_n \to \{-1,+1\}, g : \Omega_m \to \{-1,+1\}$, we let $f \otimes g$ denote the function $f \otimes g : \Omega_{nm} \to \{-1,+1\}$ defined by*

$$f \otimes g\ (x_1, \ldots, x_{nm}) = f\left(g(x_1, \ldots, x_m), \ldots, g(x_{(n-1)m+1}, \ldots, x_{nm})\right)$$

For ℓ an integer, we define $f^{\otimes^\ell} = f$ if $\ell = 1$, and $f^{\otimes^\ell} = f \otimes (f^{\otimes^{\ell-1}})$ otherwise. We let $\text{REC-MAJ-}t_\ell = \text{MAJ}_t^{\otimes^\ell}$.

The following proposition is immediate, yet useful.

Proposition 3.4. *If g is a balanced function and f is any function, then $Z(f \otimes g, \eta) = Z(f, Z(g, \eta))$.*

In this section we prove Theorem 1.2. It is easy to calculate (and well known) that for the majority function on $k = 2r+1$, MAJ_k,

$$II(\text{MAJ}_k) = \frac{2r+1}{2^{4r}}\binom{2r}{r}^2, \qquad I(\text{MAJ}_k) = \frac{2r+1}{2^{2r}}\binom{2r}{r}.$$

Note therefore that $I(\text{MAJ}_k) \to \sqrt{2/\pi}\sqrt{k}$ as $k \to \infty$. Hence Theorem 1.2 follows almost immediately from the following proposition:

Proposition 3.5. *Let $f : \Omega_k \to \{-1,+1\}$ be a balanced function, and let*

$$a = \sum_S |S| \hat{f}^2(S), \qquad b = \sum_{|S|=1} \hat{f}^2(S).$$

(Note that $a = I(f)$, and if f is monotone, $b = II(f)$.) If $a > 1$ and $b < 1$, then $Z(f^{\otimes^\ell}, 1-\delta) \le \epsilon$, for $\ell \ge \left(\log_a(1/\delta) + \log_{1/b}(1/\epsilon)\right)(1 + r(\epsilon,\delta))$, where $r(\epsilon,\delta) \to 0$ as $\epsilon \to 0$ and $\delta \to 0$.

Proof: (sketch) Let $f = \sum_S \hat{f}^2(S) u_S$ be the Fourier expansion of f. Letting $p(\eta) := Z(f,\eta) = \sum_S \hat{f}^2(S)\eta^{|S|}$, we see that $p(\eta)$ is a convex polynomial function of η which satisfies

$$p(0) = 0,\ p(1) = 1,\ p'(0) = \textstyle\sum_{|S|=1} \hat{f}^2(S) = b,\ p'(1) = \sum_S |S| \hat{f}^2(S) = a. \tag{9}$$

Proposition 3.4 implies that

$$Z(f^{\otimes^\ell}, \eta) = p^{(\ell)}(\eta) := \underbrace{p(p(\cdots p(\eta)\cdots))}_{\ell \text{ times}}. \tag{10}$$

The claim of the proposition now follows by standard arguments on iterations of convex functions (for more details, see the long version of this paper). □

4 Sensitivity to small noise

In this section we prove Theorem 1.3. We do this by proving

Theorem 4.1. *There exists an explicit infinite family of balanced monotone functions $f_n : \Omega_n \to \{-1, +1\}$ with the following property:*

$$Z(f_n, 1 - \epsilon/M) \leq 1 - \epsilon + O(\epsilon^2),$$

where $M = \sqrt{n}/\Theta(\log^t n)$, and $t = \log_2 \sqrt{\pi/2} = .3257\ldots$.

Proof of Theorem 1.3: Let f_n be the function constructed at Theorem 4.1, and let ϵ be such that $Z(f_n, 1 - \epsilon/M) \leq 1 - \epsilon + O(\epsilon^2) < 1 - \delta' + o(1)'$, where $\delta' > 0$. Let $g = \textsf{REC-MAJ-3}_\ell$ where ℓ is chosen is such a way that $Z(g, 1 - \delta'/2) \leq 1 - \delta$ (such ℓ exists by Theorem 1.2). Taking $g_n = g \otimes f_n$, we obtain the desired result. □

The construction in Theorem 4.1 again consists of recursive majorities, where now the number of inputs to the majority varies with the level. The estimates on the sensitivity of these majority functions are derived via Lemma 3.2.

Proof of Theorem 4.1: Since we are dealing with correlations close to 1, it will be more helpful to look at their difference from 1. In particular, we will prove the following equivalent formulation of the theorem: Let x be a randomly chosen input to f_n, and suppose we flip each bit of x independently with probability ϵ/M, forming y. Then the probability that $f_n(x) = f_n(y)$ is at least $\epsilon - O(\epsilon^2)$.

The function $f = f_n$ will be given by recursive majorities of increasing arity: $f_n = \textsf{MAJ}_{k_1} \otimes \textsf{MAJ}_{k_2} \otimes \cdots \otimes \textsf{MAJ}_{k_\ell}$. We will select $k_i = 3^{2^{i-1}+1}$, so "from the top down" the majorities have arity 9, 27, 243, etc. Note that $k_{i+1} = k_i^2/3$. With these choices, the number of inputs is $n = 3^{2^\ell + \ell - 1}$. Hence $\ell \leq \log_2 \log_3 n$.

Let $\delta_0 = \epsilon/M$, and recursively define δ_{i+1} to be the probability that the output of a $\textsf{MAJ}_{k_{\ell-i}}$ flips, given that each of its inputs is flipped independently with probability δ_i. Since all MAJ functions are balanced, Proposition 3.4tells us that the probability that the output of f is flipped is δ_ℓ. We will show that $\delta_\ell \geq \epsilon - O(\epsilon^2)$.

By Lemma 3.2,

$$\delta_{i+1} \geq g(k_{\ell-i}) \exp(-\delta_i k_{\ell-i}) \delta_i,$$

where:

$$g(t) := \frac{1}{\sqrt{\pi/2}} \sqrt{k_{\ell-i}} \exp(-1/3k_{\ell-i}).$$

Recursively define $\eta_0 = \eta_0' = \delta_0$, and:

$$\eta_{i+1} = g(k_{\ell-i}) \exp(-\eta_i k_{\ell-i}) \eta_i, \; \eta_{i+1}' = g(k_{\ell-i}) \eta_i'.$$

Since the probability that the output of MAJ flips is an increasing function of δ, we can conclude that $\delta_i \geq \eta_i$ for every i. But clearly $\eta_i' \geq \eta_i$ for every i. Hence, for every i, $\eta_{i+1} \geq g(k_{\ell-i}) \exp(-\eta'_i k_{\ell-i})\eta_i$. It follows immediately that:

$$\begin{aligned}\eta_\ell &\geq \Big(\prod_{i=0}^{\ell-1} g(k_{\ell-i}) \exp(-\eta'_i k_{\ell-i})\Big)\eta_0 \\ &= \Big(\frac{1}{\sqrt{\pi/2}}\Big)^\ell \prod_{j=1}^{\ell} \sqrt{k_j}\, \exp(-\frac{1}{3}\sum_{j=1}^{\ell} k_j^{-1}) \cdot \exp\Big[\sum_{i=0}^{\ell-1} -\eta'_i k_{\ell-i}\Big] \cdot \delta_0\end{aligned}$$

Defining

$$\begin{aligned}M := \prod_{m=1}^{\ell} g(k_m) = \Big(\frac{1}{\sqrt{\pi/2}}\Big)^{\log_2\log_3 n} & \exp(-\frac{1}{3}\sum_{j=1}^{\ell} k_j^{-1}) \\ &= \Big(\frac{1}{\sqrt{\pi/2}}\Big)^{\log_2\log_3 n} \sqrt{n}\exp(-O(1)),\end{aligned}$$

and $\delta_0 := \epsilon/M$, we obtain

$$\eta_\ell \geq M \cdot \exp\Big[\sum_{i=0}^{\ell-1} -\eta'_i k_{\ell-i}\Big] \cdot (\epsilon/M) = \epsilon \cdot \exp\Big[\sum_{i=0}^{\ell-1} -\eta'_i k_{\ell-i}\Big].$$

Since $\delta_\ell \geq \eta_\ell$, it remains to show:

$$\exp\Big[\sum_{i=0}^{\ell-1} -\eta'_i k_{\ell-i}\Big] \geq 1 - O(\epsilon).$$

By the recursive definition of η'_i, we immediately have $\eta'_i = (\prod_{j=0}^{i-1} g(k_{\ell-j}))\eta'_0$. Hence $\eta'_i = M(\prod_{m=1}^{\ell-i} g(k_m)^{-1})\eta'_0 = \epsilon(\prod_{m=1}^{\ell-i} g(k_m)^{-1})$. Therefore:

$$\exp\Big[\sum_{i=0}^{\ell-1} -\eta'_i k_{\ell-i}\Big] = \exp\Big[-\epsilon \sum_{m=1}^{\ell} \frac{k_m}{g(k_1)g(k_2)\cdots g(k_m)}\Big].$$

Hence if we can show $\sum_{m=1}^{\ell} k_m/g(k_1)g(k_2)\cdots g(k_m) = O(1)$ then we're done. The first term in this sum is $k_1/g(k_1) = O(1)$. The ratio of the mth term to the $(m-1)$th term is $k_m/k_{m-1}g(k_m)$. But $k_{m-1} = \sqrt{3}\sqrt{k_m}$ by definition, so this ratio is $\sqrt{k_m}/\sqrt{3}g(k_m) = \sqrt{\pi/2}/\sqrt{3}\exp(-1/3k_m) < 1$. Hence the terms in the sum decrease geometrically, so the sum is indeed $O(1)$. □

5 Talagrand's function

In [T96], Talagrand gives a randomized construction of a monotone $f_n : \Omega_n \to \{-1,+1\}$ with the following property: at least an $\Omega(1)$ fraction of points x in Ω_n satisfy both $f_n(x) = -1$, and $\#\{x' : \Delta(x,x') = 1 \text{ and } f(x) = +1\} \geq \Omega(n^{1/2})$, where Δ denotes Hamming distance. It is natural to conjecture that this function is sensitive to slight $n^{-1/2}$ noise, as we prove below.

Talagrand's function $f = f_n$ is a random CNF on its n inputs. Specifically, f is the $2^{\sqrt{n}}$-wise AND of $\sqrt{n}$-wise ORs, where each OR's inputs are selected independently and uniformly at random (with replacement) from $[n]$. To prove Theorem 1.4, it suffices to prove that if we pick f, x, and $x' := N_\epsilon(x)$ at random (where $\epsilon = n^{-1/2}$), then:

$$\mathbf{E}_f\left[\mathbf{P}[f(x) \neq f(N_\epsilon(x))]\right] \geq \Omega(1).$$

Proof of Theorem 1.4: (sketch)

$$\begin{aligned}\mathbf{E}_f\left[\mathbf{P}[f(x) \neq f(N_\epsilon(x))]\right] &= \mathbf{E}_{x,\,x'}\left[\mathbf{P}_f[f(x) \neq f(x')]\right] \\ &= 2\mathbf{E}_{x,\,x'}\left[\mathbf{P}_f[f(x) = -1, f(x') = +1]\right], \quad (11)\end{aligned}$$

by symmetry, since x and x' have the same distribution. We want to show that (11) $\geq \Omega(1)$.

Fix x and x'. Let $n_{+\star}$ denote the number of indices on which x is $+1$, let $n_{\star+}$ denote the number of indices on which x' is $+1$, and let n_{++} denote the number of indices on which *both* x and x' are $+1$.

Since f has a fairly simple form — the AND of ORs, where the ORs' inputs are completely independent — it is easy to write $\mathbf{P}_f[f(x) = -1, f(x') = +1]$ explicitly in terms of $n_{+\star}$, $n_{\star+}$, and n_{++}:

$$\mathbf{P}_f[f(x) = -1, f(x') = +1] = p_{\star-}^{2^{\sqrt{n}}} - p_{--}^{2^{\sqrt{n}}}, \tag{12}$$

where

$$\begin{aligned}p_{\star-} &= 1 - \left(\frac{n_{\star+}}{n}\right)^{\sqrt{n}}, \\ p_{--} &= 1 - \left(\frac{n_{\star+}}{n}\right)^{\sqrt{n}} - \left(\frac{n_{+\star}}{n}\right)^{\sqrt{n}} + \left(\frac{n_{++}}{n}\right)^{\sqrt{n}}.\end{aligned}$$

By the mean value theorem, (12) is bounded from below by:

$$2^{\sqrt{n}}(p_{\star-} - p_{--})p_{--}^{2^{\sqrt{n}}}. \tag{13}$$

Now $n_{+\star} \sim$ Binomial$(n, 1/2)$, and similarly for $n_{\star+}$. Hence for sufficiently large n, both quantities are in the range $[n/2 - \sqrt{n}, n/2 + \sqrt{n}]$, except with probability .05. Also, $n_{++} \sim$ Binomial$(n_{+\star}, 1 - \epsilon)$, so for sufficiently large n and if $\epsilon \geq n^{-1/2}$, n_{++} is no larger than $(1 - \epsilon + 2\sqrt{\epsilon/n_{+\star}})n_{+\star}$, except with probability .05. Taking all these facts together via a union bound, we may conclude that except with probability .15,

$$n_{+\star} \in \left[\frac{n}{2} - \sqrt{n}, \frac{n}{2} + \sqrt{n}\right], \quad n_{\star+} \in \left[\frac{n}{2} - \sqrt{n}, \frac{n}{2} + \sqrt{n}\right], \quad \frac{n_{++}}{n_{\star+}} \leq 1 - \epsilon + 3\sqrt{\frac{\epsilon}{n}}. \tag{14}$$

We would like to show that $\mathbf{E}_{x,x'}[(13)] \geq \Omega(1)$. Since (14) happen with probability at least .85, it suffices to prove $\mathbf{E}_{x,x'}[(13)] \geq \Omega(1)$ *conditioned on* these three events holding. But in this case,

$$\begin{aligned}
\mathbf{E}_{x,x'}[(13)] &= 2^{\sqrt{n}}(p_{\star-} - p_{--})\Big[1 - \big(\tfrac{n_{\star+}}{n}\big)^{\sqrt{n}} - \big(\tfrac{n_{+\star}}{n}\big)^{\sqrt{n}} + \big(\tfrac{n_{++}}{n}\big)^{\sqrt{n}}\Big]^{2^{\sqrt{n}}} \\
&\geq 2^{\sqrt{n}}(p_{\star-} - p_{--})\Big[1 - \big(\tfrac{n_{\star+}}{n}\big)^{\sqrt{n}} - \big(\tfrac{n_{+\star}}{n}\big)^{\sqrt{n}}\Big]^{2^{\sqrt{n}}} \\
&\geq 2^{\sqrt{n}}(p_{\star-} - p_{--})\Big[1 - (1/2 + n^{-1/2})^{\sqrt{n}} - (1/2 + n^{-1/2})^{\sqrt{n}}\Big]^{2^{\sqrt{n}}} \\
&\geq 2^{\sqrt{n}}(p_{\star-} - p_{--})[1 - 2e/2^{\sqrt{n}}]^{2^{\sqrt{n}}} \\
&\geq e^{-2e}2^{\sqrt{n}}(p_{\star-} - p_{--}) \\
&= e^{-2e}\Big(2\frac{n_{\star+}}{n}\Big)^{\sqrt{n}}\Big(1 - \Big(\frac{n_{++}}{n_{\star+}}\Big)^{\sqrt{n}}\Big) \\
&\geq e^{-2e}(1 - 2n^{-1/2})^{\sqrt{n}}\Big(1 - \Big(\frac{n_{++}}{n_{\star+}}\Big)^{\sqrt{n}}\Big) \\
&\geq e^{-2e-2}\Big(1 - \Big(\frac{n_{++}}{n_{\star+}}\Big)^{\sqrt{n}}\Big) \\
&\geq e^{-2e-2}\Big(1 - (1 - \epsilon + 2\sqrt{\epsilon/n})^{\sqrt{n}}\Big). \qquad (15)
\end{aligned}$$

When $\epsilon = n^{-1/2}$, the quantity $(1 - \epsilon + 2\sqrt{\epsilon/n})^{\sqrt{n}}$ exceeds e^{-1}. Hence (15) is at least $e^{-2e-2} \geq \Omega(1)$, and we're done. □

6 Tribes and high sensitivity

We have mostly settled the question of how small ϵ can be, such that there is a monotone function f satisfying $Z(f, 1-\epsilon) \leq 1 - \Omega(1)$. At the other end of the spectrum, one might ask: given an initial correlation $\delta < 1 - \Omega(1)$, which monotone function f makes $Z(f, \delta)$ as close to 0 as possible? A nearly optimal function for this problem (which is tight to within a constant factor if the initial correlation δ is small enough) is the so-called tribes function of Ben-Or and Linial [BL90].

Let AND_k denote the And function on k bits (i.e., $\mathsf{AND}_k(x) = -1$ iff $x_i = -1$ for all $1 \leq i \leq k$), and let OR_k denote the Or function on k bits. For each $b \in \mathbf{N}$, define $n = n_b$ to be the smallest integral multiple of b such that $(1 - 2^{-b})^{n/b} \leq 1/2$, so n is very roughly $(\ln 2)b2^b$, and $b = \lg n - \lg\ln n + o(1)$. (Here $\lg n$ denotes $\log_2 n$.) Now define the tribes function T_n to be $\mathsf{OR}_{n/b} \otimes \mathsf{AND}_b$. This function is monotone, and by construction it's near-balanced; it's easy to see that $\mathbf{P}[T_n = +1] = (1 - 2^{-b})^{n/b} = 1/2 - O(\log n/n)$.

One can calculate $Z(T_n, \eta)$ directly and exactly:

Proposition 6.1. $Z(T_n, \eta) = 1 - 4\big[(1-2^{-b})^{n/b} - (1 - (2 - (\frac{1}{2} + \frac{1}{2}\eta)^b)2^{-b})^{n/b}\big]$.

Corollary 6.2. $Z(T_n, \eta) \leq (1 + o(1))\frac{\lg^2 n}{n}\eta(1+\eta)^b + O(\log^2 n/n^2)$.
Therefore if $\eta \leq O(1/\log n)$, then $Z(T_n, \eta) \leq O(\eta \log^2 n/n)$.

We omit the proofs of these results from this extended abstract. A similar result to Corollary 6.2 appears in [O02], with a more complicated proof.

Now we give a monotone function for which $Z(f,\delta)$ is small when $\delta \le 1-\Omega(1)$.

Theorem 6.3. *Let $\delta \le 1-\Omega(1)$. Then there is an infinite family of monotone functions $\{g_n\}$ satisfying:*

$$Z(g_n,\delta) \le \frac{\log^{1+u'} n}{n},$$

where u' is any number exceeding $u = \log_{4/3} 3 = 3.818\ldots$.

Proof: The idea is to first use REC-MAJ-3 to reduce δ to $\eta := 1/\log n$; then, apply a tribes function.

Let T_n be any tribes function. We will construct $g_{n'}$ on $n' := n\log^{u'} n$ inputs. Let ℓ be the REC-MAJ-3 depth necessary from Theorem 1.2 to reduce δ correlation down to $1/\log n$ correlation. Hence $\ell = (1+o(1))\log_{4/3}(\log n)$ (since $1-\delta \ge \Omega(1)$). Put $h = \text{REC-MAJ-3}_\ell$, so h is a function on $3^\ell = \log^{u'} n$ inputs. Let $g_{n'} = T_n \otimes h$. By construction, $Z(h,\delta) \le 1/\log n$. By Corollary 6.2, $Z(T_n, 1/\log n) \le O(\log n/n)$. Since h is balanced, by Proposition 3.4 we get $Z(g_{n'},\delta) \le O(\log n/n)$. The result follows, since as a function of n', $O(\log n/n)$ is $\log^{1+u'} n'/n'$ (taking u' slightly larger to kill any constant factors). □

As we can see from the following proposition, when the initial correlation $0 < \delta < 1$ is a constant, the above result is tight up to a factor of $\log^{2.818} n$:

Proposition 6.4. *If $f : \Omega_n \to \{-1,+1\}$ is monotone, then $Z(f,\eta) \ge \Omega(\eta \log^2 n/n)$.*

Proof:

$$Z(f,\eta) = \sum_S \eta^{|S|} \hat{f}^2(S) \ge \eta \sum_{|S|=1} \hat{f}^2(S) \ge \Omega(\eta \log^2 n/n),$$

by a result of [KKL88] (using the fact that f is monotone). □

It also follows from this proposition and Corollary 6.2 that when the initial correlation η is $O(1/\log n)$, the tribes function by itself is maximally sensitive among monotone functions, to within a constant factor.

7 High sensitivity to small noise, and Fourier concentration around $\sqrt{n}$

It seems natural to combine the functions from Theorems 1.3 and 6.3, via Proposition 3.4. One gets:

Theorem 7.1. *There exists an explicit infinite family of monotone functions $f_n : \Omega_n \to \{-1,+1\}$ with the following property: $Z(f_n, 1-1/Q) \le \epsilon$, where:*

$$Q = \frac{\sqrt{n\epsilon}}{(\log(n\epsilon))^t \log(1/\epsilon)^{(1+u')/2}},$$

$t = .3257\ldots$, *and* $(1+u')/2 = 2.409\ldots$.

Using the relationship $Z(f,\eta) = \sum_S \eta^{|S|} \hat{f}^2(S)$, it's easy to conclude:

Corollary 7.2. *There exists an explicit infinite family of monotone functions $f_n : \Omega_n \to \{-1,+1\}$ satisfying:*

$$\sum_{|S| \le Q} \hat{f_n}^2(S) \le \epsilon,$$

where $Q = \tilde{\Omega}(\sqrt{n\epsilon})$ is the quantity from Theorem 7.1.

From (3), one can easily derive the well-known fact that for all monotone $f : \Omega_n \to \{-1,+1\}$, $\sum_{|S| \le \epsilon^{-1}\sqrt{n}} \hat{f}^2(S) \ge 1-\epsilon$. That is, every monotone function has almost all the ℓ_2 mass of its Fourier spectrum concentrated on coefficients of degree up to $O(\sqrt{n})$. Corollary 7.2 demonstrates that this bound is tight up to polylog factors.

Acknowledgment: We would like to thank Gil Kalai for providing encouragement to write this result and Yuval Peres for interesting discussions.

References

[AN72] K. Athreya, P. Ney. Branching processes. Springer-Verlag, New York-Heidelberg, 1972.

[BDG88] J. Balcázar, J. Díaz, J. Gabarró. Structural Complexity I, II. Springer-Verlag, Heidelberg, 1988.

[BJT99] N. Bshouty, J. Jackson, T. Tamon. Uniform-distribution attribute noise learnability. *Workshop on Computational Learning Theory*, 1999.

[BKS98] I. Benjamini, G. Kalai, O. Schramm. Noise sensitivity of boolean functions and applications to percolation. Preprint.

[BL90] M. Ben-Or, N. Linial. Collective coin flipping. In *Randomness and Computation*, S. Micali ed. Academic Press, New York, 1990.

[BT96] N. Bshouty, C. Tamon. On the Fourier spectrum of monotone functions. *Journal of the ACM* 43(4), 1996.

[DK00] D.-Z. Du, K.-I Ko. Theory of Computational Complexity. Wiley Interscience, New York, 2000.

[F98] E. Friedgut. Boolean functions with low average sensitivity depend on few coordinates. *Combinatorica* 18(1), 1998, 27–36.

[FK96] E. Friedgut, G. Kalai. Every monotone graph property has a sharp threshold. *Proc. Amer. Math. Soc.* 124, 1996, 2993–3002.

[H99] S. Haykin. Neural Networks, 2nd Edition. Prentice Hall, 1999.

[J97] J. Jackson. An efficient membership-query algorithm for learning DNF with respect to the uniform distribution. *Journal of Computer and System Sciences*, 55(3), 1997.

[KKL88] J. Kahn, G. Kalai, N. Linial. The influence of variables on boolean functions. *Foundations of Computer Science*, 1988.

[KOS02] A. Klivans, R. O'Donnell, R. Servedio. Learning intersections and thresholds of halfspaces. To appear.

[LMN93] N. Linial, Y. Mansour, N. Nisan. Constant depth circuits, Fourier transform, and learnability. *J. Assoc. Comput. Mach.* 40, 1993, 607–620.

[M94] Y. Mansour. Learning boolean functions via the Fourier transform. *Theoretical Advances in Neural Computing and Learning*, Kluwer Acad. Publ., Dordrecht (1994), 391–424.

[M98] E. Mossel. Recursive reconstruction on periodic trees. *Random Structures Algorithms*, 13, 1998, no. 1, 81–97.

[O02] R. O'Donnell. Hardness amplification within NP. Symposium on the Theory Of Computation, 2002.

[Pa93] C. Papadimitriou. Computational Complexity. Addison Wesley, Reading, MA, 1993.

[Pe98] Y. Peres. Personal communication, 1998.

[T96] M. Talagrand. How much are increasing sets positively correlated? *Combinatorica* 16, 1996, no. 2, 243–258.

[V84] L. Valiant. A theory of the learnable. *Communications of the ACM*, 40, 1994, no. 2, 445–474.

Elchanan Mossel

Hebrew University of Jerusalem and Microsoft Research
mossel@microsoft.com

Ryan O'Donnell

MIT Mathematics Department
odonnell@theory.lcs.mit.edu

Trends in Mathematics, © 2002 Birkhäuser Verlag Basel/Switzerland

Apprentissage de Séquences Non-Indépendantes d'Exemples

Olivier Teytaud

ABSTRACT: *Beaucoup de travaux récent considérent les applications pratiques des réseaux neuronaux (ou d'autres algorithmes proches) pour la modélisation de séries temporelles, par exemple chaotiques. Quelques papiers seulement (dont les résultats principaux sont rappelés ici) ont été consacrés aux applications de la partie théorique de l'apprentissage en la matière. Cet article fournit des rappels des résultats basés sur des propriétés d'ergodicité en matière d'apprentissage de suites non-indépendantes d'exemples, puis développe quelques nouveaux résultats.*

1 Introduction

La théorie de l'apprentissage, notamment théorie VC, est une grande aire de recherche, basée sur des théorèmes statistiques de convergence uniforme des moyennes empiriques vers les espérances. [10, 56] sont des états de l'art complets dans ces domaines. Alors que ces résultats viennent de la communauté de l'intelligence artificielle, les mathématiciens, dans la communauté du processus empirique, ont prouvé de nombreux théorèmes centraux généralisés, uniformes sur des espaces de fonctions et sur des espaces de distributions, résumés dans [57] notamment. [19, 25] fournissent des essais d'extensions dans la direction Markovienne. Un but de ce travail est une extension dans la même direction. Bien que l'article soit axé sur la théorie, on rappelle les principaux paradigmes usuellement mis en œuvre en contrôle et des cas concrets sont présentés. [21, 24] étudient la possibilité de prédire des systèmes chaotiques avec des outils de régression, éventuellement des réseaux neuronaux. [36] vérifie la validité pratique de la théorie de l'apprentissage dans le cas de telles séries temporelles et conclut que la théorie VC est validée. Néanmoins, les hypothèses classiques ne sont pas vérifiées dans un tel cas: des points consécutifs fournis par un système chaotique ne sont évidemment pas indépendants identiquement distribués. [17, 52, 18] ont souligné ce manque de résultat théorique et ont proposé une application de ces prédictions. L'idée consiste à transformer une prédiction en stabilisation. La section 2 présente des résultats venus du processus empirique. La section 3 présente une adaptation de la théorie VC dans le cas d'exemples distribués markoviennement. Dans l'ensemble du papier, F désigne un espace de fonctions tel que toutes les quantités en jeu soient mesurables et $\mathcal{F}$ un espace de fonctions continues.

2 Résultats asymptotiques: classes de Donsker pour des processus "ergodiques"

La présence de guillemets autour d' "ergodiques" est due au fait que nous travaillons sur différentes sortes de suites, dont certaines complètement déterministes, et nous ne demandons pas dans tous les cas une ergodicité *stricto sensu.* Comme expliqué ci-dessous, des dynamiques stochastiques peuvent apparaitre dans des systèmes complètement déterministes. Cette section est principalement basée sur [55] et [2]. On considère dans cette partie des conditions sous lesquelles les exemples générés par un processus approximent une loi asymptotique. Ceci est fait en deux étapes: 1) on montre les exemples X_n distribués par un système "ergodique" assurent la convergence des moyennes empiriques vers les espérances pour la distribution asymptotique, avec le même ordre de convergence que dans le théorème central limite. Ceci est vérifié par l'utilisation des théorèmes B.1 (systèmes dynamiques déterministes) ou C.2 (deuxième partie, chaînes de Markov stationnaires), 2) on généralise en direction de l'uniformité, ce qui se fait grâce au théorème 2.1 (mettant en jeu des résultats sur le processus empirique). On considère des X_n distribués, selon le cas, par un modèle de Markov (ie une probabilité conditionnelle $P(X_n|X_{n-1} = t) = p(t))$ (notez que des modèles de Markov de plus haut degré peuvent être utilisés de même), ou par une fonction déterministe g telle que $X_n = g(X_{n-1})$. Le premier cas est plus facile sous certaines jolies hypothèses sur p. Le second cas requiert de très beaux résultats sur les systèmes dynamiques. On utilisera pour généraliser à l'uniformité le résultat suivant de [2]:

Théorème 2.1. *Considérons $A_n(f)$ et $A(f)$ des processus stochastiques pour $f \in \mathcal{F}$. Supposons que (1) $\sup_f |A_n(f)|$ est presque sûrement fini pour tout n, que (2) $\sup_f |A(f)|$ est fini presque sûrement, que (3) les distributions de dimension finie de $\{A_n(f) : f \in \mathcal{F}\}$ convergent vers celles de $\{A(f)|f \in \mathcal{F}\}$, et enfin que (4) pour tout entier positif q il existe une application $\pi_q : \mathcal{F} \mapsto \mathcal{F}$ telle que le cardinal de $\{\pi_q f : f \in \mathcal{F}\}$ est fini et pour tout η $\lim_{q\to\infty} \lim_{n\to\infty} \sup Pr^*\{\sup_{f\in\mathcal{F}} |A_n(f) - A_n(\pi_q f)| \geq \eta\} = 0$. Alors A_n converge faiblement vers A dans $l^\infty(\mathcal{F})$.*

Ce résultat se trouve ailleurs dans la littérature. On utilise [2] pour référence car beaucoup de résultats (nouveaux) liés à notre propos peuvent être trouvés dedans, particulièrement dans le cas stationnaire. Le corollaire suivant sera utile par la suite:

Corollaire 2.2. *Considérons $A_n(f) = \frac{1}{\sqrt{n}} \sum_{i=1}^{n} (f(X_i) - Ef(X_i))$ pour une certaine suite de variables aléatoires X_i. Supposons que (1) $\mathcal{F}$ est Donsker pour une mesure de probabilité π, avec $\pi f = \lim \frac{1}{\sqrt{n}} A_n(f)$, que (2) $\sup_f |A_n(f)|$ est fini presque sûrement, que (3) il existe des bracketing nombres de couverture pour tout ϵ,* avec *des brackets inclus dans $\mathcal{F}$ (il est probable que cette hypothèse pourrait être réduite), et que (4) les distributions de dimension finie de $\{A_n(f) : f \in \mathcal{F}\}$ convergent vers celles de $\{A(f) : f \in \mathcal{F}\}$. Alors A_n converge faiblement vers A dans $l^\infty(\mathcal{F})$.*

Preuve:La condition 2 (aisément vérifiée dans beaucoup de cas pratiques) est la condition 1 du théorème 2.1. Le caractère Donsker de $\mathcal{F}$ assure que la condition 2 dans le théorème 2.1 est vérifiée, grâce au théorème A.1 (avec $Z_{n,i}(f) = \frac{1}{\sqrt{n}} f(X_{n,i})$, les $X_{n,i}$ variables aléatoires indépendantes ne dépendant que de i et de même loi π). La condition 4 garantit la condition 3 du théorème 2.1. En utilisant π_q les projections successives sur les bornes supérieures dans les brackets garantit la condition 4 du théorème 2.1.□

Grâce à ce simple corollaire, on peut utiliser des résultats forts comme le théorème A.1 pour prouver la finitude presque sûre de $\sup_f |Z_f|$, et alors utiliser Z_n simplement garantissant la convergence point à point, telle que les suites déterministes du théorème B.1, ou les suites markoviennes avec théorèmes centraux limites (théorème C.2, deuxième partie). Malheureusement, dans le dernier cas des suites markoviennes, de tels théorèmes sont seulement disponibles, pour autant que nous sachions, dans le cas de chaînes de Markov **réversibles** (ie pour tout A et B sous-ensembles de X $\int_{x\in A} P(x,B)d\pi(x) = \int_{y\in B} P(y,A)d\pi(y)$) ou des chaînes de Markov **stationnaires**. Cependant, des travaux sont en cours dans cette direction, selon [41, remark 2.4]. [55] explique les mécanismes par lesquels des systèmes entièrement déterministes (pour autant que de tels systèmes existent dans un monde quantique ...). Dans le cadre déterministe, la probabilité est définie pour X_0, point initial du système dynamique. Le théorème B.1 (voir [55]) justifie les comportements stochastiques observés en pratique. Finalement, on peut résumer ces résultats (convergence de chaines de Markov ou dynamiques stochastiques dans des systèmes déterministes, plus uniformité de la convergence faible sous des conditions de bracketing-entropie) dans le théorème suivant:

Théorème 2.3 (Convergence uniforme, suites déterministes ou markoviennes). *Si l'un des faits suivants a lieu et si $\mathcal{F}$ est Donsker pour la loi limite (en particulier si $\mathcal{F}$ est universellement Donsker): (1) les X_n sont distribués selon $X_{n+1} = g(X_n)$ et les hypothèses du théorème B.1 sont vérifiées, (2) les X_n sont distribués selon un modèle de Markov vérifiant les conditions du théorème C.2 (second cas). Alors, $\{\frac{1}{\sqrt{n}}\sum_{i=1}^{n} f(X_i) : f \in \mathcal{F}\}$ converge faiblement dans $l^\infty(\mathcal{F})$.*

Des extensions basées sur des hypothèses plus faibles que l'ergodicité aux sens définis ici existent, comme la Harris-récurence et les techniques introduites dans [11, 44] pour évaluer le temps d'oubli du point initial (en particulier pour une chaine avec de fortes symétries). On pourra consulter aussi [3, 13]. On peut alléger l'hypothèse de stationarité (voir [6]). Des convergences plus rapides que $1/\sqrt{n}$ peuvent aussi être proposées (comme dans la section qui suit, de manière non-asymptotique, ou dans [30]).

3 Bornes non-asymptotiques

Dans cette section, on rappelle que résultats classiques, et fournissant une nouvelle (pour autant que nous sachions) borne non-asymptotique basée sur la condition de Doeblin. Deux paradigmes d'apprentissage en sont déduits, l'un d'eux étant la minimisation du risque empirique et l'autre proche de la minimisation du risque empirique, et bénéficiant d'une borne un peu meilleure (nous prouvons seulement une meilleure borne mais ne prouvons pas une stricte supériorité). L'optimalité à des facteurs logarithmiques près est prouvée.

Espaces d'états finis ou dénombrables Pour autant que nous sachions, les résultats les plus généraux dans le cas d'espaces d'états finis ou dénombrables sont ceux de [19]. Leur résultat principal est résumé comme suit: dans le cas d'espaces d'états finis (avec N états) et une distribution stationnaire uniforme, la complexité d'échantillon requise est bornée par $\frac{s}{1-\lambda_2}\ln(\frac{sN}{\delta})$, avec s la complexité d'échantillon dans le cas iid, δ le risque, λ_2 la seconde plus grande valeur propre de la matrice de transition. Des résultats liés peuvent être trouvés dans [19].

Cas général On reformule le cadre PAC (Probablement Approximativement Correct) de Valiant de la façon suivante: un algorithme d'apprentissage sur une famille F de fonctions et sur une suite markovienne avec probabilité conditionnelle $p(t) = U \mapsto P(X_n \in U|X_{n-1} = t)$ est PAC pour une complexité d'échantillon m avec précision ϵ et confiance $1 - \delta$ si et seulement si pour tout $n \geq m$ et pour tout X_0 la probabilité d'une différence $> \epsilon$ entre l'erreur en généralisation pour la distribution asymptotique et l'erreur optimale dans F est $\leq \delta$ pour un échantillon de taille n. L'erreur en généralisation est définie comme l'erreur limite moyenne sur une suite Markovienne finie $X_0, \ldots, X_k$ (d'autres définitions sont possibles de même, prenant en compte la longueur de X_k; les résultats seraient similaires, à ceci près que k serait demandé plus grand qu'une quantité linéaire en $D(1 - \ln(\delta))/\epsilon^2$ avec D défini plus loin). Une famille F de fonctions est dite PAC s'il existe un algorithme choisissant $f \in F$ satisfaisant la condition ci-dessous pour m polynomial en ϵ, $\ln(1/\delta)$. L'algorithme est dit PAC. La même définition vaut uniformément pour $p \in \mathcal{P}$ si la condition ci-dessus a lieu uniformément en $p \in \mathcal{P}$ (ie m doit être indépendant de $p \in \mathcal{P}$).

La nécessité de bornes non-asymptotiques apparait en contrôle adaptatif, ie quand l'environnement varie avec le temps (notez que ceci est peut-être moins fondamental en "contrôle par apprentissage", comme les variations de l'environnement sont supposées rapides, ce qui implique que l'on utilise les passages précédents dans la même aire et donc qu'on peut utiliser des résultats asymptotiques). Alors, utiliser des milliers de points n'est plus possible, et le contrôle doit être dynamique.

On restreint notre attention à une suite $X_0, \ldots, X_n, \ldots$ de variables aléatoires, avec X_n dépendant seulement de X_{n-1}; en outre, $X_n|X_{n-1}$ est indépendant de n. Des chaînes de plus haut degré peuvent être considérées très similairement, simplement en considérant $Y_i = (X_i, X_{i+1}, \ldots, X_{i+k})$. On considère par la suite la convergence uniforme de $\frac{1}{n}\sum_{i=0}^{n} f(X_i)$; uniformément en $f \in F$ et en X_0. Ceci peut aisément être étendu à la régression $X_i \mapsto X_{i+1}$ en considérant $g(Y_i) = |f(X_i) - X_{i+1}|$ par exemple. On suppose que les X_i et les $f \in F$ sont bornés et $\in [0, 1]$. On suppose dans la suite que la loi μ_n de $X_n|X_0$ converge uniformément vers une mesure donnée μ_∞, dans le sens suivant: $\lim_{n\to\infty} \sup_{X_0} \int |\mu_n - \mu_\infty| \to 0$ La section C fournit des conditions suffisantes pour cela. On suppose, en outre, que $X_n|X_0 = t$ est continu par rapport à la mesure de Lebesgue, de loi de probabilité $f_n(t, .)$. Ceci va seulement être utilisé ci-dessous pour prouver une extension exponentielle uniforme de l'équation ci-dessus; aussi cette hypothèse peut être relâchée, en utilisant le théorème C.2 (première partie). L'intérêt est seulement de fournir ici une démonstration simple du résultat désiré, avec des prérequis aussi réduits que possible. L'idée de la preuve, détaillée dans les sous-sections suivantes, est comme suit: (1) En un certain sens, les mesures dans les séquences markoviennes convergent uniformément et rapidement vers la mesure asymptotique. La vitesse de convergence est mesurée par une dimension entière. (2) Ceci implique que pour des sous-suites bien choisies de la suite initiale, les mesures sont "presque" indépendantes. Plus précisément, la loi de cette sous-suite est proche de la loi du produit indépendant de la mesure asymptotique. (3) La convergence uniforme des moyennes empiriques dans le cas indépendant implique la convergence uniforme des moyennes empiriques dans le cas original.

Première partie: convergence Markovienne. Définissons $\Delta_n^{X_0}(P) = \mu_n^{X_0}(P) - \mu_\infty(P)$. $\mu_n^{X_0}$ est μ_n, conditionnellement à X_0. Alors: $\Delta_n^{X_0}(P) = \int_t \Delta_{n-k}^{X_0}(t) f_k(t, P)dt$ et

$$\Delta_n^{X_0}(P) = \int_t \Delta_{n-k}^{X_0}(t)(f_k(t,P) - f_\infty(t,P))dt \ (1)$$

$$|\Delta_n^{X_0}(P)| \leq \int |\Delta_{n-k}^{X_0}|(t)|f_k(t,P) - f_\infty(t,P)|dt, \ |\Delta_n^{X_0}(P)| \leq \frac{1}{2}\int_t |\Delta_{n-k}^{X_0}|(t)dt \ (2)$$

La ligne (1) vient du fait que $\Delta_{n-k}^{X_0}$ a masse 0 et est indépendant de P. La ligne (2, droite) est basée sur k suffisamment grand pour garantir $\| f_k(t,P) - f_\infty(t,P) \| < \frac{1}{2}$ pour tout t. Le plus petit tel k sera noté, par la suite, $D(X)$ (notation abusive pour $D(X_n|X_{n-1})$), et appelé la **dimension** de la chaîne de Markov X. Le D est là pour Doeblin ou pour dimension, selon les préférences du lecteur.

Lemme 3.1 (Convergence rapide non-asymptotique, chaines de Markov (Ueno, 1961)). *L'équation 2 implique que* $\Delta elta_n^{X_0}(P) \leq \frac{1}{2^{\lfloor \frac{n+1}{D(X)} \rfloor}} \Delta_1^{X_0}(P)$.

Seconde partie: apprentissage bruité. Considérons $X_N, X_{2N}, \ldots, X_{lN}$, des sous-suites finies de X, avec $N = kD(X)$. Alors, $\mu_N^{X_0}$ conditionnellement à X_0 a une loi à distance $\leq \eta$ de μ_∞, $\mu_{X_{2N}}^{X_N}$ idem, et ainsi de suite,
avec $\eta = \frac{1}{2^k} \sup_{X_0,P}$ masses de Dirac $\Delta_1^{X_0}(P) \leq \frac{1}{2^k}$ et avec la distance suivante $d(\mu_1, \mu_2) = \sup_P |\mu_1 - \mu_2|(P)$ (intuitivement, l'intégrale maximale d'une fonction bornée par 1, pour la loi $\mu_1 - \mu_2$) On a alors besoin du lemme qui suit:

Lemme 3.2 (D'une loi à plusieurs). *Soient* $Z_1, \ldots, Z_n$ *des variables aléatoires (non nécessairement indépendantes!), chaque* Z_{i+1} *ayant une loi conditionnellement à* Z_i *à distance* $< \eta$ *de la loi de* Z. *Alors, la loi de* $(Z_1, Z_2, \ldots, Z_n)$ *est à distance* $< n\eta$ *de la loi du produit de* n *variables indépendantes avec la même loi que* Z.

Preuve:Ceci se prouve par récurrence. La propriété pour $n = 1$ est claire. La récurrence est fait en intégrant la propriété au rang $n - 1$. □Ceci implique que notre sous-suite a, à une précision explicite près, la loi du produit de n lois indépendantes.
Troisième partie: conclure. Les résultats usuels de VC-théorie fournissent des bornes sur la complexité d'échantillon dans le cadre iid en $O(V - \ln(\delta)/\epsilon^2)$ (ou $O(\frac{V\ln(1/\epsilon) - \ln(\delta)}{\epsilon})$, dans le cas d'un taux d'erreur minimal nul), avec ϵ la précision, $1 - \delta$ la confiance, V la VC-dimension (on considère ici le cas de la catégorisation deux-classes - on peut directement considérer la régression de même ou utiliser des fonctions d'égalié ϵ-insensibles en régression pour garder le même cadre qu'en classification). Considérons maintenant la complexité d'échantillon nécessaire pour garantir une différence bornée par ϵ entre les moyennes empiriques et les espérances; P est la probabilité sous l'hypothèse iid, alors que P_M est la probabilité dans le cadre markovien, pour la suite suite ci-dessus (la somme dans l'équation qui suit est faite sur des copies indépendantes de μ_∞, qui sont distinguées par des indices (i)): $P(\exists f/|\frac{1}{k}\sum_{i=1}^k \mu_\infty^{(i)}(f) - \mu_\infty(f)| > \epsilon) \leq f(V,k,\epsilon)$, $P_M(\exists f, X_0/|\frac{1}{k}\sum_{i=1}^k \mu_{X_{iN}}^{X_0}(f) - \mu_\infty(f)| > \epsilon) \leq f(V,k,\epsilon) + k\eta$

Dans le théorème suivant on considère ERM (empirical risk minimization), consistant à minimiser l'erreur empirique sur tous les exempes, et ERM_S consistant à minimiser l'erreur sur les points X_N,X_{2N},...,X_{kN} avec $N = D\ln(2m/\delta)$.

Théorème 3.1. *Quand $D(X)$ et V sont finis, et pour $\delta \leq \frac{1}{2}$, alors ERM_S a une complexité d'échantillon majorée par*

$$O\left(\frac{D(V+\ln(1/\delta))\ln(1/\delta)}{\epsilon^2}\left[\ln(DV)+\ln(\ln(1/\delta))+\ln(1/\epsilon)\right]\right) \tag{3}$$

Quand l'erreur minimale est nulle, alors

$$O\left(\frac{D(V\ln(1/\epsilon)+\ln(1/\delta))\ln(1/\delta)}{\epsilon}\left[\ln(DV)+\ln(\ln(1/\delta))+\ln(1/\epsilon)\right]\right) \tag{4}$$

En outre, l'équation (4) est valable pour ERM dans le cas d'une erreur minimale nulle, aussi. Pour ERM, dans le cas général, les mêmes bornes sont vraies, à des facteurs logarithmiques près.

[19] fournit un résultat partiel dans le cas d'un espace d'état fini, pour apprendre avec une famille de fontions dont une qui a une erreur nulle en généralisation. En outre, [19], comme rappelé plus tôt, fournit des bornes explicites de convergence uniforme de distributions vers la distribution asymptotiques.

Preuve:Tout d'abord, considérons ERM_S. Grâce au lemme 3.2, les probabilités dans le cas Markovien et les probabilités dans le cas iid avec la loi μ_∞ sont à distance au plus $\delta_1 = O(\frac{k}{2^{\frac{N}{D}}})$, ce qui satisfait en particulier $\delta_1 = O(\delta)$ si $N = D\ln(m/\delta)$. Dans le cas iid, avec confiance $1-\delta_2$, la précision est bornée par $\epsilon = O(\sqrt{\frac{V-\ln(\delta_2)}{k}})$, qui avec $\delta_2 = Theta(\delta)$, conduit à $\epsilon = O(\sqrt{\frac{V-\ln(\delta)}{k}})$, vérifié avec probabilité $\Theta(\delta)$ dans le cas iid, et $\delta_1+\delta_2 = \Theta(\delta)$ aussi dans le cas markovien. Ceci conduit à la précision globale comme suit, avec confiance $1-O(\delta)$:

$$\epsilon^2 = O(\frac{D(V+\ln(1/\delta))\ln(1/\delta)\ln(m)}{m}) \tag{5}$$

Ceci conduit à une complexité d'échantillon comme dans l'équation (3). Le cas d'un taux d'erreur nul prend simplement en compte les bornes de complexité d'échantillon de la forme $(V\ln(1/e)+\ln(1/\delta))/\epsilon$. Ceci montre qu'apprendre est possible avec complexité d'échantillon polynomiale en D, V, $1/\epsilon$, $\ln(1/\delta)$. Maintenant, considérons ERM. ERM en fait consiste en gros à utiliser N algorithmes ERM_S différents, qui ne sont pas indépendants. Dans le cas d'un taux d'erreur nul, ERM inclut ERM_S et donc est aussi efficace. Ainsi, l'équation 4 a lieu. Considérons maintenant le cas général. Avec confiance $\geq 1-\delta' = 1-N\delta$, chacun des N apprentissages sur $(X_0, X_N, X_{2N}, \dots)$, $(X_1, X_{N+1}, X_{2N+1}, \dots)$,... a la précision ci-dessus (équation 5). Ainsi, on doit remplacer δ par $\delta' = O(\delta/N) = O(\frac{\delta}{D(\ln(m)+\ln(1/\delta))})$ dans l'équation 5. Ainsi il nous faut $\delta = O(\frac{\delta'}{D\ln(m)\ln(1/\delta)})$, $\delta\ln(1/\delta) = O(\frac{\delta'}{D\ln(m)})$, $\frac{1}{\delta\ln(\frac{1}{\delta})} = \Omega(\frac{D\ln(m)}{\delta'})$. Ceci est en particulier vérifié avec $\frac{1}{\delta} = (\frac{D\ln(m)}{\delta'}\times\ln(\frac{D\ln(m)}{\delta'}))$. Remplacer $\frac{1}{\delta}$ par cette expression dans l'équation 5 conduit au résultat souhaité.□

Notez que nous avons proposé ci-dessus des bornes sur la différence entre l'erreur en généralisation et l'erreur empirique uniformément pour tout classifieur. Ceci n'est pas limité à la précision du classifieur empiriquement optimal. Comme toujours en apprentissage ce théorème peut conduire à des algorithmes pratiques qui sont universellement consistent. Si V augmente suffisamment lentement (en tant que fonction de m), alors $\sqrt{DV/m}$ décroit vers 0. La minimisation du risque empirique est alors universellement consistente, pourvu que la séquence "emboitée" de classes de fonctions (chacune de VC-dimension finie), est un approximateur universel.

Optimalité On peut montrer que les dépendances linéaires en $\frac{D(X)V}{\epsilon^2}$ ne peut être supprimée, comme expliqué ci-dessous:

Théorème 3.2. *La dépendance linéaire en VD/ϵ^2 ne peut être supprimée. Précisément, pour tous V, D, δ, il existe une chaine de Markov de dimension D et une famille de VC-dimension D telle qu'avec probabilité au moins δ la précision ϵ est $\Omega(\sqrt{\frac{VD}{m}}) + O(D^{\frac{3}{2}}\ln(\frac{m}{D})/\sqrt{m} + D^3\ln(\frac{m}{D})^2/m)$.*

Preuve:

Considérez une famille F de fonctions sur $]0,1]$ avec VC-dimension V, une variable aléatoire μ telle que la complexité d'échantillon de F pour la distribution de μ soit la pire possible, Z_n chaine de Markov avec $Z_0 = 0$ et $Z_{n+1} = 1 - Z_n$ avec probabilité p et Z_n sinon, $X_n = Z_n \times \mu_{k(n)}$ avec les μ_n des copies indépendantes de μ et $k(n) = \sup([0,n] \cap \{i/Z_i \neq Z_{i-1}\})$. Alors (1) $D(X)$ est $\theta(1/p)$, (2) La complexité d'échantillon $\Omega(VD(X)/\epsilon^2) + O(D^{\frac{3}{2}}\ln(\frac{m}{D})/\sqrt{m} + D^3\ln(\frac{m}{D})^2/m)$. Le premier point est prouvé par l'évaluation de la loi de $(X_n - X_\infty)$, avec X_∞ la loi asymptotique. $X_n(0) - \frac{1}{2} = (1-2p) \times (X_{n-1}(0) - \frac{1}{2})$; la même relation est vraie pour tout sous-ensemble de $]0,1]$. Considérons maintenant l'évaluation empirique de $Eg(X_n)$ jusqu'à la $k+1^e$ occurence de $Z_n - Z_{n-1} = -1$. Ceci est une variable aléatoire N/P, avec $N = \sum_{i=1}^k \lambda_i g(A_i)$ et $P = \sum_{i=1}^k (\lambda_i + \lambda'_i)$, avec λ_i et λ'_i des variables aléatoires indépendantes égales à $k > 0$ avec probabilité $(1-p)^{k-1}p$, et A_i des variables aléatoires indépendantes avec loi commune μ. • Tout d'abord, fixons les μ_i. Les espérances et probabilités ci-dessous sont calculées conditionnellement aux μ_i. • La probabilité de $\lambda_i > K$ est bornée par $O((1-p)^K)$, et la probabilité d'avoir au moins un λ_i ou λ'_i plus grand que K est bornée par $O(k(1-p)^K)$. Pour un seuil de confiance fixé, on a $K = O(\ln(k)/p) = O(D\ln(k))$. La suite est faite conditionnellement à cela. • Sous l'hypothèse $\forall i \max(\lambda_i, \lambda'_i) \leq K$, on peut conclure qu'avec un seuil de confiance, N et P sont tous deux en gros égaux à leurs espérances (toujours conditionnellement aux μ_i), avec précision $O(K\sqrt{k}) = O(D\ln(k)\sqrt{k})$, par l'inégalité de Hoeffding. N/P est alors, avec précision $O(D\ln(k)/\sqrt{k} + D^2\ln(k)^2/k) = O(D^{\frac{3}{2}}\ln(\frac{m}{D})/\sqrt{m} + D^3\ln(\frac{m}{D})^2/m)$, l'évaluation empirique de $Eg(A_n)$, qui est connu (grâce aux bornes inférieures de VC-dimension) $\theta\left(\sqrt{(VD/m)}\right)$ pour le pire choix a posteriori de g. D'où le résultat souhaité.□

Notez que nous avons montré une un peu meilleure complexité d'échantillon de ERM_S, mais n'avons pas réussi à montrer qu'ERM_S était en un sens meilleur qu'ERM. Une remarque important est le fait que ceci amène à un algorithme qui est universellement consistant: comme dans le cas d'échantillons iid, on peut utiliser ERM avec les modifications suivantes: (1) accroissement (suffisamment lent) de la "taille" de la famille de fonction (en termes de VC-dimension), comme le nombre d'exemples augmente, (2) Eventuellement, augmentation de l'"arité" (= nombre de pas mémorisés - la largeur de la fenêtre. Ceci est usuellement la dimension de plongement, dans le cas de systèmes chaotiques), dans le cas de, disons, la prédiction (contrôle et stabilisation déduites de la prédiction de même). Pourvu que l'augmentation de VC-dimension résultante de cela est suffisamment lente (même combinée avec l'augmentation décrite ci-dessus). Ceci est universellement consistant au sens où si la condition de Doeblin a lieu et si un degré fini est suffisant, alors l'erreur converge vers la plus petite possible. Néanmoins, une forte différence avec le cas iid persiste: l'erreur décroit comme dans le cas iid, mais si D est inconnu, alors ***la validation est impossible.*** On ne peut jamais être sûr que l'erreur en généralisation est ce qu'elle a l'air d'être sur un ensemble empirique d'exemples, sans borne sur D. Ainsi, des conditions générales sous lesquelles D peut être borné sont d'une importance cruciale. Des résultats liés peuvent être trouvés dans [33].

4 Conclusion, remarques et problèmes ouverts

En conclusion nous énumérons les manques qui nous apparaissent émerger de l'état de l'art: (1) On montre dans le "beau" cas markovien une extension de la VC-théorie (bornes non-asymptotiques). On montre dans le cas déterministe, en utilisant des résultats de [55] et [2], des résultats asymptotiques. Un progrès intéressant serait une extension déterministe de la VC-théorie, qui apparait comme le résultat manquant le plus important dans cet article, (2) Comment fournir des bornes générales sur la condition de Doeblin ? Ceci semble un problème très important car dans les applications concrètes, cette constante est inconnue, et comme montré dans les résultats précédents, des bornes sur la dimension sont nécessaire pour construire des algorithmes capables de décider s'ils ont assez d'exemples. Utiliser des mots tels que "quantique" ou "effets chaotiques sur la précision du système" semble parfois une sorte de sorcellerie; toutefois, on peut penser que chercher de tels arguments est un problème concret. Aussi nous formulons ce problème ouvert: l'incertitude quantique (ou un autre argument) peut-elle justifier une condition générale sous laquelle la dimension ne peut être très haute ? (3) En outre, de multiples problèmes supplémentaires apparaissent lorsque l'on cherche à étendre ces résultats au cas non de l'identification de système, mais au contrôle.

References

[1] N.-T. Andersen, E. Giné, M. Ossiander, J. Zinn, *The central limit theorem and the law of iterated logarithm for empirical processes under local conditions, Probability theory and related fields 77, 1988.*

[2] M.A. Arcones, *Weak Convergence of stochastic processes indexed by smooth functions, Stochastic Processes and their Applications 62 115-138, 1996.*

[3] M.A. Arcones, B. Yu, *Central limit theorems for empirical and U-processes of stationary mixing sequences. J. Theoret. Probab. 7 47-71, 1994.*

[4] R. Bowen, D. Ruelle, *The ergodic theory of Axiom A flows, Invent. Math. 29, pp181-202, 1975.*

[5] K. CHELLAPILLA, *Evolving nonlinear controllers for backing up a Truck-and-Trailer using evolutionary Programming, EPS 1998.*

[6] *X. Chen, A. Guillin, The functional moderate deviations for Harris recurrent Markov chains and applications, preprint, Université Paris IX Dauphine, 2002.*

[7] P. COULLET, C. TRESSER, *Iterations d'endomorphismes et groupes de renormalisation, C.R. Acad. Sci. Paris 287, 577-580, 1978.*

[8] S.-E. DECATUR, *PAC learning with Constant-Partition Classification Noise and Applications to Decision Tree induction, Proceedings of ICML 1997.*

[9] R. DEVANEY, *Introduction to Chaotic Dynamical Systems: Theory and Experiments. Addison-Wesley, 1992.*

[10] L. DEVROYE, L. GYORFI, G. LUGOSI, *A Probabilistic Theory of Pattern Recognition, Springer, 1996.*

[11] P. DIACONIS, *The cutoff phenomenon in finite Markov chains, Proc. Natl. Acad. Sci., USA, 93:1659-1664, 1996.*

[12] W. DOEBLIN, *Eléments d'une théorie générale des chaînes simples constantes de Markov. Annales scientifiques de l'École Normale Supérieure 57 (III) 61-111, 1940.*

[13] P. DOUKHAN, P. MASSART, E. RIO, *The functional central limit theorem for strongly mixing processes, Ann. Inst. Henri Poincare, 30, 63-82, 1994.*

[14] R.-M. DUDLEY, *Central limit theorems for empirical measures, Annals of probability 6, 1978. Correction: Annals of probability 7, 1978.*

[15] R.-M. DUDLEY, *A course on empirical processes (Ecole d'été de Probabilité de Saint-Flour XII-1982), Lecture notes in Mathematics 1097, 2-141 (ed P.L. Hennequin), Springer-Verlag, New-York, 1984.*

[16] M. FEIGENBAUM, *Qualitative universality for a class of nonlinear transformations, J. Stat. Phys. 19,25-52, 1978.*

[17] J.-M. FRIEDT, O. TEYTAUD, D. GILLET, M. PLANAT, *Simultaneous amplitude and frequency noise analysis in Chua's circuit - neural network based prediction and analysis, VIII Van Der Ziel Symposium on Quantum 1/f Noise and Other Low Frequency Fluctuations in Electronic Devices, 2000.*

[18] J.-M. FRIEDT, O. TEYTAUD, M. PLANAT, *Learning from noise in Chua's oscillator, accepted in ICNF, 2001.*

[19] D. GAMARNIK, *Extension of the PAC Framework to Finite and Countable Markov Chains, Proceedings of the Workshop on Computational Learning Theory, Morgan Kaufmann Publishers, 1999.*

[20] S. HAYASHI, *Connecting invariant manifolds and the solution of C^1 stability and Ω-stability conjectures for flows, Annals of Math. 145, 81-137, 1997.*

[21] S. HAYKIN, J.PRINCIPLE, *Using Neural Networks to Dynamically model Chaotic events such as sea clutter; making sense of a complex world, IEEE Signal Processing Magazine 66:81, 1998.*

[22] M. HÉNON, *A two-dimensional mapping with a strange attractor, Comm. Math. Phys. 50, 69-77, 1976.*

[23] R.A. HOLMGREN, *A first course in discrete dynamical systems. Springer, 1996.*

[24] W. HONG, *The application of Radial Basis Function Networks to the prediction of Chaotic Time Series, Term Project, Course 2.160, Intelligent Control and Sensing, 1993.*

[25] T.A. JOHANSEN, E. WEYER, *On convergence Proofs in System Identification - A general principle using ideas from Learning Theory. Preprint, 1997: submitted to Elsevier Preprint.*

[26] V.-I. KOLCINSKI *On the central limit theorem for empirical measures. Theory of probability and mathematical statistics 24, 1981.*

[27] H. KOUL, *Some convergence theorems for ranks and weighted empirical cumulatives. Annals of Mathematical Statistics 41, 1768-1773, 1970.*

[28] E.N. LORENZ, *Deterministic nonperiodic flow, J. Atmosph. Sci. 20, 130-141, 1963.*

[29] R. MANÉ, *A proof of the C^1-stability conjecture, publ. math. IHES 66 (161-210), 1988.*

[30] E. MAMMEN, A.B. TSYBAKOV, *Smooth discrimination analysis. Ann. Statist. 27, 1808 - 1829, 1999.*

[31] M. MARCUS, J. ZINN, *The bounded law of the iterated logarithm for the weighted empirical distribution in the non-iid case. Annals of Probability 12, 334-360, 1984.*

[32] S.P. MEYN, R.L. TWEEDIE, *Markov chains and stochastic stability. Springer-Verlag, 1993.*

[33] S.P. MEYN, R.L. TWEEDIE, *Computable bounds for geometric convergence rates of Markov chains, Ann. Appl. Probab., 4(4):981-1011, 1994.*

[34] O. MICHEL, *Notes sur la prédiction de séries temporelles issues de systèmes chaotiques, Ecole des Techniques Avancées en Signal-Image-Parole de Grenoble, Signal et non-linéaire. 1998.*

[35] M. MOONEN, I. PROUDLER, *An Introduction to Adaptive Signal Processing, HC63-64 Adaptive Signal Processing, Course Notes 1998-1999.*

[36] S. MUKHERJEE, E. OSUNA, F. GIROSI, *Non-linear Prediction of Chaotic Time Series Using Support Vector Machines, Proc. of IEEE NNSP'97, Amelia Island, FL, 1997.*

[37] S. NEWHOUSE, *The abundance of wild hyperbolic sets and nonsmooth stable sets for diffeomorphisms, Publ. Math. IHES 50,101-151, 1979.*

[38] M. OSSIANDER, *A central limit theorem under metric entropy with L_2 bracketing, Annals of probability 15, 1987.*

[39] D. POLLARD, *A central limit theorem for empirical processes. Journal of the australian mathematical society, A33, 1982.*

[40] L. RALAIVOLA, F. D'ALCHÉ-BUC, *Incremental Support Vector Machine learning: a local approach. Accepted in Icann, 2001.*

[41] G.O. ROBERTS, J.S. ROSENTHAL, *Geometric Ergodicity and Hybrid Markov chains, Electronic Communications in Probability, 2, 13-25, 1997.*

[42] J.S. ROSENTHAL, *A review of asymptotic convergence for general state space Markov chains. Expository paper. Far East J. Theor. Stat. 5:37-50, 2001..*

[43] D. RUELLE, *A measure associated with Axiom A attractors, Amer. J. Math. 98, 619-654, 1976.*

[44] L. SALOFF-COSTE, *Lectures on finite Markov chains. In Ecole d"et'e de probabilit'es de Saint-Flour XXVI, P. Bernard (ed.), L.N. in Math. 1664, 301–413. SpringerVerlag, New York, 1997.*

[45] G.R. SHORACK, *Convergence of reduced empirical and quantile processes with application to functions of order statistics in the non-iid case. Annals of Statistics 1, 146-152, 1973.*

[46] G.R. SHORACK, *The weighted empirical process of row independent random variables with arbitrary distribution functions. Statistica Neerlandica 33,169-189 (1980).*

[47] G.R. SHORACK, J.A. WELLNER, *Empirical Processes with applications to statistics, Wiley, New-York, 1986.*

[48] G.R. SHORACK, J. BEIRLANT, *The appropriate reduction of the weighted empirical process. Statistica Neerlandica 40, 123-12, 1986.*

[49] Y. SINAI, *Gibbs measure in ergodic theory, Russian Math. Surveys 27, 21-69, 1972.*

[50] S. SMALE, *Differentiable dynamical systems, Bull. Amer. Math. Soc., 73, 747-817, 1967.*

[51] F. TAKENS, *Detecting Strange Attractors in Turbulence, Lecture Notes in Math., 898:366-381, 1981.*

[52] O. TEYTAUD, J.-M. FRIEDT, M. PLANAT, *Neural network based prediction of Chua's chaotic circuit frequency - possible extension to quartz oscillators, in proceedings of the EFTF, 2001*

[53] E. TRENTIN, M. GORI, *Continuous Speech Recognition with a Robust Connectionist/Markovian Hybrid Model, in Proceedings of Icann 2001.*

[54] V.N. VAPNIK, *The Nature of Statistical Learning, Springer, 1995.*

[55] M. Viana, *Stochastic dynamics of deterministic systems, http://www.impa.br/~ viana/.*

[56] M. Vidyasagar, *A theory of learning and generalization, Springer 1997.*

[57] A.-W. van der Vaart, J.-A. Wellner, *Weak convergence and Empirical Processes, Springer, 1996.*

[58] M. Van Zujilen, *Properties of the empirical distribution function for independent not identically distributed random variables. Annals of Probability 6, 250-266, 1978.*

A Processus empirique

On a besoin parfois d'une forme de nombres de couverture ou de bracketing-nombre de couverture différente de la définition classique. [57] fournit des résultats basés sur ces nombres, donnant comme référence [27, 45, 46, 58, 48, 31, 47]. Etant donné $Z_{n,i}$ des processus stochastiques indépendants pour $1 \leq i \leq n$, $N'_{[\,]}(\epsilon, F, n)$ est le cardinal (dépendant de n) de la plus petite famille (si elle est finie) de $(F_i)_{i\in I}$ tel que $\forall f \in F \exists i \in I/f \in F_i$ et $\forall i/\sum_{i=1}^{n} E(\sup_{f,g\in F_i^2}(Z_{n,i}(f) - Z_{n,i}(g))^2) \leq \epsilon^2$. Des théorèmes centraux limite généralisés (avec entropie uniforme) sont dus à [14, 39, 26], et les équivalents bracketings sont dus à [14, 15, 38, 1]. Un état de l'art général de tels résultats peut être trouvé dans [57]. Le résultat suivant sera utile avec $Z_{n,i}(f) = f(X_{n,i})/\sqrt{n}$ principalement.

Théorème A.1 (Convergence rapide). *Pour tout $n \in \mathbb{N}$, on considère $Z_{n,i}$, pour $i \in [1, n]$, des processus stochastiques indépendants de seconds moments finis, et on suppose F totalement borné pour la semimétrique d. On suppose que l'hypothèse suivante a lieu:*

$$\sum_{i=1}^{n} E \parallel Z_{n,i} \parallel_F \{\parallel Z_{n,i} \parallel_F > \eta\} \to 0 \textit{ pour tout } \eta > 0$$

$$\sup_{(f,g)\in F, d(f,g)<\delta_n} \sum_{i=1}^{n} E(Z_{n,i}(f) - Z_{n,i}(g))^2 \to 0 \textit{ pour tout } \delta_n \textit{ décroissant vers } 0 \quad (6)$$

$$\int_0^{\delta_n} \sqrt{\log N'_{[\,]}(\epsilon, F, n)} d\epsilon \to 0 \textit{ pour tout } \delta_n \textit{ décroissant vers } 0$$

Alors $E_n = \sum_{i=1}^{n}(Z_{n,i} - E(Z_{n,i}))$ est asymptotiquement tendu dans l'espace des fonctions totalement bornées de F dans $\mathbb{R}$. Ceci signifie que pour tout $\epsilon > 0$ il existe un ensemble compact K tel que $\liminf P(E_n \in K^\delta) \geq 1 - \epsilon$ pour tout $\delta > 0$, avec $K^\delta = \{y/d(y, K) < \delta\}$ le δ-agrandissment de K. Il converge en distribution, pourvu qu'il converge pour les lois marginales, et le processus limite T est centré (moyenne zéro), gaussien, avec covariance $E(TfTg) = E(Tfg) - E(Tf) \times E(Tg)$. La condition 6 peut être supprimée si la partition dans $N'_{[\,]}$ peut être choisie indépendamment de n. (La finitude des bracketing-nombres est en fait suffisante pour des théorèmes plus faibles que nous n'utiliserons pas ici.)

B Dynamiques stochastiques dans des systèmes déterministes

L'idée de dynamiques compliquées apparaissant spontanément dans des systèmes naturels provient de Landau-Lifschitz. Dans les années 60, Smale découvre que les flots réguliers et des transformations régulières peuvent entrainer une infinité de mouvements périodiques. Ces mouvement pouvaient provenir de perturbations arbitrairement petites. Smale a alors introduit la notion d'**hyperbolicité** ([50]). Dans le début des années 70, Ruelle-Takens a développé l'idée dans l'esprit de la présence d'attracteurs **"étranges"** dans l'espace des états. La notion d'hyperbolicité a été développée par un grand nombre de chercheurs dans le cadre de la théorie ergodique: [49, 4, 43]. [29, 20] ont montré l'importance de l'hyperbolicité pour la **stabilité structurelle** (un système étant structurellement stable si ses orbites sont en bijection avec les orbites des systèmes proches). D'un autre côté, [37] montre que dans de nombreux cas, les systèmes dynamiques étaient non hyperboliques - [55] fournit une liste d'exemples ([28, 22, 7, 16]). Lorenz a souligné l'importance de la sensibilité aux conditions initiales. Ceci a amené de nombreux chercheurs à étudier des systèmes *faiblement* hyperboliques. Le théorème B.1 donne des conditions générales sous lesquelles des systèmes déterministes se comportent "presque" comme des systèmes aléatoires.

Théorème B.1. *On suppose qu'au moins l'un des ensembles d'hypothèses suivants est vérifié: (A) Applications uniformément étendantes: (1) $g : M \to M$ est $C^{1+\mu_0}$ pour un certain $\mu_0 \in]0,1]$, (2) M est une variété compacte connexe, (3) g est une application étendante, au sens où il existe $\sigma > 1$ tel que $\| Dg(x).v \| \geq \sigma \| v \|$ pour tout x et v. (B) Attracteurs uniformément hyperboliques: (1) $g : M \to M$ est un difféomorphisme sur la variété M et $Q \subset M$ est un certain ensemble ouvert positivement invariant, au sens où $g(fermeture(Q)) \subset Q$, (2) $\mathcal{L} = \cap_{n\in\mathbb{N}} g^n(Q)$ est* **transitif** *(ie contient des orbites denses) et* **hyperbolique** *pour g, ie il existe une séparation du faisceau tangent de M $T_{\mathcal{L}}M = E^s_{\mathcal{L}} \oplus E^u_{\mathcal{L}}$ et un certain $\lambda_0 < 1$ tel que: (a) $Df(x)E^s_x = E^s_{g(x)}$ et $Dg^{-1}(x)E^u_x = E^u_{g^{-1}(x)}$, (b) $\| Dg(x)E^s_x \| \leq \lambda_0$ et $\| Dg^{-1}(x)|E^u_x \| \leq \lambda_0$ pour tout $x \in \mathcal{L}$. Alors, on a: (1) il existe une unique* **mesure SRB** *μ de support sur $\mathcal{L}$. Cette mesure est* ergodique, *et son bassin a une mesure de Lebesgue > 0. Le fait que μ soit SRB (pour Sinai-Ruelle-Bowen) signifie que μ est* **invariant** *(ie $\int f(t,P)d\mu(t) = \mu(P)$) et il y a un ensemble de mesure positive x tel que pour tout ϕ continu, et x dans le bassin $\int \phi d\mu = \lim_{n\to\infty} \frac{1}{n}\sum_{i=0}^{n-1} \phi \circ f^i(x)$, (2) la chaîne est* **exponentiellement mixante** *et satisfait le* **théorème central limite** *dans l'espace de Banach des fonctions continues μ-Hölder, pour tout $\mu \in]0,\mu_1]$. Le caractère "exponentiellement mixant" signifie qu'il existe $r < 1$ tel que pour tout couple de fonctions (ϕ,ψ) μ-Hölder, il existe C tel que $E((\phi \circ f^n(t,.) - E\phi \circ f^n(t,.)) \times (\psi \circ f^n(t,.) - E\psi \circ f^n(t,.))) \leq Cr^n$ (décroissante exponentielle des corrélations). Le théorème central limite signifie que pour tout ϕ μ-Hölder, il y a σ tel que pour tout interval A $\mu(\{x|\frac{1}{\sqrt{n}}\sum_{i=0}^{n-1}(\phi \circ f^j(x) - \int \phi d\mu) \in A\}) \to \frac{1}{\sqrt{2\pi\sigma}}\int_A \exp(-\frac{-t^2}{2\sigma^2})$, (3) la chaîne est* **stochastiquement stable sous de petites perturbations aléatoires**. *Voir [55] pour plus d'informations sur cette notion.*

Ce théorème a été prouvé ailleurs sous diverses formes mais nous utilisons la référence [55] en tant que joli état de l'art de résultats, incluant de nombreuses améliorations dans un cadre commun. En particulier, les applications non-uniformément hyperboliques sont traitées dans [55, section 5]. Une classe importante (mais non-exhaustive) de systèmes déterministes est la classe des **systèmes chaotiques**. De nombreuses définitions du chaos existent. Une définition classique est la suivante ([9, 23]): soit X_n un système dynamique défini par $X_{n+1} = f(X_n)$ avec $X_0 \in D$ et $f \in D^D$. Il est dit **chaotique** si (1) Les points périodiques de f sont denses dans D (un point est dit **périodique** de période k si $f^{(k)} = x$), (2) f est **topologiquement transitif**. Ceci signifie que pour tous ensembles ouverts U et V qui intersectent D, il y a $x \in U \cap D$ et un nombre entier n tel que $f^n(z)$ est dans V. Ceci est équivalent au fait que pour tout x et y dans D et $\epsilon > 0$, il y a $z \in D$ tel que $d(x,z) < \epsilon$, $d(f^{(n)}(z), y) < \epsilon$ pour un certain n, (3) f présente une **dépendance sensible aux conditions initiales**. Ceci signifie qu'il existe $\delta > 0$ tel que pour tout $x \in D$ et $\epsilon > 0$, il y a un $y \in D$ et un $n \in \mathbb{N}$ tel que $d(x,y) < \epsilon$ et $d(f^{(n)}(x), f^{(n)}(y)) > \delta$. Une classe particulière de systèmes chaotiques est issue d'équations différentielles grâce au fameux théorème de Takens. Considérons les équations différentielles suivantes: $\frac{dY}{dt} = G(Y(t))$ (évolution du système), $x(t) = H(Y(t)) + \epsilon(t)$ (mesure avec bruit indépendant ϵ), $x_n = x(nT)$ (discrétisation), $X_n = (x_n, x_{n-\tau}, x_{n-2\tau}, \ldots, x_{n-(d-1)\tau})$ (fenêtre). Selon le théorème de Takens (voir [51] pour un énoncé précis), sous des hypothèses légères sur G et H, si $d \geq 2D+1$, avec D la dimension de l'attracteur du système, alors il y a un difféomorphisme qui associe X_n et $Y(nT)$. Ceci implique, en particulier, le fait que $x_{n+1} = f(X_n)$ pour un certain f. La prédiction de séries temporelles chaotiques est un exercice classique pour les algorithmes de prédiction. Dans le cas général, les systèmes chaotiques sont beaucoup trop compliqués pour des prédictions basées sur Arma, d'où le besoin d'algorithmes intensifs comme les réseaux de neurones, les plus proches voisins ou les réseaux de fonctions à bases radiales. Voir [34] pour un résumé des points importants de la prédiction de systèmes chaotiques.

C Convergences uniformes et chaînes de Markov

Une chaîne de Markov est une suite de variables aléatoires X_n, définies par un état initial X_0, ou une distribution de probabilité pour X_0, et une probabilité de transition $P(X_n|X_{n-1})$, supposée constante pour $n \geq 1$: $P(X_n \in E|X_{n-1} = t) = f(t, E)$. Des modèles de Markov de degré plus élevé peuvent être définis, avec des probabilités de transition $P(X_n|X_{n-1}, X_{n-2}, \ldots, X_{n-k})$, avec une distribution de probabilité pour $(X_0, X_1, \ldots, X_{k-1})$. On restreint notre attention au cas ci-dessus, qui peut inclure d'autres cas par simple adaptation. $f^n(t, E)$ est défini par induction par $f^1 = f$ et $f^{n+1}(t, E) = \int f^n(u, E) f(t, du)$. Une aire de recherche importante à propos des chaînes de Markov est leur comportement asymptotique. De nombreux livres fournissent des résultats tels l'existence d'une distribution stationnaire et la convergence vers cette distribution, sous des hypothèses raisonnables, dans le cas d'espaces d'états finis. L'extension à des espaces d'état dénombrables existe, mais pour beaucoup d'applications, on a besoin de convergence rapide dans des chaines de Markov non dénombrables. [12] a prouvé un premier résultat dans cette direction, et [42] survole les résultats récents, dont:

Théorème C.1. *On suppose que (1) f admet une* **distribution stationnaire** *π, ie une distribution π telle que $\forall A$ mesurable $\pi(A) = \int f(y, A) d\pi(y)$, (2) f est* **apériodique**, *ie il n'existe pas une partition finie en $d \geq 2$ ensembles $\mathcal{X}_1, \ldots, \mathcal{X}_d$, telle que $\forall t \in X_i$ $f(t, \mathcal{X}_{i+1}) = 1$, avec $\mathcal{X}_{d+1} = \mathcal{X}_1$, (3) f est* **ϕ-irréductible**, *ie $\exists \phi$ mesure non-triviale telle que $\forall P$ mesurable $\phi(P) > 0 \Rightarrow \exists n f^n(t, P) >$*

0 *presque sûrement en* t *pour* π. *Alors, presque sûrement en* t *(pour* π*)* $\lim_{n\to\infty}\sup_P |f(t,P)-\pi(P)| = 0$.

On a besoin en fait, dans certaines applications ci-dessous, la convergence *uniforme* (en t), ou du moins des bornes explicites sur la dépendance en t. Ceci est géré dans le théorème (multiple) suivant (adapté de [32] pour le premier, [2, Theorem 4.1] pour le second):

Théorème C.2 (Ergodicité uniforme). *Premier cas: supposons qu'il existe* $m \in \mathbb{N}$, μ *une mesure de probabilité et* $\delta > 0$ *tel que* $\forall t f^m(t,.) \geq \delta\mu$; *alors* $\| f^n(t,.) - \pi \| \leq (1-\delta)^{\frac{n}{m}}$. *Cette condition est appelée la* **condition de Doeblin,** *ou* ergodicité uniforme. *Second cas: supposons que (1) les* X_n *sont un* **processus strictement stationnaire** *(ie la loi des* X_n *est indépendante de* n*), (2)* $\mathcal{F}$ *est un* α*-Hölder espace de fonctions sur* X, *sous-ensemble borné de* $\mathbb{R}^d$, *(3) on a une condition de* **mélange***: avec* $\alpha_k = \sup_{A,B\ measurable, l\geq 1}\{|Pr(AB) - Pr(A)Pr(B)|/A \in \sigma_1^l, B \in \sigma_{k+l}^\infty\}$, *avec* σ_a^b *la* σ*-algèbre générée par* $X_a,\ldots,X_{a+b-1}$, *pour un* $p > 2$, $\sum \alpha_n n^{2/(p-2)} < \infty$ *et* $\frac{d(p-1)}{p} < \alpha$. *Alors (dans les deux cas!)* $\{\frac{1}{\sqrt{n}} D_n f | f \in \mathcal{F}\} \to Df$ *dans* $l^\infty(\mathcal{F})$, *avec* D *dans le second cas un processus gaussien centré avec covariances définies par* $Df_1f_2 = Cov(f_1(X_1), f_2(X_1)) + \sum_{k=1}^\infty Cov(f_1(X_1), f_2(X_{1+k})) + Cov(f_2(X_1), f_1(X_{1+k}))$.

Notez que des comportements déterministes pour des applications uniformément dilatantes ou hyperboliques mènent à des meilleurs équivalent que ces chaînes de Markov. Voir [41] dans le cas réversible. [33] fournit des résultats proches très intéressants: on peut avoir convergence géométrique avec dépendance en le point initial (dans le théorème ci-dessus, la première partie considère la convergence *uniforme en le point de départ*) et avec des bornes (presque) explicites sur les constantes.

Olivier Teytaud

ISC, 67 Bd Pinel, 69675 Bron Cedex

Trends in Mathematics, © 2002 Birkhäuser Verlag Basel/Switzerland

Entropy Reduction Strategies on Tree Structured Retrieval Spaces

Alain Trouvé, Yong Yu

ABSTRACT: *In this paper, we study the performance of exact retrieval strategies in the case of tree structured retrieval spaces. We assume that the database $\mathcal{B}$ is indexed by the leaves of a hierarchical partitioning tree T. We study retrieval processes based on interaction with the user through simple questions attached to the nodes of T as follows: for each node b, the system can display a summary of the subset attached to b (e.g. some typical images) and get the user answer according to the target. We consider retrieval strategies based on step-wise entropy reduction, built on a user model where the answers are independent given the target. We prove an upper bound for the expected number of questions which appears to be nearly optimal in an interesting case. Moreover, we show that at each step, the next question can be found among an adaptive subset of nodes of size $\log(|\mathcal{B}|)$. Finally, the overall complexity of the algorithm (for the computer) per retrieval is $O(\log(|\mathcal{B}|)^3)$ whereas the average number of questions (for the user) is $O(\log(|\mathcal{B}|))$.*

1 Introduction

In the context of image retrieval in a large database, the use of interactive scenarios to get information from the user on her/his target appears as an interesting way to try to overcome the so-called "semantic gap" which is responsible to the limitations of "one shot" retrieval strategies working on a unique but complex query submitted by the user to the system such as an image or a sketch of what she/his is looking for (see [9]). Among these interactive scenarios, an interesting approach introduced in [4], is based on simple queries or questions submitted by the system to the user. For instance, assume that the user is looking for a given image, called the target, then two images are presented to the user and the user selects the closest one to her/his target. According to this selection, another couple of images is presented and so on until the target is found or the user simply gives up. The selection mechanism of couples of images defining the query is built on a probabilistic model connecting the user answers and her/his target. Given this user model, the system selects the most informative questions according to the criterion of the expected (given the model) uncertainty reduction of the distribution of the target given the answer. The probability model itself is built on a distance or a set of distances describing the dissimilarity between the features extracted from the images in a preliminary indexation process [3, 7, 11]. The performance of these retrieval strategies reported in [3, 7] is encouraging when tested on a moderate size database. However, it is not clear what could happen on a large and heterogeneous database since the discrimination through distances is more questionable in that case.

Instead of using distances, we propose here a coarse-to-fine approach as introduced in [1, 5] in the context of face detection. We assume that the database is organized in a tree T coding a hierarchical partitioning. This tree can have been built off-line, according to the content of the database using adapted clustering algorithms [2, 12] or even manually according to the semantic content of the images.

The paper is organized as follows. In section 2, we precise the mathematical framework. Then, in section 3, we define the entropy reduction algorithm and show that expected information gain for each new question can be efficiently and recursively computed throughout the tree T. In section 4, we prove an upper bound for the performance of this strategy and we study the tightness of this bound through a comparison with a lower bound proposed in [10] and simulations. In section 5, we propose a very fast version of the retrieval strategy by entropy reduction choosing the best question among an adaptively selected small set of nodes. We show theoretically as well on simulations that the performance does not seem to be affected. We end this paper with some concluding remarks.

2 Model description

2.1 Tree structured retrieval spaces

Let us first consider a database $\mathcal{B}$ organized in a tree T i.e. $\partial T = \mathcal{B}$ where ∂T denotes the set of the leaves of T. We will denote by λ the root node of the tree, and for any $s, t \in T$, we say that $s \preceq t$ if t is an ancestor of s. With this definition, for any $s \in T$, $s \preceq \lambda$. For any $b \in T$, we define $T_b = \{s \in T \mid s \preceq b\}$ as the subtree formed by the descendants of b . For any $b \in T$, C_b will denote the set of the children of b. Note that for any $b \in T \setminus \partial T$, the family $(T_s)_{s \in C_b}$ defines a partition of T_b.

Now, we consider that the user is looking for a target Y in the database. Assume that for any node b, the system can display to the user some kind of summary of the subset ∂T_b of the database (for instance, the system might presents random samples of 20 images drawn from ∂T_b). From that summary, the user is asked to say if the target Y could be in ∂T_b. In a simple formulation of the problem, the user's answer can be "no", "yes" or "I have seen my target" (so that the process stops). More generally, the user's answer is a label l in a finite set L of labels. Assume that the answers to different questions are stored in a family $X = (X_b)_{b \in T}$, where, for each $b \in T$, $X_b \in L$ is the answer of the user to question b. Note here however, that there is no reason why a question b should be asked only once. Indeed, the system can get interesting information about the probability to find the target below a given node by repeating the same question (showing a different summary each time). In this case, we generalize the previous framework, and we denote a question by $q = (b_q, i_q)$, where $b_q \in T$ and i_q is a positive integer saying that the question is asked for the ith time. However, in the extreme case where $b \in \partial T$, then the question attached to b is "$Y = b$?" to which the answer is deterministic and non repeatable. Let us recall the formal definition of a retrieval space as introduced in [10]:

Definition 2.1. *We say that $\mathcal{R} = (\mathcal{B}, \mathcal{T}, \mathcal{Q}, \Omega_\mathcal{Q}, \mathcal{F}_\mathcal{Q})$ is a retrieval space with targets set $\mathcal{T}$, database $\mathcal{B}$ (with $\mathcal{B} \subset \mathcal{T}$) and set of possible questions $\mathcal{Q}$ if*

1. *There exists a family of finite sets $L = (L_q)_{q \in \mathcal{Q}}$ called the family of* possible answers *such that $\Omega_\mathcal{Q}$ is defined by*

$$\Omega_\mathcal{Q} \doteq \{ \omega = (y, x) \mid x = (x_q)_{q \in \mathcal{Q}},\ x_q \in L_q,\ y \in \mathcal{T} \},$$

and $\mathcal{F}_\mathcal{Q} \doteq \sigma(Y, X)$ *where* $X = (X_q)_{q\in\mathcal{Q}}$ *and* Y *are the canonical projections;*

2. *For any* $t \in \mathcal{B}$, $t \in \mathcal{Q}$ *and* $X_t = \mathbf{1}_{Y=t}$ *(that is we can check with no error that* $Y = t$*, for any* $t \in \mathcal{T}$*). Such a question will be called a* basic question.

We will say that $\mathcal{R}$ *is a* simple *retrieval space if* $\mathcal{T} = \mathcal{B}$*, i.e. all the targets are available in the database.*

In our framework, *a tree structured retrieval space* will be a retrieval space $\mathcal{R}$ on a database $\mathcal{B}$ equipped with a tree T such that $\partial T = \mathcal{B}$ and for which the set of possible questions is given by $\mathcal{Q} \doteq \{\ (b,i) \mid b \in T \setminus \partial T,\ i \geq 1\ \} \cup \partial T$. Moreover, we will assume that for $q = (b,i)$, L_q depends only on b and not on i. We will consider essentially simple retrieval space ($\mathcal{T} = \mathcal{B}$) or retrieval spaces for which $\mathcal{T} = \mathcal{B} \cup \{o\}$ where o is an extra point representing the outside of the database.

2.2 User models

A user model will be a probability model P on $(\Omega_\mathcal{Q}, \mathcal{F}_\mathcal{Q})$. Many user models are possible and the selection of a good user model is a challenging issue. Among all the possible user models, we want to distinguish an important finite dimensional parametric family of models based on three assumptions on the user's behavior.

The *first* assumption is the conditional independence of the X_q's given Y. This assumption is widely used in this context in order to simplify the definition of the model. Hence, in that case, for any finite set $\mathcal{H} \subset \mathcal{Q}$

$$P(\ X_\mathcal{H} = x_\mathcal{H} \mid Y = t\ ) = \prod_{q\in\mathcal{H}} P(\ X_q = x_q \mid Y = t\ ). \tag{1}$$

The second assumption is that

$$P(\ X_q = l \mid Y = t\ ) \ = \ P(\ X_{q'} = l \mid Y = t\ ) \text{ if } b_q = b_{q'} \tag{2}$$

that is the conditional law of the answers of questions at the same position in the tree is independent and identically distributed. This last assumption is maybe more doubtful. However, we think that this model can capture the basic fact that by repeatedly showing random samples drawn in a given ∂T_b, the user can have a perfect representation of this subset. Finally, the *third* assumption is that

$$P(\ X_q = l \mid Y = t\ ) \ = \ P(\ X_q = l \mid Y = t'\ ) \text{ if } b(t) = b(t'), \tag{3}$$

where $b(t) = \mathbf{1}_{t\in\partial T_b}$ and $b = b_q$. Hence there exists a family $(\mu_{b,\epsilon})_{b\in T,\epsilon\in\{0,1\}}$ of probability distributions such that for any finite subset $\mathcal{H} \subset \mathcal{Q}$

$$P(\ X_\mathcal{H} = x_\mathcal{H} \mid Y = t\ ) = \prod_{b\in T}\prod_{l\in L_b} \left(\mu_{b,b(t)}(l)\right)^{n_{b,l}}. \tag{4}$$

where for any $b \in T$ and any $l \in L_b$, $n_{b,l} = \sum_{q\in\mathcal{H}} \mathbf{1}_{x_q=l, b_q=b}$.

Definition 2.2. *A user model* P *satisfying (1) to (4) will be called a* conditional i.i.d. user model *with associated answers distributions* $(\mu_{b,\epsilon})_{b\in T,\ \epsilon\in\{0,1\}}$.

3 Retrieval strategies

3.1 Definitions

In words, a strategy is a sequence of questions such that each new question is based on the answers of the previous one, which terminates with a decision (for instance the prediction of the target). A strategy can be defined by a decision tree where the internal node are the questions, the searching process defining a path from the root to some leaf where a decision is taken. To each possible answer to a question attached to a node, it corresponds a child on which a new question can be asked. We recall below the more formal definition given in [10]:

Let $\mathcal{R}$ be a retrieval space and let $S = (Q, D)$ be a random process on $(\Omega_{\mathcal{Q}}, \mathcal{F}_{\mathcal{Q}})$ where Q is a variable length random sequence $Q = (Q_n)_{1\leq n\leq \tau_S}$ with value in $\mathcal{Q}$ (the sequence of questions) and D is a random variable defined on the event $(\tau_S < +\infty)$ with value in $\mathcal{D}$ (defining the possible decision at the end of the sequence of questions). Let us consider the associated history sequence $(\mathcal{H}_n)_{n\geq 0}$ defined by $\mathcal{H}_0 = \emptyset$ and for any $n > 0$, we have $\mathcal{H}_{n+1} = \mathcal{H}_n \cup \{Q_{n+1}\}$ if $\tau_S > n$ and $\mathcal{H}_{n+1} = \mathcal{H}_n$ otherwise.

Definition 3.1. *- We say that S is a* retrieval strategy *on $\mathcal{R}$ if for any $n \geq 0$, any $\mathcal{H} \subset \mathcal{Q}$, any $q \in \mathcal{Q}$ and any $d \in \mathcal{D}$, the two events $(D = d) \cap (\tau_S = n) \cap (\mathcal{H}_n = \mathcal{H})$ and $(\tau_S > n) \cap (\mathcal{H}_n = \mathcal{H}) \cap (Q_{n+1} = q)$ belong to $\sigma(X_{\mathcal{H}}) \doteq \sigma(X_q | q \in \mathcal{H})$.*

- We say that a retrieval strategy $S = (Q, D)$ is exact for the user model P if $\mathcal{D} = \mathcal{T}$ and $P(D = Y) = 1$.

3.2 Example of the entropy reduction retrieval strategy

The entropy reduction strategy is built according to a chosen user model P on a retrieval space $\mathcal{R}$ (not necessarily tree structured) and defined as follows:

Entropy reduction strategy built on P (Complete scan)
`Let` $n \geq 0$. `Assume that` $\tau_S \geq n$ `and let` $\mathcal{H} = \mathcal{H}_n$.
`If` $P(Y = t_* | X_{\mathcal{H}}) = 1$ `for some` $t_* \in \mathcal{T}$, `then` $\tau_S = n$ `and` $D = t_*$.
`Otherwise, choose` $Q_{n+1} \in \operatorname{argmin}_{q\in\mathcal{Q}} H_{\mathcal{H}}(Y|X_q)$ `where`

$$H_{\mathcal{H}}(Y|X_q) = -\sum_{t\in\mathcal{T}}\sum_{l\in L_q} P(Y = t, X_q = l | X_{\mathcal{H}}) \log(P(Y = t | X_q = l, X_{\mathcal{H}})). \quad (5)$$

Remark 3.2. 1. Note that usual property on the conditional entropy gives

$$H_{\mathcal{H}}(Y|X_q) = H_{\mathcal{H}}(Y, X_q) - H_{\mathcal{H}}(X_q) = H_{\mathcal{H}}(X_q|Y) + H_{\mathcal{H}}(Y) - H_{\mathcal{H}}(X_q) \quad (6)$$

so that $Q_{n+1} \in \operatorname{argmax}_{q\in\mathcal{Q}} (H_{\mathcal{H}}(X_q) - H_{\mathcal{H}}(X_q|Y))$. The last quantity is exactly the conditional mutual information in [8, 7] denoted by $I_{\mathcal{H}}(Y, X_q)$.

2. In the case of tree structured retrieval space and conditional i.i.d. user model P, $I_{\mathcal{H}}(Y|X_q)$ depends only on b_q (and not on i_q) as soon as $q \notin \mathcal{H}$. This yields that it is sufficient to compute the mutual information gain on a new question asked somewhere in the tree $\mathcal{T}$. The terminology "complete scan" emphasizes that we have to scan the complete tree to get the best question. This limitation for very large database will be fixed in section 5.

3.3 Recursive computation for conditional i.i.d. user model

When the user model is a conditional i.i.d. model, the computation of the expected information gain given the past has the following straightforward simpler formulation.

Proposition 3.3. *Let $\mathcal{R}$ be a tree structured retrieval space. Assume that P is a conditional i.i.d. user model on $\mathcal{R}$. Let $\mathcal{H} \subset \mathcal{Q}$ and $q \doteq (b,i) \notin \mathcal{H}$. Then,*

$$I_{\mathcal{H}}(Y, X_q) = h\big(p\mu_{b,1} + (1-p)\mu_{b,0}\big) - \big(ph(\mu_{b,1}) + (1-p)h(\mu_{b,0})\big), \tag{7}$$

where $p \doteq P(Y \in \partial T_b | X_{\mathcal{H}})$ and $h(\mu) \doteq -\sum_l \mu(l)\log(\mu(l))$ is the usual entropy.

From proposition 3.3, we see that the key quantity is $P(Y \in \partial T_b | X_{\mathcal{H}})$ for $\mathcal{H} = \mathcal{H}_n$. From the tree structure underlying the user model, we get interesting factorization properties so that the computations can be done recursively as follows. We will assume throughout this section that the set of targets $\mathcal{T} = \mathcal{B} \cup \{o\}$ where o is an extra point representing the outside of the database. Let

$$\begin{cases} w_b(X_{\mathcal{H}}) \doteq \prod_{s\in T_b} \prod_{l\in L_s} (\mu_{s,0}(l))^{n_{s,l}(X_{\mathcal{H}})} \\ c_b(X_{\mathcal{H}}) \doteq \sum_{t\in\partial T_b} P(Y=t) \prod_{s\in T_b} \prod_{l\in L_s} (\mu_{s,s(t)}(l))^{n_{s,l}(X_{\mathcal{H}})} \end{cases}$$

and where as previously $n_{s,l}(X_{\mathcal{H}}) = \sum_{q\in\mathcal{H}} \mathbf{1}_{X_q=l,\ b_q=s}$. One notices immediately that $c_\lambda(X_{\mathcal{H}}) = P(X_{\mathcal{H}},\ Y \in \mathcal{B})$. Moreover, one proves easily:

$$w_b(X_{\mathcal{H}}) = \prod_{l\in L_b} (\mu_{b,0}(l))^{n_{b,l}(X_{\mathcal{H}})} \prod_{b'\in C_b} w_{b'}(X_{\mathcal{H}}), \tag{8}$$

$$c_b(X_{\mathcal{H}}) = \prod_{l\in L_b} (\mu_{b,1}(l))^{n_{b,l}(X_{\mathcal{H}})} \sum_{b'\in C_b} c_{b'}(X_{\mathcal{H}}) \prod_{b''\in C_b\setminus\{b'\}} w_{b''}(X_{\mathcal{H}}). \tag{9}$$

Now, if we notice that for $t \in \partial T$, we have $w_t(X_{\mathcal{H}}) = \prod_{l\in L}(\mu_{t,0}(l))^{n_{t,l}(X_{\mathcal{H}})}$, and $c_t(X_{\mathcal{H}}) = P(Y=t)\prod_{l\in L}(\mu_{t,1}(l))^{n_{t,l}(X_{\mathcal{H}})}$, we deduce from (8) and (9) a bottom-up computation of the w_b's and the c_b's starting from the leaves of T. Let

$$c_{b,r}(X_{\mathcal{H}}) \doteq \sum_{t\in\partial T_b} P(Y=t) \prod_{s\in T_r} \prod_{l\in L_s} (\mu_{s,s(t)}(l))^{n_{s,l}(X_{\mathcal{H}})} \tag{10}$$

for any $b \preceq r$, so that $c_{b,b} = c_b$. One gets for $b \preceq r'_0 \preceq r$ and $r'_0 \in C_r$ that

$$c_{b,r}(X_{\mathcal{H}}) = \left(\prod_{l\in L_r} (\mu_{r,1}(l))^{n_{r,l}(X_{\mathcal{H}})}\right) c_{b,r'_0}(X_{\mathcal{H}}) \prod_{r'\in C_r, r'\neq r'_0} w_{r'}(X_{\mathcal{H}}). \tag{11}$$

Note that $c_{b,\lambda}(X_{\mathcal{H}}) = P(X_{\mathcal{H}},\ Y \in \partial T_b)$ and given $c_{b,b} = c_b$, the value of $c_{b,\lambda}$ can be computed following the path from b to the root λ. Since $P(Y \in \partial T_b \mid X_{\mathcal{H}}) = \frac{P(X_{\mathcal{H}},\ Y\in\partial T_b)}{P(X_{\mathcal{H}},\ Y=o)+P(X_{\mathcal{H}},\ Y\in\mathcal{B})}$, we deduce

$$P(Y \in \partial T_b \mid X_{\mathcal{H}}) = \frac{c_{b,\lambda}(X_{\mathcal{H}})}{P(Y=o)w_\lambda(X_{\mathcal{H}}) + c_\lambda(X_{\mathcal{H}})}. \tag{12}$$

For simple retrieval space (i.e. $P(Y=o)=0$) we have $P(Y \in \partial T_b \mid X_{\mathcal{H}}) = c_{b,\lambda}(X_{\mathcal{H}})/c_\lambda(X_{\mathcal{H}})$.

Remark 3.4. Note that if the tree T is not too irregular then the depth of the tree is $O(\log(|\mathcal{B}|))$ where $|\mathcal{B}|$ is the size of the database. Let us consider a node $b_0 \in T$. Assume that the $c_b(X_{\mathcal{H}_n})$ has been computed previously and stored in the tree. To compute $c_{b_0,\lambda}$, we only need to compute recursively the $c_{b_0,b}$ for all the ancestors b of b_0 so that the complexity is $O(\log(|\mathcal{B}|)$. Moreover, given the user's answer at node b_0, we need to update the c_b's and the w_b's for the ancestor b of b_0. Hence the updating complexity is again $O(\log(|\mathcal{B}|)$.

4 Performance of the entropy reduction strategy

From now on we consider only *simple* tree structured retrieval space (i.e. $\mathcal{T} = \mathcal{B}$).

4.1 Theoretical upper bound for the performance

Let us extend the sequence $(Q_n)_{1\leq n\leq \tau_S}$ by $Q_n \doteq Q_{\tau_S}$ if $n \geq \tau_S$ and denote for any $n \geq 1$, $X_n \doteq X_{Q_n}$. We denote by $\mathcal{F}_n$ the σ-algebra of past before time n defined by $\mathcal{F}_n \doteq \sigma(X_i, Q_i \mid 1 \leq i \leq n)$. Moreover, for any $n \geq 1$, we define

$$H_n(Y) = H_{\mathcal{H}}(Y)_{|\mathcal{H}=\mathcal{H}_n} \text{ and } \eta_n \doteq I_{\mathcal{H}}(Y, X_q)_{|\mathcal{H}=\mathcal{H}_{n-1}, q=Q_n} \mathbf{1}_{\tau_S \geq n} \tag{13}$$

which is the expected information gain of the target given by the answer to the nth question as proved by the next simple proposition (see [10] for a proof):

Proposition 4.1. *For any $n \geq 1$ we have $\eta_n = E(H_{n-1}(Y) - H_n(Y)|\mathcal{F}_{n-1})$*

Theorem 4.2. *Let $\mathcal{R}$ be a simple tree structured retrieval space and let $S = (Q, D)$ be a reduction entropy retrieval strategy with complete scan on $\mathcal{R}$. Let τ_S be the associated stopping time and let P be a conditional i.i.d. user model. Assume that*

$$\underline{\alpha} \doteq \inf_{b\in T} \left[h\left(\frac{1}{2}\mu_{b,0} + \frac{1}{2}\mu_{b,1}\right) - \left(\frac{1}{2}h(\mu_{b,0}) + \frac{1}{2}h(\mu_{b,1})\right)\right] > 0.$$

Then, we have $E(\tau_S) \leq \frac{p+1}{2\underline{\alpha}} H(Y) + 1$, where p is the maximum number of children for any node in T.

Proof: Let $b \in T$. Since $\psi_b : \rho \to h\left(\rho\mu_{b,0} + (1-\rho)\mu_{b,1}\right) - (\rho h(\mu_{b,0}) + (1-\rho)h(\mu_{b,1}))$ is a concave function, we get from the hypothesis that

$$\psi_b(\rho) \geq 2\underline{\alpha}(\rho \wedge (1-\rho)). \tag{14}$$

Now, let $n \geq 0$ and assume that $\tau_S \geq n$. We will denote P_n the conditional law of Y given $\mathcal{F}_n$. If $\eta_n = 0$, we deduce from the definition of η_n and Proposition 3.3 that $\psi_b(P_n(Y \in \partial T_b)) = 0$ so that $P_n(Y \in \partial T_b) \in \{0,1\}$ and the position of the target is known. Hence in that case, $\tau_S = n$. Assume now that $\eta_n > 0$. We deduce that for any $b \in T$, we have $\psi_b(P_n(Y \in \partial T_b)) \leq \eta_n$ so that, using (14), we get

$$P_n(Y \in \partial T_b) \wedge (1 - P_n(Y \in \partial T_b)) \leq \frac{\eta_n}{2\underline{\alpha}}. \tag{15}$$

Lemma 4.3. *Assume that $0 < \eta_n < 2\underline{\alpha}/(p+1)$. Then, there exists $t_n \in \partial T$ such that $P_n(Y = t_n) \geq 1 - \frac{\eta_n}{2\underline{\alpha}}$.*

Proof: We first prove by induction that there exists a sequence $\lambda = b_0 \succeq \cdots \succeq b_r \in \partial T$ such that $P_n(Y \in \partial T_{b_i}) \geq 1-\eta_n/(2\underline{\alpha})$ for all $0 \leq i \leq r$. We start the induction noting that $P_n(Y \in \partial T_\lambda) = P_n(Y \in \mathcal{B}) = 1 \geq 1 - \eta_n/(2\underline{\alpha})$. Now, assume that we have constructed $\lambda = b_0 \succeq \cdots \succeq b_m$. If $b_m \in \partial T$, the construction is complete. Otherwise, there exists $b_{m+1} \in C_{b_m}$, the set of the children of b_m, such that $P_n(Y \in \partial T_{b_{m+1}}) \geq 1 - \eta_n/(2\underline{\alpha})$. Indeed, using (15), if $P_n(Y \in \partial T_b) \leq \eta_n/(2\underline{\alpha})$ for all $b \in C_{q_m}$, since $\partial T_{b_m} = \cup_{b \in C_{b_m}} \partial T_b$, we get $P_n(Y \in \partial T_{b_m}) \leq p\eta_n/(2\underline{\alpha})$ since $|C_{b_m}| \leq p$. However, by induction hypothesis, we have $P_n(Y \in \partial T_{b_m}) \geq 1 - \eta_n/(2\underline{\alpha})$ so that $1 - \eta_n/(2\underline{\alpha}) \leq p\eta_n/(2\underline{\alpha})$, and finally $\eta_n \geq 2\underline{\alpha}/(p+1)$ which is a contradiction. □

Lemma 4.4. *Assume that* $0 < \eta_n < 2\underline{\alpha}/(p+1)$. *Then* $P(\ \tau_S > n+1 \mid \mathcal{F}_n\ ) \leq P_n(Y \neq t_n)$ *where* t_n *is defined in the previous lemma.*

Proof: For any $q \in \mathcal{Q}$, if $\mathcal{H}_n = \mathcal{H}$ and $b = b_q$, we have $I_{\mathcal{H}}(Y, X_q) = H_{\mathcal{H}}(X_q) - H_{\mathcal{H}}(X_q|Y) = H_{\mathcal{H}}(X_q) - H_{\mathcal{H}}(X_q|b(Y))$ where as defined previously $b(Y) = \mathbf{1}_{Y \in \partial T_b}$. Hence, $I_{\mathcal{H}}(Y, X_q) = I_{\mathcal{H}}(b(Y), X_q) = H_{\mathcal{H}}(b(Y)) - H_{\mathcal{H}}(b(Y)|X_q)$, and

$$I_{\mathcal{H}}(Y, X_q) \leq H_{\mathcal{H}}(b(Y)), \tag{16}$$

with equality iff $H_{\mathcal{H}}(b(Y)|X_q) = 0$.

Now, denote for any $\rho \in [0,1]$, $h(\rho) \doteq -\rho\log(\rho) - (1-\rho)\log(1-\rho)$. Let $q \in \mathcal{Q} \setminus \mathcal{H}$ and $b \doteq b_q$. If $t_n \in \partial T_b$, we deduce from (16), that $I_{\mathcal{H}}(Y, X_q) \leq h(P_n(Y \in \partial T_b)) \leq h(P_n(Y = t_n))$ since $h(\rho)$ is decreasing for $\rho \geq 1/2$, and $2/3 \leq 1 - \eta_n/(2\underline{\alpha}) \leq P_n(Y = t_n) \leq P_n(Y \in \partial T_b)$. If $t_n \notin \partial T_b$, then one gets

$$I_{\mathcal{H}}(Y, X_q) \leq h(P_n(Y \in \partial T_b)) = h(1 - P_n(Y \in \partial T_b)) \leq h(P_n(Y = t_n))$$

so that for all $q \in \mathcal{Q} \setminus \mathcal{H}$, $I_{\mathcal{H}}(Y, X_q) \leq h(P_n(Y = t_n))$. Now, on basic questions, we have $H_{\mathcal{H}}(t(Y)|X_t) = 0$ for any $t \in \partial T$ so that $I_{\mathcal{H}}(Y, X_{t_n}) = h(P_n(Y = t_n))$ and for all $q \in \mathcal{Q}$

$$I_{\mathcal{H}}(Y, X_q) \leq I_{\mathcal{H}}(Y, X_{t_n}) = \eta_n. \tag{17}$$

If $Q_{n+1} = t_n$, then we deduce that $\tau_S = n+1$ if $Y = t_n$. Indeed, $\tau_S > n$ since $\eta_n > 0$ implies that $P_n(Y = t_n) < 1$ and given the answer to question t_n, the target will be found if $Y = t_n$. Hence, in that case, $P(\ \tau_S > n+1 \mid \mathcal{F}_n\ ) \leq P_n(Y \neq t_n)$.

Assume now that $Q_{n+1} = q \neq t_n$ and let $b \doteq b_q$. From (16) and (17), we get $I_{\mathcal{H}}(Y, X_q) = I_{\mathcal{H}}(Y, X_{t_n}) = H_{\mathcal{H}}(b(Y))$ so that $H_{\mathcal{H}}(b(Y)|X_q) = 0$. This means that

$$P_{n+1}(b(Y) = f) \in \{0,1\}, \tag{18}$$

for any $f \in \{0,1\}$. Now, assume that $t_n \in \partial T_b$. Then $I_{\mathcal{H}}(Y, X_q) = h(P_n(b(Y) = 1) = h(P_n(Y = t_n))$ implies that $t_n(Y) = b(Y)$ so that we get from (18) that $P_{n+1}(Y = t_n) \in \{0,1\}$ and the target will be found if $Y = t_n$. Thus, $P(\ \tau_S > n+1 \mid \mathcal{F}_n\ ) \leq P_n(Y \neq t_n)$.

Now, assume that $t_n \notin \partial T_b$. We have $h(1 - P_n(Y \in \partial T_b)) = h(P_n(Y = t_n))$ so that $t_n(Y) = 1 - b(Y)$ a.s. and similarly, we get from (18) that $P_{n+1}(Y = t_n) \in \{0,1\}$ and the target will be found if $Y = t_n$. Hence, in any case, we have $P(\ \tau_S > n+1 \mid \mathcal{F}_n\ ) \leq P_n(Y \neq t_n)$, and the proof of the lemma is ended. □

We come back here to the proof of the theorem. We first assume that $\eta_n < 2\underline{\alpha}/(p+1)$. From the lemma, we deduce that $P(\tau_S > n+1 | \mathcal{F}_n) \leq P_n(Y \neq t_n) \leq \eta_n/(2\underline{\alpha}) \leq (p+1)\eta_n/(2\underline{\alpha})$.

Now, note that if $\eta_n \geq 2\underline{\alpha}/(p+1)$, we have $(p+1)\eta_n/(2\underline{\alpha}) \geq 1$. Hence, whatever the value of η_n is, we get $P(\ \tau_S > n+1 \mid \mathcal{F}_n\ ) \leq \frac{(p+1)\eta_n}{2\underline{\alpha}}$. Now, we have

$$E(\tau_S) = P(\tau_S \geq 1) + \sum_{n=0}^{\infty} P(\tau_S > n+1) \leq 1 + \sum_{n=0}^{\infty} E(\frac{(p+1)\eta_n}{2\underline{\alpha}}), \tag{19}$$

so that using proposition 4.1, we get $E(\tau_S) \leq 1 + \frac{(p+1)}{2\underline{\alpha}} H(Y)$. □

Remark 4.5. The assumption $\underline{\alpha} > 0$ says that $\mu_{b,1} \neq \mu_{t,0}$. In other terms, the answer to question q should be informative. Moreover, this quantity $\underline{\alpha}$ is connected to some measure of randomness of the answer model introduced in [7].

4.2 Theoretical lower bound for the performance

In [10], we propose in a general framework a lower bound on the performance of exact retrieval strategies on simple retrieval spaces. We want here to recall briefly theses results in the case of our particular framework for a comparison with our upper bound.

In the case of i.i.d. user model P on a simple tree structured retrieval space, let us define

$$\overline{\alpha} = \sup_{b \in T \setminus \partial T} \sup_{\nu \in [0,1]} h\left(\nu \mu_{b,0} + (1-\nu)\mu_{b,1}\right) - \left(\nu h(\mu_{b,0}) + (1-\nu) h(\mu_{b,1})\right). \tag{20}$$

Since for any $b \in T \setminus \partial T$, we have $h\left(\frac{1}{2}\mu_{b,0} + \frac{1}{2}\mu_{b,1}\right) - \left(\frac{1}{2}h(\mu_{b,0}) + \frac{1}{2}h(\mu_{b,1})\right) \leq 1$, we deduce that $\underline{\alpha} \leq \overline{\alpha}$. Then we deduce from Corollary 3.1. in [10].

Theorem 4.6. *Let $S = (Q, D)$ be an exact retrieval strategy for P. Then we have $E(\tau_S) \geq W_{N,\overline{\alpha}}(H(Y)) - \frac{1}{N}$ with $W_{N,\alpha}(H) = \inf_{t \leq H} V_N(H - t\alpha) + t$ and $V_N(H') = \sup\{\sum_{i=1}^{N-1} i m_i + (N-1) m_N \mid m \in \mathcal{M}_1(\{1, \cdots, N\}),\ h(m) = H'\ \}$ where $\mathcal{M}_1(\{1, \cdots, N\})$ denotes the set of probability distribution on $\{1, \cdots, N\}$ and $h(m)$ is the entropy of m.*

Useful properties of V_N and $W_{N,\alpha}$ have been given in [10]. Moreover, we have the inequality $W_{N,\alpha}(H(Y)) - \frac{1}{N} \geq \frac{H(Y) - \epsilon(\alpha)}{\alpha}$ where $\epsilon(\alpha)$ is bounded and tends to 0 when $\alpha \to +\infty$ (uniformly in $H(Y)$). Thus, we get for the entropy reduction strategy $S = (Q, D)$ built on a i.i.d. user model P, that

$$\frac{H(Y) - \epsilon(\overline{\alpha})}{\overline{\alpha}} \leq E(\tau_S) \leq \frac{(p+1)H(Y)}{2\underline{\alpha}} + 1. \tag{21}$$

We can go further in the simple case where for any $b \in T \setminus \partial T$, we have $\mu_{b,1} = B_\rho$, $\mu_{b,0} = B_{1-\rho}$ and for $t \in \partial T$ we have $\mu_{t,1} = B_1$, $\mu_{t,0} = B_0$. Assume moreover that T is a complete binary tree. Then, in that case we have $p = 2$ and

$$\alpha(\rho) \doteq \underline{\alpha}(\rho) = \overline{\alpha}(\rho) = 1 + (\rho \log(\rho) + (1-\rho)\log(1-\rho)), \tag{22}$$

so that for the entropy reduction retrieval strategy we have

$$\frac{H(Y) - \epsilon(\alpha(\rho))}{\alpha(\rho)} \leq E(\tau_S) \leq \frac{3H(Y)}{2\alpha(\rho)} + 1. \tag{23}$$

Hence we see that our bound given in theorem 4.2 is quite tight in this configuration and that the entropy reduction strategy is not far from optimal performance.

4.3 Experimental evaluation

The upper bound given in Theorem 4.2 shows the performance of the algorithm, in terms of the significance $\underline{\alpha}$ of the user's answer and the entropy of the prior on the target. We want to verify it in a simple case on simulations. For that, we consider the situation described previously, where there exists $\rho \in [0,1]$ such that for any $b \in T \setminus \partial T$, we have $\mu_{b,1} = B_\rho$, $\mu_{b,0} = B_{1-\rho}$. For the leaves, the answer is deterministic i.e. we have $\mu_{t,1} = B_1$, $\mu_{t,0} = B_0$. We consider a complete binary tree with various depths d ($|\mathcal{B}| = 2^d$). Moreover, we assume a uniform prior on Y ($H(Y) = d$). As previously, we denote by $\alpha(\rho)$ the quantity defined by (22).

In our first experiment, we check the linear behavior of $E(\tau_S)$ has a function of the log size of the database for various values of the significance $\underline{\alpha}$ (see Fig. 1).

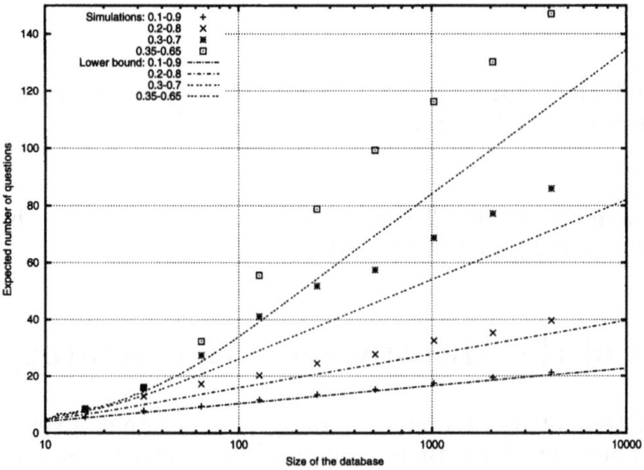

Figure 1: Average value (1000 simulations) of $E(\tau_S)$ versus $|B|$ for various values of the significance α: $(1-\rho, \rho) \in \{(0.1, 0.9), (0.2, 0.8), (0.3, 0.7), (0.35, 0.65)\}$. With lines, the corresponding lower bound given by Theorem 4.6

.

The linear dependence versus the entropy of the prior is particularly obvious in our experiment. Moreover, we see that the slope is increasing with decreasing values of the significance $\underline{\alpha}$. Note also that as the performance is quite close to the lower bound so that the entropy reduction strategy is not far from optimality in that case and seems particularly suited to tree structured retrieval spaces.

To check the accuracy of the functional form of our upper bound, we display in figure 2, the value of $C = \frac{\alpha(E(\tau_S)-1)}{H(Y)}$ which should be almost constant (the upper-bound say that this constant is upper bounded by 3/2).

Figure 2 shows that $C = \frac{\alpha(E(\tau_S)-1)}{H(Y)}$ is quite stable as soon as the database is not too small (of the order of several hundreds). Indeed, for $\rho = 0.65$ one has

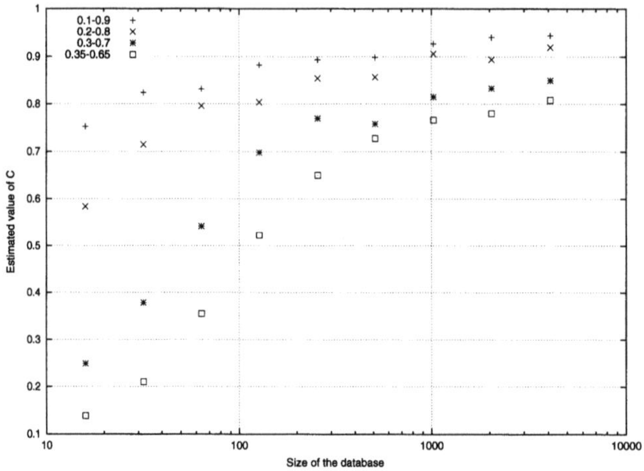

Figure 2: Average value (1000 simulations) of $\alpha(E(\tau_S)-1)/H(Y)$ versus $|\mathcal{B}|$ for various values of α: $(1-\rho,\rho) \in \{(0.1,0.9),(0.2,0.8),(0.3,0.7),(0.35,0.65)\}$.

$\alpha = 0.0659$ and for $\rho = 0.9$ one has $\alpha = 0.531$ hence $\alpha(0.9)/\alpha(0.65) = 8.05$. In the same time, we get from the figure 2 that $C(0.9)/C(0.65) \simeq 1.2$.

5 A fast version of the entropy reduction strategy

With the usual version of the entropy reduction strategy, we get that given a binary tree T, the expected number of questions is bounded by $\frac{3}{2\alpha}\log(|\mathcal{B}|)$ so that the searching time is $O(\log(|\mathcal{B}|))$. However, between questions, the algorithm has to scan the tree T for the question maximizing the mutual information. From remarks 3.4, we deduce that the complexity for the system per asked question is $O(|\mathcal{B}|\log(|\mathcal{B}|))$ so that the expected total complexity of the retrieval process for the system is $O(|\mathcal{B}|\log(|\mathcal{B}|)^2)$. Even if for the user the "burden" is not too high, the complexity for the system can be quite important for a large database or a large number of simultaneous users.

5.1 Definition and theoretical performance

We propose in this section to select a restricted subset of the database to be scanned whose size will be much smaller than the size of the database. This idea is quite natural and a usual solution is to randomly select a restricted subset $\mathcal{S}$ (via uniform sampling) of size $\gamma|\mathcal{B}|$ with $0 < \gamma < |B|$ and to scan $\mathcal{S}$ for for the best question in terms of uncertainty reduction. For this choice however, we can not prove that the overall complexity will decrease.

Our proposal is based on the selection of a scan set of size $\log(|\mathcal{B}|)$ and allows a provable bound for the overall complexity. This improved strategy is described as following (P_n denotes the conditional law of Y given $\mathcal{F}_n$):

Entropy reduction strategy (Fast Partial Scan):
`Assume that` $\tau_S \geq n$ `and let` $\mathcal{H} = \mathcal{H}_n$.

1. `Select recursively` $b_0 = \lambda$ `and` $b_{i+1} \in \text{argmax}_{b \in C_{b_i}} P_n(Y \in \partial T_b)$ `for any` $0 \leq i < r$. `Let` $\eta_n = \max_{1 \leq i \leq r} I_{\mathcal{H}}(Y, X_{b_i, n_{b_i}+1})$ $(n_b = \sum_{q \in \mathcal{H}} \mathbf{1}_{b_q = b})$.

2. `If` $\eta_n = 0$ `then` $\tau_S = n$ `and` $D = b_r = Y$. `Otherwise,` $\tau_S \geq n+1$ `and` $Q_{n+1} = (b_{i^*}, n_{b_{i^*}} + 1)$ `where` $i^* = \max\{1 \leq i \leq r |\ I_{\mathcal{H}}(Y, X_{b_i, n_{b_i}+1}) = \eta_n\ \}$.

With the fast scan version, the scanning set is of the order of the depth of the tree T. So for a balanced tree, it is of the order of $\log(|\mathcal{B}|)$ as announced. Moreover:

Theorem 5.1. *Theorem 4.2 is true for the fast partial scan version.*

Proof: The proof of this theorem is very similar to the proof of theorem 4.2 and even simpler. First lemma 4.3 still holds and we get that if $0 \leq \eta_n < 2\underline{\alpha}/(p+1)$, then for $t_n = Q_{n+1}$ we have $P_n(Y = t_n) \geq 1 - \eta_n/(2\underline{\alpha})$. Thus, for $0 \leq \eta_n < 2\underline{\alpha}/(p+1)$, $\tau_S > n+1$ implies that $Y \neq t_n$ and $P(\tau_S > n+1 \mid \mathcal{F}_n\ ) \leq P_n(Y \neq t_n) \leq \eta_n/(2\underline{\alpha})$. The proof is ended as for theorem 4.2. □

5.2 Experimental evaluation

We get from the previous theorem the same upper bound as in the case of the complete scan for an overall complexity of $O(\log(|\mathcal{B}|)^3)$ instead of the $O(|\mathcal{B}|\log(|\mathcal{B}|)^2)$ previously obtained in the complete scan version.

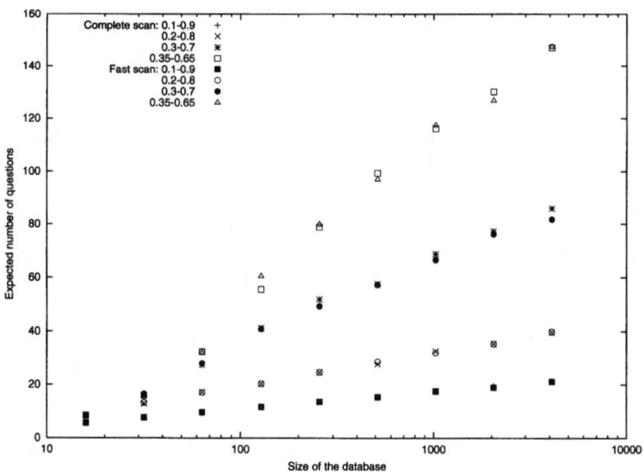

Figure 3: Complete/partial scan comparison. Average value (1000 simulations) of $E(\tau_S)$ versus $|\mathcal{B}|$ for various values of the significance: $(1-\rho, \rho) \in \{(0.1, 0.9), (0.2, 0.8), (0.3, 0.7), (0.35, 0.65)\}$.

We want to test here on simulation that the performances in terms of the expected number of questions are still the same for the fast partial scan as for the complete scan. We use the same framework where $\mu_{b,1} = B_\rho$ and $\mu_{b,0} = B_{1-\rho}$ for $b \in T \setminus \partial T$. The experiments are reported in figure 3. The most important fact is that the performances are nearly identical for all the tested values of α. This shows that our fast scan version of the entropy reduction strategy allows a considerable speed-up in the CPU time without any degradation in the performance.

6 Conclusion

It seems that, despite there is no global optimization of the number of queries from the system, the performance are not far from the optimum for conditional i.i.d user model. Hence the limitations of this greedy design, pointed out in [6] in the context of model based classification trees, does not seem to play an important role here. However, this approach needs to be tested on a real situation. The main problem will be to know what can be the value of α on a real situation. Using our toy ρ-model it seems that for a database of the order 10^5 to 10^6, one should have $\alpha \simeq 0.7$ corresponding to $\rho = 0.95$ which looks pretty high if we want an expected number of questions smaller than 20. This implies quite good trees or good summaries at each node or also patient users! Moreover, the performance of the retrieval strategy based on entropy reduction is not well understood when the user behavior is different from the user model P used to build the retrieval strategy i.e. when there is an important de-synchronization as introduced in [7]. This shows that more experimental and theoretical investigations are still needed.

References

[1] Y. Amit, D. Geman, and B. Jedynak. Efficient focusing and face detection. In H. Wechsler and J. Phillips, editors, *Face Recognition: From Theory to Applications*, NATO ASI Series G. Springer Verlag, 1998.

[2] J. Buhmann. Data clustering and learning. In *The Hand-book of Brain Theory and Neural Networks*, pages 278–281. MIT Press, 1995.

[3] I. Cox, M. Miller, T. Minka, T. Papathomas, and P. Yianilos. The bayesian image retrieval system, pichunter: Theory, implementation and psychophysical experiments. *IEEE Trans. Image Processing*, 9:20–37, 2000.

[4] I. Cox, M. Miller, T. Minka, and P. Yianilos. An optimized interaction strategy for bayesian relevance feedback. In *IEEE Conf. on Comp. Vis. and Pattern Recognition*, pages 553–558, 1998.

[5] F. Fleuret and D. Geman. Coarse-to-fine visual selection. *International Journal of Computer Vision*, 2001. To appear.

[6] D. Geman and B. Jedynak. Model-based classification trees. To appear in IEEE Information Theory, 2000.

[7] D. Geman and R. Moquet. Q & A models for interactive search. Preprint, December, 2000.

[8] M. S. Pinsker. *Information and Information Stability of Random Variables and Processes.* San-Fransisco: Holden Day, 1964.

[9] A. Smeulders, M. Worring, S. Santini, A. Gupta, and R. Jain. Content-based image retrieval at the end of the early years. *IEEE Trans. PAMI*, 22:1348–1375, 2000.

[10] A. Trouvé and Y. Yu. A lower bound on performance of exact retrieval strategies on simple retrieval spaces. Technical Report 2002-07, Université Paris 13, 2002.

[11] A. Trouvé and Y. Yu. Metric similarities learning through examples: An application to shape retieval. In *Proceeding of the Third International Workshop, EMMCVPR'01, Sophia Antipolis, September 3-5 2001*, Lecture Notes in Computer Science. Springer Verlag, 2001.

[12] A. Trouvé and Y. Yu. Unsupervised clustering trees by non-linear principal component analysis. *Pattern Recognition and Image Analysis*, II:108–112, 2001.

Alain Trouvé
Université Paris 13, LAGA et L2TI
Av. J.-B. Clément
93430 Villetaneuse, France
trouve@math.univ-paris13.fr

Yong Yu
ENST, Département TSI
46 rue Barrault
75013 Paris - FRANCE
yuyong@tsi.enst.fr

Trends in Mathematics, © 2002 Birkhäuser Verlag Basel/Switzerland

Zero-One Law Characterizations of ε_0

Andreas Weiermann

ABSTRACT: *Using methods and results from finite model theory and real analysis it is shown that the ordinal ε_0 can be characterized as the first additive principal number for which certain zero-one laws for infinitary structures do not hold. As a contribution to problem 4.17 and problem 10.6 in Burris 2001 [5] we show that additive principal numbers below ε_0 yield additive number systems in* RT_1 *and multiplicative number systems in* RV_0.

1 Introduction and motivation

In this paper we characterize via analytic combinatorics the segment of ordinals below ε_0 in terms of zero one laws. Since this goal is based on a somewhat unexpected interplay between set theory, in particular the theory of ordinals, finite model theory and analytic combinatorics it might be useful to provide in the introduction some informal explanations about the applicability of analytic combinatorics to ordinals and finite model theory.

1.1 Ordinals below ε_0 and analytic combinatorics

Ordinals appear naively in the process of counting into the transfinite. Starting from $0, 1, 2, \ldots$ we arrive at the first level of infinity and we declare ω to be the first limit number. Then we start counting from ω onwards and get a sequence $\omega, \omega+1, \omega+2, \ldots$ and we arrive at the second limit point $\omega+\omega =: \omega\cdot 2$ of this process. Clearly this can be iterated and after n steps we have arrived at the n-th limit point $\omega\cdot n$. After completing this counting process with respect to n we reach the first limit point of limit points $\omega\cdot\omega = \omega^2$. By a suitable iteration we obtain for each n the n-iterated limit point ω^n.

Up to now nothing special has happened since we can identify these ordinals with nth-tuples of natural numbers ordered with respect to the lexicographic ordering. If we iterate with respect to n again we reach ω^ω the limit of all ω^n. This process of counting can be iterated further and further and one can imagine a process of counting up to $\omega^{\omega^\omega}, \omega^{\omega^{\omega^\omega}}$ etcetera. Let $\omega_1 := \omega$ and $\omega_{n+1} := \omega^{\omega_n}$. Then ε_0 appears as the limit point of the ω_n. In set theory this intuitive approach can be formalized rigorously using the principle of transfinite recursion and of course there is no reason to stop at ε_0 but we will do so in this paper.

At this stage it does not seem plausible that analytic combinatorics can be applied to the structure of these ordinals. But using the Cantor normal form theorem from set theory it can easily been seen that each ordinal below ε_0 can be denoted uniquely by a term over a certain finite signature. Thus combinatorial enumeration applies to ordinals via counting the set of corresponding denotations. We will not make this approach more precise. Instead we will directly write down an ordering on the natural numbers such that the natural numbers represent ε_0 with respect to $\prec$ as the n-tupels of natural numbers represent ω^n with respect to the lexicographic ordering. The advantage of this approach is that one can work with these denotations for ordinals without prior knowledge of set theory. Let $(pn_i)_{i\geq 1}$ be the standard enumeration of the prime numbers and let $gcd(m,n)$ denote the greatest common divisor of m and n.

Let $m \prec n$ iff m is not equal to n and ($m = 1$ or $n = 0$ or $[\frac{m}{gcd(m,n)} = pn_{m_1} \cdot \ldots \cdot pn_{m_k}$ & $\frac{n}{gcd(m,n)} = pn_{n_1} \cdot \ldots \cdot pn_{n_l}$ & $\forall i \leq k \exists j \leq l(m_i \prec n_j)]$). With respect to $\prec$ the natural numbers start with $1 \prec pn_1^1 \prec pn_1^2 \prec \ldots \prec pn_1^n \prec \ldots \prec pn_2 \prec pn_2 \cdot pn_1 \prec \ldots \prec pn_2 \cdot pn_1^n \prec \ldots \prec pn_2^2 \prec \ldots \prec pn_2^n \prec \ldots \prec pn_3 \prec \ldots 0$. Thus pn_2^n corresponds to ω^n. The top element 0 corresponds to ε_0 itself and we have that $\langle \{\alpha : \alpha \leq \varepsilon_0\}, < \rangle$ is order isomorphic to $\langle \mathbb{N}, \prec \rangle$. If we let $q_1 := pn_2$ and $q_{n+1} := pn_{q_n}$ then the order type of q_n with respect to $\prec$ is just ω_n. Hence 0 is the limit of the q_n with respect to $\prec$ as ε_0 is the limit of the ω_n. A natural size function N on $\{\alpha : \alpha \leq \varepsilon_0\}$ can be introduced via isomorphism and the following size function $\tilde{N}$ defined on $\mathbb{N}$. $\tilde{N}(1) := 0$, $\tilde{N}(0) := 1$ and $\tilde{N}(pn_{m_1} \cdot \ldots \cdot pn_{m_k}) = k + \tilde{N}(m_1) + \cdots + \tilde{N}(m_k)$. For any $k \in \mathbb{N}$ let

$$\tilde{c}_k(n) = \#\{m \prec k : \tilde{N}(m) = n\}$$

and

$$\tilde{C}_k(x) = \#\{m \prec k : m \leq x\}.$$

As usual in combinatorics one might be interested in the asymptotics of $\tilde{c}_k$ and $\tilde{C}_k$ for various k. It turns out that this should be a quite interesting problem since $\tilde{c}_{q_2}(n) = p(n)$ is the partition function and thus $\tilde{c}_{q_2}(n) \sim \frac{\exp(\sqrt{\frac{2}{3}n})}{4\sqrt{3}n}$ by the classical Hardy Ramanujan result. Further we have $\ln \tilde{c}_{q_k}(n) \sim \frac{\pi^2}{6} \frac{n}{\ln_{k-2}(n)}$ for $k \geq 3$ by a result of Yamashita and, since c_0 is the tree enumeration function, $c_0(n) \sim D \cdot \frac{\alpha^n}{\sqrt{n^3}}$, where α is Otter's tree constant.

For certain specific further values of k we have obtained bounds for the asymptotics of $\tilde{c}_k$ and $\tilde{C}_k$ but a general classification seems to be difficult. In the appendix we include an asymptotic estimate for $\tilde{C}_5$. We conjecture that the methdods of the proof extend to the case $\tilde{C}_{q_k}$ for $k \geq 3$.

The analytical main results of this paper are as follows.

1. If $k \neq 0$ then $\tilde{c}_k$ satisfies $\lim_{n\to\infty} \frac{\tilde{c}_k(n-1)}{\tilde{c}_k(n)} = 1$.

2. If $k \neq 0$ then $\tilde{C}_k$ satisfies $\lim_{t\to\infty} \frac{\tilde{C}_k(tx)}{\tilde{C}_k(t)} = 1$.

Thus, for $k \neq 0$, $\sum_{n=0}^{\infty} \tilde{c}_k(n) z^n$ has radius of convergence 1 by the ratio test, which in Burris' book [5] is denoted by $\tilde{c}_k \in RT_1$. Moreover $\tilde{C}_k$ is slowly varying at infinity in the sense of Karamata. In Burris' book this property of $\tilde{C}_k$ is denoted by $\tilde{C}_k \in RV_0$. This is justified by the fact that slowly varying at infinity is the same as having regular variation of index 0. As a corollary of this purely analytical result and the theory developed by Compton we obtain the desired characterization of ε_0 in terms of zero one laws.

1.2 Logical limit laws and analytic combinatorics

The investigations on logical limit laws for the ordinal segment below ε_0 is inspired from related problems in finite model theory and in particular by the pioneering work of Compton in this area. For an exposition on general finite model theory we refer to the text-book by Ebbinghaus and Flum [8]. In this paper we concentrate only on a specific aspect, i.e. logical limit laws, of finite model theory. A fundamental and basic question of finite model theory is as follows: What is the probability that a given property holds in a randomly chosen large finite structure? This rather vague formulation can be put into mathematical terms as follows. Assume that we have given a class $\mathcal{K}$ of finite structures, such that for every natural number n there are at most finitely many members in $\mathcal{K}$ which have a universe of cardinality n. For a given property $\mathcal{P}$ let r_n be the proportion of structures in $\mathcal{K}$ of size n that satisfy $\mathcal{P}$. As the probability that $\mathcal{P}$ holds for a randomly chosen structure of large size from $\mathcal{K}$ we consider the limiting distribution $\lim_{n\to\infty} r_n$ if this limit exists. This limit is called the labeled asymptotic density of $\mathcal{P}$. Alternatively we may as well consider s_n, the proportion of all isomorphism types of structures in $\mathcal{K}$ of size n that satisfy $\mathcal{P}$. Here $\lim_{n\to\infty} s_n$ is called the unlabeled asymptotic density of $\mathcal{P}$. In both cases it is then an obvious problem to figure out for which classes of structures and which classes of properties the labeled and unlabeled asymptotic probabilities exist.

Classical results of Fagin [9] and Glebskij et al. [11] show that if the underlying first order language L contains only relation symbols but no function symbols and no constants then the labeled and unlabeled asymptotic densities exist for properties expressible by L-formulas. Moreover these densities are either 0 or 1 and therefore a zero one law holds for first order logic. Moreover their work show that zero one laws hold in the labeled and unlabeled case for the class of finite (as well as for directed as for undirected) graphs.

In pushing these results further to very general classes of finite structures Compton applied methods and results from analytic combinatorics to prove logical limit laws i.e. the existence of asymptotic densities properties expressed by formulas of languages of first order and even of monadic second order logic [6, 7]. For a given language L he considered adequat classes $\mathcal{K}$ of structures which means that

1. $\mathcal{K}$ is closed under disjoint unions,
2. members of $\mathcal{K}$ with non empty universe can be uniquely decomposed into (a disjoint union of) $\mathcal{K}$ indecomposable structures and
3. the L-structure with empty domain is in $\mathcal{K}$.

For adequat classes $\mathcal{K}$ he was able to reduce the limit law question to purely analytical problems on the count function for $\mathcal{K}$. In the unlabeled case, on which we concentrate from now on, the count function $c_{\mathcal{K}}$ for $\mathcal{K}$ is defined as follows: $c_{\mathcal{K}}(n)$ is the number of isomorphism types of structures in $\mathcal{K}$ which have a universe of cardinality n. Then, according to Compton's results, the condition $\lim_{n\to\infty} \frac{\mathcal{K}(n-1)}{\mathcal{K}(n)} = 1$ is sufficient to yield a zero one law for $\mathcal{K}$.

The underlying idea is as follows. If $\mathcal{K}$ is adequate then the isomorphism types of $\mathcal{K}$ give rise to an additive number system (or additive arithmetical semigroup) by considering the (isomorphism types of) indecomposables as indecomposable elements and by interpreting disjoint union as addition. The norm of a finite structure is given by the size of its universe. The condition $\lim_{n\to\infty} \frac{c_{\mathcal{K}}(n-1)}{c_{\mathcal{K}}(n)} = 1$ is sufficient to yield that every partition set (in the sense of [5]) has asymptotic density 0 or 1. Using Ehrenfeucht-Fraïssé games one can then show that for any first order (monadic second order) sentence ϕ the set of isomorphism types of members $K \in \mathcal{K}$ in which ϕ holds is a finite union of disjoint partition sets and thus ϕ has unlabeled asymptotic density 0 or 1.

A typical way for proving a condition like $\lim_{n\to\infty} \frac{c_{\mathcal{K}}(n-1)}{c_{\mathcal{K}}(n)} = 1$ is to consider the fundamental identity for the additive number system for $\mathcal{K}$. Let $c_{\mathcal{I}}(n)$ be the number of isomorphism types of indecomposables from $\mathcal{K}$ which have a universe of cardinality n. Then we have the following fundamental identity

$$\sum_{n=0}^{\infty} c_{\mathcal{K}}(n)z^n = \prod_{n=1}^{\infty} (1 - z^n)^{-c_{\mathcal{I}}(n)}$$

and extracting information on $c_{\mathcal{K}}$ from $c_{\mathcal{I}}$ and vice versa is a classical topic in analytic combinatorics. A similar phenomenon occurs when we consider appropriate $\mathcal{K}$ as multiplicative number systems but here we would like to refer to the literature [5].

In contrast to finite model theory we consider in this paper infinite structures but it turns out that for first order languages the general theory of Compton (as for example presented in Burris' book [5]) applies mutatis mutandis to this situation as well. The basic observation is that we can replace the size function of finite structures by a size function for the denotation of an infinite structure. The crucial fact for applying analytic combinatorics is that there are only finitely many structures which have a representation of a given finite size.

2 Basic definitions

Throughout the paper we denote (if not stated otherwise) ordinals less than $\varepsilon_0 = \min\{\xi : \xi = \omega^\xi\}$ by small Greek letters. If not stated otherwise small Latin numbers range over natural numbers and t ranges over reals. Disjoint union is denoted by $\uplus$ and isomorphism of structures is denoted by $\simeq$. We write $\alpha =_{NF} \omega^{\alpha_1} + \cdots + \alpha_n$ if $\alpha = \omega^{\alpha_1} + \cdots + \alpha_n > \alpha_1 \geq \ldots \geq \alpha_n$. (This means α is in normal form. The existence of this representation follows from the Cantor normal form theorem.) The (commutative) natural sum of ordinals $\oplus$ is defined as usual, i.e. for $\alpha =_{NF} \omega^{\alpha_1} + \cdots + \omega^{\alpha_m}$ and $\beta =_{NF} \omega^{\alpha_{m+1}} + \cdots + \omega^{\alpha_{m+n}}$ we have $\alpha \oplus \beta = \omega^{\alpha_{p(1)}} + \cdots + \omega^{\alpha_{p(m+n)}}$ where p is a permutation of $\{1, \ldots, m+n\}$ such that $\alpha_{p(1)} \geq \cdots \geq \alpha_{p(m+n)}$.

We define the (additive) norm $N\alpha$ (or length) of an ordinal α as follows.

$$N0 := 0$$

and

$$N\alpha := n + N\alpha_1 + \ldots + N\alpha_n$$

if $\alpha =_{NF} \omega^{\alpha_1} + \cdots + \omega^{\alpha_n}$. Let $pn_1, pn_2, \ldots$ be the enumeration of prime numbers. Inspired by Schütte's (1977) [14] Gödel numbering of ordinals below Γ_0 we define the (multiplicative) norm $M\alpha$ (or Gödel number) of an ordinal α as follows.

$$M0 := 1$$

and

$$M\alpha := pn_{M\alpha_1} \cdot \ldots \cdot pn_{M\alpha_n}$$

if $\alpha =_{NF} \omega^{\alpha_1} + \cdots + \omega^{\alpha_n}$.

Then $\{\alpha : N\alpha \leq n\}$ and $\{\alpha : M\alpha \leq n\}$ are always finite for any n. We equip any ordinal α with a relation r_α as follows. $r_0 := \emptyset$ and if $\alpha =_{NF} \omega^{\alpha_1} + \cdots + \omega^{\alpha_n}$ then we put for $\beta, \gamma < \alpha$: $\beta r_\alpha \gamma$ if there are $m < n$ and $\xi, \eta < \omega^{\alpha_{m+1}}$ with $\beta = \omega^{\alpha_1} + \cdots + \omega^{\alpha_m} + \xi$ and $\gamma = \omega^{\alpha_1} + \cdots + \omega^{\alpha_m} + \eta$ and $\xi < \eta$. Then $(\alpha, r_\alpha) \simeq (\omega^{\alpha_1}, <) \uplus \ldots \uplus (\omega^{\alpha_n}, <)$. With $\mathcal{K}_\beta$ we denote for $\beta \leq \varepsilon_0$ the class of structures $\{(\alpha, r_\alpha) : \alpha < \omega^\beta\}$. Then $\mathcal{K}_\beta$ is closed under finite disjoint unions since $(\gamma, r_\gamma) \uplus (\delta, r_\delta) \simeq (\gamma \oplus \delta, r_{\gamma \oplus \delta})$. Further the empty structure is in $\mathcal{K}_\beta$. Moreover elements of $\mathcal{K}_\beta$ with nonempty universe can be decomposed uniquely into a disjoint union of $\mathcal{K}_\beta$ indecomposable structures. Here a structure (α, r_α) is called $\mathcal{K}_\beta$ indecomposable if it is not the finite disjoint union of structures (α_i, r_{α_i}) with $\alpha_i < \alpha$. Of course then α has to be additive principal. Moreover if $(\alpha, r_\alpha) \simeq (\beta, r_\beta)$ then $\alpha = \beta$. This rigidity property leads in our situation to the fact that the labeled and unlabeled densities coincide.

We denote the cardinality of a finite set M by $\#M$. Let $\mathcal{L}_<$ be the language of orders which contains a binary relation symbol $<$ as only non logical symbol. For an $\mathcal{L}_<$ symbol ϕ we write $\alpha \models \phi$ if $(\alpha, r_\alpha) \models \phi$ where r_α is the interpretation of $<$. Following the terminology of Burris' book [5] we say that a subset $A \subseteq \beta$ has an additive (local) density $\delta_\beta(A)$ if

$$\delta_\beta(A) := \lim_{n \to \infty} \frac{\#\{\alpha \in A : N\alpha = n\}}{\#\{\alpha < \beta : N\alpha = n\}}$$

exists and a multiplicative (global) density $\Delta_\beta(A)$ if

$$\Delta_\beta(A) := \lim_{n \to \infty} \frac{\#\{\alpha \in A : M\alpha \leq n\}}{\#\{\alpha < \beta : M\alpha \leq n\}}$$

exists.

We say that β satisfies an additive limit law for $\mathcal{L}_<$ if

$$\delta_\beta(\{\alpha < \beta : \alpha \models \phi\})$$

exists for any $\phi \in \mathcal{L}_<$ and if this is the case we say that β satisfies the additive zero-one law for $\mathcal{L}_<$ if

$$\delta_\beta(\{\alpha < \beta : \alpha \models \phi\}) \in \{0, 1\}$$

for any $\phi \in \mathcal{L}_<$. We say that β satisfies a multiplicative limit law for $\mathcal{L}_<$ if

$$\Delta_\beta(\{\alpha < \beta : \alpha \models \phi\})$$

exists for any $\phi \in \mathcal{L}_<$ and if this is the case we say that β satisfies the multiplicative zero-one law for $\mathcal{L}_<$ if

$$\Delta_\beta(\{\alpha < \beta : \alpha \models \phi\}) \in \{0, 1\}$$

for any $\phi \in \mathcal{L}_<$.
Our aim is to show that any additive principal number $\beta < \varepsilon_0$ satisfies the additive and multiplicative zero-one laws for $\mathcal{L}_<$ but ε_0 does not. For this purpose it proves useful to introduce machinery used in [5] for proving zero-one laws in finite model theory. For a function $f : [0,\infty[\to [0,\infty[\cap\mathbb{N}$ say $f \in RT_1$ if $\lim_{n\to\infty} \frac{f(n-1)}{f(n)} = 1$ and $f \in RV_0$ if $\lim_{t\to\infty} \frac{f(tx)}{f(t)} = 1$ for any real $x > 0$. Let

$$c_\beta(n) := \#\{\alpha < \beta : N\alpha = n\}$$

and

$$C_\beta(t) := \#\{\alpha < \beta : M\alpha \leq t\}.$$

For proving that $c_\beta \in RT_1$ and $C_\beta \in RV_0$ we relate the concept of reduced additive number systems and multiplicative number systems to additive principal numbers. Given $\beta \leq \varepsilon_0$ let

$$\mathcal{A}_\beta := \{\alpha : \alpha < \omega^\beta\}$$

and

$$\mathcal{P}_\beta := \{\omega^\gamma : \gamma < \beta\}.$$

Then $\mathcal{A}_\beta$ is an additive number system with respect to $\oplus$ (when restricted to $\mathcal{A}_\beta$) and 0. Any element of $\mathcal{A}_\beta$ can be written (modulo commutativity) uniquely as a natural sum of elements from $\mathcal{P}_\beta$. Furthermore let

$$a_\beta(n) := c_{\omega^\beta}(n)$$

and

$$p_\beta(n) := \#\{\omega^\gamma < \omega^\beta : N\omega^\gamma = n\}$$

Then the structure $(\mathcal{A}_\beta, \mathcal{P}_\beta, \oplus \restriction \mathcal{A}_\beta, 0, N)$ is a reduced additive number system with local counting functions a_β and p_β. See, for example, Burris [5] for a definition. Moreover let

$$A_\beta(n) := C_{\omega^\beta}(n)$$

and

$$P_\beta(n) := \#\{\omega^\gamma < \omega^\beta : M\omega^\gamma \leq n\}$$

Then the structure $(\mathcal{A}_\beta, \mathcal{P}_\beta, \oplus \restriction \mathcal{A}_\beta, 0, M)$ is a multiplicative number system with global counting functions A_β and P_β when we consider $\oplus$ as multiplication with neutral element 0 considered as 1.

3 Proof of the additive results

Lemma 3.1. *If $\omega \leq \beta < \varepsilon_0$ then $c_\beta \in \mathrm{RT}_1$.*

Proof. By induction on β. If $\beta = \omega^k$ then $\mathcal{P}_k = \{\omega^0, \ldots, \omega^{k-1}\}$ consists of k indecomposables. Hence $a_k \in RT_1$ by Theorem 2.46 in Burris [5] or by Theorem 1.3 in Bell [2]. Assume now that $\beta = \omega^\gamma$ where $\omega \leq \gamma$. The induction hypothesis yields $c_\gamma \in RT_1$. We have

$$p_\gamma(n) = \#\{\omega^\delta < \omega^\gamma : N\omega^\delta = n\} = c_\gamma(n-1).$$

Thus $p_\gamma \in RT_1$, hence $a_\gamma \in RT_1$ by Bell and Burris [4]. This gives $c_{\omega^\gamma} = a_\gamma \in RT_1$ for $0 \leq \gamma < \varepsilon_0$. Finally assume that $\beta =_{NF} \omega^\gamma \cdot m + \omega^{\delta_1} + \ldots + \omega^{\delta_k}$ with $m + k > 1$, $\gamma > \delta_1 \geq \ldots \geq \delta_k$ and $m > 0$. The induction hypothesis yields $c_{\omega^\gamma}, c_{\omega^{\delta_i}} \in RT_1$ for $1 \leq i \leq k$. Then $c_{\omega^\gamma}(n) \geq c_{\omega^{\delta_1+1}}(n)$. Let us consider the additive number system provided by $\mathcal{A}_{\delta_1+1}$. This is built from the system for $\mathcal{A}_{\delta_1}$ by adding a new indecomposable ω^{δ_1} to $\mathcal{P}_{\delta_1}$ and the norm function satisfies $N(\omega^{\delta_1} \cdot m + \alpha) = m \cdot N\omega^{\delta_1} + N\alpha$ for any $\alpha \in \mathcal{A}_{\delta_1}$. The proof of Lemma 3.58 in Burris [5] (confer claim 2 or line 7 on page 72) yields

$$\lim_{n\to\infty} \frac{c_{\omega^{\delta_1}}(n)}{c_{\omega^{\delta_1+1}}(n)} = 1 - 1^{N\omega^{\delta_1+1}} = 0. \tag{1}$$

Hence $\lim_{n\to\infty} \frac{c_{\omega^{\delta_i}}(n)}{c_{\omega^\gamma}(n)} = 0$ for $1 \leq i \leq k$. In particular

$$\lim_{n\to\infty} \frac{c_{\omega^{\delta_i}}(n-l)}{c_{\omega^\gamma}(n)} = 0 \tag{2}$$

for $1 \leq i \leq k$ and any fixed l. Moreover $c_{\omega^\gamma} \in RT_1$ yields $\lim_{n\to\infty} \frac{c_{\omega^\gamma}(n-l)}{c_{\omega^\gamma}(n)} = 1$ for any fixed l. We have

$$\begin{aligned} c_\beta(n-1) &= c_{\omega^\gamma}(n-1) + \ldots + c_{\omega^\gamma}(n-1-N(\omega^\gamma \cdot (m-1))) \\ &\quad + c_{\omega^{\delta_1}}(n-1-N(\omega^\gamma \cdot m)) \\ &\quad + \ldots + c_{\omega^{\delta_k}}(n-1-N(\omega^\gamma \cdot m + \omega^{\delta_1} + \cdots + \omega^{\delta_{k-1}})). \end{aligned}$$

The equations (1) and (2) yield $\lim_{n\to\infty} \frac{c_\beta(n-1)}{c_{\omega^\gamma}(n)} = m$ and similarly we obtain $\lim_{n\to\infty} \frac{c_\beta(n)}{c_{\omega^\gamma}(n)} = m$, hence $\lim_{n\to\infty} \frac{c_\beta(n-1)}{c_\beta(n)} = 1$.

Theorem 3.2. *If* $\beta < \varepsilon_0$ *then* ω^β *satisfies the additive zero-one law for* $\mathcal{L}_<$.

Proof. This follows from Lemma 3.1 and the proof of theorem 6.29 in Burris [5].

Theorem 3.3. *The ordinal* ε_0 *does not satisfy the additive zero-one law for* $\mathcal{L}_<$.

Proof. Let $\phi := \exists x \forall y (y \neq x \implies \neg y < x \wedge \neg x < y)$ Then $(\alpha, r_\alpha) \models \phi$ iff α is a successor ordinal. We claim that $\delta_{\varepsilon_0}(\{\alpha < \varepsilon_0 : \alpha \models \phi\})$ is the radius of convergence ρ of the tree generating function. By Otter 1948 [13] we know that $\frac{1}{\rho} = 2.95576\ldots$, hence $0 < \rho < 1$. Further [13] yields that there exists a constant D such that $c_{\varepsilon_0}(n) \sim \frac{D}{\rho^n \sqrt{n^3}}$ for $n \to \infty$. Hence $\lim_{n\to\infty} \frac{c_{\varepsilon_0}(n-1)}{c_{\varepsilon_0}(n)} = \rho$. Now the result follows from observing that $\#\{\alpha < \varepsilon_0 : (\exists\beta)[\alpha = \beta + 1] \;\&\; N\alpha = n\} = c_{\varepsilon_0}(n-1)$. Similarly we can use an $\mathcal{L}_<$ sentence ψ (with the same limit density) describing that the universe is linearly ordered.

4 Proof of the multiplicative results

Lemma 4.1. *If* $\omega \leq \beta < \varepsilon_0$ *then* $C_\beta \in \mathrm{RV}_0$.

Proof. By induction on β. If $\beta = \omega^k$ then $\mathcal{P}_k = \{\omega^0, \ldots, \omega^{k-1}\}$ consists of k many indecomposables. Hence $A_k \in RV_0$ by Theorem 8.30 in Burris [5]) or by Theorem 1.8 in Bell [2]. Assume now that $\beta = \omega^\gamma$ where $\omega \leq \gamma$. The induction hypothesis yields $C_\gamma \in RV_0$. We have $P_\gamma(t) = \#\{\omega^\delta < \omega^\gamma : M\omega^\delta \leq t\} = \#\{\delta < \gamma : pn_{M\delta} \leq t\}$. We claim that $P_\gamma \in RV_0$. Here we proceed with some elementary intermediate steps. Let $D_\gamma(t) = \#\{\delta < \gamma : M\delta \cdot \ln M\delta \leq t\}$. We first claim that

$$\lim_{t\to\infty} \frac{D_\gamma(t)}{C_\gamma(\frac{t}{\ln t})} = 1. \tag{3}$$

Proof of the claim: Assume $M\delta \leq \frac{t}{\ln t}$. Then $\ln M\delta \leq \ln t$ hence $M\delta \ln M\delta \leq \ln t M\delta \leq t$. This discussion shows

$$\frac{D_\gamma(t)}{C_\gamma(\frac{t}{\ln t})} \geq 1.$$

Further, for t large enough,

$$\frac{D_\gamma(t)}{C_\gamma(\frac{t}{\ln t})} \leq \frac{C_\gamma(2\frac{t}{\ln t})}{C_\gamma(\frac{t}{\ln t})}.$$

$C_\gamma \in RV_0$ yields

$$\lim_{t\to\infty} \frac{C_\gamma(2\frac{t}{\ln t})}{C_\gamma(\frac{t}{\ln t})} = 1.$$

Hence (3) follows. We further have for $x > 0$ that

$$\lim_{t\to\infty} \frac{C_\gamma(\frac{tx}{\ln(tx)})}{C_\gamma(\frac{t}{\ln t})}. \tag{4}$$

Proof of the claim: First note that

$$\frac{C_\gamma(\frac{tx}{\ln(tx)})}{C_\gamma(\frac{t}{\ln t})} \leq \frac{C_\gamma(\frac{t}{\ln t}x)}{C_\gamma(\frac{t}{\ln t})} \rightarrow_{t\to\infty} 1$$

since $C_\gamma \in RV_0$.
Further, for t large enough,

$$\frac{C_\gamma(\frac{tx}{\ln(tx)})}{C_\gamma(\frac{t}{\ln t})} \geq \frac{C_\gamma(\frac{1}{2}x\frac{t}{\ln t})}{C_\gamma(\frac{t}{\ln t})} \rightarrow_{t\to\infty} 1$$

since $C_\gamma \in RV_0$.

We now claim

$$D_\gamma \in RV_0. \tag{5}$$

Indeed

$$\frac{D_\gamma(tx)}{D_\gamma(t)} = \frac{D_\gamma(tx)}{C_\gamma(\frac{tx}{\ln(tx)})} \frac{C_\gamma(\frac{tx}{\ln(tx)})}{C_\gamma(\frac{t}{\ln t})} \frac{C_\gamma(\frac{t}{\ln t})}{D_\gamma(t)} \rightarrow_{t\rightarrow\infty} 1$$

by (3) and (4).
Next we claim that

$$\lim_{t\rightarrow\infty} \frac{P_\gamma(t)}{D_\gamma(t)} = 1. \tag{6}$$

Elementary number theory yields the existence of constants E and F such that $0 < E < F$ and

$$En\ln(n) \leq p_n \leq Fn\ln(n) \tag{7}$$

for all $n > 1$. (See, for example, [1] theorem 4.7, for a proof. The full strength of the classical prime number theorem is not required here.) The inequality (7) gives

$$\frac{D_\gamma(\frac{1}{F}t)}{D_\gamma(t)} \leq \frac{P_\gamma(t)}{D_\gamma(t)} \leq \frac{D_\gamma(\frac{1}{E}t)}{D_\gamma(t)}.$$

Thus (6) follows since $D_\gamma \in RV_0$ by (5). Now we can show that

$$P_\gamma \in RV_0. \tag{8}$$

Indeed we have

$$\lim_{t\rightarrow\infty} \frac{P_\gamma(tx)}{P_\gamma(t)} = \frac{P_\gamma(tx)}{D_\gamma(tx)} \frac{D_\gamma(tx)}{D_\gamma(t)} \frac{D_\gamma(t)}{P_\gamma(t)} \rightarrow_{t\rightarrow\infty} 1$$

by (5) and (6). Now $C_\beta = A_\gamma \in RV_0$ follows by Theorem 1 in Bell [3].
Finally assume that $\beta =_{NF} \omega^{\delta_1} + \ldots + \omega^{\delta_k}$ with $k > 1$. The induction hypothesis yields $C_{\omega^{\delta_i}} \in RV_0$ for $1 \leq i \leq k$. Hence $t \mapsto C_{\omega^{\delta_i}}(\frac{t}{pn_{M\delta_1} \cdot \ldots \cdot pn_{M\delta_{i-1}}}) \in RV_0$ for $1 \leq i \leq k$. We have

$$C_\beta(t) = C_{\omega^{\delta_1}}(t) + \cdots + C_{\omega^{\delta_k}}(\frac{t}{pn_{M\delta_1} \cdot \ldots \cdot pn_{M\delta_{k-1}}}).$$

Assertion (2) of Theorem 1.3.7 in Geluk and de Haan [10] yields $C_\beta \in RV_0$.

Theorem 4.2. *If $\beta < \varepsilon_0$ then ω^β satisfies the multiplicative zero-one law for $\mathcal{L}_<$.*

Proof. This follows from Lemma 2 and the proof of theorem 6.29 and theorem 10.2 in Burris [5].

Theorem 4.3. *The ordinal ε_0 does not satisfy the multiplicative zero-one law for $\mathcal{L}_<$.*

Proof. Again consider $\phi := \exists x \forall y (y \neq x \implies \neg y < x \wedge \neg x < y)$. Then $(\alpha, r_\alpha) \models \phi$ iff α is a successor ordinal. We claim that $\Delta_{\varepsilon_0}(\{\alpha < \varepsilon_0 : \alpha \models \phi\}) = \frac{1}{2}$. Indeed

$$\frac{\#\{\alpha < \varepsilon_0 : (\exists \beta)\alpha = \beta + 1 \;\&\; M\alpha \leq n\}}{\#\{\alpha < \varepsilon_0 : M\alpha \leq n\}} = \frac{\#\{m : 0 < m \leq \frac{n}{2}\}}{\#\{m : 0 < m \leq n\}} \rightarrow_{n \to \infty} \frac{1}{2}.$$

If we use an $\mathcal{L}_<$ sentence ψ (with the same limit density) describing that the universe is linearly ordered then its limit probability is 0 in contrast to the additive case.

Questions: Is it possible to replace in Theorems 1 and 3 the relations r_α by the more natural less than relation or other relations? Is it possible to give a complete asymptotics for c_β and C_β for $\beta \geq \omega^{\omega^\omega}$? Do general (additive and multiplicative) limit laws hold for ordinals above ε_0? The results in [15] indicate a positive answer to the first and last question in the additive case.

5 Appendix

5.1 An asymptotic bound for C_{ω^ω}

In this section we show via a Hardy Ramanujan style Tauberian argument from [12] the following theorem which seems to be of interest in its own since the additive variant c_{ω^ω} of C_{ω^ω} is the partition function.

Theorem 5.1.

$$C_{\omega^\omega}(n) = \exp((1 + o(1))\frac{\pi\sqrt{2}}{\sqrt{3\ln(2)}}\sqrt{\ln(n)}).$$

For complex z let $\Gamma_{\omega^\omega}(z) := \sum_{n=1}^{\infty} \gamma_{\omega^\omega}(n) n^{-z}$ where $\gamma_{\omega^\omega}(n) = 1$ if $n = gn(\alpha)$ for some $\alpha < \omega^\omega$ and $\gamma_{\omega^\omega}(n) = 0$ otherwise. Moreover let $\Pi_{\omega^\omega}(z) := \sum_{n=1}^{\infty} \pi_{\omega^\omega}(n) n^{-z}$ where $\pi_{\omega^\omega}(n) = 1$ if $n = gn(\omega^i)$ for some $i < \omega$ and $\pi_{\omega^\omega}(n) = 0$ otherwise. Then the fundamental identity (cf. [5] Theorem 8.13) yields

$$\Gamma_{\omega^\omega}(z) = \sum_{n=1}^{\infty} \gamma_{\omega^\omega}(n) n^{-z} = \prod_{i=2}^{\infty} (1 - n^{-z})^{-\pi_{\omega^\omega}(n)}$$

Moreover we have $\Pi_{\omega^\omega}(z) = \sum_{l=0}^{\infty} pn_{2^l}^{-z}$ and $C_{\omega^\omega}(n) = \sum_{i \leq n} \gamma_{\omega^\omega}(i)$.

Lemma 5.2.

$$\Pi_{\omega^\omega}(x) \sim \frac{1}{x \cdot \ln(2)}$$

for real $x \downarrow 0$.

Proof. Since $\lim_{x \downarrow 0} \int_0^{x \ln(2)} e^{-v} v^{-x} dv = 0$ we have by continutiy of the Γ function

$$\lim_{x \downarrow 0} \int_{x \ln(2)}^{\infty} e^{-v} v^{-x} dv = \Gamma(1) = 1. \tag{9}$$

Fix real numbers A, B with $A, B > 0$ such that $A \cdot l \cdot \ln(l) < p_l < B \cdot l \cdot \ln(l)$ for $l \geq 1$. Then for $x > 0$

$$\Pi_{\omega^\omega}(x) \leq \sum_{l=1}^{\infty} \frac{1}{(A 2^l l \ln(2))^x} \sim \frac{1}{\ln(2)x}$$

since

$$\int_1^\infty \frac{1}{(A2^y y \ln(2))^x} dy = \int_x^\infty \frac{1}{2^u (\frac{u}{x})^x x (\ln(2)A)^x} du = \frac{1}{x} (\frac{1}{x \ln(2) A})^x \int_x^\infty \frac{1}{e^{u \ln(2)} u^x} du$$
$$= \frac{1}{x} \frac{\ln(2)^x}{x^x \ln(2)^x A^x \ln(2)} \int_{x \ln(2)}^\infty e^{-v} v^{-x} dv \sim \frac{1}{\ln(2)x}.$$

by (9). Similarly we obtain

$$\Pi_{\omega^\omega}(x) \geq \sum_{l=1}^{\infty} \frac{1}{(B 2^l l \ln(2))^x} \geq \frac{1}{x^{x+1} \ln(2)^x B^x \ln(2)} \int_{x \ln(2)}^\infty e^{-v} v^{-x} dv$$
$$\sim \frac{1}{x} \frac{1}{x^x B^x \ln(2)} \sim \frac{1}{\ln(2)x}.$$

Lemma 5.3.

$$\Gamma_{\omega^\omega}(x) \sim \frac{\pi^2}{6 \ln(2)} \frac{1}{x}$$

Proof. Let $\epsilon > 0$. Then $\sum_{N<n} \frac{1}{n^2} < \epsilon$ for some large N. Prop. 8.22 in [5] yields

$$\ln(\Gamma_{\omega^\omega}(x)) = \sum_{n=1}^{\infty} \frac{1}{n} \Pi_{\omega^\omega}(nx) = \Phi_1(x) + \Phi_2(x) + \Phi_3(x) + \Phi_4(x)$$

where $\Phi_1(x) = \sum_{n \leq N} \frac{1}{n} \Pi_{\omega^\omega}(nx)$, $\Phi_2(x) = \sum_{N<n<\frac{1}{\sqrt{x}}} \frac{1}{n} \Pi_{\omega^\omega}(nx)$,
$\Phi_3(x) = \sum_{\frac{1}{\sqrt{x}} \leq n \leq \frac{1}{x}} \frac{1}{n} \Pi_{\omega^\omega}(nx)$, $\Phi_4(x) = \sum_{\frac{1}{x}<n} \frac{1}{n} \Pi_{\omega^\omega}(nx)$.
Ad Φ_1: We have for x small enough

$$\Phi_1(x) = \sum_{n \leq N} \frac{1}{n} \Pi_{\omega^\omega}(nx) \sim \sum_{n \leq N} \frac{1}{n} \frac{1}{nx \ln(2)}$$

since $\lim_{x \downarrow 0} nx \Pi_{\omega^\omega}(nx) = 1$. Moreover $\frac{1}{\ln(2)x}(\frac{\pi^2}{6} - \epsilon) \leq \sum_{n \leq N} \frac{1}{n} \frac{1}{nx \ln(2)} \leq \frac{1}{\ln(2)x} \frac{\pi^2}{6}$.
Ad Φ_2: For $N < n < \frac{1}{\sqrt{x}}$ we have $\Pi_{\omega^\omega}(nx) \leq \frac{K}{nx}$ for some K since for $x \downarrow 0$ we have $nx \downarrow 0$ and $\Pi_{\omega^\omega} \sim \frac{1}{x \ln(2)}$. Hence $0 < \Phi_2(x) = \sum_{N \leq n \leq \frac{1}{\sqrt{x}}} \frac{1}{n} \Pi_{\omega^\omega}(nx) \leq \sum_{N<n} \frac{K}{n^2 x} \leq \frac{K\epsilon}{x}$.
Ad $\Phi_3(x)$: There is a constant L such that for $\sqrt{x} \leq nx \leq 1$ we have $\Pi_{\omega^\omega}(nx) \leq \frac{L}{\sqrt{x}}$. Lemma 5.2 yields the existence of L with $\Pi_{\omega^\omega}(\sqrt{x}) \leq \frac{L}{\sqrt{x}}$ and the claim follows by monotonicity of Π_{ω^ω}. This gives for a certain L' independent of x

$0 < \Phi_3(x) \le \frac{L}{\sqrt{x}} \sum_{n=1}^{\frac{1}{x}} \frac{1}{n} \le L\frac{\ln(\frac{1}{x})}{\sqrt{x}} \le L'\frac{\sqrt[6]{x}}{\sqrt{x}} \sim 0$ for $x \downarrow 0$.
Ad Φ_4: Here $nx > 1$ and for x small enough

$$\begin{aligned}\Pi_{\omega^\omega}(nx) &= \sum_{l=0}^{\infty} pn_{2^l}^{-nx} \le 2^{-nx} + \sum_{l=1}^{\infty} 2^{-nxl} \\ &\le 2^{-nx} + 2^{-nx}\sum_{l=1}^{\infty} 2^{-l} \le \frac{M}{2^{nx}}\end{aligned}$$

Thus $\Phi_4(x) < M\sum_{\frac{1}{x}<n} \frac{1}{2^{nx}n} < xM\sum_{\frac{1}{x}<n} 2^{-nx} < \frac{xM}{1-2^{-x}} \le M'$ for a suitable M'. Putting things together we obtain $\ln(\Pi_{\omega^\omega}(x)) = \frac{1}{\ln(2)x}(\frac{\pi^2}{6} \pm o(1)) + o(x) + O(x^{-\frac{1}{3}}) + O(1)$ and the assertion follows.

Theorem 5.4 (Hardy and Ramanujan [12]). *Suppose that* $\lambda_1 \ge 0$, $\lambda_n > \lambda_{n-1}$, $\lambda_n \to \infty$; $\frac{\lambda_n}{\lambda_{n-1}} \to 1$, $a_n \ge 0$, $A > 0$, $a > 0$, $\sum a_n \exp(-\lambda_n s)$ *is convergent for* $s > 0$ *and* $f(s) = \sum a_n \exp(-\lambda_n s) = \exp((1+o(1))As^{-\alpha}\ln(\frac{1}{s})^{-\beta})$. *Then* $A_n = a_1 + a_2 + \cdots + a_n = \exp((1+o(1))B\lambda_n^{\frac{\alpha}{1+\alpha}}(\ln(\lambda_n)^{\frac{-\beta}{1+\alpha}}))$ *where* $B = A^{\frac{1}{1+\alpha}}\alpha^{\frac{-\alpha}{1+\alpha}}(1+\alpha)^{1+\frac{\beta}{1+\alpha}}$ *when* $n \to \infty$.

Proof of Theorem 5.1. Put $\lambda_n := \ln(n)$, $A := \frac{\pi^2}{6\ln(2)}$, $\alpha = 1$, $\beta = 1$. Then Lemma 5.3 yields $\Gamma_{\omega^\omega}(s) = \sum_{l=1}^{\infty} \gamma_2(l)\exp(-\ln(n)s) = \exp((1+o(1))As^{-1})$. Thus Theorem 5.4 yields $C_{\omega^\omega}(n) = \gamma_{\omega^\omega}(1) + \cdots + \gamma_{\omega^\omega}(n) = \exp((1+o(1))B\sqrt{\ln(n)})$ where $B = \frac{\pi\sqrt{2}}{\ln(2)\sqrt{3}}$.

Remark: Suitable extensions of the Theorem 5.4 yield bounds on $C_{\omega_n(d)}$ for $n \ge 3$ where $\omega_0(d) := d$ and $\omega_{n+1}(d) := \omega^{\omega_n(d)}$. These will be included in a subsequent paper.

References

[1] T.M. Apostol: *Introduction to Analytic Number Theory.* Springer, 1976.

[2] J.P. Bell: *Sufficient conditions for zero-one laws.* Preprint, 2000.

[3] J.P. Bell: *The second and third conjectures of number theoretic density and logical limit laws.* Preprint, 2000.

[4] J.P. Bell and S.N. Burris: *Asymptotics for logical limit laws. When the growth of the components is in an RT class.* Preprint, 2000.

[5] S.N. Burris: *Number Theoretic Density and Logical Limit Laws.* Mathematical Surveys and Monographs 86. American Mathematical Society 2001.

[6] K.J. Compton. *A logical approach to asymptotic combinatorics. I. First order properties.* Advances in Mathematics 65 (1987), 65-96.

[7] K.J. Compton. *A logical approach to asymptotic combinatorics. II. Monadic second order properties.* Journal of Combinatorial Theory Series A 50 (1989), 110-131.

[8] E. Ebbinghaus and J. Flum. *Finite model theory.* Springer 1995.

[9] R. Fagin. *Probabilities on finite models.* The Journal of Symbolic Logic 41 (1976), 50-58.

[10] J.L. Geluk and L. de Haan: *Regular Variation, Extensions and Tauberian Theorems.* CWI Tract 40, 1980.

[11] Y.V. Glebskij, D.I. Kogan, M.I. Liogonkij and V.A. Talanov. *Range and degree of realizability of formulas in the restricted predicate calculus.* Cybernetics 5 (1969), 142-154.

[12] G.H. Hardy and S. Ramanujan. *Asymptotic formulae for the distribution of integers of various types.* Proceedings of the London Mathematical Society 2 (1917), 112-132.

[13] R. Otter: *The number of trees.* Annals of Mathematics 49 (1948), 583-599.

[14] K. Schütte: *Proof Theory.* Springer 1977.

[15] A.R. Woods: *Coloring rules for finite trees, and probabilities of monadic second order sentences.* Random Structures and Algorithms, 10 (1997) 453-485.

Andreas Weiermann

Institut für Mathematische Logik und Grundlagenforschung
der Westfälischen Wilhelms-Universität Münster
Einsteinstr. 62, 48149 Münster, Germany

Trends in Mathematics, © 2002 Birkhäuser Verlag Basel/Switzerland

Further Applications of Chebyshev Polynomials in the Derivation of Spanning Tree Formulas for Circulant Graphs

Yuanping Zhang, Mordecai J. Golin

ABSTRACT: *Kirchhoff's* Matrix Tree Theorem *permits the calculation of the number of spanning trees in any given graph G through the evaluation of the determinant of an associated matrix. Boesch and Prodinger [6] have shown how to use Chebyshev polynomials to evaluate the associated determinants and derive closed formulas for the number of spanning trees of graphs in certain special classes.*

In this paper we extend this work to describe two further applications of Chebyshev polynomials in the evaluation of the numbers of spanning trees of Circulant *Graphs.*

Note: In this extended abstract some proofs are omitted.

1 Introduction

An undirected graph G is a pair (V, E), where V is its vertex set and E its edge set. All graphs considered in this paper are finite and undirected with self-loops and multiple copies of the same edge permitted.

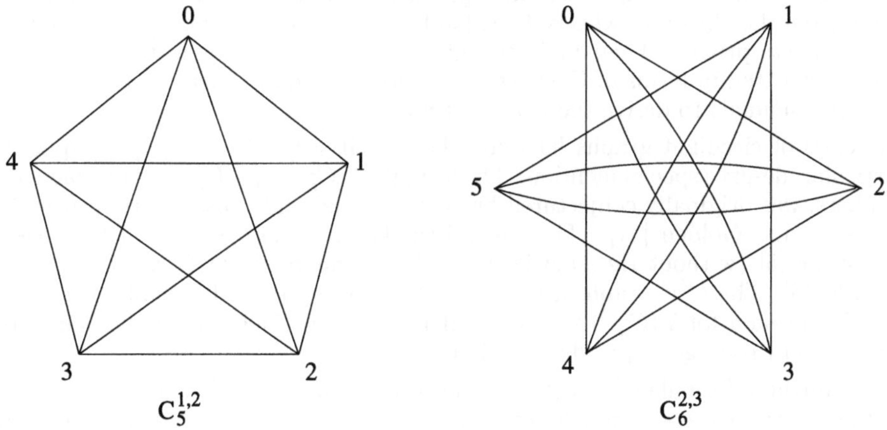

Figure 1: Two examples of circulant graphs. Note that $C_6^{2,3}$ has multiple edges.

Let $1 \le s_1 < s_2 < \cdots < s_k$, the s_j positive integers. The *undirected circulant graph*, $C_n^{s_1,s_2,\cdots,s_k}$, has n vertices labeled $0,\ 1,\ 2,\ \cdots,\ n-1$, with each vertex

*This work was supported by Hong Kong CERG grants HKUST6137/98E, HKUST6162/00E and HKUST DAG98/99.EG23.

i $(0 \le i \le n-1)$ adjacent to $2k$ vertices $i \pm s_1,\ i \pm s_2,\ \cdots,\ i \pm s_k \pmod n$. The simplest circulant graph is the n vertex cycle C_n^1. Figure 1 illustrates two circulant graphs. We note that our definition here specifically forces the graph to be $2k$ regular so, if $i \pm s_i \equiv i \pm s_j \pmod n$ for some i, j then the graph would have repeated edges. See, for example, $C_6^{2,3}$ in Figure 1.

For graph G, a spanning tree in G is a tree which has the same vertex set as G. The number of spanning trees in G denoted by $T(G)$, is a well studied quantity, being interesting both for its own sake and because it has practical implications for network reliability, e.g., [7, 8].

Let $A(G)$, or simply A be the adjacency matrix of G. If G has vertex set $V = \{v_1,\ v_2,\ \cdots,\ v_n\}$, the number d_i of the edges adjacent to vertex v_i is called the degree of v_i $(1 \le i \le n)$. Let B denote the diagonal matrix with $\{d_1,\ d_2,\ \cdots,\ d_n\}$ as diagonal entries. The classic result known as the *Matrix Tree Theorem* [10] states that, the Kirchhoff matrix H defined as $H = B - A$ has all its co-factors equal to $T(G)$, providing a method for calculating $T(G)$ for any particular given graph.

The number of spanning trees in graph G also can be calculated from the eigenvalues of the Kirchhoff matrix H. Let $\mu_1 \ge \mu_2 \ge \cdots \ge \mu_n (= 0)$ denote all eigenvalues of H. Kel'mans and Chelnokov [9] have shown that the *Matrix Tree Theorem* implies

$$T(G) = \frac{1}{n} \prod_{j=1}^{n-1} \mu_j. \tag{1}$$

For special classes of graphs it is possible to show that their Kirchhoff matrices have special structures and then bootstrap off of Kel'mans and Chelnokov's formula to get formulae for $T(G)$ when G is in those classes.

In [6] Boesch and Prodinger use this approach to derive closed formulae when G comes from the classes of wheels, fans, ladders, Moebius ladders, squares of cycles and complete prisms. Their main technique was to show that in these cases (1) can be rewritten in terms of Chebyshev polynomials and to then use properties of these polynomials to derive the closed formulae.

The class of circulant graphs have also been well studied. The $C_n^{1,2}$ graphs, in particular, deserve special mention. The formula $T(C_n^{1,2}) = nF_n^2$, F_n the *Fibonacci* numbers, was originally conjectured by Bedrosian [2] and subsequently proven by Kleitman and Golden [11]. The same formula was also conjectured by Boesch and Wang [5] (without the knowledge of [11]). Different proofs can been found in [1, 6, 13]. The $C_n^{1,2}$ graphs are actually the squares of cycles mentioned above and the formula for $T(C_n^{1,2})$ was also rederived using Chebyshev polynomials by Boesch and Prodinger [6] as described above.

Going further, formulae for $T(C_n^{1,3})$ and $T(C_n^{1,4})$ are provided in [12]. A connection between these formulae was given in [14] by showing that, for any *fixed* $s_1, s_2, \cdots, s_k$,

$$T(C_n^{s_1, s_2, \cdots, s_k}) = n a_n^2,$$

where the a_n satisfy a recurrence relation of the form

$$\forall n > 2^{s_k - 1},\ a_n = \sum_{i=1}^{2^{s_k-1}} b_i a_{n-i}$$

with the b_i are reals (but not necessarily nonnegative). Recall that the *Matrix Tree Theorem* gives us a method of calculating $T(C_n^{s_1,s_2,\cdots,s_k}) = na_n^2$ for any arbitrary n by building the Kirchhoff matrix and evaluating a determinant. This means that we can find the b_i by calculating all of the a_i for $i \le 2^{s_k}$ and then solving for the b_i. The asymptotics of $T(C_n^{s_1,s_2,\cdots,s_k})$ could then be found by solving for the minimum modulus root of the characteristic polynomial of the recurrence relation. This was done in [14] for all circulant graphs with $s_k \le 5$.

In this note we extend the ideas in [6] in two directions. In the first we show how to use the Chebyshev polynomial technique to derive a much simpler proof that $T(C_n^{s_1,s_2,\cdots,s_k}) = na_n^2$ where the a_n satisfy a linear recurrence relation of order 2^{s_k-1}. This new proof will have the added advantage of providing a method of deriving the minimum modulus root of the characteristic polynomial of the recurrence relation ***without having to construct the recurrence relation***, thus obviating the need to calculate the determinants (it will only require finding the roots of a particular polynomial of order $s_k - 1$).

In the second we describe how to use the Chebyshev polynomial technique for deriving closed formulae for some circulant graphs with *non-constant* jumps. More specifically, the technique will permit the derivation of formulae for circulant graphs of the form $C_n^{s1,\cdots,s_k,\frac{n}{a_1},\cdots,\frac{n}{a_l}}$ where $s_1, \cdots, s_k$ are constant integers, $a_1, \cdots, a_l \in \{2,3,4,6\}$ and $\forall i \le l$, $a_i | n$. As examples, we will derive formulae for $T(C_{2n}^{1,n})$, $T(C_{3n}^{1,n})$, $T(C_{4n}^{1,n})$, $T(C_{6n}^{1,n})$ and $T(C_{6n}^{1,2n,3n})$.

The rest of the paper is structured as follows. In section 2 we briefly review the basic facts we will need. In section 3 we rederive $T(C_n^{s_1,s_2,\cdots,s_k}) = na_n^2$ and describe how to efficiently calculate its asymptotics. Finally, in section 4, we discuss non-constant jumps.

2 Basic concepts and lemmas

An $n \times n$ matrix C is said to be a *circulant matrix* [4] if its entries satisfy $c_{ij} = c_{1,j-i+1}$, where the subscripts are reduced modulo n and lie in the set $\{1, 2, \cdots, n\}$. In the other words, i-th row of C is obtained from the first row of C by a cyclic shift of $i - 1$ steps, and so any circulant matrix is determined by its first row. Let W denote the circulant matrix whose first row is $(0, 1, 0, \cdots, 0)$, and let C denote a general circulant matrix whose first row is $(c_1, c_2, \cdots, c_n)$. Then a straightforward calculation shows that

$$C = \sum_{i=1}^{n} c_i W^{i-1}.$$

Since the eigenvalues of W are $1, \varepsilon, \varepsilon^2, \cdots, \varepsilon^{n-1}$, where $\varepsilon = e^{\frac{2\pi i}{n}}$, it follows that the eigenvalues of C are

$$\lambda_j = \sum_{i=1}^{n} c_i \varepsilon^{(i-1)j}, \; j = 0, 1, \cdots, n-1. \tag{2}$$

It's clear that the adjacency matrix of the circulant graph $C_n^{s_1,s_2,\cdots,s_k}$ is a circulant matrix. The first row $(c_1, c_2, \cdots, c_n)$ of the adjacency matrix is determined by the connection jumps $s_1, s_2, \cdots, s_k$. More specifically, an edge $(1, i)$ is in the graph if and only if $i \equiv (1 \pm s_j) \pmod{n}$ for some s_j, $1 \le j \le k$. (Note that it is possible for the $c_i > 1$. This happens if $(1 \pm s_j) \equiv (1 \pm s_{j'}) \pmod{n}$ for some $j \ne j'$. In this case the graph is a multigraph and c_i is the *number* of different edges connecting 1 and i. This can only happen when n is small, though.) From the adjacency matrix of $C_n^{s_1,s_2,\cdots,s_k}$ and the definition of the Kirchhoff matrix it's easy to see that the Kirchhoff matrix of $C_n^{s_1,s_2,\cdots,s_k}$ is also a circulant matrix. This can be used to find its eigenvalues.

The starting point of our calculations is actually the following lemma which is a direct application of Proposition 3.5 on page 16 of Biggs [4]:

Lemma 1. The *Kirchhoff* matrix of the circulant graph $C_n^{s_1,s_2,\cdots,s_k}$ has n eigenvalues. They are 0 and, $\forall j$, $1 \le j \le n-1$ the values $2k - \varepsilon^{-s_1 j} - \cdots - \varepsilon^{-s_k j} - \varepsilon^{s_1 j} - \cdots - \varepsilon^{s_k j}$, where $\varepsilon = e^{\frac{2\pi i}{n}}$.

Plugging this into (1) yields the following well known corollary:

Corollary 1. Set $\varepsilon = e^{\frac{2\pi i}{n}}$. Then

$$\begin{aligned} T(C_n^{s_1,s_2,\cdots,s_k}) &= \frac{1}{n}\prod_{j=1}^{n-1}(2k - \varepsilon^{-s_1 j} - \varepsilon^{-s_2 j} - \cdots - \varepsilon^{-s_k j} \\ &\qquad -\varepsilon^{s_1 j} - \varepsilon^{s_2 j} - \cdots - \varepsilon^{s_k j}), \\ &= \frac{1}{n}\prod_{j=1}^{n-1}\left(\sum_{i=1}^{k}\left(2 - 2\cos\frac{2js_i\pi}{n}\right)\right). \end{aligned}$$

An important special case of this occurs when we examine the cycle C_n^1. Clearly C_n^1 has exactly n spanning trees. Applying the corollary therefore yields ([6]) the nonobvious

$$n = T(C_n^1) = \frac{1}{n}\prod_{j=1}^{n-1}\left(2 - 2\cos\frac{2j\pi}{n}\right) = \frac{1}{n}\prod_{j=1}^{n-1}(4\sin^2\frac{j\pi}{n}), \tag{3}$$

which will be useful to us later.

The other main tools we use are various standard properties of Chebyshev polynomials of the second kind. For reference we quickly review them here. The following definitions and derivations (with the exception of (11)) follow [6].

For positive integer m, the Chebyshev polynomials of the first kind are defined by

$$T_m(x) = \cos(m \arccos x). \tag{4}$$

The Chebyshev polynomials of the second kind are defined by

$$U_{m-1}(x) = \frac{1}{m}\frac{d}{dx}T_m(x) = \frac{\sin(m \arccos x)}{\sin(\arccos x)}. \tag{5}$$

It is easily verified that

$$U_m(x) - 2xU_{m-1}(x) + U_{m-2}(x) = 0. \tag{6}$$

Solving this recursion by using standard methods yields

$$U_m(x) = \frac{1}{2\sqrt{x^2-1}}[(x+\sqrt{x^2-1})^{m+1} - (x-\sqrt{x^2-1})^{m+1}], \tag{7}$$

where the identity is true for all complex x (except at $x = \pm 1$ where the function can be taken as the limit).

The definition of $U_m(x)$ easily yields its zeros and it can therefore be verified that

$$U_{m-1}(x) = 2^{m-1} \prod_{j=1}^{m-1} (x - \cos \frac{j\pi}{m}). \tag{8}$$

One further notes that

$$U_{m-1}(-x) = (-1)^{m-1} U_{m-1}(x). \tag{9}$$

These two results yield another formula for $U_m(x)$,

$$U^2_{m-1}(x) = 4^{m-1} \prod_{j=1}^{m-1} \left(x^2 - \cos^2 \frac{j\pi}{m} \right). \tag{10}$$

Finally, simple manipulation of the above formula yields the following, which will also be very useful to us later:

$$U^2_{m-1}\left(\sqrt{\frac{x+2}{4}}\right) = \prod_{j=1}^{m-1} \left(x - 2\cos \frac{2\pi j}{m} \right), \tag{11}$$

where the identity is true for all complex x.

3 Recurrence relations for fixed step circulant graphs

In this section we assume that $s_1, s_2, \cdots, s_k$ are *fixed* positive integers with $1 \le s_1 < s_2 < \cdots < s_k$ and use the properties of Chebyshev polynomials to reprove the main result in [14], i.e, that there exist reals $b_1, b_2, \ldots, b_{2^{s_k-1}}$ such that

$$T(C_n^{s_1,s_2,\cdots,s_k}) = na_n^2, \quad \text{where} \quad \forall n > 2^{s_k-1},\ a_n = \sum_{i=1}^{2^{s_k-1}} b_i a_{n-i}. \tag{12}$$

We start with a basic lemma on trigonometric polynomials; its proof is quite tedious but straightforward so we omit it in this extended abstract:

Lemma 2. Let $k > 0$ be any integer. Then $2 - 2\cos(2kx)$ can be rewritten in the form $4^k f_k(\cos^2 x)\sin^2 x$, where f_k is a polynomial of order $k-1$ with leading coefficient 1 that does not have 1 as a root.

Combining this with Corollary 1 and some manipulation yields

Lemma 3. The number of spanning trees $T(C_n^{s_1,s_2,\cdots,s_k})$ satisfies

$$T(C_n^{s_1,s_2,\cdots,s_k}) = \frac{1}{n}\prod_{j=1}^{n-1} 4^{s_k} f(\cos^2\frac{j\pi}{n})\sin^2\left(\frac{j\pi}{n}\right),$$

where f is a polynomial of order $s_k - 1$ with leading coefficient 1 that does not have 1 as a root.

Now let $x_1, x_2, \cdots, x_{s_k-1}$ be the roots of $f(x)$. Then

$$f(x) = (-1)^{s_k-1}\prod_{i=1}^{s_k-1}(x_i - x).$$

Plugging this into Lemma 3 and using formulae (3) and (10) gives

$$\begin{aligned} &T(C_n^{s_1,s_2,\cdots,s_k}) \\ &= \frac{1}{n}\prod_{j=1}^{n-1} 4^{s_k}(-1)^{s_k-1}\left(\prod_{i=1}^{s_k-1}(x_i - \cos^2\frac{j\pi}{n})\right)\sin^2\left(\frac{j\pi}{n}\right) \\ &= (-1)^{(n-1)(s_k-1)}\frac{1}{n}\prod_{i=1}^{s_k-1}\left(4^{n-1}\prod_{j=1}^{n-1}(x_i - \cos^2\frac{j\pi}{n})\right)\cdot 4^{n-1}\prod_{j=1}^{n-1}\sin^2\frac{j\pi}{n} \\ &= (-1)^{(n-1)(s_k-1)} n\prod_{i=1}^{s_k-1} U_{n-1}^2(\sqrt{x_i}). \end{aligned}$$

Using formula (7) to rewrite $U_{n-1}^2(\sqrt{x_i})$ gives

$$T(C_n^{s_1,s_2,\cdots,s_k}) = n\left[\prod_{i=1}^{s_k-1}\frac{1}{2\sqrt{1-x_i}}\left((\sqrt{-x_i}+\sqrt{1-x_i})^n - (\sqrt{-x_i}-\sqrt{1-x_i})^n\right)\right]^2.$$

This actually provides a 'closed formula' for $T(C_n^{s_1,s_2,\cdots,s_k})$, albeit, not a particularly satisfying one. We now continue by, for all i, $1 \le i \le s_k - 1$, set $y_{i,0} = \sqrt{-x_i} + \sqrt{1-x_i}$ and $y_{i,1} = \sqrt{-x_i} - \sqrt{1-x_i}$. For $(\delta_1, \delta_2, \ldots, \delta_{s_k-1}) \in \{0,1\}^{s_k-1}$ set

$$R_{\delta_1,\delta_2,\ldots,\delta_{s_k-1}} = (-1)^{\sum_{i=1}^{s_k-1}\delta_i}\cdot\prod_{i=1}^{s_k-1} y_{i,\delta_i}.$$

Also set $c = \prod_{i=1}^{s_k-1}\frac{1}{2\sqrt{1-x_i}}$. If a_n is defined so that $T(C_n^{s_1,s_2,\cdots,s_k}) = na_n^2$, then

$$a_n = c\sum_{(\delta_1,\delta_2,\ldots,\delta_{s_k-1})\in\{0,1\}^{s_k-1}} R^n_{\delta_1,\delta_2,\ldots,\delta_{s_k-1}}.$$

Since there are at most 2^{s_k-1} different values $R_{\delta_1,\delta_2,\ldots,\delta_{s_k-1}}$ this immediately implies (12) and we have proved what we claimed.

As noted in [14] one way to find the b_i is to simply use the *Matrix Tree Theorem* to calculate the value of $T(C_n^{s_1,s_2,\cdots,s_k})$ for all $n \le 2^{s_k}$ yielding all of the values of a_n and then solve for the b_n. Once the b_n are known the asymptotics of a_n (and therefore $T(C_n^{s_1,s_2,\cdots,s_k})$) could be found by standard generating function techniques, i.e., by calculating the roots of the characteristic equation of the a_n. This is what was done in [14]. That paper actually proved a stronger result; that is, if $\gcd(s_1, s_2, \cdots, s_k) = 1$, then ϕ, the smallest modulus root of the generating function of the a_n, is unique and real so $a_n \sim c\phi^n$ for some c, and $T(C_n^{s_1,s_2,\cdots,s_k}) \sim nc^2\phi^{2n}$. The asymptotics of $T(C_n^{s_1,s_2,\cdots,s_k})$ could therefore be found by calculating the smallest modulus root of the generating function. If $\gcd(s_1, s_2, \cdots, s_k) = d \ne 1$ it is described in [14] how this case can be reduced down to evaluating $T(C_n^{s_1/d,s_2/d,\cdots,s_k/d})$. Since $\gcd(s_1/d, s_2/d, \cdots, s_k/d) = 1$ we may always restrict ourselves to assuming that $\gcd(s_1, s_2, \cdots, s_k) = 1$.

The difficulty with this technique is that, in order to derive the generating function, it was necessary to apply the *Matrix Tree theorem* 2^{s_k} times, evaluating a determinant each time. Since the underlying matrices get quite large this can be very calculation intensive.

Our new proof of (12) immediately yields a much more efficient method of deriving the asymptotics. Note that the roots of the generating function are exactly $\frac{1}{R_{\delta_1,\delta_2,\ldots,\delta_{s_k-1}}}$. Finding the smallest modulus root is therefore the same as finding $R_{\max}$, the $R_{\delta_1,\delta_2,\ldots,\delta_{s_k-1}}$ with maximum modulus since the smallest modulus root is real, $R_{\max}$ is real as well. We can therefore easily find. $R_{\max}$ by setting $y_i = \max(|y_{i,0}|, |y_{i,1}|)$ for all $i \le s_k - 1$ and then noting that $|R_{\max}| = \prod_{i=1}^{s_k-1} y_i$. This technique yields the asymptotics of $T(C_n^{s_1,s_2,\cdots,s_k})$ without requiring the evaluation of any determinants; all that is needed is the calculation of all of the roots of a degree $s_k - 1$ polynomial. Note that it is not a-priori obvious that $R_{\max}$ is positive but, since we are only interested in na_n^2 and not a_n, knowing $|R_{\max}|$ suffices.

As an example we work through the process for $T(C_n^{1,2,3})$:

$$\begin{aligned}
T(C_n^{1,2,3}) &= \frac{1}{n}\prod_{j=1}^{n-1}(6 - e^{\frac{2\pi j}{n}} - e^{\frac{4\pi j}{n}} - e^{\frac{6\pi j}{n}} - e^{\frac{-2\pi j}{n}} - e^{\frac{-4\pi j}{n}} - e^{\frac{-6\pi j}{n}})\\
&= \frac{1}{n}\prod_{j=1}^{n-1}(6 - 2\cos\frac{2\pi j}{n} - 2\cos\frac{4\pi j}{n} - 2\cos\frac{6\pi j}{n})\\
&= \frac{1}{n}\prod_{j=1}^{n-1} 64(\cos^4\frac{\pi j}{n} - \frac{1}{4}\cos^2\frac{\pi j}{n} + \frac{1}{8})\sin^2\frac{\pi j}{n}\\
&= n\prod_{j=1}^{n-1} 16(\cos^4\frac{\pi j}{n} - \frac{1}{4}\cos^2\frac{\pi j}{n} + \frac{1}{8}).
\end{aligned}$$

The roots of the polynomial $x^2 - \frac{1}{4}x + \frac{1}{8}$ are

$$x_1 = \frac{1}{8} - \frac{\sqrt{7}}{8}i \quad\text{and}\quad x_2 = \frac{1}{8} + \frac{\sqrt{7}}{8}i.$$

Thus

$$\begin{aligned}
y_{1,0} &= \sqrt{-x_1} + \sqrt{1-x_1} = \frac{1}{4}\sqrt{-2+2\sqrt{7}i} + \frac{1}{4}\sqrt{14+2\sqrt{7}i},\\
y_{1,1} &= \sqrt{-x_1} - \sqrt{1-x_1} = \frac{1}{4}\sqrt{-2+2\sqrt{7}i} - \frac{1}{4}\sqrt{14+2\sqrt{7}i},\\
y_{2,0} &= \sqrt{-x_2} + \sqrt{1-x_2} = \frac{1}{4}\sqrt{-2-2\sqrt{7}i} + \frac{1}{4}\sqrt{14-2\sqrt{7}i},\\
y_{2,1} &= \sqrt{-x_2} - \sqrt{1-x_2} = \frac{1}{4}\sqrt{-2-2\sqrt{7}i} - \frac{1}{4}\sqrt{14-2\sqrt{7}i}.
\end{aligned}$$

Therefore, $T(C_n^{1,2,3}) = na_n^2$, $a_n \sim c\phi^n$ where $c = \frac{1}{2\sqrt{1-x_1}} \cdot \frac{1}{2\sqrt{1-x_2}} = \frac{1}{\sqrt{14}} \doteq 0.2672612$ and $\phi = y_{1,0} \cdot y_{2,0} = \frac{1}{16}(\sqrt{32} + \sqrt{224} + \sqrt{64\sqrt{7}}) \doteq 2.102256$. These are exactly the same values c and ϕ derived in [14] using the longer method.

4 The number of spanning trees in some non fixed-jump circulant graphs

In the previous section we examined the spanning tree number for circulant graphs in which the ***steps***, i.e., the s_i, were fixed and the number of nodes, i.e., n, changed. In this section we derive formulae for some graphs in which the step sizes can be functions of n. Our approach is, as before, to expand $T(G)$ for some circulant graph G as a product of trigonometric polynomials and then express it in terms of Chebyshev polynomials, in this case, ratios of such polynomials. We will see though, that this technique is not totally general and only works for particular values of jumps.

We illustrate the technique via three examples; starting from a easy one, $T(C_{2n}^{1,n})$, then seeing $T(C_{3n}^{1,n})$, which is more complicated and ending at $T(C_{4n}^{1,n})$ which reveals where the difficulties can lie.

We start by calculating $T(C_{2n}^{1,n})$. Recall that, according to our definition of circulant graphs, $C_{2n}^{1,n}$ is the *four*-regular graph with $2n$ vertices $0, 1, \ldots, 2n-1$ such that node i has one edge connecting it to $(i+1) \pmod{2n}$ one edge connecting it to $(i-1) \pmod{2n}$ and *two* edges connecting it to $(i+n) \pmod{2n}$. We should note that this is not the same graph as the Moebius ladder which is a *three*-regular graph on the same vertex set in which node i has one edge connecting it to each of $(i+1) \pmod{2n}$, $(i-1) \pmod{2n}$ and $(i+n) \pmod{2n}$. The techniques described here, though, could be used to rederive closed formulae for the spanning tree numbers of Moebius ladders and similar graphs (see [6] for such a derivation).

Theorem 4.

$$T(C_{2n}^{1,n}) = \frac{n}{2}[(\sqrt{2}+1)^n + (\sqrt{2}-1)^n]^2.$$

Proof. Let $\varepsilon_2 = e^{\frac{2\pi i}{2n}}$. By Lemma 1, we have

$$\begin{aligned}
T(C_{2n}^{1,n}) &= \frac{1}{2n} \prod_{j=1}^{2n-1} (4 - \varepsilon_2^j - \varepsilon_2^{-j} - \varepsilon_2^{nj} - \varepsilon_2^{-nj}) \\
&= \frac{1}{2n} \prod_{j=1}^{2n-1} \left(4 - 2\cos\frac{2\pi j}{2n} - 2\cos(\pi j)\right) \\
&= \frac{1}{2n} \prod_{\substack{j=1 \\ 2 \nmid j}}^{2n-1} \left(6 - 2\cos\frac{2\pi j}{2n}\right) \prod_{\substack{j=1 \\ 2 \mid j}}^{2n-1} \left(2 - 2\cos\frac{2\pi j}{2n}\right).
\end{aligned}$$

Noting that if $j = 2j'$ for some integer j' then $\cos\frac{2\pi j}{2n} = \cos\frac{2\pi j'}{n}$ gives

$$\begin{aligned}
T(C_{2n}^{1,n}) &= \frac{1}{2n} \prod_{j=1}^{2n-1} \left(6 - 2\cos\frac{2\pi j}{2n}\right) \prod_{j=1}^{n-1} \frac{2 - 2\cos\frac{2\pi j}{n}}{6 - 2\cos\frac{2\pi j}{n}} \\
&= \frac{1}{2n} U_{2n-1}^2(\sqrt{2}) \cdot \frac{n^2}{U_{n-1}^2(\sqrt{2})} \\
&= \frac{n}{2}[(\sqrt{2}+1)^n + (\sqrt{2}-1)^n]^2.
\end{aligned}$$

Where (3), (7) and (11) are used to derive the last two steps. □

We now go to

Theorem 5.

$$T(C_{3n}^{1,n}) = \frac{n}{3}\left[\left(\frac{\sqrt{7}}{2} + \frac{\sqrt{3}}{2}\right)^{2n} + \left(\frac{\sqrt{7}}{2} - \frac{\sqrt{3}}{2}\right)^{2n} + 1\right]^2.$$

Proof. The proof is similar to the previous one. Let $\varepsilon_3 = e^{\frac{2\pi i}{3n}}$. By Lemma 1, we have

$$\begin{aligned}
T(C_{3n}^{1,n}) &= \frac{1}{3n} \prod_{j=1}^{3n-1} (4 - \varepsilon_3^j - \varepsilon_3^{-j} - \varepsilon_3^{nj} - \varepsilon_3^{-nj}) \\
&= \frac{1}{3n} \prod_{j=1}^{3n-1} \left(4 - 2\cos\frac{2\pi j}{3n} - 2\cos\frac{2\pi j}{3}\right) \\
&= \frac{1}{3n} \prod_{\substack{j=1 \\ 3 \nmid j}}^{3n-1} \left(5 - 2\cos\frac{2\pi j}{3n}\right) \prod_{\substack{j=1 \\ 3 \mid j}}^{3n-1} \left(2 - 2\cos\frac{2\pi j}{3n}\right).
\end{aligned}$$

Note that if $j = 3j'$ for some integer j' then $\cos\frac{2\pi j}{3n} = \cos\frac{2\pi j'}{n}$. In this case we will need the added observation that if $3 \nmid j$ then $\cos\frac{2\pi j}{3} = -\frac{1}{2}$. This gives

$$\begin{aligned} T(C_{3n}^{1,n}) &= \frac{1}{3n}\prod_{j=1}^{3n-1}\left(5 - 2\cos\frac{2\pi j}{3n}\right)\prod_{j=1}^{n-1}\frac{2-2\cos\frac{2\pi j}{n}}{5-2\cos\frac{2\pi j}{n}} \\ &= \frac{1}{3n}U_{3n-1}^2\left(\sqrt{\frac{7}{4}}\right)\cdot\frac{n^2}{U_{n-1}^2\left(\sqrt{\frac{7}{4}}\right)} \\ &= \frac{n}{3}\left[\left(\frac{\sqrt{7}}{2}+\frac{\sqrt{3}}{2}\right)^{2n}+\left(\frac{\sqrt{7}}{2}-\frac{\sqrt{3}}{2}\right)^{2n}+1\right]^2. \end{aligned}$$

□

We next see

Theorem 6.

$$T(C_{4n}^{1,n}) = \frac{n}{4}\left[\left(\sqrt{\frac{3}{2}}+\sqrt{\frac{1}{2}}\right)^{2n}+\left(\sqrt{\frac{3}{2}}-\sqrt{\frac{1}{2}}\right)^{2n}\right]^2\left[\left(\sqrt{2}+1\right)^n+\left(\sqrt{2}-1\right)^n\right]^2.$$

Proof. The proof again starts similarly to the previous ones. Let $\varepsilon_4 = e^{\frac{2\pi i}{4n}}$. We have

$$\begin{aligned} T(C_{4n}^{1,n}) &= \frac{1}{4n}\prod_{j=1}^{4n-1}(4-\varepsilon_4^j-\varepsilon_4^{-j}-\varepsilon_4^{nj}-\varepsilon_4^{-nj}) \\ &= \frac{1}{4n}\prod_{j=1}^{4n-1}\left(4-2\cos\frac{2\pi j}{4n}-2\cos\frac{\pi j}{2}\right) \\ &= \frac{1}{4n}\prod_{\substack{j=1\\2\nmid j}}^{4n-1}\left(4-2\cos\frac{2\pi j}{4n}\right)\prod_{\substack{j=1\\2|j}}^{4n-1}\left(4-2\cos\frac{2\pi j}{4n}-2\cos\frac{\pi j}{2}\right), \end{aligned}$$

where the last derivation follows from the fact that if $2 \nmid j$ then $\cos\frac{2\pi j}{4n} = 0$. Unlike in the previous proofs, though, if $2|j$ it is not true that $\cos\frac{2\pi j}{4n}$ equals some constant, so we will have to derive further. We use the fact that if $j = 2j'$ then $\cos\frac{2\pi j}{4n} = \cos\frac{2\pi j}{2n}$ to get

$$T(C_{4n}^{1,n}) = \frac{1}{4n}\prod_{j=1}^{4n-1}\left(4-2\cos\frac{2\pi j}{4n}\right)\prod_{j=1}^{2n-1}\frac{4-2\cos\frac{2\pi j}{2n}-2\cos(\pi j)}{4-2\cos\frac{2\pi j}{2n}}.$$

At this point we can evaluate both the leftmost product and the denominator of the rightmost product in terms of Chebyshev polynomials. To evaluate the numerator of the rightmost product we will need to split it into two cases depending upon whether j is odd or even, and apply the same type of procedure again. This yields

$$\begin{aligned}
T(C_{4n}^{1,n}) &= \frac{1}{4n}\cdot\frac{U_{4n-1}^{2}\left(\sqrt{\frac{3}{2}}\right)}{U_{2n-1}^{2}\left(\sqrt{\frac{3}{2}}\right)}\prod_{j=1}^{2n-1}\left(6-2\cos\frac{2\pi j}{2n}\right)\prod_{j=1}^{n-1}\frac{2-2\cos\frac{2\pi j}{n}}{6-2\cos\frac{2\pi j}{n}}\\
&= \frac{1}{4n}\cdot\frac{U_{4n-1}^{2}\left(\sqrt{\frac{3}{2}}\right)}{U_{2n-1}^{2}\left(\sqrt{\frac{3}{2}}\right)}\cdot\frac{U_{2n-1}^{2}(\sqrt{2})}{U_{n-1}^{2}(\sqrt{2})}\cdot n^{2}\\
&= \frac{n}{4}\left[\left(\sqrt{\frac{3}{2}}+\sqrt{\frac{1}{2}}\right)^{2n}+\left(\sqrt{\frac{3}{2}}-\sqrt{\frac{1}{2}}\right)^{2n}\right]^{2}\left[\left(\sqrt{2}+1\right)^{n}+\left(\sqrt{2}-1\right)^{n}\right]^{2}.
\end{aligned}$$

□

The proofs of Theorem 4, 5 and 6 depend on certain symmetry properties of the *cosine* functions, e.g., if $3\nmid j$ then $\cos\frac{2\pi j}{3}=-\frac{1}{2}$ that permitted us to write products out as ratios that were in the proper form to express as Chebyshev polynomials. Unfortunately, this can not always be done. For example, we do not seem to be able to use this technique to derive a formula for $T(C_{5n}^{1,n})$. The most that we are currently able to push this technique is to derive closed formulae for the number of spanning trees (as a function of n) for all circulant graphs of $C_n^{s1,\cdots,s_k,\frac{n}{a_1},\cdots,\frac{n}{a_l}}$ where $s_1,\cdots,s_k$ are constant integers and all $a_1,\cdots,a_l$ are in the set $\{2,3,4,6\}$ with $a_u|n$ for any u, $1\le u\le l$. (The proof of this fact is omitted in this extended abstract).

We conclude with a few more applications (proofs omitted):

Theorem 7.

$$T(C_{6n}^{1,n}) = \frac{n}{6}\left[\left(\frac{\sqrt{5}+1}{2}\right)^{3n}+\left(\frac{\sqrt{5}-1}{2}\right)^{3n}\right]^{2}\left[\left(\sqrt{2}+1\right)^{n}+\left(\sqrt{2}-1\right)^{n}\right]^{2}\cdot\left[\left(\frac{\sqrt{5}+1}{2}\right)^{n}+\left(\frac{\sqrt{5}-1}{2}\right)^{n}\right]^{2}\left[\left(\frac{\sqrt{7}+\sqrt{3}}{2}\right)^{2n}+\left(\frac{\sqrt{7}-\sqrt{3}}{2}\right)^{2n}+1\right]^{2}.$$

Theorem 8.

$$T(C_{6n}^{1,2n,3n}) = \frac{n}{6}\left[\left(\frac{\sqrt{11}}{2}+\frac{\sqrt{7}}{2}\right)^{2n}+\left(\frac{\sqrt{11}}{2}-\frac{\sqrt{7}}{2}\right)^{2n}-1\right]^{2}\cdot\left[\left(\frac{\sqrt{7}+\sqrt{3}}{2}\right)^{2n}+\left(\frac{\sqrt{7}-\sqrt{3}}{2}\right)^{2n}+1\right]^{2}\left[\left(\sqrt{2}+1\right)^{n}+\left(\sqrt{2}-1\right)^{n}\right]^{2}.$$

A major open problem still remaining is to devise a technique that would work to derive closed formulae for $T(C_n^{s1,\cdots,s_k,\frac{n}{a_1},\cdots,\frac{n}{a_l}})$ where the a_i could be arbitrary.

Acknowledgment. The authors would like to thank Yong Xuerong for his suggestions and comments. We would also like to thank the anonymous referees for their suggestions.

References

[1] G. Baron, H. Prodinger, R. F. Tichy, F. T. Boesch and J. F. Wang, The Number of Spanning Trees in the Square of a Cycle, Fibonacci Quarterly 23.3(1985), 258-264.

[2] S. Bedrosian, The Fibonacci Numbers via Trigonometric Expressions, J. Franklin Inst. 295(1973), 175-177.

[3] S. D. Bedrosian, Tree counting polynomials for labelled graphs, J. Franklin Inst., 312 (1981) 417-430.

[4] N. Biggs, Algebraic Graph Theory, London: Cambridge University Press, Second Edition, 1993.

[5] F. T. Boesch and J. F. Wang, A Conjecture on the Number of Spanning Trees in the Square of a Cycle, In: Notes from New York Graph Theory Day V, p. 16, New York: New York Academy Sciences, 1982.

[6] F. T. Boesch, H. Prodinger, Spanning Tree Formulas and Chebyshev Polynomials, Graph and Combinatorics 2 (1986) 191-200.

[7] C. J. Colbourn, The combinatorics of network reliability, Oxford University Press, New York, 1987.

[8] D. Cvetkovivc, M. Doob, H. Sachs, Spectra of Graphs: Theory and Applications, Third Edition, Johann Ambrosius Barth, Heidelberg, 1995.

[9] A. K. Kel'mans, V. M. Chelnokov, A certain polynomial of a graph and graphs with an extremal number of trees, J. Comb. Theory (B) 16(1974) 197-214.

[10] G. Kirchhoff, Über die Auflösung der Gleichungen, auf welche man bei der Untersuchung der linearen Verteilung galvanischer Ströme geführt wird, Ann. Phys. Chem. 72 (1847) 497-508.

[11] D. J. Kleitman and B. Golden, Counting Trees in a Certain Class of Graphs, Amer. Math. Monthly 82(1975), 40-44.

[12] X. Yong, Talip and Acenjian, The Numbers of Spanning Trees of the Cubic Cycle C_N^3 and the Quadruple Cycle C_N^4, Discrete Math. 169(1997), 293-298.

[13] X. Yong and F. J. Zhang, A simple proof for the complexity of square cycle C_p^2, J. Xinjiang Univ. 11(1994), 12-16.

[14] Y. P. Zhang, X. Yong, M. J. Golin, The number of spanning trees in circulant graphs, Discrete Math. 223 (2000) 337-350.

Yuanping Zhang

Department of Computer Science
Hong Kong U.S.T.
Clear Water Bay
Kowloon, Hong Kong
(Also at
The Department of Mathematics
Hunan Normal University
Changsha, 410081, PRC)
ypzhang@cs.ust.hk

Mordecai J. Golin

Department of Computer Science
Hong Kong U.S.T.
Clear Water Bay
Kowloon, Hong Kong
golin@cs.ust.hk

Key Words

List of Authors

MIX
Papier aus verantwortungsvollen Quellen
Paper from responsible sources
FSC® C105338

If you have any concerns about our products,
you can contact us on
ProductSafety@springernature.com

In case Publisher is established outside the EU,
the EU authorized representative is:
Springer Nature Customer Service Center GmbH
Europaplatz 3, 69115 Heidelberg, Germany

Printed by Libri Plureos GmbH
in Hamburg, Germany